# AEROSOL PROCESSING OF MATERIALS

# AEROSOL PROCESSING OF MATERIALS

**Toivo T. Kodas and Mark J. Hampden-Smith**

Superior Micropowders, Albuquerque, NM

WILEY-VCH

**New York • Chichester • Weinheim • Brisbane • Singapore • Toronto**

This book is printed on acid-free paper. ♾



Published simultaneously in Canada.



***Library of Congress Cataloging in Publication Data:***

Kodas, Toivo T. (Toivo Tarmo), 1958–
Aerosol processing of materials / by Toivo T. Kodas and Mark J. Hampden-Smith
p. cm.
Includes index.
ISBN 0-471-24669-7 (cloth : alk. paper)
1. Thin films. 2. Aerosols. 3. Chemical vapor deposition.
I. Hampden-Smith, Mark J. II. Title.
TA418.9.T45K63 1999
621.3815'2—dc21 98-3623

Printed in the United States of America

10 9 8 7 6 5 4 3 2 1

# CONTENTS

# PREFACE

There is currently considerable commercial and scientific interest in the production of powders and the deposition of films employing aerosol-based methods. Existing applications that employ aerosol techniques range from the production of highly dispersed metal oxide powders for use in the pigment industry to the deposition of thin films of materials, which modify the optical characteristics of glass windows and display screens.

This book describes the principles and practice of aerosol science and engineering along with the relevant aspects of materials science and engineering. The book is organized in the following manner: the early chapters deal with the basic principles of aerosols. After a general introduction in Chapter 1, Chapters 2–7 describe the fundamental aspects of aerosols such as particle size, transport, growth, evaporation, and collision and coalescence. Chapter 8 describes the chemistry relevant to aerosol processes to materials and Chapter 9 describes the characteristics of nanostructured materials because aerosol methods are among the best synthetic routes to prepare this class of materials. This is followed by a discussion of the overall qualitative behavior and technology of gas-to-particle conversion; liquid-to-solid, and solid-to-solid conversion; and film formation in Chapters 10–15. These chapters combine the principles described in Chapters 1–9 to better understand the practicalities of these systems. Equipment that can be used to produce materials by aerosol methods is found in Chapter 16 and Chapter 17 describes the measurement techniques that have been used to characterize materials produced by aerosol methods.

# ACKNOWLEDGMENTS

We are indebted to a number of people who have contributed in direct and indirect ways to this work. We appreciate the members of our research groups for their patience in reading, correcting and making suggestions to the drafts of the manuscript and we want to acknowledge Quint Powell, Dave Dericotte, James Brewster, Dr. Sanjeev Jain, Dr. Abhijit Gurav, and Dr. Paolina Atanasova, who were particularly insightful.

We thank the employees of Nanochem Research, Inc. and Superior MicroPowders, LLC for their patience and input over the last 5 years and their dedication to making many of the aerosol routes described in this book a manufacturing reality.

We are grateful to Professor Sotiris Pratsinis for his energy, enthusiasm and assistance in coming up with the idea for this book, the basic outline and detailed input on many chapters of this book, particularly Chapters 1–7.

We thank Judith Binder for her tireless work throughout the preparation of this manuscript. She more than anyone is responsible for pulling the details together into a useful form. In addition we thank Nena Davis and Kelley Burns for various editorial contributions.

Finally we thank our families for their love and patience during the time this work has taken. Toivo thanks Cynthia, Aidan, and Maila for enduring the countless hours spent away from them. Mark thanks Tessa, Emily, and Harry.

# NOMENCLATURE

| | |
|---|---|
| $A$ | total particle surface area per volume of gas, $cm^2/cm^3$ |
| $a$ | surface area of agglomerates or particles, $cm^2$ |
| $A_A$ | Avogadro's number, number/mole |
| $a'$ | interatomic distance, cm |
| $a_a$ | water activity in solution droplet, dimensionless |
| $A_c$ | constant |
| $a_c$ and $a_{sol}$ | solute activity, dimensionless |
| $a_o$ | monomer surface area, $cm^2$ |
| $A_r$ | rate constant |
| $A_s$ | substrate surface area or heated area over which deposition takes place, $cm^2$ |
| $a_s$ | surface area of sphere, $cm^2$ |
| $b$ | number of moles of solid required for reaction with one mole of gaseous reactant |
| $B_1$, $B_2$, $B_3$, etc. | constants |
| $b_{th}$ | deposition efficiency, dimensionless |
| $C$ | concentration, $moles/cm^3$ or $molecules/cm^3$ |
| $\mathbf{c}$ | velocity vector, cm/sec |
| $C_{A_g}$ | concentration of species A in gas, $moles/cm^3$ |
| $C_b$ | percentage of atoms in grain boundary, dimensionless |
| $C_c$ | Cunningham correction factor, dimensionless |
| $C_D$ | drag coefficient, dimensionless |
| $C_L$ | concentration of reactant in liquid, $g/cm^3$ or $moles/cm^3$ |
| $C_o$ | concentration of solute at center of drop, $mole/cm^3$; or solubility over a flat surface, $g/cm^3$ |
| $C_p$ | heat capacity at constant pressure, erg/K mol |
| $C_{pa}$ | heat capacity of wet air, erg/K mol |
| $C_r$ | molar concentration of gaseous reactant, $moles/cm^3$ |
| $C_s$ | concentration at surface of particle or immediately above substrate, moles or $molecules/cm^3$ |
| $C_{sat}$ | saturation concentration, $moles/cm^3$ or $molecules/cm^3$ |
| $C_T$ | pre-exponential for temperature-dependent form of precursor vapor pressure, $moles/cm^3$ |
| $c_x$ | velocity in $x$-direction, cm/sec |

| | |
|---|---|
| $C_\infty$ | vapor concentration far from droplet, moles/cm$^3$ or molecules/cm$^3$ |
| $\bar{c}$ | mean molecular velocity, cm/sec |
| $D$ | particle diffusion coefficient or diffusivity of gas or vapor, cm$^2$/sec |
| $D_{cr}$ | effective diffusivity of vapor in crust, cm$^2$/sec |
| $D_e$ | effective diffusion coefficient in film or porous layer, cm$^2$/sec |
| $d_e$ | equivalent volume diameter, diameter of sphere corresponding to same volume as irregular particle, cm |
| $D_f$ | mass fractal dimension, dimensionless |
| $D_g$ | diffusion coefficient in gas phase, cm$^2$/sec |
| $d_g$ | average particle diameter for a lognormal distribution, cm |
| $d_{gm}$ | geometric mass average diameter, cm |
| $d_{gn}$ | geometric number average diameter, cm |
| $d_{gr}$ | average grain diameter, cm |
| $D_L$ or $D_l$ | liquid phase diffusion coefficient of solute, cm$^2$/sec |
| $D_o$ | diffusion coefficient at reference conditions, cm$^2$/sec |
| $d_o$ | diameter of primary particle, cm |
| $d_p$, $d_{p1}$, $d_{p2}$ | droplet or particle diameter, cm |
| $d_p^*$ | critical particle diameter, cm |
| $d_{pi}$, $d_{pj}$ | particle diameter, cm |
| $d_{pm}$ | mass average diameter, cm |
| $d_{pn}$ | number (count) average diameter, cm |
| $d_{po}$ | initial particle diameter, cm |
| $d_{pp}$ | average primary particle diameter, cm |
| $d_{ps}$ | equivalent sphere diameter of completely dried droplet, cm |
| $d_{p\bar{m}}$ | diameter of particle with average mass, cm |
| $D_s$ | surface fractal dimension, dimensionless |
| $d_s$ | Sauter diameter, cm |
| $d_t$ | tube or pipe diameter, cm |
| $D_v$ | vapor diffusion coefficient in air, cm$^2$/sec |
| $d_{\bar{m}}$ | mass mean diameter, cm |
| $E$ | electric field |
| $e$ | charge of an electron, statcoulomb |
| $E_I$ | efficiency of impaction, dimensionless |
| $E_A$ | activation energy, erg/mol K |
| $E_d$ | deposition efficiency, fraction of particles entering that deposit inside tube or other channel or body, dimensionless |
| $E_L$ | surface field strength for spontaneous emission of electrons or ions, statvolt/cm |
| $F$ | rate at which molecules enter gas phase by diffusion from single droplet, molecules/sec |
| $f$ | Fanning friction factor, dimensionless |

| | |
|---|---|
| $f_0$ | fraction of material $A$ in mixture, dimensionless |
| $F_1$, $F_2$ | Fuchs–Sutugin correction factor, dimensionless |
| $F_a$ | air molar flow rate, moles/sec |
| $F_E$ | electrophoretic force exerted on particle, dyne |
| $F_G$ | force of gravity, dyne |
| $F_i$ | force acting on a particle, dyne |
| $F_{i,j}$ | free energy, erg/cm$^3$ |
| $F_R$ | gas drag, dyne |
| $F_r$ | feed rate of reactant, moles/sec |
| $F_S$ | gravitational force exerted on particle, dyne |
| $f_t$ | transducer frequency, 1/sec |
| $F_{\text{th}}$ | thermophoretic force exerted on particle, dyne |
| $G$ | particle growth law, cm$^3$/sec |
| $g$ | acceleration due to gravity, cm/sec$^2$ |
| $G_c$ | constant, cm$^2$/sec |
| $H$ | Henry's constant |
| $h_g$ | heat transfer coefficient in gas surrounding particle, erg/cm$^2$ sec K |
| $H_L$ | latent heat of water evaporation, erg/mol |
| $H_r$ | heat of reaction, erg/mol |
| $h_s$ | heat transfer coefficient around droplet, erg/cm$^2$ sec K |
| $h_w$ | heat transfer coefficient at reactor wall, erg/cm$^2$ sec K |
| $I$ | nucleation rate or particle current in volume space, molecules/cm$^3$ sec |
| $i^*$ | number of monomer units in particle of critical size, dimensionless |
| $i$ | number of monomer units in particle, dimensionless |
| $J$ | particle flux, number/cm$^2$ sec |
| $J_m$ | molecular flux, moles/cm$^2$ sec |
| $K$ or $K_{\text{th}}$ | thermophoretic coefficient, dimensionless |
| $k$ | thermal conductivity, erg/cm sec K |
| $K'$ | volume compressibility coefficient |
| $k^*$ | number of monomers in critical-size particle, dimensionless |
| $k_1$ | first-order reaction rate constant, 1/sec |
| $k_B$ | Boltzmann's constant, erg/molecule, K |
| $K_C$ | coagulation coefficient in continuum regime, cm$^3$/sec |
| $K_{\text{eq}}$ | equilibrium constant, dimensionless |
| $K_F$ | coagulation coefficient in free molecule regime, cm$^{5/2}$/sec |
| $k_g$ | mass transfer coefficient, cm/sec |
| $k_H$ | Hall–Petch intensity parameter, units vary with definition [see Equation (9.14)] |
| $k_L$ | first-order reaction rate constant in particle, 1/sec |
| $k_m$ | vapor mass transfer coefficient for laminar tube flow |
| Kn | Knudsen number, dimensionless |
| $k_o$ | pre-exponential constant, cm/sec |
| $k_p$ | thermal conductivity of particle |

| | |
|---|---|
| $k_s$ | surface reaction rate constant, cm/sec |
| $K_T$ | turbulent coagulation coefficient, $cm^2$/sec |
| $L$ | dimension of system, cross section of flow, characteristic length, substrate diameter, tube length, thickness of coating, or thickness of concentration boundary layer, cm |
| $L_f$ | latent heat of fusion, erg/mol |
| $L_v$ | latent heat of vaporization, erg/g or erg/mol |
| $M$ | molecular weight, g/mole |
| $m'$ | exponent that depends on sintering mechanism, dimensionless |
| $M(d_p)$ | cumulative mass distribution function with respect to diameter, g/$cm^3$ |
| $m(d_p)$ | continuous mass distribution function with respect to diameter, g/$cm^4$ |
| $M_0, M_1, M_2$ | moments of particle size distribution [see Equation (2.16)] |
| $M_1$ | molecular weight, g/mole |
| $m_1$ | mass of a molecule or atom, g |
| $m_d$ | mass concentration of solvent at droplet surface, g/$cm^3$ |
| $m_g$ | mass concentration of solvent far from particle, g/$cm^3$ |
| $M_i$ | mass of particles within an interval, g |
| $M_i(d_{pi})$ | discrete mass size distribution [see Equation (2.6)] |
| $M_j$ | $j$th moment of distribution [see Equation (2.16)] |
| $M_m$ | atomic weight of depositing species, g/mole |
| $m_m$ | solution molality |
| $m_p$ | particle mass, g |
| $M_s$ | solute molecular weight, g/mole |
| $M_s'$ | saturation magnetization, emu/g |
| $M_\infty$ | total mass concentration, g/$cm^3$ |
| $\overline{m}$ | average mass of a particle, g |
| $N$, $N_1$, or $N_2$ | number concentration of aggregate, number/$cm^3$ |
| $N(d_p)$ | cumulative number distribution function with respect to diameter, number/$cm^3$ |
| $n(d_p)$ | continuous number distribution with respect to diameter, number/$cm^4$ |
| $n(s)$ | particle size distribution function with respect to surface area, number/$cm^5$ |
| $n(v)$ | particle size distribution with respect to volume, number/$cm^6$ |
| $n, n_1, n_g$ | atomic or molecular concentration, molecules/$cm^3$ |
| $n_{1s}$ | reactant concentration in gas phase at particle surface, molecules/$cm^3$ |
| $N_A$ | Avogadro's number, moles/molecule |
| $N_f$ | number of primary particles, dimensionless |
| $N_i$ | number of particles in an interval in an agglomerate, dimensionless |

| | |
|---|---|
| $n_i$ | particle concentration for discrete size $i$, number/cm$^3$ |
| $N_i(d_{pi})$ | number discrete size distribution [see Equation (2.6)] |
| $n_{\text{in}}$ | particle concentration entering bend, number/cm$^3$ |
| $n_l$ | lognormal distribution, number/cm$^4$ |
| $n_{\text{out}}$ | particle concentration exiting bend, number/cm$^3$ |
| $N_o$ | initial or inlet particle concentration, number/cm$^3$ |
| $n_p$ | number of primary particles in an aggregate, dimensionless |
| $n_s$ | saturation concentration of monomer, molecules/cm$^3$ |
| $\text{Nu}_M$ | Nusselt number for mass transport, dimensionless |
| $n_w$ | vapor concentration at reactor wall, molecules/cm$^3$ |
| $N_\infty$ | total particle or aggregate concentration, number/cm$^3$ |
| $p$ | gas pressure, dyne/cm$^2$ or cm Hg |
| $p_1$ | partial pressure of species 1, dyne/cm$^2$ |
| $p_{1s}$ | reactant partial pressure at particle surface, dyne/cm$^2$ |
| $P_d$ | aerosol penetration, dimensionless |
| $p_d$ | vapor or partial pressure at particle surface, dyne/cm$^2$ |
| $p_{\text{ds}}$ | solvent vapor pressure at droplet surface, dyne/cm$^2$ |
| Pe | Peclet number, dimensionless |
| $p_e$ | equilibrium vapor pressure over flat surface, dyne/cm$^2$ |
| $p_i$ | partial pressure of species $i$ corresponding to complete evaporation, dyne/cm$^2$ |
| $p_o$ | total pressure at reference conditions, dyne/cm$^2$ |
| Pr | Prandtl number, dimensionless |
| $p_r$ | reactant partial pressure outside concentration boundary, dyne/cm$^2$ |
| $p_s$ | vapor pressure at droplet surface at $T_s$, dyne/cm$^2$ |
| $p_{\text{sat}}$ | vapor pressure for pure solvent, dyne/cm$^2$ |
| $p_s^o$ | equilibrium water vapor pressure at $T_o$, dyne/cm$^2$ |
| $p_\infty$ | partial pressure of solvent far from droplet, dyne/cm$^2$ |
| $Q$ | aerosol or gas flow rate, cm$^3$/sec |
| $q$ | number of charges on a particle |
| $q_{\text{rms}}$ | rms charge on a particle |
| $r$ | radial coordinate, cm |
| $R$, $R_G$, $R_g$ | gas constant, cm$^3$ atm/mol K |
| $r_1$ or $r_2$ | monomer radius or radius of curvature, cm |
| $R_a$ | characteristic length of aggregate, cm |
| $R_c$ | collision radius, cm |
| Re | Reynolds number, dimensionless |
| $\text{Re}_j$ | jet Reynolds number, dimensionless |
| $\text{Re}_L$ | Reynolds number based on length $L$, dimensionless |
| $\text{Re}_p$ | Reynolds number based on particle diameter, dimensionless |
| $\text{Re}_t$ | tube Reynolds number, dimensionless |
| $r_i$, $r_j$ | particle radius, cm |
| $R_p$ | droplet or particle radius, cm |

| | |
|---|---|
| $r_p$ | radius of primary particle, cm |
| $R_r$ | reaction rate, molecules/cm$^3$ sec |
| $r_s$ | surface reaction rate, molecules/cm$^2$ sec |
| $R_t$ | tube radius, cm |
| $r_v$ | rate of reaction per unit volume, moles/cm$^3$ sec |
| $S$ | saturation ratio, dimensionless |
| $S$ | stopping distance, cm |
| $s$ | surface area of a particle, cm$^2$ |
| $S^*$ | critical saturation ratio, dimensionless |
| $S_1$ or $S_2$ | surface area, cm$^2$ |
| Sc | Schmidt number, dimensionless |
| Sh | Sherwood number, dimensionless |
| St | Stokes number, dimensionless |
| $T$ | temperature, K or °C |
| $t$ | time, sec |
| $T_c$ | Curie temperature, K |
| $T_d$ | droplet temperature, K or °C |
| $T_e$ | temperature at entrance or far from surface, K or °C |
| $T_g$ | gas temperature at entrance or far from surface, K or °C |
| $T_o$ | reference temperature for gas above substrate or inlet temperature, K |
| $T_p$ | temperature of particle, K or °C |
| $T_{pre}$ | preheater temperature, K or °C |
| $T_s$ | droplet temperature, K or °C |
| $T_w$ | temperature at surface or wall, K or °C |
| $T_\infty$ | temperature far from particle, K or °C |
| $U$ | average gas velocity, cm/sec |
| $u$ | particle velocity, cm/sec |
| $U_o$ or $u_e$ | velocity of gas far from surface (free-stream velocity), cm/sec |
| $V$ | total aerosol volume, cm$^3$/cm$^3$ |
| $V_1$ and $V_2$ | volumes of particles and coating on particle, respectively, cm$^3$ |
| $v$ and $v'$ | particle volume, cm$^3$ |
| $\mathbf{v}$ | gas velocity vector, cm/sec |
| $v^*$ | critical particle size, cm |
| $v_0$ | monomer volume, cm$^3$ |
| $v_1$ | molecular volume, cm$^3$ |
| $v_a$ | atomic volume, cm$^3$ |
| $V_{dif}$ | diffusion velocity, cm/sec |
| $V_e$ | electrical migration velocity, cm/sec |
| $v_g$ | geometric mean volume, cm$^3$ |
| $v_i$ and $v_j$ | particle volume, cm$^3$ |
| $V_M$ | particle deposition velocity, cm$^3$ |

| | |
|---|---|
| $v_m$ | volume of deposited species, $cm^3$ |
| $v_p$ | volume of a primary particle, $cm^3$ |
| $V_s$ | sedimentation settling velocity, cm |
| $V_t$ | total or terminal velocity, cm/sec |
| $V_{th}$ | thermophoretic velocity, cm/sec |
| $v_{tr}$ | thermophoretic velocity in radial direction, cm/sec |
| $v_{tz}$ | thermophoretic velocity in $z$ direction, cm/sec |
| $v_x$ | gas velocity in $x$ direction, cm/sec |
| $\bar{v}$ | average particle volume, $cm^3$ |
| $\bar{v}$ | variable of integration, $cm^3$ |
| $W$ | spread of the particle size distribution, dimensionless |
| $W_{ij}$ | factor for coagulation, dimensionless |
| $X$ | residence time, dimensionless |
| $x$ | reactor axial coordinate, cm |
| $x_A$ | mole fraction in liquid, dimensionless |
| $x_{rms}$ | root mean square diffusion distance, cm |
| $x_s$ | mole fraction of solvent, dimensionless |
| $y_i$ | concentrations of particles of size $i$, dimensionless |
| $Y_k$ | concentrations of particles in sections $k$, dimensionless |
| $y_w$ | water vapor mole fraction in air, dimensionless |
| $Z$ | number of charges per particle, dimensionless |
| $\bar{Z}$ | average number of charges per particle, dimensionless |
| $z$ | coordinate in axial direction, cm |
| $z_i$ | number of charges on particle, dimensionless |
| $Z_{s,\max}$ | maximum number of acquired charges, dimensionless |

## GREEK

| | |
|---|---|
| $\alpha$ | thermophoretic coefficient, dimensionless |
| $\alpha_c$ | ratio of characteristic times for collision and fusion, dimensionless |
| $\alpha_g$ | thermal diffusivity of gas, $cm^2/sec$ |
| $\alpha_p$ | thermal diffusivity of particle, $cm^2/sec$ |
| $\alpha_s$ | sticking coefficient, dimensionless |
| $\beta_C$ | collision frequency in continuum regime, $cm^3/sec$ |
| $\beta_{Ch}$ | collision frequency function for charged particles, $cm^3/sec$ |
| $\beta_F$ | collision frequency in free-molecule regime, $cm^3/sec$ |
| $\beta_{FA}$ | coagulation frequency function of agglomerates, $cm^3/sec$ |
| $\beta_{i,j}$ | collision frequency function, $cm^3/sec$ |
| $\beta_T$ | collision frequency for turbulent coagulation, $cm^3/sec$ |
| $\beta(v, v')$ | collision frequency of particles of size $v$ and $v'$, $cm^3/sec$ |
| $\gamma$ | surface tension, $erg/cm^2$ |
| $\gamma_A$ | activity coefficient, dimensionless |

| | |
|---|---|
| $\gamma_{i,j}$ | dimensionless function |
| $\gamma_s$ | activity coefficient of solution, dimensionless |
| $\gamma_{sol}$ and $\gamma_c$ | activity coefficient of solute, dimensionless |
| $\Delta H_{vap}$ | enthalpy of evaporation of precursor, erg/mole K |
| $\delta$ | delta function, dimensionless |
| $\delta_c$ | crust thickness, cm |
| $\varepsilon$ | dielectric constant, dimensionless |
| $\varepsilon_c$ | rate of gas molecule collisions, 1/sec |
| $\varepsilon_d$ | rate of energy dissipation, $cm^2/sec^2$ |
| $\epsilon$ | total number of collisions with particle surface per unit time, number/sec |
| $\epsilon_p$ | prefactor, dimensionless |
| $\eta$ | particle volume, dimensionless |
| $\eta_r$ | size of smallest eddies, cm |
| $\theta$ | reaction time, dimensionless |
| $\theta_1$ | characteristic time for reaction, dimensionless |
| $\theta_2$ | numerical coefficient, dimensionless |
| $\theta^*$ | temperature, dimensionless |
| $\kappa$ | mass transfer coefficient (or deposition velocity) by turbulent diffusion, cm/sec |
| $\lambda$ | gas mean free path, cm |
| $\lambda_p$ | heat of phase transition, $erg/cm^3$ |
| $\mu$ | gas viscosity, g/cm sec |
| $\nu$ | kinematic viscosity of gas, $cm^2/sec$ |
| $\rho$ | density, $g/cm^3$ |
| $\rho_1$ | droplet density, $g/cm^3$ |
| $\rho_b$ | molar density, $mole/cm^3$ |
| $\rho_f$ | gas density, $g/cm^3$ |
| $\rho_L$ | density of pure water, $g/cm^3$ |
| $\rho_{liq}$ | density of liquid, $g/cm^3$ |
| $\rho_p$ | particle density, $g/cm^3$ |
| $\rho_s$ | solute density, $g/cm^3$ |
| $\rho_{sol}$ | density of solid, $g/cm^3$ |
| $\sigma$ | surface tension, $erg/cm^2$ |
| $\sigma_d$ | hard-sphere molecular diameter, cm |
| $\sigma_g$ or $\sigma_{gs}$ | geometric standard deviation of size distribution, dimensionless |
| $\sigma_s$ | surface energy of $TiO_2$, dyne/cm or $erg/cm^2$ |
| $\sigma_{ys}$ | yield strength, $dyne/cm^2$ |
| $\tau_{as}$ | characteristic time for reaction in particle when ash diffusion controls, sec |
| $\tau_C$ | characteristic time for particle collisions, sec |
| $\tau_{CC}$ | characteristic time for coagulation in continuum regime, sec |

| | |
|---|---|
| $\tau_{CF}$ | characteristic time for coagulation in free-molecule regime, sec |
| $\tau_D$ | characteristic time for diffusion in particles, sec |
| $\tau_{dg}$ | characteristic time for concentration profile to achieve steady state, sec |
| $\tau_e$ | evaporation time, sec |
| $\tau_f$ | characteristic time for sintering or fusion, sec |
| $\tau_{fs}$ | coagulation parameter, dimensionless |
| $\tau_{GC}$ | characteristic time for particle growth in continuum regime, sec |
| $\tau_{GF}$ | characteristic time for particle growth in free-molecule regime, sec |
| $\tau_{gs}$ | characteristic time for reaction in particle when gas phase transport controls, sec |
| $\tau_{GSR}$ | characteristic time for particle growth by surface reaction, sec |
| $\tau_{GV}$ | characteristic time for particle growth by volume reaction, sec |
| $\tau_L$ | characteristic time for diffusion in coating, sec |
| $\tau_p$ | characteristic time for particle formation, sec |
| $\tau_{pd}$ | time between particle collisions with same region on surface, sec |
| $\tau_r$ | reactor residence time, sec |
| $\tau_{rs}$ | characteristic time for reaction in particle when chemical reaction controls, sec |
| $\tau_S$ | time needed for an aerosol to reach a self-preserving distribution, sec |
| $\tau_s$ | settling time, sec |
| $\tau_{sat}$ | characteristic time to saturate gas with vapor from evaporating droplets, sec |
| $\tau_{SC}$ | time needed to reach self-preserving distribution in continuum regime, sec |
| $\tau_{SF}$ | time needed to reach self-preserving distribution in free-molecule regime, sec |
| $\tau_{Tg}$ | characteristic time for heat transfer in gas phase around a particle, sec |
| $\tau_{th}$ | time for particle transport by thermophoresis, sec |
| $\tau_{Tp}$ | characteristic time for heat transfer in solid particle or droplet, sec |
| $\phi$ | water osmotic coefficient |
| $\chi$ | dynamic shape factor, dimensionless |
| $\psi$ | normalized charge $= q/q_{rms}$ |
| $\psi(\eta)$ | self-preserving size distribution, dimensionless |

## ACRONYMS

| | |
|---|---|
| AACVD | aerosol-assisted chemical vapor deposition |
| AES | Auger electron spectroscopy |
| BET | Brauner–Emmett–Teller |
| CSTAR | continuous stirred tank aerosol reactor |
| CVD | chemical vapor deposition |
| DTA | differential thermal analysis |
| FTIR | Fourier transform infrared |
| GDE | general dynamic equation |
| LFAR | laminar-flow aerosol reactor |
| MLCC | multilayer ceramic capacitor |
| NMR | nuclear magnetic resonance |
| PCB | printed circuit board |
| PE | plasma enhanced |
| PFAR | plug-flow aerosol reactor |
| PSD | particle size distribution |
| RF | radio frequency |
| SEM | scanning electron microscope |
| SPSD | self-preserving size distribution, dimensionless |
| SSA | specific surface area |
| TEM | transmission electron microscope |
| TGA | thermogravimetric analysis |
| TMEDA | tetramethylethylenediamine |
| UHV | ultrahigh vacuum |

# AEROSOL PROCESSING OF MATERIALS

# 1 Introduction

## 1.1 BACKGROUND

This text describes the science, engineering, and technology involved in the synthesis and processing of powders and films by gas-phase, or aerosol, techniques. These routes to producing materials exploit aerosols—solid or liquid particles in gases—to make films and powders with special characteristics (Fig. 1.1). In what follows, *aerosols* are considered to be solid or liquid particles in a gas, where the particle sizes range from molecular dimensions to roughly 100 micrometers. This definition is somewhat different from that used in other books.[1–8] An outgrowth of aerosol and materials research, the discipline of aerosol processing of materials is quite young.

As shown in Figure 1.1, the expression *aerosol process* denotes systems in which the presence of particles in the gas phase is an integral part of the materials synthesis or processing scheme. Included in this definition are many types of systems in which (1) aerosols are formed in the gas phase by chemical or physical processes; (2) particles or droplets suspended in gases are deposited onto surfaces, in some cases simultaneously with vapor, to form films; (3) particles are directly ejected from surfaces into a gas; and (4) volatile species evaporate from an aerosol and carry out either chemical vapor deposition or etching of films.

Historically, aerosol science has focused on the harmful effects of aerosols, e.g., the effects of particles formed by combustion of coal and other fossil fuels in power plants, burning of gasoline in automobiles, and photochemical reactions of hydrocarbons and other molecules in the atmosphere, as well as dust suspension from the ground and other routes that can degrade air quality and visibility. In addition, toxic aerosols are formed in many manufacturing processes, such as welding. At the same time, vast sums of money have been spent to eliminate asbestos from buildings without exposing workers to the potential health hazards associated with the inhalation of asbestos aerosols created during removal of the mineral. The destruction of large quantities of grain and loss of life and property are sometimes caused by dust explosions in grain elevators. Contamination of semiconductor wafers in clean rooms by aerosol particle deposition is a major contributor to reduced yields during the manufacture of microelectronic components, which leads to higher prices for goods incorporating these items. Human-derived aerosols, the result of sneezing and coughing, are the main mode of transfer of the

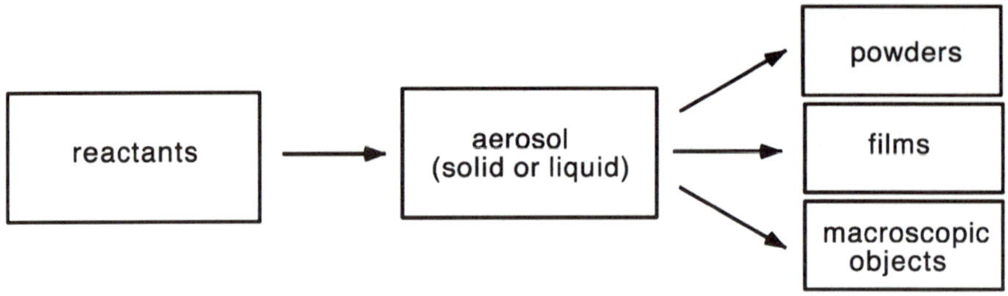


Fig. 1.1. Aerosol routes to powders and films utilize aerosols as an integral part of the process.

viruses and bacteria that cause many diseases. Nuclear reactor accidents such as the one in Chernobyl, Ukraine, form radioactive aerosols. Nature also contributes undesirable aerosols from volcanic eruptions or the impact of meteors, and such aerosols have been proposed to have played a major role in the elimination of many species, including dinosaurs. Models for global warming incorporate the role of aerosols to predict future global climate changes. Seemingly innocuous sources of indoor air pollution in the form of aerosols emitted from burning candles and fireplaces are also of major concern. Of scientific curiosity is interstellar dust, which can be considered to be made up of particles in a very low pressure gas.

Beneficial aerosols have also been studied, although to a lesser extent. In addition to introducing compounds into instruments such as mass spectrometers, other examples include humidifying air, delivering drugs by inhalation,[9] and spraying pesticides on crops, as well as a variety of household applications in which sprays are used to deliver materials to surfaces (paints, cleaners, disinfectants, deodorants, etc.).

Until recently, however, only a few aerosol routes for the formation of materials were developed for use in industrial practices, even though virtually all types of materials have been generated in the laboratory, including carbon (diamond, diamondlike carbon, fullerenes, graphitic carbon), other elemental materials (metals, semiconductors), nonoxide ceramics, metal oxides, organics, polymers, and composites. These industrial routes were developed long before the basic science and engineering of aerosols were understood. Examples are the formation of carbon black particles by hydrocarbon combustion for applications such as pigments and reinforcing agents in materials for car tires; the generation of titania pigment by the reaction of titanium tetrachloride ($TiCl_4$) and small amounts of aluminum trichloride ($AlCl_3$) with oxygen ($O_2$) for use in paints, plastics, and other materials (Fig. 1.2); and the formation of fumed silica and titania by the chemical reaction of silicon tetrachloride ($SiCl_4$) and titanium tetrachloride ($TiCl_4$) in flames for a variety of applications, such as free flow agents in other powdered materials and photocatalysis. Optical fibers are manufactured by aerosol routes in

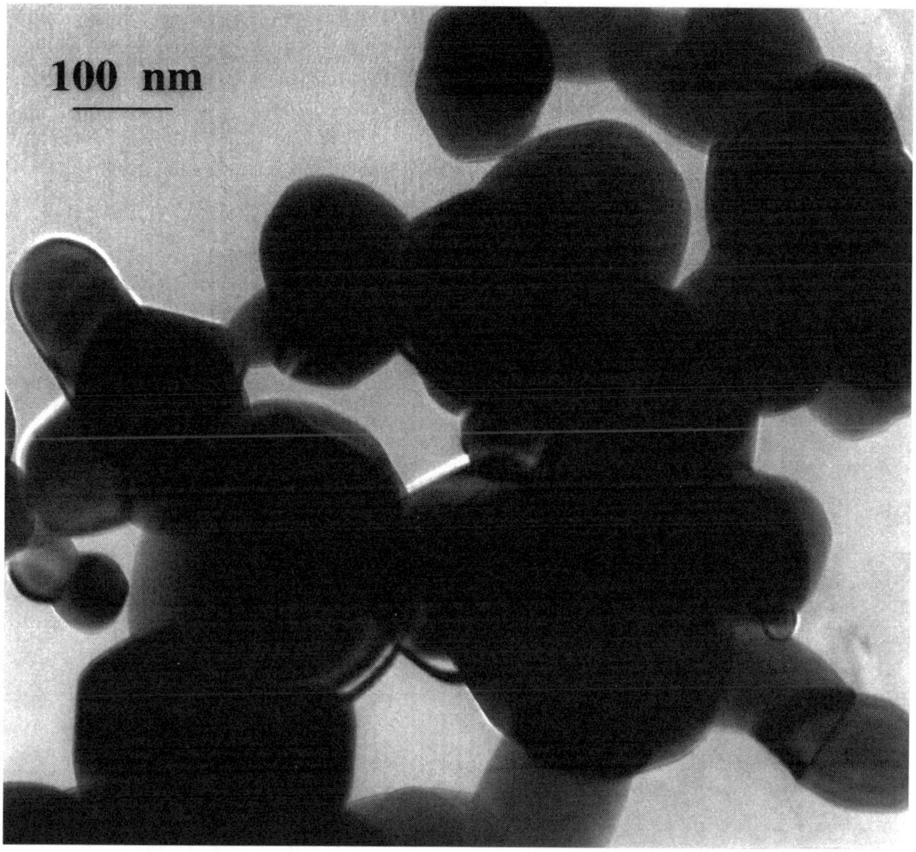


**Fig. 1.2.** Titanium dioxide pigment particles produced by gas-to-particle conversion with $TiCl_4$ and $O_2$ as reactants. (Courtesy of Kemira Pigments.)

which particles are formed in the gas phase and then deposited to form rods that are eventually heated and drawn out to make fibers.

These industrial approaches have been used for only a very limited number of materials, such as simple metal oxides like $SiO_2$, $TiO_2$, and $Al_2O_3$. Thus, tremendous opportunities exist to generate powders and films of electroceramics (piezoelectrics, dielectrics, etc., used in varistors, resistors, and electronics in general), optical materials (phosphors, electrochromics, and reflectors used in flat panel displays and windows), precious metals (Ag, Pd, Au, and Pt used to fabricate metal features on circuit boards),[10] and numerous other materials.[11-14]

Several new aerosol methods for the generation of materials have been commercialized recently. These routes were developed to meet the need for producing particles and films with controlled composition and particles with a controlled size distribution and other properties that cannot be obtained by alternative routes. One example utilizing the conversion of gaseous species to particles is the formation of nanophase particles (materials in which the particle or grain size is in the nanometer range) by sputtering or evaporation and condensation of metals, metal oxides, and other materials (Fig. 1.3). Ultrafine titanium dioxide produced by this route is now available commercially by the ton and has applications in sunscreens and cosmetics. Another driving force for examining nanophase materials is the possibility of superplastic processing of these powders (Fig. 1.4). In this technique, the small particle size allows the material to be deformed much like a metal, dramatically simplifying the processing of ceramic materials.

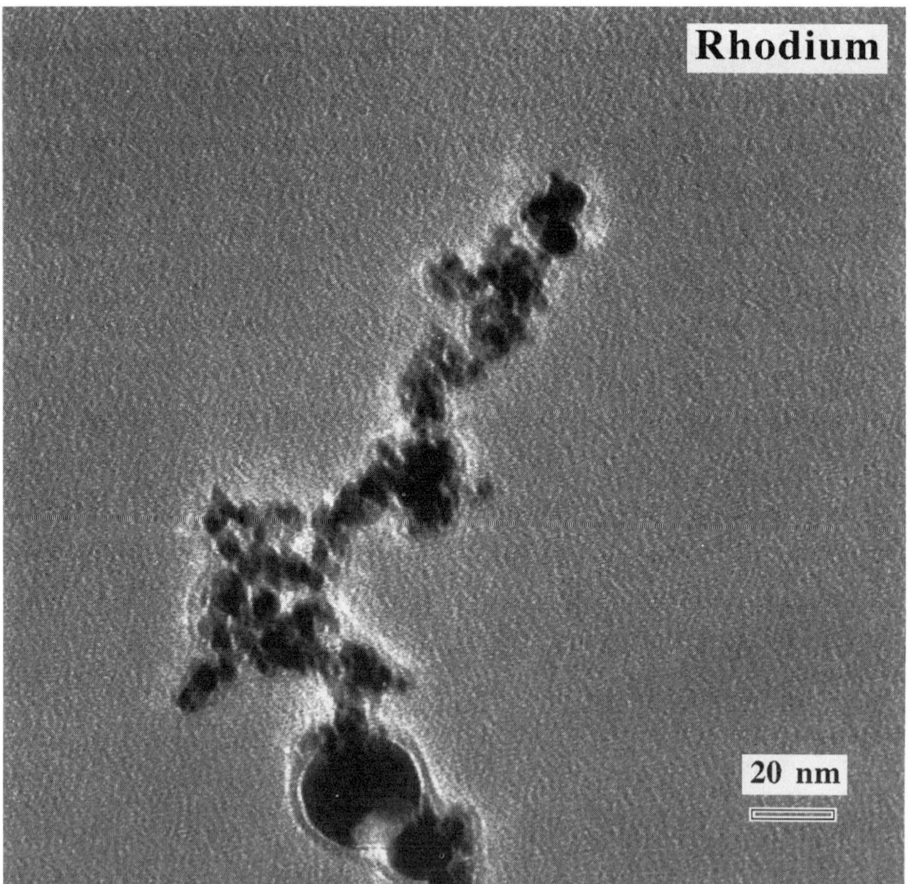


**Fig. 1.3.** Nanophase particles produced by laser ablation. (Courtesy of S. Jain, T.T. Kodas.)

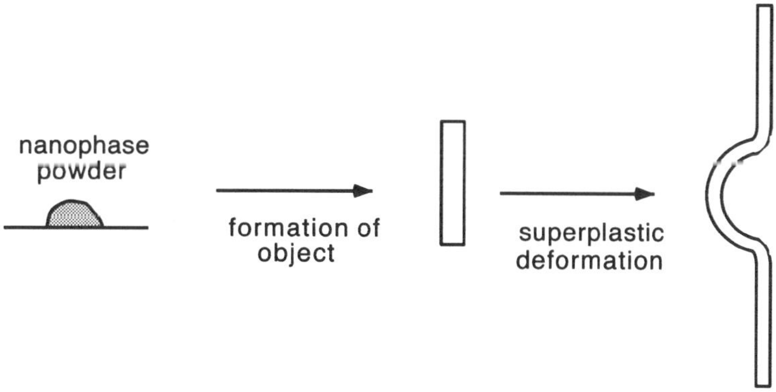


**Fig. 1.4.** Superplastic deformation exploits nanophase powder to allow the formation of objects without sintering.

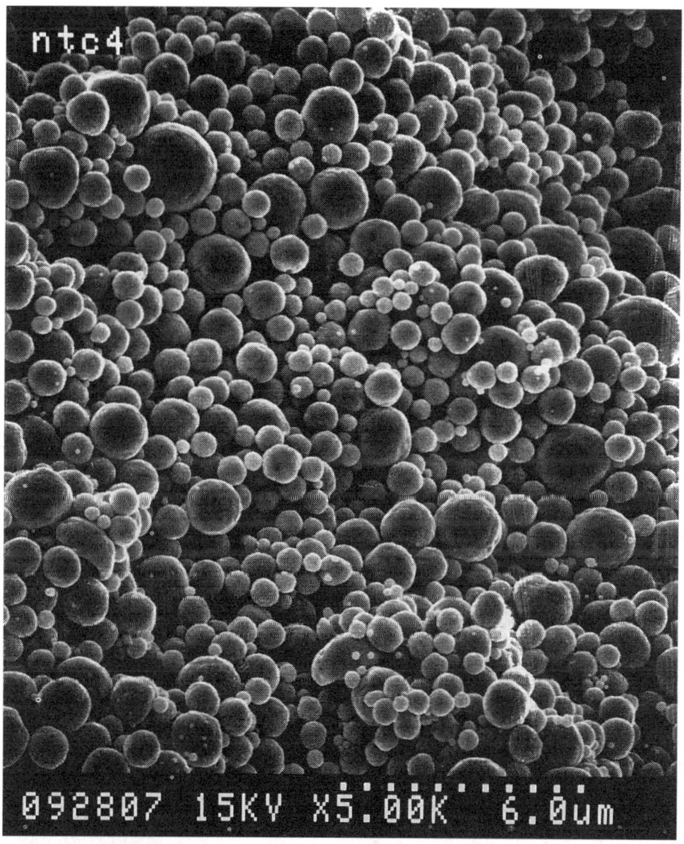


**Fig. 1.5.** Palladium particles produced by spray pyrolysis for application in thick-film technology for microelectronics. (Courtesy of Q. Powell, Superior Micro Powders, LLC.)

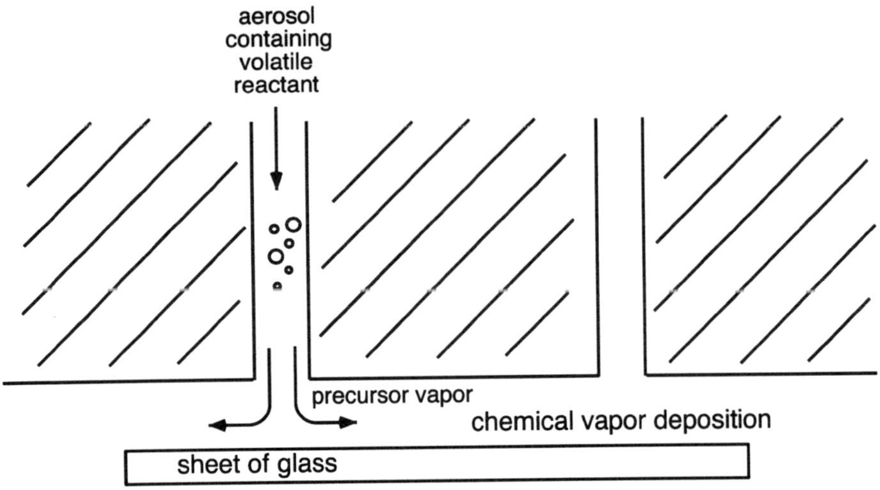


**Fig. 1.6.** Aerosol delivery of volatile reactants allows the coating of glass panels by chemical vapor deposition.

An example exploiting the conversion of sprayed solutions into particles is the formation of precious metal and metal-oxide powders for use in pastes to make features on circuit boards and electrodes in capacitors (Fig. 1.5). Powders of various precious metals and metal oxides, including superconducting ceramics, produced by this route are now sold commercially by several companies. An example of film formation is the fabrication of coatings on large panes of glass, which is carried out by aerosol delivery of volatile reactants to regions near surfaces, followed by evaporation and chemical vapor deposition (CVD) of the volatile species on the surface (Fig. 1.6). Film formation is also carried out via the deposition of droplets on surfaces. This process is used commercially on a massive scale for coating glass for various applications, where the coatings may provide desirable optical properties such as colors and ultraviolet (UV) and infrared (IR) reflectance. Solar cells have also been produced by spray processes. Aerosols are now playing an increasingly critical role in fundamental research that utilizes clusters (small numbers of particles, often 1 to 100 atoms) to understand the basic physics and chemistry of materials. Other recent examples of fundamental advances in science and engineering involving aerosols are the production of various new forms of carbon ($C_{60}$ and related materials) and diamond particles by high-temperature aerosol processes.

The preceeding examples illustrate the development of new aerosol processes by industry to meet the demands of the marketplace. And in recent years, great progress has also been achieved in understanding and describing the operation of established industrial processes for manufacturing $TiO_2$ pigment, fumed $SiO_2$, carbon black, and optical fibers that rely on aerosol formation and processing of particles.

By contrast, aerosols may be undesirable during the synthesis and processing of some materials. Examples are particle formation during the coating of glass windows by CVD and particle formation during CVD to form films for microelectronics applications. In these approaches, the particles may deposit on the substrate, thereby degrading optical properties or causing defects, or they may actually clog the tubing used in processing and lead to labor-intensive cleaning procedures. In both cases, an understanding of particle formation and deposition is required to develop strategies to minimize or eliminate the problems, and industry is moving in this direction.

Finally, the basic science and engineering of aerosol approaches in synthesizing and processing materials has been studied extensively in recent years, both theoretically and experimentally, and has matured into a self-contained discipline. The major classes of systems for powder and film formation have been demonstrated, and the theory in most cases is now well developed.

## 1.2 DEFINITIONS

Historically, the term *aerosol* has been used to describe small particles (solid or liquid) suspended in a gas. The term is analogous to the term *hydrosol* (or simply *sol*) used to describe particles suspended in a liquid, most commonly water. For the purposes of this book, particle sizes range from molecular clusters with dimensions of several angstroms up to roughly 100 microns. This upper limit is higher than that conventionally used for aerosols in atmospheric processes. The reason is that, in atmospheric processes, a 100-micron droplet would rapidly settle out of the air relative to natural time scales, which are usually hours or days. However, most aerosol routes for the synthesis and processing of materials involve time scales of a few seconds or less. Over these time scales, even particles that are 100 $\mu$m in diameter remain in the gas phase and can be considered aerosols in the sense that standard concepts and equations for aerosols are applicable. The lower size limit for which a cluster or particle can be considered an aerosol is not strictly defined, but even molecular-scale clusters can be considered aerosols.

A large number of expressions describe particles suspended in a gas, and the use of these expressions varies somewhat, depending on the field. In this volume, an *aerosol* is a gas containing particles (typically, the particle volume fraction does not exceed 0.01%) that can be either solid or liquid (or both). Several expressions are used to describe aerosols consisting of solid particles. We will refer to them simply as *solid particles*, which can have a variety of shapes and are not necessarily spherical. *Powder* is used to describe a group of particles after collection from the gas phase. *Mist* refers to a cloud of droplets and is most often employed to describe the aerosol formed by atomization of a liquid. The term *droplet* is synonymous with *liquid particle*.

The term *cluster* has a variety of uses in the scientific literature, depending on the context and discipline. For the purposes of this book, a *cluster* refers to particles smaller than 10 nm that have an appreciable fraction ($> 10\%$) of surface atoms or molecules. A cluster may consist of atoms or molecules and may be either a solid or a liquid. The term *cluster* is applied to particles that are both stable and unstable with respect to evaporation, sublimation, or other processes that result in the removal of atoms from a surface. Clusters that are stable in this sense are often called *supercritical clusters*, while unstable ones are called *subcritical clusters*.[1, 8] It is worthwhile to note that the term *cluster* has another meaning in chemistry as well: It is also used to describe molecules containing two or more metal or semiconductor atoms supported by metal–metal bonds and the necessary ligands.[14]

Several other words are used in the literature to describe aerosols. *Smoke* refers to aerosols consisting of residual materials resulting from combustion processes or from materials formed by condensation processes. *Fume* is a solid aerosol that results from the condensation of vapors, usually in high-temperature processes such as those involving flames. *Haze* refers to aerosols with sufficiently high atmospheric concentrations to result in a degradation of visibility. *Smog* refers to the aerosols observed in urban areas caused by airborne pollutants. *Fog* is a visible mist and is usually used to describe natural atmospheric aerosols. *Dust* is a suspension of particles produced by the mechanical disintegration of material that is then suspended in a gas.

Aerosols can be formed as the result of a wide variety of processes. In this text, it will be advantageous to classify these as physical or chemical processes. Mechanical processes are considered physical. *Physical routes* form particles by processes such as condensation of a hot vapor during cooling; the composition of the material is the same before and after processing, but the form of the material is changed from bulk to particles. Similarly, the ejection of particles from a surface can be considered a physical process. *Chemical processes* form particles as the result of chemical reactions; the composition of the precursor is different from that of the final product. Chemical reactions can occur in the gas phase, on surfaces, or inside particles (as in the solid state).

## 1.3 IMPORTANT PHYSICOCHEMICAL PHENOMENA

Aerosol methods for generating powder can involve a variety of fundamental physicochemical phenomena (Fig. 1.7), described in Chapters 1 through 8:

- Gas-phase chemical reactions
- Single-particle chemical processes such as surface and intraparticle chemical reactions
- Single-particle physical processes such as nucleation, evaporation, and condensation
- Single-particle physical processes such as solid-state diffusion and phase transformations

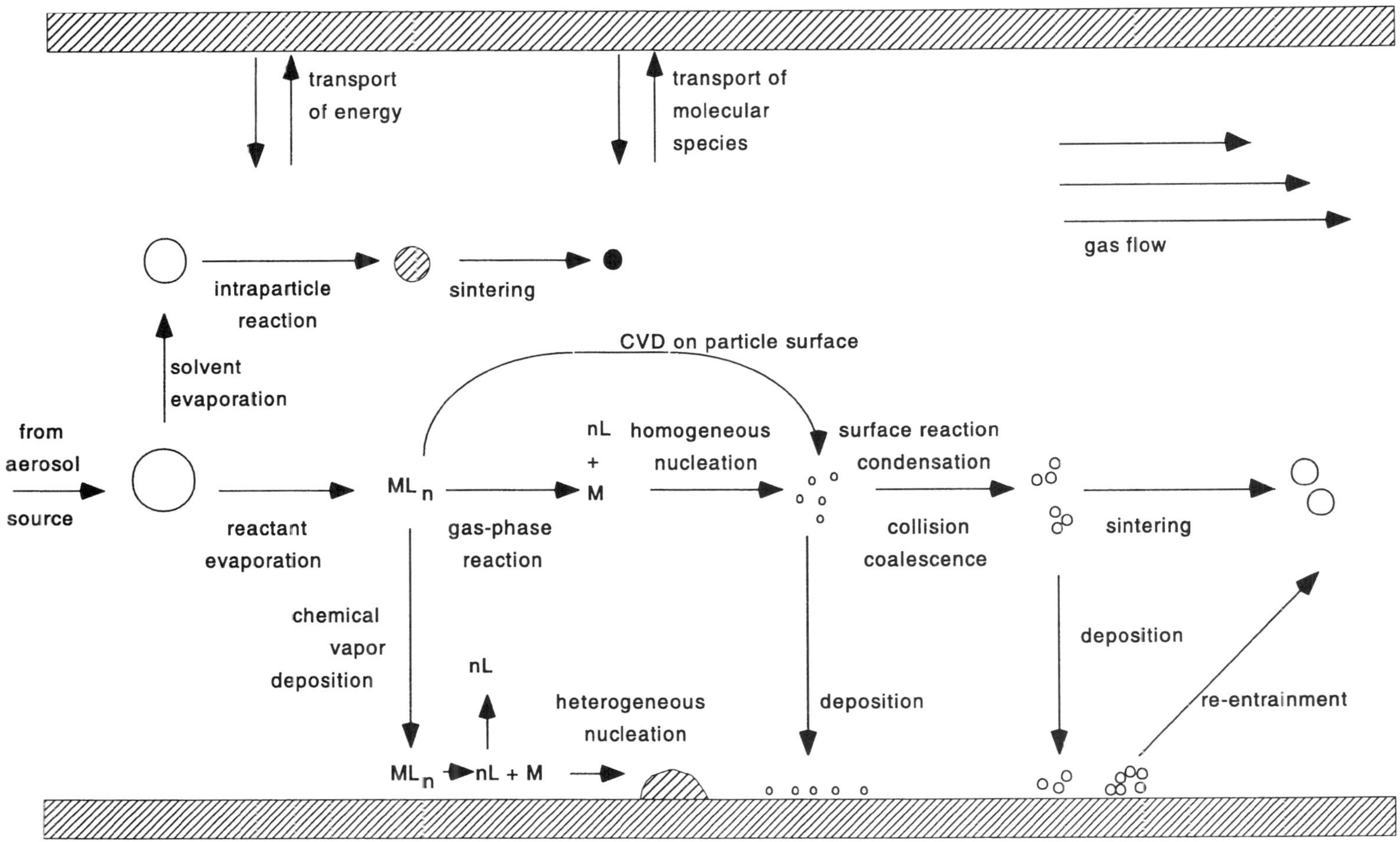


**Fig. 1.7.** Physicochemical phenomena occurring during aerosol synthesis of materials.

- Multiparticle physical processes such as collision, coalescence, and agglomeration
- Transport and deposition of reactant and product vapors and particles by a variety of mechanisms
- Reactor-scale mass transport, heat transport, and fluid flow leading to nonuniform temperature, mass, and velocity fields
- Effects associated with plasmas, photons, electric fields, and supercritical fluids
- Particle ejection or entrainment from surfaces

The description of these phenomena requires combining information about gas-phase temperatures, velocities, and species concentrations with the aerosol aspects of the problem. For example, the sintering rate of particles depends strongly on temperature, which usually varies dramatically with position in a system; particle formation rates depend very strongly on the gas-phase concentration of the nucleating species; and velocities in a system determine the amount of time particles spend in the system. This overall integration of physical and chemical phenomena has been carried out in a highly effective manner for CVD processes,[15, 16] but is more difficult for aerosol reactions due to the complexity of describing particle formation and growth problems, often in addition to simultaneous CVD.

The relative importance of the preceding phenomena depends on the mode of aerosol reactions. One of the simplest cases, *intraparticle reaction*, *aerosol decomposition*, or *spray pyrolysis*, can sometimes involve only intraparticle processes for particles in nonuniform temperature and velocity fields. In such cases, multiparticle processes such as collision, coalescence, and agglomeration often do not have to be considered. *Gas-to-particle conversion* processes can, however, involve many of the aforementioned phenomena. Similarly, intraparticle reaction processes with one or more volatile species can involve many of these phenomena. Because of the variety of processes that exist, a complete description of aerosol methods for film and particle generation can be given only by an interdisciplinary approach.

In recent years, a large number of theoretical studies have examined the behavior of aerosol processes for film and particle generation. Models for single-component systems incorporating gas-phase reactions, particle formation and growth, nonuniform velocity profiles, and diffusion of gaseous reactants, as well as products in flow systems, have been developed. The complexities of coagulation and transport of particles have been added to these models, and the aerosol dynamics in a variety of different types of reactors has been examined. Detailed studies of the roles of particle nucleation, growth, and coagulation in determining particle characteristics have provided a basis for understanding aerosol dynamics in a variety of systems for powder and film generation.

**TABLE 1.1. Aerosol processes used for powder and film generation**

| Process | Films | Powders |
|---|---|---|
| Gas-to-Particle Conversion (physically induced) | Ionized and neutral cluster beam; Supercritical fluid expansion; Nozzle expansion | |
| Gas-to-Particle Conversion (chemically induced) | Laser evaporation/deposition | Laser-driven reactors |
| | Hot-wall reactors | Hot-wall reactors |
| | Plasma spraying | Plasma reactors |
| | Flame spraying | Flame reactors |
| Intraparticle Reaction | Droplet or particle deposition | Spray pyrolysis |
| Hybrid Processes | Aerosol-assisted CVD | Condensation followed by intraparticle reaction |
| Direct Ejection | Laser ablation; Sputtering | |

## 1.4 TYPES OF AEROSOL PROCESSES

A variety of technologies for generating powders and films are discussed in three main sections of this book: *gas-to-particle conversion*, *intraparticle reaction* (also called *liquid/solid-to-solid conversion* or *particle-to-particle conversion*), *and films*. Table 1.1 lists the major aerosol processes that have been used for generating particles and films. These aerosol routes to materials are now well-developed, and the processes can be grouped according to how the final product is produced: gas-to-particle conversion, intraparticle reaction, and bulk-to-particle direct ejection or entrainment processes.

In *gas-to-particle conversion* processes (Fig. 1.8), a supersaturated vapor of a condensable gaseous species is formed either as the result of chemical reactions that create new species or as the result of physical processes such as cooling that reduce the vapor pressure of condensable species. A *supersaturated vapor* consists of atoms or molecules of given species that are present at a partial pressure higher than the vapor pressure of that species. At sufficiently high supersaturation, new particles form by *homogeneous nucleation*, defined as the formation of a new phase, either solid or liquid, without the aid of any foreign surfaces. Following nucleation, particle growth can occur by physical condensation of vapor onto particles, by chemical reaction on the particle surface, or by diffusion of the reactant into a particle, followed by reaction within the bulk of the particle. Collisions occur at high particle concentrations. After collision, *coalescence* (the fusing of two or

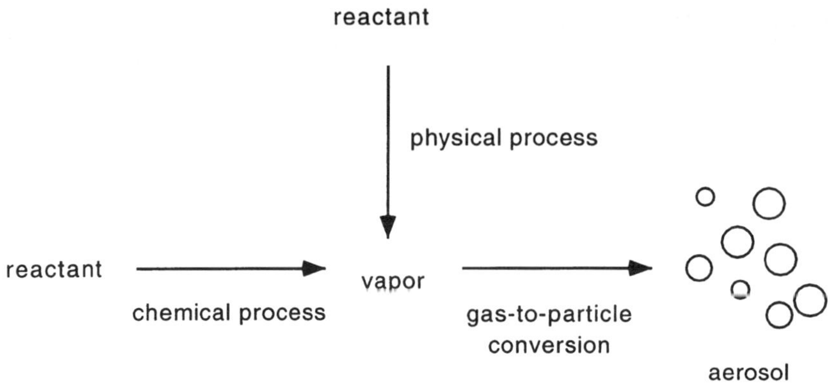


**Fig. 1.8.** Gas-to-particle conversion processes rely on the formation of a vapor with a low vapor pressure that nucleates to form particles.

more particles) can occur to form a single spherical particle. Coalescence is often nearly instantaneous for liquids (except for highly viscous materials) and may be slow or fast for solids, depending on the conditions, process, and material properties associated with the solids. Coalescence normally takes place at high temperatures. Finally, *agglomeration* results from collisions between particles to form collections of particles that do not completely coalesce (and thus, that are not spherical). Gas-to-particle conversion processes can be coupled with particle deposition from the gas phase to provide methods for fabricating films. Particle deposition from the gas phase onto surfaces can take place by mechanisms that include impaction, thermophoresis, electrophoresis, and diffusion. These phenomena are all discussed in Chapters 1 through 8.

In *intraparticle reaction* processes, some of which are called *spray pyrolysis*, *evaporative decomposition*, *spray roasting*, *spray calcination*, or *aerosol decomposition*, particles containing precursors are formed mechanically using a droplet or particle generator (Fig. 1.9). The precursors in each solid particle

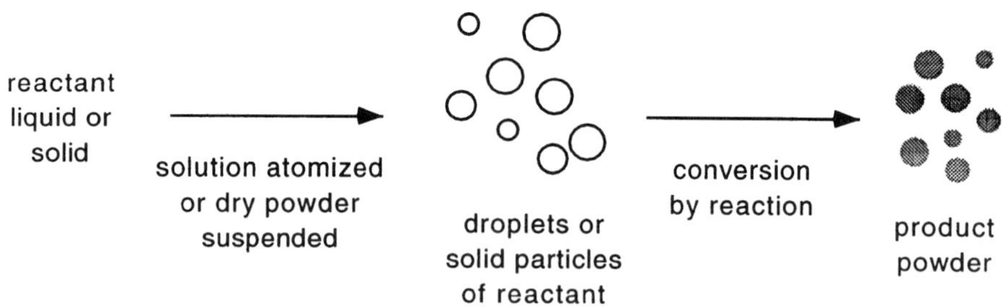


**Fig. 1.9.** Intraparticle reaction routes rely on chemical reactions and/or phase changes in particles to form the product.

or droplet are then reacted with a gas or pyrolyzed in the particle to form the product. This can be accomplished using a hot-wall reactor, a flame, or another energy source that heats the particles. In some cases, no reaction occurs, and the droplets are just dried to form a product (spray drying).

*Spray pyrolysis* also can be used for film generation (Fig. 1.10). Instead of the droplets drying and reacting while they are suspended in a gas, the liquid droplets are deposited onto a heated substrate, where the precursors decompose. Further heating results in the formation of the product film. A variation on this process is to first form solid particles in the gas phase, then deposit them onto a substrate, and, finally, sinter them to form the solid ceramic film. Another variation, *aerosol-assisted chemical vapor deposition*, uses volatile reactants that evaporate from the solid particles or droplets into the gas phase and deposit materials onto the substrate by CVD. This approach has numerous names in the literature, including *spray CVD* and *spray pyrolysis*.

Another class of aerosol processes invovles systems in which particles are ejected directly from surfaces. Examples of these processes are *laser ablation* and *sputtering* in certain modes of operation.

Although gas-to-particle conversion and intraparticle reaction processes provide convenient groups for classifying aerosol processes, some methods incorporate elements of both. For example, in the *hybrid process* cited in Table 1.1, droplets of an organometallic species are formed by nucleation and condensation and are then converted to a ceramic species by intraparticle chemical reaction. Metal particles can be formed by gas-to-particle conversion and then reacted with sources of oxygen or carbon to form metal oxides or carbides. Numerous other examples exist.

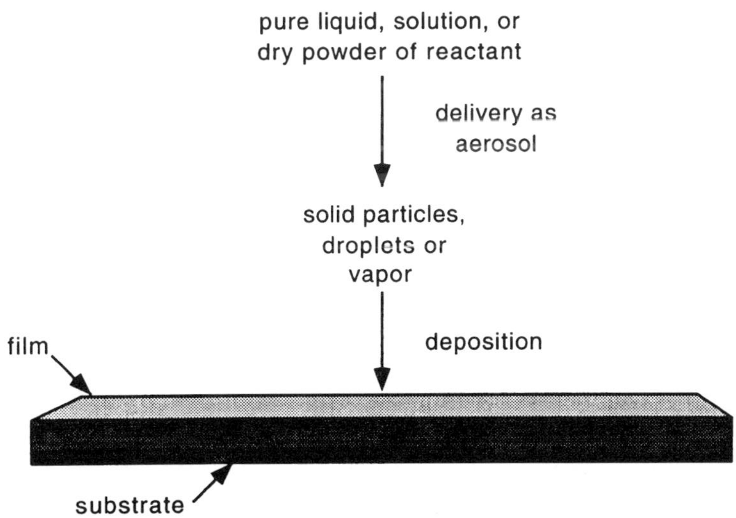


**Fig. 1.10.** Aerosol routes to films rely on deposition of droplets, solid particles, or vapor.

## 1.5 ADVANTAGES AND DISADVANTAGES OF AEROSOL PROCESSES

Aerosol routes to manufacturing powders and films have both advantages and disadvantages with respect to both the materials they produce and the processes themselves. Each aspect must be evaluated cautiously, because the specific application usually determines the criteria, which may vary for each application.

Aerosol processes allow the generation of particles with a unique combination of properties that in many cases cannot be obtained by other processes. Among these properties are:

- High purity
- Unagglomerated, soft or hard agglomerates
- Porous or nonporous structure
- Hollow to dense material
- Multicomponent materials
- Coated and/or multilayer, chemically homogeneous, or composites
- Controlled spread of the size distribution of the particles
- Size from less than 1 nm to 100 microns; nanophase materials
- Single-crystal, polycrystalline, amorphous material, or a mixture thereof

Aerosol processes also provide several advantages for the powder generation process:

- Energy efficient in many cases
- Dry process
- Continuous process
- Environmentally safer than most competing processes
- Either a high- or a low-temperature process
- Simple unit operations

Disadvantages of the powders (for some, but not all, cases) produced by aerosol routes include:

- Hard agglomerates
- Contamination from solvent
- Hollow particles
- Porous particles
- Broad particle size distribution
- Difficult to control stoichiometry if volatile intermediates or reagents are used

Among the disadvantages of aerosol routes (again, for some, but not all, cases) for powder generation are:

- Dilute process (large volumes of gas per quantity of product)
- High cost of reactants

The desirable film properties that can be obtained by aerosol dcposition for film fabrication include:

- High purity
- Controlled film thickness
- Multicomponent materials
- Porous or nonporous structure
- Single-crystal or polycrystalline material
- Composites

Aerosol processes used to form coatings have some useful features:

- High deposition rates
- Simplicity
- Reproducibility
- Low-volatility precursors can be used
- Low-thermal-stability precursors can be used

Aerosol processes for coating formation also have some disadvantages:

- Contamination by solvent in some cases
- Residual particles may remain after droplet evaporation during aerosol delivery of volatile reactants
- Particle formation and deposition lead to degraded properties

Relative to solution routes to powders, aerosol routes have the advantage that their particles are formed at high temperatures, which allows the formation of highly crystalline materials. Liquid-phase routes, in contrast, are usually carried out at ambient temperature and may result in poorly crystallinc matcrials, unless the particles are heated—a process that can lead to agglomeration of particles. An exception is hydrothermal synthesis, in which high temperatures and pressures are used to produce highly crystalline materials. Here, however, control over the size of the particle is a problem. The costs of the precursors are often similar for both aerosol and liquid-phase routes. Compared to solid-state routes, some aerosol processes, such as spray pyrolysis, can provide powders with greater homogeneity than those produced by solid-state routes. At the same time, the major disadvantage of aerosol routes compared to solid-state routes is the lower cost of precursors for the solid-state methods. This, however, is often compensated for by the absence of the need to mill and classify materials, steps that lead to increased costs and degraded powder properties.

An advantage of aerosol routes to films, compared to dip coating and other liquid-phase routes, is that many aerosol processes are dry, and only vapor or solid particles contact the surface. This can be important for applications in which exceptionally high purity films are required. An important example is optical fibers, in which the high purity of the gaseous reactants allows high-purity materials to be formed.

## 1.6 RECOMMENDED BOOKS BY TOPIC

### Transport Processes

Bird, R., Stewart, W.E., & Lightfoot, E.N. *Transport Phenomena*. New York: John Wiley and Sons, 1962.

### General Aerosol

Friedlander, S.K. *Smoke Dust and Haze*. New York: Wiley Interscience, 1977.

Fuchs, N.A. *The Mechanics of Aerosols*. New York: Dover Publications, 1964.

Hidy, G. *Aerosols: An Industrial and Environmental Science*. New York: Academic Press, 1984.

Hidy, G.M. & Brock, J.R. *The Dynamics of Aerocolloidal Systems*. New York: Pergamon, 1970.

Hinds, W.C. *Aerosol Technology*. New York: Wiley Interscience, 1982.

Meyer, R.E. (Ed.). *Theory of Dispersed Multiphase Flow*. New York: Academic Press, 1983.

Reist, P.C. *Introduction to Aerosol Science*. New York: MacMillan Publishing Co, 1984.

Seinfeld, J. *Atmospheric Chemistry and Physics of Air Pollution*. New York: Wiley Interscience, 1986.

Williams, M.M.R. & Loyalka, S.K. *Aerosol Science: Theory and Practice*. New York: Pergamon Press, 1991.

### Materials

Brook, R.J. (Ed.). *Concise Encyclopedia of Advanced Ceramic Materials*. New York: Pergamon Press, 1991.

Edelstein, A.S. & Cammarata, R.C. (Eds.). *Nanomaterials: Synthesis, Properties and Applications*. Philadelphia and Bristol, PA: Institute of Physics Publishing, 1996.

Kingery, W.D., Bowen, A.K., & Uhlmann, D.R. *Introduction to Ceramics*. New York: John Wiley and Sons, 1976.

Reed, J.S. *Introduction to the Principles of Ceramic Processing*. New York: John Wiley and Sons, 1988.

## Chemistry

Arai, Y. *Chemistry of Powder Production.* New York: Chapman and Hall, 1996.

## Clusters

Schmidt, G. (Ed.). *Clusters and Colloids.* New York: VCH, 1994.

## Sprays

Bayvel, L. & Orzechowski, Z. *Liquid Atomization.* Washington: Taylor and Francis, 1993.

Lefebvre, A.H. *Atomization and Sprays.* New York: Hemisphere Publishing Co., 1989.

## Chemical Vapor Deposition

Hitchman, M.L. & Jensen, K.F. *Chemical Vapor Deposition.* New York: Academic Press, 1993.

Kodas, T.T. & Hampden-Smith, M. (Eds.). *The Chemistry of Metal CVD.* New York: VCH, 1994.

## Pharmaceuticals

Chulia, D., Deleuil, M., & Pourcelot, Y. (Eds.). *Powder Technology and Pharmaceutical Processes.* New York: Elsevier, 1994.

Hickey, A.J. (Ed.). *Inhalation Aerosols.* New York: Marcel Dekker, 1996.

## Measurements

Vincent, J.H. *Aerosol Sampling.* New York: John Wiley and Sons, 1989.

Willeke, K. & Baron, P.A. (Eds.). *Aerosol Measurement.* New York: Van Nostrand Reinhold, 1993.

## Sintering

German, R.M. *Sintering Theory and Practice.* New York: John Wiley and Sons, 1996.

## Turbulence

Tennekes, H. & Lumley, J.L. *A First Course in Turbulence.* Cambridge, MA: MIT Press, 1972.

## Reactor Design

Froment, G.F. & Bischoff, K.B. *Chemical Reactor Analysis.* New York: Wiley, 1990.

Levenspiel, O. *Chemical Reaction Engineering.* New York: John Wiley and Sons, 1972.

## 1.7 REFERENCES


1. Friedlander, S.K. *Smoke, Dust and Haze*. New York: Wiley Interscience, 1977.
2. Fuchs, N.A. *The Mechanics of Aerosols*. New York: Dover Publications, 1964.
3. Hidy, G. *Aerosols: An Industrial and Environmental Science*. New York: Academic Press, 1984.
4. Hidy, G.M. & Brock, J.R. *The Dynamics of Aerocolloidal Systems*. New York: Pergamon, 1970.
5. Hinds, W.C. *Aerosol Technology*. New York: Wiley Interscience, 1982, p. 424.
6. Meyer, R.E. (Ed.). *Theory of Dispersed Multiphase Flow*. New York: Academic Press, 1983.
7. Reist, P.C. *Introduction to Aerosol Science*. New York: MacMillan Publishing Co., 1984.
8. Seinfeld, J.H. *Atmospheric Chemistry and Physics of Air Pollution*. New York: Wiley Interscience, 1986.
9. Hickey, A.J. (Ed.). *Inhalation Aerosols*. New York: Marcel Dekker, 1996.
10. Eck, R. *Intern. J. Powder Metallurgy & Powder Tech. 17*(3), 1981, 201–211.
11. Arai, Y. *Chemistry of Powder Production*. New York: Chapman and Hall, 1996.
12. Brook, R.J. (Ed.). *Concise Encyclopedia of Advanced Ceramic Materials*. New York: Pergamon Press, 1991.
13. Kingery, W.D., Bowen, A.K., & Uhlmann, D.R. *Introduction to Ceramics*. New York: John Wiley and Sons, 1976.
14. Schmidt, G. (Ed.). *Clusters and Colloids*. New York: VCH, 1994.
15. Hitchman, M.L. & Jensen, K.F. *Chemical Vapor Deposition*. New York: Academic Press, 1993.
16. Kodas, T.T. & Hampden-Smith, M.J. (Eds.). *The Chemistry of Metal CVD*. Weinheim, Germany: VCH, Verlagsgesellschaft, 1994

# 2 Particle Size Distribution and Physical and Chemical Characteristics

## 2.1 INTRODUCTION

Particles with specific physicochemical properties are required in applications of the aerosol synthesis of films and powders. To describe aerosol processes theoretically, several process variables, such as precursor and product characteristics, must be considered. For particles, average size, spread of the size distribution, concentration, surface area, morphology, density, chemical composition, extent of agglomeration, optical properties, and other parameters are reported in the literature. However, these properties are not sufficient to describe materials. As a result, many other properties (crystallinity, phase composition, elemental and phase distribution within a particle, etc.) are required.[1-4] For aerosol routes to films, such properties are also relevant to understanding the process of film formation, although additional quantities are necessary to describe the films themselves.[5, 6]

More specifically, for droplets, particle size distribution and composition (which determine vapor pressure, for example) are the most important parameters. For solid particles, particle size distribution, morphology, chemical composition, phase composition, and microstructure are relevant properties. These parameters enter into aerosol processes in several ways. High delivery rates of precursors (in the form of droplets that evaporate to give a gaseous precursor) for CVD require high gas-phase concentrations of particles of sufficiently large size and sufficiently high concentration of the precursor in the droplets.[6] Similarly, when the product is a powder, high production rates are often desired and require high aerosol mass concentrations in the process streams.

Control of aerosol processes for the production of materials with the foregoing properties also requires understanding the interaction between the two aerosol components: gas and particles. The nature of this interaction is largely determined by the particle size and the mean free path of the gas. Accordingly, we open our discussion of particle characteristics in this chapter with the latter important parameter.

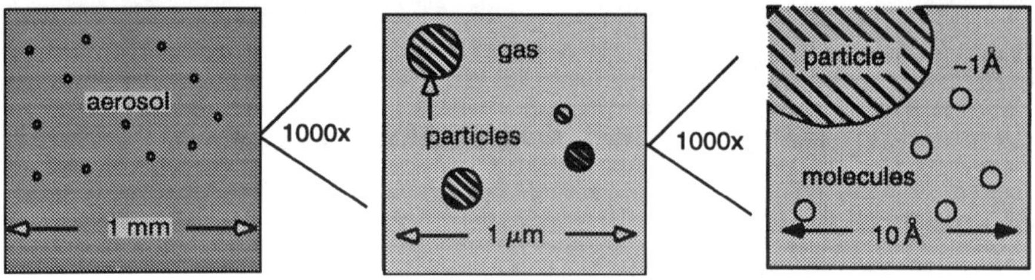


**Fig. 2.1.** Relative sizes of molecules, gas mean free path, and intermolecular spacing. At standard conditions for air, the molecular diameter is 0.0004 $\mu$m, the molecular spacing is 0.004 $\mu$m, and the mean free path is 0.067 $\mu$m. For a particle concentration of $10^{12}/cm^3$ the particles are roughly 1 $\mu$m apart.

## 2.2 MEAN FREE PATH OF THE GAS

A major difference between gases and liquids is that the molecules and atoms in liquids are separated by a distance roughly equal to molecular or atomic diameters, while the molecules and atoms in gases are separated by distances that are much larger than an atomic diameter. As a consequence, gas molecules travel relatively long distances, the *gas mean free path* ($\lambda$), before colliding with each other (Fig. 2.1). Furthermore, species that are suspended or dissolved in liquids carry several "spheres" of coordinated solvent which often dominate their chemical reactivity.

The mean free path of a gas—the average distance that a molecule or atom of the gas travels before colliding with another atom, another molecule, or the walls—is one of the most important characteristics associated with an aerosol. Along with other dimensions such as particle size, the mean free path determines the form of the theoretical expressions describing particle growth and collisions, as well as various transport processes. For example, nanophase particle formation can occur under conditions in which the particles grow no larger than 10 nm, while many spray processes involve particles that are larger than 1000 nm; these dimensions are much smaller and much larger, respectively, than the gas mean free path. The most important consequence of this distinction is that the theoretical expressions used to describe collision, condensation, evaporation, and transport rates differ vastly in form.

The expression for the gas mean free path is derived from kinetic theory by assuming that a gas consists of molecules resembling hard spheres that collide with each other.[7-10] These collisions take place with high frequency and determine the macroscopic and measurable properties of a gas—its temperature, pressure, viscosity, thermal conductivity, diffusivity, etc. The average distance traveled by the molecules between collisions is the mean free path of the gas (cm), given by

$$\lambda = \frac{\bar{c}}{\varepsilon_c}, \tag{2.1}$$

where $\bar{c}$ (cm/sec) is the arithmetic average (the mean) of the velocities of all the gas molecules. The quantity

$$\bar{c} = \left( \frac{8k_B T}{\pi m_1} \right)^{1/2} \tag{2.2}$$

where $k_B$ (erg/K) is the Boltzmann constant, $T$(K) is the temperature of the gas, and $m_1$ (g) is the mass of a gas atom or molecule. Values of $\bar{c}$ at 20° C are 46,300 cm/sec for air and 176,000 cm/sec for $H_2$.[8] The quantity $\varepsilon_c$ (1/sec) in Equation (2.1) is the rate of collisions of gas molecules and is given by

$$\varepsilon_c = \left( \frac{16k_B T}{\pi m_1} \right)^{1/2} \pi \sigma_d^2 n_g, \tag{2.3}$$

where $\sigma_d$ (cm) is the hard-sphere molecular diameter and $n_g$ (number molecules/cm$^3$) is the molecular concentration. Substituting Equations (2.2) and (2.3) into Equation (2.1) and invoking the ideal gas law, we find that the mean free path of the gas is

$$\lambda = \frac{k_B T}{\sqrt{2}\, \pi p \sigma_d^2}, \tag{2.4}$$

where $p$ is the pressure in dynes/cm$^2$.

A slightly more accurate expression for $\lambda$ that relies not on the molecular diameter $\sigma_d$, but on experimentally measured gas properties, is[9]

$$\lambda = 2.00\mu/\rho\bar{c}, \tag{2.5}$$

where $\mu$ (g/cm sec) and $\rho$ (g/cm$^3$) are the gas viscosity and density, respectively, $\bar{c}$ is in cm/sec, and $\lambda$ is in cm. As a point of reference, the mean free path at 500 Torr (atmospheric pressure at sea level is 760 Torr) is on the order of 0.1 $\mu$m, or 100 nm; that is, molecules travel a distance of roughly a thousand atomic diameters before colliding with one another. It is worth noting that the approximate molecular spacing in air at standard conditions is 0.004 $\mu$m, while the molecular diameter is 0.00037 $\mu$m, and the mean free path is 0.067 $\mu$m (Fig. 2.1).[8]

The physical situations encountered for large and small particle sizes relative to the gas mean free path are quite different. When the particle diameter is much larger than the mean free path ($d_p \gg \lambda$), the particle "sees" the gas as a continuous medium. The situation is analogous to moving your hand through water or air, in that your hand feels a continuous medium and does not feel the individual impacts of the gas molecules. For this case, continuum theories (e.g., Navier–Stokes equations and diffusion equations for heat and mass transfer)[7] can be used to describe transport processes.

For example, the movement of the gas around a particle can be described by treating the gas as a continuous fluid. When $d_p \ll \lambda$, the particle is effectively a large molecule. In this case, a collision is not unlike having your hand struck by individual tennis balls: Your hand feels the separate collisions with each ball that strikes it. Under these conditions, each particle behaves as if it were in a rarefied medium in which the particles do not, as a first approximation, influence the velocity distribution of surrounding molecules and atoms. When the particle size is approximately the same as the gas mean free path, a rigorous description of the transport processes requires solution of the Boltzmann equation, a very complicated approach that is rarely used in practice.[11, 12] Because particles often grow from nanometer size to microme-

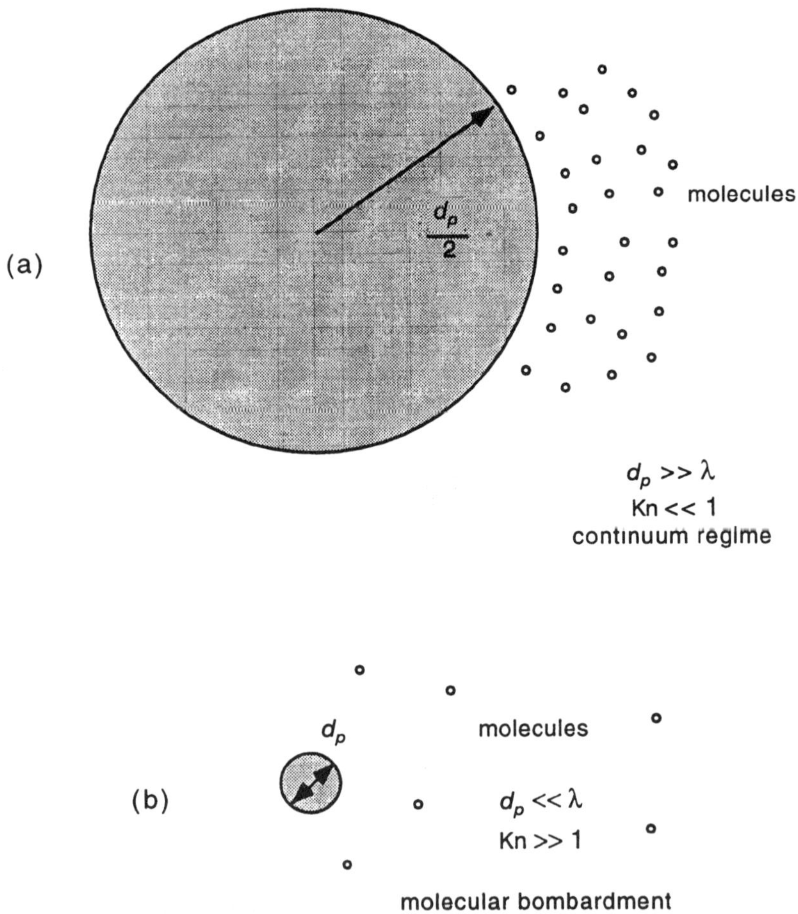


**Fig. 2.2.** Relative sizes of particles and molecules and mean free path for (a) continuum and (b) free molecular regimes.

ter size, the change in $\lambda/d_p$ requires complicated expressions to describe particle growth and collisions over the evolving size range. This situation is depicted schematically in Figure 2.2.

It is worth noting that the gas mean free path is not the same as the *particle mean free path*. The latter quantity, discussed by Hinds and Seinfeld,[8, 10] is a measure of how far a particle travels before it changes direction.

To simplify analyses involving particle size and gas mean free path, the *Knudsen number* $\mathrm{Kn} = 2\lambda/d_p$ is defined as the ratio of the mean free path of the gas to the particle radius. This quantity also emerges naturally when theoretical descriptions of aerosol processes are derived. When $\mathrm{Kn} \ll 1$, the particles are said to be in the *continuum regime*. When $\mathrm{Kn} \gg 1$, the particles are in the *free molecular regime*. The intermediate regime $(0.1\lambda < d_p < 10\lambda)$ is called the *transition-slip regime*, since the particles effectively "slip" between the gas molecules.

## 2.3 MASS AND PARTICLE CONCENTRATION

An important aspect of aerosols is their dilute nature relative to liquids. Even in the most heavily loaded gas streams, the volume fraction of particles relative to the gas is much less than 1%, and the overall density of the aerosol (gas and particles) is two to three orders of magnitude lower than for a liquid. From a practical perspective, this means that much larger volumes of gas have to be processed to form a given mass of material than is the case for a liquid containing the same mass of product.

These considerations suggest that one of the most significant aerosol properties with respect to the material production rate is the *total mass concentration* $M_\infty$ (g/cm$^3$), the mass of particles per unit volume of gas. Multiplying the value of $M_\infty$ at the system outlet by the aerosol or gas flow rate (cm$^3$/s) gives the powder production rate: mass per unit time (g/s). Typical industrial production rates for $TiO_2$ are roughly 10 Kg/sec, while those in the laboratory are roughly 10 mg/sec.

The *total particle concentration* ($N_\infty$, particles/cm$^3$) is of interest primarily because it plays a major role in determining the importance of collisions between particles. Particle concentrations in aerosol systems range from very low levels, on the order of 1 particle/cubic foot, or about 30 particles/m$^3$, up to $10^{16}$ particles/m$^3$ ($\sim 3 \times 10^{14}$/ft$^3$) or more. The low particle concentrations are found in clean rooms used in the electronics industry and in some hospitals where great precautions are taken to remove particles from the air. High particle concentrations are found in industrial processes, such as the manufacture of particles in flames. Although the particle concentration in aerosol processes can reach these very high values for short times, the final concentration of the aerosol is usually much lower because collisions between particles reduce the particle concentration in a very short time. Concentrations, therefore, can vary 10 to 15 orders of magnitude or more, often within a single process. This variation is, on an area basis, analogous to having one person on the earth versus having 1 million people on each square kilometer of our planet. A consequence is that theories describing aerosol behavior in systems used to generate particles or films can be different, depending on the particle concentration. Another consequence is that it is not possible to measure the concentration or composition of particles over a broad range without resorting to the use of several instruments whose operations depend on very different principles.

In the absence of particle losses and chemical reactions, or after the reactions have gone to completion, the aerosol mass concentration is constant. For processes in which the concentration of particles is high enough for collisions to occur, the concentration decreases and the average particle size increases with increasing process residence time. Collisions between particles are very effective in reducing high particle concentrations even for short residence times. For example, it takes about 1 second to reduce the concentration of 0.1-$\mu$m particles from $10^{10}/\text{cm}^3$ to $10^9/\text{cm}^3$ by coagulation even at room temperature.[8, 9] For processes in which the concentration of particles is low enough that collisions do not occur, the concentration remains constant.

Finally, the aerosol concentration is a statistical property,[9] and as such, it must be measured using sufficient sampling volumes to assure consistent and accurate data. This is an important issue in clean-room technology, where extremely low particle concentrations are found. Typical values are 1 particle in one cubic foot (a class-1 clean room), which suggests that many cubic feet of air must be sampled to obtain statistically significant values for the particle concentration. At the other extreme, because common aerosol instruments cannot measure very high concentrations (many are restricted to a range below $10^7$ particles/$\text{cm}^3$), dilution of the aerosol is often required in sampling from industrial systems.

## 2.4 PARTICLE SIZE DISTRIBUTION

### 2.4.1 Introduction

Aerosols include particles with sizes ranging from molecular clusters up to roughly 100 microns. Examples of processes in which large (100-micron) particles or droplets are present are molten metal atomization, spray pyrolysis (a term used to describe both droplet deposition to form films and spraying and reaction of solutions to form powders), freeze drying, and spray drying. At the other extreme, clusters play a role in all aerosol routes to materials in which the particles are built up from the vapor, illustrating a unique aspect of aerosol routes to materials, namely, the broad range of sizes encountered, which can cover five or six orders of magnitude. To provide some perspective, the relative sizes of a 1-nm particle and a 100-$\mu$m particle (a difference of five orders of magnitude) are the same as for the head of a needle and a hot-air balloon.

For films, the dimensions of interest range from a monolayer with thickness of roughly 0.1 nm to greater than millimeters; the length scale is even broader than for aerosols. This implies that the aerosol methods used to form films must provide deposition rates that are slow for ultrathin films (angstroms per minute) and rapid for thick films (microns per minute or more), to allow the formation of films with controlled thicknesses. There are also applications in which thin films must be deposited at high rates of up to microns per second—in, for example, the glass-coating industry.

The broad range of particle sizes suggests that another important parameter is the spread of the particle size distribution, a property that often determines the usefulness of particles in the final product or as intermediates to films. A *monodisperse* aerosol consists of particles of exactly the same size. A *polydisperse* aerosol contains particles of different sizes. A nearly monodisperse aerosol can be compared (on the basis of height) to a classroom of children of the same age and roughly the same size where the heights differ by less than 10%. A polydisperse aerosol is analogous to a playground with children of different ages where their sizes differ by up to a factor of two. An aerosol with a very broad distribution cannot be described by this analogy, because the range in sizes of humans is far smaller than that of aerosol particles; for aerosols, this difference can be a factor of 100 or more. The description is also complicated by the occurrence of *multi-modal* distributions, in which two or more groups of particles exist with similar sizes in each group, but with very different average sizes for the groups.

The particle size distribution of an aerosol plays a major role in determining how particles behave as they form and grow or as films form and grow and in how a product powder behaves during the processing of parts or coatings. While a powder or film is being synthesized, the particle size distribution determines the average size, surface area, and spread of the distribution, which together influence the rates of particle growth and deposition and, eventually, the yield of material from the process. For subsequent processing of powders into bulk materials and coatings, broad size distributions are generally undesirable. The tolerance of the presence of undesirable large or small particles depends on the application. Typically, large particles act as sites where flaws can form during the fabrication of objects.[3] Very small particles can sinter (fuse together) faster than larger ones, resulting in nonuniform sintering and defects in a material. For $TiO_2$ pigment, deviations of only 5 to 10 nm in average size and small changes in the spread of the distribution can have a large influence on the perceived color of the pigment and the amount of pigment required in a paint to completely obscure a surface.[13]

The size of aerosol particles is characterized by a *particle size distribution function*. An aerosol can be described by several types of size distributions, in

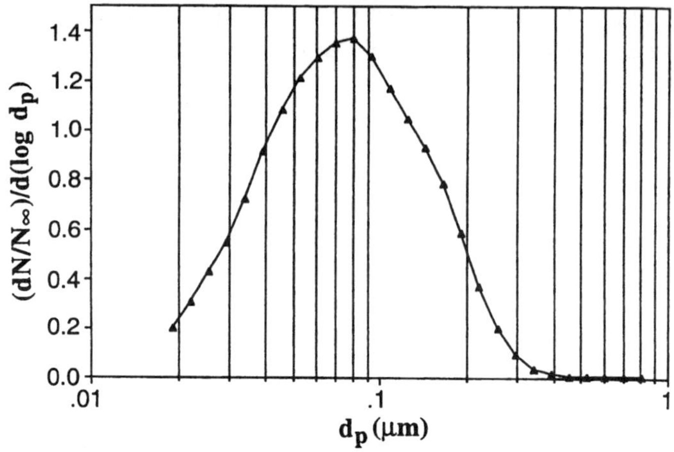


**Fig. 2.3.** A discrete gas-phase particle size distribution during the synthesis of (Bi, Pb)–Sr–Ca–Cu–O by spray pyrolysis at 900° C, as measured by the differential mobility particle sizer (DMPS). (Courtesy Gurav et al.)[14]

which "size" denotes the diameter $d_p$ of a spherical particle when the particle size is continuously distributed or $d_{pi}$ when the particle size is discretely distributed. In principle, counting and determining the sizes of a large number of individual particles can give a complete picture of the particle size distribution. Such information however, is often neither practical to use nor easy to obtain. Frequently, particles are grouped in size ranges with a significant width, 0.1–0.3 $\mu$m, for example. The number $N_i$ or the mass $M_i$ of the particles within each range $i$ is measured. All particles in the range are assumed to have a size equal to the midpoint of the size range, 0.2 $\mu$m in this example. Listing all the size ranges and the corresponding $N_i$ or $M_i$ gives a *discrete particle size distribution* (Fig. 2.3), because discrete groups of particles are considered.[14] As the size range narrows, this description becomes more accurate, but, unfortunately, more cumbersome and time consuming.

For spherical particles, the preceding quantities can be related by

$$M_i = \rho_p N_i \pi d_{pi}^3 / 6, \tag{2.6}$$

where $\rho_p$ is the particle density (g/cm$^3$). The description in terms of a listing of $N_i(d_{pi})$ is a discrete number size distribution, while the description in terms of $M_i(d_{pi})$ is a discrete mass size distribution. This discrete approach is the one used to count and determine the sizes of particles either manually or with the use of a computer.

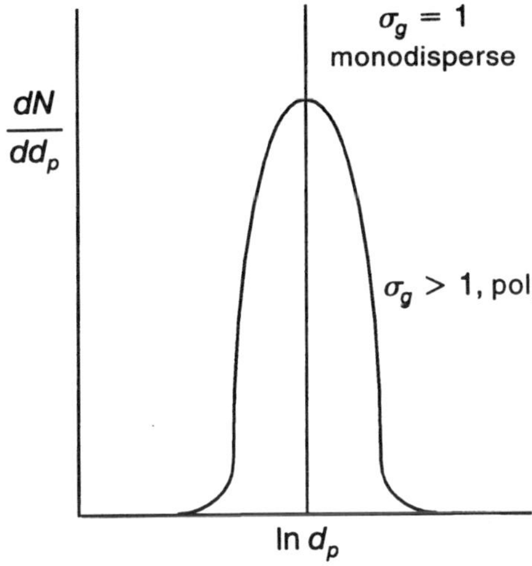


**Fig. 2.4.** Continuous size-distribution function.

In order to circumvent the difficulties associated with keeping track of a large number of particle counts and sizes, a discrete size distribution can be converted into a *continuous size distribution* (Fig. 2.4). This is accomplished by keeping in mind that, for the huge number of particles present in practice, the particles are continuously spread in the size spectrum, and the particle size distribution can be mathematically described as a continuous function. The continuous number distribution is defined as the number of particles per unit gas volume and per particle size interval and is given mathematically by $n(d_p) = N_i/\Delta d_{pi}$, for $\Delta d_{pi} \longrightarrow 0$, where $\Delta d_{pi}$ is the difference of the upper and lower limits of the $i$th size interval and $N_i$ is the number of particles in the interval per volume of gas. Similarly, the continuous mass distribution is expressed mathematically as $m(d_p) = M_i/\Delta d_{pi}$ for $\Delta d_{pi} \longrightarrow 0$. In both cases, the differences are converted to differentials: $n(d_p) = dN/dd_p$ and $m(d_p) = dM/dd_p$, and $N(d_p)$ and $M(d_p)$ are the cumulative number- and cumulative mass-distribution functions, respectively.[9] Hinds, Reist, and Seinfeld give introductions to this and other aspects of particle size-distribution functions.[8, 10, 15]

The size of aspherical particles is usually based on characteristics other than a geometric dimension. An aspherical particle (an agglomerate, for example) will settle at a different rate, usually more slowly, than a spherical particle. This difference manifests itself in measurements of particle size that, in liquids, rely on sedimentation velocity. A measure of particle size that accounts for the difference is useful. For this reason, a variety of so-called spherical-equivalent diameters are used to characterize particles. Two such measures are the *Stokes diameter* and *aerodynamic diameter* (discussed in later sections). The *Stokes diameter* is the diameter of a spherical particle with the same settling velocity and density as the actual particle and is measured during characterization of the aerosol in multistage impactors. The *aerodynamic diameter* is the diameter of a spherical particle with a density of 1 $gm/cm^3$ and the same settling velocity as an aspherical particle. It is important to keep in mind that measured particle size distributions are often plotted not in terms of the actual physical diameters of the particles, but in terms of the Stokes, aerodynamic, and other equivalent diameters.[9]

### 2.4.2 Average Particle Size

Although a detailed knowledge of the size distribution is desirable, an estimate of the average size and the spread of the distribution is usually sufficient to characterize particle sizes. Commonly, definitions of "average" particle size are determined by the diameter, which may be defined by the mean, median, or mode. *Mean diameters* weight the diameters of the particles in a size interval by some property such as the number or mass fraction of particles and then average the results. A *median diameter* provides the size above which 50% of the particles, by number, mass, or some other measure, exist. The *modal diameter* is the size for which the largest number of particles are found. The mean and the median measurements are considered most important.

One of the most common definitions of "average" particle size is the number (also called count) average (mean) diameter,

$$d_{pn} = \frac{\Sigma N_i d_{pi}}{N_\infty} = \frac{\int_0^\infty n(d_p) d_p dd_p}{N_\infty}, \tag{2.7}$$

where

$$N_\infty = \Sigma N_i = \int_0^\infty n(d_p) dd_p \tag{2.8}$$

is the total particle concentration.

Another common quantity is the mass average (mean) diameter

$$d_{pm} = \frac{\Sigma M_i d_{pi}}{M_\infty} = \frac{\int_0^\infty m(d_p) d_p dd_p}{M_\infty}, \tag{2.9}$$

where

$$M_\infty = \Sigma M_i = \int_0^\infty m(d_p) dd_p. \tag{2.10}$$

Recall that $d_{pi}$ refers to the discrete distribution and $d_p$ to the continuous distribution. In this case, the number or mass of particles in an interval is weighted by the diameter. (Note that the summations and integrals are all from 0 to $\infty$.)

Several other types of diameters are used, such as the diameter of the particle with the average mass $\overline{m}$, given by

$$d_{p\overline{m}} = \left( \frac{6M_\infty}{\pi \rho_p N_\infty} \right)^{1/3}, \tag{2.11}$$

where $\overline{m} = \dfrac{\Sigma M_i}{N_\infty}$ and $\rho_p$ is the particle density (g/cm$^3$).

A different class of, or a different way of thinking about, average diameters can be obtained by weighting the number of particles in an interval by the diameter to some power. An average diameter that is frequently considered in characterizing droplets and powders is the *Sauter diameter* $d_s$, which is proportional to the ratio of the aerosol mass to its area:

$$d_s = \frac{\Sigma N_i d_{pi}^3}{\Sigma N_i d_{pi}^2} = \frac{6}{\rho_p} \frac{M_\infty}{A}. \tag{2.12}$$

Here $A = \pi \Sigma N_i d_{pi}^2$ is the total particle surface area per volume of gas of the aerosol or powder.

The preceding averages were obtained directly from the diameter. In other cases, mainly for distributions that are close to lognormal in shape, it is useful to compute averages using the natural logarithm of the diameter. This leads to other useful averages, viz., the *geometric number* ($d_{\text{gn}}$) and *geometric mass* ($d_{\text{gm}}$) *average diameters*, given respectively by

$$\ln d_{\text{gn}} = \frac{\Sigma N_i \ln d_{pi}}{N_\infty} \quad \text{and} \quad \ln d_{\text{gm}} = \frac{\Sigma M_i \ln d_{pi}}{M_\infty}. \tag{2.13}$$

The geometric number and mass diameters are used extensively in size characterizations of aerosols. The geometric average diameter is equal to the median diameter for a lognormal size distribution. These diameters are frequently considered for materials synthesis in the aerosol literature.

The two averages for a lognormal distribution have very different physical meanings and different numerical values for the same size distribution. Table 2.1 illustrates this point for a lognormal distribution with a 1-$\mu$m number median diameter and a geometric standard deviation (see shortly) of $\sigma_g = 1.5$ The mass mean diameter is often more important than the number mean diameter, because it is more closely related to the mass of material produced; small numbers of very large particles would manifest themselves as a large increase in mass mean diameter, but only a small change in number mean diameter.

**TABLE 2.1. Different average and median diameters for a lognormal distribution with 1-μm number median and $\sigma_g = 1.5$**

| Type of Diameter | Value, μm |
|---|---|
| Mode | 0.85 |
| Number median diameter | 1.0 |
| Number mean diameter | 1.1 |
| Diameter of average mass | 1.3 |
| Mass median diameter | 1.6 |
| Mass mean diameter | 1.8 |

In many cases, average diameters do not provide the complete picture. For example, in making powders for structural ceramics, the absence of a coarse tail in the size distribution (a few large particles in an otherwise finely distributed powder) is crucial for the synthesis of defect-free sintered products. This coarse tail may contain a negligible fraction of the mass and number of particles and, as a result, may not influence any of the average diameters in a measurable way.

There are many other types of averages in addition to those mentioned here. Furthermore, simple equations exist that allow the conversion of one type of average into another for the case of distributions with certain properties (e.g., a lognormal distribution). The texts by Hinds and Reist give a detailed discussion of these topics.[8, 15]

### 2.4.3 Spread of the Distribution

The spread of a particle size distribution is a number that provides a measure of how large and small particles are relative to some average size and how many of these particles there are. As there are many ways to quantify the average size, there are also many ways to define the spread of the distribution. The geometric standard deviation of the number size distribution, however, is most commonly used:

$$\ln^2 \sigma_g = \frac{\Sigma N_i \left[\ln\left(d_{pi}/d_{gn}\right)\right]^2}{N_\infty}. \tag{2.14}$$

This quantity multiplies the number of particles $N_i$ in an interval $d_{pi}$ by the square of the natural logarithm of the ratio of $d_{pi}$ to the geometric number average diameter, to obtain a measure of the spread of the distribution. Although the equation is derived for a lognormal distribution, it is frequently used for any distribution, because particle size distributions from several different types of aerosol processes are roughly lognormal and justify the use of such an approach. Note that measured particle size distributions cannot ever be monodisperse. For this reason, it is common to use the term *monodisperse* to describe a size distribution that is *close* to being monodisperse, with the acceptable deviation from a truly monodisperse distribution being smaller or larger, depending on the application. The details of lognormal distributions are discussed in Hinds.[8]

Knowing $N_\infty$, $d_{gn}$, and $\sigma_g$, we define the lognormal distribution $n_l$, as

$$\frac{dN}{dd_p} = n_l(d_i) = \frac{N_\infty}{\sqrt{2\pi}\, d_i \ln\sigma_g} \exp\left[-\frac{\ln^2(d_i/d_{gn})}{2\ln^2\sigma_g}\right]. \tag{2.15}$$

Interesting examples of lognormal distributions are the annual personal income distribution or the number of publications there arc per researcher. A lognormal distribution is normally distributed with respect to $\ln d_p$. For such a distribution, 67% of the particles are between $d_{gn}/\sigma_g$ and $d_{gn}\sigma_g$, while 95% of the particles are in the range $d_{gn}/\sigma_g^2$ to $d_{gn}\sigma_g^2$. For example, for a lognormally shaped aerosol size distribution with geometric number average diameter of $d_{gn} = 1$ $\mu$m and $\sigma_{gn} = 2$, 67% of the particles are between 0.5 and 2 $\mu$m, and 95% are between 0.25 and 4 $\mu$m.

Aerosol distributions are commonly lognormal or near lognormal, so numerical values of $\sigma_g$ give a quantitative and tangible estimate of the shape of the distribution. Furthermore, $\sigma_g$ can be used to calculate other average diameters from $d_{gn}$ for a lognormal distribution through the *Hatch–Choate* conversion equations.[8] These equations are simple expressions that are mathematically exact and very useful for converting between, for example, number and mass average diameters. For a monodisperse aerosol, $\sigma_g$ is 1 and increases as the spread of the distribution widens. A value of roughly 1.5 is considered polydisperse and is often found for $TiO_2$ pigment powders. For most applications, a $\sigma_g$ value of 1.1 or 1.2 can be considered monodisperse, depending on the end application.

### 2.4.4 Moments of Particle Size Distribution

The overall physical and chemical properties of an aerosol or powder are often more important than the detailed information given by the full particle size distribution and are the basis for calculating certain types of average particle sizes. Examples of overall properties are the powder production rate in materials synthesis (g/sec), the total aerosol number or mass concentration in air pollution (particles/cm$^3$ or $\mu$g/cm$^3$), and the specific surface area of the powder (m$^2$/g) in catalysis. These properties are proportional to a power of the particle size; for example, the mass is proportional to the third power, while the area is proportional to the second power, of particle diameter. The properties are obtained by integration of the particle size distribution, weighted by the appropriate power of the particle size, with or without some other parameter over the entire size range.[9] These integral forms are the $j$th moments ($M_j$) of the size distribution [not to be confused with total mass and mass in an interval ($M_\infty$ and $M_i$, respectively)]:

$$M_j = \int_o^\infty n(d_p) d_p^j dd_p = \sum N_i d_{pi}^j. \tag{2.16}$$

Some of the moments have special physical meaning. The zeroth moment ($j = 0$) corresponds to the particle number concentration $N_\infty = M_0$; the second moment with respect to particle diameter ($j = 2$) is proportional to the particle surface area $A = \pi M_2$; the third moment with respect to particle diameter is proportional to the total aerosol volume $V = \pi M_3/6$. The moments also are useful in defining production criteria and average particle sizes obtained from weighted distributions. For example, the powder production rate is $\pi \rho_p M_3 Q/6$, where $Q$ is the aerosol or gas flow rate, and the specific surface area is $\text{SSA} = A/\rho_p V = 6M_2/\rho_p M_3$, while the *mass mean* and *Sauter* (*surface mean* or *mean volume–surface*) diameters are $d_{\overline{m}} = M_4/M_3$ and $d_s = M_3/M_2$, respectively.

Moments can also be defined with respect to any measure of particle size —diameter, area, or volume. For example, the moments with respect to particle volume are

$$M_j = \int_0^\infty n(v) v^j dv. \tag{2.17}$$

It is not uncommon to employ noninteger moments for many models describing aerosol system behavior. In this case, the zeroth moment is still the total particle concentration, the two-thirds moment is proportional to the particle area, and the first moment with respect to particle volume is the total aerosol volume. Hence, care should be taken in defining the basis of each size distribution function and moment (e.g., particle volume or diameter).

Particle size distributions with respect to any size measure are related to each other by

$$n(d_p) dd_p = n(v) dv = n(s) ds \tag{2.18}$$

where $s$ is the surface area of a particle of volume $v$.[9]

Finally, the term *moment* can be confusing because it has nothing to do with the inertia or other mechanical properties of the particles. Rather, it is just a convenient way to define integral properties of the particle size distribution.

## 2.5 DENSITY AND SURFACE AREA

The apparent density of aerosol particles ranges from much less than 1 $gm/cm^3$ for hollow or porous particles to the densities of more than 10 $gm/cm^3$ associated with the heavier metals. This is another important contrast with atmospheric aerosols, in which particles usually have densities of a few grams per cubic centimeter. At the same time, films formed by aerosol routes can have densities of 0.01 $g/cm^3$ and have applications as low-dielectric-constant materials for microelectronics. At the other extreme, they can have densities corresponding to those of solid materials.

Aerosols can have a high surface area per unit mass, or *specific surface area*, which can range from less than 1 $m^2/g$ (roughly micron-size particles) up to many hundreds of square meters per gram (nanometer-size particles). As an example, 10 grams of a powder with a surface area of 1000 $m^2/g$ has enough cumulative surface area to cover approximately four football fields. This large surface area is the chief reason for the attraction of aerosol routes for the generation of catalysts, adsorbents, and nanophase materials. A useful, yet simple, expression is

$$SSA = \frac{6}{\rho_p d_p}, \tag{2.19}$$

which relates the specific surface area (SSA) to particle size and density for monodisperse particles.

Films can have surface areas that are as high as those of aerosol particles on a mass basis. An obvious example is a film formed by depositing particles. But films can also be formed from molecular precursors (as in the aerosol delivery of volatile reactants for CVD) to provide highly interconnected porous films with similarly high surface areas.

## 2.6 PARTICLE MORPHOLOGY

Aerosol particles are either solid or liquid. Droplets are spheres, while solid particles come in various shapes: spheres, rods, cubes, agglomerates, polyhedra, and even fibers. Several types of particles that have been formed are shown in Figure 2.5; the materials include several simple metal oxides, as well as the more complex material $CaMoO_4$.[16, 17] These materials were produced at high temperatures by condensation from a gas exiting a plasma reactor.

Because of the broad range of particle shapes, a common practice in characterizing aspherical particles is to measure a property (for example, drag force or electrical mobility) of a given particle and obtain an "equivalent" diameter by matching the behavior of the aspherical particle to that of a spherical one. The ratio of this diameter to that of a sphere with the same volume is the *shape factor* of the particle. Hinds has tabulated various shape factors from drag force measurements.[8] The shape factor concept applies best to particles with well-defined geometries and to fibers, for which an aspect ratio is relevant. Shape factors are used in a variety of theoretical descriptions of aerosol phenomena, such as transport and coagulation, and are described in more detail in Chapter 3.

Aspherical particles are frequently agglomerates consisting of many smaller particles. *Soft agglomerates* are composed of several small (primary) particles loosely attached to each other, usually by van der Waals forces. The term *floc* has come to be synonymous with *soft agglomerates*. *Hard agglomerates* have structures similar to those of soft agglomerates, but the bonds between

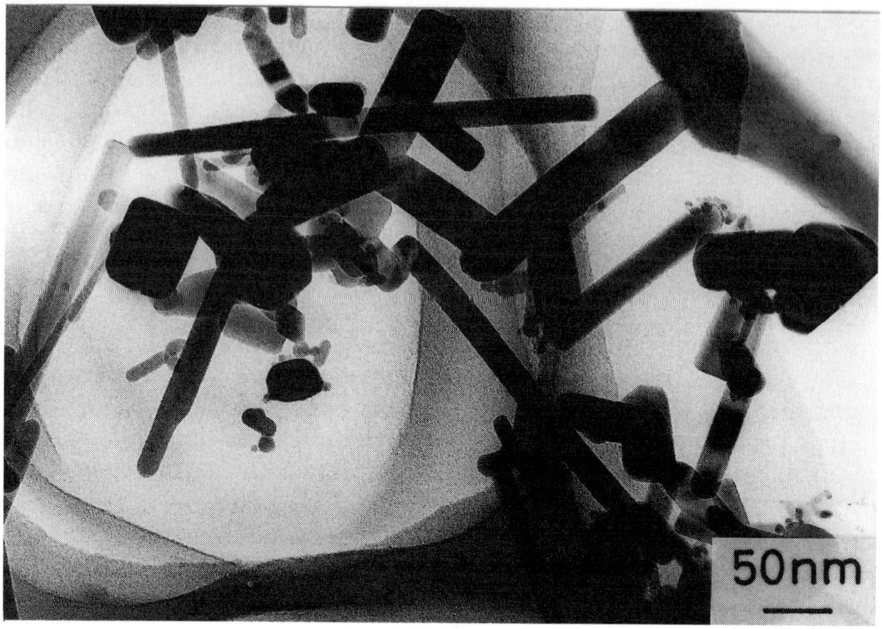


*(a)*

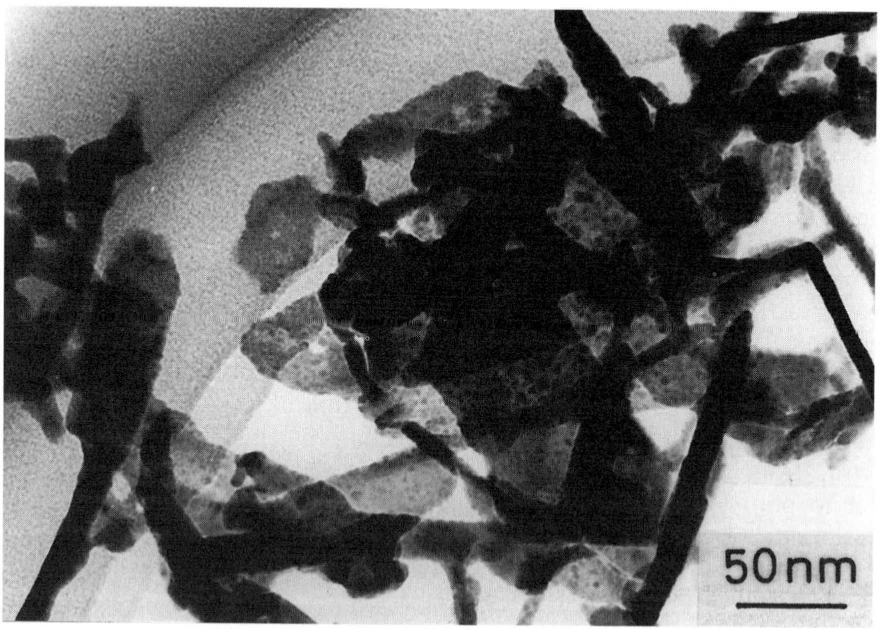


*(b)*

**Fig. 2.5.** TEM photographs of (a) ZnO, (b) PbO, (c) $CaMoO_4$, (d) MgO, (e) $Nd_2O_3$, and (f) $Fe_2O_3$ particles produced by spray inductively coupled plasma (ICP). (Courtesy of Suzuki et al.)[16, 17] (a) and (b) Reprinted from *Journal of Material Science* *27*(3), 679–684, 1992, Synthesis of ultrafine single-component particles by the spray-ICP technique, Suzuki et al., Chapman & Hall.

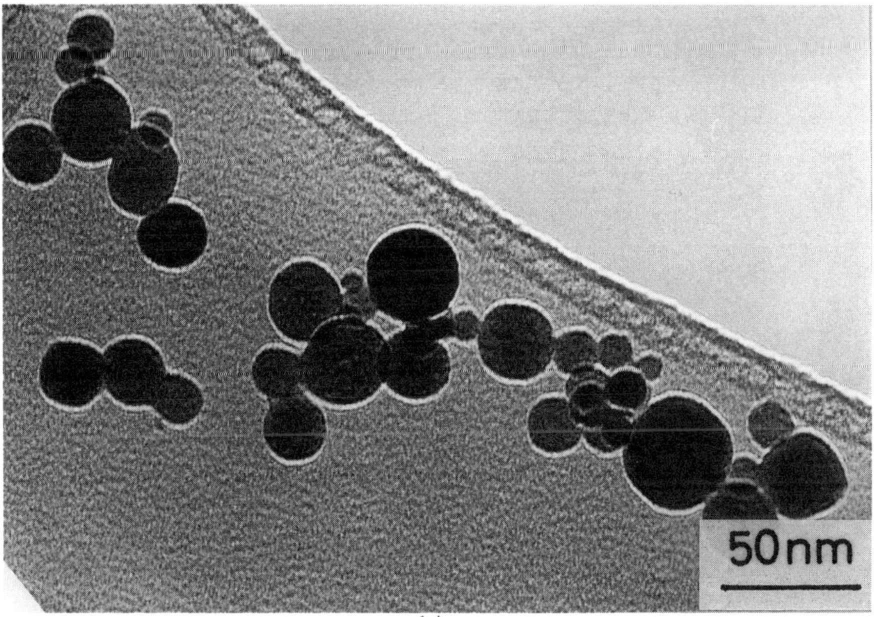


(c)

(d)

**Fig. 2.5.** (*Continued*) (c)–(d) Reprinted with permission of The American Ceramic Society, Post Office Box 6136, Westerville, OH 43086-6136. 

(f)

**Fig. 2.5.** (*Continued*) (e)–(f) Reprinted with permission of The American Ceramic Society, Post Office Box 6136, Westerville, OH 43086-6136. 

(primary) particles may be stronger, as is the case when the particles sinter (fuse). As a result, de-agglomeration can be easily achieved for soft agglomerates, but is considerably more difficult (requiring more energy input) or even impossible for hard agglomerates. In the literature, hard agglomerates are frequently termed *aggregates*. These structures are encountered frequently in aerosol processing of materials because they are formed by the collision and partial sintering of particles in the gas phase. Agglomerates can have very complicated, branched structures with large variations in open area. A significant consequence of agglomerate formation is that the transport properties (diffusion, settling, etc.) of agglomerates are very different from those of the spherical primary particles. This difference has motivated the use of fractal concepts to describe the irregular structure of agglomerate particles commonly encountered in aerosol synthesis of materials (Fig. 2.6). Specifi-

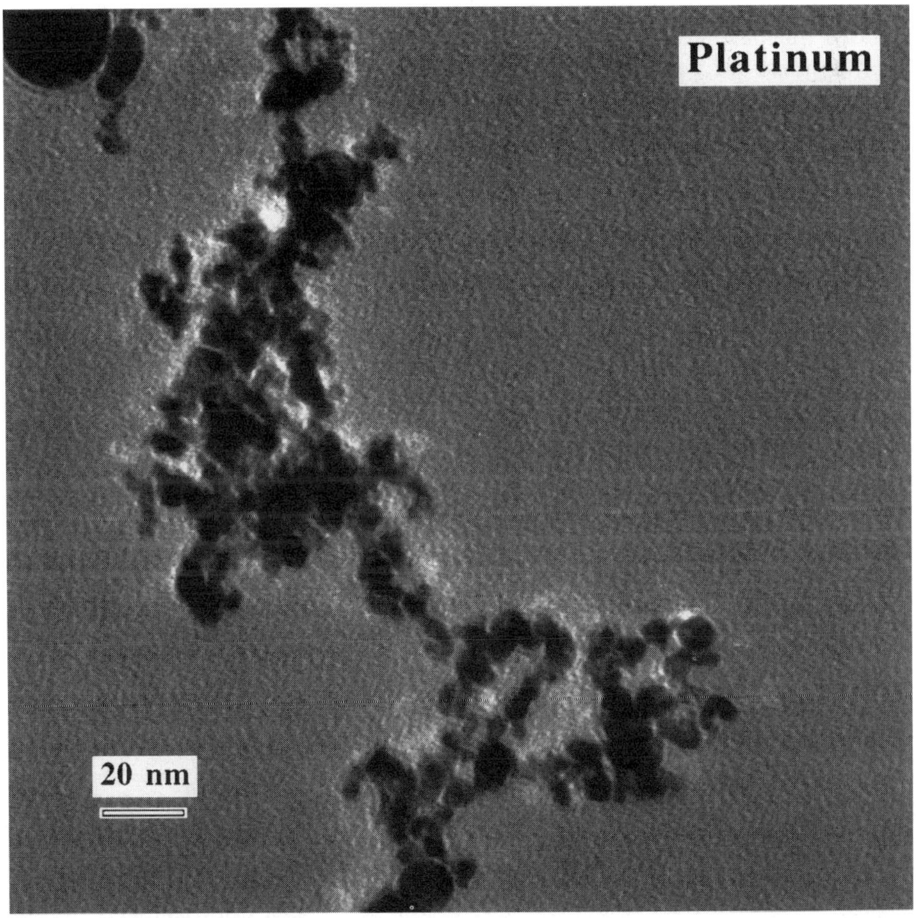


**Fig. 2.6.** TEM photograph of an agglomerate formed by pulsed laser ablation. (Courtesy of S. Jain and T. Kodas.)

cally, fractal or fractallike aggregates composed of $N_f$ primary particles follow the law[18]

$$N_f = \epsilon_p(2R_a/d_o)^{D_f}, \tag{2.20}$$

where $R_a$ is a characteristic length of the aggregate (the radius of gyration or the square root of the product of the length and width of the two-dimensional particle image; $d_o$ is the diameter of a primary particle; $\epsilon_p$ is a dimensionless prefactor, usually assigned unit value; and $D_f$ is the *mass fractal dimension*. For a sphere or a compact particle (a cube or rectangle), $D_f = 3$; for a thin platelet, $D_f = 2$; and for a fiber, $D_f = 1$. Typically, $D_f$ values for aerosol-generated materials are nonintegral and cover the entire range from 1 to 3. Fractal values may also vary from particle to particle in a given sample. The fractal concept is useful in correlating particle characteristics with particle generation conditions.

The mass fractal dimension gives a quantitative estimate of the compactness of the particles. A similar estimate can be obtained for the smoothness or roughness of a particle or film surface from the *surface fractal dimension*, $D_s$ instead of $D_f$. For a smooth sphere, $D_s = 2$, while for an infinitely porous one (in which every molecule is exposed), $D_s = 3$.

Fractal dimension is a useful concept for many theoretical descriptions of aerosol phenomena, including collision theory.[19–31] It is usually obtained by detailed imaging of particles, which allows routine characterization of the fractal dimensions of both films and particles. Although the concept of fractals is useful for describing the shapes of some particles, the theory of fractals does not apply when only a few particles are included in an agglomerate. In that case, fractal concepts provide convenient correlations between the number of primary particles in the agglomerate, the size of the agglomerate, and the fractal dimension. Frequently, aspherical particles are called *fractallike* to circumvent the rigor of fractal theory. It is useful to note that shape factors are also employed to account for the fact that the agglomerates are not spherical.[8] Finally, the concept of fractals, as applied to agglomerates, is used primarily for particles made by gas-phase, and not spray, routes.

## 2.7 CHEMICAL AND PHASE COMPOSITION

Particles and films often do not contain a uniform distribution of the elements that make up the material, either while it is being formed or processed or even when it is collected. This nonuniform distribution may be intentional or unintentional. Examples of intentional inhomogeneity are coated particles, as when titanium dioxide pigment particles are coated with silica and alumina, and multilayer films used in solar cells. Another example is functionally graded films that rely on a gradient in composition to obtain unique properties (e.g., optical fibers), join metals to ceramics, or obtain graded interfaces between two phases. In these cases, to characterize the material, it is necessary to know the location of the elements and phases within the particles or films.

There are a variety of cases in which unintentional inhomogeneity occurs. In some of them, aerosol-derived materials involve elemental particles such as metals (e.g., Pd, Ag, Au, etc.) and semiconductors (e.g., Si and Ge). But more commonly, these materials are doped with other elements, or more complicated materials are formed in which two, three, or more metallic elements are present, in addition to oxygen, nitrogen, or sulfur. Several scenarios are possible: The variation in composition from particle to particle may be completely random, particles of a given size may have different compositions, or the inhomogeneity can be associated with particular sizes (i.e., particles of similar size could have the same composition, but larger particles could have a different composition from smaller particles). A knowledge of the distribution of the compositional inhomogeneity as a function of particle size often provides an insight into the mechanism of particle formation and the origin of the inhomogeneity. The same effects may result in crystalline-phase nonuniformity with respect to particle size: The observed phase may depend on the particle size or may vary from particle to particle without being correlated with the particle size. An example in which the distribution of elements with respect to particle size is nonuniform is the Pb–Bi–Sr–Ca–Cu–O superconductor materials. (See Fig. 2.7.)[32]

In most cases, the lack of a constant composition from particle to particle is undesirable, and strategies to eliminate this problem are required. For example, in aerosol synthesis of Pb–Ru–O materials, which have applications as resistors (e.g., lead ruthenate), preferential loss of the more volatile ruthenium component as ruthenium oxides ($RuO_3$, $RuO_4$) can lead to large Ru-deficient particles and the formation of much smaller ruthenium oxide particles by gas-to-particle conversion.[33] The electrical properties of the nonvolatile ruthenium oxide ($RuO_2$) are different from those of the lead ruthenate ($Pb_2Ru_2O_x$) and may lead to problems in processing during the formation of resistors. Therefore, the chemical and phase composition of the particles must be characterized and measured.

When both the number and composition of particles vary with size, aerosols must be characterized by *multidimensional distribution functions*. For example, the particle mass and elemental composition distribution for two-component particles requires knowledge of both the distribution of the number or mass of particles with size $d_p$ and the fraction $x_1$ of component 1 in the material as a function of $d_p$. The distribution function $m(d_p, x_1)$ has two variables, $d_p$ and $x_1$, (size and composition), so it is two dimensional. Although this type of distribution function is conceptually advantageous it is rarely used in practice because of the difficulties in obtaining the requisite information experimentally and in developing theoretical descriptions.

Solid particles or films can consist of phases that may be either amorphous or crystalline.[1] In this book, a phase is a region over which a material has a

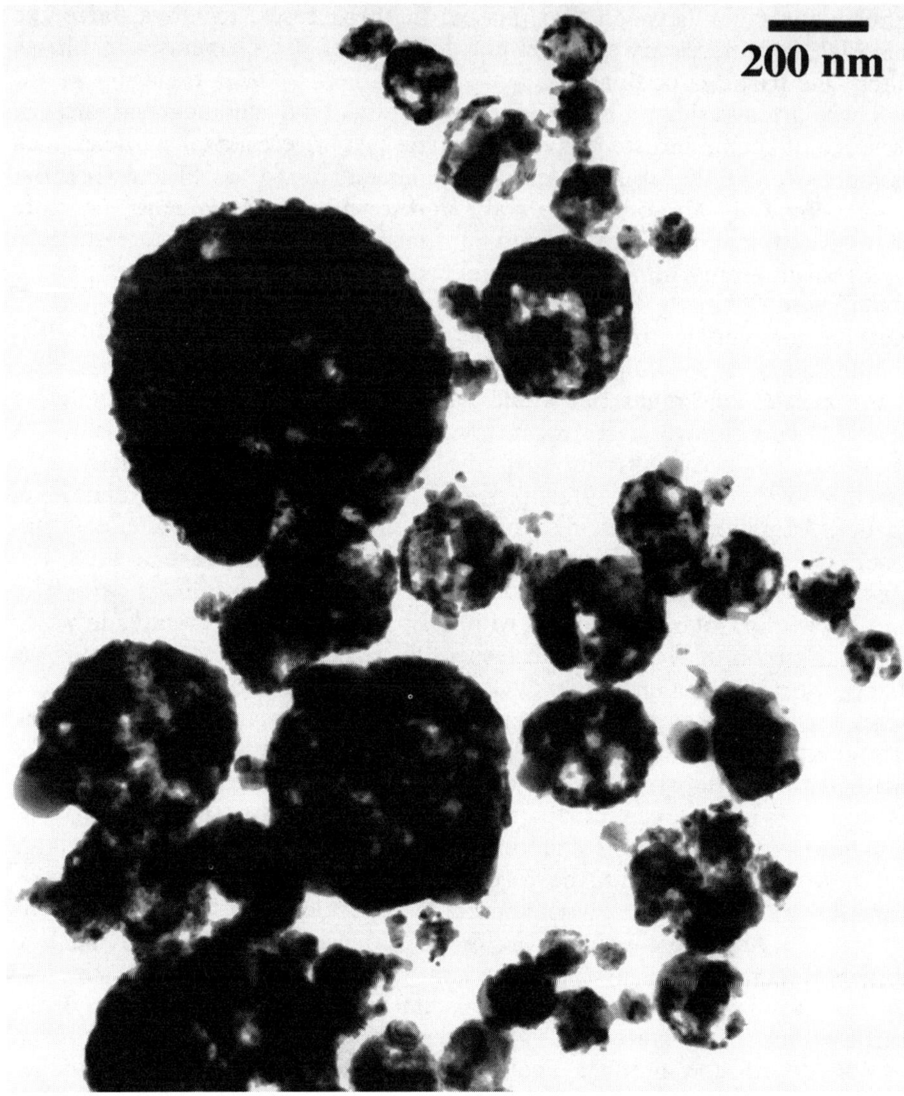


**Fig. 2.7.** TEM photograph of Pb-Bi-Sr-Ca-Cu-O particles produced by spray pyrolysis. (Courtesy of Gurav and Kodas.)

well-defined, uniform composition and structure; a phase may be crystalline or *glassy* (amorphous). Although the composition of the material can be the same in either phase, the crystalline structure may be different. This phenomenon is termed *polymorphism*. An example of a polymorphic system (two or more phases with the same composition but different structures) is anatase and rutile, two forms of $TiO_2$ in which the elemental compositions are identical, but the crystal structures of the Ti and O atoms are different, which leads to different properties. Another possibility is that the compositions of two phases are not the same. An example of two phases containing the same elements, but in different ratios (stoichiometries), is $BaTiO_3$ and $Ba_2TiO_4$; the latter phase is undesirable for most applications. A more subtle example is $PbTiO_3$, which is the perovskite phase of lead titanate, while $Pb_2Ti_2O_{7-x}$, where $x = 1–0$, is the pyrochlore phase. More extreme examples of two or more phases of a material are those in which each phase contains different elements, such as particles or films consisting of regions (phases) of $TiO_2$ and $Ba_2TiO_4$ where the overall composition averaged over both phases is $BaTiO_3$. Depending on the system, these phases may pose a difficult characterization problem. As with crystallinity, phase composition plays a role in the dynamics (particle formation, growth, interdiffusion, etc.) of materials synthesis and processing.

The description of phase composition as a function of particle size is similar to that of elemental composition. The use of distribution functions to account for phase distribution is even rarer than the use of such functions to describe size and elemental composition. However, one motivation for considering this type of distribution function is that phase evolution in particles can be dependent on particle size and is relatively common. This kind of dependence has been shown for certain materials in which the thermodynamically stable phase changes below a certain particle size. (See Chapter 9.) For example, $BaTiO_3$, which is used as a temperature sensor, exists in tetragonal crystal form at room temperature when the crystallite size is larger than 10 nm, but exists in cubic crystal form below this size. The effect on the particle properties is severe, because the higher symmetry results in a loss of properties (such as ferro-, pyro-, and piezo-electricity) associated with the lower symmetry (polar) crystal form. Similarly, smaller particles are often formed near the outlet of a reactor towards the end of the residence time, leading to less time available for phase transformation and crystal growth. This difference means that the smaller particles can have a different phase composition than larger particles. One situation in which this can occur is during the conversion of anatase to rutile.[34]

## 2.8 CRYSTALLINITY

In addition to the crystalline phase (i.e., the particular crystalline structure a material adopts), the level of crystallinity is important. Particles and coatings can be amorphous or crystalline in structure.[1] In *amorphous materials* there is no long-range order over distances greater than a few angstroms, while in *crystalline materials* there is a regular repetition of an arrangement of atoms or molecules (commonly called a *unit cell*) over a distance much larger than the size of the atom or molecule, commonly up to length scales of micrometers. Highly crystalline materials have large *crystallites*, corresponding to a long distance over which the unit cell is repeated. In contrast, poorly crystalline materials have small crystallites corresponding to a short distance over which the unit cell is repeated. Crystallinity plays a major role in determining the properties of materials and is important in determining the dynamics of both particle and coating formation. It may also determine the presence or absence of specific properties, as in the case of pigments, electroceramics, etc. *Polycrystalline* particles or films contain many crystallites, while *single-crystal* particles and films contain only a single crystal. The connections between properties and crystallite sizes for nanostructured materials are described in Chapter 9.

Materials produced by aerosol processes often exist under dramatically different conditions, depending on the method used to form the particles. Reactors can operate near room temperature or at temperatures high enough to dissociate molecules and ionize atoms, as in plasmas. These extremes can lead to large differences in the crystallinity of materials and in the size of the crystallites found in them. In extreme cases, the material may be single

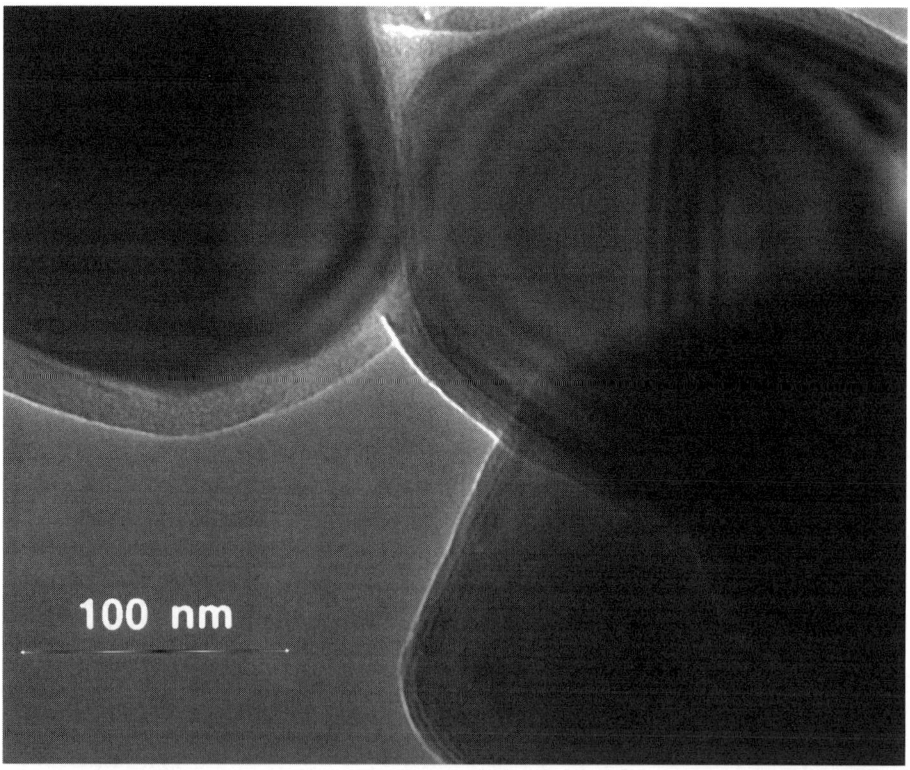


**Fig. 2.8.** TEM photograph of titanium dioxide particles produced by gas-phase reaction of $TiCl_4$ and $O_2$. (Courtesy of Gurav, Kodas, and Kemira, Inc.)

crystal, may exhibit a preferred orientation (in the case of films), or, at the opposite end, may be amorphous. However, in many cases, the crystallinity of the film or powder can be controlled through the processing conditions, particularly the temperature-vs.-time history of the process. Because crystallinity is an important factor in determining powder and film properties, methods are needed to quantify this parameter. Experimentally, crystallinity can be probed using the Scherrer equation for X-ray diffraction and transmission electron microscopy, with electron diffraction or direct imaging of the material at the atomic scale.[2] These approaches can provide the crystallite size in particles and films (Fig. 2.8) and are discussed in more detail in Chapter 17.

In principle, the average crystallite size for a particle can vary with the particle size, requiring the use of a multidimensional distribution function to describe the material. Fortunately, crystallite size does not usually depend *strongly* on particle size, because it is mainly a function of temperature. For this reason, studies usually report the crystallite size for a powder or film, averaged over the entire sample.

## 2.9 REFERENCES


1. Kingery, W.D., Bowen, A.K., & Uhlmann, D.R. *Introduction to Ceramics*. New York: John Wiley and Sons, 1976.
2. Arai, Y. *Chemistry of Powder Production*. New York: Chapman and Hall, 1996.
3. German, R.M. *Sintering Theory and Practice*. New York: John Wiley and Sons, 1996.
4. Schmidt, G. (Ed.). *Clusters and Colloids*. New York: VCH, 1994.
5. Hitchman, M.L. & Jensen, K.F. *Chemical Vapor Deposition: Principles and Applications*. New York: Academic Press, 1993.
6. Kodas, T.T. & Hampden-Smith, M.J. (Eds.). *The Chemistry of Metal CVD*. Weinheim, West Germany: VCH Verlagsgesellschaft, 1994.
7. Bird, R., Stewart, W.E., & Lightfoot, E.N. *Transport Phenomena*. New York: John Wiley and Sons, 1960.
8. Hinds, W.C. *Aerosol Technology*. New York: Wiley Interscience, 1982.
9. Friedlander, S.K. *Smoke, Dust and Haze*. New York: Wiley Interscience, 1977.
10. Seinfeld, J.H. *Atmospheric Chemistry and Physics of Air Pollution*. New York: Wiley Interscience, 1986.
11. Williams, M.M.R. & Loyalka, S.K. *Aerosol Science: Theory and Practice*. New York: Pergamon Press, 1991.
12. Hidy, G.M. & Brock, J.R. *The Dynamics of Aerocolloidal Systems*. New York: Pergamon Press, 1970.
13. Patton, T.C. (Ed.). *Pigment Handbook*. New York: John Wiley & Sons, 1972.
14. Gurav, A.S., Kodas, T.T., Kauppinen, E.I., Joutsensaari, J., & Zilliacus, R. *J. Mater. Res.* *10*(7), 1995, 1644–1652.

15. Reist, P.C. *Introduction to Aerosol Science*. New York: MacMillan Publishing Co., 1984.
16. Suzuki, M., Kagawa, M., Syono, Y., & Hirai, T. *J. Mater. Sci. 27*(3), 1992, 679–684.
17. Suzuki, M., Kagawa, M., Syono, Y., & Hirai, T. *Ceramic Powder Science IV, Ceramic Transactions 22*, 1991 147–152.
18. Mandelbrott, B. *The Fractal Geometry of Nature*. New York: W.H. Freeman, 1983.
19. Akhtar, M.K., Pratsinis, S.E., & Mastrangelo, S.V.R. *J. Mater. Res. 9*(5), 1994, 1241–1249.
20. Akhtar, M.K., Lipscomb, G.G., & Pratsinis, S.E. *Aerosol Sci. Technol. 21*, 1994, 83–93.
21. Akhtar, M.K., Vemury, S., & Pratsinis, S.E. *AIChE J. 40*(7), 1994, 1183–1192.
22. Akhtar, M.K., Vemury, S., & Pratsinis, S.E. *NanoStruct. Mater. 4*(5), 1994, 537–544.
23. Rogak, S.N., Baltensperger, U., & Flagan, R.C. *Aerosol Sci. Technol. 14*, 1991, 447–458.
24. Rogak, S.N., Flagan, R.C., & Nguyen, H.V. *Aerosol Sci. Technol. 18*(1), 1993, 25–47.
25. Xie, Y., Hopke, P.K., Casuccio, G., & Henderson, B. *Aerosol Sci. Technol. 20*, 1994, 161–168.
26. Xie, G.W., Patel, S., Wong, F.C.H., & Shaw, D.T. *Aerosol Sci. Technol. 22*, 1995, 219–235.
27. Gutsch, A., Pratsinis, S.E., & Loffler, F. *J. Aerosol Sci. 26*(2), 1995, 187–199.
28. Colbeck, I. & Wu, Z. *J. Phys. D. Appl. Phys. 27*(3), 1994, 670–675.
29. Rogak, S.N. & Flagan, R.C. *J. Aerosol Sci. 23*(7), 1992, 693–710.
30. Rogak, S.N. & Flagan, R.C. *J. Colloid Interface Sci. 151*(1), 1992, 203–224.
31. Rogak, S.N. & Flagan, R.C. *Part. Syst. Charact. 9*(1), 1992, 19–27.
32. Ward, T., Lyons, S., Kodas, T.T., Brynestad, Y., Kroeger, D., & Hsu, H. *Physica C 200*, 1992, 31–42.
33. Gurav, A.S., Kodas, T.T., Kauppinen, E.I., Joutsensaari, J., & Zilliacus, R. unpublished manuscript, 1997.
34. Akhtar, M.K., Xiong, Y., & Pratsinis, S.E. *AIChE 37*(10), 1991, 1561–1570.

# 3 Particle Transport

## 3.1 INTRODUCTION

During the aerosol synthesis of materials, particle characteristics change dramatically, resulting in a profound effect on particle transport properties. For example, in the manufacture of carbon blacks, hydrocarbon vapors and oxygen react in turbulent flame reactors to form small spherical or near-spherical soot particles (Fig. 3.1). These small particles may leave the process stream by *diffusion* to the reactor walls. As the particles grow larger and the temperature of the process drops, particle fusion slows down, and aggregates of smaller, spherical (primary) particles are formed. As the agglomerates flow along the reactor, some may become so large that they settle towards the reactor walls via *sedimentation*. Some particles—especially those near the reactor walls downstream of the flame front—experience strong temperature gradients, forcing them to deposit by *thermophoresis*. Other particles also end up at the reactor walls by means of turbulent eddies (*turbulent diffusion*) or *inertial impaction* at tube bends. Later in the process, the particles may be charged and collected by using *electric fields* in electrostatic precipitators. In other material systems, the particles may be magnetic and can be removed by the use of *magnetic fields*. In addition, some kinds of single-crystal particles or fibers can be aligned in magnetic fields. All of these physico-chemical phenomena can drastically alter the particle size distribution.

Several of these basic deposition phenomena are also common to a variety of film generation routes including solid-particle deposition, spray pyrolysis, and aerosol-assisted chemical vapor deposition (AACVD) (Fig. 3.2). The mechanisms underlying these phenomena are shown in Figure 3.3. The *resuspension* of particles also can be important during particle deposition. This complex problem was reviewed recently by Ziskind et al.[1] and others.[2, 3]

## 3.2 DIFFUSION

Particles suspended in gases are in continuous motion as a result of collisions with gas molecules, which transfer some of their translational energy to the particles, forcing them to move in a random (Brownian) motion. This motion, although random at the atomic scale, results in net particle transport from areas of high to low concentration. Diffusion of particles is also the mechanism by which collisions occur between particles larger than the gas mean

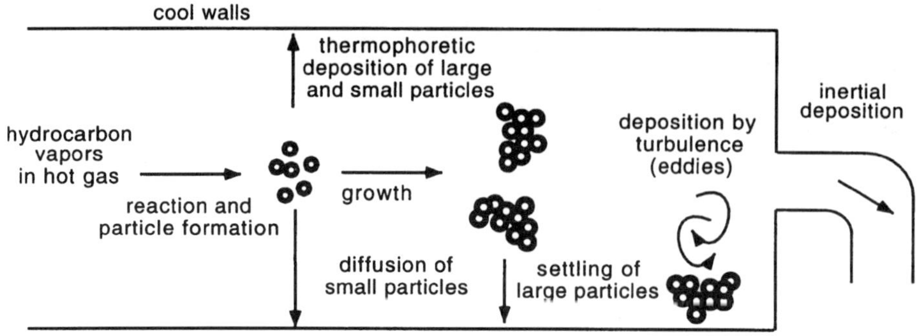


**Fig. 3.1.** Physicochemical phenomena occurring during carbon black formation, with emphasis on particle transport.

free path, and it is a mechanism of particle deposition onto surfaces when the particle size is sufficiently small and the diffusion distance sufficiently short, as in CVD processes in which particles can be formed just above the surface of a material.[4]

The process of particle diffusion in a single direction is usually quantified in terms of the flux of the particles, $J$ (number/cm$^2$ sec)—that is, the number of particles crossing a unit area of gas per unit time:[5]

$$J = -D\frac{\partial N_\infty}{\partial x}. \tag{3.1}$$

Here $D$ (cm$^2$/sec) is the diffusion coefficient, $N_\infty$ (number/cm$^3$) is the particle concentration, and $x$ (cm) is the coordinate in the direction of transport.

The diffusion coefficient or particle diffusivity is given by[5, 6]

$$D = \frac{k_B T C_c}{3\pi\mu d_p}, \tag{3.2}$$

where $k_B$ is the Boltzmann constant (erg/molecule K), T is temperature (K), $\mu$ is the gas viscosity (g/cm sec), $d_p$ is the particle diameter (cm), and $C_c$ is the Cunningham correction factor (dimensionless).

The Cunningham correction factor accounts for the reduced drag of particles when their size is comparable to or smaller than the gas mean free path $\lambda$ (say, $d_p \leq 10\lambda$):

$$C_c = 1 + \frac{2\lambda}{d_p}\left[A_1 + A_2 \exp\left(-A_3\frac{d_p}{2\lambda}\right)\right]. \tag{3.3}$$

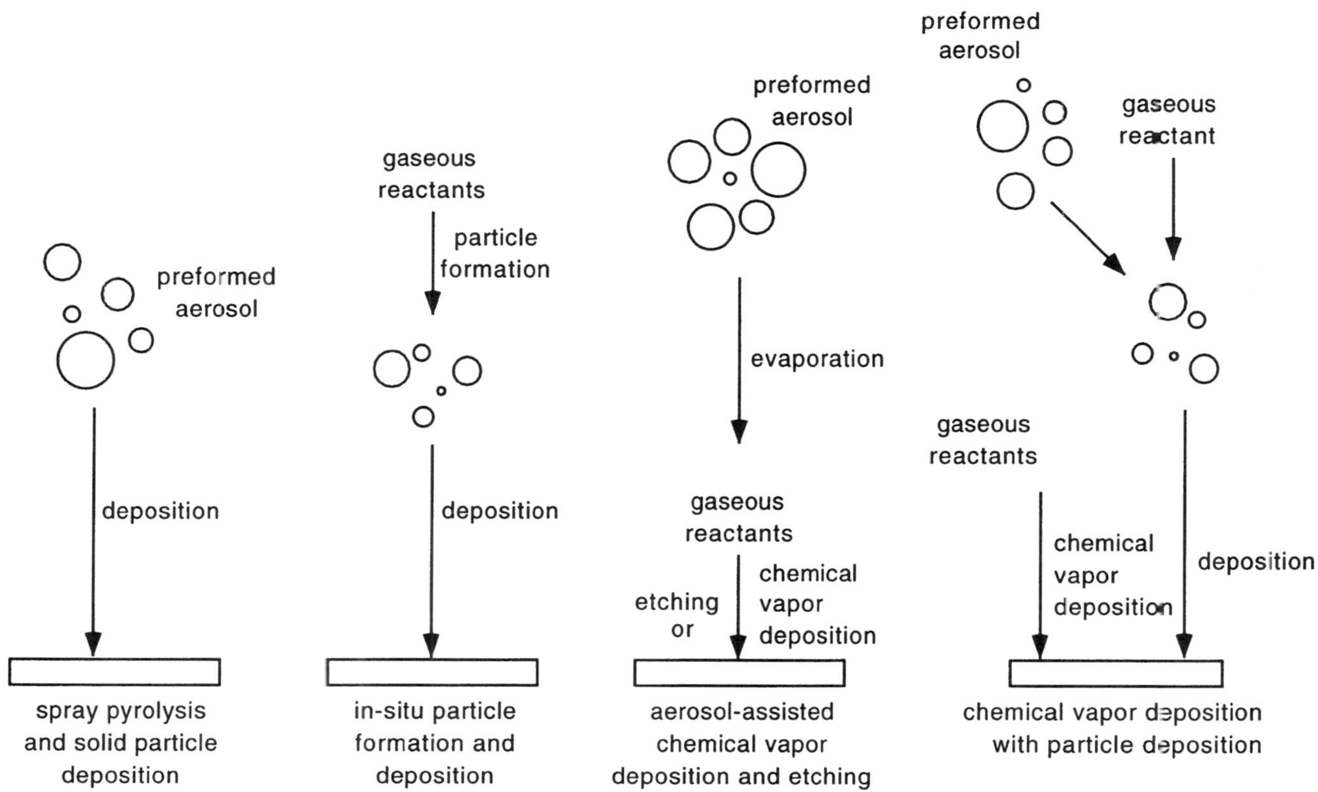


**Fig. 3.2.** Role of particle deposition in various processes for fabrication of films.

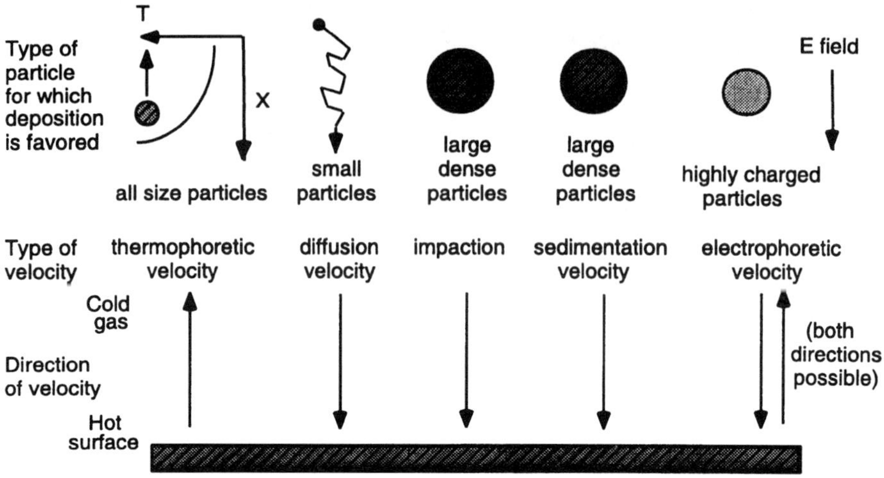


**Fig. 3.3.** Particle transport phenomena.

The coefficients $A_1$, $A_2$, and $A_3$ are empirical constants whose values have been determined by measuring the settling velocities of particles in various gases. Table 3.1 gives the values of these constants for various gases.[6, 7] The Cunningham correction factor does not change very much with different gases for constant Kn. The values of $A_1$, $A_2$, and $A_3$ for air can be used for other gases not listed in the table. (See Tables 3.1 and 3.2.) The Cunningham correction factor is high for large Kn because a large Kn corresponds to a small particle size relative to the gas mean free path. This in turn results in a greater degree of "slip."

In many aerosol routes to particles and films, it is useful to know the dependencies of the deposition velocities on pressure and particle size due to the different driving forces. The information discussed in this chapter can be used to determine what changes to make to a system to increase or decrease its deposition efficiency. In physical terms, the dependencies on pressure and particle size are primarily due to the "slip" of the particles as their size approaches the gas mean free path, which is described by the Cunningham

**TABLE 3.1. Constants for the Cunningham correction factor in various gases**

| Gas | $A_1$ | $A_2$ | $A_3$ |
|---|---|---|---|
| Air | 1.207 | 0.4 | 0.78 |
| Ar | 1.227 | 0.38 | 0.85 |
| $H_2$ | 1.141 | 0.506 | 0.85 |
| $CO_2$ | 1.208 | 0.44 | 0.92 |

**TABLE 3.2. Cunningham correction factors for various gases and Knudsen numbers**

| $\mathrm{Kn} = 2\lambda/d_p$ | Air | Ar | He |
|---|---|---|---|
| 0.01 | 1.01207 | 1.01227 | 1.01227 |
| 0.1 | 1.1207 | 1.1227 | 1.1277 |
| 1 | 2.3904 | 2.3894 | 2.3203 |
| 10 | 16.770 | 16.7608 | 16.390 |
| 100 | 161.40 | 161.38 | 160.07 |

correction factor $C_c$. In terms of pressure and particle diameter, the Cunningham correction factor is[6]

$$C_c = 1 + (2/pd_p)\left[6.32 + 2.01\exp(-0.1095pd_p)\right], \tag{3.4}$$

where $p$ is the absolute pressure (cm Hg) and $d_p$ is particle diameter ($\mu$m). Figure 3.4 is a plot of this equation as a function of particle diameter with pressure as a parameter.[8] The plot shows that $C_c$ depends only on the product $pd_p$. The correction factor increases as pressure and particle size decrease. This factor appears in the expression for nearly all of the particle velocities (thermophoretic, diffusional, impaction, electrophoretic, and settling) where it is usually the term with the strongest dependence on pressure.

Diffusion coefficients can be calculated once the Cunningham correction factors are known. Hinds has given a table of particle diffusivities as a function of particle size.[6] As an example, the diffusivity of a 1-nm-diameter

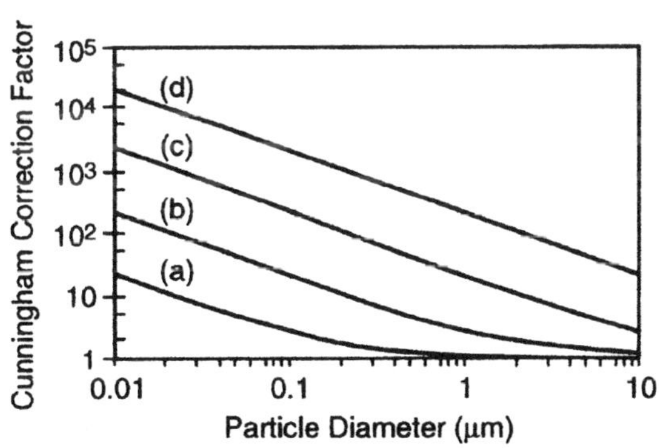


**Fig. 3.4.** Cunningham correction factor as a function of particle size and gas pressure. (a) 1013 mbar, (b) 100 mbar, (c) 10 mbar, (d) 1 mbar. Adapted from Donovan, R.P., Yamamoto, T., Periasamy, R., & Clayton, A.C., Mechanisms of particle transport in process equipment, *J. Electrochem. Soc. 140* (10), 2917–2927, 1993. Reproduced by permission of The Electrochemical Society, Inc.

particle is roughly 0.05 $cm^2/sec$, while that for a particle 1 $\mu m$ in diameter is $2.8 \times 10^{-7}$ $cm^2/sec$, in air at 20°C and 1 atm. By way of comparison, the diffusion coefficient of a small molecule such as $CO_2$ in liquid water is roughly $10^{-5}$ $cm^2/sec$ and of $CO_2$ in air is roughly 0.1 $cm^2/sec$.

Diffusion is significant only in the transport of submicron-size particles because the diffusion coefficient depends strongly on particle size. The probability for a particle to diffuse a root mean square distance $x_{rms}$ is related to the diffusion time $t$ and the diffusion coefficient $D$ by the equation[6]

$$x_{rms} = (2Dt)^{1/2} \tag{3.5}$$

This equation can be used with values of the diffusion coefficient to calculate the average distance particles diffuse during a typical reactor residence time. At standard temperature and pressure (STP), particles with diameters of 1 $\mu m$ and 0.01 $\mu m$ have diffusion coefficients of $2.8 \times 10^{-7}$ $cm^2/sec$ and $5.3 \times 10^{-4}$ $cm^2/sec$, respectively. For a 10-sec residence time, these correspond to diffusion distances of 0.02 mm and 1 mm, respectively. The numbers emphasize that diffusion is important only for small particles and short distances.

As a note of caution, several concepts exist regarding the transport of aerosols for which intuition from liquid-phase processes is not applicable. One of these concepts pertains to the diffusion coefficients of particles in gases relative to those in liquids. A particle with a diameter of 0.1 $\mu m$ in a liquid can have a diffusion coefficient of $10^{-10}$ $cm^2/sec$, while the same particle can have a diffusion coefficient of $10^{-6}$ $cm^2/sec$ in a gas. This difference is similar to that for molecular species, for which a molecule may have a diffusion coefficient of $10^{-5}$ $cm^2/sec$ in a liquid and 0.1 $cm^2/sec$ in a gas. As a consequence, diffusion is a poor method for collecting particles from liquids, while in contrast, it plays a dominant role in the collection of particles from gases when the particle size is less than 0.1 $\mu m$.

### 3.2.1 Wall Losses by Diffusion

In geometries of practical importance, such as those describing the sampling of aerosol streams and the transport of particles through conduits, diffusional losses often occur in a cylindrical tube. The fraction of particles entering such a tube that deposits inside the tube or other channel or body is called the *deposition efficiency* $E_d$, while the remaining fraction of the particles that does not deposit is called the *penetration* $P_d$. This situation has been studied for a variety of geometries.[9–14]

When the flow is laminar (the Reynolds number $Re_t$ of the tube is less than 2100; $Re_t = d_t U \rho / \mu$, where $U$ is the average gas velocity through the tube, $\rho$ is the gas density, $d_t$ is tube diameter, and $\mu$ is the gas viscosity), the fraction of particles lost by diffusion to the tube walls is obtained by solving the diffusion equation for fully developed laminar flow (i.e., flow such that the velocity profile of the gas does not change with distance along the tube) in circular conduits. Hinds has cast these solutions in an easy-to-use format for calculation of the aerosol penetration $P_d$ through tubes:[6]

$$P_d = 1 - 5.50\left(\frac{DL}{Q}\right)^{2/3} + 3.77\left(\frac{DL}{Q}\right) \quad \text{for} \quad \frac{DL}{Q} < 0.007; \tag{3.6}$$

$$P_d = 0.718\exp\left(-\frac{11.5DL}{Q}\right) + 0.0975\exp\left(-\frac{70.1DL}{Q}\right) \tag{3.7}$$

$$+0.0325\exp\left(-\frac{179DL}{Q}\right) \quad \text{for} \quad \frac{DL}{Q} > 0.007.$$

In these equations, $D$ (cm$^2$/sec) is the particle diffusivity, $Q$ (cm$^3$/sec) is the aerosol or gas flow rate, and $L$ (cm) is the tube length. It is interesting to note that there is no dependence on the tube diameter. Also the equations do not account for particle losses at the tube inlet when the flow is not fully developed. As an example, applying them to 0.1-$\mu$m particles in a gas flow of 1.0 L/min through a 1-m tube gives 0.006 fractional loss. For 0.001-$\mu$m particles, the fractional loss is 0.976.

When the flow is turbulent ($\mathrm{Re}_t > 2100$), the aerosol penetration is

$$P_d = \exp\left(-\frac{4\kappa L}{d_t U}\right), \tag{3.8}$$

where $d_t$ is the tube diameter (cm), $U$ is the average gas velocity (cm/sec), $L$ is the tube length (cm), and $\kappa$ is the mass transfer coefficient (or deposition velocity) by turbulent diffusion (cm/sec).[15] Depending on the assumptions made, this expression for the penetration can be written in various forms to obtain the mass transfer coefficient. For example,

$$P_d = \exp\left(\frac{-0.14L}{d_t \mathrm{Re}_t^{1/8} \mathrm{Sc}^{3/4}}\right) \tag{3.9}$$

for $\mathrm{Re}_t > 10{,}000$, where Sc is the Schmidt number $\mu/\rho D$.[6] This expression, as well as other forms, is for turbulent diffusion of submicron-size particles. For larger particles, their inertia should be explicitly considered. Several papers discuss these and other aspects of turbulent deposition.[16–27]

The preceding expressions for the penetration shows that, in general, Brownian diffusion is important only for smaller particles, 10 nm or less, and small tube diameters. The phenomenon is also more important in turbulent flows, for which the effective diffusion coefficients are much higher than for laminar flows. Because most systems have high flow rates and large tube diameters, diffusional losses are usually negligible. An exception is the case of ultrafine (< 10 nm) particles, for which diffusion coefficients can be high enough to result in losses.

The preceding expressions for the penetration are inadequate for more complicated situations in which particles are formed in the system. In these situations, two types of approaches are often taken to model the problem. The first is empirical and relies on a mass transfer coefficient; this approach is described in the next section. Several correlations have been developed to predict the amounts of deposition of particles in tubes and on surfaces with various geometries. The second approach is more theoretical and more complex; in it, spatial variations of particle concentration are taken into account, and particle diffusion is handled in a manner similar to that for diffusion of molecular species. This approach is also described in Section 3.4.

## 3.3 TRANSPORT COEFFICIENT APPROACH TO DIFFUSIONAL DEPOSITION OF PARTICLES

Particle deposition onto surfaces is important in methods for fabricating films. In some cases, the flow geometry is more complex than that for either a jet directed at a flat surface or flow through a tube. As with the latter, diffusion of particles is a mechanism by which only very small (usually $< 10$ nm) particles can deposit on the surface. Thus, this phenomenon is often of negligible importance in droplet deposition, molten particle deposition, and optical fiber production. It can, however, play an important role in aerosol delivery of volatile precursors for CVD, particle deposition during CVD and etching, and simultaneous particle deposition and CVD if the particles are small enough.

This empirical approach exploits a quantity called a *transport coefficient* or *deposition velocity*, which can be defined in terms of the concentration of the particles in the gas above the surface and other parameters as

$$V_M = \frac{J}{N_o} = \frac{\Delta N_\infty}{N_o A_s \Delta t}, \tag{3.10}$$

where $V_M$ is the particle deposition velocity (cm/sec), $J$ is the particle flux to the surface (number/cm$^2$ sec), $N_o$ is the particle concentration at the edge of the concentration boundary layer adjacent to the surface (number/cm$^3$), $\Delta N_\infty$ is the total particles deposited (number), $A_s$ is the substrate surface area (cm$^2$), and $\Delta t$ is the deposition time (sec). The deposition velocity varies with the mechanism of particle transport and deposition.

The use of the mass transport coefficient or deposition velocity to estimate deposition rates $J$ requires the ability to estimate values for $V_M$. It is difficult to make theoretical estimates of $V_M$ for situations in which the diffusion is coupled to the flow of the gas—that is, where there is *convective diffusion*. In this case, particle transport occurs by a combination of flow to regions near the substrate surface, followed by diffusion across a concentration boundary layer (a region over which the particle concentration changes). Therefore, an alternative approach is to use, for the transport coefficient, empirical correlations that take into account convective diffusion. The form of these correlations is supported by theory, which says that the diffusion velocity in dimensionless form (Sherwood number = Sh = $V_M L/D$) can be written in terms of other dimensionless groups (Reynolds and Schmidt numbers) describing the importance of fluid flow and diffusion of particles.

One equation that can be used to predict particle deposition flux due to convective diffusion for a jet directed at a flat plate is[8]

$$V_M = \mathrm{Sh}\, D/L, \tag{3.11}$$

where

$$\mathrm{Sh} = 1.08\left(\frac{U_o L \rho_f}{\mu}\right)^{1/2}\left(\frac{\mu}{\rho_f D}\right)^{1/3} = 1.08\,\mathrm{Re}_L^{1/2}\,\mathrm{Sc}^{1/3}, \tag{3.12}$$

in which Sh is the Sherwood number (also known as the Nusselt number $\mathrm{Nu}_M$ for mass transport, $D$ is the particle diffusion coefficient, $L$ is the substrate diameter, $\mu$ is the viscosity of the gas, $U_o$ is the free-stream velocity, and $\rho_f$ is the gas density. The quantities $\mathrm{Re}_L$ and Sc are the Reynolds number for the gas flow and the Schmidt number for the particles, respectively. The dependence of the diffusion velocity on pressure is a function of both $C_c$ (which appears in $D$) and the gas density $\rho_f$ because, for an ideal gas, $\rho_f$ depends directly on pressure.

It is useful to examine the predictions of this equation regarding deposition flux to a flat plate. The sum $V_t$ of the convective diffusion velocity and the sedimentation velocity is shown in Figure 3.5.[8] This figure uses the same

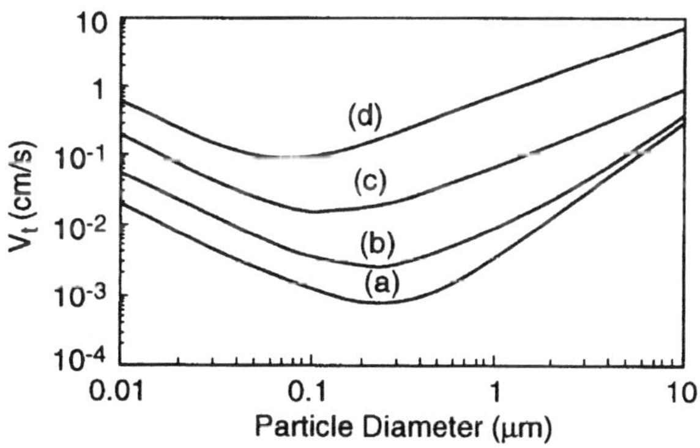


**Fig. 3.5.** The net deposition velocity due to sedimentation and diffusion as a function of particle size in a free-stream velocity of 50.8 cm/s. (a) $P = 1013$ mbar, (b) $P = 100$ mbar, (c) $P = 10$ mbar, (d) $P = 1$ mbar. Adapted from Donovan, R.P., Yamamoto, T., Periasamy, R., & Clayton, A.C., Mechanisms of particle transport in process equipment, *J. Electrochem. Soc. 140* (10), 2917–2927, 1993. Reproduced by permission of The Electrochemical Society, Inc.

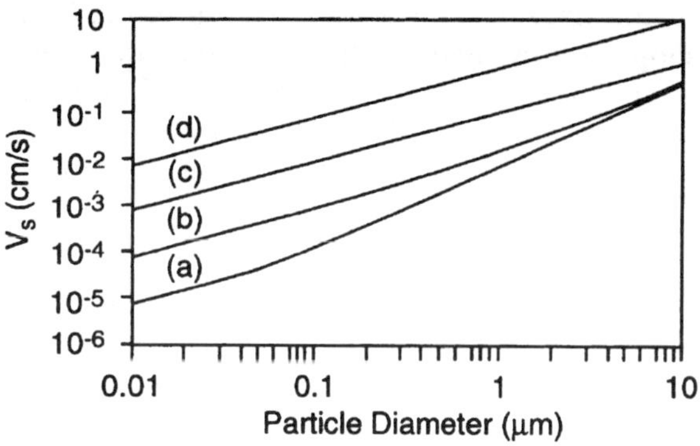


**Fig. 3.6.** Particle deposition velocity due to sedimentation as a function of particle size. (a) $P = 1013$ mbar, (b) $P = 100$ mbar, (c) $P = 10$ mbar, (d) $P = 1$ mbar. Adapted from Donovan, R.P., Yamamoto, T., Periasamy, R., & Clayton, A.C., Mechanisms of particle transport in process equipment, *J. Electrochem. Soc. 140* (10), 2917–2927, 1993. Reproduced by permission of The Electrochemical Society, Inc.

values for $V_s$ that appear in Figure 3.6: the values of $V_M$ calculated for a substrate diameter of 12.5 cm, a free-stream velocity of 50.8 cm/s, and a temperature of 20°C.[8] At atmospheric pressure, convective diffusion, $V_M$, dominates the deposition velocity for particle diameters less than 0.1 $\mu$m, while settling, $V_s$, dominates for particle diameters greater than 0.5 $\mu$m. The region between these two values is a transition region where both mechanisms are important. The influence of decreasing pressure is to decrease the diameter for which the transition region occurs, indicating that sedimentation is more important relative to diffusion as pressure decreases. However, both the diffusion velocity and the sedimentation velocity increase significantly with decreasing pressure. The minimum in deposition velocity occurs for the following reason: The diffusional deposition velocity *increases* monotonically with decreasing particle size because the diffusion coefficient of the particles increases with decreasing size, while the sedimentation velocity *increases* monotonically with increasing particle size because of the increased mass of the particles.

## 3.4 DIFFUSION EQUATION APPROACH TO PARTICLE DEPOSITION

Particle transport with negligible inertial effects can be described by equations of the same form as those describing the diffusion of molecular species.[28] This approach is often taken when modeling the deposition of particles onto surfaces. Equations of the following form must be solved:

$$\mathbf{v} \cdot \nabla n = D\nabla^2 n - \nabla \cdot \mathbf{c} n. \tag{3.13}$$

Here **v** is the fluid velocity vector, $n$ is the particle size distribution function, $D$ is the particle diffusion coefficient, and **c** is the velocity vector.

Note that this form of the diffusion equation assumes that the diffusion coefficient does not depend on position or particle size. For transport in one dimension, the equation is of the form

$$v_x \frac{\partial n}{\partial x} - D\frac{\partial^2 n}{\partial x^2} - \frac{\partial (c_x n)}{\partial x}, \tag{3.14}$$

where $v_x$ is the gas velocity in the $x$-direction and $c_x$ is the velocity in the x-direction associated with thermophoresis, electrophoresis, photophoresis, and settling. This equation couples particle transport by diffusion, convection (gas flow), and the various particle velocities due to force fields and other phenomena. Evaporation of particles enters into the equation through its influence on the diffusion coefficient $D$ and the velocities $c$. Equation (3.14) has been solved both analytically and numerically for a wide variety of physical situations.[10–13] Equations of this form can in principle be solved to give the deposition velocity of particles due to diffusion in different geometries. An alternative approach is to use empirical correlations for diffusion velocities, as discussed in Section 3.3.

## 3.5 PARTICLE MOTION IN EXTERNAL FORCE FIELDS

External force fields (gravitational, electrical, magnetic, or thermal) can influence particle motion through effects that are exploited in particle collection and separation devices such as settling chambers, centrifuges, cyclones, and electrical and thermal precipitators.[6] These effects are also important in aerosol processing of materials, not only for collecting particles, but in specific processes as well. For example, in aerosol-assisted CVD of films, sprays containing precursor droplets are directed at a substrate, above which the droplets evaporate. The volatile reactants then decompose on the surface. The particles can be charged and an electric field applied to increase the transport rate of particles to regions near the hot surface, where the precursors can evaporate.

Strong force fields such as gravitational, electrical, magnetic, and thermal fields normally overshadow the forces of Brownian motion. Particle motion in these fields rather quickly becomes steady (reaching a constant migration velocity $V_t$), because of the balance between the field forces and the aerodynamic resistance (drag) from the gas. The migration velocity $V_t$ determines the deposition flux $J$ of the aerosol according to the equation

$$J = -V_t N_\infty, \tag{3.15}$$

where $N_\infty$ is the particle concentration in the gas.

For the fraction of the particles that does not deposit during flow through a conduit or over a surface, $P_d$, the penetration, can be determined by the corresponding equation for the flow geometry. For example, in Equation (3.8), the mass transfer coefficient $\kappa$ is the same as the migration (deposition) velocity $V_t$. As a result, it is important to determine this velocity for various fields in order to calculate particle losses.

### 3.5.1 Sedimentation in a Gravitational Field

Sedimentation, or settling, refers to the steady particle motion resulting from the balance of gravity and gas resistance (drag) forces. The gravity force is important only when large, dense particles are involved. Applications include the collection of highly agglomerated particles or large particles in large chambers with long residence times. Sedimentation can also play a role in the deposition of droplets when the droplet size and density are sufficiently large.[11–13]

The gravitational force that drives the settling velocity is

$$F_G = m_p g = \frac{\pi d_p^3}{6}(\rho_p - \rho) g, \tag{3.16}$$

where $g$ is the gravitational acceleration, $\rho_p$ is the density of a spherical particle of diameter $d_p$ (g/cm$^3$), $m_p$ is the particle mass (g), and $\rho$ is the gas density (g/cm$^3$). The resistance force (drag) on the particle is

$$F_R = \frac{C_D}{C_c}\frac{\pi}{8}\rho d_p^2 V_s^2, \tag{3.17}$$

where $V_s$ is the settling velocity of the particle, $C_c$ is the Cunningham correction factor, $\rho$ is the gas density, and $C_D$ is the drag coefficient of the particle, which depends on its Reynolds number $\mathrm{Re}_p$.

Particles commonly encountered in materials synthesis are of micron and submicron sizes. For these particles, $\mathrm{Re}_p$ is usually less than 0.1 $\mu$m, for which

$$C_D = 24/\mathrm{Re}_p \tag{3.18}$$

and particle motion is in the Stokes regime. If $\mathrm{Re}_p$ is larger than 0.1, then other expressions should be used for $C_D$[29–31] (Table 3.3).

The *dynamic shape factor* is used to account for the effect of shape on particle motion. This factor, defined as the ratio of the actual drag force of the nonspherical particle to the drag force of a sphere having the same volume and velocity, is given by

$$\chi = \frac{F_R}{3\pi\mu V d_e}, \tag{3.19}$$

where $d_e$ (cm) is the equivalent volume diameter—that is, the diameter of a sphere with the same volume as the irregular particle.

**TABLE 3.3. Particle drag coefficient as a function of particle Reynolds number**[31]

| | Regimes of $Re_p$ | $C_D$ |
|---|---|---|
| Laminar Flow/ Stokes Regime | $Re_p < 0.1$ | $\dfrac{24}{Re_p}$ |
| Transition Flow | $0.1 \leq Re_p \leq 2$ | $\dfrac{24}{Re_p}\left(1 + \dfrac{3}{16}Re_p + \dfrac{9}{160}Re_p^2\right)\ln(2\,Re_p)$ |
| | $2 \leq Re_p < 500$ | $\dfrac{24}{Re_p}(1 + 0.15\,Re_p^{0.687})$ |
| Turbulent Flow | $500 < Re_p < 2 \times 10^5$ | 0.44 |

The sedimentation, or settling, velocity of particles in the Stokes regime ($Re_p < 0.1$) can be obtained by combining Equations (3.16) to (3.19) and observing that $\rho_p \gg \rho$ for aerosols:

$$V_s = \frac{\rho_p d_p^2 g C_c}{18 \mu \chi}. \tag{3.20}$$

It is worthwhile to examine the terminal settling velocities, the time required for settling, and the ratio of the settling time to the residence time in a reactor, as shown in Table 3.4. In general, particles smaller than 1 $\mu$m follow the flow of the gas and do not settle at the commonly employed residence times (< 1 min) in aerosol synthesis of materials. Note that artificial g fields (as in a spiral duct centrifuge) increase the effective gravitational field; the settling velocity equation is still valid then, but with increased g.

As another example, Table 3.5 lists settling velocities for spherical titania particles ($\rho_p = 4.26$ g/cm$^3$) in air. At 20°C and 500°C, the viscosity of the air is $1.813 \times 10^{-4}$ g/cm sec and $3.5 \times 10^{-4}$ g/cm sec, respectively, while the corresponding mean free paths are 0.065 $\mu$m and 0.172 $\mu$m, respectively. Again, settling velocities are so low that the distance traveled in typical

**TABLE 3.4. Settling velocities, settling time, and ratio of settling time ($\tau_s$) to reactor residence time ($\tau_r$ = 10 sec) for unit density particles at standard conditions with 10-cm reactor dimension in direction of settling**

| $d_p$, $\mu$m | $V_s$ cm/sec | $\tau_s$ sec, | $\tau_s/\tau_r$ |
|---|---|---|---|
| 0.1 | $8.7 \times 10^{-5}$ | $1.1 \times 10^5$ | $1.1 \times 10^4$ |
| 1 | $3.5 \times 10^{-3}$ | $2.9 \times 10^3$ | $2.9 \times 10^3$ |
| 10 | 0.3 | 33 | 3.3 |
| 100 | 25 | 0.4 | 0.04 |

**TABLE 3.5. Settling velocities (cm/sec) of $TiO_2$ particles in air at two temperatures (1 atm)**

| $d_p$, μm | 20°C | 500°C |
|---|---|---|
| 0.01 | $2.76 \times 10^{-5}$ | $3.71 \times 10^{-5}$ |
| 0.1 | $3.65 \times 10^{-4}$ | $4.14 \times 10^{-4}$ |
| 1 | 0.015 | $9.48 \times 10^{-3}$ |
| 10 | 1.3 | 0.691 |
| 100 | 128 | 66 |

residence times is negligible. For this reason, deposition by settling is only rarely incorporated into models for gas-to-particle conversion. The influence of turbulence or gravitational settling in a well-mixed system has also been studied.[32-34]

The theory for deposition velocity also provides insight into film fabrication processes that exploit droplets of volatile species. Gravitational settling of particles can be (1) the primary mechanism for the deposition of large particles in vertically oriented systems with the substrate placed at the bottom and (2) a particle loss mechanism in horizontally oriented systems. It is often the primary mechanism by which the deposition of droplets (spray pyrolysis) occurs. Deposition by settling is not significant in most other aerosol routes to films.

Some insight can be gained by considering the terminal settling velocity of evaporating aerosol droplets or particles. (See Table 3.4.)[6] Settling velocities for unit-density 10-μm particles are 0.3 cm/sec, often negligible relative to the velocity of the gas stream. At the same time, particles larger than 10 μm have appreciable settling velocities and can strike the surface before evaporation is complete. For example, particles with diameters on the order of 100, 500, and 1000 μm have settling velocities of 25, 200, and 390 cm/sec, respectively.[6] These particles have sufficiently high settling velocities and long evaporation times to allow droplets to deposit. Particles consisting of high-density liquids such as molten metals or of solids have even higher deposition velocities for the same diameters, because of the dependence of the deposition velocity on density.

Although settling is not important for submicron particles at one atmosphere, reduced pressures result in higher settling velocities. Figure 3.6 shows the settling velocity as a function of particle size, with pressure as a parameter, for unit-density particles in air.[8] The primary factor that changes with pressure in the equation for the settling velocity is $C_c$. As shown earlier, $C_c$ increases as pressure and particle size decrease. The dependence of the gas viscosity on pressure is not great and in the limit of low pressures is zero.[28] Figure 3.6 demonstrates that reducing pressure increases the settling velocity nearly inversely with pressure and that smaller particles settle more slowly than larger particles.

The discussion above suggests that particles formed by gas-to-particle conversion processes whose sizes are almost always below 1 μm would not deposit by settling unless the pressure is reduced dramatically. Thus, settling is mainly a problem in spray processes with large droplets, as when large powder particles are entrained into a gas and when particles with high densities are used.

### 3.5.2 Motion of Charged Particles in Electric Fields

Particles are frequently charged and passed through electric fields to facilitate their separation (e.g., in electrical precipitators), characterization (e.g., in electrical aerosol analyzers), or deposition for film formation (e.g., by spray pyrolysis). Or one may alter the particles' characteristics by enhancing or inhibiting coagulation. In many practical systems, such as plasma reactors or high-temperature flame reactors, particles are charged by the process itself. Particle losses due to charging of particles are also of concern.[35, 36]

The force exerted by an electric field on a particle is[35]

$$F_E = ZeE, \tag{3.21}$$

where $Z$ is the number of charges per particle, which depends on the material, particle size, and charging mechanism (field, diffusion, contact, or thermionic charging); $e$ is the elemental electron charge ($4.8 \times 10^{-10}$ statcoulomb); and $E$ is the intensity of the electric field. For an aerosol having the equilibrium bipolar (equal numbers of positively and negatively charged ions) charge distribution, often called the *Boltzman equilibrium distribution*, the average number of charges per particle is[6]

$$Z = 2.37\sqrt{d_p}\,, \tag{3.22}$$

where $d_p$ is the particle diameter in μm. For example, for a particle with a diameter of 0.2 μm, $Z = 1$. Much higher charge levels, however, can be reached by unipolar charging (inducing one type of charge, such as negative charges obtained from electrons). Hinds has plotted charge limits for various cases in which particles in the range of 0.1 to 1.0 μm can acquire 1000 to $10^6$ charges in the electron limit and $10^4$ to $10^7$ charges in the positive-ion limit.[6]

Charged particles can be formed in a variety of ways.[5, 6, 29, 30] In diffusion charging, ions are created in the gas, and Brownian motion (diffusion) of the ions and particles is the mechanism by which the ions reach the particles. The ions can be created by dissociation of species such as CsCl at high temperatures. Particles can also be charged by field charging, in which charged species are moved by the force of an electric field. When the electric field lines intersect the particles, the ions strike the particles, which results in charging. This can occur in corona discharges, a common method of producing charged species. Other modes of charging also exist.[6]

For solid particles, the maximum number of acquired charges is

$$Z_{s,\,\mathrm{max}} = \frac{d_p^2 E_L}{4e}, \tag{3.23}$$

where the surface field strength for spontaneous emission of electron or ions, $E_L$, is $3 \times 10^4$ stV/cm and $7 \times 10^5$ stV/cm for negatively and positively charged particles, respectively. This can correspond to thousands of charges per particle for many materials. For liquid droplets, the maximum number of charges per particle is the *Rayleigh limit*;

$$Z_{l,\,\mathrm{max}} = \frac{\left(2\pi\gamma d_p^3\right)^{1/2}}{e}, \tag{3.24}$$

where $\gamma$ is the surface tension.

When particle motion under the influence of the electric field force $F_E$ is balanced by the resistance of the fluid drag, $F_R$, the electrical migration velocity $V_E$ can be obtained by equating Equations (3.21) and (3.17). For the common condition of $\mathrm{Re}_p < 0.1$, so that $C_D = 24/\mathrm{Re}_p$, the electrical migration velocity in the direction of the field becomes

$$V_E = \frac{ZeEC_c}{3\pi\mu d_p}. \tag{3.25}$$

The deposition of charged particles is of greatest importance in aerosol routes to films. In this case, highly charged particles and an applied electrical field are often used to increase the deposition rate. The approach is useful for droplet deposition systems in which a significant fraction of the droplets would not strike the surface because of the mechanisms described earlier. Charged particles in electric fields are commonly exploited for deposition of droplets (spray pyrolysis), a mechanism that is the basis for particle collection in what are known as *electrostatic precipitators*.

To examine this problem further, Equation (3.25) can be written in the slightly different form,[8]

$$V_E = \frac{q_{\mathrm{rms}}\,\psi E C_c}{3\pi\mu d_p}, \tag{3.26}$$

where $E$ is the electric field, $C_c$ is the Cunningham correction factor, $q_{\mathrm{rms}}$ is the rms charge on a particle, $\psi = q/q_{\mathrm{rms}}$ is the normalized charge, and $q$ is the number of charges on a particle.

The terminal velocity of a charged particle can be much greater than the velocities due to diffusion, settling, and thermophoresis. The quantity $V_E/V_s$ is 1160 for a 0.6-$\mu$m particle with 40 charges.[6] The charging of particles has been used to dramatically increase deposition rates during spray pyrolysis.[37, 38]

The quantity $\psi E$ in Equation (3.26) contains the electrophoretic variables. With $E$ in V/cm, $\psi E$ values on the order of 10 V/cm are very low, $\psi E$'s of 100 V/cm low, $\psi E$'s of 1000 V/cm moderate, $\psi E$'s of 10,000 V/cm high, and $\psi E > 10{,}000$ V/cm very high. A key feature of electrophoresis is that even low values of $\psi E$ correspond to significant contributions to the total deposition velocity.

Figure 3.7 shows the sum of the electrophoretic, diffusional, and settling velocities.[8] These curves show that even a weak field increases the deposition velocity dramatically and that lowering pressure also increases the deposition velocity. At both pressures illustrated, the electrophoretic velocity is higher than the sedimentation or convective diffusion velocity for particles with diameters below 1 $\mu$m. Only above 1 $\mu$m does sedimentation begin to dominate.

The solid curves in Figure 3.7 assume that the electrophoretic velocity is in the direction of settling and diffusion. However, the electric field can be reversed, resulting in repulsion of particles from the surface, as shown by the dashed curves. These curves show that modest electrical fields can result in a complete absence of deposition for particles less than 1 $\mu$m in diameter and can serve as the basis for a method to avoid particle deposition during the formation of particles in CVD and etching processes if the particles can be charged appropriately.

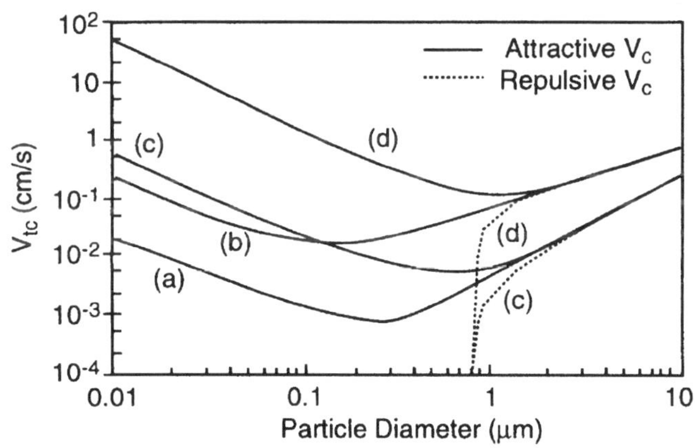


**Fig. 3.7.** Effects of electrostatics and pressure on particle deposition. (a) $P = 1013$ mbar, $\psi E = 0$ V/cm; (b) $P = 10$ mbar, $\psi E = 0$ V/cm; (c) $P = 1013$ mbar, $\psi E = 100$ V/cm; (d) $P = 10$ mbar, $\psi E = 100$ V/cm. Adapted from Donovan, R.P., Yamamoto, T., Periasamy, R., & Clayton, A.C., Mechanisms of particle transport in process equipment, *J. Electrochem Soc.*, *140* (10), 2917–2927, 1993. Reproduced by permission of The Electrochemical Society, Inc.

### 3.5.3 Particle Motion in Thermal Gradients

Temperature gradients in a gas induce particle motion by a process called *thermophoresis*, in which particles move from high- to low-temperature regions.[6] Thermophoresis is responsible for particle deposits at cold spots on reactor walls and exhaust tubes. The phenomenon is exploited in aerosol synthesis of preforms for optical fibers, but reduces deposition rates during aerosol-assisted CVD, droplet deposition, and solid-particle deposition onto heated surfaces.

The thermophoretic force $F_{th}$ on the particles is balanced by the fluid drag force (Eq. 3.17), resulting in a constant thermophoretic velocity.[39] The most commonly used expression for the thermophoretic velocity is from Talbot et al., who analyzed a large number of data sets and proposed a correlation for the thermophoretic coefficient encompassing a spectrum of particle sizes.[40] These researchers suggested the equation

$$V_{\text{th}} = \frac{-\alpha\mu}{\rho}\nabla \ln T, \tag{3.27}$$

with

$$\alpha = \frac{2.294\left[k/k_p + 2.20\,\text{Kn}\right]}{(1 + 3.438\,\text{Kn})\left[1 + 2(k/k_p) + 4.40\,\text{Kn}\right]} \times\{1 + \text{Kn}[1.2 + 0.41\exp(-0.88/\text{Kn})]\} \tag{3.28}$$

where $\alpha$ is the thermophoretic coefficient, Kn is the Knudsen number, $k$ is the thermal conductivity of the gas (ergs/cm sec K), $k_p$ is the thermal conductivity of the particle (ergs/cm sec K), $\mu$ is the gas viscosity (g/cm sec), and $\rho$ is the gas density (g/cm$^3$).[41]

For submicron particles, the thermophoretic velocity is nearly independent of the particle size. For larger particles, temperature gradients develop inside the particle, so the thermophoretic coefficient $\alpha$ depends on the thermal conductivity and other material properties of the particle. Table 3.6 gives examples of the thermophoretic velocity.[6] This velocity is often much higher than the diffusional and settling velocities, making it an effective method of particle deposition or collection. Numerous papers have addressed various aspects of thermophoretic particle motion and deposition.[39–60]

**TABLE 3.6. Thermophoretic velocity for particles in temperature gradient of 100°C/cm at 20°C for reactor dimension of 1 cm, reactor residence time $\tau_r$ of 10 seconds, and $k_p = 10k$**

| $d_p$, μm | $V_{th}$, cm/sec | $\tau_{th}$, sec | $\tau_{th}/\tau_r$ |
|---|---|---|---|
| 0.1 | $2.0 \times 10^{-2}$ | 50 | 5 |
| 1 | $1.3 \times 10^{-2}$ | 77 | 7.7 |
| 10 | $7.8 \times 10^{-3}$ | 130 | 13 |

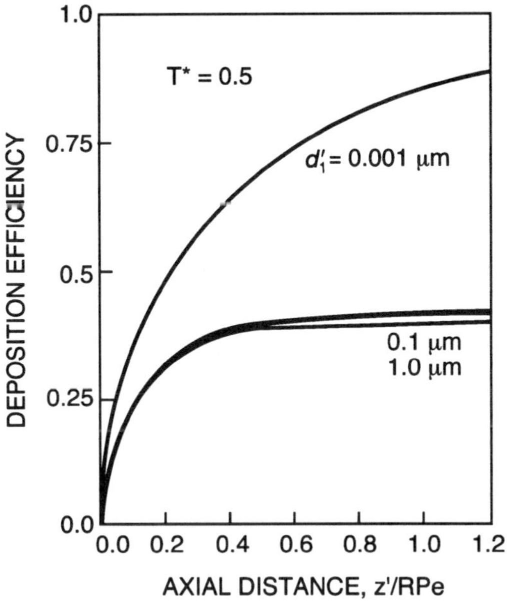


**Fig. 3.8.** Thermophoretic deposition efficiency for $SiO_2$ particles flowing through a tube with cool walls. Reprinted from *Journal of Aerosol Science*. Pratsinis, S.E. and Kim, K.-S, *20*(1), 101–111, Copyright 1989, with kind permission from Elsevier Science Ltd., The Boulevard, Langford Lane, Kidlington OX5 1GB UK.

Thermophoresis is often the most common loss mechanism for particles at both the laboratory and industrial scales because it is driven by temperature differences between a hot gas and the cool walls, a situation that is commonly encountered in aerosol reactors. This configuration is found in the industrial-scale production of $TiO_2$ pigment, carbon black, fumed silica, and other materials where a hot gas must be passed through a tube with cooled walls in order to cool the gas before particles are collected. Figure 3.8 shows the efficiency of thermophoretic deposition of $SiO_2$ particles onto the walls of a cooled tube.[61] Because deposition efficiency depends on particle size only for particles smaller than 0.01 $\mu$m and occurs in the regions with the highest temperature gradients (shortest axial distances), significant portions of the aerosol may be lost to the walls. Therefore, there is a motivation to use more sophisticated methods of cooling the gas, such as the annular introduction of cool gas or the introduction of cool gas through porous walls.

Thermophoresis plays a somewhat different role in aerosol routes to films, where it is a mechanism that provides a driving force to keep particles that are suspended in a cool gas stream from striking a hot surface. Thermophoresis is especially important for keeping undesired particles from depositing on surfaces during CVD with a hot substrate and cooler walls. It is the mechanism by which optical fibers are produced and is also the basis for

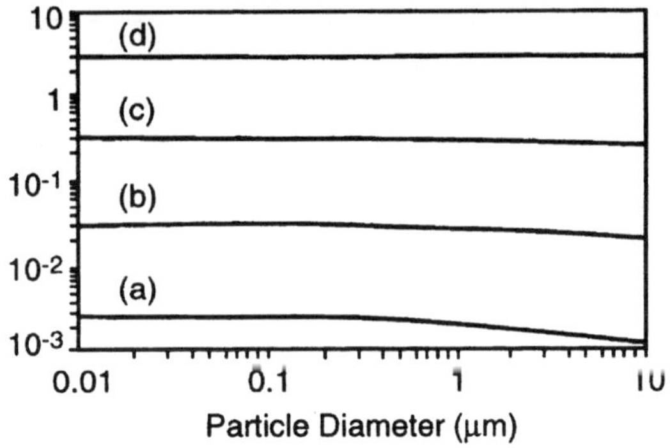


**Fig. 3.9.** Pressure-dependent particle deposition velocity due to thermophoretic force alone. $\nabla T$ is 10 K/cm. (a) 1013 mbar, (b) 100 mbar, (c) 10 mbar, (d) 1 mbar. Adapted from Donovan, R.P., Yamamoto, T., Periasamy, R., & Clayton, A.C., Mechanisms of particle transport in process equipment *140* (10), 2917–2927, 1993. Reproduced by permission of The Electrochemical Society, Inc.

particle-precipitation-aided CVD. In droplet deposition and aerosol delivery of precursors for CVD, thermophoresis drives particles away from the hot surface, keeping them from penetrating into hotter regions of the gas where their evaporation rates would be increased.

Simple calculations can be used to understand the role of thermophoresis in aerosol routes to films. The thermophoretic velocity for ambient conditions with a 100 K/cm gradient (a value commonly found in many CVD processes) is greater than the settling velocity only for particles smaller than 1.0 $\mu$m. (See Table 3.6.) Since $V_{th}$ is not a strong function of particle size, thermophoresis will not hinder the gravitational deposition of particles larger than approximately 1 $\mu$m. However, for particles less than 1 $\mu$m in diameter, the thermophoretic velocity can be much greater than the diffusion velocity, resulting in inhibition of deposition of small particles by diffusion. Nevertheless, for particles that have diameters on the order of nanometers, diffusional velocities can exceed thermophoretic velocities, a fact that can be exploited for deposition of clusters.

Figure 3.9[8] shows the thermophoretic velocity calculated using the expression of Batchelor and Shen; the main dependence of $V_{th}$ on pressure is due to $C_c$ and the Knudsen number.[62] These two terms have nearly opposite dependencies on particle diameter, so that $V_{th}$ has only a weak dependence on particle diameter. The thermophoretic velocity shows an inverse dependence on pressure, with the values of $V_{th}$ for temperature gradients of 10 K/cm being similar to those for the diffusional and settling velocities. The plot is for a temperature gradient of 10 K/cm at a temperature of 293 K, the average molecular weight of air, and a ratio of the thermal conductivities of the gas and the solid of $k/k_p = 0.1$. At atmospheric pressure, decreasing $k/k_p$ to 0.01 influences the results only for particle diameters greater than 1 $\mu$m.

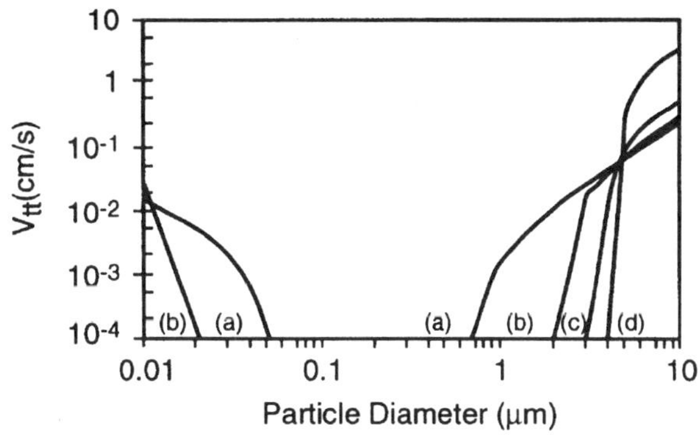


**Fig. 3.10.** Sum of the thermophoretic, settling, and diffusional velocities when the temperature gradient is 10 K/cm with $k/k_p = 0.1$. (a) 1013 mbar, (b) 100 mbar, (c) 10 mbar, (d) 1 mbar. Adapted from Donovan, R.P., Yamamoto, T., Periasamy, R., & Clayton, A.C., Mechanisms of particle transport in process equipment *140* (10), 2917–2927, 1993. Reproduced by permission of The Electrochemical Society, Inc.

Insight into the combined effects of thermophoresis, settling, and diffusion can be obtained by adding the thermophoretic, settling, and diffusional velocities (Fig. 3.10).[8] A key point is that the thermophoretic velocity is often in the direction opposite that of the settling and diffusional velocities; when the substrate temperature is greater than that of the gas, the thermophoretic effect pushes particles away from the surface. The negative sign of the thermophoretic velocity gives a net deposition velocity in the direction away from the surface for particles of roughly 0.05 to 0.7 $\mu$m in diameter at atmospheric pressure; particle deposition occurs only for larger and smaller particle sizes, for which settling and diffusion, respectively, dominate. At lower pressures, the range of particle sizes that do not deposit broadens to the point where, at 10 mbar, particles from 0.02 to 2 $\mu$m do not deposit. Situations like this would require the use of charged particles in an electric field or impaction from jets with high velocities to deposit particles.

## 3.6 IMPACTION

Particle-laden streams often flow in tubes or jets. Neglecting all other forces (electrophoretic, thermophoretic, etc.), particles follow the gas stream lines until the fluid stream changes direction, as in the case of a bend or obstruction in a pipe, a change in pipe diameter, or aerosol spraying of a substrate using a jet of gas. Large particles cannot keep up with the change in the course of the flow; inertia forces them to move in the original direction and impact. On the other hand, small particles possess much less inertia and follow the gas flow. Inertial deposition is the basis for a class of particle-sizing equipment (multistage inertial impactors). Also, the technique is often

**TABLE 3.7. Stopping distance of particles for length scale of 10 cm**

| $d_p$, μm | Stopping Distance $S$ (cm) | St |
|---|---|---|
| 0.1 | $8 \times 10^{-5}$ | $8 \times 10^{-6}$ |
| 1 | $4 \times 10^{-3}$ | $4 \times 10^{-4}$ |
| 10 | 0.23 | 0.023 |
| 100 | 13 | 1.3 |

used to form films by depositing droplets onto surfaces during spray pyrolysis, as well as in solid-particle deposition.

The likelihood of particle deposition by impaction depends on the dimensionless quantity St, the Stokes number, for the particles. The relationship is

$$\mathrm{St} = \frac{\rho_p d_p^2 U C_c}{18 \mu L} = \frac{S}{L}, \tag{3.29}$$

where $S$ is the stopping distance (the distance a particle travels in a gas until it reaches zero velocity with respect to the gas), $\rho_p$ is the particle density, $d_p$ is the particle diameter, $U$ is the velocity of the fluid, $\mu$ is the viscosity of the gas, and $L$ is a characteristic length of the flow geometry.[6] For large Stokes numbers (St $\gg$ 1), impaction occurs; for small Stokes numbers (St $\ll$ 1), it does not. For aerosol flow around cylinders of radius $R$, the threshold (critical) Stokes number for impaction is St = 0.125 with $L = R$. The Stokes number can be interpreted as the ratio of the stopping distance to a length scale over which the fluid turns. Large Stokes numbers mean that the particle cannot stop (i.e., follow the gas flow) in the distance allowed (Table 3.7).

In more complex situations, it is important to develop equations that relate the process conditions ($d_p, U$) to the fraction of particles deposited on surfaces by impaction.[63, 64] This relationship is discussed further in Chapter 7. Here we introduce the topic enough to allow discussion of various impaction processes. In principle, one needs to write the detailed fluid flow and particle motion equations for the specific process. Most of the time, the gas velocity ($U$) is affected not by the particles (unless the loadings are extremely high), but only by the flow geometry and the momentum of the fluid. The equation for particle motion states that the particle acceleration or deceleration depends on the forces acting on the particle and the drag of the fluid; mathematically,

$$m_p \frac{du}{dt} = \Sigma F_i \pm \frac{C_D}{C_c} \frac{\pi}{8} \rho d_p^2 (u - U)^2, \tag{3.30}$$

where $u$ is the particle velocity, $U$ is the gas velocity, $C_D$ is the drag coefficient, $m_p$ is the particle mass, $t$ is time, $C_c$ is the Cunningham correction factor, $F_i$ are the forces exerted on the particle, and $\rho$ is the gas density. The sign on the right side of Equation (3.30) is negative for $u > U$

and positive otherwise. This is the starting point for calculations of particle deposition by impaction using impactors to size particles and for deposition onto surfaces for film generation. A typical procedure requires calculating the velocity field for the particular flow geometry to get $U$, followed by integration of Equation (3.30) to get particle trajectories. The drag on the particle is zero when the particle does not have a relative velocity with respect to the gas ($u = U$). Particle deposition by impaction in the absence of other deposition mechanisms is investigated with the $F_i$ set equal to zero. Deposition with gravitational settling, thermophoresis, electrophoresis, photophoresis, and other external forces is described by inserting the appropriate expression in $F_i$. Evaporation of particles enters into this equation through its influence on many of the terms in the equation: $m_p$, $F_i$, $C_D$, $C_c$, and $d_p$. Crudely, Equation (3.30) is used for particles somewhat larger than 0.1 $\mu$m. For smaller particles, the diffusion rate becomes important, and a different approach must be used, as outlined in Section 3.2.

The foregoing analysis must be generalized when a polydisperse aerosol is considered. The spread of a droplet or particle size distribution is important because it plays a key role in determining the rates of processes such as droplet evaporation and the efficiencies of aerosol deposition. The settling velocity and probability of impaction depend on the particle diameter to a positive power. Charged-particle deposition has a more complex dependence on particle size because of the relationship between the maximum number of charges on a particle and the particle size. Thus, a broad size distribution will result in a broad range of deposition velocities. For this reason, a broad size distribution may provide deposition of only the fraction of the particles with diameters smaller or larger than a certain value. If most of the mass of the particles is in a size range smaller or larger than this value, the efficiency of film formation may be low, even though most of the particles may deposit.

From a fundamental aspect, calculating the deposition rate of an aerosol with a spread of particle sizes involves different approaches, depending on whether Equation (3.14) or (3.30) is appropriate. For deposition in which inertial effects are important, particle trajectories can be calculated separately for different particle sizes using Equation (3.30).[63] The results are then combined to determine the deposition efficiency of a polydisperse aerosol. For deposition in which diffusion is important (Eq. 3.14), the convective diffusion equation can be solved with the particle size distribution function and a diffusion coefficient that depends on particle size and location.[5] This requires a more complex form of the diffusion equation than Equation (3.14). There is a voluminous body of work on particle deposition to which the reader is referred for more detailed discussions.[5, 6]

### 3.6.1 Particle Deposition by Impaction from Aerosol Jets

From the perspective of materials processing, one of the most important geometries pertaining to impaction is a jet directed at a substrate. This geometry is the most common one for fabrication of films. Figure 3.11 shows

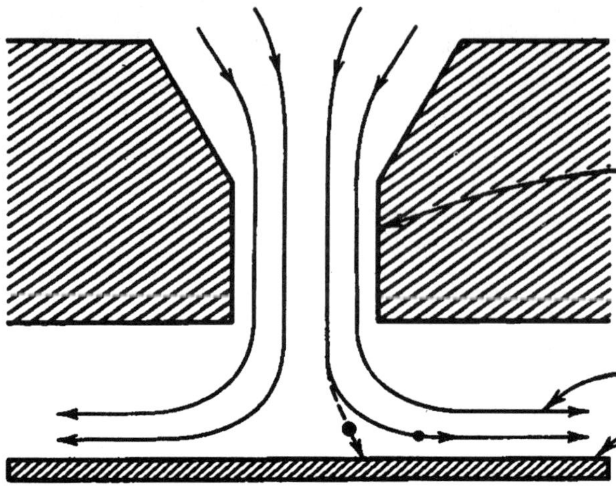


**Fig. 3.11.** Aerosol jet directed at flat plate. Adapted from *Aerosol Technology*, Hinds, W.C., Copyright 1982 Wiley Interscience, Reprinted by permission of John Wiley & Sons, Inc.

a schematic of an aerosol jet expelled from a circular nozzle and impinging on a substrate.[6] As the flow lines show, the fluid flows toward the substrate and then changes direction by 90°, away from the substrate. Particles initially follow the fluid, but as the gas changes direction, larger particles deviate from the gas flow and deposit on the plate, while smaller particles follow the gas and bypass the substrate.

Detailed calculations of the fluid flow field and the equation for particle motion (Eq. 3.30) have been carried out for aerosol jets issuing from rectangular or round nozzles that impinge on plates at cross flows.[65–67] These calculations afford an insight into the conditions required for film generation by droplet or particle deposition. The shape of the nozzle and the Reynolds number for the jet, $\mathrm{Re}_j = 2RU\rho/\mu$ [where $R$ is the radius (or half-width) of the round (rectangular) nozzle], determine the velocity profiles used in the calculations. The Stokes number determines the fraction of particles depositing on the plate (the deposition efficiency), $E_I$, for impactors. Figure 3.12 shows deposition efficiency curves as a function of particle size ($\sqrt{St}$) for different stages of a multijet impactor.[68] In this case, each stage corresponds to a smaller jet diameter ($R$ in the preceding equation for the Reynold's number) and lower pressure (which influenced the Cunningham correction factor $C_c$). The figure shows that impaction can be used to deposit particles with diameters much smaller than can be deposited by gravitational settling. Particles with diameters on the order of nanometers have been deposited by reducing pressure and increasing gas velocity, suggesting that cluster deposition can be carried out by impaction.[69]

Table 3.7 provides some useful guidelines for film formation processes. Larger particle sizes and shorter distances to the substrate favor impaction

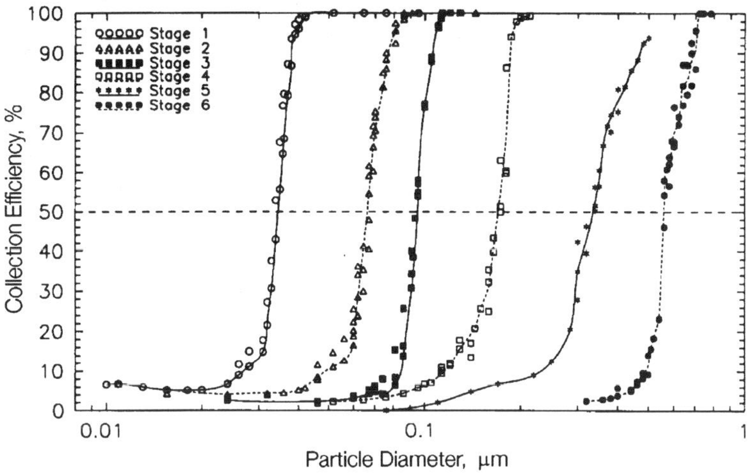


**Fig. 3.12.** Particle collection efficiencies for different stages in a multistage impactor. Reprinted with permission from *Aerosol Science and Technology*, Hillamo, R.E. and Kauppinen, E.I., "On the performance of the Bernor low pressure impactor," **14**, 33–47, 1991. Elsevier Science Inc., 655 Ave. of the Americas, New York, NY 10010-5107.

(higher Stokes numbers). Thus, impaction is favored by directing the aerosol particles at the surface at a high velocity, using particles that are sufficiently large and dense, and reducing pressure. For example, stopping distances for unit-density spheres (in the absence of gravity at 1 atmosphere) of 1, 10, and 100 $\mu$m are on the order of $10^{-3}$, 0.1, and 10 cm, respectively. Thus, particles or droplets much smaller than 10 $\mu$m cannot be deposited by impaction unless gas jets with narrow diameters and high velocities are used or the pressure is reduced. Examining evaporation times and terminal settling velocities suggests that droplets greater than 1 to 100 $\mu$m are required for droplet deposition by this mechanism using centimeter-size nozzles and typical gas flow rates.

In practical applications, impaction efficiencies can be increased by decreasing the pressure, because of the dependence of the Cunningham correction factor on pressure. This approach has been used extensively for inertial particle deposition. Particles smaller than 10 nm can be deposited at low pressures, at high gas flow rates, and with small nozzle diameters.[69]

### 3.6.2 Particle Deposition in Tube Bends

Aerosol processing of materials requires passing particle-laden streams through tubing and ducts of a variety of diameters and geometries. The role of particle losses in sampling is discussed by Vincent.[70] As an aerosol flows through pipes, particles can impact and deposit at the tube bends. Pui et al. measured particle losses at various angles of tube bends for $Re_t$ = 100 to 10,000 and Stokes numbers of 0.03 to 1.46.[71] The fractions of particles that passed through (penetrated) the tube bend was correlated by

$$P_{90°} = \frac{n_{\text{out}}}{n_{\text{in}}} = 10^{-0.963\,\text{St}} \tag{3.31}$$

for a bend angle of 90° and

$$P_{45°} = \frac{n_{\text{out}}}{n_{\text{in}}} = 10^{-0.482\,\text{St}} \tag{3.32}$$

for a bend angle of 45°, where $n_{\text{out}}$ and $n_{\text{in}}$ are the particle concentrations exiting and entering the bend, respectively, and St is the Stokes number.[71] As an example, for an aerosol of 0.5 $\mu$m with a density of 1 g/cm$^3$ in air at 20°C flowing through a 5-cm-radius tube at 1 m/s, the penetration is 0.70, meaning that 30% of the particles will deposit to the tube bend.

### 3.6.3 Turbulent Deposition in Pipe Flows

Most industrial systems utilize turbulent flow through tubes for powder production. The highly turbulent flows (with Re = $10^6$) are most important for particles larger than 1 $\mu$m, upon which inertial forces are significant, and for high Reynolds numbers, for which small eddies are present near the tube walls. Particles in *eddies*, small swirling currents of gas, are directed at walls by the flow in the eddies. The inertia of the particles causes them to depart from the flow of the eddies and to pass through the laminar sublayer near the tube wall and strike the wall.

Friedlander and Johnstone proposed that particles migrate from the turbulent core to the viscous sublayer near the tube walls with a diffusivity equal to the eddy diffusivity of the fluid.[72] Then the particles travel by inertia across the laminar sublayer if their stopping distance is greater than the thickness of the sublayer. The stopping distance, the distance a particle travels into a fluid if the particle has an initial velocity $u$, is

$$S = \frac{\rho_p d_p^2 u}{18\mu}. \tag{3.33}$$

In the case of turbulent deposition, $u$ is the mean radial fluctuating velocity of the fluid and is given by

$$u_{\text{mrf}} = 0.9U\left(\frac{f}{2}\right)^{1/2}, \tag{3.34}$$

where $f$ is the Fanning friction factor and $U$ is the main stream fluid velocity. The quantity $f$ is a standard chemical engineering concept discussed in many texts.[28] Crudely speaking, this fluctuating velocity is due to the eddies in the turbulent flow and is responsible for propelling the particles across the laminar sublayer near the tube walls.

When the stopping distance is smaller than the viscous sublayer, particles can still deposit from the presence of fluctuations in the viscous sublayer, and the deposition velocity (or mass transfer coefficient) $\kappa$ becomes

$$\frac{\kappa}{U} = \frac{\rho^2 d_p^4 \rho_p^2 U^4}{6 \times 10^5 \mu^4}\left(\frac{f}{2}\right)^{5/2} \qquad \left[\text{for } \frac{SU}{\nu}\left(\frac{f}{2}\right)^{1/2} < 5\right], \tag{3.35}$$

where $\nu$ is the kinematic viscosity. When the stopping distance is larger than the thickness of this buffer layer,

$$\frac{\kappa}{U(f/2)^{1/2}} = \frac{1 - 2S/d_t}{25\ln(d_t/S) - 1.73} \qquad \left[\text{for } \frac{S}{d_t} < 0.5 \text{ and } \frac{SU}{\nu}\left(\frac{f}{2}\right)^{1/2} > 30\right], \tag{3.36}$$

where $d_t$ is the tube diameter. The fraction of particles not deposited on the pipe walls (the fraction exiting the tube) can be determined from Equation (3.8), using Equation (3.35) or (3.36) for $\kappa$.[73]

It is worth noting that, in some cases, foreign particles are introduced into reactors purposely to remove wall deposits. For example, fine sand particles can be added to gas flows with the intent of keeping the walls clean via particle re-entrainment.

## 3.7 PHOTOPHORESIS AND RADIOMETRY

Photophoresis is particle motion induced by photons. There are two interactions: radiometric effects and radiation pressure. Radiometry is similar in some ways to thermophoresis: The photons impinging on the particle are absorbed, which results in an uneven temperature in the particle, which in turn creates a temperature gradient in the gas. This induced temperature difference in the particles results in a thermophoretic force on the particle. For materials processing, the effect is important only for laser-induced processes. Arnold and Lewittes give expressions for this radiometric force on a particle.[73]

Radiation pressure is the result of the momentum transfer that is associated with photons interacting with the particles. Expressions for the force associated with this phenomenon are given by Ashkin.[74] Again, the force is important only for laser-induced processes.

## 3.8 STRUCTURE OF DEPOSITS

Particle and vapor deposition onto surfaces can result in coatings with a wide variety of structures; however, the phenomenon remains poorly understood.[75] Structure is an important consideration in the formation of optical fibers,

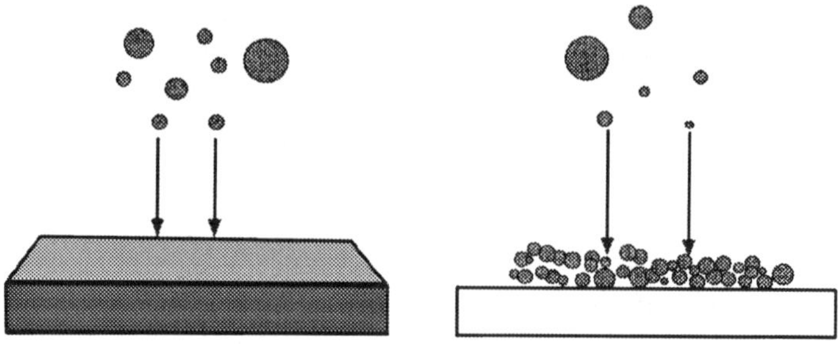

**Fig. 3.13.** Extremes of film morphology resulting from rapid and slow sintering of particles relative to the particle deposition rate.

sensors, dense and porous membranes, etc. An extensive literature relates the structure of films formed by physical and chemical vapor deposition processes to the deposition conditions.[4] This literature can be used to obtain insights into film morphology during the formation of films by aerosol delivery of volatile precursors for CVD and for deposition of clusters. Far fewer results are available for particle deposition,[75] and these are mostly from the literature on filtration, in which the structure of the deposits influences the collection efficiency.[6]

The overall features of importance for particle deposition include the structure of the depositing particles (spheroids, agglomerates, etc.), the mechanism by which particles deposit (ballistic, diffusive), reentrainment, and the sintering characteristics of the deposited particles. Several types of limiting cases can be identified. Deposition of particles onto a heated substrate can occur in a mode in which the film is smooth and dense at all times if the characteristic time for sintering or fusion ($\tau_f$) is much shorter than the time between particle collisions with the same region on the surface ($\tau_{pd}$). Physically, particles collide and sinter to the surface before another particle strikes the surface at the same location (Fig. 3.13).

The other extreme is that of slow sintering, in which particles strike the surface and do not sinter before another particle strikes at the same location. In this case, porous films can be formed with characteristics that depend on the deposition conditions.

## 3.9 REFERENCES

1. Ziskind, G., Fichman, M., & Gutfinger, C. *J. Aerosol Sci. 26*(4), 1995, 613–644.
2. Matsusaka, S. & Masuda, H. *Aerosol Sci. Tech. 24*, 1996, 69–84.
3. Punjrath, J.S. & Heldman, D.R. *J. Aerosol Science 3*, 1972, 429–440.
4. Kodas, T.T. & Hampden-Smith, M.J. (Eds.). *The Chemistry of Metal CVD*. Weinheim, West Germany: VCH, Verlagsgesellschaft, 1994.
5. Friedlander, S.K. *Smoke, Dust and Haze*. New York: Wiley Interscience, 1977.
6. Hinds, W.C. *Aerosol Technology*. New York: Wiley Interscience, 1982.

7. Cunningham, E. *Proceedings of the Royal Society*, *A-83*, 1910, 357.
8. Donovan, R.P., Yamamoto, T., Periasamy, R., & Clayton, A.C. *J. Electrochem. Soc.* *140*(10), 1993, 2917–2927.
9. Fotou, G.P. & Pratsinis, S.E. *Aerosol Sci. Technol.* *18*, 1993, 213–218.
10. Ingham, D.B. *J. Aerosol Sci.* *6*, 1975, 125–132.
11. Pich, J. *Aerosol Sci.* *3*, 1972, 351–361.
12. Taulbee, D.B. *J. Aerosol Sci.* *9*, 1978, 17–23.
13. Taulbee, D.B. & Carpenter, P.S. *J. Aerosol Sci.* *9*, 1978, 513–519.
14. Thomas, J.W. & Knuth, R.H. *Am. Industrial Hygiene Assoc.*, May/June 1967, 229–237.
15. Wells, A.C. & Chamberlain, A.C. *Brit. J. Appl. Phys.* *18*, 1967, 1793.
16. Bcal, S.K. *Nuclear Science & Engineering* *40*, 1970, 1–11.
17. Brockmann, J.E., McMurry, P.H., & Liu, B.Y.H. *Aerosol Sci. Technol.* *1*(2), 1982, 163–178.
18. El-Shobokshy, M.S. & Ismail, I.A. *Atmos. Env.* *14*(3), 1980, 297–304.
19. El-Shobokshy, M.S. *Atmos. Env.* *17*(3), 1983, 639–644.
20. Forney, L.J. & Spielman, L.A. *J. Aerosol Sci.* *5* 1974, 257–271.
21. Kader, B.A. & Yaglom, A.M. *Int. J. Heat Mass Transfer* *20*, 1977, 345–357.
22. Kubie, J. *Chem. Eng. Sci.* *35*, 1980, 1787–1993.
23. Matsui, H., Yoshida, Y., Murata, M., & Ohata, T. *J. Nucl. Sci. Technol.* *11*(7), 1974, 300–307.
24. Reeks, M.W. & Skyrme, G. *J. Aerosol Sci.* *7*(6), 1976, 485–495.
25. Schack, C.J., Jr., Pratsinis, S.E., & Friedlander, S.K. *Atmos. Env.* *19*(6), 1985, 953–960.
26. Sehmel, G.A. *J. Aerosol Sci.* *4*, 1973, 125–138.
27. Yaglom, A.M. & Kader, B.A. *J. Fluid Mech.* *62*(3), 1974, 601–623.
28. Bird, R., Stewart, W.E., & Lightfoot, E.N. *Transport Phenomena*. New York: John Wiley and Sons, 1960.
29. Flagan, R.C. *Ceram. Trans.* *1*(2, A), 1988, 229–243.
30. Flagan, R.C. *J. Aerosol Sci.* *19*(7), 1988, 783–787.
31. Seinfeld, J.H. *Atmospheric Chemistry and Physics of Air Pollution*. New York: Wiley Interscience, 1986.
32. Crump, J.G. & Seinfeld, J.H. *J. Aerosol Sci.* *12*(5), 1981, 405–415.
33. Crump, J.G., Flagan, R.C., & Seinfeld, J.H. *Aerosol Sci. Technol.* *2*, 1983, 303–309.
34. Shimada, M., Okuyama, K., Kousaka, Y., & Minamino, D. *J. Chem. Eng. Jpn.* *24*(2), 1991, 203–209.
35. McMurry, P.H. & Rader, D.J. *Aerosol Sci. Technol.* *4*, 1985, 249–268.
36. Shimada, M. & Okuyama, K. *J. Aerosol Sci.* *23*(7), 1992, 711–721.
37. Siefert, W. *Thin Solid Films* *120*, 1984, 267–274.
38. Siefert, W. *Thin Solid Films* *121*, 1984, 275–282.
39. Brock, J.R. *J. Colloid. Sci.* *17*, 1962, 768–780.
40. Talbot, L., Cheng, R.K., Schefer, R.W., & Willis, D.R. *J. Fluid Mech.* *101*(4), 1980, 737–758.

41. Shimada, M., Seto, T., & Okuyama, K. *Jpn. J. Appl. Phys.* *33*(1(2)), 1994, 1174–1181.
42. Bae, G.-N., Lee, C.S., & Park, S.O. *Aerosol Sci. Technol.* *23*, 1995, 321–330.
43. Berger, C., Horvath, H., & Schindler, W. *J. Aerosol Sci.* *26*(2), 1995, 211–217.
44. Byers, R.L. & Calvert, S. *Indust. & Eng. Chem. Fundamentals* *8*(4), 1969, 646–655.
45. Chang, Y.-C., Ranade, M.B., & Gentry, J.W. *J. Aerosol Sci.* *26*(3), 1995, 407–428.
46. Derjaguin, B.V., Rabinovich, Y.I., Storozhilova, A.I., & Shcherbina, G.I. *J. Colloid Interface Sci.* *57*(3), 1976, 451–461.
47. El-Shobokshy, M.S. *Canadian J. Chem. Eng.* *59*, 1981, 155–157.
48. Fulford, G.D., Moo-Young, M., & Babu, M. *Canadian J. Chem. Eng.* *49*, 1971, 553–556.
49. Gokoglu, S.A. & Rosner, D.E. *Int. J. Heat & Fluid Flow* *5*(1), 1984, 37–41.
50. Goren, S.L. *J. Colloid Interface Sci.* *61*(1), 1977, 77–85.
51. Homsy, G.M., Geyling, F.T., & Walker, K.L. *J. Colloid Interface Sci.* *83*(2), 1981, 495–501.
52. Keh, H.J. & Yu, J.L. *Aerosol Sci. Technol.* *22*, 1995, 250–260.
53. Lin, S.H. *Appl. Sci. Res.* *32*, 1976, 637–648.
54. Lin, S.H. *Chem. Eng. Sci.* *32*, 1977, 458.
55. Maynard, A.D. *Aerosol Sci. Technol.* *23*, 1995, 521–533.
56. Prodi, F., Santachiara, G., & Prodi, V. *J. Aerosol Sci.* *10*, 1979, 421–425.
57. Shchukin, E.R., Shulimanova, Z.L., & Yalamov, Y.I. *High Temp-Engl. Tr.* *32*(5), 1994, 679–684.
58. Shimada, M., Seto, T., & Okuyama, K. *AIChE J.* *39*(11), 1993, 1859–1869.
59. Sleicher, C.A., Jr. & Churchill, S.W. *Ind. Eng. Chem.* *48*(10), 1956, 1819–1924.
60. Weinberg, M.C. *J. Am. Ceram. Soc.* *66*(6), 1983, 439–443.
61. Pratsinis, S.E. & Kim, K.-S. *J. Aerosol Sci.* *20*(1), 1989, 101–111.
62. Batchelor, G.K. & Shen, C. *J. Colloid Inter. Sci* *107*, 1985, 21.
63. Oh, E.-K. & Kim, S.-G. *J. Aerosol Sci.* *27*(8), 1996, 1143–1154.
64. Oh, M.D., Yoon, K.H., & Mong, H.K. *Aerosol Sci. Technol.* *25*(2), 1996, 141–156.
65. Marple, V.A. *A Fundamental Study of Inertial Impactors*. Ph.D. dissertation, University of Minnesota, Minneapolis, 1970.
66. Marple, V.A. & Liu, B.Y.H. *Environ. Sci. Technol.* *8*, 1980, 648.
67. Rader, D.J. & Marple, V.A. *Aerosol Sci. Technol.* *4*, 1985, 141–156.
68. Hillamo, R.E. & Kauppinen, E.I. *Aerosol Sci. Technol.* *14*, 1991, 33–47.
69. de la Mora, F., Hering, S.V., Rao, N., & McMurry, P.H. *J. Aerosol Sci.* *21*, 1990, 169–187.
70. Vincent, J.H. *Aerosol Sampling*. New York: John Wiley and Sons, 1989.
71. Pui, D.Y.M., Romay-Novas, F., & Lui, B.Y.H. *Aerosol Sci. Technol.* *7*, 1987, 301.
72. Friedlander, S.K. & Johnstone, H.F. *Ind. Eng. Chem.* *49*, 1957, 1151.
73. Arnold, S. & Lewittes, M. *J. Appl. Phys.* *53*, 1982, 5314.
74. Ashkin, A. *Phys. Rev. Lett.* *24*, 1970, 156.
75. Tassopoulos, M., O'Brien, J.A., & Rosner, D.E. *AIChE J.* *35*, 1989, 967.

# 4 Particle Growth, Evaporation, and Nucleation

## 4.1 INTRODUCTION

Aerosol routes to materials exploit the growth of particles by a variety of phenomena, the evaporation of droplets (spray pyrolysis) and the formation of new particles (nucleation). Evaporation and condensation of gaseous species from and onto particles can occur for both droplets and solid particles. These two mechanisms together make up one of the ways particles grow and shrink in size after they are formed, often by nucleation. In the case of droplets, condensation and evaporation occur onto and from a liquid and require a change in phase from vapor to liquid or vice versa. For solid particles, atomic and molecular species can physisorb or chemisorb onto the particles. *Physisorption*, or *Physical adsorption*, involves only weak attractions such as van der Waals forces and is important only for species with partial pressures near their vapor pressures. On the other hand, *chemisorption*, or *chemical absorption*, involves the formation of chemical bonds with the surface and can occur at any temperature and partial pressure. This is the mechanism by which many solid particles grow. A more detailed discussion of this distinction is found in various texts on physical chemistry.[1, 2]

Condensation and evaporation of liquids leading to growth and shrinkage of particles is an important concept in a wide variety of aerosol processes for the synthesis and processing of materials. Film formation by the deposition of droplets onto surfaces (often called spray pyrolysis) involves directing the droplets at hot surfaces. The gas temperature rises as the droplets approach the surface, resulting in evaporation of the liquid. This in turn causes the droplets to decrease in size, leading to a lower probability of inertial impaction and gravitational sedimentation. Aerosol-assisted chemical vapor deposition (AACVD) involves using droplets to transport volatile precursors, either in solution or as a pure liquid, to a region where the solvent (if present) and the precursor(s) evaporate to form a vapor of the reactant(s). This is then followed by reaction of the volatile reactant(s) on the surface to form a film or, in the gas phase, to form particles. In spray drying, solid particles are made by evaporating the solvent and precipitating the solute.[3] Spray pyrolysis for the production of powder involves converting a liquid containing a precursor, in solution or as a suspension, into aerosol droplets.

The droplets then evaporate in a heated gas, yielding particles of the reactant, which decompose at higher temperatures to produce the final product. In metal alkoxide hydrolysis processes, aerosol streams of precursor droplets of metal alkoxides are mixed with water vapor. The water vapor is transported to the surfaces of the droplets and then into the droplets themselves, where the reaction occurs to form a solid product.

Solid particles can grow by condensation and shrink by sublimation, and they can be used to deliver precursors for CVD processes in a manner similar to that for AACVD with droplets. Solid particles of volatile reactants can be deposited onto surfaces if they do not evaporate before deposition. Surface reaction is a particle growth mechanism in which volatile reactants chemisorb on particle surfaces and react to form a product with a low vapor pressure, which then remains on the particle surface. Often, transport and adsorption of co-reactants such as oxygen are required for these reactions to occur. Numerous other examples exist, some of which are mentioned subsequently.

Transport of gaseous species to particle surfaces and, more specifically, condensation and evaporation also play central roles in the theory of gas-phase *nucleation*—that is, the formation of new particles. Here, condensation is one of the processes by which small clusters form and then grow to their final size.

In this chapter, we discuss the theory of particle growth and shrinkage by vapor transport, surface reaction, and volume reaction for both droplets and solid particles. The theories for gas-phase transport processes are nearly identical for solid and liquid particles. For this reason, much of the discussion is general and is not restricted to either solids or liquids. However, surface reaction is relevant in the case of solid particles, while volume reaction is more common in the context of liquid aerosols. Further in the chapter, we discuss the *Kelvin equation*, followed by nucleation theory.

## 4.2 QUALITATIVE DESCRIPTION OF PARTICLE GROWTH

Before addressing the details of particle formation and growth, it is useful to give a qualitative summary of how the various phenomena are related. For the example of particle formation followed by growth, the overall process begins with the formation of new particles by homogeneous nucleation.[4] In situations in which particle collisions do not occur, growth often occurs by vapor condensation. After the particles are formed, vapor diffusion takes place, with a much shorter time constant than growth. Before significant growth occurs, vapor diffusion comes to a pseudosteady state: The vapor concentration profile within many radii from the particle is at steady state, and the steady-state diffusional flux may be used to calculate the particle growth rate.

The effects of the release of latent heat due to vapor condensation at the particle surface can be described by an energy balance. Heat conduction occurs both inside and outside the particles. When the particle is much more conductive than the gas, the particle temperature is nearly uniform, and the latent heat is conducted primarily outward into the gas. The times taken to reach steady external temperature and vapor concentration profiles are short compared to the particle growth times. As a result, the steady-state heat and vapor fluxes may be used to determine the steady-state surface temperature at all times.

For most aerosol routes to materials, the particle temperature is the same as that of the gas, and, as a result, gas properties like total molar concentration can be assumed constant. The partial pressures of volatile species at the particle surface are equal to the thermodynamic values at the particle temperature for particles larger than roughly 100 nm, for which curvature effects can be neglected.

In summary, nearly all transport processes can be assumed to be at a pseudosteady state, which allows us to obtain steady-state solutions to the various equations, using a fixed radius for the particle. These equations can then be utilized in the unsteady-state equations for particle growth.

## 4.3 PARTICLE GROWTH: FREE-MOLECULE AND TRANSITION REGIMES

This section introduces the equations used to calculate rates of particle growth and shrinkage. The goal is to obtain simple equations for the times required for particle growth or shrinkage that allow comparison with other relevant times—for instance, process residence times. The starting point for any discussion of the theory of condensation, evaporation, surface reaction, volume reaction, and nucleation is the transport of atomic and molecular species to particle surfaces. The transport of atoms, molecules, and even clusters (collections of atoms or molecules) to and from particles is commonly encountered in many material-processing applications. The general features of the evaporation and growth of particles have been discussed in a variety of articles.[5–10]

The simplest case of molecular transport to a particle surface is when the particle size is much smaller than the gas mean free path ($d_p \ll \lambda$) and molecules arrive at the particle surface in a ballistic fashion (Fig. 4.1). This occurs for particles roughly 10 nm or smaller in size at standard temperature and pressure. The gas or vapor does not develop concentration or temperature gradients around the particle. Physically, the particle is so small that it fits into the spaces between gas molecules and atoms and behaves like a large molecule. According to the kinetic theory of gases, the flux of molecules

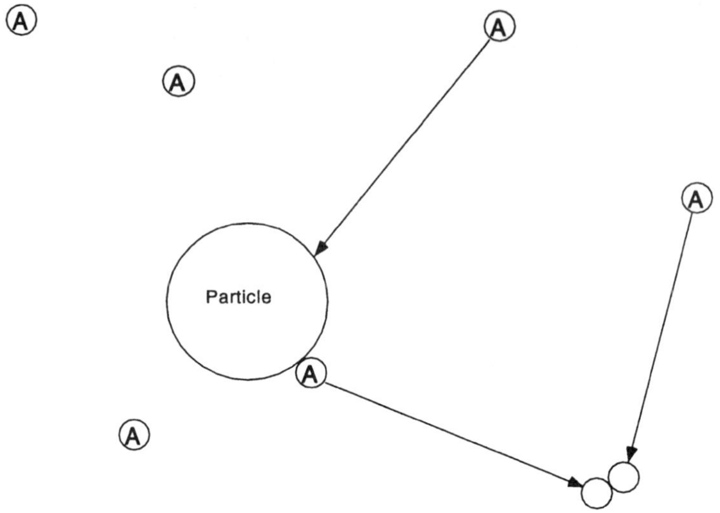


**Fig. 4.1.** Schematic illustrating how atoms and/or molecules (A) collide with particles when the particle diameter is much less than the gas mean free path.

(number of molecules/area × time) to the particle surface is[11]

$$J_m = \frac{n_1 \bar{c}}{4} = \frac{p_1}{k_B T} \frac{\bar{c}}{4}, \tag{4.1}$$

where $n_1$ is the molecular concentration (number of molecules/cm$^3$) and is related to $p_1$ by $p_1 = n_1 k_B T$, in which $p_1$ is the partial pressure of species 1. The mean molecular velocity $\bar{c} = (8k_B T/\pi m_1)^{1/2}$ is the average velocity of the gas molecules, where $m_1$ is the mass of a molecule or atom—that is, $m_1 = M/N_A$, in which $M$ is the molecular weight and $N_A$ is Avogadro's number. A useful relationship that can be calculated from Equation (4.1) is that a species at $10^{-6}$ torr partial pressure provides one monolayer of gas molecules in 1 second on a particle surface.

The total number of molecular collisions with the particle surface per unit time, $\epsilon$, is obtained by multiplying the flux $J_m$ by the area of a particle, $\pi d_p^2$:[4]

$$\epsilon = J_m \cdot \pi d_p^2 = \frac{p_1 \pi d_p^2}{(2\pi m_1 k_B T)^{1/2}}. \tag{4.2}$$

If only a fraction $\alpha_s$, the sticking coefficient, of these collisions results in sticking of the species on the particle surface, if the evaporation or sublimation is taken into account, and if a correction factor $F_1$ is added to allow use of the equation for particles corresponding to Kn ~ 1, the change in particle volume with time is

$$\frac{dv}{dt} = \epsilon\alpha_s F_1 v_1 = \frac{\alpha_s \pi d_p^2 (p_1 - p_d) F_1 v_1}{(2\pi m_1 k_B T)^{1/2}}, \tag{4.3}$$

where $v_1 = m_1/\rho_1$ ($\rho_1$ is the density of the material in the particle in the solid or liquid state and $m_1 = M/N_A$ are the volume and mass, respectively, of the transported molecule in the particle), $p_1$ is the partial pressure of the vapor far from the particle, and $p_d$ is the partial pressure at the particle surface. Note that this equation is generic and applies to both solid and liquid particles and both condensation and evaporation/sublimation, depending on the relative values of $p_1$ and $p_d$. The correction factor $F_1$ is discussed shortly.

The volume of the particle can be converted to the diameter via the expression $v = \pi d_p^3/6$. From this relationship, the rate of change of the particle diameter with time at temperature $T$ in the free-molecule regime (Kn $\gg$ 1) is given by[4]

$$\frac{dd_p}{dt} = \frac{2\alpha_s v_1 (p_1 - p_d)}{(2\pi m_1 k_B T)^{1/2}} F_1 \tag{4.4}$$

Figure 4.2 shows the symbols involved. Note that the vapor concentration profiles shown are for Kn $\ll$ 1 (discussed in Section 4.4 on transport in the continuum regime) and not for Kn $\gg$ 1. Equation (4.4) was derived for the free-molecule regime and modified by the addition of the quantity $F_1$, the Fuchs–Sutugin expression for describing the transition regime between the free-molecule (Kn $\gg$ 1) and the continuum regimes (Kn $\ll$ 1),[12] that is,

$$F_1 = \frac{1.33\text{Kn}(1 + \text{Kn})}{1 + 1.71\text{Kn} + 1.33\text{Kn}^2}. \tag{4.5}$$

The Knudsen number is $\text{Kn} = 2\lambda/d_p$. Equation (4.5) allows the use of Equation (4.4) in the transition regime, where Kn is roughly unity.

In order to use Equation (4.4) to calculate particle growth and shrinkage rates, the partial pressure of the vapor at the particle surface must be known. For a pure liquid or solid and large droplets (> 0.1 $\mu$m), $p_d$ is equal to the vapor pressure of the liquid or solid over a flat surface. For smaller particles, the vapor pressure above the surface of the particle is higher than that over a flat surface because, in a crude sense, the molecules near the particle surface have fewer nearest neighbors than they would in a flat surface. This results in less energy being required to remove an atom or molecule from the curved surface into the gas phase. It is worth noting that, because this effect is important only for particle diameters of less than 0.1 $\mu$m, most applications with standard droplet generators are in size regimes in which the effect is

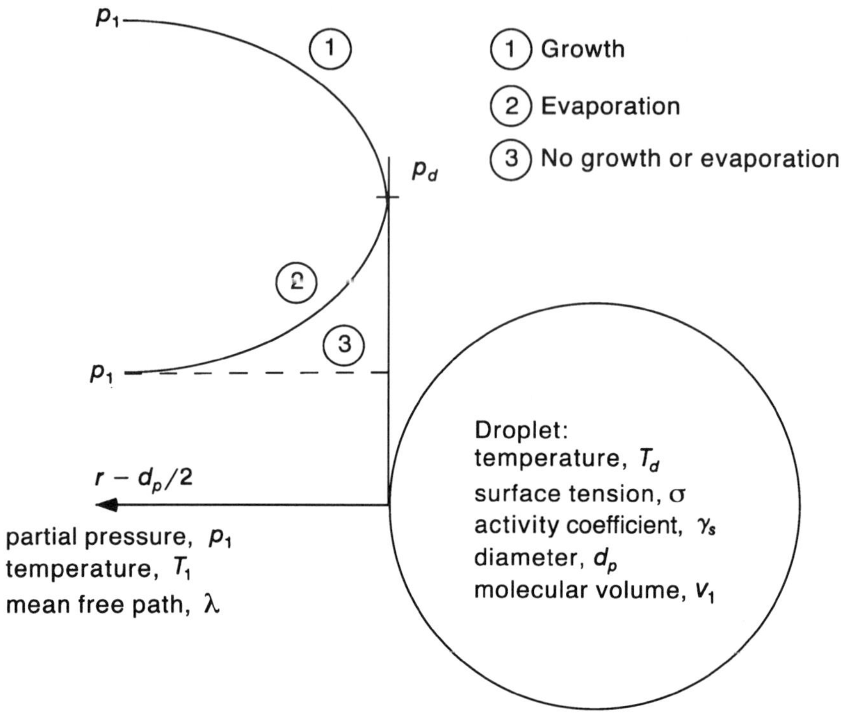


**Fig. 4.2.** Definitions of terms used in equations describing growth and shrinkage of particles.

negligible. Thus, Equation (4.4) is most important for understanding the dynamics of new-particle formation, wherein droplets and/or solid particles grow from molecular to larger dimensions.

The theory of how the vapor pressure depends on surface curvature is reasonably well understood.[4, 13, 14] The relationship between the vapor pressure over a flat surface ($p_e$), that over a curved surface ($p_d$), and other parameters is commonly called the *Kelvin equation* for aerosol particles and is given by

$$S = \frac{p_d}{p_e} \exp\left(\frac{4\sigma v_1}{d_p k_B T}\right), \tag{4.6}$$

where $S$ is the *saturation ratio* of the vapor pressures and $\sigma$ is the surface tension (erg/cm$^2$) (for liquids) or surface energy (erg/cm$^2$) (for solids) of the material in the particle. The *Kelvin diameter* is the droplet diameter defined by this equation for a given saturation ratio.

An important result derived from Equations (4.6) and (4.4) is that only for $p_1 > p_d$ can condensation take place, and the vapor must be slightly supersaturated. The aerosol literature uses the concept of the *critical droplet* or *particle diameter* to describe the particle size above which droplets grow and below which they shrink because of condensation and evaporation for a given saturation ratio. This critical diameter is often designated by $d_p^*$ and depends on the saturation ratio, as shown in Equation (4.6).

The Kelvin equation is important for an understanding of particle formation. As the saturation ratio is raised, the size of the particle that is stable with respect to evaporation and/or sublimation decreases. For most inorganic materials of interest, this critical size is often smaller than atomic dimensions. Physically, this means that the vapor pressure of the material is so low that even the very smallest particles do not evaporate, in spite of their extreme surface curvature. An example is the formation of $TiO_2$ at temperatures below roughly 1500°C, for which the critical particle size is less than the size of a $TiO_2$ unit.[15] Thus, all $TiO_2$ particles formed below 1500°C are stable and do not evaporate, a major implication for nucleation theory. (See Section 4.10.)

## 4.4 TRANSPORT: CONTINUUM REGIME

The previous section described the theory of particle growth or shrinkage for Kn ≫ 1. The theory for Kn ≪ 1 assumes that the transport mechanism by which atoms and molecules move to and from a particle's surface is diffusion, which occurs when the particle size is much larger than the gas mean free path, or about 1000 nm for standard temperature and pressure. The gaseous species being transported undergoes numerous collisions with other gas molecules on its way to or from the particle surface (Fig. 4.3). In addition,

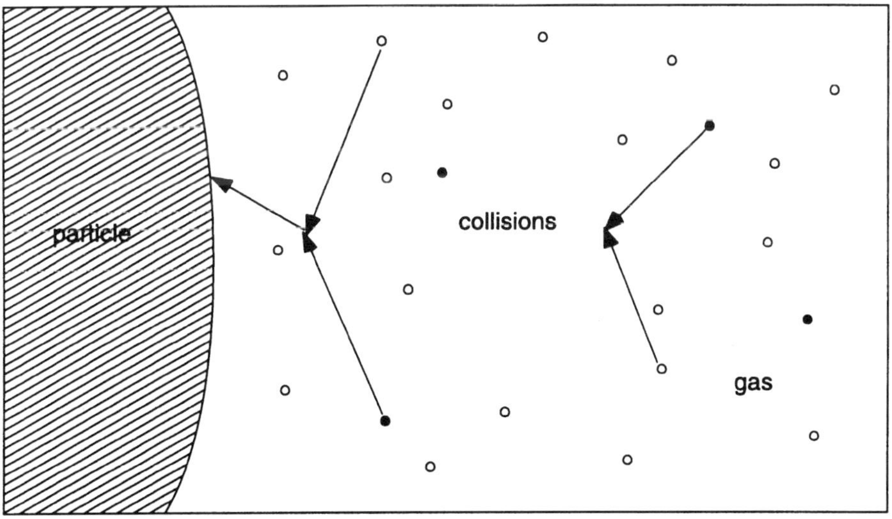


**Fig. 4.3.** Schematic illustrating how atoms and molecules are transported to a particle surface by diffusion when the gas mean free path is much less than the particle size.

during condensation, evaporation, or sublimation, a concentration gradient (a variation in concentration with distance from the particle) can be established around the particle, as shown graphically in Figure 4.2. The background theory for particle growth and shrinkage with $\mathrm{Kn} \ll 1$ is discussed in standard transport phenomena texts.[11]

The rate of change of particle diameter with time at a temperature $T$ in the continuum regime is given by[4]

$$\frac{dd_p}{dt} = \frac{4Dv_1(p_1 - p_d)}{k_B T d_p} F_2, \tag{4.7}$$

where $D$ (cm$^2$/sec) is the diffusivity (diffusion coefficient) of the gas or vapor and

$$F_2 = \frac{1 + \mathrm{Kn}}{1 + 1.71\mathrm{Kn} + 1.333\mathrm{Kn}^2} \tag{4.8}$$

is the dimensionless Fuchs–Sutugin expression for bridging the equations for the continuum and free-molecule regimes.[12] Note that the correction factor $F_2$ is different from $F_1$ (used to extrapolate the equation for growth in the free-molecular regime to $\mathrm{Kn} \sim 1$ and $\mathrm{Kn} \ll 1$).

For many applications of aerosols in materials synthesis and processing, it is not necessary to know the detailed evolution of particle size for the processes of condensation and evaporation. Rather, we are often interested in the time required for particles to grow or shrink a certain amount. Thus, it is useful to define the *characteristic time for condensation or evaporation* as the time required to halve the particle diameter by evaporation or sublimation or to double it by condensation. In the continuum limit ($\mathrm{Kn} \ll 1$), for large particles such that $p_d = p_e$, and neglecting the correction factor, this characteristic time is given by

$$\tau_{\mathrm{GC}} = \frac{3k_B T d_{po}^2}{32D(p_1 - p_e)v_1}, \tag{4.9}$$

where $d_{po}$ is the initial particle size. In the free-molecule limit ($\mathrm{Kn} \gg 1$) and with $p_d = p_e$, the equilibrium vapor pressure over a flat surface, the characteristic time for particle condensation or evaporation, neglecting the correction factor and $\alpha_s$, is

$$\tau_{\mathrm{GF}} = \frac{(2\pi m_1 k_B T)^{1/2}}{4v_1(p_i - p_e)}. \tag{4.10}$$

It is useful to note that (1) if $p_1$ is close to $p_e$, the time for evaporation is long; (2) the dependencies on particle size are different in the two regimes; and (3) the most effective way to influence the evaporation time is through the vapor pressure, which depends exponentially on the temperature.

Typical numbers for evaporation times of particles provide a useful insight into the many applications that involve the evaporation of water droplets. Water droplets (at room temperature with vapor pressure roughly 20 torr and in surroundings with relative humidities of 50%) with diameters of 0.01, 1, and 10 $\mu$m, can evaporate completely over time scales of roughly $10^{-6}$, $10^{-3}$, and 0.1 second, respectively. These times can be compared with process residence times to determine whether droplets can evaporate completely.

Equations (4.9) and (4.10) also describe the time constants for particle growth when $p_1 > p_e$ or $p_d$; they show that particle growth is not possible unless the vapor is supersaturated. Supersaturation is induced during the synthesis or processing of materials by cooling the vapor or gas-phase reaction to form species—usually solids—with low vapor pressures.

A significant aspect of particle growth by condensation in the continuum limit (Kn $\ll$ 1) is that it can narrow the particle size distribution. A close examination of Equation (4.7) reveals that the particle growth rate is inversely proportional to the particle diameter. As a result, small particles grow faster than larger ones until they "catch up" in size.

In the preceding equations, the effect of the latent heat of the phase change (vapor to solid or vapor to liquid) has been neglected. The heat released by condensation, for example, can raise the temperature of the droplet, so that the vapor pressure at the droplet surface increases, thus reducing the rate of condensation. The opposite process results in a temperature depression during evaporation of a droplet. To fully assess this effect, one has to carry out an energy balance on the gas and the particles, an approach that is discussed elsewhere with standard equations for estimating the effect.[12, 16]

Large droplets can also have significant settling velocities, which result in convective transport (due to the gas motion relative to the particle). This problem is discussed in standard texts on transport phenomena.[11]

## 4.5 DROPLET EVAPORATION: CHARACTERISTIC TIME FOR GAS-PHASE DIFFUSION TO A PARTICLE

The previous sections introduced the concept of a *characteristic time*, a measure of the time it takes for a process to occur. Because particle growth and shrinkage for Kn $\ll$ 1 occur by diffusion, it is useful to develop an expression that is a measure of the time required to establish a steady-state diffusion process. This time is the characteristic time for diffusion to a particle.[12]

The analysis begins with the unsteady-state diffusion of a species $A$ to the surface of a droplet, which is described by

$$\frac{\partial C}{\partial t} = D_g\left(\frac{\partial^2 C}{\partial r^2} + \frac{2}{r}\frac{\partial C}{\partial r}\right), \tag{4.11}$$

$$C(r,0) = C_\infty, \qquad r > R_p, \tag{4.12}$$

$$C(r,t) = C_\infty, \qquad r \to \infty, \tag{4.13}$$

$$C(R_p,t) = 0, \tag{4.14}$$

where $R_p$ is the droplet radius, $C(r,t)$ is the gas-phase concentration of $A$, $D_g$ is the molecular diffusivity of $A$ in air, $C_\infty$ is the concentration of $A$ far from the droplet, and $C_s(t)$ is the concentration of $A$ in the gas phase at the droplet surface.[11]

An expression for the characteristic time of this diffusion process can be obtained by solving Equations (4.11) to (4.14). The solution of Equation (4.11), subject to the given boundary and initial conditions, is

$$\frac{C(r,t)}{C_\infty} = 1 - \frac{R_p}{r} + \frac{2R_p}{r\sqrt{\pi}}\int_0^{(r-R_p)/2\sqrt{D_g t}} e^{-\varepsilon^2}\,d\varepsilon. \tag{4.15}$$

The characteristic time for the concentration profile to achieve the steady-state value near the particle surface is

$$\tau_{\text{dg}} = \frac{R_p^2}{4D_{\text{g}}}, \tag{4.16}$$

where the subscript dg refers to diffusion in the gas phase. Some insight can be gained by calculating this characteristic time for a typical gas-phase molecular diffusivity of $D_g = 0.1$ cm$^2$ sec$^{-1}$ and various values of $R_p$. For $R_p = 10$ $\mu$m, $\tau_{\text{dg}} = 2.5 \times 10^{-6}$ sec. Thus, the concentration profile will be at steady state for most conditions of interest. This supports the use of steady-state concentration profiles when one calculates the rate of change of particle size.

## 4.6 EVAPORATION FROM SOLUTION DROPLETS

Particle formation by evaporation from solution droplets takes place during spray drying and spray pyrolysis to form both films and powders. The equations for particle growth and shrinkage and the Kelvin equation (4.6) were developed for pure materials and must be modified to handle suspensions, particles in liquids, or solutions. The major changes in these equations are that the effective vapor pressure at the particle surface is no longer that of the pure liquid, but is reduced by the suspended or dissolved species. Another complication occurs when precipitation takes place in droplets, to give solids that form a shell around the liquid. Again, the effective vapor pressure at the particle surface is no longer that of the pure liquid.

The presence of the solute lowers the vapor pressure of the solvent.[4, 12, 16] This effect is caused, first, by the presence of nonvolatile molecules near the droplet surface, thus reducing the area for evaporation, and second, by the development of chemical (affinity) forces between solute and solvent molecules. In this case, the vapor pressure of the solvent above the solution droplet, $p_{\mathrm{ds}}$, needs to be corrected for the presence of the solute. We have

$$p_{\mathrm{ds}} = \gamma_s x_s p_d = \gamma_s x_s p_e \exp\left(\frac{4\sigma v_1}{d_p k_B T}\right), \tag{4.17}$$

where $\gamma_s$ is the activity coefficient of the solution, $x_s$ is the mole fraction of the solvent, $\sigma$ is the surface tension of the solvent, and $v_1$ is the molecular volume of the solvent in the droplet. The information required to use this equation in a quantitative manner is usually not available; in most cases, the equation is used for qualitative guidance or modeling where the parameters are varied over assumed, but realistic, ranges.[17]

During the evaporation of solution droplets, there are two competing effects: As the solvent evaporates, the solute concentration increases and the solvent vapor pressure over the droplet decreases; at the same time, evaporation of the solvent reduces the droplet size, increasing the droplet curvature and therefore increasing the solvent vapor pressure according to the Kelvin effect, but only for droplets with diameters less than or equal to approximately 0.1 $\mu$m. Thus, in materials processing, the main concern is usually the reduction of vapor pressure, which forces the use of longer residence times or higher process temperatures for complete removal of the solvent; the Kelvin equation plays a minor role. A more extreme effect is the formation of a crust, such as during powder generation by spray pyrolysis, in which evaporation rates are reduced dramatically.

## 4.7 DROPLET EVAPORATION DURING MATERIALS PROCESSING

Droplet evaporation plays a key role in droplet and solid particle deposition, as well as in the aerosol delivery of volatile precursors for CVD. If the solvent evaporates before the droplet strikes the substrate, solid precursor particle deposition will then lead to rough films. If the precursor does not evaporate fully (after evaporation of the solvent) during the aerosol delivery of precursors for CVD, deposition rates are reduced and some of the precursor is not utilized. Furthermore, partial evaporation of the precursor can lead to rather complex temperature dependencies for the deposition rate. Insight into these problems can be gained by considering the time required for evaporation of droplets described in the last section and comparing it with the appropriate residence time in the deposition system.

The most common technology for which these equations are useful is droplet deposition (also known as spray pyrolysis). Droplet deposition is accompanied by some degree of solvent (and, in some cases, solute) evaporation. If the droplet concentration is sufficiently high, this evaporation process is a non-equilibrium one in which the pseudoequilibrium state is gas with 100% relative humidity. In the case of water droplets containing small amounts of a dissolved species, droplet lifetimes accounting for the lowering of the temperature due to the latent heat of evaporation for droplet diameters of 0.1, 1, and 10 $\mu$m diameter are $5 \times 10^{-5}$, $2 \times 10^{-3}$, and 0.15 second, respectively,[16] at atmospheric pressure, 50% relative humidity, and 20°C ambient temperature. These calculations assume no increase in the relative humidity of the surrounding gas as the water evaporates, a situation that is sometimes unrealistic. The preceding lifetimes can be reduced when solvents with higher vapor pressures are used; the lifetimes for ethanol droplets of the same diameters as those just mentioned are $10^{-5}$, $3 \times 10^{-4}$, and 0.03 second, respectively, at 20°C with vapor-free air at one atmosphere.[16]

The foregoing calculations can be used as a qualitative guide to determine when a process occurs by droplet deposition or by some other mechanism, such as vapor deposition (when volatile solutes and precursors are used). The given evaporation times suggest that droplet deposition without appreciable solvent evaporation requires that the droplet diameter be greater than roughly 10 $\mu$m for droplet residence times of about 0.1 sec before impaction. Thus, initial droplet diameters of 100 $\mu$m or greater usually ensure droplet deposition before evaporation. Conversely, experiments carried out with certain classes of ultrasonic atomizers, as described in Chapter 16, on system components (using droplets less than 5 $\mu$m in diameter), will provide solvent evaporation well before a droplet reaches a surface. In these cases, deposition can occur by solid particles for nonvolatile solutes or by CVD for volatile solutes.

An effect neglected in Equations (4.9) and (4.10) is the reduction in the vapor pressure and, therefore, increase in droplet lifetime due to the presence of solutes in the droplet. This effect does not, however, greatly influence the preceding results, because crust formation often occurs before the droplet evaporates fully, and such crust must usually be avoided. A second effect that is neglected is the increase in evaporation rate due to heating of the gas above the substrate. This is a complicated problem, because thermophoresis keeps the particles from approaching the substrate too closely, and the large volume of water vapor produced by evaporation on the substrate results in a net gas velocity away from the substrate.

In contrast to the situation as regards droplet deposition (spray pyrolysis), complete solvent and precursor evaporation is required for the aerosol delivery of volatile precursors in CVD without particle or droplet deposition. In this case, two evaporation problems should be considered: evaporation of the solvent and evaporation of the precursor, both of which must occur before the droplet or particle reaches the substrate. Typically, droplet diameters of less than 10 $\mu$m are used, which ensures evaporation of the solvent within the residence time of most systems, as discussed earlier. Precursor evaporation, however, is not ensured in all cases. Evaporation of the precursors requires operating at a temperature that is sufficiently high to provide a precursor vapor pressure of roughly 10 torr, which provides precursor particle lifetimes of roughly 0.1 second, depending on the particle size. For precursors and temperatures giving lower vapor pressures of 1 and 0.1 torr, the lifetimes become roughly 1 second and 10 seconds, respectively. This contrast illustrates that precursors may not evaporate fully during aerosol delivery of CVD precursors, unless the vapor pressure is high enough and the residence time is long enough, which explains the motivation for adding a preheated section before the deposition chamber in AACVD processes, to increase both the temperature and residence time of the droplet in the gas phase.

Aerosol particles cannot evaporate once equilibrium (gas phase saturated with vapor) is achieved and the partial pressure of the evaporating species in the gas phase reaches its vapor pressure ($p_{\text{sat}}$). This failure to evaporate occurs because, as seen in Equations (4.9) and (4.10) for the characteristic times, the driving force for evaporation is the difference between the vapor pressure at the particle surface and the partial pressure far from the particle. If the vapor pressure is low, only a small amount of the precursor must evaporate before the vapor is saturated and evaporation stops. The partial pressure corresponding to complete evaporation of monodisperse particles of a precursor is

$$p_i = N_\infty\left(\pi d_p^3/6\right)\left(\rho/M\right)RT \tag{4.18}$$

where $N_\infty$ is the particle concentration (number/cm$^3$), $d_p$ is the diameter of a particle consisting of pure precursor (cm), $\rho$ is the density of the precursor (g/cm$^3$), $M$ is the molecular weight of the precursor (g/mol), $R$ is the gas constant (cm$^3$ atm/mol K), and $T$ is the temperature of the gas. If $p_i > p_e$, the particles will evaporate only partially before the vapor becomes saturated. In this case, the temperature must be raised, or the precursor must be removed from the gas by diffusion or reaction, in order for the particles or droplets to evaporate further.

The foregoing analysis takes into account the thermodynamics, but not the kinetics, of evaporation. A deeper understanding of the kinetics can be obtained by considering the problem of the evaporation of $N_\infty$ droplets/cm$^3$ with diameter $d_p$, where the diameter does not change significantly during evaporation, as is the case with large droplets. The rate at which molecules of the evaporating species enter the gas phase by diffusion from a single droplet (number of molecules/sec) in the continuum regime (Kn $\ll$ 1), which, for practical purposes, is valid for particles greater than roughly 0.1 $\mu$m, is[4]

$$F = \frac{2\pi d_p D(p_d - p_1)}{k_B T}. \tag{4.19}$$

This equation shows that the evaporation rate (also, the rate of introduction of vapor into the gas phase) is highest when the particle is the largest and decreases as the particle shrinks in size when compared on a per-particle basis as opposed to a per-mass basis. The physical reason for this relationship is that larger particles have higher surface areas from which evaporation takes place. Surface area does not appear in Equation (4.19) because the vapor diffusion rate from the particle varies inversely with the particle size, a linear dependence. The increase in vapor pressure due to surface curvature is negligible for particles larger than 0.1 $\mu$m. For $N_\infty$ particles per volume of gas, the rate at which vapor molecules enter the gas phase is $N_\infty F$ (molecules/sec). After converting $N_\infty F$ to units of pressure divided by time, we have, for the equation describing the change in $p_1$ with time,

$$\frac{dp_1}{dt} = \left(2\pi d_p D(p_d - p_1)\right) N_\infty. \tag{4.20}$$

Solving this equation for $p_1$ yields

$$p_1 = p_d\left(1 - \exp(-2\pi d_p D N_\infty t)\right). \tag{4.21}$$

The characteristic time [the time to reach $(1 - 1/e)p_d$] for this process is

$$\tau_{\text{sat}} = \frac{1}{(2\pi d_p D N_\infty)}. \tag{4.22}$$

Equation 4.22 shows that the time required to saturate the gas with the evaporating vapor does not depend on the vapor pressure of the evaporating material: Materials with lower vapor pressures evaporate more slowly, but less material must evaporate to reach the vapor pressure. The time required to saturate the gas with vapor depends on particle concentration and size. Larger particles and higher particle concentrations result in shorter times. The equation neglects the decrease in particle size as the particles evaporate but is valid if the fraction of the material that evaporates is small compared to the total amount in each particle. A simple calculation with reactant particles with diameter 0.1 $\mu$m ($10^{-5}$ cm), diffusion coefficient of 0.1 $cm^2$/sec, and droplet concentration of $10^6/cm^3$ shows that this time is 0.2 sec. For reactant particles with diameter 0.01 $\mu$m ($10^{-6}$ cm), diffusion coefficient of 0.1 $cm^2$/sec, and droplet concentration of $10^5/cm^3$, the time is 20 sec. These numbers are typical of those used during AACVD. This discussion shows that the droplet generator (which controls the droplet size and concentration) plays a key role in film formation processes based on droplet or vapor deposition. Unless sufficiently high precursor solubility in the solvent and sufficiently high droplet concentrations are used, the solute will not have sufficient time to evaporate in the deposition system and achieve the equilibrium vapor pressure.

More sophisticated calculations for particle evaporation, accounting for the shrinking particle size, must couple the equations for particle size and partial pressures in the gas phase.[16] Even more complicated approaches have been studied in which the particle mass and size change with the position of the particle in the system.[17]

## 4.8 CHEMICAL REACTIONS: SURFACE-REACTION-LIMITED GROWTH

Chemical reactions at the surface of, or inside, particles are often exploited for the generation or modification of materials. Powders can be formed by atomizing a liquid containing precursors or by suspending dry particles of a precursor in a gas and then introducing a gaseous reactant into the aerosol stream to convert the precursor to the product. Alternatively, particles can be formed by the evaporation of a metal, followed by condensation; the gaseous reactant can then be transported to the particle surface, where reactions take place to form the product. These processes are discussed here because of their mathematical similarity to growth by condensation.

One distinction that must be made in order to describe these processes theoretically is that between the relative rates of transport of gaseous reactants to the particle and chemical reactions on or in the particles. If the rate of reaction on the particle surface is able to consume the reactant faster than it can be supplied to the particle by vapor transport, then transport limits the growth rate, and the previous analyses positing growth and shrinkage rates based on transport rates describe the rate of particle growth. If the particles are larger than the gas mean free path, a concentration gradient can exist in the gas phase. If, however, the rate of consumption of the gaseous reactant is slower than the maximum potential rate at which vapor can be supplied by gas-phase transport, then the particle growth rate is determined by the chemical reaction rate, regardless of whether the reaction takes place on the surface or inside the particle. In this case, there are no gradients in the concentration of the reactant in the gas phase even for $Kn \ll 1$, and the concentration of the reactant at the surface is the same as that in the bulk of the gas. It is worth noting that this interplay between reaction and transport is common in chemical engineering, where the Damköler number, a dimensionless quantity, is used to determine which effect dominates.[1]

A second distinction that must be made is where the reaction takes place: on the surface or inside the particles. Surface reaction occurs when the

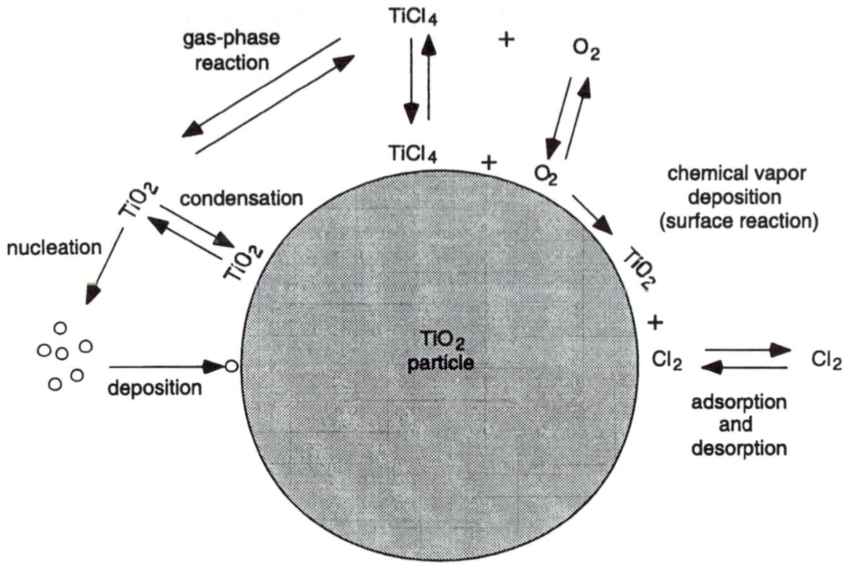


**Fig. 4.4.** Physical and chemical phenomena involved in growth of particles. Surface reaction is only one pathway that occurs in parallel with vapor condensation and cluster deposition.

particles are growing by CVD, as in the case of $TiO_2$ particle growth by the surface reaction of $TiCl_4$ and $O_2$ (Fig. 4.4). Another example is the application of coatings to particles by surface reaction, as in the case of coating $TiO_2$ particles with $SiO_2$ by CVD using $SiCl_4$ and oxygen as reactants. Sometimes it is not clear where the reaction occurs, as in the case of nitridation of aluminum particles (formed by evaporation and condensation of aluminum metal) by reaction with ammonia or nitrogen to form aluminum nitride. In this case, the reaction may occur at the interface between the AlN and Al, or it may be more complicated. Some of these possibilities are found in Chapter 6. In contrast, volume reaction occurs in cases where the particle consists of one reactant material that must be converted to a product by chemical reaction, and the reaction takes place throughout the particle, not just at an interface. This situation is also discussed in more detail in Chapter 6, as well as in the next section.

A note of caution is in order here: Surface reaction on particles without gas-phase reaction of the same reactant is not always possible. At a given temperature, the surface reaction and gas-phase reaction will provide different rates of consumption of the reactant. The total rate of reactant consumption by surface reaction (g/sec) is proportional to the product of the reactant concentration and the total aerosol surface area (the second moment of the particle size distribution with respect to diameter), while the total rate of reactant consumption by gas-phase reaction (g/sec) is proportional to the product of the gas volume and the concentration of the reactant in the gas phase. Thus, the overall rate of reactant consumption by surface reaction may be small, and the surface reaction rate cannot be calculated correctly without considering the dynamics of the entire process.

The rate of particle growth by surface reaction can be put into mathematical form for various limiting cases.[4] In one situation, slow surface reaction with respect to the potential transport rate of gaseous reactant results in reaction-limited growth. In one of the simplest cases, the surface reaction rate is first order in the gaseous reactant concentration. The product species is formed at a surface reaction rate (molecules/cm$^2$ sec) of

$$r_s = k_s n_{1s} = k_s \frac{p_{1s}}{k_B T}, \tag{4.23}$$

where $k_s$ is the surface reaction rate constant (cm/sec), $n_{1s}$ is the reactant concentration in the gas phase at the particle surface (molecules/cm$^3$ sec), and $p_{1s}$ is the reactant partial pressure at the particle surface.

The particle growth rate due to surface reaction can be obtained from the product of the particle surface area, surface reaction rate, and volume of the reaction product:

$$\frac{dv}{dt} = r_s \pi d_p^2 v_1. \tag{4.24}$$

This expression can be converted into the rate of change of the particle diameter by relating the particle volume to the particle diameter by $v = \pi d_p^3/6$, noting that

$$\frac{dv}{dt} = \frac{\pi}{2} d_p^2 \frac{dd_p}{dt} \tag{4.25}$$

and that

$$p_{1s} = n_{1s} k_B T, \tag{4.26}$$

to arrive at

$$\frac{dd_p}{dt} = \frac{2k_s p_{1s} v_1}{k_B T}. \tag{4.27}$$

Observe that this expression is zero order in particle diameter and first order in reactant concentration, making it very similar to the expression for particle growth by condensation in the free-molecule regime. As for changes in particle size by evaporation and condensation, the characteristic time for particle growth by surface reaction is

$$\tau_{\text{GSR}} = \frac{k_B T d_{po}}{2k_s p_1 v_1}, \tag{4.28}$$

namely, the time required for the particle to double in size.

It should be emphasized that surface reactions often work in series with vapor transport for particle growth and in parallel with gas-phase reactions to form clusters that can deposit onto particles. However, growth by deposition of small particles often overwhelms the surface reaction, making the theoretical description far more complex than presented here.[18]

## 4.9 CHEMICAL REACTIONS: VOLUME REACTION-LIMITED GROWTH

When the maximum potential rate of gas-phase transport to the particles is rapid relative to the reaction rate, and the reaction occurs homogeneously throughout the particles, the rate of particle growth is limited by the reaction rate inside the liquid or solid in the particles. For liquids, the reactant is transported to the particles, dissolves, diffuses, and then reacts inside the particles (Fig. 4.5). Alternatively, the reactant may move through a porous solid, where it may react on surfaces inside the particles. The reader is referred to standard texts for more extensive discussion of these and other possibilities;[1] only a simple example is given here to illustrate the principles involved.

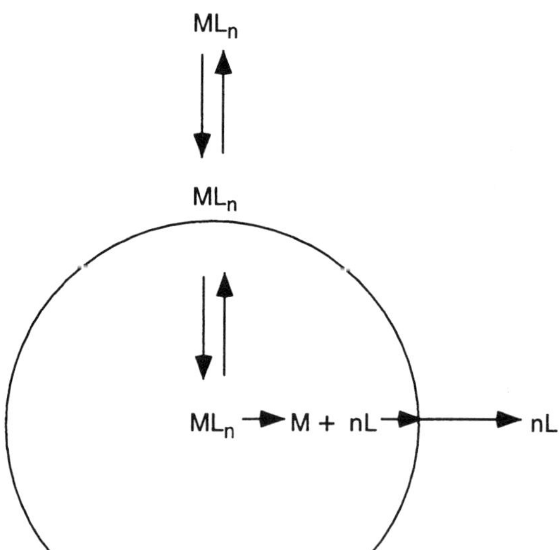


**Fig. 4.5.** Volume reaction can lead to particle growth by chemical reactions. Reaction of precursors $ML_n$ throughout a particle's volume leads to the addition of M with transport of ligands L into the gas phase.

One of the simplest cases of volume-reaction-limited growth is that of a reactant dissolved in water droplets (or droplets of some other solvent) and a gaseous coreactant. In one situation, the coreactant enters the particles and reacts with the dissolved reactant to form a product, thereby increasing the volume of the droplet. The rate of the reaction per unit volume (moles/cm$^3$ sec) in the particle for a first-order reaction can be written as

$$r_v = k_L C_L = k_L H \frac{p_1}{k_B T}, \tag{4.29}$$

where $H$ is Henry's constant, which relates the concentration in the liquid, $C_L$, to the partial pressure in the gas through the expression $C_L = Hp_1/k_BT$, in which $k_L$ is the first-order reaction rate constant in the particle. The rate of particle growth by volume reaction is

$$\frac{dv}{dt} = \frac{r_v v M}{\rho_p}, \tag{4.30}$$

where $\rho_p$ is the particle density, $M$ is the molecular weight of the material out of which the particle is made, and $v$ is the particle volume. Following the other analyses, we find that the characteristic time for particle growth by volume reaction is the time needed to double the particle diameter and is given by

$$\tau_{\mathrm{GV}} = \frac{3(\ln 2)\,\rho_p k_B T}{k_v H p_1 M}. \tag{4.31}$$

## 4.10 NUCLEATION

Gas-phase nucleation is the process by which gas atoms or molecules are transformed into liquid or solid particles and is the basis for gas-to-particle conversion processes.[4, 12] The theory is related to that of nucleation in liquids, which plays a role in spray-based aerosol processes.[19, 20] Gas-phase nucleation is discussed here because it is intimately related to the concepts described by the Kelvin equation and the equation for particle growth in the free-molecule regime. Two mechanisms with different driving forces occur—*Physical* and *Chemical nucleation*—although it is often not clear which mechanism is occurring, and there is some debate over the distinctions between the phenomena.[21, 22] An example of physical nucleation is formation of metal particles from a metal vapor. An example of chemical nucleation is the polymerization of styrene.[23, 24] Nucleation can also be homogeneous or heterogeneous. *Homogeneous nucleation* occurs without the presence of a foreign phase (only with existing particles) and is common in industrial processes, in which enormous saturation ratios frequently occur. This topic has been the subject of numerous reviews and studies.[25–32] In contrast, *heterogeneous nucleation* requires the presence of a foreign body, usually aerosol particles of some other material onto which nucleation takes place. This is most common in atmospheric processes and is usually not important in industrial processes. Several other terms are used to describe nucleation processes. *Homogeneous homomolecular* nucleation is the self-nucleation of a single species with no foreign nuclei or surfaces involved. *Homogeneous heteromolecular* nucleation is the self-nucleation of two or more species with no foreign nuclei or surfaces involved. This topic has been discussed extensively in the literature.[33–35] *Heterogeneous homomolecular* nucleation is the nucleation of a single vapor species on a foreign substance. And *heterogeneous heteromolecular* nucleation is that of two or more species on a foreign substance. Photo-induced nucleation[36–38] and ion-induced nucleation[39–43] have also been examined.

Before presenting any equations, it is useful to discuss these phenomena in qualitative terms. In physical nucleation, the particles are built up from monomer units that have the same or nearly the same form and composition, both in the gas phase and as particles. An example is the formation of metal alkoxide droplets by cooling a metal alkoxide vapor. Another example is the formation of nanophase metal particles, in which metal atom vapors are cooled to drive condensation of the metal to form particles. In general, this process does not involve chemical reactions between the monomer units to displace smaller molecules. It is worth noting, however, that bonds are formed between the atoms in the particle and the arriving atom, making the process in some way a chemical one.

An understanding of physical nucleation requires insight into the behavior of small collections of atoms or molecules called clusters. Clusters of all types of atoms and molecules exist at all pressures and temperatures; however, they exist only for an extremely short time (Fig. 4.6)[15]. Two or more molecules (or atoms) can form a cluster by random collisions, followed by

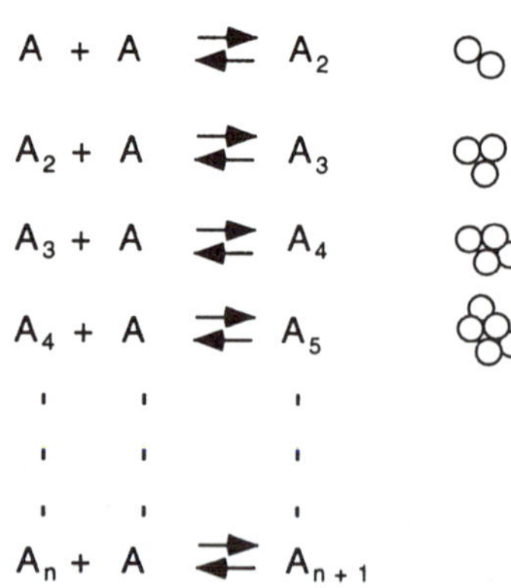


**Fig. 4.6.** Nucleation in terms of collisions between atoms and molecules of substance A. The formation of larger clusters requires a large number of collisions with monomer A.

collisions with other molecules to form larger clusters. A molecule may also leave a cluster, causing it to shrink in size. The clusters are formed and dispersed continuously in a gas even when it is not supersaturated. For a supersaturated vapor, however, the clusters also form, but most of them are still thermodynamically unstable with respect to evaporation at the particular saturation ratio; only sufficiently large clusters are thermodynamically stable. This is the result of the *Kelvin effect*, which states that clusters and particles are unstable with respect to evaporation unless the partial pressure around the particle is greater than the vapor pressure above the curved surface of the particle. Mathematically, this is expressed as

$$d_p^* = \frac{4\sigma v_1}{k_B T \ln S}, \tag{4.32}$$

where $\sigma$ (ergs/cm$^2$) is the surface tension, $v_1$ (cm$^3$) is the molecular or atomic volume, $k_B$ is Boltzmann's constant, $T$ is the temperature, and $S$ is the saturation ratio. Remembering that the vapor pressure above a curved surface is greater than that above a flat surface, we see that the saturation ratio ($S = p_1/p_e$, where $p_1$ is the actual partial pressure of species 1 and $p_e$ is the equilibrium vapor pressure over a flat surface) must be greater than unity in order for clusters to grow. Only if the driving force of the phase transition (e.g., condensation) is sufficient can aerosol nucleation take place from the vapor; saturation ratios of slightly more than unity do not result in homogeneous nucleation (Fig. 4.6). The highly supersaturated state necessary for nucleation can be achieved in practice by cooling a gas to decrease the equilibrium vapor pressure or by driving chemical reactions that form a product at a concentration corresponding to a supersaturated state. Once stable clusters are formed, they grow by the addition of monomer units.

Two limiting cases are of interest for physical nucleation theory in this context: (1) that wherein the dimers, trimers, etc., are not thermodynamically stable with respect to evaporation and (2) that in which the dimers, trimers, etc., are all thermodynamically stable. In (1) nucleation occurs with balanced evaporation and condensation for the clusters, while in (2) evaporation or sublimation does not occur, and nucleation becomes collision limited. These cases are discussed separately.

Chemical nucleation can be considered the result of chemical reactions between a molecule and a growing oligomer (several monomers bonded to each other). An example is the formation of particles during CVD, where the precursor may react in the gas phase to form oligomers with low vapor pressures that in turn nucleate to form particles. Two monomer units may react to form a dimer by eliminating part of the molecules. A dimer may react with another monomer to form a trimer. As the process is repeated, the particle grows. The key difference from physical nucleation is that the clusters grow by chemical reactions and not by physical processes such as condensation. This is important because there may be only a few sites on the cluster (oligomer) where further reaction can take place. In contrast, a monomer can add to any location on a cluster during physical nucleation.

A situation intermediate between chemical nucleation and physical nucleation is the formation of oligomers by the reaction of several monomer units. The oligomers that are formed may have such low vapor pressures, that they nucleate to form particles. Thus, the particles are not composed of unreacted monomer units, but consist of a partially reacted species.

Several other aspects of nucleation have also been studied, including the effect of turbulence on nucleation rates,[11] the effect of surface tension,[45] the control of nucleation to suppress agglomerate formation,[46] and the influence of a growing aerosol on nucleation rates,[47] as well as other features.

### 4.10.1 Classical Homogeneous Nucleation

Many varieties of nucleation theory can be found in the literature. Because the qualitative description given in the preceding section does not adequately convey some features of nucleation, we present a brief quantitative introduction to theoretical, or "classical," nucleation.

*Classical nucleation theory* is one of the simplest theories used to describe the homogeneous nucleation of solid or liquid materials from a gas.[4, 12] The approach, although not necessarily numerically accurate, is useful because it illustrates the dependence of the nucleation rate on various parameters, including the saturation ratio and surface tension. One or another variation of classical nucleation has been used in a wide variety of models to describe the formation of new particles. (See Chapter 7.) *Nucleation theory in the form given here is valid only when the critical particle size is much larger than the size of a monomer.*

The theory presupposes a dilute mixture of a vapor in a carrier gas. Clusters are formed by random collisions and are held together by van der Waals forces (Fig. 4.6). A snapshot of the gas would show mostly monomers, with increasingly smaller concentrations of higher order clusters. Some clusters are able to grow larger than the critical size $d_p^*$ given by the Kelvin equation, become thermodynamically stable, and continue to grow. Classical nucleation theory provides an expression for the rate at which these stable particles are formed (number of nuclei/cm$^3$ sec), namely,

$$I = \left( \frac{p_1}{(2\pi m_1 k_B T)^{1/2}} \right) \left( \frac{2 v_1 \sigma^{1/2}}{(k_B T)^{1/2}} \right) n_1 \exp\left( \frac{-16\pi v_1^2 \sigma^3}{3(k_B T)^3 (\ln S)^2} \right), \quad (4.33)$$

where the parameters are defined in the expressions (Eq. 4.3 and 4.6) for the growth rate of particles and the Kelvin equation (4.32).

Perhaps the most interesting aspect of Equation (4.33) is the extremely strong dependence it posits of the nucleation rate on the saturation ratio $S$,

**TABLE 4.1. Nucleation rate as a function of saturation ratio, where $\sigma$ = 72.75 dyne/cm, $v_1 = 2.99 \times 10^{-23}$ cm$^3$/molecule, $N_1$ = (17.525 torr/kT) $S$, and $m_1 = 2.99 \times 10^{-23}$ g/molecule. (Adapted from Seinfeld)[12]**

| $S$ | $J$ number/cm$^3$ sec |
|---|---|
| 2 | $4 \times 10^{-54}$ |
| 3 | $3.7 \times 10^{-6}$ |
| 3.5 | 94 |
| 4 | $3.4 \times 10^{6}$ |
| 5 | $6.7 \times 10^{11}$ |
| 6 | $6.5 \times 10^{14}$ |
| 7 | $5.6 \times 10^{16}$ |
| 10 | $8.5 \times 10^{19}$ |



as shown in Table 4.1. This comes about because of the $(\ln S)^2$ term in the exponent. As an example, the nucleation rate $I$ varies over 70 orders of magnitude for $S$ varying from 2 to 10 for nucleation of water vapor.[12] The nucleation rate also depends very strongly on the value of the surface tension through the $\sigma^3$ term in the exponential. Because the surface tension can vary by more than an order of magnitude from material to material,[48–51] this results in tremendous changes in nucleation rates as different materials are considered, all other parameters being kept constant. This example illustrates the idea that nucleation is often an all-or-nothing event: Small increases in the saturation ratio result in the nucleation rate increasing from nearly imperceptible levels to high levels.

### 4.10.2 Collision-Controlled Nucleation

Homogeneous, collision-controlled nucleation occurs when there is no thermodynamic barrier to nucleation. Whether this is true or not can be determined in a crude manner by using the vapor pressure and partial pressure of the nucleating species, surface energy, and temperature to evaluate the critical particle size given by the Kelvin equation (4.32). If the critical particle size $d_p^*$ is smaller than the monomer size (on the order of 0.1 nm, the size of the atom or molecule nucleating to form the particle), the monomer is thermodynamically stable, and there is no thermodynamic barrier to the formation of particles. In this case, each atom or molecule is, in a thermodynamic sense, a particle. The implication is that the growth of clusters is not inhibited by the thermodynamics and becomes collision limited. In cases where chemical reaction occurs to form the monomer, and when the critical size is smaller than the monomer size, nucleation is equivalent to reaction in the sense that each reactant molecule which forms a product monomer in essence forms a thermodynamically stable particle. This is an important distinction, because classical nucleation theory is based on the concept of a critical particle size much larger than the monomer size and does not apply to situations in which there is no thermodynamic barrier to the formation of particles. Instead, the rate of particle formation must be replaced by the rate of the chemical reaction that forms the product atoms or molecules. This situation is common and occurs for many materials produced at low temperatures relative to their melting points.

For the vast majority of cases of aerosol material synthesis in the literature, the nucleation mechanism responsible for particle formation is not discussed, and the parameters required to evaluate the critical particle size are not available. Only for well-studied systems such as $TiO_2$ and many elements is this information available. Critical particle sizes have been estimated for the generation of Si in furnace reactors, $SiO_2$ in flame reactors, $TiO_2$ in furnace reactors, $Al_2O_3$ in flame reactors, and other materials for which it has been shown that the critical particle size can be much less than the size of the monomer. (See Chapter 7.) In general, the formation of most metal-oxide particles such as $TiO_2$, $SiO_2$, $Al_2O_3$, etc., with high melting points from a volatile reactant takes place under conditions in which nucleation is indistinguishable from chemical reaction and growth occurs by collisions and coalescence as long as the reactor temperature is below roughly 1500°C. The mode of nucleation depends on the temperature of the reactor: Very high temperature processes such as those occurring in plasma reactors may allow nucleation with a thermodynamic barrier, while lower temperature furnace reactors may provide reaction-limited nucleation.

Evaluation of the critical particle size requires the ability to estimate the saturation ratio and surface energy of the nucleating species. For a pure vapor containing no other species except a carrier gas, estimation of the saturation ratio requires a knowledge of the vapor pressure and partial pressure of the material. Several studies have summarized surface-energy data,[48–51] some of which are presented in Table 4.2. Vapor pressures of various species are given in Chapter 8.

In the case of chemical transformations of reactants to products (as discussed in Section 4.10.4 for dissociative sublimation), the saturation ratio can be written in terms of the equilibrium constant and the partial pressures of the reactants and products. As an example, the saturation ratio for $TiO_2$ can be written as

$$S = K_{eq}(T)\frac{p_{TiCl_4}p_{O_2}}{(p_{Cl_2})^2}, \tag{4.34}$$

where $K_{eq}$ is the equilibrium constant; $p_{TiCl_4}$, $p_{O_2}$, and $p_{Cl_2}$ are the partial pressures of the various species; and $S$ is the saturation ratio. Substituting

**TABLE 4.2. Surface energy of solid metals at melting point. (Adapted from Jones)[51]**

| Metal | Solid–Vapor Surface Energy at Melting Point ($erg/cm^2$) | Liquid–Vapor Surface Tension ($erg/cm^2$, temperature, °C) | Melting Point (°C) |
|---|---|---|---|
| Ag | 1120 | 800 (970) | 961 |
| Al | 1080 | 840 (700) | 660 |
| Au | 1390 | 580–1000 (1070) | 1063 |
| Be | 900 | — | 1287 |
| Co | 1950 | — | 1495 |
| Cr | 2300 | — | 1857 |
| Cu | 1720 | 1034 (1150) | 1084 |
| Fe | 2080 | 1360 (1600) | 1536 |
| Mo | 1960 | — | 2617 |
| Nb | 2260 | — | 2467 |
| Ni | 1810 | — | 1453 |
| Pt | 2280 | 1819 (2000) | 1769 |
| Sn | 680 | 526 (253) | 232 |
| Ta | 2500 | | 2977 |
| Ti | 1690 | — | 1670 |
| W | 2650 | — | 3407 |
| Zn | 830 | — | 420 |

this expression into the Kelvin equation provides a criterion for when the critical particle size is less than the monomer size and when particle formation is identical to chemical reaction.[15] We have

$$\frac{2\sigma_s v_1}{k_B T r_1 \ln\left[K_{eq}(T) p_{TiCl_4} p_{O_2}/(p_{Cl_2})^2\right]} \leq 1, \tag{4.35}$$

where $\sigma_s$ is the surface energy of $TiO_2$ (dyne/cm or ergs/$cm^2$), $k_B$ is Boltzmann's constant, $v_1$ is the volume of the $TiO_2$ monomer ($cm^3$), $r_1$ is the radius of the $TiO_2$ monomer (cm), and $T$ is the temperature (K). This equation is presented graphically in Figure 4.7,[15] which shows that, for almost all conditions of practical importance, particle formation and growth occur by chemical reaction and collisions (region labeled "coagulation"). Several general features of this type of problem are evident from Equation (4.35) and Figure 4.7. Increasing the temperature tends to favor conversion by classical nucleation; the higher temperature tends to lower the saturation ratio. Increasing conversion of the reactants tends to favor conversion by classical nucleation because this lowers the reactant concentrations and shifts the equilibrium away from the formation of $TiO_2$. Increasing the surface energy favors classical nucleation because it results in a higher energy barrier to the formation of the new solid phase from the gas.

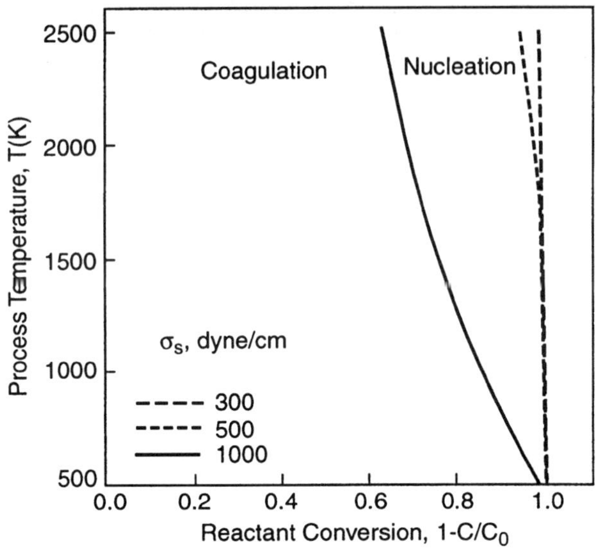


**Fig. 4.7.** Regions in which nucleation is coagulation controlled and in which classical nucleation theories are valid for $TiO_2$ formation from $TiCl_4$ and $O_2$. Reprinted from *Journal of Aerosol Science 22* (5), 1991, 637–655, Xiong, Y. and Pratsinis, S.E., Copyright 1991, with kind permission from Elsevier Science Ltd., The Boulevard, Langford Lane, Kidlington OX5 1GB UK.

### 4.10.3 Chemical Nucleation

Chemical nucleation is best explained via an example after giving a qualitative overview. A typical reactant consists of a metal center with one or more attached ligands. Chemical reaction with or without a co-reactant resulting in the elimination of some of the supporting ligands can lead to the formation of molecular species that react to form, in turn, species of higher molecular weight. In purely chemical nucleation, the formation and growth of particles occurs only by chemical reaction. However, species formed by the reaction of monomer units are likely to exhibit lower vapor pressures than the starting material, which can also result in particle formation by classical nucleation of the higher-molecular-weight species. In this case, there is no clearly defined condensing species, as there is when a single species such as water condenses. Rather, the condensing species may be molecules with a distribution of molecular weights, as discussed shortly. Finally, if the reaction of a monomer results in the formation of a single molecular species with a vapor pressure low enough to have a critical particle size (as defined by the Kelvin equation) less than its molecular size, nucleation becomes indistinguishable from chemical reaction and is collision controlled, as discussed in Section 4.10.2. The product can be a droplet or a solid particle, depending on the chemical and physical characteristics of the species involved.

As a specific example of chemical nucleation, a variety of different species with low vapor pressure can be formed from precursors intended to form pure TiN particles or films. The precursor $Ti(NMe_2)_4$ (a common volatile precursor) can form TiN in the presence of ammonia according to the overall reaction

$$Ti(NMe_2)_4(g) + \text{excess } NH_3(g) \longrightarrow TiN(g) + 4\,HNMe_2(g) + N_2(g). \tag{4.36}$$

However, depending on the temperature, other reactions can occur, leading to the formation of particles with different compositions.

Stable molecular species can form with lower vapor pressures than the precursor. For example, the stoichiometric reaction of $Ti(NMe_2)_4$ with ammonia to form TiN can occur via steps that involve the elimination of $HNMe_2$ ligands. This elimination reaction probably results in the formation of species with a less sterically encumbered metal center than the starting material. As a consequence, the products are likely to react, yielding a species with higher molecular weight and therefore lower vapor pressure, as is presumed in the following equations:

$$Ti(NMe_2)_4(g) + NH_3(g) \longrightarrow Ti(NMe_2)_3NH_2(g) + HNMe_2(g), \tag{4.37}$$

$$Ti(NMe_2)_3NH_2(g) + Ti(NMe_2)_4(g) \longrightarrow [Ti(NMe_2)_3]_2NH(g) + HNMe_2(g), \tag{4.38}$$

$$2\,[Ti(NMe_2)_3]_2NH(g) \longrightarrow Ti_4(NMe_2)_{11}N_2H(g) + HNMe_2(g). \tag{4.39}$$

Higher-molecular-weight species can also be formed. Chemical nucleation would be the formation of particles via reactions similar to these. Analysis of the chemical composition of the particles would show molecular species with significant amounts of contaminants from the ligands. Alternatively, the species containing four Ti atoms—for example, $Ti_4(NMe_2)_{11}N_2H$—has a much lower vapor pressure than the reactant, $Ti(NMe_2)_4$, and can nucleate to form particles or grow to even larger sizes by chemical reactions analogous to those in Equation (4.39). This example illustrates the potential complexity of nucleation phenomena when chemical reactions occur.

### 4.10.4 Materials That Sublime Dissociatively

Nucleation theory, in the context of aerosols, was developed primarily for simple liquids. This approach is reflected in the large number of studies of the nucleation of water, ethanol, hydrocarbons, and other simple species. Aerosol routes to materials, however, usually involve inorganic substances. Some examples are elements, metal oxides, nitrides, carbides, and borides, as well as more complex substances. Many of these materials exhibit behavior that is unique to binary (two-element) or ternary (three-element) systems and is not found in elements. One of the most important examples is dissociative sublimation.

Dissociative sublimation involves processes of the type

$$AB\ (\text{solid}) \rightleftarrows A\ (\text{gas}) + \frac{1}{m} B_m\ (\text{gas}). \tag{4.40}$$

A specific example is

$$2\text{BN}\ (\text{solid}) \rightleftarrows 2\text{B}(s) + \text{N}_2\ (\text{gas}). \tag{4.41}$$

In these cases the monomer in the solid state does not exist in the gas phase, and alternative methods are required to describe the nucleation process. As a starting point, the saturation ratio can be expressed in terms of the equilibrium constant, analogous to the approach outlined for $TiO_2$. (See Eq. 4.35.) A variety of sources provide useful background material for dealing with this problem.[52–58] References for specific materials include SiC,[59, 60] CdS and CdSe,[61–64] metal nitrides,[65, 66] and $TiB_2$.[67, 68]

## 4.11 REFERENCES

1. Froment, G.F. & Bischoff, K.B. *Chemical Reactor Analysis*. New York: Wiley, 1990.
2. Arai, Y. *Chemistry of Powder Production*. New York: Chapman and Hall, 1996.
3. Masters, K. *Spray Drying*. New York: John Wiley and Sons, 1972.
4. Friedlander, S.K. *Smoke, Dust and Haze*. New York: Wiley Interscience, 1977.
5. Heidenreich, S. *J. Aerosol Science 25*(1), 1994, 49–59.
6. Kreidenweis, S.M., Flagan, R.C., & Seinfeld, J.H. *Aerosol Sci. Technol. 6*, 1987, 1–14.
7. Kulmala, M. *Aerosol Sci. Technol. 19*, 1993, 381–388.
8. Rader, D.J., McMurry, P.H., & Smith, S. *Aerosol Sci. Technol. 6*, 1987, 247–260.
9. Tsang, T.H., Cook, S.M., & Marra, M.E. *Aerosol Sci. Technol. 12*, 1990, 386–398.
10. Williams, M.M.R. *J. Colloid and Interface Sci. 93*(1), 1983, 252–263.
11. Bird, R., Stewart, W.E., & Lightfoot, E.N. *Transport Phenomena*. New York: John Wiley and Sons, 1960.
12. Seinfeld, J.H. *Atmospheric Chemistry and Physics of Air Pollution*. New York: Wiley Interscience, 1986.
13. Heidenreich, S. & Büttner, H. *J. Aerosol Sci. 26*(2), 1995, 335–339.
14. Crouzet, Y. & Marlow, W.H. *Aerosol Sci. Technol. 22*, 1995, 43–59.
15. Xiong, Y. & Pratsinis, S.E. *J. Aerosol Sci. 22*(5), 1991, 637–655.
16. Hinds, W.C. *Aerosol Technology*. New York: Wiley Interscience, 1982.
17. Xiong, Y. & Kodas, T. *J. Aerosol Sci. 24*, 1993, 893.

18. Jain, S., Fotou, G.P., & Kodas, T.T. *J. Colloid Interface Sci. 185*, 1997, 26–38.
19. Chiang, P.-P. & Donohue, M.D. *J. Colloid Interface Sci. 122*(1), 1988, 230–250.
20. Chiang, P.-P., Donohue, M.D., & Katz, J.L. *J. Colloid Interface Sci. 122*(1), 1988, 251–265.
21. Frenklach, M. *Ceramic Powder Science III 12*, 1990, 239–250.
22. Katz, J.L. & Donohue, M.D. *J. Colloid Interface Sci. 85*(1), 1982, 267–277.
23. El-Shall, M.S., Bahta, A., Rabeony, H., & Reiss, H. *J. Ceram. Phys. 87*(2), 1987, 1329–1345.
24. Reiss, H., Rabeony, H., El-Shall, M.S., & Bahta, A. *J. Chem. Phys. 87*(2), 1987, 1315–1328.
25. Reiss, H. *Ind. Eng. Chem. 44*(6), 1952, 1284–1287.
26. Burton, J.J. *Surface Sci. 26*, 1971, 1–13.
27. McDonald, J.E. *Am. J. Physics 30*(12), 1962, 870–877.
28. McDonald, J.E. *Am. J. Physics 31*(1), 1962, 31–41.
29. Uhlman, D.R. & Chalmers, B. *Ind. Eng. Chem 57*(9), 1965, 19–31.
30. Abraham, F.F. *Homogeneous Nucleation Theory*. New York: Academic Press, 1974.
31. Becker, C. & Reiss, H. *J. Chem. Phys. 65*(6), 1976, 2066–2070.
32. Wegener, P.P. & Mirabel, P. *Natur wissenchaften 74*, 1987, 111.
33. Reiss, H. *J. Chem. Phys. 18*(6), 1950, 840–848.
34. Shugard, W.J., Heist, R.H., & Reiss, H. *J. Chem. Phys. 61*(12), 1974, 5298–5305.
35. Mirabel, P. & Katz, J.L. *J. Chem. Phys. 60*, 1974, 1138.
36. Reiss, H., Marvin, D.C., & Heist, R.H. *J. Colloid Interface Sci. 58*(1), 1977, 125–140.
37. Marvin, D.C. & Reiss, H. *J. Chem. Phys. 61*, 1978, 1897–1918.
38. Gertler, A.W., Almeida, B., El-Sayed, M.A., & Reiss, H. *Chem. Phys. 42*, 1979, 429–439.
39. Adachi, M., Okuyama, K., & Seinfeld, J.H. *J. Aerosol Sci. 23*(4), 1992, 327–337.
40. Raes, F. & Janssens, A. *J. Aerosol Sci. 16*(3), 1985, 217–227.
41. Raes, F. *Health Phys.*, Fall–Winter, 1985, 1177–1187.
42. Raes, F., Janssens, A., & Eggermont, G. *Atmos. Env. 19*(7), 1985, 1069–1073.
43. Raes, F., Janssens, A., & Vanmarcke, H. *Science of the Total Environment 45*, 1985, 205–218.
44. Lesniewski, T. & Friedlander, S.K. *Aerosol Sci. Technol. 23*, 1995, 174–182.
45. Rao, N.P. & McMurry, P.H. *Aerosol Sci. Technol. 13*, 1990, 183–195.
46. Zachariah, M.R. & Dimitriou, P. *Aerosol Sci. Technol. 13*(4), 1990, 413–425.
47. Alam, M.K. & Flagan, R.C. *J. Colloid Interface Sci. 97*(2), 1984, 232–246.
48. McLachlan, D., Jr. *Acta Metallurgica 5*(2), 1957, 111–112.
49. Benson, G.C. & Yun, K.S. Surface energy and surface tension of crystalline solids. In Flood, E.A. (Ed.), *The Solid–Gas Interface*. New York: Marcel Dekker, Inc., 1967, 203–269.
50. Anand, M.K., Solanki, S.K., Jain, K.C., & Shastry, V.R. *J. Indian Chem. Soc. 47*(12), 1970, 1177–1181.

51. Jones, H. *Metal Sci. J. 5*, 1971, 15–18.

52. Colin, R. & Goldfinger, P. Correlation of dissociation energies of gaseous molecules and of heats of vaporization of solids. Intrnl. Symp. Condensation and Evaporation of Solids, Dayton, OH, 1962, 165–179.

53. Drowart, J. Mass spectrometric studies of the vaporization of inorganic substances at high temperatures. Intrnl. Symp. Condensation and Evaporation of Solids, Dayton, OH, 1962, 256–309.

54. Hirschwald, W. & Stranski, I.N. Theoretical considerations and experiments on evaporation of solids. Internl. Symp. Condensation and Evaporation of Solids, Dayton, OH, 1962, 59–84.

55. Trulson, O.C. & Schissel, P.O. Mass spectrometric study of vaporization coefficients. Internl. Symp. Condensation and Evaporation of Solids, Dayton, OH, 1962, 313–317.

56. Cabera, N. & Coleman, R.V., Theory of crystal growth from the vapor. In Gilman, J.J. (Ed.), *The Art and Science of Growing Crystals*. New York: John Wiley & Sons, 1963, 3–28.

57. Strickland-Constable, R.F. *Kinetics and Mechanism of Crystallization*. London: Academic Press, 1968, p. 347.

58. Battat, D., Faktor, M.M., & Garrett, I. A model of vapor transport and its interplay with surface reactions. In Wilcox, W.R. (Ed.), *Preparation and Properties of Solid State Materials*, New York: Marcel Dekker, Inc., 1976, 43–128.

59. O'Connor, J.R. Silicon carbide. In Gilman, J.J. (Ed.), *The Art and Science of Growing Crystals*, New York: John Wiley & Sons, Inc., 1963, 93–118.

60. Lyon, T.F. Vapor pressures of several carbides and hafnium metal by the Langmuir method, Intrnl. Symp. Condensation and Evaporation of Solids, Dayton, OH, 1962, 435–449.

61. Somorjai, G.A. The evaporation rate and mechanism of CdS and CdSe, Internl. Symp. Condensation and Evaporation of Solids, Dayton, OH, 1962, 417–433.

62. Kaldis, E. *J. Crystal Growth 5*, 1969, 376–390.

63. Kaldis, E. Principles of the vapour growth of single crystals. In Goodman, C.H.L. (Ed.), *Crystal Growth Theory and Techniques*, London: Plenum Press, 1974.

64. Hariharan, A.V. & Eick, H.A., *High Temp. Sci. 3*, 1971, 123–129.

65. Hildenbrand, The vaporization behavior of boron nitride and aluminum nitride, Internl. Symp. Condensation and Evaporation of Solids, Dayton, Ohio, 1962, 399–415.

66. Glasson, D.R. & Jayaweera, S.A.A. *J. Appl. Chem. 18*(March), 1968, 65–77.

67. Schissel, P.O. & Trulson, O.C. Mass spectrometric study of the vaporization of the Ti–B system. Internl. Symp. Condensation and Evaporation of Solids, Dayton, OH, 1962, 321–322.

68. Schissel, P.O. & Trulson, O.C. *J. Phys. Chem. 66*, 1962, 1492–1496.

# 5 Collision and Coalescence

## 5.1 INTRODUCTION

*Coagulation* is the process by means of which two particles collide and stick together to form a new, larger particle. One of the most common particle growth mechanisms in industrial processes, coagulation increases the average particle size, decreases the total particle concentration, and conserves the total aerosol mass concentration and volume (Fig. 5.1). Coagulation is usually not important in spray processes for powders and films, because the particle concentration is not high enough to result in particle collisions during a typical process residence time. However, coagulation *is* the method by which particles of fumed silica ($SiO_2$) and the pigments titanium dioxide ($TiO_2$) and carbon black grow, and it determines the final spread of the particle size distribution and average particle size. This is because the concentrations of particles are sufficiently high to allow numerous collisions, even in the short residence times found in these processes. Hinds and Reist provide an introduction to the topic,[1, 2] and more advanced treatments can be found elsewhere.[3–5]

Because particle collisions and fusion always occur together, but with different relative rates, coagulation can result in particles with varying morphologies (Fig. 5.1). If two droplets collide, they can immediately coalesce to form a larger spherical droplet. However, if two solid particles collide, the result may be an agglomerate or a spherical particle, depending on the relative rates of fusion and collision. However, the total concentration of individual particles in the gas decreases regardless of the rate of fusion, because colliding particles always stick and form one particle.

Coagulation usually takes place by the Brownian motion of particles. Electrical, gravitational, shear, and other forces that affect particle motion can also affect the coagulation rate of aerosols. In general, the rate of change of the particle size distribution function $n(v)$ (which can be viewed as the rate of change of the concentration of particles of size $v$) by coagulation is given by[3]

$$\frac{dn(v)}{dt} = \underbrace{\frac{1}{2}\int_0^v \beta(v', v - v')n(v')n(v - v')dv'}_{(A)} - \underbrace{n(v)\int_0^\infty \beta(v, v')n(v')dv'}_{(B)}, \quad (5.1)$$

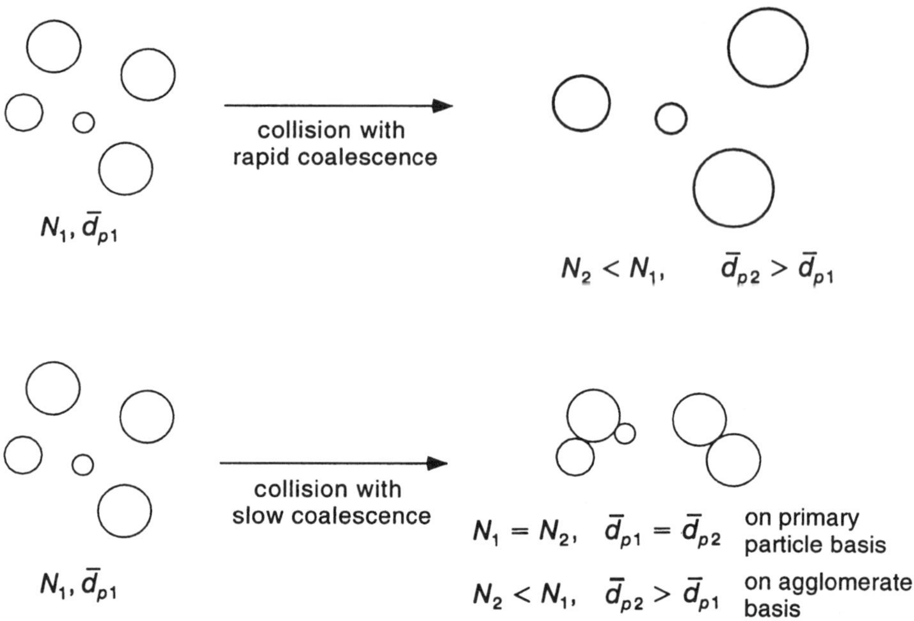


**Fig. 5.1.** Collision and coalescence of particles for slow (bottom) and rapid (top) particle fusion relative to collision rate, illustrating influence on particle morphology. $N_1$ and $N_2$ are particle concentrations.

where $n(v)$ is the particle size distribution function (number/cm$^6$) and $\beta(v, v')$ is the collision frequency of particles of size $v$ and $v'$ (cm$^3$/sec) ($v$ and $v'$ are particle volumes). This equation is one of the most general forms of the theory describing coagulation; simpler expressions are given shortly. It is more convenient to work in terms of the particle volume than diameter, because two particles colliding always result in the formation of a particle with a volume equal to the sum of the volumes of the original two particles. Volumes are also useful because, when sintering of particles after collision is not rapid, agglomerates are formed, and it is not simple to define a size or area. In Equation (5.1), the first term ($A$) accounts for the formation of particles of volume $v$ by coagulation of smaller particles. (The factor 1/2 is necessary to avoid double-counting collisions.) The second term ($B$) accounts for the loss of particles of volume $v$ by coagulation with all other particles.

The evolution of the particle size distribution function by coagulation can be obtained by using the appropriate collision frequency functions $\beta(v, v')$ and solving Equation (5.1). This equation is used in numerical modeling and more complicated coagulation problems; simpler forms are given in Section 5.2. The goal of this chapter is to examine simple equations that provide the time scales associated with coagulation and that allow comparison with other relevant times, including process residence time.

The chapter first presents Brownian coagulation and provides the characteristic time for coagulation. Other relevant types of coagulation (turbulent, electrically induced) are then introduced. The concept of a self-preserving size distribution is discussed. Finally, the chapter concludes with a discussion of particle coalescence.

## 5.2 BROWNIAN COAGULATION

In the absence of any external field forces, the coagulation of spherical particles is governed by the Brownian motion of these particles, which is induced by the random thermal motion of gas molecules. This is the most common type of coagulation and is included in most models of aerosol dynamics in reactors. As with many other aerosol processes, the frequency function describing the collision takes different forms, corresponding to conditions where the particles are much smaller and much larger than the gas mean free path.[3, 5] The development of the theory has been reviewed elsewhere.[6]

When the particle size is much larger than the mean free path of the gas ($\mathrm{Kn} \ll 1, d_p \gg \lambda$), coagulation takes place in the continuum regime such that

$$\beta_C(v_i, v_j) = K_C\left(v_i^{1/3} + v_j^{1/3}\right)\left(\frac{1}{v_i^{1/3}} + \frac{1}{v_j^{1/3}}\right), \tag{5.2}$$

where $v_i$ and $v_j$ are particle volumes and

$$K_C = \frac{2k_BT}{3\mu}, \tag{5.3}$$

in which $k_B$ is Boltzmann's constant, $T$ is temperature (K), and $\mu$ is the gas viscosity (g/cm sec). Coagulation occurs by the diffusion of the particles through the continuous gas, as shown in Figure 5.2. Consistent with this thinking, the derivation of Equation (5.2) examines the Brownian diffusion of particles towards each other, followed by collision. Equations (5.2) and (5.3) represent the form of the coagulation coefficient $\beta_C$ most commonly used in modeling aerosol dynamics during the synthesis of materials. Note that these equations hold only for spherical particles.

Equation (5.1) can be rearranged into a simpler form by neglecting the spread of the particle size distribution and assuming a monodisperse aerosol. The rate of change in particle concentration in the continuum regime is obtained from Equations (5.1) and (5.2) by setting $v_i = v_j = v$, so that

$$\frac{dN_\infty}{dt} = -2K_C N_\infty^2, \tag{5.4}$$

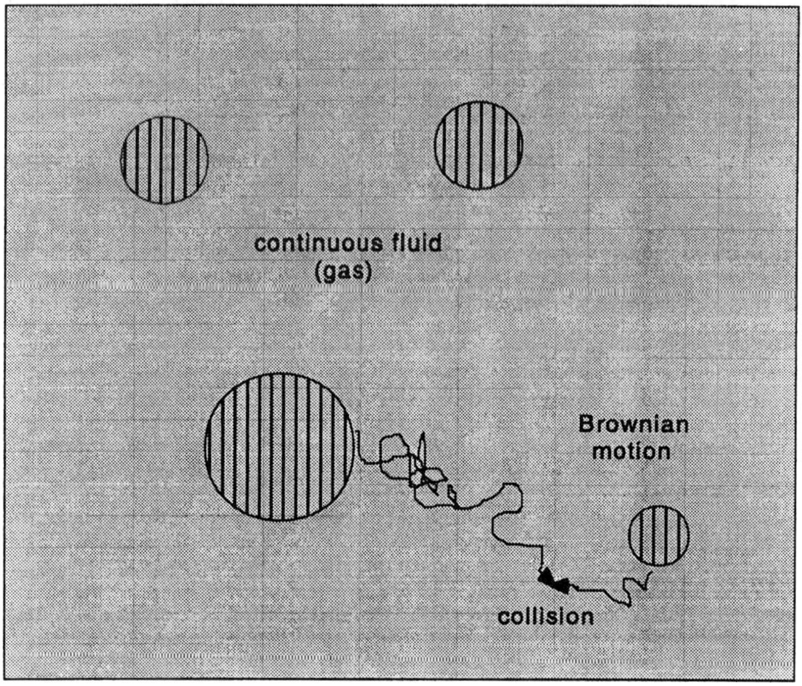


**Fig. 5.2.** Particle collisions for particle diameters much larger than the gas mean free path occur by diffusion.

where $N_\infty$ is the total particle concentration. Coagulation is a second-order process analogous to a second-order chemical reaction. The corresponding evolution of the particle diameter with time for a monodisperse aerosol is[1]

$$d_p = d_{p0}(1 + 2K_C N_0 t)^{1/3}, \tag{5.5}$$

where $d_{p0}$ and $N_0$ are the initial particle diameter and total particle concentration, respectively. This equation shows that the diameter increases linearly for short times. For longer times, the assumption of a monodisperse aerosol is not valid, and the increase is no longer linear. Hinds has tabulated values of $K_C$ for various particle sizes and spreads of the distribution.[1] His values are not the same as those used here because of the difference in the definition of $K_C$; instead, we take Friedlander's approach.[3] Equation (5.4) will also be used to derive a characteristic time for coagulation in the continuum regime.

When the particle size is much smaller than the mean free path of the gas (Kn $\gg$ 1, $d_p \ll \lambda$), coagulation takes place in the free-molecule regime, and

$$\beta_F = K_F \left(\frac{3}{4\pi}\right)^{1/6} \left(\frac{1}{v_i} + \frac{1}{v_j}\right)^{1/2} \left(v_i^{1/3} + v_j^{1/3}\right)^2, \tag{5.6}$$

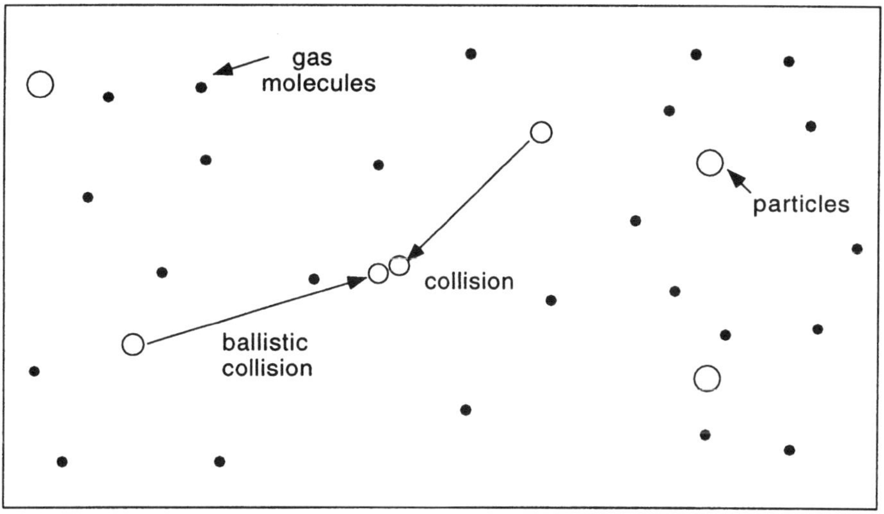


**Fig. 5.3.** Particle collisions for particle diameters much less than the gas mean free path occur by a ballistic process in which the particles behave like large molecules.

where $K_F = (6k_BT/\rho_p)^{1/2}$, in which $\rho_p$ is the particle density.[3, 5] Collisions take place by a ballistic process in which the particles can be treated as large molecules, as shown in Figure 5.3.

In material processing, where submicron particles are produced at high temperatures, particle sizes are often comparable to the gas mean free path. In that case, particle growth by coagulation takes place in the transition regime (Kn $\sim$ 1, $d_p \sim \lambda$). Then the most appropriate function for the coagulation rate is the Fuchs interpolation formula, which converges to the free-molecule and continuum limits. The Fuchs equation is discussed in more detail elsewhere.[5] It is useful to note that Equations 5.1 through 5.6 are for single-component systems (i.e., systems of one material); the equations for two-component coagulation are more complex.[7]

## 5.3 CHARACTERISTIC TIME FOR COAGULATION

A qualitative understanding of aerosol behavior in reactors almost always requires knowledge of the importance of coagulation. For that reason, it is useful to have a simple method for determining its contribution to a reduction in particle concentration and an increase in particle size during the reactor residence time. This is conveniently done by computing the reactor residence time and comparing it with the time required for an appreciable fraction of the particles to collide with other particles, a time commonly called the *characteristic time for coagulation.*

The characteristic time for coagulation can be defined as the half-life of the aerosol by coagulation, $\tau_C$, the time needed to reduce the particle concentration to half of its initial value. For spherical monodisperse particles, this characteristic time for coagulation in the continuum limit is

$$\tau_{CC} = \frac{1}{2K_C N_0}. \tag{5.7}$$

Equation (5.7) can be obtained from Equation (5.4) by a procedure outlined by Friedlander.[3]

The characteristic time in the free-molecule regime is

$$\tau_{CF} \approx \frac{1}{\left(\frac{3V}{4\pi}\right)^{1/6} K_F N_0^{5/6}}, \tag{5.8}$$

where $V$ is the total particle volume ($cm^3/cm^3$). An example is an aerosol containing $10^{10}$ particles/$cm^3$ of 0.1 $\mu$m diameter, coagulating at 25°C and atmospheric pressure ($K_C = 4 \times 10^{-9}$ $cm^3$/sec), for which $\tau_{CC} \approx 0.1$ sec. This means that only 0.1 second is required for a significant fraction of the particles to collide with each other. The value of $\tau_{CC}$ is very important for determining the role of coagulation in a given process. For example, if the reactor residence time is 10 seconds, then coagulation is important and reduces the particle concentration and increases the particle size. Because the residence times in most industrial reactors are much less than 10 seconds and particle concentrations are often much higher than these values, coagulation will modify the particle concentration and control not only the coalescence rate, but also the growth of particles in the majority of systems. Thus, the effects of coagulation have to be considered for most gas-to-particle conversion processes.

## 5.4 SHEAR-INDUCED AND TURBULENT COAGULATION

In addition to thermal or Brownian coagulation, other types of coagulation (laminar shear, turbulent, gravitational, and others) may occur under certain circumstances.[3–5] Here, we consider the qualitative behavior of turbulent coagulation because of its importance in industrial flow systems. Manufacturing particles at large production rates requires the use of high aerosol flow rates and subsequent turbulent flows (with high Reynolds numbers, often $10^6$). This turbulence results in the formation of small eddies (swirling currents of fluid). The more turbulent the flow, the smaller is the size and the higher is the velocity of the eddies. If a particle (more specifically, its stopping distance, which is related to the Stokes number) is much smaller in size than the smallest eddies, it is not affected by the large-scale motion of

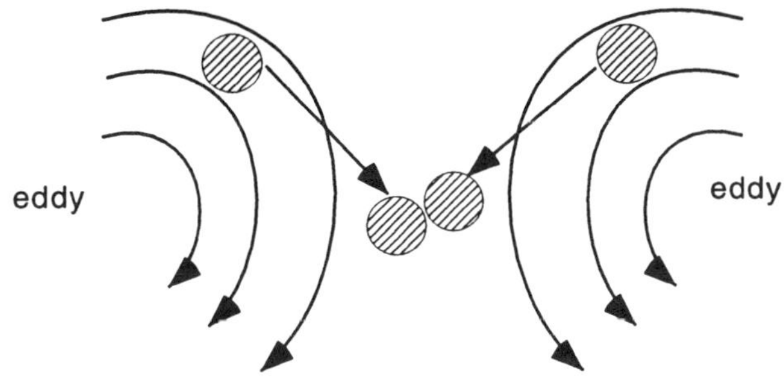


**Fig. 5.4.** Turbulent coagulation occurs as a result of particle inertia and velocity gradients in eddies.

the eddies, but is instead transported along with the fluid inside the eddies. If, however, the particle is larger (with a stopping distance on the order of, or larger than, the eddy size), it experiences rapid changes in speed and direction. The eddies impart a velocity to the particles, so, as the fluid changes direction, the particles cannot change direction because of inertia and, as a consequence, cross fluid flow lines. This displacement results in collisions of particles as they are thrown around by the eddies (Fig. 5.4).

Turbulence can increase the collision rate of particles only when the particle size is near or exceeds the size of the smallest eddies, which is also called the *Kolmogorov microscale of turbulence*[8] and is given by

$$\eta_r = \left(\frac{UL}{\nu}\right)^{3/4} L, \tag{5.9}$$

where $\nu$ is the kinematic viscosity of the gas (cm$^2$/sec), $L$ is the cross section of the flow (cm), and $U$ is the average gas velocity (cm/sec). Typical values of the size of the smallest eddies are $1 \times 10^{-3}$ cm, or 10 $\mu$m, for Re = 100,000 and a tube diameter ($L$ in Equation (5.9)) of 10 cm. Comparing this with typical particle sizes, one sees that the effect is important only for large particles (with long stopping distances) and highly turbulent flows. It is at these conditions alone that particles are influenced by the forces associated with eddies in the fluid.

Following the reasoning just presented allows the determination of the collision frequency for turbulence (shear) induced coagulation,[5, 9, 10] namely,

$$\beta_T = K_T(d_{pi} + d_{pj})^3, \tag{5.10}$$

where $d_{pi}$ and $d_{pj}$ are particle diameters (cm) and

$$K_T = \left(\frac{\pi^2 \varepsilon_d}{120\nu}\right)^{1/2}, \tag{5.11}$$

in which $K_T$ is the turbulent coagulation coefficient ($\text{cm}^3/\text{sec}$), $\nu$ is the kinematic viscosity ($\text{cm}^2/\text{sec}$), and $\varepsilon_d$ ($\text{cm}^2/\text{sec}^2$) is the rate of energy dissipation for the core region of turbulent pipe flow (the central part of the tube where the flow is the most turbulent)[3] and is given by

$$\varepsilon_d = \frac{4}{d_t}\left(\frac{f}{2}\right)^{3/2} U^3. \tag{5.12}$$

Here, $d_t$ is the pipe diameter (cm), $f$ is the Fanning friction factor (dimensionless),[11] and $U$ is the average fluid velocity (cm/sec). The Fanning friction factor is a function of the flow Reynolds number and can be estimated from simple plots given in standard texts on particle transport.[11] As an example, for a smooth pipe 10 cm in diameter with air at 20°C and a Reynolds number of 50,000, $\varepsilon_d = 2 \times 10^4$ $\text{cm}^2/\text{sec}^3$.[3]

In practice, turbulent coagulation is not important in laboratory studies, where the Reynolds numbers are low and the eddies are very large relative to the particle size. In the case of laminar flow, the eddies do not even exist. It is only in the highly turbulent flows encountered in industry that this effect is important, and even then, only for sufficiently large particle sizes.[12]

## 5.5 COAGULATION BY ELECTRIC FORCES

Electrical charges can play a significant role in aerosol coagulation.[3–5] Their effect is important when the particles are intentionally charged in electrostatic precipitators or when the process is operated at high temperatures, where ionized species can form. The latter can be obtained when species like KCl and CsCl are used in flames and high-temperature processes. The ionization temperatures for these species are low enough to allow an appreciable fraction of the salts to ionize. The salt ions then collide with the aerosol particles, resulting in charged particles. When aerosol particles carry charges, (ionic) Coulomb forces enhance or retard the coagulation rate of particles with opposite or similar charges, respectively. The standard Brownian coagulation frequency function (Equation (5.2) or Equation (5.6)) needs to be corrected to account for charging. The correction is usually accomplished by correcting the standard frequency function by a dimensionless factor $W_{ij}$,[13] yielding

$$\beta_{\text{Ch}}(v_i, v_j) = \frac{\beta(v_i, v_j)}{W_{ij}}, \tag{5.13}$$

where

$$W_{ij} = \frac{1}{\omega}(e^{\omega} - 1), \tag{5.14}$$

in which

$$\omega = \frac{z_i z_j e^2}{\varepsilon(r_i + r_j)k_B T}, \tag{5.15}$$

where $z_i$ and $z_j$ are the number of charges on particles of radius $r_i$ and $r_j$, and $\varepsilon$ is the dielectric constant for the material of which the particles are constituents. Equations (5.13) and (5.14) are strictly valid only for spherical particles.

The correction for charged particles is usually not significant, unless the particles are intentionally charged in a corona discharge[3] or by collisions with high concentrations of ions created by molecular dissociation at high temperatures.

The foregoing discussion applies to the case where charged particles exist, but an electric field is not present, such as particles charged by ions formed by dissociation at high temperatures. A different physical situation occurs, however, when charged aerosols flow through electric fields; particle motion is then affected by the potential lines of the field. The time for coagulation to occur is a strong function of the field intensity. This problem is more specialized than those just described and is analyzed in detail elsewhere.[14]

## 5.6 SELF-PRESERVING SIZE DISTRIBUTIONS: UNCHARGED SPHERICAL PARTICLES

For aerosols undergoing coagulation, the particle size distribution changes continuously: The particle size increases monotonically, and the particle concentration decreases monotonically. However, when plotted in a certain dimensionless form, particle size distributions can become invariant with respect to time.[3] This situation can occur if there are no losses or additions of particles to the system. The approach requires scaling the particle volume with respect to the average particle volume; that is,

$$\eta = \frac{v}{\bar{v}} = \frac{N_\infty v}{V}, \tag{5.16}$$

where $N_\infty$ is the total particle concentration (number/cm$^3$), $v$ is the particle volume, $\bar{v}$ is the average particle volume, and $V$ is the total aerosol volume (cm$^3$/cm$^3$), so that the particle size distribution is a function of the dimensionless particle volume only. Mathematically, we have

$$\psi(\eta) = \frac{n(v,t)V}{N_\infty^2}, \tag{5.17}$$

where $\psi(\eta)$ is the self-preserving size distribution (SPSD).[3] These time-invariant *self-preserving distributions* are reached after a finite time, depending on the coagulation rate and the initial shape of the distribution. They are independent of the initial particle size distribution.

The usefulness of self-preserving size distributions is that all coagulation processes which form particles by a mechanism that reaches a self-preserving distribution will give the same self-preserving size distribution if sufficient residence time is provided. Thus, plots of data can show when the size distribution from a process is consistent with a certain physical mechanism, such as Brownian coagulation in the continuum regime, and exclude other mechanisms as possibilities. *Furthermore, the self-preserving distribution is the narrowest the particle size distribution can become, in the absence of a change in the mechanism of evolution of particle size.* Thus, this approach can indicate when it is no longer advantageous to increase the reactor residence time in order to decrease the spread in the size distribution.[15]

It is important to understand what types of coagulation processes lead to SPSDs. Fortunately, the most commonly encountered situations, from a practical point of view, fall into this category. Aerosols undergoing Brownian coagulation in the continuum[16] and free-molecule regimes[17] reach self-preserving size distributions. Aerosols in the transition regime reach quasi-self-preserving distributions.[18]

Figure 5.5 shows that self-preserving distributions are nearly lognormal in shape.[3] The geometric standard deviations $\sigma_g$ for the number and mass SPSDs are about 1.45 and 1.3, for both the continuum and free-molecule limits, respectively.[12] As a result, the SPSD places a lower limit on the spread of the size distributions for Brownian coagulation. If the aerosol is

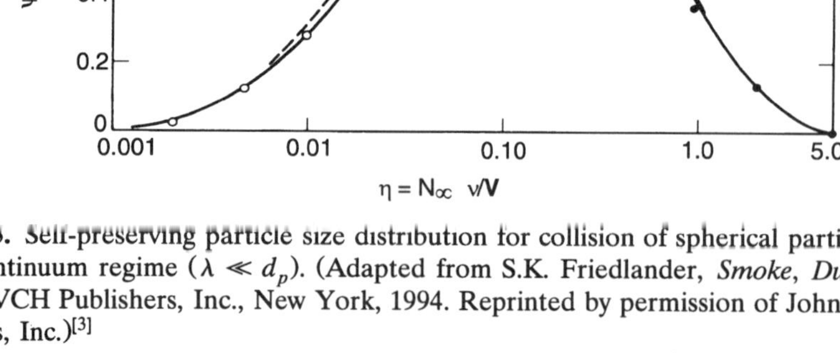


**Fig. 5.5.** Self-preserving particle size distribution for collision of spherical particles in the continuum regime ($\lambda \ll d_p$). (Adapted from S.K. Friedlander, *Smoke, Dust and Haze*, VCH Publishers, Inc., New York, 1994. Reprinted by permission of John Wiley & Sons, Inc.)[3]

initially monodisperse, the spread of the distribution (quantified by $\sigma_g$) initially increases; the SPSD then provides an upper limit for the spread.

Not all types of coagulation processes give self-preserving distributions. Shear-induced coagulation does not develop an SPSD, except for a special, rather unrealistic case.[19] The same is true for coagulation in the transition regime, where quasi-self-preserving distributions can be obtained.[18]

Because an SPSD is often the narrowest distribution that can be obtained, it is useful to know the time it takes to reach such a distribution and to compare it to the reactor residence time. The time needed for an aerosol to reach a self-preserving distribution, $\tau_S$, can be defined as the time needed to develop a distribution with $\sigma_g$ within 1% of the listed values. Landgrebe and Pratsinis calculated the time needed to reach the self-preserving distribution in the free-molecular regime, $\tau_{SF}$, given an initially monodisperse aerosol.[20] The value they obtained was

$$\tau_{SF} = 10\tau_{CF}. \tag{5.18}$$

For an initially monodisperse aerosol in the continuum regime, this time is[3]

$$\tau_{SC} = 12\tau_{CC}. \tag{5.19}$$

These times are extremely short for the aerosol concentrations employed in material synthesis by gas-phase routes. For example, in the synthesis of $B_4C$ particles by carbothermal reduction of soot–boron oxide powders,[21] $\tau_{SF}$ was half a microsecond, while the reactor residence time was on the order of one second! This difference suggested that a self-preserving distribution would be obtained rapidly relative to the reactor residence time. Clearly, in material processing, self-preserving conditions can be easily obtained in short periods of time when extremely high particle concentrations are employed.

The time required to reach an SPSD for particles in the continuum regime, $\tau_{SC}$, has been calculated in detail[15] and is shown in Figure 5.6. The dimensionless parameter $\tau_C$ is given by the expression

$$\tau_C = \frac{2k_BT}{3\mu}N_0t = K_CN_0t. \tag{5.20}$$

This corresponds to a time

$$t = \frac{3\mu}{2k_BTN_0}\tau_C = \frac{1}{K_CN_0}\tau_C = 2\tau_{CC}\tau_C, \tag{5.21}$$

expressed in multiples (given by $2\tau_C$) of the characteristic time for collision in the continuum regime, $\tau_{CC}$. As an example, a size distribution with an initial $\sigma_g$ value of 2.2 requires a $\tau_C$ of roughly 1000. For a particle concentration of $10^{10}/\text{cm}^3$, $T = 300$ K, $\mu = 0.00018$ g/cm sec$^2$, $k_B = 1.38 \times 10^{-16}$ g cm$^2$/sec$^2$

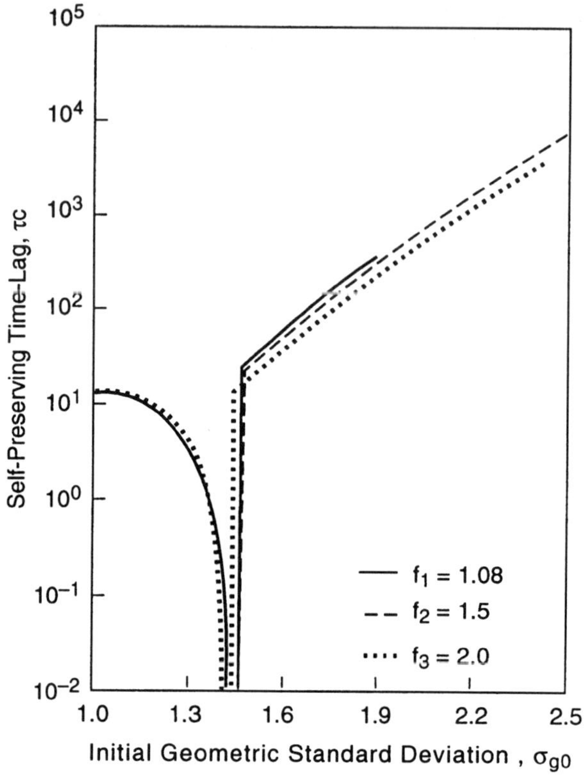


**Fig. 5.6.** Time to reach self-preserving distribution in the continuum regime for different initial geometric standard deviations. Values of f correspond to different modeling approaches. (Adapted from Vemury et al.[15] Reproduced with permission from Academic Press, Inc.)

molecule K, and $t = 65$ seconds. For a concentration of $10^{12}$ molecules/cm$^3$, $t = 0.6$ second. These values illustrate that the time required to reach an SPSD is highly dependent upon the initial spread of the distribution, residence time, and particle concentration.

If there is not enough time to reach an SPSD, the size distribution may be broad. Figure 5.7 shows how the actual particle size distribution can deviate from an SPSD for different deviations of the spread of the distribution from the value for the SPSD. Even a 5% deviation in the value of the spread of the distribution will manifest itself as large deviations in the shape of the distribution. It is important to note, however, that these distributions may have only a small fraction of the mass in the portion of the distribution that does not follow the SPSD.

The coagulation rate of an aerosol at self preserving conditions does not significantly differ from that of a monodisperse aerosol. The change in particle concentration with time in the continuum regime (Kn $\ll$ 1) of a

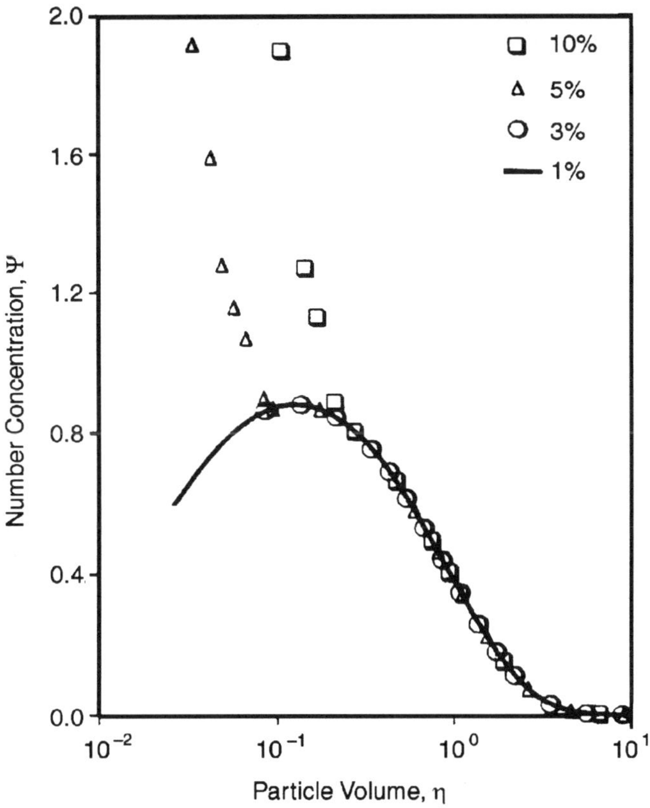


**Fig. 5.7.** Particle size distribution for various deviations of $\sigma_g$ from self-preserving value in continuum regime. (Adapted from Vemury, et al.[15] Reproduced with permission from Academic Press, Inc.)

self-preserving aerosol is[3]

$$\frac{dN_\infty}{dt} = -2.12 K_C N_\infty^2. \tag{5.22}$$

This is 6% faster than for a monodisperse aerosol. Similarly, the coagulation rate in the free-molecule regime (Kn $\gg$ 1) is

$$\frac{dN_\infty}{dt} = -\frac{6.67}{2}\left(\frac{3V}{4\pi}\right)^{1/6} K_F N_\infty^{11/6}, \tag{5.23}$$

where $V$ is the total aerosol volume ($cm^3/cm^3$).

These equations allow the change in particle concentration with time to be estimated if the distribution is self-preserving. Given the rapid attainment of the self-preserving state when high particle concentrations are employed, Equations (5.22) and (5.23) often give a good description of the coagulation of spherical particles. The foregoing values of particle concentration, when combined with the detailed self-preserving solutions, give the full particle size distribution as a function of time. Other expressions apply when agglomerates are present, as will be discussed in Section 5.7.

Self-preserving distributions are roughly lognormal in shape, a fact that is often used to justify the use of lognormal models and size distributions for analyzing data. An example of measured particle size distributions that are

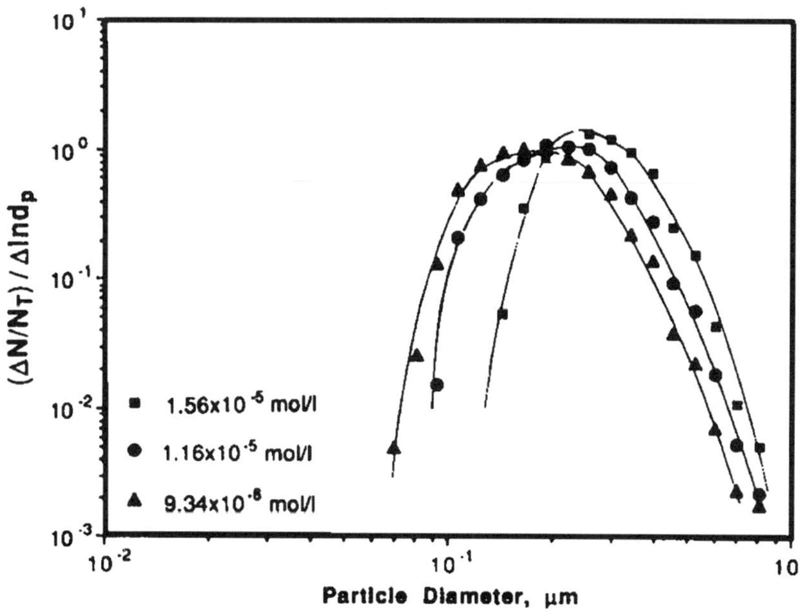


| $TiCl_4$ Concentration (mol/l) | Geometric Volume Mean Particle Size (μm) | Geometric Standard Deviation | Total Number Concentration (#/cm³) |
|---|---|---|---|
| $1.56\times10^{-5}$ | 0.26 | 1.34 | $1.2\times10^{7}$ |
| $1.16\times10^{-5}$ | 0.21 | 1.42 | $1.1\times10^{7}$ |
| $9.34\times10^{-6}$ | 0.18 | 1.44 | $1.0\times10^{7}$ |

**Fig. 5.8.** Effect of initial $TiCl_4$ concentration on average particle size at temperature of 1400 K and reactor residence time of 1.6 sec. (Adapted from Akhtar, M.K., Xiong, Y., Pratsinis, S.E., Vapor synthesis of titania powder by titanium tetrachloride oxidation, *37* (10), 1561–1570, 1991. Reproduced with permission of the American Institute of Chemical Engineers. Copyright © 1991 AIChE. All rights reserved.)[22]

roughly lognormal is shown in Figure 5.8.[22] These distributions also have spreads of the particle size distribution, quantified by the geometric standard deviation of 1.34 to 1.44 in the range expected for a self-preserving distribution.

## 5.7 SELF-PRESERVING SIZE DISTRIBUTIONS: UNCHARGED AGGLOMERATES

Particle growth by collision and coalescence does not always result in the formation of spherical particles. When sintering of the particles is slow relative to the rate at which they collide (i.e., the characteristic time for sintering is longer than the characteristic time for collision), the particles can be aggregates similar to those in the simulation presented in Figure 5.9.[23-26] Sketch (a) shows the case for particles colliding by diffusion with no sintering, while sketch (b) shows the same situation for collisions with some sintering of the particles.

The existence of aggregates during particle growth results in different self-preserving distributions than for spherical particles.[27] Figure 5.10 shows

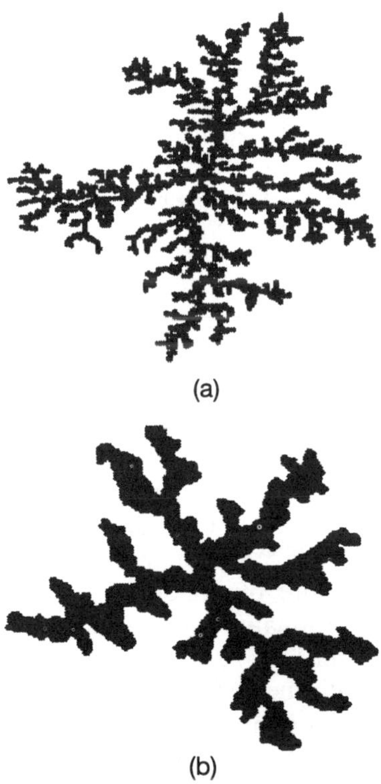


**Fig. 5.9.** (a) Unsintered aggregate. (b) Sintered aggregate. (Reprinted with permission from *Aerosol Science and Technology 21*, Akhtar, M.K., Lipscomb, G.G. and Pratsinis, S.E., Monte Carlo simulation of particle coagulation and sintering, 83–93, 1994. Elsevier Science Inc., 655 Ave. of the Americas, New York, NY 10010-5107.)[25]

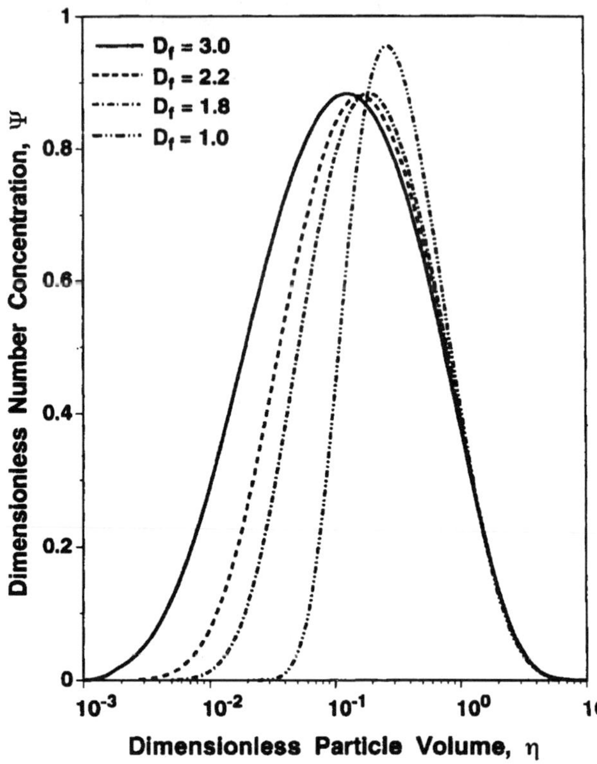


**Fig. 5.10.** Self-preserving size distributions for different fractal dimensions. On a volume basis, the distributions are narrower for higher fractal dimensions (more spherical particles). (Reprinted from *Journal of Aerosol Science 26*, Vemury, S. and Pratsinis, S.E., 175–185, Copyright 1995, with kind permission from Elsevier Science Ltd., The Boulevard, Langford Lane, Kidlington OX5 1GB UK.)[27]

self-preserving distributions for collision in the continuum regime with different fractal dimensions. The defining equation for fractal dimension is given in Chapter 2. The $D_f = 3$ curve in the figure corresponds to spherical particles, while smaller values correspond to particles with progressively more open structures. The broadest distribution that can be formed is for the case of spherical particles. This dependence of the spread $\sigma_{gs}$ of the distribution on fractal dimension is quantified in Figure 5.11.[12, 16, 27] There has been extensive discussion in the literature of the structure of agglomerates and their relation to fractal dimension,[28–35] and the formation of agglomerates by coagulation has been modeled extensively as well.[36–41]

As with other SPSDs, the time required to reach the SPSD for agglomerates must be shorter than the reactor residence time in order to reach the SPSD. Figure 5.12 shows the time to reach these distributions.[27] The

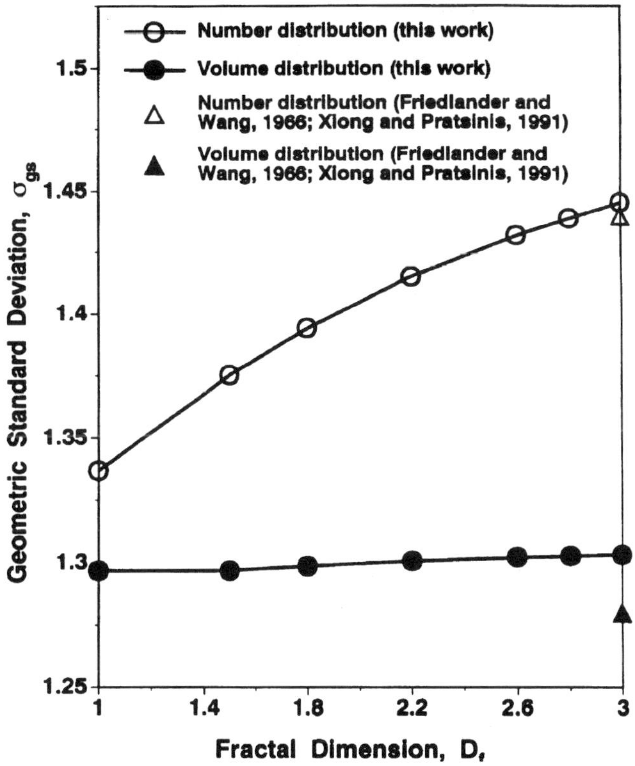


**Fig. 5.11.** Geometric standard deviation of self-preserving distributions obtained for different fractal dimensions. (Reprinted from *Journal of Aerosol Science 26*, Vemury, S. and Pratsinis, S.E., 175–185, Copyright 1995, with kind permission from Elsevier Science Ltd., The Boulevard, Langford Lane, Kidlington OX5 1GB UK.)[27] (See also Xiong and Pratsinis[12] and Friedlander and Wang.)[16]

parameter $\tau_{FS}$ is given by

$$\tau_{FS} = \left(\frac{6k_B T r_p}{\rho_p}\right)^{1/2} \left(\frac{3}{4\pi}\right)^{1/6} N_0 t, \tag{5.24}$$

where $r_p$ is the primary particle radius, $\rho_p$ is the particle density, and the other terms are the same as before. The quantity $\tau_{CS}$ is defined similarly and is closely related to the characteristic time for coagulation in the free-molecule regime. The plot can be used in the same manner as that for spherical particles.

The time required for agglomerates to reach an SPSD is usually within a factor of two of the time required for spherical particles. Also, the spreads of the distributions are narrower for agglomerates because the spread of the distribution is quantified on the basis of the volume of the agglomerates.

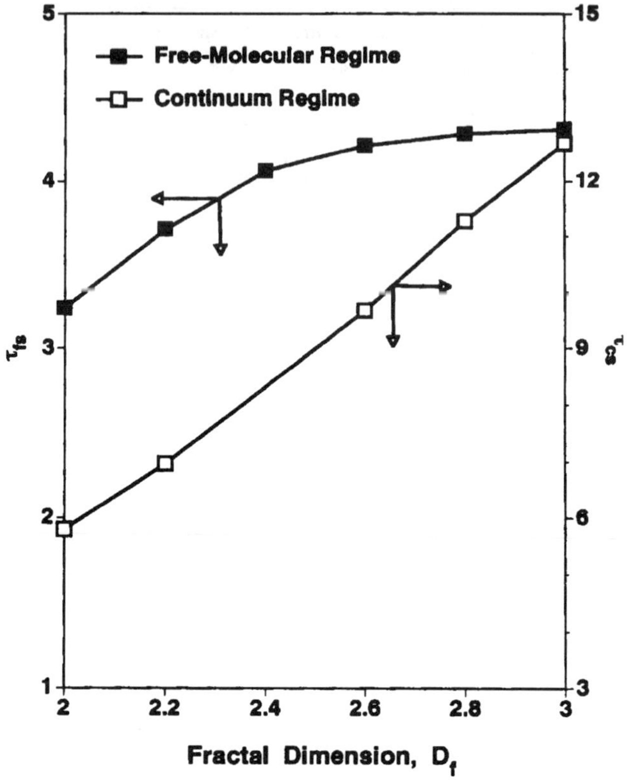


**Fig. 5.12.** Time to reach self-preserving distribution as a function of fractal dimension. (Reprinted from *Journal of Aerosol Science 26*, Vemury, S. and Pratsinis, S.E., 175–185, Copyright 1995, with kind permission from Elsevier Science Ltd., The Boulevard, Langford Lane, Kidlington OX5 1GB UK.)[27]

## 5.8 SELF-PRESERVING SIZE DISTRIBUTIONS: CHARGED PARTICLES

In many systems, additives such as KCl or CsCl can create charged particles at high temperatures (Fig. 5.13).[42] These particles can interact with each other through the electric fields created by the charges. As a result, the asymptotic particle size distribution can be different from the self-preserving distributions discussed. Figure 5.14 shows how the distributions obtained with charged particles can be much narrower than those obtained with uncharged particles.[42] The interpretation of this distribution is the same as that for the other SPSDs.

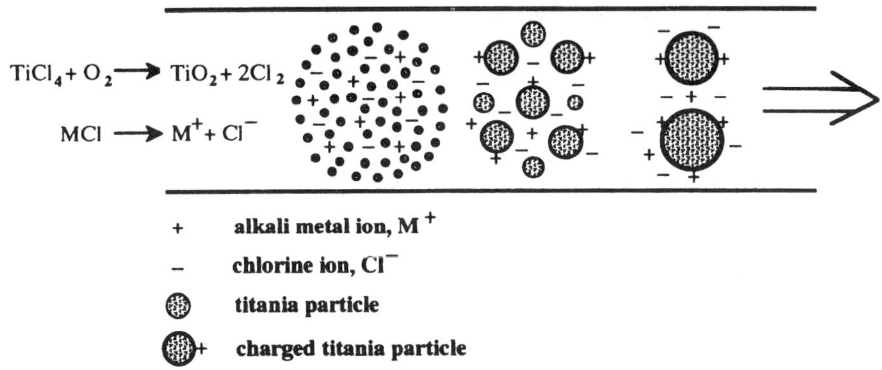


**Fig. 5.13.** Idealized representation of charged-particle coagulation during particle synthesis of $TiO_2$. (Adapted from Xiong et al.[42] Reproduced with permission from Academic Press, Inc.)

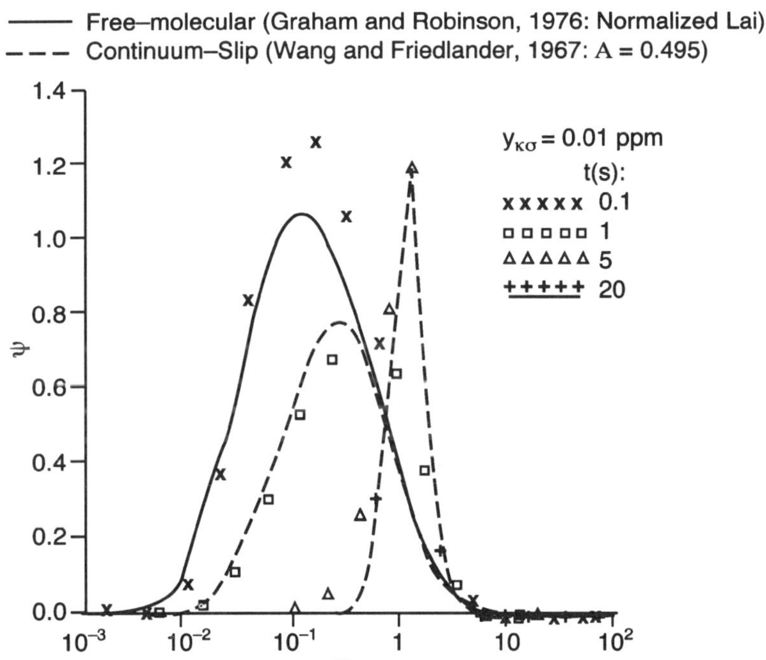


**Fig. 5.14.** Self-preserving distributions for charged particles. Narrower distributions can be obtained than for uncharged particles. (Adapted from Xiong et al.[42] Reproduced with permission from Academic Press, Inc.)

## 5.9 COALESCENCE OF PARTICLES

In material processing, product powders such as commercial $TiO_2$ pigment, fumed $SiO_2$, and carbon black made by coagulation frequently consist of agglomerates or aggregates of fine primary particles. Thus, there is a motivation for describing the rate of coalescence of particles.

Particle coalescence (or fusion) occurs by sintering, the rate of which is a strong function of temperature, particle size, and material properties. A driving force for sintering is the excess surface area an agglomerate possesses over the area $a_s$ of a spherical particle with the same mass as the agglomerate.[43] Sintering is also driven by the reduction in the radius of curvature at the neck between particles.[44, 45] The process may proceed by various mechanisms: viscous sintering; surface, bulk or grain boundary diffusion; and evaporation and condensation. Viscous sintering is common for glasses and materials like $SiO_2$, where the particles behave like fluids with high viscosities. Sintering by solid-state diffusion involves the transport of atoms by diffusion along the particle surfaces, across grain boundaries, or through the bulk of the solid. Evaporation and condensation sintering occurs when volatile species are present, enter the gas phase, and condense in the neck region between particles. All of these are discussed in standard texts on materials science.[44, 45] During the course of particle coalescence, one or more sintering mechanisms can be important.

The overall conversion of two or more spherical particles into a single spherical particle is a complicated process that is difficult to describe exactly. An approximate expression that has proven useful for simple calculations can be written in terms of the surface area of the particles[46 – 51] as

$$\frac{da}{dt} = \frac{1}{\tau_f}(a - a_s), \tag{5.25}$$

where $\tau_f$ is the characteristic time of fusion (sintering) for a specific mechanism and material, $a$ is the surface area of the agglomerate, and $a_s$ is the surface area of a sphere with the same volume as the agglomerate. The characteristic time for fusion can be defined as the time needed to reduce the excess surface area of the agglomerate over that of a sphere of the same mass by 63% (1/2.718 . . . ).

The characteristic time for sintering (fusion) can be written as

$$\tau_f = A d_p^m \exp\left(\frac{E}{R_g T}\right), \tag{5.26}$$

where $d_p$ is the primary particle diameter, $E$ is the activation energy for diffusion, $R_g$ is the gas constant, $A$ is a constant, and the exponent $m$ depends on the sintering mechanism.[11]

As an example, for viscous-flow sintering of silica,[44]

$$\tau_f = 6.5 \times 10^{-15} d_{pp} \exp\left(\frac{83{,}000}{T}\right), \tag{5.27}$$

where $d_{pp}$ is the primary particle diameter (cm) and $T$ is the temperature (K). For sintering of $TiO_2$ by grain boundary diffusion,[52]

$$\tau_f = 7.4 \times 10^{8} T d_{pp}^4 \exp\left(\frac{31{,}000}{T}\right). \tag{5.28}$$

For water droplets, $\tau_f$ is approximately zero, so a spherical droplet is instantaneously formed upon collision.

Aside from its merit for providing a quantitative measure of the rate of particle coalescence, $\tau_f$ can be used to estimate the role of agglomerates in the particle collision rate. This is important in the free-molecular regime, where agglomerates collide faster than spheres of the same volume because of the enhanced area of collision. The effect is less important in the continuum regime, where the enhanced collision area of agglomerates is balanced by the higher drag of agglomerates over that of spheres with the same mass. However, when the characteristic time for collision, $\tau_C$, is much

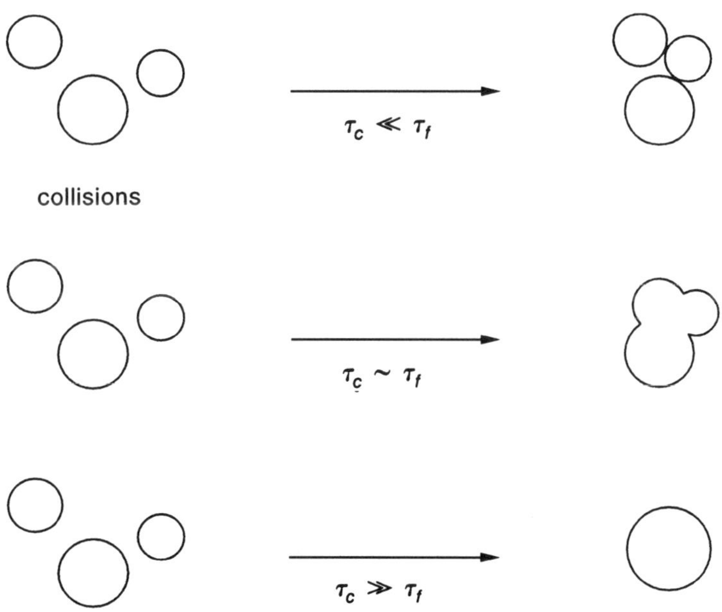


**Fig. 5.15.** Particle morphologies with slow (top) and fast (bottom) coalescence relative to particle collision rate.

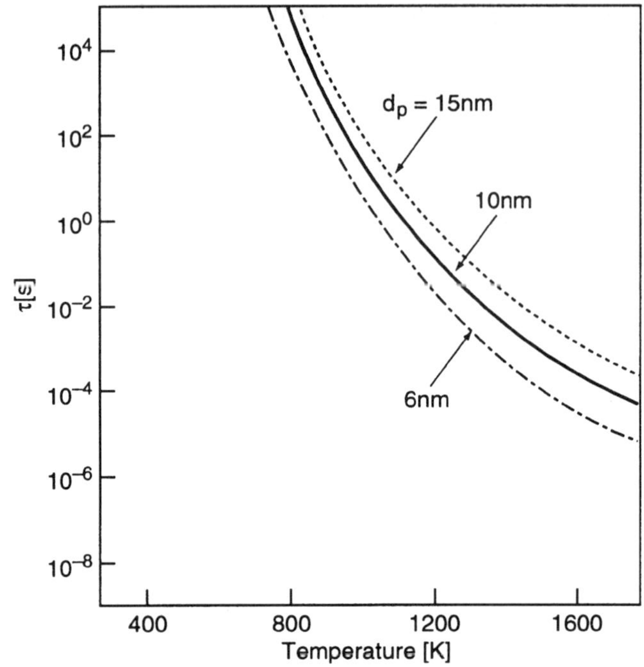


**Fig. 5.16.** Characteristic time for fusion of $TiO_2$ particles. (Reprinted with permission from *Aerosol Science and Technology*, *23*, Seto, T., Shimada, M., & Okuyama, K., Evaluation of sintering of nanometer-sized titania using aerosol method, 183–200, 1995. Elsevier Science Inc., 655 Ave. of the Americas, New York, NY 10010-5107.)[53]

longer than the characteristic time for fusion (e.g., $\tau_f/\tau_C < 0.1$), spherical particles are formed after collision, but before the next collision. When $\tau_f/\tau_C > 10$, fusion is slow and results in agglomerate particles. In the latter case, the coagulation rate needs to be corrected to account for the irregular shape of the particles. This situation is shown in Figure 5.15.

Figure 5.16 shows the characteristic time for sintering of $TiO_2$, plotted as a function of the reactor temperature with particle diameter as a parameter.[53] For a temperature of 1000 °C (1273 K), the sintering time constant is roughly 0.01 second. This demonstrates that the characteristic time for collision must be greater than 0.01 second for the particles to be able to fuse together before other collisions occur.

Quantification of this problem requires an expression for the collision frequency function of agglomerates that have a fractallike structure. In the free-molecule regime, the coagulation frequency function of agglomerates having fractal dimension $D_f$ is given by

$$\beta_{\mathrm{FA}} = K_F \left(\frac{3}{4\pi}\right)^{1/6} \left(\frac{1}{v_i} + \frac{1}{v_j}\right)^{1/2} \left(v_i^{\frac{1}{D_f}} + v_j^{\frac{1}{D_f}}\right)^2, \tag{5.29}$$

where $K_F$ must be suitably modified.[54] In the continuum and transition regimes, similar expressions have been obtained by Mountain et al.[37] In these expressions, $D_F$ is a measure of the openness of the agglomerate, with $D_F = 3$ corresponding to a solid sphere and lower values corresponding to agglomerates. This equation 5.29 is, however, limited to values of $D_f \geq 2$.

## 5.10 REFERENCES


1. Hinds, W.C. *Aerosol Technology*. New York: Wiley Interscience, 1982.
2. Reist, P.C. *Introduction to Aerosol Science*. New York: MacMillan Publishing Co., 1984.
3. Friedlander, S.K. *Smoke, Dust and Haze*. New York: Wiley Interscience, 1977.
4. Williams, M.M.R. & Loyalka, S.K. *Aerosol Science: Theory and Practice*. New York: Pergamon Press, 1991.
5. Seinfeld, J.H. *Atmospheric Chemistry and Physics of Air Pollution*. New York: Wiley Interscience, 1986.
6. Hidy, G.M. & Brock, J.R. *J. Colloid Science 20*, 1965, 477–491.
7. Okuyama, K., Kousaka, Y., & Hayashi, K. *J. Colloid Interface Sci. 113*(1), 1986, 42–54.
8. Tennekes, H. & Lumley, J.L. *A First Course in Turbulence*. Cambridge, MA: MIT Press, 1972.
9. Saffman, P.G. & Turner, D.S. *J. Fluid Mech. 1*, 1956, 16.
10. Okuyama, K. *J. Aerosol Sci. 9*, 1978, 399–410.
11. Bird, R., Stewart, W.E., & Lightfoot, E.N. *Transport Phenomena*. New York: John Wiley and Sons, 1960.
12. Xiong, Y. & Pratsinis, S.E. *J. Aerosol Sci. 22*(5), 1991, 637–655.
13. Fuchs, N.A. *The Mechanics of Aerosols*. New York: Dover Publications, 1964.
14. Lehtinen, K.E.J., Jokiniemi, J.K., & Kauppinen, E.I. *Aerosol Sci. Technol. 23*, 1995, 422–430.
15. Vemury, S., Kusters, K.A., & Pratsinis, S.E. *J. Colloid Interface Sci. 165*, 1994, 53–59.
16. Friedlander, S.K. & Wang, C.S. *J. Colloid Interface Sci. 22*, 1966, 126–132.
17. Lai, F.S., Friedlander, S.K., Pich, J., & Hidy, G.M. *J. Colloid Interface Sci. 39*(2), 1972, 395–405.
18. Otto, E., Stratmann, F., Fissan, H., Vemury, S., & Pratsinis, S.E. *J. Aerosol Sci. 25*(1), 1994, S383–S384.
19. Wang, C.S. & Friedlander, S.K. *J. Colloid Interface Sci. 24*, 1967, 170–179.
20. Landgrebe, J.D. & Pratsinis, S.E. *J. Colloid Interface Sci. 139*(1), 1990, 63–86.
21. Weimer, A.W., Moore, W.G., Roach, R.P., Haney, C.N., & Rafaniello, W. *AIChE J. 37*, 1991, 759–768.
22. Akhtar, M.K., Xiong, Y., & Pratsinis, S.E. *AIChE 37*(10), 1991, 1561–1570.
23. Akhtar, M.K., Vemury, S., & Pratsinis, S.E. *NanoStruct. Mater. 4*(5), 1994, 537–544.
24. Akhtar, M.K., Vemury, S., & Pratsinis, S.E. *AIChE J. 40*(7), 1994, 1183–1192.

25. Akhtar, M.K., Lipscomb, G.G., & Pratsinis, S.E. *Aerosol Sci. Technol. 21*, 1994, 83–93.
26. Akhtar, M.K., Pratsinis, S.E., & Mastrangelo, S.V.R. *J. Mater. Res. 9*(5), 1994, 1241–1249.
27. Vemury, S. & Pratsinis, S.E. *J. Aerosol Sci. 26*, 1995, 175–185.
28. Wu, M.K. & Friedlander, S.K. *J. Aerosol Sci. 23*(1), 1992, S71–S74.
29. Wu, M.K. & Friedlander, S.K. *J. Colloid Interface Sci. 159*(1), 1993, 246–248.
30. Rogak, S.N., Flagan, R.C., & Nguyen, H.V. *Aerosol Sci. Technol. 18*(1), 1993, 25–47.
31. Wu, M.K. & Friedlander, S.K. *J. Aerosol Sci. 24*(3), 1993, 273–282.
32. Rosner, D.E. & Tandon, P. *AIChE J. 40*(7), 1994, 1167–1182.
33. Weber, A.P. & Friedlander, S.K. *Mater. Res. Soc. Symp. Proc., 351*, 1994, 331–336.
34. Tandon, P. & Rosner, D.E. *Ind. Eng. Chem. Res. 34*, 1995, 3265–3277.
35. Köylü, Ü.Ö., Xing, Y., & Rosner, D.E. *Köylü, Ü.Ö. 11*, 1995, 4848–4854.
36. Rogak, S.N. & Flagan, R.C. *J. Colloid Interface Sci. 151*(1), 1992, 203–224.
37. Mountain, R.D., Mulholland, G.W., & Baum, H. *J. Colloid Interface Sci. 114*(1), 1986, 67–81.
38. Matsoukas, T. & Friedlander, S.K. *J. Colloid Interface Sci. 146*(2), 1991, 495–506.
39. Xiong, Y. & Pratsinis, S.E. *J. Aerosol Sci. 24*(3), 1993, 283–300.
40. Xiong, Y., Akhtar, M.K., & Pratsinis, S.E. *J. Aerosol Sci. 24*(3), 1993, 301–313.
41. Xiong, Y., Pratsinis, S.E., & Weimer, A.W. *AIChE 38*(11), 1992, 1685–1692.
42. Xiong, Y., Pratsinis, S., & Mastrangelo, S.V.R. *J. Colloid Interface Sci. 153*, 1992, 106–117.
43. Friedlander, S.K. & Wu, M.K. *Physical Review B 49*(5), 1994, 3622–3624.
44. Kingery, W.D., Bowen, A.K., & Uhlmann, D.R. *Introduction to Ceramics*. New York: John Wiley and Sons, 1976.
45. German, R.M. *Sintering Theory and Practice*. New York: John Wiley and Sons, 1996.
46. Koch, W. & Friedlander, S.K. *J. Aerosol Sci. 21*(1), 1990, S73–S76.
47. Koch, W. & Friedlander, S.K. *J. Colloid Interface Sci. 140*(2), 1990, 419–427.
48. Koch, W. & Friedlander, S.K. *J. Aerosol Sci. 20*(8), 1990, 891–894.
49. Koch, W. & Friedlander, S.K. *Part. Part. Syst. Charact. 3*, 1991, 35–39.
50. Koch, W. & Friedlander, S.K. *Part. Part. Syst. Charact. 8*, 1991, 86–89.
51. Lehtinen, K.E.J., Windeler, R.S., & Friedlander, S.K. *J. Aerosol Sci. 27*(6), 1996, 883–896.
52. Kobata, A., Kusakabe, K., & Morooka, S. *AIChE 37*(3), 1991, 347–359.
53. Seto, T., Shimada, M., & Okuyama, K. *Aerosol Sci. Technol. 23*, 1995, 183–200.
54. Kaplan, C.R. & Gentry, J.W. *Aerosol Sci. Technol. 8*, 1988, 11.

# 6 Intraparticle Transport Processes

## 6.1 INTRODUCTION

Intraparticle processes are important during powder formation by spray pyrolysis, vacuum atomization of molten metals, droplet-phase reactions to form liquid or solid products, and gas–solid reactions to form solid products. Film formation processes such as AACVD also can involve intraparticle reactions when the volatile precursor in each droplet reacts before evaporating.

There are several important classes of intraparticle processes: gas–solid reactions at interfaces, gas–liquid reactions at interfaces, solid-state reactions, liquid-phase reactions, solid-state diffusion, transport of gaseous reactants and products into and out of particles, phase transformations, and crystal growth. Gas–solid reactions can occur at the interface of particles with a gas, or the gaseous species can diffuse into a porous solid, where the reaction occurs. Some examples are the calcination of colloidal materials and the conversion of inorganic metal compounds to metal oxides.[1] Other, specific examples are the conversion of boehmite (AlOOH) to alumina by the loss of water;[2] the conversion of PdO to Pd by the loss of oxygen;[3] oxidation and nitridation of metals, such as particles of Ti or Al, to form $TiO_2$ and AlN;[4] and the formation of polymer particles by intraparticle polymerization.[5]

Gas–liquid reactions can occur at or near the droplet–gas interface, as well as within a droplet. Examples are the conversion of liquid reactant aerosol droplets to solid products during spray pyrolysis and the decomposition of volatile precursors in aerosol droplets during aerosol delivery of precursors for CVD. Metal alkoxide droplets can be converted to hydrous metal oxides by reaction with water.[6] A specific example for CVD is the reaction of $Pd(hfac)_2$ (where hfac = hexafluoroacetylacetonate) with hydrogen in toluene droplets during AACVD; the $Pd(hfac)_2$ is reacted to form Pd and hfacH.[7, 8]

Examples of intraparticle transport processes are the transport of a variety of species inside particles to the surface and into the gas phase: volatile components such as PbO in solids, volatile molecules such as organometallics in precursor particles, and metal organics and inorganic compounds in solvents. A specific example is the loss of PbO during the synthesis of materials such as $PbTiO_3$ and Pb–Bi–Sr–Ca–Cu–O[9] by spray pyrolysis. Two-phase problems also occur: The reaction of metal alkoxide droplets with water results in the formation of a crust through which the water must diffuse to react further with the metal alkoxide inside the crust. A specific example is the reaction of aluminum-sec-butoxide droplets with water vapor.[6]

Phase transformations involve materials going from liquid to solid, amorphous to crystalline, and crystalline to crystalline phases. Specific examples are the anatase-to-rutile transformation (crystalline to crystalline)[10] and the solidification of molten metal droplets (liquid to solid).[11] The precipitation of solutes during spray pyrolysis is followed by crystal growth.[12, 13] Crystallites in solids grow at high temperatures, leading in some cases to the formation of solid single-crystal particles. A specific example is the formation of single-crystal precious-metal particles during spray pyrolysis.[14–18] Phase transformations in solids and liquids are covered in Chapters 8 and 13.

Our goal in this chapter is to discuss the basic theory only at a level sufficient to introduce the reader to the topics and to connect this theory to the behavior of aerosol reactors. The chapter is divided into a discussion of diffusion in solid and liquid particles, heat transfer within particles, gas–solid reactions, and gas–liquid reactions. The chemical aspects of gas–solid and gas–liquid reactions are also discussed in Chapter 8.

## 6.2 SOLID-STATE AND LIQUID-PHASE DIFFUSION

### 6.2.1 Introduction

Diffusive transport—the movement of species due to their random thermal motion in a concentration gradient—occurs in solids, liquids, and gases. Reactants are transported into the interior of particles and reaction products are removed from the interior of particles by diffusion. Diffusion is also the process that enables crystal growth inside the particles after solute precipitation in droplets. Sintering of particles and the interdiffusion of species to provide solid-state reactions after collision also rely on diffusion.

Atomic and molecular diffusion processes are often driven either by gradients in species concentrations that are caused by reactions inside particles or droplets or by transport processes outside the particles. They can also be driven by other phenomena, such as surface curvature.[19] Figure 6.1 shows a gaseous reactant transported through a gas, through a reacted layer surrounding a core of unreacted material, and, finally, into the unreacted material, where reaction takes place. This mechanism illustrates that diffusion inside particles can be in series with gas-phase transport and reaction. Gas-phase transport in the region surrounding the particles almost never limits the overall rate when intraparticle transport occurs, because the transport rates in these regions are proportional to the diffusion coefficients.

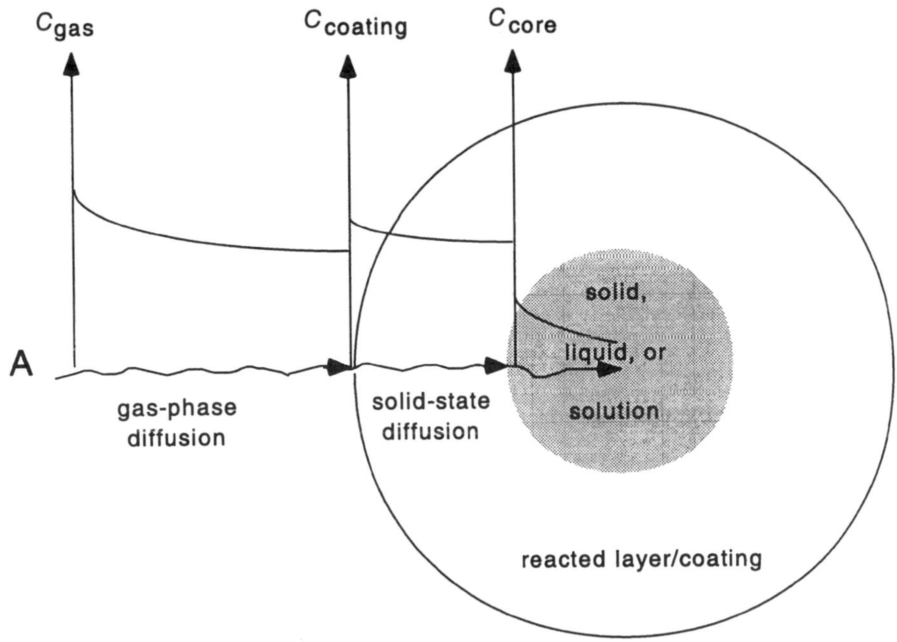


**Fig. 6.1.** Qualitative depiction of transport of reactant into a particle containing a core of unreacted material surrounded by a reacted layer. Concentrations in the various phases are in different units; hence, equilibrium relationships are required at the interfaces to relate the quantities.

Diffusion coefficients in gases ($\sim$ 0.1 $cm^2/sec$) are higher than the effective diffusion coefficients in porous materials, and diffusion coefficients in solids ($\sim 10^{-7}$ to $10^{-15}$ $cm^2/sec$) and liquids ($\sim 10^{-5}$ $cm^2/sec$) have even lower values. The rate of diffusion through the shell, however, can be slow relative to the rate of reaction inside the particle and can limit the conversion rate. This is known as shell diffusion-limited or transport-limited conversion. If the transport rate is sufficiently high relative to the reaction rate, it may not limit the conversion rate, which is then given by the reaction rate. Thus, a key goal for any analysis of problems of this type is identifying the slow step in the series.[20]

Diffusion also plays a critical role in determining the time required to convert nonuniformly mixed solid reactants or intermediates into a final product. Figure 6.2 shows how two species can be distributed inside particles or agglomerates. The examples with large diffusion distances (large regions of each material) require longer times for conversion to the final product by solid-state diffusion. The agglomerate structures can be formed by the reaction of two or more volatile species in the gas phase, generating separate particles that then collide. An example is the nonuniform species distribution

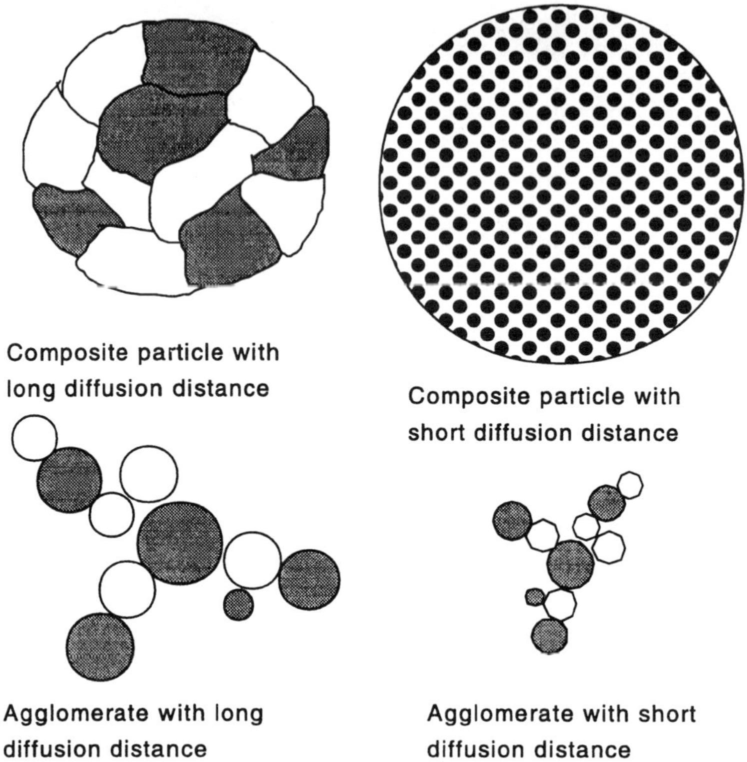


**Fig. 6.2.** Schematic illustration of spatial distribution of two components in spherical particles, as could be formed by spray pyrolysis, and in agglomerates, as could be formed by gas-to-particle conversion. Shorter diffusion distances (a smaller scale of segregation) on the right side of the figure favor more rapid conversion to homogeneous material, while larger distances (on the left side of the figure) favor longer conversion times.

that can occur when particles of $Al_2O_3$ and $TiO_2$ collide in attempts to form Al-doped $TiO_2$.[10] For spray processes, two or more precursors can precipitate separately and react, followed by incomplete interdiffusion resulting in a multiphase material, as is the case for yttrium and aluminum nitrates and their incomplete conversion to yttrium aluminum garnet.[21] This segregation is of critical importance, because the time for conversion can depend on the square of the scale of the segregation, as discussed subsequently. Several other factors that affect the chemical homogeneity in these systems are described in Chapter 8.

### 6.2.2 Theory

The theoretical approach to this class of problems relies on solving equations that describe the diffusion of species in and around solid particles and droplets.[22, 23] The equations describing diffusion can be put into the same form for gases, liquids, and solids. Atoms and molecules, however, move through gases, liquids, solids, and pores by different mechanisms, resulting in different diffusion coefficients.[24] Transport through gases occurs by Brownian (thermal) motion with a long gas mean free path between collisions of the molecules relative to their size. In liquids, the atoms and molecules are close together at all times, resulting in more resistance to the motion of species than that which is found in gases. In solids, the diffusing species—atoms, not molecules—encounter an even higher resistance to motion because of the relatively rigid structure of solids.[25] Diffusion through porous materials is actually diffusion through a gas when the pore diameter is larger than the gas mean free path. However, when the gas mean free path is *much* larger than the pore diameter, the transporting species interacts more with the walls of the pore and not with other gaseous species. This is known as *Knudsen diffusion*.[26] Even though there is little resistance to the motion of species through the pores relative to liquids, the reduced cross-sectional area and the tortuosity of the pores result in an effective diffusion coefficient that is much lower than in gases.

Because most aerosol problems involve spherical solid particles or droplets, it is convenient to assume spherical symmetry. Diffusion of species in the radial direction in and around particles of liquids and solids can be described by Fick's second law in spherical coordinates,[24] namely,

$$\frac{\partial C}{\partial t} = \frac{D}{r}\frac{\partial}{\partial r}\left(r\frac{\partial C}{\partial r}\right), \tag{6.1}$$

where $C$ is the concentration of a gaseous species (mole/cm$^3$), $r$ is the radial coordinate (cm), $D$ is the diffusion coefficient (cm$^2$/sec), and $t$ is time (sec). This highly simplified equation is not valid in the preceding form for all cases, but it serves to illustrate the relevant principles.

The diffusion equation (6.1) can be solved to obtain the detailed species concentration profiles, fluxes, and time for completion of transport. Here, however, we would like to obtain characteristic times for the transport processes. These times can be compared with other characteristic times for reactor residence time or other transport or reaction steps in series with this step. The approach begins with solutions of various forms of the equation, which can be obtained for a wide variety of geometries and boundary conditions, as described by Carslaw and Jaeger[22] and Crank.[23] The characteristic time for diffusion can then be obtained from these solutions. In order to proceed further, specific situations must be defined.

One of the most common problems is the diffusion of a species into or out of the bulk of a particle. An example is $H_2$ diffusion into a droplet, where the time of interest is that required to reach a uniform equilibrium concentration of $H_2$ in the droplet. This rather involved problem is discussed in Carslaw and Jaeger[22] and Crank.[23]

The derivation of the characteristic time for diffusion into a particle begins with Equation (6.1). Boundary and initial conditions are required to solve the problem. Here we use a fixed concentration of the diffusing species at the particle surface with no initial dissolved species. The equations are

$$C = C_S \qquad @\, r = R_p, t > 0, \tag{6.2}$$

$$\frac{\partial c}{\partial r} = 0 \qquad @\, r = 0, t > 0, \tag{6.3}$$

$$C = 0 \qquad @\, 0 \le r \le R_p, t = 0. \tag{6.4}$$

The solution of this equation is given by[22, 23]

$$\frac{C(r,t)}{C_S} = 1 + \frac{R_p}{r} \sum_{n=1}^{\infty} (-1)^n \frac{2}{n\pi} \sin\left(\frac{n\pi r}{R_p}\right) \exp\left(-\frac{n^2\pi^2 Dt}{R_p^2}\right). \tag{6.5}$$

Examination of Equation (6.5) for long times (the most persistent term corresponds to $n = 1$) shows that the characteristic time is

$$\tau_D = \frac{R_p^2}{\pi^2 D}. \tag{6.6}$$

In physical terms, this time is roughly the time required for a species to diffuse into or out of a particle. For times much longer than $\tau_D$, the droplet or particle will be at equilibrium with the surface concentration of the diffusing species. This shows that the characteristic time scales as the square of the diameter; larger particles require much longer times to reach equilibrium. A typical time is $10^{-2}$ second for a 10-$\mu$m water droplet with a diffusion coefficient for the gaseous species in the droplet of $10^{-5}$ cm$^2$/sec. For a 1-$\mu$m droplet, the time is $10^{-4}$ sec. This illustrates the short times required for diffusion into droplets.

A similar analysis for a 1-$\mu$m solid particle gives 10 seconds for a diffusion coefficient of $10^{-10}$ cm$^2$/sec. For a 10-nm particle, the time is 0.001 sec, or 1 msec. This illustrates that diffusion into solids can be rapid or slow relative to reactor residence times, which ultimately depend on system parameters.

Diffusion through a coating on a spherical particle is also of interest in a number of problems. As an example, a porous layer can be formed when a metal alkoxide droplet is reacted with water vapor. After the porous coating is formed, the water vapor must diffuse through the coating to reach the metal alkoxide. A mathematical analysis similar to that just presented leads to an expression for the characteristic time for diffusion through a coating. The equation is

$$\tau_L = L^2/D_e, \tag{6.7}$$

where $L$ is the thickness of the coating and $D_e$ is the effective diffusion coefficient in the porous coating (cm$^2$/sec). The physical interpretation of

Equation (6.7) is that $\tau_L$ is the time to diffuse through the coating or the time to establish a constant concentration gradient in the coating. Examples of the use of these times will be given in the sections that follow.

A second useful type of equation is one giving the rate of diffusion into a particle. This information allows calculation of the *extent* of conversion, as opposed to the time required for conversion. The diffusion rate is usually discussed in terms of a flux $J_m$ [molecules/(cm$^2$sec)] and is obtained by differentiating the concentration profile:

$$J_m = -D\frac{\partial C}{\partial r}\bigg|_{R_p}. \tag{6.8}$$

The flux times the area of the particle gives the total flow of diffusing species into the particle. As an example, the species uptake rate into the sphere as a function of time for Equation (6.5) is

$$4\pi R_p^2 D\left(\frac{\partial C}{\partial r}\right)_{R_p} = 8\pi RDC_S \sum_{n=1}^{\infty} \exp\left(-\frac{n^2\pi^2 Dt}{R_p^2}\right). \tag{6.9}$$

As another example, the flux [molecules/(cm$^2$sec)] in the radial direction through a shell with thickness much less than the particle radius can be written as

$$J_m \equiv D_e \Delta C/L, \tag{6.10}$$

where $D_e$ is the effective diffusion coefficient in the shell, $\Delta C$ is the concentration difference across the shell, and $L$ is the thickness of the shell. The flux into the particle depends on the driving force (concentration difference), diffusion coefficient, and diffusion distance.

In describing the rate of species diffusion into particles, these expressions are useful because the temperature dependence of the transport rate can be inferred. The latter information can be employed in determining the rate-limiting step in a process. The primary influence of temperature comes from the diffusion coefficient. Diffusion is an activated process (the rate increases exponentially with temperature) for solids and has a weak temperature dependence for gases in pores. This temperature dependence can be compared to any experimentally determined dependencies to assist in identifying the rate-limiting step.

### 6.2.3 Liquid Phase

The preceding analysis of characteristic times was general and was not restricted to a certain type of material. In the sections that follow, we discuss how to use the results of the analysis for several cases of specific interest.

Diffusion in droplets occurs in several contexts. To form particles by spray pyrolysis, the solvent must diffuse through the droplet to the surface, where it then evaporates and diffuses into the gas phase. At the same time, the solute must diffuse towards the center of the particle. This results in concentration gradients of the solute in the droplet that are the cause of precipitation of species at the droplet surface, which can lead to formation of a shell.[27] The diffusing species must then also diffuse through this shell.

It is useful to consider some order-of-magnitude numbers for this type of process. Diffusion coefficients are approximately $10^{-5}$ $cm^2/sec$ for most small molecules in water, a value that corresponds to a characteristic time for diffusion of 0.001 to $10^{-4}$ sec for a 1-$\mu$m droplet. This means that the time to completely saturate a 1-$\mu$m water droplet with a gaseous species is approximately 0.1 to 1 msec, which shows that gaseous reactants such as hydrogen and oxygen can diffuse into droplets very rapidly relative to typical reactor residence times. It also shows that gaseous products can diffuse out of droplets rapidly relative to most residence times. As a specific example, consider the transport of $H_2$ gas into a particle, followed by chemical reaction. The characteristic time for diffusion of a gaseous species into a droplet is roughly $R_p^2/\pi^2 D$. For 1-$\mu$m and 10-$\mu$m droplets and a liquid-phase diffusion coefficient of $10^{-5}$ $cm^2/sec$, the characteristic time is $10^{-4}$ sec and 0.01 sec, respectively. This indicates that the gaseous coreactant has ample time to diffuse into the droplet and react and that the rate-limiting step in the conversion can be the rate of the reaction. In this example, the characteristic time for reaction must be compared with the residence time to determine whether the reaction will consume much of the precursor. If the

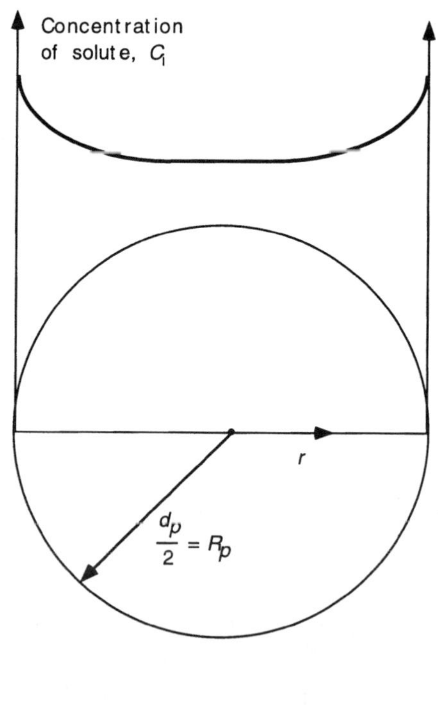


**Fig. 6.3.** Solute concentration profiles in an evaporating droplet. The solute concentration is highest at the droplet surface from which the solvent evaporates.

kinetics of the liquid-phase reaction are sufficiently slow, no appreciable reaction may be observed. This problem has been examined qualitatively, and it has been shown that the rate of the liquid-phase reaction can be rapid enough to result in reaction of an appreciable fraction of the precursor.[7, 8] In cases where no coreactant is needed for reaction to take place and the precursor decomposes thermally by a unimolecular process, the time constant for reaction must be compared with the residence time in the system. If the time for reaction in the particle is shorter than the residence time, then the reaction will convert an appreciable fraction of the precursor to product(s).

Analysis of characteristic times also provides insight into the problem of solute precipitation at the droplet surfaces. This problem can be approached by comparing the characteristic time for solute diffusion in droplets with the characteristic time for droplet evaporation. Comparing the evaporation time for a 1-$\mu$m droplet of water, roughly 1 msec, to the characteristic time for diffusion in the droplet, roughly 1 msec, shows that (at room temperature) nonuniform concentration profiles of the solute will be maintained in the droplet during evaporation. This occurs because the solvent evaporates from the surface of the droplet faster than solute can diffuse into the interior of the droplet and maintain a uniform concentration profile (Fig. 6.3).

### 6.2.4 Solid State

Problems involving solid-state diffusion in particles are, in general, complex; only the simplest of analyses in terms of characteristic times is presented here. The reader is referred to materials science texts for background in this subject.[19, 25]

Several examples are relevant. Although interdiffusion of a coating and the core material in a particle can occur under some conditions, it is often not desired; for example, $TiO_2$ particles have been coated with $SiO_2$, $Al_2O_3$, and $ZrO_2$ at temperatures well above 1000°C.[28–31] Solid-state diffusion can also be the result of a decomposition reaction, as in the decomposition of PdO to form Pd metal and oxygen:

$$2\,PdO(s) \longrightarrow 2\,Pd(s) + O_2(g) \tag{6.11}$$

Here, atomic oxygen diffuses to the surface, where it forms $O_2$.[32] The loss of volatile species can also occur in this manner, as when PbO is lost from a variety of materials, including the superconductor Pb–Bi–Sr–Ca–Cu–O by Pb and O diffusion in the particle and PbO diffusion in the gas surrounding the particle.[9]

The characteristic times given earlier offer a useful insight into these problems. The diffusion coefficients of species like oxygen and various cations in solids are usually $10^{-7}$ to $10^{-15}$ $cm^2/sec$ at elevated temperatures, giving a time scale for diffusion into a 1-$\mu$m particle of roughly 0.1 to $10^7$ sec. This demonstrates that appreciable solid-state diffusion often cannot occur during normal reactor residence times, unless the particles are sufficiently small or the temperature is raised to very high levels such that diffusion coefficients are sufficiently high. Also, other gaseous reactant species often do not have enough time to penetrate into a solid particle during typical reactor residence times. Some examples are oxidation and nitridation of metal particles to form metal oxides and metal nitrides.[4, 33] Furthermore, these slow transport rates mean that diffusion through solids is often much slower than chemical reaction and can limit the rate of the reaction. A variation of this problem can occur when a shell is formed around a core of reactant and a species must diffuse into the particle in order for reaction to occur.

A note of caution is in order, however. The preceding analysis assumes a spherically symmetric homogeneous particle. By contrast, in reality many particles are porous, leading to diffusion through pores followed by diffusion into the solid, which has the effect of drastically reducing the diffusion distances in the solid, thereby reducing the time required for diffusion. Furthermore, solid-state diffusion and reaction is a complicated phenomenon in many other respects.[19, 25]

## 6.3 HEAT TRANSFER WITHIN PARTICLES

Heat transfer within a particle is of interest in nonisothermal systems, where the gas temperature changes rapidly, or when chemical reactions or phase changes occur in the particle or at the surface. Often, the main question is, Does the particle remain in thermal equilibrium with the gas? Thermal equilibrium occurs only when the rate of heat transfer inside the particle and in the gas surrounding the particle are more rapid than the rate of heat evolution within the particle and the rate of change of gas temperature surrounding the particle. In this case, any heat produced in the particle is, for example, transferred out before it leads to a temperature rise in the particle. Probably, the most common situation is the reduction in temperature of an evaporating droplet, in which the heat transfer from the gas into the droplet is not fast enough relative to the rate of heat consumption by evaporation.[34]

The basic equation describing heat transfer (Fourier's law) in spherical coordinates[24] is similar in form to Fick's second law for mass transfer by diffusion:

$$\frac{\partial T}{\partial t} = \frac{\alpha_p}{r}\frac{\partial}{\partial r}\left(r\frac{\partial T}{\partial r}\right). \qquad (6.12)$$

In this equation, in the particle, $T$ is the temperature (K) at radial position $r$ and $\alpha_p$ is the thermal diffusivity (cm$^2$/sec) $k/\rho C_p$, where $k$ is the thermal conductivity (erg/cm sec K), $\rho$ is the density (g/cm$^3$), and $C_p$ is the heat capacity of the material. For heat transfer in the gas surrounding the particle, the same equation is valid, but with the properties of the gas substituted for those of the particle.

As with Fick's second law, solutions for the temperature profiles have been obtained for a wide variety of geometries and boundary conditions, as described by Carslaw and Jaeger[22] and Crank.[23] Again, however, the main concern here is determining how fast heat transfer takes place relative to other transport and heat evolution processes; full detailed solutions of the equation are of less interest. The time for heat transfer is often quantified as a characteristic time, which can be obtained by solving the heat transfer equation with appropriate initial and boundary conditions. For example, if a particle is introduced rapidly into a heated gas, the time for the interior of the particle to reach the temperature of the gas is approximately the characteristic time for heat transfer. The mathematical analysis is exactly the same as in the example given for mass transfer (Section 6.2.2). This leads to the following approximate expression for the characteristic time for heat transfer in the solid particle or droplet:

$$\tau_{Tp} = d_p^2/\alpha_p. \tag{6.13}$$

A similar analysis for heat transfer in the gas surrounding the particle shows that the characteristic time for heat transfer in the gas phase around the particle is

$$\tau_{\mathrm{Tg}} = d_p^2/\alpha_g, \tag{6.14}$$

where $\alpha_g$ ($cm^2$/sec) is the thermal diffusivity of the gas and $d_p$ is the particle diameter. Here, $\tau_{\mathrm{Tg}}$ is the time necessary to establish a steady-state heat flux from the particle. The length scale, $d_p$, occurs in this semi-infinite system because most of the temperature change in the gas extends only a few particle diameters away from the surface of the particle.

These characteristic times allow us to examine the conditions required for a solid particle to remain at the same temperature as the surrounding gas. The relevant times are those for heat transfer in the particle and in the gas around the particle, as well as that for the temperature change in the gas around the particle. For times much greater than $\tau_{\mathrm{Tg}}$, the particle will have reached thermal equilibrium with the gas if heat transfer inside the particle is rapid. The time required to heat the particle itself is $\tau_{\mathrm{Tp}}$. Because a typical thermal diffusivity for a solid material is usually much higher than for a gas, heat transfer in the particle will often be rapid relative to heat transfer in the gas phase. This means that the temperature inside the particle is normally uniform and will be determined by the rate of heat transfer to or from the gas phase. Because the thermal diffusivity of air is roughly 0.1 $cm^2$/sec and particle diameters are roughly $10^{-3}$ cm (10 $\mu$m) or less for most applications, the time taken for the particle to reach thermal equilibrium with the gas is approximately $10^{-5}$ sec or less. This value can be compared with time constants associated with heating rates. A time constant shorter than $10^{-5}$ sec is required for a change in the temperature of the gas before the particle is not in thermal equilibrium with the gas. As an example, typical reactor temperatures are 1000°C and suggest that gas heating rates must be greater than $10^8$ °C/sec [1000°C/($10^8$ °C/sec) = $10^{-5}$ sec] before the temperature of the particle will not be equal to that of the surrounding gas. Such an extremely rapid heating rate is not found under most conditions. A similar analysis can be performed for heat evolution in the particle by chemical reaction; the conclusion is similar: *For most situations of practical interest, the particles remain in thermal equilibrium with the gas*. The only cases where nonevaporating particles are not in thermal equilibrium with a gas are large particles and extremely rapid heating or cooling rates in nozzle expansions or contractions.

It is worth emphasizing that the order-of-magnitude analysis presented here was for solid particles with negligible internal heat generation. For evaporating droplets, a completely different situation occurs for which the droplet temperature can be more than 10°C lower than that of the surrounding gas.[34] Here, every mole of evaporating liquid requires a certain amount of energy, the latent heat of evaporation. In this situation, the temperature is determined by a balance between heat transfer to the sphere and heat consumption by evaporation, the rate of which is determined by the vapor transport rate from the droplet. Hinds discusses the mechanism in detail.[34]

## 6.4 GAS–SOLID REACTIONS

### 6.4.1 Introduction

Gas–solid reactions play important roles in the formation of materials. They are especially important for nanometer-sized metal oxide particles, which are typically formed by reaction of oxygen with metal particles. Larger particles require conversion times that are far too long for practical purposes. Examples are the conversion of Al to $Al_2O_3$ and Ti to $TiO_2$ by reaction with oxygen. Gas–solid reactions are the mechanism by which various species (metal oxalates, hydroxides, carbonates, nitrates, sulfates, acetates, etc.) in the form of particles form metal oxides and volatile products ($SO_2$, $NO_x$, $CO_2$), as shown in Figure 6.4. Reaction of the metal-containing reactant requires the transport of co-reactants such as oxygen into the particle and reaction products ($SO_2$, $NO_x$, $CO_2$) out of the particles. In the figure, the reaction occurs in the interior of the particle.

The concept of a rate-limiting step is critical to the understanding of gas–solid reactions. The overall conversion process consists of several steps in series: transport of gaseous species in the gas to or from the particle (often called gas or "film" diffusion), transport through the reacted layer to or from the gas (often called the shell or "ash" layer), and transport in the bulk of the unreacted material. Not all of these steps occur in a given situation. Rapid reaction can result in reactions confined to an interface between the reacted

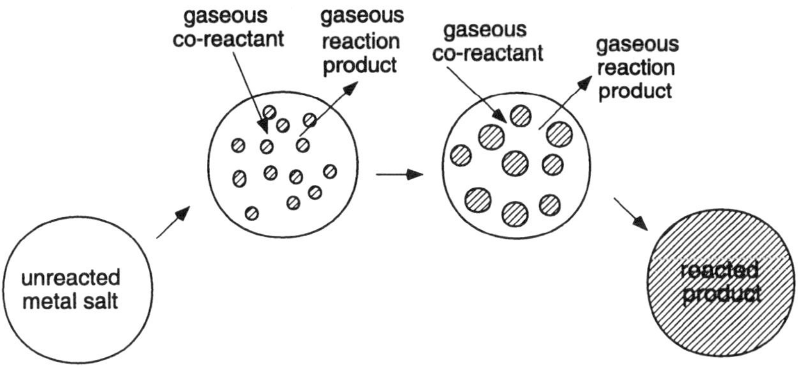


**Fig. 6.4.** Metal salt reaction in particle showing directions of transport of reactants and products.

and unreacted materials. This idea leads to that of a shrinking-core model, described next. Several other limiting cases can also occur, as discussed elsewhere.[20, 26] Rapid transport in the gas and solids relative to the reaction rate can give rise to a conversion controlled by the rate of the solid-state reaction.

### 6.4.2 Shrinking-Core Model

The shrinking-core model is a classical chemical engineering concept and refers to situations in which a clearly defined interface exists between the reacted and unreacted materials (Fig. 6.5).[20, 26] This is in contrast to the situation in Figure 6.4, where the reaction is shown to occur in the interior of the particle in an inhomogeneous manner.

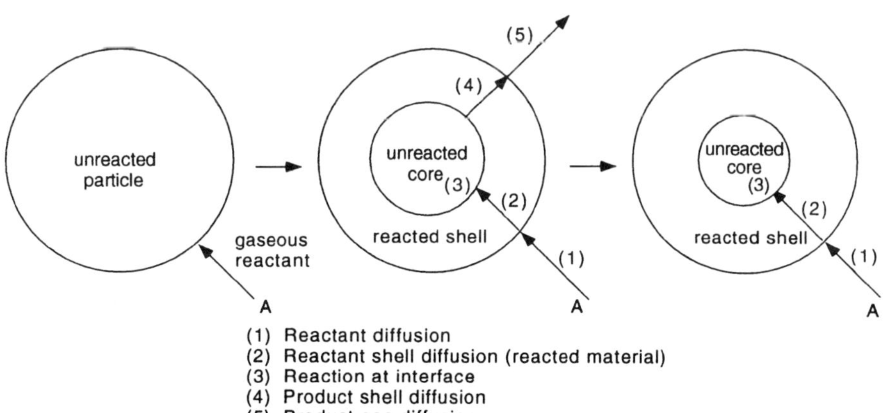


**Fig. 6.5.** Shrinking-core concept for reaction of material in a particle.

If gas-phase transport is rate limiting for the conversion of reactant to product, the time for complete reaction of the particle is[20]

$$\tau_{gs} = \frac{\rho_b d_p}{6bk_g C_{Ag}}, \tag{6.15}$$

where $\rho_b$ is the molar density of material in the unreacted core (moles/cm$^3$), $d_p$ is the diameter of the particle (including unreacted and reacted materials) (cm), $b$ is the number of moles of solid required to react with gaseous reactant (unitless), $k_g$ is the mass transfer coefficient of the gaseous reactant $A$ in the gas phase around the particle (cm/sec), and $C_{Ag}$ is the concentration of gaseous reactant in the gas phase (moles/cm$^3$). The way to increase the rate of conversion is to increase the concentration of the gaseous reactant (oxygen, for example) or the rate of mass transfer in the gas phase ($k_g$). The latter quantity is a function of particle size and gas properties and cannot be easily manipulated; it is relatively insensitive to temperature and, if the particle size is fixed, leaves only the concentration of the gaseous reactant as a control parameter. The reaction rate does not influence the time for conversion because it is fast enough not to be rate limiting. The temperature dependence of this reaction time is weak.

If both the gas-phase transport and the solid-state reaction are rapid, ash diffusion (diffusion through the reacted layer) controls the rate of the conversion of reactant to product. The time for complete reaction of the particle is then[20]

$$\tau_{as} = \frac{\rho_b d_p^2}{24bD_e C_{Ag}}, \tag{6.16}$$

where $D_e$ is the effective diffusion coefficient in the reacted shell of species $A$ (cm$^2$/sec). Note that $D_e$ depends on the permeability of the reacted layer, which cannot be predicted in most cases. For this reason, the most common approach to obtaining values for the effective diffusion coefficient is to use experimental results from similar systems. Equation (6.16) shows that little can be done to increase the conversion rate, except to increase the concentration of the gaseous reactant; the other parameters in the equation cannot be controlled. As before, the reaction rate does not influence the time for conversion because it is rapid enough not to be rate limiting. This equation does show, however, that larger particles will have much longer conversion times ($\tau_{as} \sim d_p^2$) than will smaller particles. The temperature dependence of $D_e$ is weak, leading to a weak temperature dependence for $\tau_{as}$.

When chemical reaction at the reacted–unreacted solid interface is slow and controls the conversion rate, the expression for the time for complete reaction becomes[20]

$$\tau_{rs} = \frac{\rho_b d_p}{2bk_s C_{Ag}}, \tag{6.17}$$

where $k_s$ is the surface reaction rate constant and the expression assumes a first-order reaction. This is the only case ($\tau_{rs}$, $\tau_{as}$, $\tau_{gs}$) where the reaction rate influences the conversion time; the parameter $k_s$ follows an Arrhenius behavior and can provide a strong temperature dependence.[20] Therefore, increasing temperature results in much shorter times for conversion. As with the limiting cases of Equations (6.15) and (6.16), increasing the concentration of the gaseous reactant can also increase the conversion rate.

Each of the foregoing cases provides different functions for conversion versus time. As a result, measurements of conversion as a function of time can be used to determine which mechanism is controlling. Also, the temperature dependence provides insights into the controlling mechanism, because only reaction-controlled conversion has a strong temperature dependence for the rate. Details on determining the rate-controlling steps are given in many books.[20, 24, 26]

## 6.5 GAS–LIQUID REACTIONS

Gas–liquid reactions occur during aerosol processing of both films and powders. For aerosol delivery of CVD precursors, volatile precursors in droplets can react with gaseous species before the solvent evaporates.[35] An example is the reaction

$$H_2(g) + Pd(hfac)_2(l) \longrightarrow 2hfacH(g) + Pd(s), \tag{6.18}$$

where the $H_2$ gas diffuses into a solvent droplet containing $Pd(hfac)_2$ and reacts to form Pd metal and volatile hfacH, which diffuses out of the droplet and into the gas phase. Liquid reactant droplets can react with gaseous reactants to form solid products (Fig. 6.6). The hydrolysis of metal alkoxide

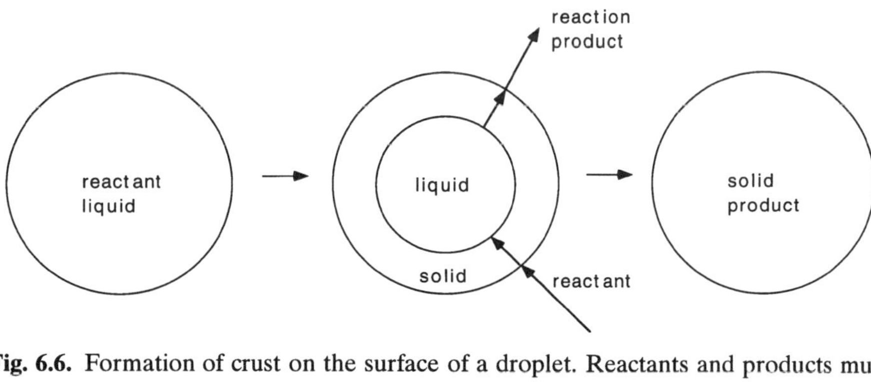


**Fig. 6.6.** Formation of crust on the surface of a droplet. Reactants and products must diffuse through the solid shell to reach the liquid and the gas.

droplets by reaction with water vapor is an example:[6]

$$Al(OR)_3(l) + 3H_2O(g) \longrightarrow Al(OH)_3(s) + 3\,ROH(l). \qquad (6.19)$$

In many cases, crusts of the hydrolyzed alkoxide are formed, leading to an even more complex problem.

As with gas–solid reactions, the overall conversion process in gas–liquid reactions consists of several steps in series. The situation is slightly more complicated, however, because a solid crust often forms at the surface of the particle if the reaction forms a solid product (Fig. 6.6). The steps in the series are film diffusion in the gas surrounding the droplet, diffusion through a solid reaction-product layer (when it exists), and diffusion through the liquid core, or, in some cases, reaction at the liquid–solid interface. If the reaction in the liquid is rapid enough, the reaction occurs at the interface between the reacted layer and the liquid core. The limiting cases are film diffusion (diffusion through the gas surrounding the droplet), diffusion through the reacted layer, diffusion through the unreacted liquid, and reaction either at the solid–liquid interface or in the bulk of the liquid.

Cases where film diffusion, diffusion through the reacted layer, or reaction at the interface limits the conversion give the same time for conversion of the droplet to a solid product as in gas–solid reactions. [See Equations (6.14) through (6.16).] The presence of the liquid inside the solid shell does not influence the conversion, as long as the reaction of the gaseous reactant with the liquid is sufficiently rapid. For gas–liquid reactions, the conversion rate can be controlled by the rate of reaction in the bulk of the fluid. Levenspiel[20] and Friedlander[36] give a more detailed discussion.

## 6.6 REFERENCES


1. Arai, Y. *Chemistry of Powder Production.* New York: Chapman and Hall, 1996.
2. Moore, K.A., Cesarano, J.I., Smith, D.M., & Kodas, T.T. *J. Amer. Cer. Soc.* *75*(1), 1992, 213–215.
3. Jain, S., Atanasova, P., Kodas, T.T., & Hampden-Smith, M.J. *J. Electrochem. Soc.* *143*(11), 1996, 3762–3770.
4. Pratsinis, S.E., Wang, G.Z., Panda, S., Guiton, T., & Weimer, A.W. *J. Mater. Res.* *10*(3), 1995, 512–520.
5. Shin, D.-S., Oh, E.-K., & Kim, S.-G. *Aerosol Sci. Technol.* *24*(4), 1996, 243–254.
6. Kodas, T.T., Sood, A., & Pratsinis, S.E. *Powder Technology* *50*, 1987, 47–53.
7. Xu, C., Hampden-Smith, M.J., & Kodas, T.T. *Chem. Mater.* *7*, 1995, 1539–1546.
8. Xu, C., Hampden-Smith, M., Kodas, T.T., Duesler, E.N., Rheingold, A.L., & Yap, G. *Inorganic Chem.* *34*, 1995, 4767–4773.
9. Gurav, A., Kodas, T.T., Kauppinen, E., Joutsensaari, J., & Zilliacus, R. *Nano-Struct. Mater.* *4*(5), 1994, 583–589.
10. Akhtar, M.K., Pratsinis, S.E., & Mastrangelo, S.V.R. *J. Mater. Res.* *9*(5), 1994, 1241–1249.

11. Bayvel, L. & Orzechowski, Z. *Liquid Atomization*. Washington: Taylor and Francis, 1993.
12. Tang, I.N., Fung, K.H., Imre, D.G., & Munkelwitz, H.R. *Aerosol Sci. Technol. 23*, 1995, 443–453.
13. Rubel, G.O. & Gentry, J.W. *J. Aerosol Sci. 15*(6), 1984, 661–671.
14. Kodas, T.T., Datye, A., Lee, V., & Engler, E. *J. Appl. Phys. 65*, 1989, 2149–2151.
15. Pluym, T.C., Lyons, S.W., Powell, Q.H., Gurav, A.S., Kodas, T.T., Wang, L.M., & Glicksman, H.D. *Mat. Res. Bull. 28*, 1993, 369–376.
16. Pluym, T.C., Powell, Q.H., Gurav, A.S., Ward, T.L., Kodas, T.T., Wang, L.M., & Glicksman, H.D. *J. Aerosol Sci. 24*(3), 1993, 383–392.
17. Pluym, T., Kodas, T.T., Wang, L.-M., & Glicksman, H. *J. Mater. Res. 10*(7), 1995, 1661–1673.
18. Gurav, A., Pluym, T., Kodas, T.T., & Pratsinis, S.E. *Chem. Eng. Commun. 151*, 1996, 211–226.
19. German, R.M. *Sintering Theory and Practice*. New York: John Wiley and Sons, 1996.
20. Levenspiel, O. *Chemical Reaction Engineering*. New York: John Wiley and Sons, 1972.
21. Nyman, M., Caruso, J., Hampden-Smith, M.J., & Kodas, T.T. *J. Am. Cer. Soc. 80*(5), 1997, 1231–1238.
22. Carslaw, H.S. & Jaeger, J.C. *Conduction of Heat in Solids*. London: Clarendon Press, 1959.
23. Crank, J. *The Mathematics of Diffusion*. London: Clarendon Press, 1957.
24. Bird, R., Stewart, W.E., & Lightfoot, E.N. *Transport Phenomena*. New York: John Wiley and Sons, 1960.
25. Kingery, W.D., Bowen, A.K., & Uhlmann, D.R. *Introduction to Ceramics*. New York: John Wiley and Sons, 1976.
26. Froment, G.F. & Bischoff, K.B. *Chemical Reactor Analysis*. New York: Wiley, 1990.
27. Foss, W.R. & Davis, E.J. *J. Aerosol Sci. 25*(1), 1994, 33–48.
28. Powell, Q.H., Kodas, T.T., & Anderson, B.M. *J. Chem. Vapor Deposition 2*, 1996, 179–181.
29. Powell, Q.H., Fotou, G.P., & Kodas, T.T. *Chem. Mater. 9*(3), 1996, 685–693.
30. Powell, Q.H., Fotou, G.P., Kodas, T.T., Anderson, B.M., & Guo, Y. *J. Mater. Res. 22*(2), 1997, 552–559.
31. Fotou, G.P. & Kodas, T.T. *Adv. Mater. 9*(5), 1997, 420–423.
32. Jain, S., Kodas, T.T., & Hampden-Smith, M.J. *J. Electrochem. Soc. 14*(4), 1997, 1222–1231.
33. Panda, S. & Pratsinis, S.E. *NanoStruct. Mater. 5*(7–8), 1995, 755–767.
34. Hinds, W.C. *Aerosol Technology*. New York: Wiley Interscience, 1982.
35. Kodas, T.T. & Hampden-Smith, M.J. (Eds.). *The Chemistry of Metal CVD*. Weinheim, West Germany: VCH, Verlagsgesellschaft, 1994.
36. Friedlander, S.K. *Smoke, Dust and Haze*. New York: Wiley Interscience, 1977.

# 7 Models Based on the General Dynamic Equation

## 7.1 INTRODUCTION

In flow reactors used for particle generation, many of the separate phenomena described in Chapters 2 through 6 occur simultaneously over at least part of the reactor. Although the separate equations describing the individual phenomena are useful for crude calculations, there is a need for more sophisticated models that describe the overall behavior of these systems, taking into account the interplay between all the relevant phenomena. For industry, such models provide a method for the design, modification, and operation of reactors. Part of the motivation for creating the models is that industrial systems are expensive, making any modifications to them also expensive. Further, it is very cumbersome to probe for characteristics such as particle size in industrial systems. Models allow numerical experiments that can predict, at least qualitatively, the effects of various system modifications on the characteristics of the particles produced. For this reason, various types of models have been developed to describe the behavior of aerosol flow reactors for powder generation and aerosol routes to films. These models build on the concepts presented in previous chapters and combine them into mathematical descriptions of the aerosol dynamics.

The starting point for developing most models is the *general dynamic equation* (GDE),[1] a complete mathematical description of the aerosol dynamics in terms of a particle size distribution function. This function is usually cast in terms of volume, $n(v)$. There are several forms of the GDE, depending on the physical and chemical phenomena under consideration. The most useful and common form includes spatially one-dimensional systems with particle formation by chemical reaction or nucleation, collisions and coalescence between particles, and particle growth. The more complicated versions are spatially two-dimensional systems that take into account particle transport and deposition by diffusion, thermophoresis, and other mechanisms. These models are also computationally intensive. A number of papers discuss and review various aspects of models for gas-to-particle conversion processes.[2–11]

In this chapter, we review the types of models used to describe aerosol routes to powders and film. The GDE is introduced first. The various models

—discrete, sectional, moment, and monodisperse—used to describe gas-to-particle conversion processes are presented together with examples of how quantities such as primary particle size, agglomerate size, and spread of the size distribution evolve along a reactor and how this evolution depends on operating parameters. Finally, models for film formation are discussed. Models for spray processes are discussed in Chapter 13.

## 7.2 GENERAL DYNAMIC EQUATION

The starting point for most models is the general dynamic equation for simultaneous aerosol nucleation, condensation, and coagulation for a continuous distribution function.[1] This equation is

$$\frac{\partial n}{\partial t} + \frac{\partial (Gn)}{\partial v} - I(v^*)\delta(v - v^*)$$
$$= \frac{1}{2}\int_0^v \beta(v - \bar{v}, \bar{v}) n(v - \bar{v}, t) n(\bar{v}, t)\, d\bar{v}$$
$$- n(v,t)\int_0^\infty \beta(v, \bar{v}) n(\bar{v}, t)\, d\bar{v}, \tag{7.1}$$

where $n$ is the particle size distribution function (number/cm$^6$), $G$ is the particle growth law (cm$^3$/sec), $v$ is the particle volume (cm$^3$), $I$ is the particle nucleation rate (number/cm$^3$sec), $\delta$ is the Dirac delta function (unitless), and $\beta$ is the collision frequency function (cm$^3$/sec). The first term on the left side describes the rate of change of the particle size distribution function with respect to time, which, for a particle volume interval $[v, v + dv]$, is the change in particle concentration. The second term on the left accounts for the gain or loss of particles from a given size range by condensation at a rate $G = dv/dt$, where $dv/dt$ is given by the appropriate particle growth law[1] described in Chapter 4. The third term on the left accounts for the formation of new particles of critical size $v^*$ (given by the Kelvin equation) by nucleation at rate $I$. The two terms on the right side account for the gain and loss of particles in a volume interval $[v, v + dv]$ by Brownian coagulation. Equation (7.1) is restricted to spherical particles with no velocities due to external forces on them and no diffusion of particles. Equation (7.1) has also been generalized to include the finite sintering rate of particles for the case where the sintering rate can be written in terms of the surface area of an aggregate, the surface area of the volume-equivalent sphere, and a sintering time constant. (See Koch and Friedlander[12, 13] and the material presented later in the chapter.)

In cases where the critical particle size $v^*$ is larger than the monomer size (as in the formation of droplets of many metal-organic and organometallic compounds), the nucleation rate can be assumed to follow that postulated in the classical Becker–Doring[1, 14] or other nucleation theory. (See Chapter 4.) Classical nucleation theory is often used because it is simple and has wider experimental support than any other model for homogeneous nucleation. For critical particle sizes less than the monomer size, as is usually the case when materials with low vapor pressures are formed, the expression for the nucleation rate is replaced by an expression for the rate of the reaction forming the monomer.

The GDE cannot be solved analytically except in highly simplified cases; therefore, it is usually converted into forms that allow a solution using numerical techniques. The most common forms include the discrete, sectional, lognormal, and monodisperse models. Each of these models has advantages and disadvantages. Although more sophisticated models based on two-dimensional size distributions accounting for agglomerate formation are not discussed here,[15, 16] a variety of specialized numerical schemes have been developed to solve equations of this type.[17, 18]

## 7.3 DISCRETE MODELS

The class of discrete models divides the particle size distribution into discrete sizes corresponding to the monomer (single atom or molecule), dimer, trimer, etc.[19–26] The major application of these models is to nanoparticle formation, where the particle size is limited to hundreds or thousands of atoms or molecules. This approach is attractive for dealing with nanoparticles because it is simple both conceptually and computationally. The approach is computationally impossible for larger particle sizes because one equation is required for each particle size and adjacent particle sizes differ by one atom or one molecule.

Most formulations of discrete models presuppose that the particles are formed by homogeneous chemical reaction of a precursor gas at an initial concentration $C(0)$ in a plug-flow reactor (with a flat velocity profile) at constant temperature and pressure. Newly formed particles are assumed to grow by coagulation and coalescence of spherical particles with a unity accommodation coefficient. The discrete GDE without growth by condensation is[1]

$$\frac{dn_i}{dt} - I(i^*)\,\delta(i - i^*) = \frac{1}{2}\sum_{j=1}^{i-1} \beta_{j,i-j} n_j n_{i-j} - n_i \sum_{j=1}^{\infty} \beta_{i,j} n_j, \qquad (7.2)$$

where $n_i$ is the particle concentration for size $i$ (molecules/cm$^3$), $I$ is the nucleation rate of critical-size particles (number/cm$^3$sec), $\beta_{i,j}$ is the collision frequency function (cm$^3$/sec), and $\delta$ is the Dirac delta function, which mathematically accounts for the introduction of particles with size $i^*$, the number of monomer units in a particle of critical size. The first term on the left side is the rate of change of the number concentration of $i$-mers (particles consisting of $i$ atoms or molecules) with respect to time. The second term accounts for the formation of $i$-mers of critical size by nucleation or reaction. The two terms on the right side respectively account for the formation of $i$-mers by collision and the loss of $i$-mers by coalescence.

For species with low equilibrium vapor pressure (metal oxides, borides, nitrides, carbides, and many metals), even single atoms or molecules can be thermodynamically stable particles.[27] In these species, there is no free-energy barrier to particle nucleation, and the thermodynamic critical particle size is equal to the size of the product monomers. In this case, $i^*$ equals unity, and the nucleation rate is equal to the reaction rate [molecules/(cm$^3$sec)] that is necessary to form the monomer. Any reaction scheme can be used to describe new-particle formation, but most studies have considered, for convenience, a first-order irreversible reaction scheme. A first-order or pseudo-first-order mechanism with respect to precursor vapor concentration is encountered in the high-temperature oxidation of $TiCl_4$ and $SiCl_4$ in excess $O_2$ to form $TiO_2$ and $SiO_2$, respectively.[28-30]

The discrete model can only describe particle dynamics in the free-molecule regime, because the number of equations needed for particles large enough to be in the continuum regime is enormous. In the free-molecule regime, and with a critical size of $i^* = 1$, the discrete GDE can be written for the monomers and all other $i$-mers ($i = 2, 3, 4, \ldots, i_{\max}$) as[1, 22]

$$\frac{dn_1}{dt} = C(0)\,A_A k_1 \exp(-k_1 t) - n_1 \sum_{j=1}^{\infty} \beta_{1,j} n_j, \tag{7.3}$$

and

$$\frac{dn_i}{dt} = \frac{1}{2} \sum_{j=1}^{i-1} \beta_{j,i-j} n_j n_{i-j} - n_i \sum_{j=1}^{\infty} \beta_{i,j} n_j, \tag{7.4}$$

where $C(0)$ is the reactant concentration (moles/cm$^3$), $A_A$ is Avogadro's number, and $k_1$ is the first-order reaction rate constant (1/sec).

In the monomer balance (Equation 7.3), the first term on the right side is the rate of monomer formation by chemical reaction, and the second is the loss of monomer to coagulation with all other particles. In the equations for dimers and larger particles (Equation 7.4), the terms on the right side are the gain and loss of particles of size $i$ by coagulation. The collision frequency function for Kn $\gg$ 1 can be written in terms of $i$ and $j$ as[22]

$$\beta_{i,j} = \left[\frac{6k_B T r_1}{\rho_p}\right]^{1/2} \left[\frac{1}{i} + \frac{1}{j}\right]^{1/2} \left(i^{1/3} + j^{1/3}\right)^2, \tag{7.5}$$

where $k_B$ is Boltzmann's constant, $r_1$ is the monomer radius (cm), and $\rho_p$ is the particle density (g/cm$^3$).

These equations are often placed into dimensionless form to facilitate calculations and interpretation of the results. Regardless of the order of the reaction and the conditions of the process, two dimensionless groups suffice to determine the particle size distribution at isothermal conditions: (1) the ratio of the characteristic time for chemical reaction to the characteristic time for monomer coagulation and (2) the ratio of the process residence time to the characteristic time for monomer coagulation. This conclusion has been used to produce nomographs that give a summary of the reactor behavior on simple plots.[22, 24]

## 7.4 SECTIONAL MODELS

Sectional models rely on dividing the particle size distribution into sections.[19–26] The characteristics of the particles in each section are described by average values for that section. Sectional models are often combined with discrete models to give a hybrid in which the smallest particles are described on a discrete basis and the larger particles are described using the sectional approach. The main advantage of this dual approach is that no assumptions are made about the shape of the particle size distribution. The main disadvantage is the complexity, which makes handling spatially nonuniform systems difficult and computationally intensive. The size spectrum is divided into discrete sizes for the lower end of the distribution and into sections for the remainder. Population balance equations are written for both the discrete and sectional size regimes.

For simplicity, these equations are presented here in nondimensional form; the details are found elsewhere.[19–26] The $y_i$ are the dimensionless concentrations of particles of size $i$, while the $Y_k$ are the dimensionless concentrations in the sections $k$. The $\gamma_{i,j}$ are functions that include the collision frequency functions. The quantity $X$ is the dimensionless residence time in the system, while $\theta$ is the dimensionless reaction time. The following equations are quite detailed, but simple in the sense that they are a set of coupled ordinary differential equations:

$$\frac{dy_1}{dX} = \frac{\exp(-X/\theta)}{\theta} - y_1 \sum_{j=1}^{i_{\max}} \gamma_{1,j}^{*} y_j - y_1 \sum_{k=1}^{m} {}^4\bar{\gamma}_{1,k}^{D} Y_k, \tag{7.6}$$

$$\frac{dy_i}{dX} = \frac{1}{2} \sum_{j=1}^{i-1} \gamma_{j,i-j}^{*} y_j y_{i-j} - y_i \sum_{j=1}^{i_{\max}} \gamma_{i,j}^{*} y_j - y_1 \sum_{k=1}^{m} {}^4\bar{\gamma}_{i,k}^{D} Y_k, \tag{7.7}$$

$$\frac{dY_1}{dX} = \frac{1}{2} \sum_{i=1}^{i_{\max}} \sum_{j=1}^{i_{\max}} {}^1\bar{\gamma}_{i,j,1}^{DD} y_i y_j - Y_1 \sum_{i=1}^{i_{\max}} {}^2\bar{\gamma}_{i,1}^{D} y_i + Y_1 \sum_{i=1}^{i_{\max}} {}^5\bar{\gamma}_{i,1}^{D} y_i - \frac{1}{2} {}^3\bar{\gamma}_1 Y_1^2 - Y_1 \sum_{i=2}^{m} {}^4\bar{\gamma}_{1,i} Y_i, \tag{7.8}$$

$$\begin{aligned}\frac{dY_k}{dX} = {} & \frac{1}{2} \sum_{i=1}^{i_{\max}} \sum_{j=1}^{i_{\max}} {}^1\bar{\gamma}_{i,j,k}^{DD} y_i y_j + \sum_{i=1}^{i_{\max}} \sum_{p=1}^{k-1} {}^1\bar{\gamma}_{i,p,k}^{D} y_p Y_k \\ & + \frac{1}{2} \sum_{i=1}^{k-1} \sum_{j=1}^{k-1} {}^1\bar{\gamma}_{i,j,k} Y_i Y_j - Y_k \sum_{i=1}^{i_{\max}} {}^2\bar{\gamma}_{i,k}^{D} y_i \\ & - Y_k \sum_{i=1}^{k-1} {}^2\bar{\gamma}_{i,k} Y_i + Y_k \sum_{i=1}^{i_{\max}} {}^5\bar{\gamma}_{i,k}^{D} y_i \\ & + Y_k \sum_{i=1}^{k-1} {}^5\bar{\gamma}_{i,k} Y_i - \frac{1}{2} {}^3\bar{\gamma}_k Y_k^2 \\ & - Y_k \sum_{i=k+1}^{m} {}^4\bar{\gamma}_{i,k} Y_i.\end{aligned} \tag{7.9}$$

Although there are several variations on sectional models in which different quantities are preserved, those models that preserve the highest moments of the particle size distribution are the most accurate.[23]

The first term on the right side in the monomer balance, (Equation 7.6), accounts for the production of monomers by chemical reaction, while the second and third terms account for the loss of monomers by coagulation with particles in the discrete size regime and the sectional size regime, respectively. For the balance equations for particle (molecular cluster) sizes $i = 2$ to $i_{max}$ (Equation 7.7), the first term on the right is the rate of production of $y_i$ from coagulation of smaller particles, while the last two terms account for the loss of $y_i$ by coagulation with particles in the discrete and sectional size regimes, respectively. In the balance equation for the first section of the distribution (Equation 7.8), the first term on the right side gives the rate of production of $Y_1$ by the coagulation of two particles in the discrete size regime. The second and third terms give the loss and production rates, respectively, of $Y_1$ from coagulation of one particle in the first section and one particle in the discrete size regime. The fourth term on the right accounts for the loss of $Y_1$ by coagulation of two first-section particles. The last term gives the rate of loss of $Y_1$ from coagulation of first-section particles with particles in larger sections. Finally, the balance equation for section $k$ (where $k = 2$ to $m$), (Equation 7.9) includes the first three terms on the right side for the rates of production of $Y_k$ from coagulation of, respectively, two particles in the discrete size regimes, one particle in the discrete size regime and one particle in the sectional size regime from a section smaller than the $k$th section, and two particles in the sectional size regime from sections smaller than the $k$th sections. The fourth and sixth terms on the right account for the loss and gain, respectively, of $Y_k$ from coagulation of one particle from the discrete size regime with one particle from section $k$ of the sectional size regime. The fifth and seventh terms account for the loss and gain, respectively, of $Y_k$ from coagulation of particles in the sectional size regime from a section smaller than the $k$th section with particles from the $k$th section of that regime. The eighth term on the right gives the loss of $Y_k$ from coagulation of two k particles from section $k$ in the sectional size regime, and the ninth term on the right is the rate of loss of $Y_k$ from coagulation of particles from section $k$ in the sectional size regime with particles from larger sections in that regime. Expressions are required for the DSM collision integrals $\gamma$ in discrete-sectional models and are given in the cited references.[19-26]

The dimensionless groups $Y$ and $X$ and the dimensionless collision integrals $\gamma_{i,j}$ determine the evolution of the size distribution for any aerosol system. The group $X$ (the dimensionless residence time) is the ratio of the process residence time to the characteristic time for aerosol coagulation. The group $Y$ (a dimensionless time for chemical reaction) is the ratio of the characteristic time for chemical reaction to the characteristic time for aerosol coagulation. The collision integrals $\gamma_{i,j}$ are constant numerical values for coagulation in the free-molecular regime[23] that depend on temperature, the gas medium, and particle properties in the transition and continuum regimes.

Equations (7.6) through (7.9) are a closed system of $i_{max} + m$ ordinary differential equations. The system can be integrated numerically, providing the evolution of the size distribution in the parameter space $X$ and $Y$.

Commonly, fewer than 100 discrete sizes and fewer than 100 sections are used.[23] Although the equations can be employed to describe nonisothermal powder production, the computation time for their integration can become a serious issue. Specifically, at fixed gas properties and isothermal conditions, the $\gamma_{i,j}$ need to be calculated only once and can be stored for subsequent integrations, while at nonisothermal conditions, the elaborate expressions for $y$ need to be recalculated at each integration step. The problem can be mitigated by calculating the $y$ values at an average temperature.

## 7.5 LOGNORMAL AND OTHER MOMENT MODELS

The lognormal and other types of moment models rely on converting the integro-differential general dynamic equation (7.1) to ordinary differential equations by making assumptions about the shape of the particle size distribution function. The main disadvantage of this approach is the assumption of a lognormal size distribution, an assumption that is often not valid. The main advantage is that the simplicity of the approach allows the inclusion of particle transport and the relatively simple description of two-dimensional systems. The earliest of this class of models assumed that the particles remained smaller than the mean free path and that no collisions occurred between particles. This assumption allowed the GDE to be converted into a closed set of ordinary differential equations for the particle concentration, average diameter, and surface area without making explicit assumptions about the shape of the particle size distribution.[31-42] Unfortunately, such an approach cannot describe systems in which particles collide and coalesce or systems in which the particles grow into the continuum regime, because the set of moment equations cannot be closed under either of those conditions. This shortcoming motivated the development of lognormal models,[43, 44] which are discussed here because of their greater usefulness.

Lognormal models rely on the assumption that the particle size distribution has a lognormal shape throughout the entire system. Making this assumption allows the GDE to be converted to ordinary differential equations while taking into account most of the important physical and chemical phenomena, including particle collisions, coalescence, and deposition for all particle sizes. The assumption also allows convenient expressions for the geometric mean volume $\upsilon_g$ and geometric standard deviation $\sigma_g$ of the size distribution. The lognormal distribution is defined as[43]

$$n(\upsilon, t) = \frac{1}{3\sqrt{2}\,\pi \ln \sigma_g} \exp\left[\frac{\ln^2(\upsilon/\upsilon_g)}{18 \ln^2 \sigma_g}\right] \frac{1}{\upsilon}. \tag{7.10}$$

The geometric mean particle size, in terms of volume, is given as

$$v_g = \frac{M_1^2}{M_0^{3/2} M_2^{1/2}}, \tag{7.11}$$

The geometric standard deviation, $\sigma_g$, is given by

$$\ln^2 \sigma_g = \frac{1}{9} \ln\left(\frac{M_0 M_2}{M_1^2}\right), \tag{7.12}$$

where $M_0$, $M_1$, and $M_2$ are the zeroth, first, and second moments with respect to the particle volume $v$.

A closed set of moment equations can be obtained by first substituting the expression for the lognormal size distribution (Equation 7.10) into the GDE (Equation 7.1). In the example that follows, this is done, ignoring particle transport. Integrating over all particle sizes gives the rate of change of the zeroth moment with respect to the particle volume. Multiplying both sides of the GDE by that volume $v$ and integrating over all particle sizes gives the rate of change of the first moment. Multiplying both sides by $v^2$ and integrating over all particle sizes gives the rate of change of the second moment. Friedlander outlines this procedure, while Hinds discusses various features of lognormal distributions.[1, 45] A balance on the monomer for particle growth is also required.

The equations for a constant rate of monomer production are

$$\frac{dM_0}{dt} I - B_4\Big[M_0^2 + M_{1/3} M_{-1/3} + B_5 \lambda (4\pi/3)^{1/3} \left(M_0 M_{-1/3} + M_{1/3} M_{-2/3}\right)\Big], \tag{7.13}$$

$$\frac{dM_1}{dt} = I v^* + B_3 (S - 1) M_{1/3}, \tag{7.14}$$

$$\frac{dM_2}{dt} = I v^{*2} + 2B_3 (S - 1) M_{4/3} + 2B_4\Big[M_1^2 + M_{4/3} M_{2/3} + B_5 \lambda (4\pi/3)^{1/3} \times \left(M_1 M_{2/3} + M_{1/3} M_{4/3}\right)\Big], \tag{7.15}$$

and

$$\frac{dS}{dt} = \frac{R_r}{n_S} - \frac{Ik^*}{n_S} - \frac{B_3}{n_S v_1}(S - 1)M_{1/3}, \tag{7.16}$$

where the integral moments are $M_0$, $M_1$, and $M_2$; the fractional moments $M_{-1/3}$, $M_{1/3}$, etc.; the nucleation rate $I$ (number/cm$^3$ sec); the reaction rate $R_r$ (number/cm$^3$ sec); the critical particle size $v^*$ and $k^*$, in terms of volume (cm$^3$) and number of monomers, respectively; the saturation concentration of the monomer $n_s$ (number/cm$^3$); the saturation ratio $S$; the monomer volume $v_1$ (cm$^3$); the constants $B_1$, $B_2$, $B_3$, and $B_4$ incorporating properties of the material; Boltzmann's constant $k_B$; the temperature $T$; and the mean free path $\lambda$ (cm). A variety of auxiliary equations not listed here are required to solve the differential equations. Typically, the terms for particle growth and coagulation are obtained from the harmonic average of the rates in the free-molecule and continuum regimes.[43] The dynamic behavior of a lognormally preserving aerosol can be described using these equations for the first three moments.

The GDE can also include terms to account for particle transport in two-dimensional systems.[46] The following equation gives the GDE accounting for laminar flow in a tubular reactor, diffusion of particles in the radial direction, thermophoretic transport in the radial direction, coagulation, and particle formation by reaction:

$$\begin{aligned} 2U\left[1 - \left(\frac{r}{R}\right)^2\right]\frac{\partial n}{\partial z} = {} & \frac{1}{r}\frac{\partial}{\partial r}\left[r\frac{\partial (Dn)}{\partial r}\right] + \frac{k\nu}{r}\frac{\partial}{\partial r}\left(nr\frac{\partial \ln T}{\partial r}\right) \\ & + \frac{1}{2}\int_0^v \beta(v, v - \bar{v})n(v)n(v - \bar{v})\, d\bar{v} \\ & - \int_0^\infty \beta(v, \bar{v})n(v)n(\bar{v})\, d\bar{v} + R_r\delta(v - v_1). \end{aligned} \tag{7.17}$$

Here, $r$ is the radial coordinate (cm), $R$ is the tube radius (cm), $U$ is the average gas velocity (cm/sec), $n$ is the particle size distribution function (number/cm$^6$), $z$ is the axial coordinate (cm), $D$ is the particle diffusion coefficient (cm$^3$/sec), $K$ is the thermophoretic constant, and where $R_r$ is the chemical reaction rate (molecules/cm$^3$ sec); the delta function indicates that the reaction is equivalent to nucleation.

The term on the left describes the change in the size distribution function with the residence time along the streamlines at each radial position. The first term on the right accounts for particle losses by radial Brownian diffusion, with the particle diffusivity related to particle size.[1] The second term on the right accounts for particle losses by thermophoresis.[47] The thermophoretic coefficient is related to the particle size by a correlation over the entire particle size spectrum.[48] The third and fourth terms account for particle gain and loss, respectively, by Brownian coagulation. The last term on the right-hand side accounts for the formation of new particles by chemical reaction. Equation (7.17) is the formulation used when materials such as silica particles are instantaneously formed because nucleation of silica by reaction of $SiCl_4$ cannot be distinguished from chemical reaction;[49] the size of the newly formed particles is less than or equal to the equivalent molecular diameter of $SiO_2$ (4.35 Å). The rate of particle formation is then equal to the reaction rate of $SiCl_4$.

As with the equations that apply in the absence of transport processes, the evolution of the $q$th aerosol moment is obtained by substituting the expression for the lognormal size distribution (Equation 7.10) into the GDE, multiplying both sides by $v$ to the zeroth, first, and second powers, and integrating over all particle sizes. This procedure requires converting the functions describing the basic particulate processes (the collision frequency function for the free-molecule regime and particle diffusivity) into forms that allow closure of the moment model. Such conversion can be done relatively easily; the various fractional moments that arise can be easily related to the integer moments through standard equations.

As an example, following this procedure, we can write the evolution of the zeroth, first, and second aerosol moments with time as[46]

$$2U\left[1-\left(\frac{r}{R}\right)^2\right]\frac{\partial M_0}{\partial z} = \frac{B_1}{r}\frac{\partial}{\partial r}\left[r\frac{\partial}{\partial r}(M_{-1/3}+B_2M_{-2/3})\right] + \frac{K\nu}{r}\frac{\partial}{\partial r}\left(M_0 r\frac{\partial \ln T}{\partial r}\right) + \xi M_0^2 + k_0 N_A \exp\left(\frac{-E_A}{R_G T}\right)C, \tag{7.18}$$

$$2U\left[1-\left(\frac{r}{R}\right)^2\right]\frac{\partial M_1}{\partial z} = \frac{B_1}{r}\frac{\partial}{\partial r}\left[r\frac{\partial}{\partial r}(M_{2/3}+B_2M_{1/3})\right] + \frac{K\nu}{r}\frac{\partial}{\partial r}\left(M_1 r\frac{\partial \ln T}{\partial r}\right) + k_0 N_A v_1 \exp\left(\frac{-E_A}{R_G T}\right)C, \tag{7.19}$$

$$2U\left[1-\left(\frac{r}{R}\right)^2\right]\frac{\partial M_2}{\partial z} = \frac{B_1}{r}\frac{\partial}{\partial r}\left[r\frac{\partial}{\partial r}(M_{5/3}+B_2M_{4/3})\right] + \frac{K\nu}{r}\frac{\partial}{\partial r}\left(M_2 r\frac{\partial \ln T}{\partial r}\right) + \zeta M_1^2 + k_0 N_A v_1^2 \exp\left(\frac{-E_A}{R_G T}\right)C, \tag{7.20}$$

where $B_1$, $B_2$, etc., are constants, $\xi$ and $\zeta$ are functions, $N_A$ is Avogadro's number, $R_G$ is the gas constant, $C$ is the reactant concentration, $k_0$ is the reaction-rate pre-exponential constant, and $E_A$ is the activation energy for the gas phase reaction. A variation of this approach has also been used to model the aerosol dynamics in turbulent flow.[50]

## 7.6 MONODISPERSE WITH SINTERING

The assumption that particles are monodisperse allows the aerosol dynamics to be modeled using simple equations, to the point where the finite rate of particle sintering can be incorporated easily. The main advantage of this model is its simplicity of implementation, while the main disadvantage is the assumption of a monodisperse size distribution, a constraint that is often unrealistic. This section develops a monodisperse model accounting for particle sintering.

For many materials, the product molecules or atoms are formed by chemical reaction and grow to molecular clusters and macroscopic particles by successive collisions and coalescence. The particle coalescence rate is initially much higher than the collision rate, so spherical particles are found. As the particle size increases, the sintering rate decreases, and aggregates of spherical primary particles form. Assuming monodisperse aggregates and that all aggregates contain the same number of equally sized primary particles, we find that the rate of change of the total aggregate number concentration with respect to time is

$$\frac{dN_\infty}{dt} = -\frac{1}{2}\beta N_\infty^2, \tag{7.21}$$

where $N_\infty$ is the number concentration of the aggregates and $\beta$ is the frequency function for collisions between aggregates.

Sintering of the particles is taken into account through the surface area of an aggregate particle, $a$, which increases by coagulation and decreases by sintering.[12, 13, 51] We have

$$\frac{da}{dt} = \frac{1}{N_\infty}\frac{dN_\infty}{dt}a - \frac{1}{\tau_f}(a - a_s), \tag{7.22}$$

where $\tau_f$ is the characteristic sintering time (the time needed to reduce excess aggregate surface area over that of a sphere of equal mass by about 63%) and $a_s$ is the surface area of the completely fused (spherical) aggregate. The aggregate and primary particle surface areas are related through the expression

$$a_s = \left(\frac{v}{v_0}\right)^{2/3} a_0, \tag{7.23}$$

with $a_0$ and $v_0$ denoting monomer surface area and volume, respectively. The volume of an aggregate, $v$, changes with time because of coagulation and not sintering:

$$\frac{dv}{dt} = \frac{1}{N_\infty}\frac{dN_\infty}{dt}v. \tag{7.24}$$

Finally, the various diameters and volumes are related through the equations

$$d_{\mathrm{pp}} = \frac{6v}{\alpha}, \tag{7.25}$$

$$n_p = \frac{v}{v_p}, \tag{7.26}$$

where $d_{\mathrm{pp}}$ is the primary particle size, $n_p$ is the number of primary particles in an aggregate, and $v_p$ is the volume of a primary particle $(= \pi d_p^3/6)$.

Aggregates are often fractallike and exhibit a power law relationship between volume and a characteristic aggregate radius, such as the radius of gyration or the so-called collision radius.[52, 53] The latter relationship follows from the equivalent spherical collision cross section of fractallike aggregates. A relationship between the aggregate volume and collision radius is[54]

$$r_c = r_p\left(\frac{v}{v_p}\right)^{1/D_f} = r_p(n_p)^{1/D_f}, \tag{7.27}$$

where $r_c$ is the collision radius, $r_p$ is the radius of the primary particle, and $D_f$ is the mass fractallike dimension. This function suggests that if the number of primary particles in the aggregate reduces to unity, the collision radius becomes equal to the primary particle radius. For more than one primary particle per aggregate, the collision radius approaches the radius of gyration of the aggregate. Calculation of the collision frequency function $\beta$ for these irregular-shaped aggregates can be accomplished by using the collision radius $r_c$ to replace the solid-sphere radius in expressions for Brownian coagulation in the free-molecule and continuum regimes, as given in Chapters 3 and 5.[14] It must be assumed, however, that the fractallike dimension of the aggregates is not altered by sintering, which is reasonable only if the particles remain as aggregates. This is perhaps the weakest assumption in existing models.

In addition to affecting the coagulation rate, the characteristic time for sintering influences the evolution of the particle morphology. Several primary particles in contact can sinter to form a single compact particle by surface diffusion, grain boundary diffusion, evaporation–condensation, and/or lattice diffusion. (See Chapter 5.) Each of these sintering mechanisms may contribute to the reduction in surface area at a different rate. However, the driving force for all of these processes is the reduction in surface energy, which is related to the particle surface energy and curvature. Expressions relating the characteristic time for sintering to temperature and the primary particle size are derived from the so-called initial-stage sintering models, which describe the center-to-center approach and neck growth between two equal-sized spheres.[55] Unfortunately, the lack of reliable data on properties of the material in the particles and the supposition of different mechanisms to determine the rate-controlling mechanism can frustrate the interpretation of experimental results. Ashby sintering diagrams have been used to identify the dominant mechanism for different materials.[55] Ashby diagrams are useful because they show the main transport mechanisms for various particle and neck sizes at a constant temperature.

## 7.7 COMPARISON OF MODELS

Each of the models described in the preceding sections has certain weaknesses and strengths that make it useful in some situations and not in others. Figure 7.1 shows a comparison of the predicted particle concentration, average diameter, and geometric standard deviation for the discrete–sectional, lognormal, and monodisperse models for noninstantaneous chemical reaction and for a self-preserving distribution.[23] In general, the models agree only after a sufficiently long time has elapsed to reach a self-preserving distribution. During particle formation, the models disagree to significant extents, depending on the residence time and nature of the model. For the geometric standard deviation, only the lognormal and discrete–sectional models are shown, because the other two models have fixed spreads of the size distribution that do not change with time. Overall, the discrete–sectional model is the most accurate, followed by the lognormal and the monodisperse models.

Different models are better suited to different situations. The discrete and sectional models are the most computer intensive; the lognormal and monodisperse models make the most assumptions. These assumptions may be unrealistic for the particular situation; for example, most size distributions are not monodisperse. Similarly, many reactors do not produce particles with lognormal distributions until sufficiently long residence times. From the point of view of increasing complexity, the models are monodisperse, lognormal, discrete, sectional, and discrete–sectional.

Finally, it is worth adding that various other models have been developed for a variety of physical situations. Coating of particles by CVD[56] and formation of aggregates with charged particles[57] are but a few of the situations that have been modeled.

## 7.8 MODELING OF PARTICLE DEPOSITION PROCESSES

Models for processing of films using aerosols can be developed utilizing both Lagrangian and Eulerian formulations for particle transport. To describe the overall behavior of a system, these approaches rely on obtaining the fluid velocity profiles from measurements, analytical solutions, or commercially available codes such as those using a finite-element approach. Flow patterns

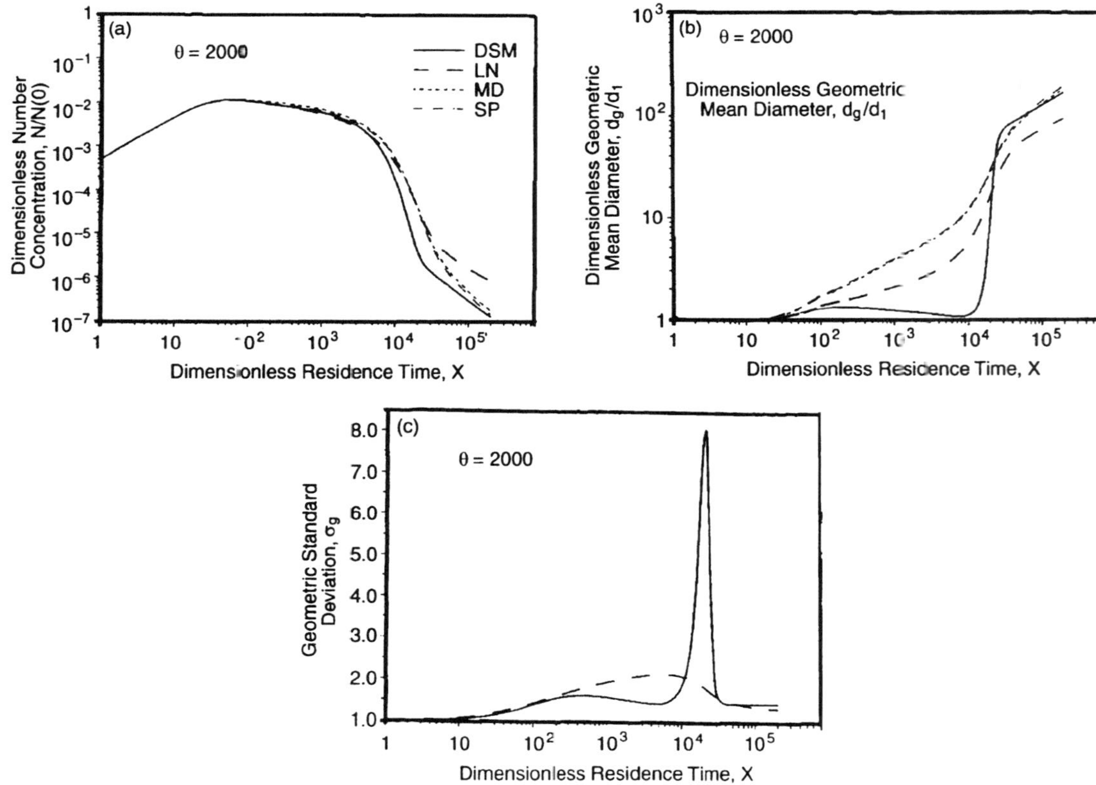


**Fig. 7.1.** Aerosol properties predicted by discrete–sectional (DSM), lognormal (LN), monodisperse (MD), and self-preserving (SP) models for noninstantaneous chemical reaction ($\theta = 2000$): (a) total number concentration, (b) geometric mean diameter, and (c) geometric standard deviation. Poor agreement is obtained until $X = 30{,}000$, after which a self-preserving distribution is established. (Adapted from Landgrebe and Pratsinis.[23] Reprinted with permission from Academic Press, Inc.)

have been measured and calculated for a variety of geometries; the most important results have been obtained for flow impinging on a flat plate. The temperature profiles must also be obtained and are often measured or calculated at the same time as the velocity profiles. It is assumed in most cases that the presence of the particles does not influence the velocities and temperatures. The velocity and temperature profiles are then used with the equations describing particle motion within the fluid to give deposition rates to surfaces. The Lagrangian approach is employed when particle inertia and applied forces are important, while the Eulerian approach is used when particle Brownian motion and applied forces are important.

For the Lagrangian approach, the equation describing particle motion is

$$m_p \frac{du}{dt} = \pm \frac{C_D}{C_c} \frac{\pi}{8} \rho d_p^2 (u - U)^2 + F_E + F_{\mathrm{Th}} + F_S \tag{7.28}$$

(as described in Chapter 3), where $m_p$ is the particle mass (g), $u$ is the particle velocity (cm/sec), $t$ is time (sec), $F_E$ is the electrophoretic force exerted on the particle (dynes), $F_{\mathrm{Th}}$ is the thermophoretic force exerted on the particle (dynes), $F_S$ is the gravitational force exerted on the particle (dynes), $C_D$ is the drag coefficient (dimensionless), $C_c$ is the Cunningham correction factor (dimensionless), $\rho$ is the gas density (g/cm$^3$), and $U$ is the gas velocity (cm/sec). The gas velocity profiles enter into this equation directly through $U$, while the temperature enters through $F_{\mathrm{Th}}$ and through its effect on parameters, including viscosity and density. Systems in which new particles are formed cannot be described by this approach in most cases because the small particles have high diffusion coefficients and negligible inertia. Droplet evaporation is described by having the mass of the particle vary with position. An example of the approach is given by Oh and Kim.[58]

The Eulerian approach is used when the particles have negligible inertia and when new particles are formed in a system. Particle transport occurs by the flow of the gas (convection), diffusion, thermophoresis, gravitational settling, electrophoresis, and other phenomena. The equation describing particle transport is the general dynamic equation, including terms describing those effects:[1]

$$\nabla \cdot n\mathbf{v} + \frac{\partial I}{\partial v} = \nabla \cdot D \nabla n + \frac{1}{2}\int_0^v \beta(\bar{v}, v - \bar{v}) n(\bar{v}) n(v - \bar{v})\, d\bar{v} - \int_0^\infty \beta(v, \bar{v}) n(v) n(\bar{v})\, d\bar{v} - \nabla \cdot \mathbf{c} n \qquad (v \gg v_m) \tag{7.29}$$

Here, $\mathbf{v}$ is the fluid velocity vector (cm/sec), $v$ is the particle volume (cm$^3$), $\bar{v}$ is the variable of integration (cm$^3$), $n$ is the particle size distribution function (number/cm$^6$), $I$ is the particle current in volume space (number/cm$^3$ sec), $D$ is the particle diffusion coefficient (cm$^2$/sec), $\beta$ is the coagulation coefficient (cm$^3$/sec), and $\mathbf{c}$ is the velocity vector associated with thermophoresis, electrophoresis, photophoresis, and settling (cm/sec). This equation is also called the convective diffusion equation in cases where the velocities due to force fields are negligible. The first term on the left side describes transport by convection, while the second accounts for the formation and growth of particles. The first term on the right describes diffusion of particles, while the integral terms describe collisions between particles. The last term in the equation accounts for the velocities, **c**, due to thermophoresis, electrophoresis, settling, etc. As with the Lagrangian approach, the gas velocities and temperatures in the system must be measured or calculated using separate equations. The gas velocities enter into the GDE through the first term in Equation (7.29). Temperature enters through its influence on parameters, including particle diffusion coefficients, and through the thermophoretic velocity, which appears in **c**. This equation is the general starting point for many particle deposition problems when new particles are formed in a system by gas-to-particle conversion or when small particles are formed ex situ and then introduced into a deposition system. As a crude rule of thumb, the approach is used for atmospheric pressures when the particle size is less than 100 nm; when the particle size is greater than 1000 nm, the Lagrangian approach is used. An example of the Eulerian approach is given by Oh et al.[59]

Equation (7.29) can be simplified for flow through a tube, where there is no velocity or particle transport in the theta (angular) direction and the problem is effectively two-dimensional. We have

$$\frac{1}{r}\frac{\partial}{\partial r}[r\rho(u + v_{Tr})n] + \frac{\partial}{\partial z}[\rho(v + v_{Tz})n] = \frac{1}{r}\frac{\partial}{\partial r}\left(r\rho D\frac{\partial n}{\partial r}\right) + \frac{\partial}{\partial z}\left(\rho D\frac{\partial n}{\partial z}\right), \tag{7.30}$$

where $r$ is the radial coordinate (cm), $z$ is the coordinate in the axial direction (cm), $\rho$ is the fluid density (g/cm$^3$), $v$ is the fluid velocity (cm/sec), $v_{Tr}$ is the thermophoretic velocity in the radial direction (cm/sec), and $v_{Tz}$ is the thermophoretic velocity in the $z$-direction (cm/sec). The terms on the left side account for both the transport by flow of the gas and thermophoresis. This is done by bringing the thermophoresis terms to the left side of the equation. The terms on the right side account for diffusion in the radial and axial directions, respectively. The most important auxiliary equations are those relating the thermophoretic velocities in the radial and axial directions to the temperature gradients:

$$v_{Tr} = -\alpha\frac{\mu}{\rho}\frac{\partial \ln T}{\partial r}, \quad v_{Tz} = -\alpha\frac{\mu}{\rho}\frac{\partial \ln T}{\partial z}. \tag{7.31}$$

The thermophoretic coefficient is given by

$$\alpha = \frac{2.294\left[(k/k_p) + 2.20\mathrm{Kn}\right]}{(1 + 3.438\mathrm{Kn})\left[1 + 2(k/k_p) + 4.40\mathrm{Kn}\right]} \times \{1 + \mathrm{Kn}[1.2 + 0.41\exp(-0.88/\mathrm{Kn})]\}, \qquad (7.32)$$

where Kn is the Knudsen number, $k$ is the thermal conductivity of the gas, and $k_p$ is the thermal conductivity of a particle. A wide variety of other situations have also been described, and numerous comparisons with measurements are available. Variations of this approach have been used to describe the formation of optical fibers by particle formation and deposition.[46, 60, 61]

## 7.9 REFERENCES


1. Friedlander, S.K. *Smoke, Dust and Haze*. New York: Wiley Interscience, 1977.
2. Flagan, R.C. *J. Aerosol Sci. 19*(7), 1988, 783–787.
3. Flagan, R.C. *Ceram. Trans. 1*(2, A), 1988, 229–243.
4. Flagan, R.C., Nguyen, H.V., & Rogak, S.N. Formation of refractory aerosol particles. *Conf. Combust. Plasma Synthesis High-Temp. Mater.*, 1990, 357–365.
5. Flagan, R.C. *Fuel Processing Technol. 39*(1–3), 1994, 319–336.
6. Friedlander, S.K. *Aerosol Sci. Technol. 1*, 1982, 3–13.
7. Friedlander, S.K. Dynamics of aerosol formation by chemical reaction. In Pfeffer, R. (Ed.), *Fourth Intrnl. Conf. Physiochemical Hydrodynamics*. New York: Annals of the New York Academy of Sciences, 1983, 355–364.
8. Friedlander, S.K., Koch, W., & Main, H.H. *J. Aerosol Sci. 22*(1), 1991, 1–8.
9. Rao, N.P. & McMurry, P.H. *Aerosol Sci. Technol. 11*(2), 1989, 120–132.
10. Shaw, B.D. & Lawman, J. *Aerosol Sci. Technol. 20*, 1994, 363–374.
11. Zachariah, M.R. & Dimitriou, P. *Aerosol Sci. Technol. 13*(4), 1990, 413–425.
12. Koch, W. & Friedlander, S.K. *J. Colloid Interface Sci. 140*(2), 1990, 419–427.
13. Koch, W. & Friedlander, S.K. *J. Aerosol Sci. 21*(1), 1990, S73–S76.
14. Seinfeld, J.H. *Atmospheric Chemistry and Physics of Air Pollution*. New York: Wiley Interscience, 1986.
15. Xiong, Y. & Pratsinis, S.E. *J. Aerosol Sci. 24*(3), 1993, 283–300.
16. Xiong, Y., Akhtar, M.K., & Pratsinis, S.E. *J. Aerosol Sci. 24*(3), 1993, 301–313.
17. Gelbard, F. & Seinfeld, J.H. *J. Computational Physics 28*, 1978, 357–375.
18. Jacobson, M.Z. & Turco, R.P. *Aerosol Sci. Technol. 22*, 1995, 73–92.
19. Gelbard, F. & Seinfeld, J.H. *J. Colloid Interface Sci. 68*(2), 1979, 363–382.
20. Gelbard, F., Tambour, Y., & Seinfeld, J.H. *J. Colloid Interface Sci. 76*(2), 1980, 541–556.
21. Gelbard, F. *Aerosol Science and Technology 12*, 1990, 399–412.

22. Landgrebe, J.D. & Pratsinis, S.E. *Ind. Eng. Chem. Res. 28*(10), 1989, 1474–1481.
23. Landgrebe, J.D. & Pratsinis, S.E. *J. Colloid Interface Sci. 139*(1), 1990, 63–86.
24. Landgrebe, J.D., Pratsinis, S.E., & Mastrangelo, S.V.R. *Chem. Eng. Sci. 45*, 1990, 2931–2941.
25. Wu, J.J. & Flagan, R.C. Modeling particle formation and growth in aerosol reactors. *Aerosols: Formation and Reactivity*. Berlin: 2nd Int. Aerosol Conf., 1986, 9914–9917.
26. Wu, J.J. & Flagan, R.C. *J. Colloid Interface Sci. 123*(2), 1988, 339–352.
27. Ulrich, G.D. *Combustion Sci. Technol. 4*, 1971, 47–57.
28. French, W.G., Pace, L.J., & Foertmeyer, V.A. *J. Phys. Chem. 82*(20), 1978, 2191–2194.
29. Powers, D.R. *J. Am. Ceram. Soc. 61*(7–8), 1978, 295–297.
30. Suyama, Y. & Kato, A. *J. Am. Ceram. Soc. 59*(3–4), 1976, 146–149.
31. Barrett, J.C. & Jheeta, J.S. *J. Aerosol Sci. 27*(8), 1996, 1135–1142.
32. Kodas, T.T., Pratsinis, S.E., & Friedlander, S.K. *J. Colloid Interface Sci. 111*(1), 1986, 102–111.
33. Kodas, T.T. & Friedlander, S.K. *Ind. Eng. Chem. Res. 26*, 1987, 1999–2007.
34. Okuyama, K., Kousaka, Y., Tohge, N., & Adachi, M. *J. Mater. Sci. Lett. 6*(12), 1987, 1466–1468.
35. Okuyama, K., Kousaka, Y., Warren, D.R., Flagan, R.C., & Seinfeld, J.H. *Aerosol Sci. Technol. 6*(1), 1987, 15–27.
36. Pratsinis, S.E., Kodas, T.T., Dudukovic, M.P., & Friedlander, S.K. *Powder Tech. 47*, 1986, 17–23.
37. Pratsinis, S.E., Kodas, T.T., Dudukovic, M.P., & Friedlander, S.K. *Ind. Eng. Chem. Process Des. Dev. 25*(3), 1986, 634–642.
38. Fotou, G.P. & Kodas, T.T., personal comm.
39. Pratsinis, S.E., Friedlander, S.K., & Pearlstein, A.J. *AIChE J. 32*(2), 1986, 177–185.
40. Stern, J.E., Wu, J.J., Flagan, R.C., & Seinfeld, J.H. *J. Colloid Interface Sci. 110*(2), 1986, 533–543.
41. Warren, D.R. & Seinfeld, J.H. *Aerosol Sci. Technol. 3*, 1984, 135–153.
42. Warren, D.R. & Seinfeld, J.H. *J. Colloid Interface Sci. 105*(1), 1985, 136–142.
43. Pratsinis, S.E. *J. Colloid Interface Sci. 124*(2), 1988, 416–427.
44. Heintzenberg, J. *Aerosol Sci. Tech. 21*, 1994, 46–48.
45. Hinds, W.C. *Aerosol Technology*. New York: Wiley Interscience, 1982.
46. Kim, K.-S. & Pratsinis, S.E. *AIChE J. 34*(6), 1988, 912–921.
47. Walker, K.L., Geyling, F.T., & Nagel, S.R. *J. Am. Ceram. Soc. 63*(9–10), 1980, 552–558.
48. Talbot, L., Cheng, R.K., Schefer, R.W., & Willis, D.R. *J. Fluid Mech. 101*(4), 1980, 737–758.
49. Ulrich, L., Cheng, R.K., Schefer, R.W., & Willis, D.R. *J. Fluid Mech. 101*, 1980, 737.
50. Xiong, Y. & Pratsinis, S.E. *J. Aerosol Sci. 22*(5), 1991, 637–655.

51. Koch, W. & Friedlander, S.K. *J. Aerosol Sci. 20*(8), 1990, 891–894.
52. Megaridis, C.M. & Dobbins, R.A. *Combust. Sci. and Tech. 71*, 1990, 95–109.
53. Schmidt-Ott, A., Baltenberger, U., Gaeggeler, H.W., & Jost, D.T. *J. Aerosol Sci. 21*, 1990, 711–717.
54. Matsoukas, T. & Friedlander, S.K. *J. Colloid Interface Sci. 146*(2), 1991, 495–506.
55. Coblenz, W.S., Dynys, J.M., Cannon, R.M., & Coble, R.L. In Kuczynski, G.C. (Ed.), *Sintering Processes, Material Science Research. New York: Plenum*, 1980, 141–157.
56. Jain, S., Fotou, G.P., & Kodas, T.T. *J. Colloid Interface Sci. 185*, 1997, 26–38.
57. Xiong, Y., Pratsinis, S., & Mastrangelo, S.V.R. *J. Colloid Interface Sci. 153*, 1992, 106–117.
58. Oh, E.-K. & Kim, S.-G. *J. Aerosol Sci. 27*(8), 1996, 1143–1154.
59. Oh, M.D., Yoon, K.H., & Mong, H.K. *Aerosol Sci. Technol. 25*(2), 1996, 141–156.
60. Kim, K.-S. & Pratsinis, S.E. *Chem. Eng. Sci. 44*(11), 1989, 2475–2482.
61. Kim, K.S. & Pratsinis, S.E. *Intl. J. Heat Mass Trans. 33*, 1990, 1977–1986.

# 8 Chemistry

## 8.1 INTRODUCTION

The goal of this chapter is to describe the chemistry that is associated with the aerosol processing of materials. Our approach is to dissect powder and film formation processes so that the fundamental chemical steps can be identified. These steps are then discussed individually in order to describe the underlying principles and to bring them together with physical aspects of aerosol processing described in Chapters 1 through 7. Fundamental aspects of physical processes that occur during aerosol processing have been described in previous sections and have an intimate relationship with chemical processes. During aerosol processes, many chemical and physical phenomena take place simultaneously, which leads to a high level of complexity in the systems involved.[1] In order to describe the processes that *can* occur, it is valuable to separate them artificially so that we can discuss them individually, even though this may not represent the actual situation.

Chemical processes that occur during the aerosol synthesis of materials can be classified in a number of different ways. One classification method is to describe the processes in terms of the final product (i.e., films vs. powder), while another is to classify the chemical processes according to the phase in which they occur (i.e., solid, liquid, or gas). Whichever method is chosen, there is a significant amount of overlap between each classification. Furthermore, it is often difficult to determine from data in the literature which chemical processes are most important, such as which step is rate limiting and whether gas-phase, liquid-phase, or solid-state reactions occur during the aerosol synthesis of materials. There are relatively few reports in which chemical aspects of aerosol processing have been studied and the key processes identified unambiguously. For example, film deposition can occur by simultaneous particle deposition or CVD. More generally, literature reports often do not determine or distinguish the mechanism of powder or film formation. As a result, we will describe an overview of all the processes that can occur, and a number of which *may* occur, during a particular process.

In the aerosol processing of materials, chemistry plays a crucial role and has a profound influence on the overall behavior of the system in a number of ways, illustrated schematically in Figures 8.1 through 8.4. It is important to understand the nature of these processes in order to identify the key steps

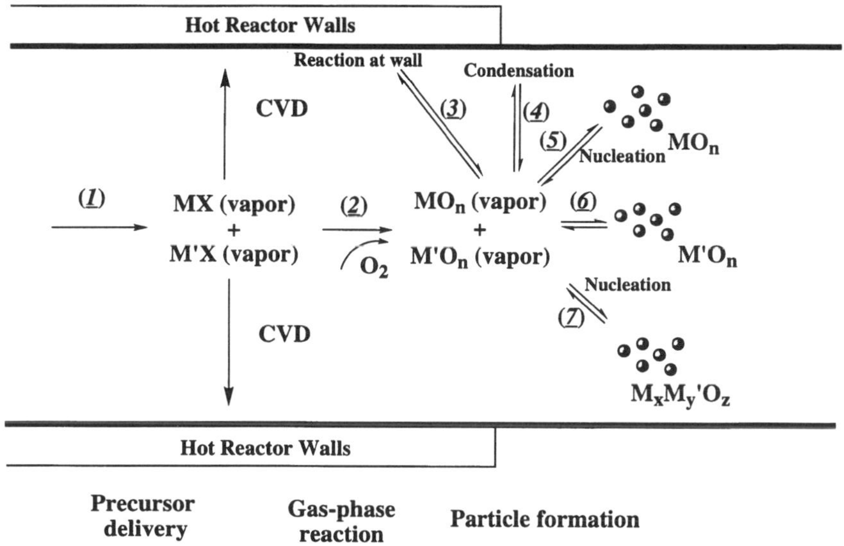


**Fig. 8.1.** Overview of gas-to-particle conversion.

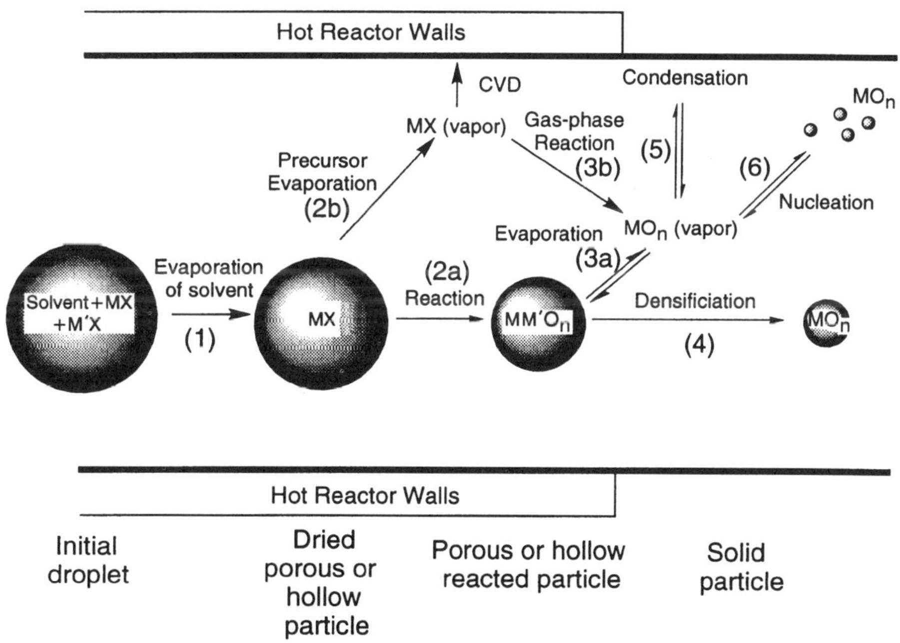


**Fig. 8.2.** Overview of particle formation by spray pyrolysis.

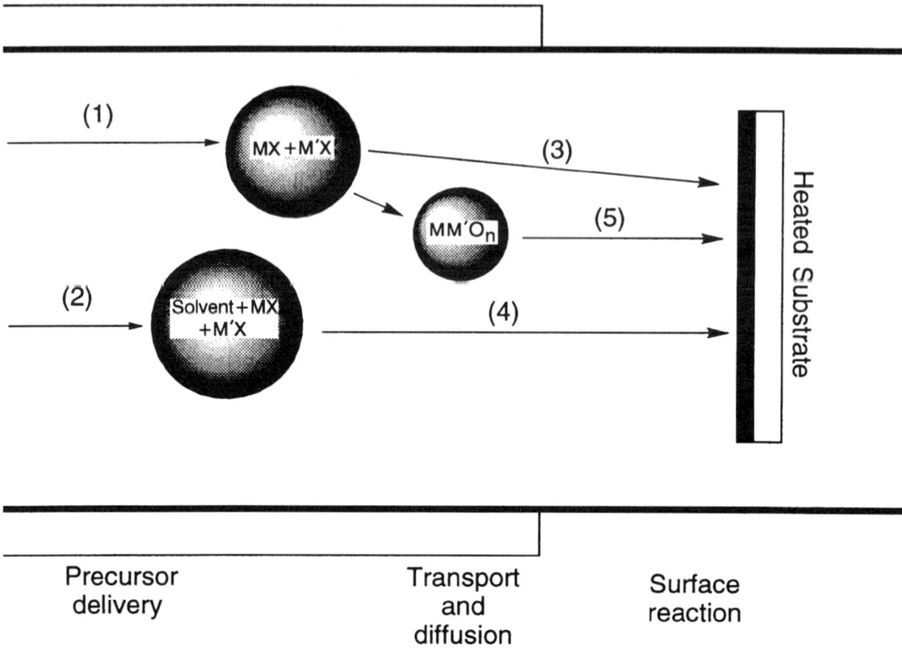


**Fig. 8.3.** Overview of film deposition by spray pyrolysis and solid-particle deposition.

that control the overall behavior of the system. Accordingly, we next describe the chemical aspects of these processes.

Powder formation by gas-to-particle conversion can occur via a number of chemical and physical routes. Figure 8.1 shows a chemical gas-to-particle conversion process in which molecular metal-containing precursors (MX, M′X) to metal-oxide products are (1) delivered to a reactor as gaseous species, where (2) they undergo chemical reactions and are converted to a metal oxide using a variety of forms of energy, including thermally-induced, laser-induced, or plasma-induced chemical reactions, to name but a few. Then, depending on the conditions, the vapor-phase species $MO_n$ and $M'O_n$ may (3) react further on a hot surface or (4) condense on a cool surface or nucleate and, finally (5, 6, 7), grow to form particles. If a more complex material, such as $M_xM'_yO_z$, is desired, then the relative rates of chemical reaction between $MO_n$ and $M'O_n$ versus nucleation and growth of separate $MO_n$ and $M'O_n$ particles must be considered. For complex materials, the volatile metal-containing precursor(s) should exhibit reasonable vapor pressures and undergo chemical reactions that result in the elimination or reaction of gaseous by-products to give the desired product with high purity. The volatile precursors may also react with the hot walls of the reactor (in CVD) to form a coating. A complete understanding of the chemical and physical properties of the reagents and intermediates is clearly crucial for

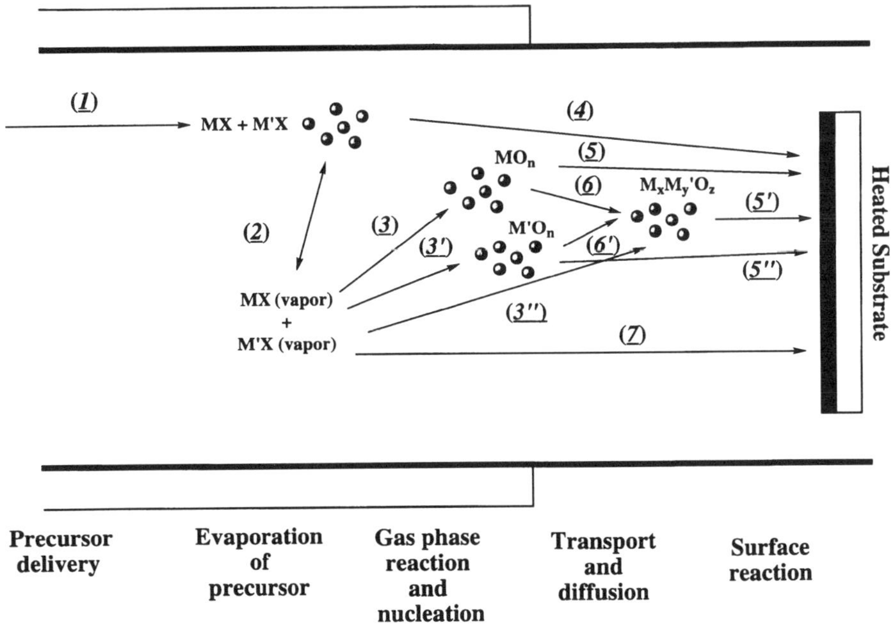


**Fig. 8.4.** Overview of film deposition by chemical vapor deposition.

controlling the characteristics (composition, purity, phase purity, stoichiometry, particle size and distribution, etc.) and properties of the product. The by-products that evolve should take *no further part in the reaction* and should be removed from the system.

Powder (i.e., particle) formation by spray pyrolysis, summarized in Figure 8.2, generally involves the delivery of droplets of one or more metal-containing precursors in a solvent to a reactor. The solvent (1) generally evaporates, leaving a dried particle containing the precursor(s) that (2a) can react in the particle to form the final product, (4) followed by densification. Alternatively, a volatile precursor, intermediate, or product (2b and 3b) can evaporate, which is often detrimental to the process, since it leads to loss of control over particle size through (3b) gas-phase decomposition reactions followed by (6) nucleation and growth of undesired small particles or (5) a loss of material to the system by condensation or reaction on the reactor walls. Among the chemical aspects of spray pyrolysis, it is important to understand and control the solubility, volatility, and reactivities of the precursors, as well as the volatility of the intermediates. In more complex materials, it is important to understand the relative reactivities of the precursors and intermediates, which, if not correctly matched, can lead to phase segregation in the product due to any or all of three mechanisms: preferential precipitation of one reagent in the droplet, premature reaction of one precursor in the dry particle, or preferential loss of one precursor or intermediate due to evaporation.

Film deposition can occur by spray pyrolysis (droplet deposition) solid particle deposition, and CVD. In spray pyrolysis (Fig. 8.3) a droplet (2) composed of the solvent and the metal-containing precursor (4) is delivered to a heated substrate, where the solvent evaporates and the precursors react to form a film of the desired material. In this case, the solubilities and volatilities of the metal-containing precursors are important, as are their reactivities individually and with each other on the heated substrate. The solvent may also play a role in determining the purity and composition of the film produced.

In solid particle deposition (Fig. 8.3), solid metal-containing precursor particles (1) are delivered to the substrate (3), where they react to form the desired film. Depending on the conditions, the solid particle (5) may also undergo thermally induced reactions in the gas phase to form the product. An understanding of the thermal decomposition pathway of the precursor to produce a compositionally controlled, pure film and of the relative reaction rates of the precursors to produce a stoichiometric, phase-pure film is crucial. Again, it is desirable that the precursors be nonvolatile, to avoid gas-phase processes and retain control over the deposition process.

Chemical vapor deposition[2–5] (Fig. 8.4) involves (1) the delivery of vapor-phase precursors, in this context using an aerosol delivery process[6] to the reactor, followed by (7) transport to the surface where reaction to deposit a film occurs. Gas-phase reactions (3, 3′, 3″) can lead to the formation of undesired particles, which may be lost to the system or be co-deposited during CVD, both of which are often detrimental to the deposition process. Solid precursor particles (2) may also be present (in the case of aerosol precursor delivery), which (4) may be transported to the surface and react or be deposited there. In CVD, it is important that the precursors be volatile and capable of reacting on the substrate surface, the result of which is the formation of high-purity films. When complex materials are to be deposited, the relative vapor pressures and reactivities of the precursors are again important. Compositional or phase inhomogeneities or loss of control over stoichiometry could arise in a number of different steps, including those in which the reaction and transport rates are different for each precursor and/or its derivatives. Such steps include feed-rate, transport-rate, and surface-reaction-rate-limited deposition. Various gas-phase reaction rates (3, 3′, 3″, 5, 5′, 5″, 6, 6′) for which particle formation occurs in the gas phase may also be limiting. These aspects will be discussed in more detail later in the chapter.

It is clear from this introductory discussion that there are a number of underlying chemical steps, the understanding of which is required to successfully synthesize powders and films by aerosol means. First, the three most important properties of precursors and intermediates are *solubility*, *volatility*, and *reactivity*. For example, gas-phase particle formation is generally detrimental to a CVD process because it can lead to nonuniformity in the film. It is especially detrimental in applications where a thin film on the order of 100 Å is desired, because particles larger than this will destroy the properties of the film.

Second, there is a distinction between the chemistry that occurs in the gas, liquid, and solid phases, because of the different diffusion coefficients and different nature of the nearest neighbors in their phases. In the gas phase, a reactive intermediate is likely to travel a considerable distance before it encounters another species and will undergo many vibrations, which often favors intramolecular rearrangement. At the same time, intermediates in the liquid phase, either in a pure liquid or a solvent, are likely to encounter another species on a timescale similar to that of a molecular vibration, more strongly favoring intermolecular reactions. In the solid state, interactions between neighbors are even more important.

Finally, a number of other considerations, more practical in nature, occur when the chemical design of the system plays a crucial role. For example, the choice of solvent in spray pyrolysis is important because the solvent can take part in chemical reactions. The reactivities of precursors and their reaction products towards each other also are important. In the case of powder formation, the temperature of the collection apparatus, the collection method, and the materials used are likely to be important to ensure that the conditions are not detrimental to the powder collected. The by-products formed during chemical reaction can react with collected powder or during the growth of a film and be detrimental to the purity and other properties of the product. On a larger scale, a thorough analysis of the by-products is also crucial for the operation of any powder or film formation process in which waste separation or the conversion of waste streams to environmentally benign products is necessary. In an industrial context, the cost of the precursor may be a crucial factor and may even determine the choice of processing method. In general, volatile metal-containing compounds are likely to be more expensive than nonvolatile ones, because volatile metal-containing compounds (with the exception of metal halides) generally contain more complex ligands and require more steps to prepare.

*The chemistry associated with aerosol synthesis of materials is classified here according to chemical reactions that can occur in the gas, liquid, or solid phase (Sections 8.4 through 8.7), surface chemical reaction processes (Section 8.9), and solid-state processes (Section 8.10).* Any or all of these processes may occur simultaneously during powder or film formation, and their relative importance in a given situation will depend on the conditions employed. For this reason, this method of classification has been chosen, rather than describing the chemistry relevant to gas-to-particle conversion, liquid-to-solid conversion, solid-to-solid conversion, and film formation. For example, the discussion of chemical reaction processes is relevant to homogeneous *intra*molecular gas-phase reactions of a volatile molecule to form a solid particle and *inter*molecular reactions of a solid particle containing a molecular precursor. Surface processes are important especially in the deposition of films and gas-to-particle conversion and may also be important in other particle formation processes, such as the formation of coated particles. In the section on solid-state processes, key aspects of chemical reactions in the solid state are described. However, solid-state chemical reactions of molecular species that are central to control of the compositional homogeneity and crystal-phase evolution in films and powders are described in Section 8.4, on molecular processes. This aspect is often crucial in aerosol synthesis of powders, due to the high heating rates (~ 100°C/min) and short residence times (~ 1 sec) associated with that mechanism. In a given system, it is desirable to understand which of these processes is likely to be most important and, as a result, how to control the overall system to produce the optimum product.

## 8.2 GENERAL FEATURES OF METAL-CONTAINING COMPOUNDS

The general features of metal-containing compounds are described in a number of texts.[7–9] Here, we encapsulate these features in some definitions, the nomenclature used in the sections that follow, and a discussion of some typical properties of metal-containing compounds. The latter includes the distinction between empirical and molecular formulas and the origin of different degrees of oligomerization and provides a background for the subsequent discussion of molecular, surface, and solid-state processes.

It is first important to define the general nomenclature used in this section. The term *inorganic compound* refers to a compound that does not contain organic ligands. Typical examples of such compounds are metal nitrates, carbonates, sulfates, or halides [e.g., $Fe(NO_3)_3$, $BaCO_3$, $Al_2(SO_4)_3$, $BaCl_2$, and $SnCl_4$]. *Metal-organic* compounds are compounds that contain organic ligands, but no metal–carbon bond [e.g., $Ba(OAc)_2$, $Ca(C_2O_4)$, $Zr(O\text{-}t\text{-}Bu)_4$, $Al(OR)_3$, $Cu(hfac)_2$, and $Zr(NMe_2)_4$]. *Organometallic* compounds are compounds that contain organic ligands *and* a metal carbon bond, such as $Sn(CH_3)_4$ and $Al(\text{-}i\text{-}Bu)_3$. Frequently, it may not be possible to classify molecular compounds using these terms, because a single compound can contain a number of different types of ligand [e.g., $Me_2Sn(OAc)_2$]. Accordingly, *metal-containing* compound refers to any compound that contains a metal.

It is worth remembering that the *empirical formula* given for an inorganic compound (and often metal-organic and organometallic compounds) such as a metal nitrate or metal chloride is seldom also the *molecular formula*. This is because most metal atoms have an affinity for *coordination numbers* (the number of bonds subtended between the metal center and the ligands) greater than their ligands can provide. As a result, these compounds tend to *oligomerize* (aggregate by forming covalent bonds utilizing vacant coordination sites at the metal and unused electrons of certain elements in the

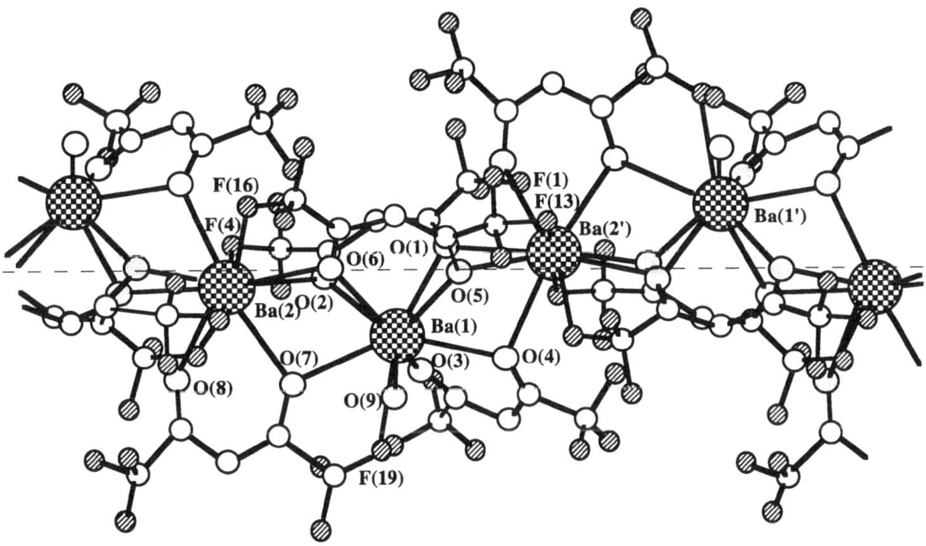


**Fig. 8.5.** Example of an oligomeric metal-containing compound, $\{[Ba(hfac)_2]_2 \cdot Et_2O]\}_\infty$. (Adapted from Drozdov and Troyanov.[10] Permission granted by The Royal Society of Chemistry.)

ligands, thereby creating bridging ligands) in the solid state. Oligomerization can render them both involatile and perhaps insoluble, as shown by the example given in Figure 8.5.[10]

Where soluble and volatile compounds are required, this problem can be overcome in a number of different ways. Inorganic compounds that are oligomeric or polymeric in the solid state can be de-oligomerized by reaction with a *donor ligand* (a ligand that donates electron density, usually in the form of a lone pair of electrons, to a vacant (acceptor) orbital at the metal center). The donor ligand replaces a *bridging ligand* (a ligand that coordinates to two or more metal centers), improving the solubility of the inorganic compound. This is often the case when inorganic compounds are dissolved in water. The oxygen lone pair in water acts as a donor to the metal center, displacing a bridging ligand, and this process frequently occurs to the extent that the ligands are completely displaced from the metal atom environment, giving an ionic compound. Ligand displacement can have a profound effect on the reactivity of a particular species, depending on the nature of the solvent. The solid-state structure of $Al(NO_3)_3 \cdot 9H_2O$ is an example of a species in which the anionic ligand ($NO_3$) is not bonded to the metal. (See Fig. 8.6)[11] This species is used in the preparation of aluminum-oxide powders by spray pyrolysis.[12]

A similar strategy can be employed to deoligomerize metal-containing compounds in order to improve their volatility for applications in AACVD

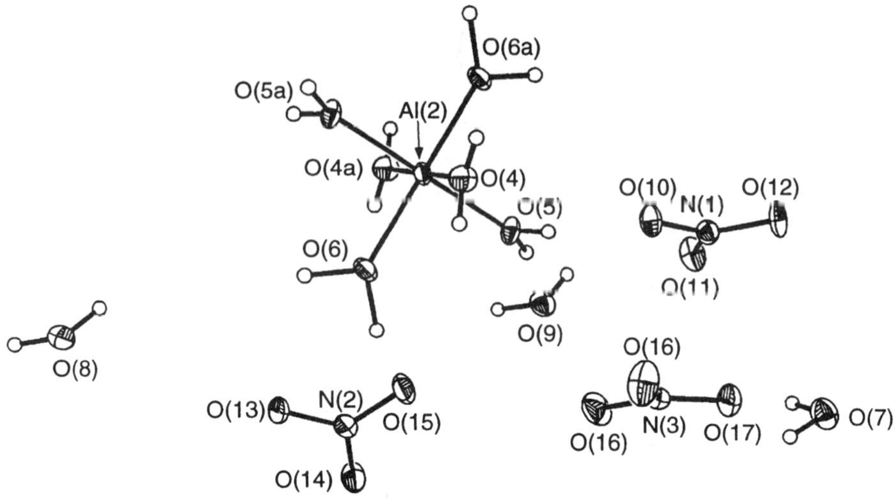


**Fig. 8.6.** Solid-state structure of $Al(NO_3)_3 \cdot 9H_2O$, as determined by single-crystal X-ray diffraction. (Courtesy of Nyman et al., unpublished results.)[11]

and in gas-to-particle conversion. A donor (or coordinating) ligand can be used to occupy a coordination site at the metal center, preventing oligomerization, as is evidenced by the solid-state structures of $\{[Ba(hfac)_2]_2 \cdot Et_2O\}_\infty$[10] (Fig. 8.5) vs. $Ba(hfac)_2 \cdot$(18-C-6)[13] (Fig. 8.7), and $Ba(hfac)_2 \cdot$(tetraglyme)[14] vs. $[Ba(tmhd)_2]_4$[15] (Fig. 8.8a, b.)[16]

In some cases, the donor group has been incorporated into the ligand, which also imparts volatility. This occurs for the triki (2,2-dimethyl-5-N-(2-(2-

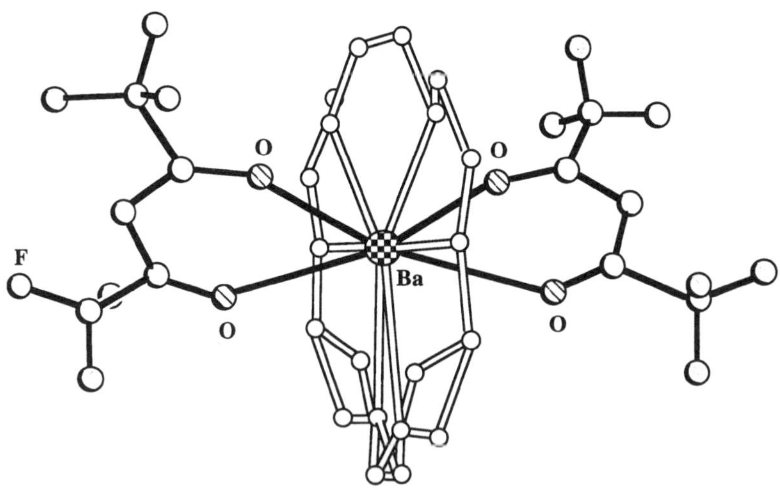


**Fig. 8.7.** Solid-state structure of $Ba(hfac)_2 \cdot$ 18-C-6. (Adapted from Norman and Pez.[13] Permission granted by The Royal Society of Chemistry.)

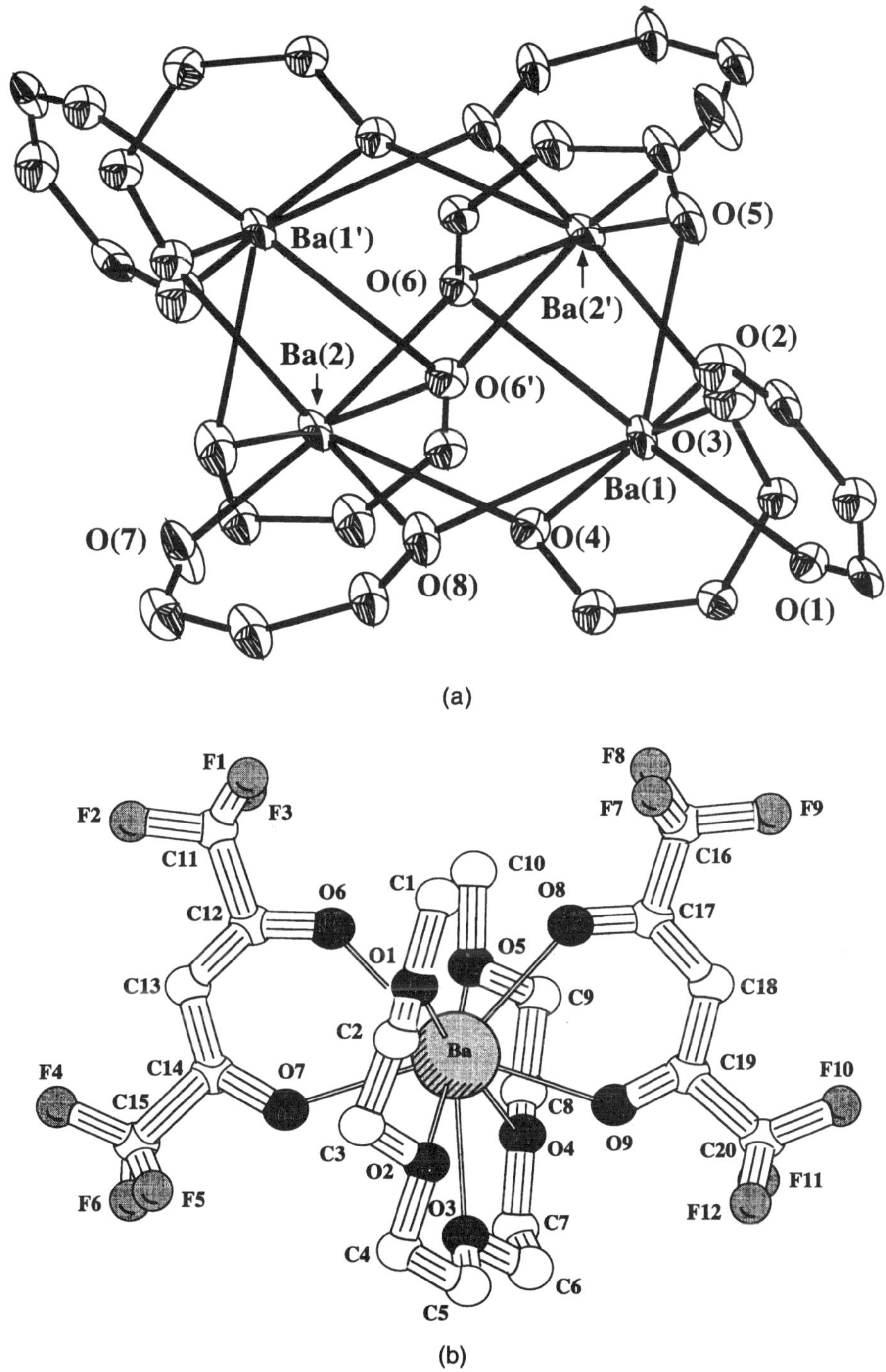


**Fig. 8.8.** Comparison of solid state structures of (a) $[Ba(tmhd)_2]_4$ (adapted from Sluis and Spek[14] and Drake et al.[15] Permission granted by The Royal Society of Chemistry) vs. (b) $Ba(hfac)_2 \cdot$ tetraglyme.

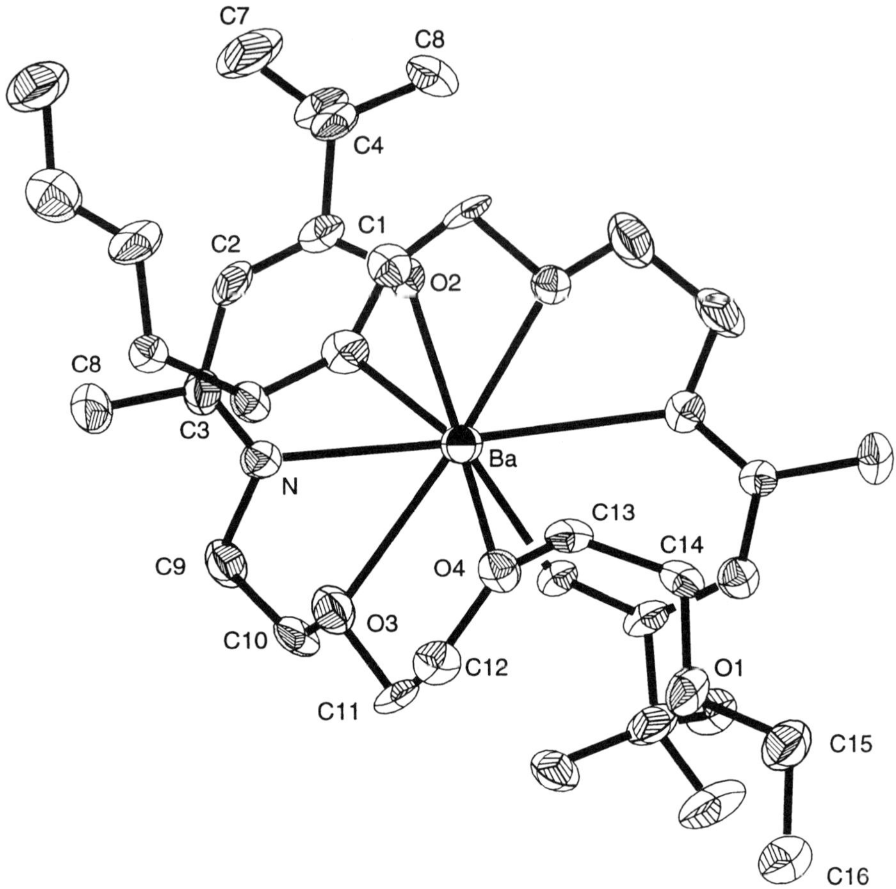


**Fig. 8.9.** Solid-state structure of $Ba(triki)_2$. (Reprinted with permission from Schulz, Hinds, Neumayer, Stern, and Marks.[17] *Chem. Mater.* *5*, 1993, 1605. Copyright 1993, American Chemical Society.)

(2-ethoxy)ethoxy)-ethoxyamino)-3-hexanone) systems, as exemplified by $Ba(triki)_2$ in Figure 8.9.[17, 18] These Ba-containing compounds have been used for CVD of Ba-containing films, especially $BaTiO_3$.[16]

Thermally induced deoligomerization can occur if the metal-containing compound is thermally robust. This frequently happens with metal halides (e.g., $[AlCl_3]_\infty$ and $[SnCl_2]_\infty$): At room temperature the halide is polymeric, but at elevated temperatures where the thermal energy available can overcome the energy of the intermolecular attraction, it undergoes thermally induced deoligomerization.[9] However, in some cases, covalent bonds are broken at these energies and cause decomposition rather than deoligomerization. The solid-state and gas-phase structures of "$SnCl_2$" and "$AlCl_3$" (shown in Figs, 8.10 and 8.11) provide an example in which $[AlCl_3]_\infty$ and $[SnC_2]_\infty$ are

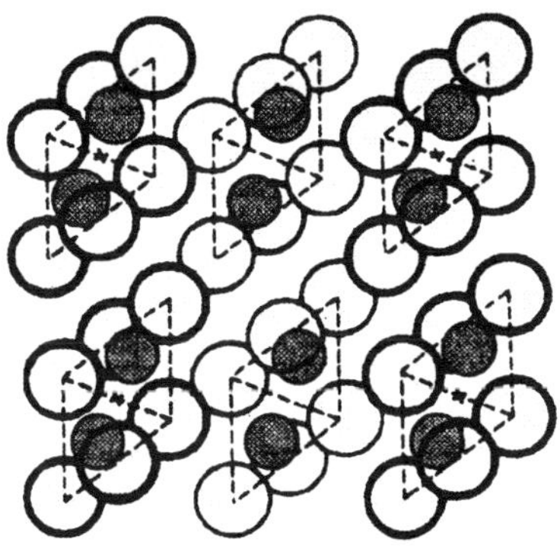

a.

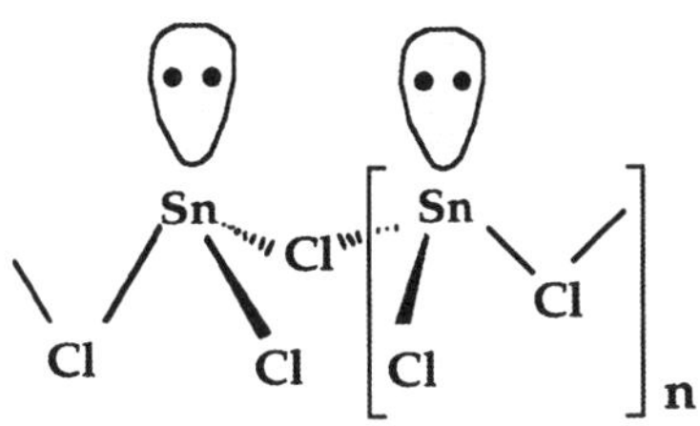


b.

**Fig. 8.10.** (a) Solid-state structure of $AlCl_3$. Unshaded spheres are chloride atoms and shaded spheres are aluminum atoms. (b) Solid-state structure of $SnCl_2$. (Adapted from Wells.)[9]

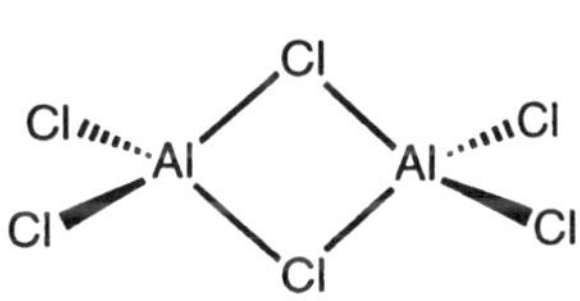


**a.**

**b.**

**Fig. 8.11.** Gas-phase structure of (a) $Al_2Cl_6$ (a) and (b) $SnCl_2$.

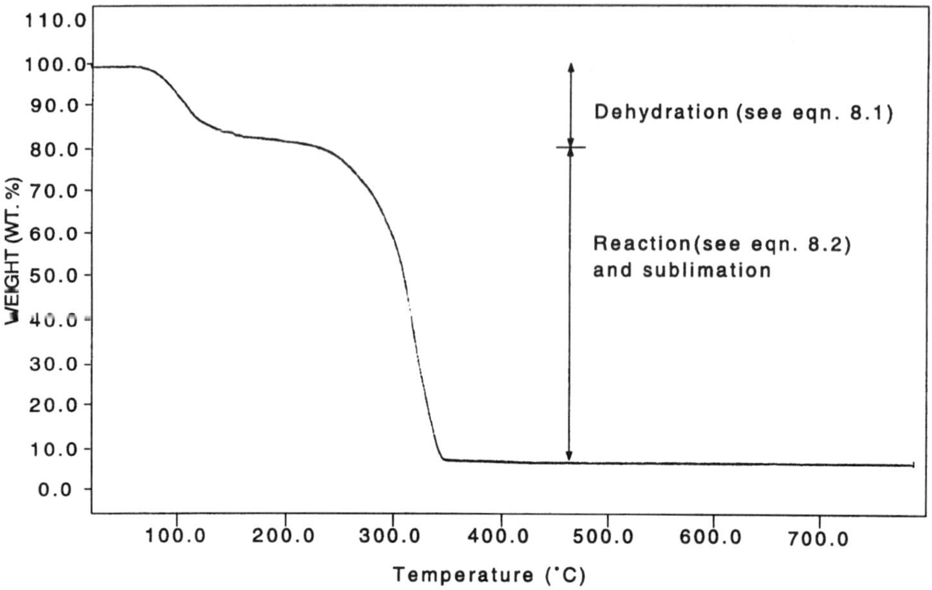


**Fig. 8.12.** Data from thermogravimetric analysis of $Zn(OAc)_2 \cdot 2H_2O$.

infinite polymers in the solid state at room temperature, but on heating (e.g., to 600°C for $[AlCl_3]_\infty$), a volatile compound ($Al_2Cl_6$) is evolved in the gas phase, with a monomeric substance ($AlCl_3$) formed at a higher temperature.

In some cases, apparent thermally induced deoligomerization is actually accompanied by thermal decomposition. For example, $Zn(OAc)_2 \cdot 2H_2O$ thermally dehydrates at approximately 100°C to form $Zn(OAc)_2$ and, on further heating to roughly 250°C, reacts to form $Zn_4O(OAc)_6$, which is monomeric and volatile, as described in the following equations. (See also Figure 8.12.)[19–21]

$$Zn(OAc)_2 \cdot 2H_2O \longrightarrow Zn(OAc)_2 + 2H_2O, \tag{8.1}$$

$$4\,Zn(OAc)_2 \longrightarrow Zn_4O(OAc)_6 + (Ac)_2O. \tag{8.2}$$

As a result, a potentially involatile precursor can be converted to a different, volatile precursor during aerosol synthesis of ZnO powders or films.

All of the features described in this section should be taken into account when one chooses metal-containing compounds for use in aerosol processes, because they may determine the presence or absence of gas-phase reactions.

## 8.3 SELECTION OF CHEMICAL REAGENTS AS PRECURSORS: IMPORTANT CHEMICAL PROPERTIES RELATED TO AEROSOL PROCESSING

Among the features of chemical species that are important to consider during aerosol processing are solubility, volatility, and reactivity.

### 8.3.1 Solubility

The solubility of a solute in a solvent is important in aerosol processes in which the solute (or metal-containing precursor) is first dissolved in a solvent in order to deliver the solute to the system, often using various aerosol generators. Two such processes are spray pyrolysis and AACVD. Solubility can affect particle morphology through different solubilities of the reagent, compositional homogeneity during powder formation, and different transport rates during film deposition. The affect of solubility on morphology has been demonstrated in the formation of $ZrO_2$.[22-24] Similarly, the effect of solubility on the compositional control of iron-oxide powders[25] using a number of different metal-containing precursors has been demonstrated. The choice of both the solute and the solvent generally determines the outcome of the process.

The general description, "like dissolves like," is a good rule of thumb in considering a solvent for a metal-containing compound. For example, inorganic compounds such as metal nitrates generally dissolve in water (an inorganic solvent), while metal-organic and organometallic compounds generally dissolve in organic solvents and do not dissolve in water. In choosing a solvent, one should consider various solvent properties. The *donor or coordinating ability* of the solvent describes its ability to coordinate to a particular metal (acceptor) center, which depends on both the nature of the solvent and the metal center to which electrons are being donated and is an important aspect of dissolving a metal-containing compound by de-oligomerization. The donor/acceptor ability of solvents has been quantified through defining *donor/acceptor numbers*. The donor number is the negative of the standard reaction enthalpy for the reaction between the donor in question with the acceptor $SbCl_5$ in 1,2-dichloroethane. The acceptor number is an arbitrary number derived from the change in $^{31}P$ NMR chemical shift when trimethylphosphine oxide ($Me_3PO$) is used as a reference base; a value of 0

**TABLE 8.1. Donor numbers for various solvents at 298 K**

| Solvent | Donor Number | Acceptor Number |
|---|---|---|
| Acetic acid | — | 52.9 |
| Acetone | 17.0 | 12.5 |
| Benzene | 0.1 | 8.2 |
| Carbon tetrachloride | — | 8.6 |
| Diethyl ether | 19.2 | 3.9 |
| Dimethylsulfoxide | 29.8 | 19.3 |
| Ethanol | 19.0 | 37.1 |
| Pyridine | 33.1 | 14.2 |
| Tetrahydrofuran | 20.0 | 8.0 |
| Water | 18 | 54.8 |

Values unavailable:—

**TABLE 8.2. Dielectric constants of important solvents at 298 K. (Superscript indicates temperature in °C)**

| Name of Solvent | Formula | M.P./°C | b.p./°C | Dielectric Const. |
|---|---|---|---|---|
| Water | $H_2O$ | 0 | 100 | 78.5 |
| Chloroform | $CHCl_3$ | −63.5 | 61.7 | 1.7 |
| Nitromethane | $CH_3NO_2$ | −28.5 | 101 | 38.6 |
| Methanol | $CH_3OH$ | −97.5 | 64.5 | 32.6 |
| Acetonitrile | $CH_3CN$ | −45.7 | 81.6 | 36.2 |
| Acetic acid | $CH_3CO_2H$ | 16.6 | 118 | 6.2 |
| Ethanol | $C_2H_5OH$ | −114.5 | 78.3 | 24.3 |
| Dimethylsulfoxide | $(CH_3)_2SO$ | 18.4 | 189 (d) | 49 |
| Hexafluoroacetone | $(CF_3)_2CO$ | −125 | −27.4 | 2 |
| Acetone | $(CH_3)_2CO$ | −95.4 | 56.2 | 20.7 |
| Trichloroacetic acid | $Cl_3CCO_2H$ | 58 | 198 | $4.6^{60}$ |
| Trifluoroacetic acid | $F_3CCO_2H$ | −15 | 72.4 | $39^{20}$ |
| N,N-dimethylformamide | $(CH_3)_2NCHO$ | −61 | 152 | 36.7 |
| 2-methoxyethanol | $MeOCH_2CH_2OH$ | −85 | 124 | $16^{30}$ |
| Tetrahydrofuran | $(CH_2)_4O$ | −65 | 66 | 7.3 |
| 1,4-dioxane | $(CH_2)_4O_2$ | 11.8 | 102 | 2.2 |
| Diethyl ether | $(C_2H_5)_2O$ | −116 | 34.5 | $4.3^{20}$ |
| Pyridine | $C_5H_5N$ | −41.8 | 115.6 | 12.3 |
| Hexane | $C_6H_{14}$ | −95 | 69 | $1.9^{20}$ |
| Toluene | $C_7H_8$ | −95 | 110 | 2.4 |
| Hexamethylphosphoramide | $OP(NMe_2)_3$ | — | 66 | — |
| 1,1-dimethoxyethane | $MeOCH_2CH_2OMe$ | −113 | 64.5 | $3.5^{20}$ |
| Ethylacetate | $C_2H_5OAc$ | −83.6 | 77.1 | 6.0 |

(d) = decomposes; — = data unavailable

corresponds to a noninteracting base such as hexane, while a value of 100 is given to the acceptor $SbCl_5$. Some examples of donor and acceptor numbers are given in Table 8.1.[26] In each case, a higher numerical value indicates a stronger donor or acceptor. This trend can be used as a guide to help dissolve reagents by choosing a better donor solvent over a weaker donor solvent to improve solubility.

The *dielectric constant* is also an important property of a solvent, describing its ability to separate charges. Solvents with a high dielectric constant are good for dissolving ionic species. The dissolution of metal-containing compounds by polar solvents with a low dielectric constant can generally be attributed to the donor ability of the solvent. Some examples of the range of dielectric constants of various solvents are given in Table 8.2.[27] Water exhibits a high dielectric constant and is often a solvent of choice for spray pyrolysis. The solubility of common reagents in water is given in Table 8.3.[27]

Finally, it is important to note that the solvent may participate in the chemistry of the reaction prior to or during aerosol processing. For example,

**TABLE 8.3. Solubility of some metal-containing compounds at 298 K. (Superscript indicates temperature in °C)**

| Name of Compound | Formula | Solubility in Water (g/100 g of water) |
|---|---|---|
| Aluminum (III) chloride | $AlCl_3$ | 46.6 |
| Aluminum (III) nitrate | $Al(NO_3)_3$ | 81.8 |
| Aluminum (III) sulfate | $Al_2(SO_4)_3$ | 40.4 |
| Barium acetate | $Ba(OAc)_2 \cdot 3H_2O$ | 75 |
| Barium chloride | $BaCl_2 \cdot 2H_2O$ | 38.1 |
| Barium formate | $Ba(O_2CH)_2$ | 31.9 |
| Barium hydroxide | $Ba(OH)_2$ | 5.6 |
| Barium nitrate | $Ba(NO_3)_2$ | 11.5 |
| Barium sulfate | $Ba(SO_4)$ | insoluble |
| Cadmium chloride | $CdCl_2 \cdot H_2O$ | 135 |
| Cadmium nitrate | $Cd(NO_3)_2$ | 167 |
| Cadmium sulfate | $CdSO_4$ | 77 |
| Calcium acetate | $Ca(OAc)_2 \cdot 2H_2O$ | 33.8 |
| Calcium chloride | $CaCl_2 \cdot 6H_2O$ | 100 |
| Calcium nitrate | $Ca(NO_3)_2 \cdot 4H_2O$ | 152 |
| Calcium sulfate | $CaSO_4 \cdot 2H_2O$ | 0.26 |
| Cerium(III) sulfate | $Ce_2(SO_4)_3 \cdot 9H_2O$ | 7.3 |
| Chromium (III) nitrate | $Cr(NO_3)_3$ | 152.35 |
| Cobalt (II) chloride | $CoCl_2$ | 59.7 |
| Cobalt (II) nitrate | $Co(NO_3)_2$ | 111 |
| Cobalt (II) sulfate | $CoSO_4$ | 42 |
| Copper (II) chloride | $CuCl_2$ | 77.3 |
| Copper (II) nitrate | $Cu(NO_3)_2$ | 156 |
| Copper (II) sulfate | $CuSO_4 \cdot 5H_2O$ | 37.8 |
| Iron (II) chloride | $FeCl_2$ | 66.7 |
| Iron (III) chloride | $FeCl_3 \cdot 6H_2O$ | 106.8 |
| Iron (II) nitrate | $Fe(NO_3)_2 \cdot 6H_2O$ | $266^{60}$ |
| Iron (III) nitrate | $Fe(NO_3)_3 \cdot 9H_2O$ | $137.7^{20}$ |
| Iron (II) sulfate | $FeSO_4 \cdot 7H_2O$ | 60 |
| Lanthanum nitrate | $La(NO_3)_3$ | $168^{40}$ |
| Lead (II) acetate | $Pb(OAc)_2$ | 69.8 |
| Lead (II) nitrate | $Pb(NO_3)_2$ | 64.4 |
| Magnesium acetate | $Mg(OAc)_2$ | 68.6 |
| Magnesium chloride | $MgCl_2$ | 55.8 |
| Magnesium nitrate | $Mg(NO_3)_2$ | 73.6 |
| Magnesium sulfate | $MgSO_4$ | 38.9 |
| Nickel chloride | $NiCl_2$ | 70.6 |
| Nickel nitrate | $Ni(NO_3)_2$ | 105 |
| Nickel sulfate | $NiSO_4 \cdot 6H_2O$ | 44 |
| Silver acetate | $AgOAc$ | 1.2 |
| Silver chloride | $AgCl$ | insoluble |
| Silver nitrate | $AgNO_3$ | $216^{20}$ |
| Silver sulfate | $Ag_2SO_4$ | 0.9 |
| Strontium acetate | $Sr(OAc)_2$ | 39.5 |
| Strontium chloride | $SrCl_2$ | 58.7 |
| Strontium nitrate | $Sr(NO_3)_2$ | 88.7 |
| Strontium sulfate | $SrSO_4$ | 0.0138 |
| Zinc chloride | $ZnCl_2$ | 437 |
| Zinc nitrate | $Zn(NO_3)_2$ | 138 |
| Zinc sulfate | $ZnSO_4$ | 66.5 |

many metal-containing compounds are hydrolyzed by water, and many metal-organic compounds react with alcohol solvents at room temperature, as the following equation illustrates:[28, 29]

$$Ti(OR)_4 + 4R'OH \longrightarrow Ti(OR')_4 + 4HOR. \qquad (8.3)$$

This type of reaction, such as ligand replacement by reaction with the solvent, can result in different reactivities and various degrees of oligomerization of the ligands, possibly leading to preferential precipitation or vaporization of one component of the solution. At higher temperatures, the solvents themselves may react, in the absence or presence of the precursor or product, to produce other species, but the reaction may be detrimental if impurities are formed. However, in some cases the reaction is advantageous. For example, alcohols dehydrate according to the following equation, a reaction that is catalyzed by certain metal-oxide surfaces such as $TiO_2$:

$$RCH_2CH_2OH \xrightarrow{[MO_x]} RCH = CH + H_2O. \qquad (8.4)$$

This reaction serves as a method of introducing water at a surface, avoiding the presence of water in the vapor phase. In a similar fashion, other reagents can be added that react at elevated temperatures in the gas phase or on a surface to produce other reagents *in situ*, which, in turn, react with metal-containing precursors to effect conversion to the final product as a method of controlled delivery of that reagent. An example is thiourea as a source of sulfur.

### 8.3.2 Volatility

The selection of metal-containing compounds with suitable volatility is necessary for gas-to-particle conversion and for CVD. The *volatility*, or *vapor pressure*, $p_e$, of a given species increases with increasing temperature and generally increases with decreasing molecular weight. The relationship between vapor pressure and temperature is given by the Clausius–Clapeyron equation,

$$\frac{d \ln p_e}{dT} = \frac{\Delta H_{\text{vap}}}{RT^2}. \qquad (8.5)$$

where $p_e$ is the vapor pressure, $T$ is temperature in Kelvin, $R$ is the gas constant and $\Delta H_{\text{vap}}$ is the enthalpy of vaporization. Using this equation, Table 8.4 shows the temperatures at which various vapor pressures have been determined.[27] Fluorinated metal-organic compounds generally tend to be more volatile than their nonfluorinated counterparts, as is described in recent reviews.[16, 30–32] Lower degrees of oligomerization result in lower

**TABLE 8.4. Vapor pressures of some metal-containing compounds. (s = solid, temperatures in °C)**

| Name | Formula | 1 mm | 10 mm | 40 mm | 100 mm | 400 mm | 760 mm | Melting Point |
|---|---|---|---|---|---|---|---|---|
| Aluminum chloride | $AlCl_3$ | 100.0s | 123.8s | 139.9s | 152.0s | 171.6s | 180.2s | 192.4 |
| Antimony trichloride | $SbCl_3$ | 49.2s | 85.2 | 117.8 | 143.3 | 192.2 | 219.0 | 73.4 |
| Antimony pentachloride | $SbCl_5$ | 22.7 | 61.8 | 91.0 | 114.1 | — | — | 2.8 |
| Arsenic trichloride | $AsCl_3$ | −11.4 | +23.5 | 50.0 | 70.9 | 109.7 | 130.4 | −18 |
| Arsenic trifluoride | $AsF_3$ | s | s | −2.5 | +13.2 | 41.5 | 56.3 | −5.9 |
| Arsenic pentafluoride | $AsF_5$ | −117.9s | −103.1s | −92.4s | −84.3s | −64.0 | −52.8 | −79.8 |
| Bismuth trichloride | $BiCl_3$ | s | 264 | 311 | 343 | 405 | 441 | 230 |
| Cadmium chloride | $CdCl_2$ | s | 656 | 736 | 797 | 908 | 967 | 568 |
| Chromium carbonyl | $Cr(CO)_6$ | 36.0 | 68.3 | 91.2 | 108.0 | 137.2 | 151.0 | — |
| Cobalt nitrosyl tricarbonyl | $Co(CO)_3NO$ | s | s | +11.0 | 29.0 | 62.0 | 80.0 | −11 |
| Cuprous chloride | $CuCl_2$ | 546 | 702 | 838 | 960 | 1249 | 1490 | 422 |
| Ferric chloride | $FeCl_3$ | 194.0s | 235.5s | 256.8s | 272.5s | 298.0s | 319.0 | 304 |
| Ferrous chloride | $FeCl_2$ | — | 700 | 779 | 842 | 961 | 1026 | — |
| Gallium trichloride | $GaCl_3$ | 48.0s | 76.5s | 107.5 | 132.0 | 176.3 | 200.0 | 77.0 |
| Iron pentacarbonyl | $Fe(CO)_5$ | — | +4.6 | 30.3 | 50.3 | 86.1 | 105.0 | −21 |
| Lead chloride | $PbCl_2$ | 547 | 648 | 725 | 784 | 893 | 954 | 501 |
| Magnesium chloride | $MgCl_2$ | 778 | 930 | 1050 | 1142 | 1316 | 1418 | 712 |
| Mercuric chloride | $HgCl_2$ | 136.2s | 180.2s | 212.5s | 237.0s | 275.0s | 304.0 | 277 |
| Molybdenum hexafluoride | $MoF_6$ | −65.5s | −40.8s | −22.1s | −8.0s | +17.2 | 36.0 | 17 |
| Nickel chloride | $NiCl_2$ | 671 s | 759 s | 821 s | 866 s | 945 s | 987 s | 1001 |
| Nickel carbonyl | $Ni(CO)_4$ | s | s | −23.0 | −6.0 | +25.8 | 42.5 | −25 |
| Phosphorus trichloride | $PCl_3$ | −51.6 | −21.3 | +2.3 | 21.0 | 56.9 | 74.2 | −111.8 |
| Phosphorus pentachloride | $PCl_5$ | 55.5s | 83.2s | 102.5s | 117.0s | 147.2s | 162.0s | — |
| Phosphorus hydride (phosphine) | $PH_3$ | s | s | −129.4 | 118.8 | −98.3 | −87.5 | −132.5 |
| Silane | $SiH_4$ | −179.3 | −163.0 | −150.3 | −140.5 | −122.0 | −111.5 | −185 |
| Chlorosilane | $SiH_3Cl$ | −117.8 | −97.8 | −81.8 | −68.5 | −44.5 | −30.4 | — |
| Disilane | $Si_2H_6$ | −114.8 | −91.4 | −72.8 | −57.5 | −29.0 | −14.3 | −132.6 |
| Silver chloride | $AgCl$ | 912 | 1074 | 1200 | 1297 | 1467 | 1564 | 455 |
| Stannic chloride | $SnCl_4$ | −22.7 | +10.0 | 35.2 | 54.7 | 92.1 | 113.0 | −30.2 |
| Stannous chloride | $SnCl_2$ | 316 | 391 | 450 | 493 | 577 | 623 | 246.8 |
| Tantalum pentafluoride | $TaF_5$ | s | s | s | 130.0 | 194.0 | 230.0 | 96.8 |
| Thallium chloride | $TlCl$ | s | 517 | 589 | 645 | 748 | 807 | 430 |
| Titanium tetrachloride | $TiCl_4$ | −13.9 | 21.3 | 48.4 | 71.0 | 112.7 | 136.0 | −30 |
| Tungsten hexafluoride | $WF_6$ | −71.4s | −49.2s | −33.0s | −20.3s | +1.2 | 17.3 | −0.5 |
| Uranium hexafluoride | $UF_6$ | −38.8s | −13.8s | +4.4s | 18.2s | 42.7s | 55.7s | 69.2 |
| Vanadyl trichloride | $VOCl_3$ | −23.2 | +12.2 | 40.0 | 62.5 | 103.5 | 127.2 | — |
| Zinc chloride | $ZnCl_2$ | 428 | 508 | 566 | 610 | 689 | 732 | 365 |
| Zirconium tetrachloride | $ZrCl_4$ | 190 s | 230 s | 259 s | 279 s | 312 s | 331 s | 437 |

molecular weights and therefore higher volatilities, with monomeric metal-containing compounds generally being optimum. It is important not to confuse the empirical formula with the molecular formula. For example, "Al(-i-Bu)$_2$H" is less volatile than Al(-i-Bu)$_3$ because "Al(-i-Bu)$_2$H" is trimeric with a molecular formula [Al(-i-Bu)$_2$H]$_3$.[33] Thermally induced de-oligomerization can occur to form higher volatility species, as in the case of [Al(-i-Bu)$_2$H]$_3$,[33] or thermally induced reactions can occur to form more volatile species, as described previously for $Zn_4O(OAc)_6$.

In the CVD of films and in gas-to-particle formation of powders using conventional precursor delivery methods such as a heated bubbler, the long-term thermal stability of the metal-containing compound at low temperatures is important. Thermal decomposition of the precursor in the delivery system is detrimental to the deposition process and should be avoided.[16] This creates a dichotomy, however, in that a relatively low precursor deposition (surface reaction) temperature is often desirable, but the precursor must have sufficient thermal stability that it can be heated for long periods at elevated temperatures where it exhibits a reasonable vapor pressure so that it can be transported to the system at an acceptable rate. In addition, many metal-containing compounds used for CVD are solids so that, during evaporation, the surface area can change with time, leading to changes in evaporation rates. These problems can be alleviated by using alternative precursor delivery methods, such as aerosol precursor delivery and liquid delivery.[6]

Finally, a number of inorganic materials are volatile. These species may be formed as intermediates during aerosol processing, either intentionally or unintentionally. For example, some metal oxides, such as PbO, $RuO_3$, $RuO_4$, $V_2O_5$, and $MoO_3$, are volatile at elevated temperatures (see Table 8.5),[27] either because they are thermally de-oligomerized or because they can react

**TABLE 8.5. Vapor pressures of some metal oxides. (s = solid, temperatures in °C)**

| Name of Compound | Formula | 1 mm | 10 mm | 40 mm | 100 mm | 400 mm | 760 mm | Melting Point |
|---|---|---|---|---|---|---|---|---|
| Aluminum oxide | $Al_2O_3$ | 2148 | 2385 | 2549 | 2665 | 2874 | 2977 | 2050 |
| Antimony trioxide | $Sb_2O_3$ | 574 s | 666 | 812 | 957 | 1242 | 1425 | 656 |
| Arsenic trioxide | $As_2O_3$ | 212.5s | 259.7s | 299.2s | 332.5 | 412.2 | 457.2 | 312.8 |
| Cadmium oxide | CdO | 1000 s | 1149 s | 1257 s | 1341 s | 1484 s | 1559 s | — |
| Lead oxide | PbO | 943 | 1085 | 1189 | 1265 | 1402 | 1472 | 890 |
| Selenium dioxide | $SeO_2$ | 157.0s | 202.5s | 234.1s | 258.0s | 297.7s | 317.0s | 340 |
| Silicon dioxide | $SiO_2$ | s | 1732 | 1867 | 1969 | 2141 | 2227 | 1710 |
| Strontium oxide | SrO | 2068 s | 2262 s | 2410 s | — | — | — | 2430 |
| Molybdenum trioxide | $MoO_3$ | 734 s | 814 | 892 | 955 | 1082 | 1151 | 795 |
| Rhenium heptoxide | $Re_2O_7$ | 213 s | 248 s | 272 s | 289 s | 336 | 362 | 296 |

— = no data available

with an ambient species (such as $O_2$) to form volatile species. An example of this reactivity to produce high-volatility species is that of ruthenium dioxide, which reacts with oxygen to form more volatile oxides such as $RuO_3$ and $RuO_4$ according to the following equations:

$$RuO_2 + \tfrac{1}{2}O_2 \longrightarrow RuO_3, \tag{8.6}$$

$$RuO_3 + \tfrac{1}{2}O_2 \longrightarrow RuO_4. \tag{8.7}$$

For gas-to-particle conversion, evaporation/condensation processes can be beneficial, resulting in the formation of extremely small particles (on the order of nanometers). However, these processes can also be detrimental if small particles are not desired, such as when they form as intermediates and evaporate unexpectedly. In addition, during the formation of a more complex material, evaporation can lead to the selective loss of one component. This has been demonstrated in the formation of perovskite materials such as $PbTiO_3$[34] and Pb-Bi-Sr-Ca-Cu-O,[35] wherein lead-deficient phases can be formed, and in the formation of pyrochlore materials such as $Bi_2Ru_2O_7$, in which Ru-deficient phases are formed.[36]

### 8.3.3 Reactivity

It is important to understand the reactivity of the metal-containing precursor used in a particular aerosol process. The rate of reaction and the reaction chemistry are crucial to the purity, composition, and value of the product of the process; therefore, it is desirable to have an understanding of the reaction pathway, reaction rate as a function of temperature, role of other potential reagents (solvents, carrier gases) in the system, relative reactivities of the precursors if two or more are used, reactivities of the precursors towards each other, and reactivity of a precursor on different surfaces during film deposition, to name but a few items. Other aspects of chemical reactivity, such as sintering and crystallization behavior, are also important during aerosol processing and are discussed in Section 8.10.

In considering which metal-containing compounds to choose when preparing a particular material, one generally examines precursor characteristics, such as solubility, volatility, and reactivity, as well as commercial availability and cost. From a manufacturing viewpoint, it is also important to consider the nature of the by-products and other effluents evolved during the process, the potential for recycling materials, and the efficiency of production. For example, in the formation of $TiO_2$ pigment by aerosol routes, a large part of the manufacturing process is devoted to synthesis of the metal-containing precursor, $TiCl_4$, and to recycling of the $Cl_2$ evolved during the aerosol process for reaction with the $TiO_2$ ore to form $TiCl_4$.

## 8.4 THERMALLY INDUCED REACTIONS OF METAL-CONTAINING COMPOUNDS

The goal of this section is to review the reaction pathways that can occur during aerosol processing, in order to gain an insight into these processes, which may then lead to a more rational choice of metal-containing precursors. An overview of precursor characteristics as a function of the nature of the ligand (e.g., halide, alkyl, etc.) has been presented elsewhere,[37] primarily with respect to their use as precursors for CVD, and details of the reactivity of individual precursors can be found in standard texts and other reference works. Here, we survey the reaction pathways according to the nature of the source of energy and then the pathway itself. These reactions are likely to occur primarily in the gas phase or in the solid state, either homogeneously or heterogeneously (i.e., on surfaces). In many laboratory-scale studies, commercial availability has been the major criterion for choosing precursors, and little attention has been paid to precursor reactivity.

We define *thermally induced reactions* as reactions that are initiated by thermal energy, such as heating of a solid, liquid, or gaseous metal-containing precursor in a gas stream. Direct combustion and flame processes are not included in this classification, due to the chemical difference of reactions in a flame. Although chemistry similar to that of thermally induced reactions can occur in flames as a result of the high temperatures, there is often a distinct effect of the presence of the oxidant and reductant, as well as of species derived from them. Thermally induced reactions have been used extensively in aerosol processing of materials; in this section, we specifically describe the thermal reactivity of metal-containing compounds.

### 8.4.1 Thermal Decomposition of Inorganic Compounds

Inorganic compounds such as metal nitrates, carbonates, sulfates, and halides have been used for the formation of metal-containing materials by aerosol processes mainly because they are generally inexpensive and commercially available. In most cases, metal nitrates, metal carbonates, and metal sulfates have been employed for the formation of metal-oxide-based materials by spray pyrolysis, while metal halides have been utilized in a wide variety of aerosol processes for the formation of a wider variety of materials, due to the ability of metal halides to react with co-reagents and eliminate hydrogen halides, thereby forming metal oxides, sulfides, nitrides, borides, and other materials. Metal nitrates, carbonates, and sulfates are generally nonvolatile, so they have been used mainly in aerosol processes, such as spray pyrolysis, that do not involve gas-phase reactions, although intermediates produced during the reactions, (e.g., certain metal oxides) may be volatile. These reagents have been used mainly as precursors to metal-oxide powders formed by spray pyrolysis. This method also takes advantage of the relatively high solubility of these metal complexes in water.

In contrast, the vapor pressures of metal halides range from extremely high ($WF_6$) to extremely low ($YCl_3$). As a result, metal halides have been used in all the different aerosol routes to materials that require a metal-containing precursor, including spray pyrolysis, gas-to-particle conversion, and film formation.

The thermal decomposition behavior of metal nitrates, carbonates, sulfates, and halides is described next in some detail, because these inorganic reagents have been employed extensively in aerosol routes to materials. The discussion is also aimed at illustrating trends that might be expected as a function of the substituent (nitrate, carbonate, sulfate, or halide) and the nature of the metal center (noble (Ag), electropositive (Ca) or intermediate (Ni)).

*8.4.1.1* ***Metal Nitrates*** Metal nitrates have been used extensively to form metal oxides by spray pyrolysis, although the reaction mechanisms have not been studied in detail during aerosol-processing experiments. However, a great deal is known about the thermal decomposition of metal nitrates in the solid state. In general, metal nitrates decompose to form the corresponding metal oxides, together with $NO_2$ or NO and $O_2$, in various stoichiometries:

$$M(NO_3)_2 \longrightarrow MO + xNO_2 + yNO + zO_2 \qquad (8.8)$$

$$MO \longrightarrow M + \tfrac{1}{2}\,O_2 \qquad (8.9)$$

The noble metals, such as Ag and Pd, tend to react further and lose $O_2$ at higher temperatures, forming the corresponding metals.

We illustrate the types of reactions expected for metal nitrates by choosing a highly electropositive metal nitrate, $Ca(NO_3)_2$, a less electropositive metal nitrate, $Ni(NO_3)_2$, and a noble metal, $Ag(NO_3)$. Calcium nitrate thermally decomposes to CaO over the temperature range 350 to 550°C. The thermal decomposition reaction has been studied in the solid state by means of a variety of techniques, including gas chromatography, gas pressure measurements, IR spectroscopy, thermogravimetric and differential thermal analysis, and X-ray diffraction. Based on the interpretation of these data and thermodynamic calculations, the following reaction scheme was proposed:[38]

$$Ca(NO_3)_2 \longrightarrow Ca(NO_2)_2 + O_2, \qquad (8.10)$$

$$Ca(NO_3)_2 \longrightarrow CaO + 2NO_2 + \tfrac{1}{2}O_2, \qquad (8.11)$$

$$Ca(NO_2)_2 \longrightarrow CaO + NO_2 + NO, \qquad (8.12)$$

$$Ca(NO_2)_2 \longrightarrow CaO + 2NO + \tfrac{1}{2}O_2, \qquad (8.13)$$

$$Ca(NO_3)_2 \longrightarrow CaO + 2NO + \tfrac{3}{2}O_2, \qquad (8.14)$$

$$Ca(NO_3)_2 \longrightarrow CaO + NO_2 + NO + O_2, \qquad (8.15)$$

$$Ca(NO_2)_2 + 2NO_2 \rightleftarrows Ca(NO_3)_2 + 2NO, \qquad (8.16)$$

$$Ca(NO_2)_2 + 2NO \rightleftarrows Ca(NO_3)_2 + N_2. \qquad (8.17)$$

Although some of these reactions may be disfavored under aerosol synthesis conditions, this example serves to illustrate the potential complexity of the systems. Calcium nitrate has been used as a source of calcium in calcium-containing metal-oxide materials, such as $CaTiO_3$. Nickel nitrate, $Ni(NO_3)_2$, thermally decomposes in a single process to form $NO_2$ and $O_2$ exclusively, with an activation energy of approximately 21 Kcal/mol:[39]

$$Ni(NO_3)_2 \longrightarrow NiO + 2NO_2 + 0.5O_2. \tag{8.18}$$

It is interesting to compare the preceding results with an analogous study of the thermal decomposition of the corresponding hydrate, $Ni(NO_3)_2 \cdot 2H_2O$, which occurs in a two-step process. In the first step, a mixture of $Ni(NO_3)_2$ and turbostratic $Ni(OH)_2$ in a molar ratio of 1:1 was isolated, and in the second step, dehydration and loss of $NO_2$ and $O_2$ resulted in the formation of NiO.[40] These observations illustrate the potential importance of the solvent in a thermal decomposition reaction: The solvent, such as water, may play a significant role in the composition and phase evolution of the system. Indeed, spray pyrolysis of aqueous nickel nitrate solution under conditions of fast heating to 450°C produced powders consisting of a mixture of tiny crystallites of NiO, hydrated nitrate, and other noncrystalline phases. After heating above 480°C, the powder comprised only crystalline NiO.[41]

The thermal decomposition of liquid $Ag(NO_3)$ has been studied in vacuo over the temperature range 275 to 575°C using evolved gas analysis–Fourier transform IR spectroscopy and mass loss measurements. The reaction followed second-order kinetics and produced $NO_2$ as the only nitrogen-containing product. Thermal decomposition occurred in a single step yielding $Ag_2O$, with no evidence of the formation of an intermediate, as reported previously. Further heating above 550°C resulted in a reduction of the oxide to Ag metal. A kinetic analysis of the data was consistent with catalysis of the reaction by the products formed.[42] Silver nitrate has been used as a source of silver in the formation of silver and silver–palladium alloy particles by spray pyrolysis.

One further example warrants mention because it illustrates the diversity in chemical reactivity of the system: mixed hydroxynitrates of Zn and Cd. The thermal decomposition of $Cd(OH)(NO_3) \cdot H_2O$ shows three major stages: At 80 to 130°C, one equivalent of $H_2O$ is removed, leaving $Cd(OH)(NO_3)$ (with $\Delta H$ of dehydration = ~ 16 Kcal/mol); at 170°C, further $H_2O$ is lost (through condensation between hydroxyl groups), followed by NO at 200°C; finally, at 270°C $NO_2$ and $O_2$ are lost, and at 390°C only CdO remains. There was evidence of the transient formation of $Cd(NO_3)_2$.[43] In contrast, the thermal decomposition of the brucite-structured Zn compound showed an initial loss of $H_2O$ and $HNO_3$ and no formation of $Zn(NO_3)_2$.

These examples illustrate the variety of reactions and products that can be formed. Since oxides of nitrogen have a significant environmental impact, any industrial process that involves the use of metal-nitrate species as precursors has to be regulated to control emissions.

In common with other metal-containing precursors, many factors can add complexity to these systems. The influence of the solvent, the presence of another metal complex—especially if it has different substituents—and the presence of a mixture of substituents on the same metal center, as seen in some of the examples presented, can have a dramatic effect on the chemical reactions.

A simple factor that should be considered in more complex systems is the relative thermal reactivities of the different reagents. The case of unmatched reactivities for metal nitrates used for the formation of mixed metal-oxide powders by spray pyrolysis can be illustrated by a comparison between the thermal decomposition temperatures of strontium and iron nitrates, as shown by data from a thermogravimetric analysis in Figure 8.13. Iron nitrate, $Fe(NO_3)_3$, thermally decomposes at approximately 400°C lower than $Sr(NO_3)_2$. As a result, phase segregation might be expected in the formation of strontium–iron oxides such as $SrFe_{12}O_{19}$ during spray pyrolysis, as was in fact recently demonstrated.[44]

It is interesting to note that the thermal decomposition temperature of $Sr(NO_3)_2$ is lower when the compound is in the presence of $Fe(NO_3)_3$. For comparison purposes, the thermal decomposition temperatures of a number of metal nitrates are presented in Table 8.6.[27] The thermal decomposition temperature is likely to be higher in spray pyrolysis as a result of the high heating rate, compared to methods with lower heating rates (e.g., in a thermogravimetric analysis pan).

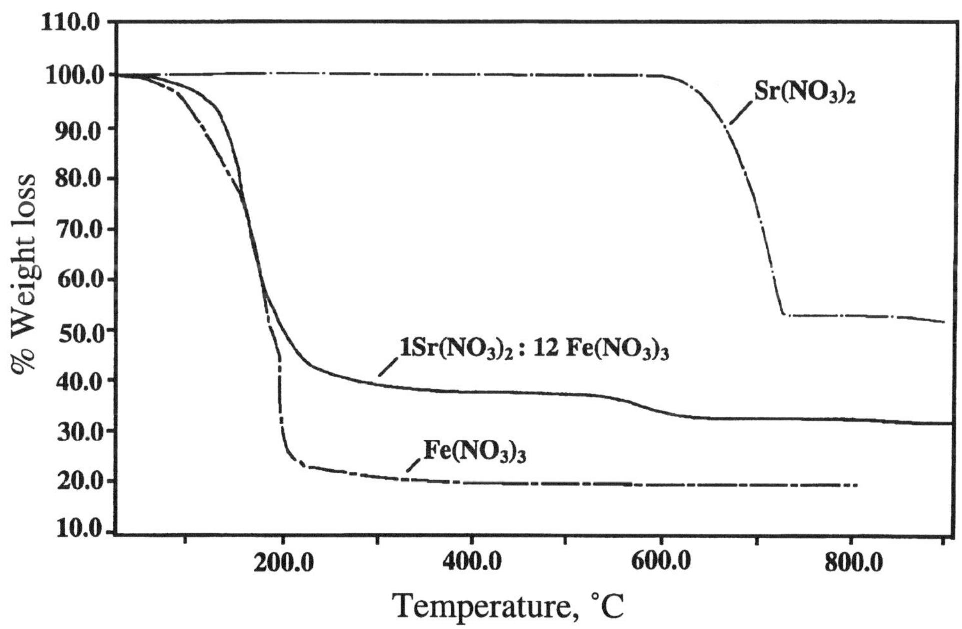


**Fig. 8.13.** Data from thermogravimetric analysis of $Fe(NO_3)_3$, $Sr(NO_3)_2$, and a 12:1 mixture of $Fe(NO_3)_3$ and $Sr(NO_3)_2$. (Adapted from Senzaki et al.)[44]

**TABLE 8.6. Thermal decomposition temperature of a number of metal nitrates[27]**

| Metal Nitrate | Melting Point, °C | Thermal Decomposition Temperature, °C |
|---|---|---|
| $Al(NO_3)_3$ | 73 | 150 |
| $Ba(NO_3)_2$ | 592 | N/A |
| $Cd(NO_3)_2$ | 350 | N/A |
| $Ca(NO_3)_2$ | 561 | N/A |
| $Cu(NO_3)_2 \cdot 2H_2O$ | 114.5 | 170 |
| $Fe(NO_3)_3 \cdot 9H_2O$ | 47.2 | 125 |
| $Pb(NO_3)_2$ | 470 | 470 |
| $Mg(NO_3)_2$ | 95 | 129 |
| $AgNO_3$ | 210 | 440 |
| $Sr(NO_3)_2$ | 570 | 1100 |

N/A = Not available

*8.4.1.2 **Metal Carbonates*** Metal carbonates[45] have been used as precursors for the formation of metal oxides by spray pyrolysis and are generally the intermediates formed when metal carboxylates, which have also been used as precursors in aerosol routes to materials, are thermally decomposed to form metal oxides. Metal carbonates generally decompose thermally via the elimination of $CO_2$ according to the equation.[26]

$$M(CO_3) \longrightarrow MO + CO_2. \tag{8.19}$$

However, in some cases, a number of steps are involved that can influence composition and phase behavior in the system. To illustrate these points and to provide a comparison with the discussion on the thermal decomposition of metal nitrates in Section 8.4.1.1, we briefly describe some analogous examples of the thermal decomposition of metal carbonates.

Silver carbonate, $Ag_2(CO_3)$, decomposes thermally in three stages during heating at a constant rate in a number of different atmospheres, including $CO_2$, He, and $O_2$.[46-49] In the first step, the compound undergoes a reversible crystallographic phase change from a low-temperature $\beta$-form to a high-temperature $\alpha$-form between 100 and 200°C. Above 200°C, $CO_2$ gas evolves, while at 400°C, $CO_2$ and $O_2$ simultaneously evolve. Finally, above 400°C, reduction to Ag is observed. These reactions are summarized in the following equations:

$$\beta\text{-}Ag_2(CO_3) \longrightarrow \alpha\text{-}Ag_2(CO_3), \tag{8.20}$$

$$\alpha\text{-}Ag_2(CO_3) \longrightarrow Ag_2O + CO_2, \tag{8.21}$$

$$Ag_2O \longrightarrow Ag + 0.5O_2. \tag{8.22}$$

The thermal decomposition of $Ag_2(CO_3)$ was accelerated by the presence of water, as well as metal dopants such as Y and Gd. Analysis of the kinetic data revealed a mechanism consistent with decomposition via an interface mechanism.[50]

The thermal decomposition of carbonates of the more electropositive elements has been studied in some detail, especially the alkaline earth elements Mg, Ca, Sr, and Ba. The thermal decomposition of each of these species has been examined separately as well as in mixtures. The individual carbonates decompose thermally, according to the stoichiometry of Equation 8.19, to liberate $CO_2$ and the corresponding metal oxide. Initial reports of the thermal decomposition of binary mixtures of these metal carbonates provided evidence of the existence of two separate steps, which were interpreted, for example, in the case of a 1:1 $SrCO_3$:$BaCO_3$ solid solution, as the result of phase segregation between the metal carbonates, followed by individual thermal decomposition of $SrCO_3$ and then $BaCO_3$.[51] This finding is a little unexpected, because the thermal decomposition temperatures of $SrCO_3$ and $BaCO_3$ are quite similar. (See Table 8.7.)[27] However, in a more recent study involving constant-rate thermal analysis (CRTA), Criado reports that the thermal decomposition of solid solutions of $Sr_xBa_{1-x}(CO_3)$ results in a single-step loss of $CO_2$, to form a solid solution of the corresponding $Sr_xBa_{1-x}O$.[52–54] These data are supported by the results and interpretation of the thermal decomposition of the ternary system (Ca, Sr, Ba)($CO_3$), which show that the mixture initially thermally decarbonylates to form an amorphous metal oxide that, upon further heat treatment, yields the crystalline metal oxide (Ca, Sr, Ba)O as a solid solution. Hydration experiments on the X-ray diffraction amorphous intermediate were interpreted as being consistent with a microcrystalline or structurally disordered solid solution.[55] It is noteworthy that the thermally induced decarbonylation is reversible, as shown by reaction between the metal-oxide products MO with gaseous $CO_2$

**TABLE 8.7. Thermal decomposition temperature of some metal carbonates**

| Metal Carbonate | Thermal Decomposition Temperature, °C |
|---|---|
| $BaCO_3$ | 1360 |
| $CdCO_3$ | 500 |
| $CaCO_3$ | 900 |
| $PbCO_3$ | 340 |
| $MgCO_3$ | 402 |
| $Ag_2CO_3$ | 218 |
| $SrCO_3$ | 1172 |
| $ZnCO_3$ | 300 |

to re-form metal carbonates. This phenomenon is important in aerosol synthesis because, in many particle collection systems, the gaseous by-products pass over the collected powder.

Dolomite, $CaMg(CO_3)_2$, is reported to decompose in a two-stage process corresponding to the formation of MgO, followed by the decomposition of $CaCO_3$.[56] The hypothesis is reasonable for this mixture, compared to the controversy surrounding thermal decomposition data for the mixtures of other metal carbonates of Group 2 as a result of the substantially lower thermal decomposition temperature of $MgCO_3$. (See Table 8.7.)[27] These results illustrate the need to consider the relative reactivities of the metal-containing precursors when attempting to form materials with controlled composition and homogeneity.

The decomposition of $PbCO_3$ was investigated in the solid state and serves to illustrate the chemical diversity in an apparently chemically simple system. The decomposition proceeds in three stages:

$$6PbCO_3 \longrightarrow 3(PbO \cdot PbCO_3) + 3CO_2, \tag{8.23}$$

$$3(PbO \cdot PbCO_3) \longrightarrow 2(2PbO \cdot PbCO_3) + CO_2, \tag{8.24}$$

$$2(2PbO \cdot PbCO_3) \longrightarrow 6PbO + 2CO_2. \tag{8.25}$$

The results were explained in terms of kinetic rather than thermodynamic control over the reactions.[57] The formation of so-called basic salts—i.e., oxo species such as $(2PbO \cdot PbCO_3)$ or $Pb_3O(CO_3)_2$—is common to many inorganic and metal-organic systems and is analogous to the formation of $Zn_4O(OAc)_6$ described earlier. Oxidation of the metal center in the presence of an oxygen ambient is also likely to occur in these systems, where the metal has an accessible and stable higher oxidation state. This is the case for the thermal decomposition of $CoCO_3$, which, in the presence of an oxygen atmosphere, follows the overall reaction:[58]

$$3\ CoCO_3 + \tfrac{1}{2}\ O_2 \rightleftarrows Co_3O_4 + 3\ CO_2. \tag{8.26}$$

*8.4.1.3 **Metal Sulfates*** Metal sulfates also have been used in aerosol synthesis (primarily spray pyrolysis) of materials and generally decompose according to the stoichiometry of the equation

$$M(SO_4) \longrightarrow MO + SO_3. \tag{8.27}$$

The $SO_3$ produced often undergoes further reactions at high temperatures to produce $SO_2$ and $O_2$ according to the equation

$$SO_3 \longrightarrow SO_2 + 0.5O_2. \tag{8.28}$$

For example, aluminum sulfate, $Al_2(SO_4)_3$, which has been used for the formation of alumina powders by spray pyrolysis,[59] thermally decomposes via the formation of an oxosulfate intermediate according to the equation

$$Al_2(SO_4)_3 \longrightarrow Al_2O(SO_4)_2 + SO_3. \tag{8.29}$$

This reaction is followed by further thermal decomposition to form amorphous alumina and crystallization into various phases of alumina according to the equation

$$Al_2O(SO_4)_2 \longrightarrow Al_2O_3 + 2SO_3. \tag{8.30}$$

These reactions have been studied under a variety of conditions. The rate of conversion of $SO_3$ to $SO_2$ and $O_2$ is slow, so that $SO_3$ is the major S-containing product in a flow reactor. Even in the presence of a reducing atmosphere such as 95%Ar/5%$H_2$, the formation of the intermediate $Al_2O(SO_4)_2$ was observed and the reaction mechanism proposed to be unchanged. The thermal decomposition of mixtures of metal sulfates containing aluminum has also been studied.[60] For example, mixtures of $MgSO_4$ and $Al_2(SO_4)_3$ were freeze-dried from equimolar mixed aqueous solutions of the sulfates and were calcined at 900 to 1400°C for 1 h. The freeze-dried sulfate was amorphous with composition $MgAl_2(SO_4)_4 \cdot (8\text{-}9)H_2O$, which corresponded to partially dehydrated $MgAl_2(SO_4)_4 \cdot 22H_2O$. The sulfate completely decomposed to spinel above 1000°C. Powders obtained from the calcination of sulfate at 1100 to 1300°C were nearly perfect spinel crystals.[60]

The thermal decomposition temperatures of $Fe_2(SO_4)_3$ and $MSO_4$ (M = Cu, Ni, Zn, Co, Pb) were investigated by thermogravimetric analysis and differential thermal analysis. The mechanism of decomposition of these sulfates has been described, making use of additional information obtained from isothermal studies and X-ray diffraction measurements.[61] The activation energies of the reactions were calculated, and it was shown that they increased, with almost the same increments, in the order Zn < Fe < Co < Ni < Cu.[61] These data provide guidelines which aid in the selection of reagents with reactivities that are matched to form single-phase, rather than phase-segregated, materials.

As with the metal carbonates and metal nitrates, the noble-metal salts are generally reduced at high temperatures to form the corresponding metals, and the more electropositive metal sulfates form oxosulfate intermediates before decomposing to form the metal oxide. Redox chemistry is also important where accessible oxidation states are available, and this has been shown for the thermal decomposition of Ce(IV) and Co(II) sulfates.[62, 63] The highly oxidative nature of $SO_3$ can play an important role in oxidation reactions.

For industrial processes based on the use of metal-sulfate materials, a knowledge of the thermal reaction chemistry to determine the relative proportions of the sulfur-containing products in the effluent stream is important because of the environmental impact of these gases. The conversion of

**TABLE 8.8. Thermal decomposition temperature of a number of metal sulfates**

| Metal Sulfate | Melting Point, °C | Thermal Decomposition Temperature, °C |
|---|---|---|
| $Al_2(SO_4)_3$ | 770 | 770 |
| $BaSO_4$ | 1580 | N/A |
| $CdSO_4$ | 1,000 | N/A |
| $CaSO_4$ | 1,400 | N/A |
| $CoSO_4$ | 735 | 735 |
| $CuSO_4$ | 200 | 650 |
| $Fe_2(SO_4)_3$ | 480 | 480 |
| $PbSO_4$ | 1,090 | N/A |
| $MgSO_4$ | 1124 | 1124 |
| $NiSO_4$ | 848 | 848 |
| $Ag_2SO_4$ | 660 | 1085 |
| $SrSO_4$ | 1,600 | N/A |
| $ZnSO_4$ | 1,200 | N/A |

N/A = Not available

sulfur oxides to sulfuric acid under the conditions of aerosol processes in the presence of water is also a concern and can be detrimental to the properties of the final material. Table 8.8 provides some information on the thermal decomposition temperature for a number of metal sulfates.[27] Finally, it is worth noting that under some conditions metal sulfates can form metal sulfides rather than metal oxides.[64]

*8.4.1.4 Metal Halides* The use of metal halides for aerosol routes to materials is chemically distinct from that of the other inorganic precursors discussed thus far, for two reasons: First, many examples of volatile metal halides exist, and a variety of aerosol synthesis techniques are available for these compounds; second, unless metal-halide products such as $BaF_2$ films are required, the metal-halide precursor has to completely eliminate the halide ligand, either by thermal reduction to form a metal or by reaction with another, usually protic reagent such as $H_2O$, $NH_3$, $H_2S$, $PH_3$, $CH_4$, $SiH_4$, or $BH_3$ to form the corresponding desired products—oxides, nitrides, sulfides, phosphides, carbides, silicides, or borides, respectively. The gross features of the overall pathway are likely to be similar, regardless of the methods (solid state, gas phase, surface) by which either powders or films are formed. These pathways are illustrated in the following equations, which, it should be noted, show only overall stoichiometries for hypothetical systems, so it is likely that the detailed reaction mechanisms and products could vary, depending on the conditions employed:

$$MX_2 \longrightarrow M + X_2, \tag{8.31}$$

$$MX_2 + H_2 \longrightarrow M + 2HX, \tag{8.32}$$

$$MX_2 + H_2O \longrightarrow M(OH)_2 + 2HX, \tag{8.33}$$

$$M(OH)_2 \longrightarrow MO + H_2O, \tag{8.34}$$

$$MX_3 + NH_3 \longrightarrow MN + 3HX, \tag{8.35}$$

$$MX_2 + H_2S \longrightarrow MS + 2HX, \tag{8.36}$$

$$MX_3 + PH_3 \longrightarrow MP + 3HX, \tag{8.37}$$

$$MX_4 + CH_4 \longrightarrow MC + 4HX, \tag{8.38}$$

$$MX_4 + SiH_4 \longrightarrow MSi + 4HX, \tag{8.39}$$

$$MX_3 + BH_3 \longrightarrow MB + 3HX. \tag{8.40}$$

In some actual cases, the overall reaction has been determined as illustrated by the following equation:[65-67]

$$6\,TiCl_4 + 8\,NH_3 \longrightarrow 6\,TiN + 24\,HCl + N_2. \tag{8.41}$$

One of the potential problems associated with the use of a metal halide is the possibility of reaction of the HX liberated with the product and the possibility of halide contamination in the final material. Halide contamination of electronic materials is a particular problem, because halide ions often act as charge carriers in these systems. There is also the problem that the halide-containing waste stream must be treated.

Other, nonprotic regeants have been used in reactions involving metal halides to form metal-containing materials. The reaction with $N_2$ at high temperatures is

$$2\,TiCl_4 + 4\,H_2 + N2 \longrightarrow 2\,TiN + 8\,HCl, \tag{8.42}$$

which should be compared with Equation (8.41). In the deposition of films by CVD, sacrificial reactions can also occur by reaction of the metal halide with the substrate surface. The classic example is the deposition of tungsten through the reduction of $WF_6$ by Si according to the stoichiometry of the equation[68]

$$2WF_6 + 3Si \longrightarrow 2W + 3SiF_4. \tag{8.43}$$

This reaction is the basis of the selective deposition of W on a Si surface in the presence of other, unreactive surfaces such as $SiO_2$.

The reaction of volatile metal halides with oxygen has a significant technological importance, because this reaction is the basis of the large-scale production of pigments (especially $TiO_2$), the formation of metal-oxide thin films (particularly $SiO_2$), and the formation of optical fibers (especially oxides of Si, Ge, P, and B).[69] Some detailed analyses of the kinetics and mechanisms of these oxidation reactions are available in the literature. The overall reaction stoichiometries of the most important materials are summarized in the following equations, using chlorides as a representative example of the halides in general:

$$TiCl_4 + O_2 \longrightarrow TiO_2 + 2Cl_2, \tag{8.44}$$

$$SiCl_4 + O_2 \longrightarrow SiO_2 + 2Cl_2, \tag{8.45}$$

$$GeCl_4 + O_2 \longrightarrow GeO_2 + 2Cl_2, \tag{8.46}$$

$$BCl_3 + 3/2O_2 \longrightarrow B_2O_3 + 3Cl_2, \tag{8.47}$$

$$POCl_3 + 3/2O_2 \longrightarrow P_2O_5 + 3Cl_2. \tag{8.48}$$

Equation (8.44) has been examined in most detail. A study of the oxidation of $TiCl_4$ with $O_2$ by Fourier transform infrared (FTIR) spectroscopy in the temperature range 700 to 1000°C showed that the reaction rate was first order in $TiCl_4$ concentration and nearly zero order for $O_2$ concentration, up to a ten-fold excess of $O_2$.[70] At higher $O_2$ concentrations, the reaction rate was half order in $O_2$ concentration. The activation energies and pre-exponential factors of this and other reactions are given in Table 8.9,[69–71] and the form of the rate expression for Equation (8.44) is

$$-d[TiCl_4]/dt = \left(k + k'[O_2]^{1/2}\right)[TiCl_4], \tag{8.49}$$

where $k$ and $k'$ are constants.

**TABLE 8.9. Activation energies**

| $MX_n$ | $E_a$/kcal/mol | Temp. Range/°C | Bond dissociation energy (M-Cl) kcal/mol | Ref. |
|---|---|---|---|---|
| $TiCl_4$ | 22 | 700–1000 | 118 | [70] |
| $SiCl_4$ | 98 | 1050–1250 | 96 | [69] |
| $SiCl_4$ | 96 | 1100–1300 | 96 | [71] |
| $SiBr_4$ | 67 | 950–1050 | 88 | [69] |
| $GeCl_4$ | 64 | 1050–1250 | N/A | [69] |
| $POCl_3$ | 62 | 850–1000 | 69 | [69] |
| $BCl_3$ | N/A | 1050–1200 | 128 | [69] |

N/A = Not Available

These observations in the last paragraph were interpreted as follows.[70] The reaction of $TiCl_4$ involves a radical mechanism that is initiated by an active site, M, according to the equation

$$TiCl_4 + M \longrightarrow Ti^{III}Cl_3 + Cl^{\bullet} + M. \quad (8.50)$$

This homolytic Ti–Cl bond cleavage is formally a reduction process that produces a chlorine radical ($Cl^{\bullet}$) and reduced (neutral) $Ti^{III}Cl_3$. The $Cl^{\bullet}$ radical can then abstract another chlorine radical from $TiCl_4$, according to the reaction

$$TiCl_4 + Cl^{\bullet} \longrightarrow Ti^{III}Cl_3 + Cl_2, \quad (8.51)$$

a process that takes place at lower energies than does Eq. (8.50). In addition, $Ti^{III}Cl_3$ species can undergo further reduction or chlorine radical abstraction reactions according to the equations

$$Ti^{III}Cl_3 + M \longrightarrow Ti^{II}Cl_2 + Cl^{\bullet} + M \quad (8.52)$$

and

$$Ti^{III}Cl_3 + Cl^{\bullet} \longrightarrow Ti^{II}Cl_2 + Cl_2, \quad (8.53)$$

respectively.

The reduced $TiCl_{4-x}$ can undergo a number of reactions, including reduction–oxidation (redox) reactions, which are likely to be particularly significant when $TiCl_4$ is reduced, as, for example, in the reaction

$$TiCl_4 + TiCl_x \longrightarrow Ti^{III}Cl_3 + TiCl_{x+1} \quad (8.54)$$

or, more generally,

$$TiCl_y + TiCl_x \longrightarrow TiCl_{y-1} + TiCl_{x+1}. \quad (8.55)$$

The reduced $TiCl_x$ species can then oxidize via reaction with $O_2$ to form oxyhalide species according to the equation

$$TiCl_x + O_2 \longrightarrow TiOCl_{x-1} + {}^{\bullet}OCl. \quad (8.56)$$

The following reactions then occur between oxyhalides, causing coalescence and the buildup of small clusters and, upon more collisions, forming particles that are increasingly oxide rich and halide poor:

$$TiO_kCl_x + TiO_kCl_x \longrightarrow (TiO_k)_2Cl_{x-n} + n/2Cl_2, \quad (8.57)$$

$$(TiO_k)_iCl_x + (TiO_k)_jCl_x \longrightarrow (TiO_k)_{i+j}Cl_{x-n} + n/2Cl_2. \quad (8.58)$$

Species such as $TiOCl_2$ are highly abundant in these reactions, as is demonstrated by Fourier transform infrared (FTIR) measurements.[70] Finally, the halide is completely removed (or it may still exist on the surface of the particles), and titania particles are formed.

The foregoing reactions proceed according to the equations

$$TiCl_4 \longrightarrow R + Cl, \tag{8.59}$$

$$TiCl_4 + Cl \longrightarrow R + Cl^{\bullet}, \tag{8.60}$$

$$R + O_2 \longrightarrow P + Cl^{\bullet}, \tag{8.61}$$

$$P + P \longrightarrow T, \tag{8.62}$$

and can be classified as follows: reaction of $TiCl_4$ to form reduced titanium-halide species, $TiCl_x$, denoted R; oxidation of the reduced titanium-halide species to form titanium-oxyhalide species, denoted P; and collision of the oxyhalides to form titania particles, denoted T.

When rate expressions are derived for the concentrations of P and R using the steady-state approximation, it can be shown that at low $O_2$ concentrations, the reaction is dependent on the $O_2$ concentration, whereas at intermediate $O_2$-to-$TiCl_4$ ratios the reaction rate is independent of the $O_2$ concentration.[70] At higher $O_2$ concentrations (i.e., when $O_2$ is in large excess), the formation of O–Cl species and oxygen radicals ($O^{\bullet}$) according to the equations

$$O_2 + Cl^{\bullet} \longrightarrow {}^{\bullet}OCl + O^{\bullet}, \tag{8.63}$$

$$O_2 + R \longrightarrow RO^{\bullet} + O^{\bullet}, \tag{8.64}$$

becomes important, and this helps explain the half-order dependence of the reaction rate on $O_2$ concentration under these conditions.

In the case of other species, it has been demonstrated that the reactions of $SiCl_4$, $SiBr_4$, $GeCl_4$, and $POCl_3$ with $O_2$ are first order in the halide concentration of the *p*-block element.[69] The activation energies are shown in Table 8.9. Under different conditions, also shown in the table, the reaction of $SiCl_4$ with $O_2$ was first order in $SiCl_4$ concentration and zero order in $O_2$ concentration, up to a twentyfold excess of $O_2$, and first order in $O_2$ concentration at higher concentrations.[69] The reaction of $BCl_3$ with $O_2$ does not fit a first-order rate expression under the conditions studied, and it has been proposed that the reaction is third order in $O_2$ concentration, although no mechanism to interpret this observation has been offered.[69]

A comparison of the activation energies and thermochemical data for the halides allows some speculation about the importance of various steps as potentially rate limiting. For example, if the thermally induced homolytic cleavage of an M–Cl bond [e.g., Equation (8.50)] is rate limiting, then one might expect an activation energy close to a bond dissociation energy. This appears to be the case for the reactions of $SiCl_4$, $POCl_3$, and perhaps $SiBr_4$. However, the significantly lower values of the activation energies measured for the reactions of $TiCl_4$ and $GeCl_4$, compared with their respective bond dissociation energies, suggest that other steps, such as $Cl^{\bullet}$ abstraction, may be rate limiting.

### 8.4.2 Thermal Decomposition of Metal-Organic Compounds

A wide variety of metal-organic compounds have been used as precursors to a number of materials via thermally induced aerosol processes. These compounds include metal derivatives such as carboxylates, alkoxides, β-diketonates, and amides, which are most frequently used. The mechanistic details of the thermal decomposition of metal carboxylates and metal alkoxides have been studied to a limited extent, but the thermal decomposition of metal β-diketonates and amides has received little attention. Metal carboxylates and metal alkoxides can thermally decompose to form metal oxides, although an added reagent such as water often plays a role, particularly in the case of metal alkoxides. Metal β-diketonates and amides are often used as sources of metal-containing materials and frequently require reaction with an added reagent. In general, metal carboxylates have not been used in gas-to-particle conversion processes because they exhibit low volatility and exist as oligomers or infinite polymers in the solid state. In contrast, a number of metal alkoxides, β-diketonates, and amides are volatile, so these species have been used in gas-to-particle conversion and AACVD. In particular, fluorinated metal β-diketonates are frequently used sources of metals for CVD of films.[4, 16]

***8.4.2.1 Metal Carboxylates*** Metal carboxylates[29, 72] are often utilized as sources of metals during the formation of metal oxides by spray pyrolysis.[73] Simple metal organic carboxylates, such as acetates of the electropositive elements, typically decompose thermally in air to form the corresponding metal carbonate as an intermediate, which then requires heating to high temperatures to decarbonylate and produce the metal oxide. This process is particularly well known for the Group 2 elements (especially Ca, Sr, and Ba), and proceeds according to the equation

$$Ba(OAc)_2 \text{ — (air) } \longrightarrow Ba(CO_3) \text{ — } \Delta \longrightarrow BaO + CO_2. \qquad (8.65)$$

The fate of the organic substituents depends upon the conditions of the reactions, and under most common conditions (air or $O_2$ as ambient and water as solvent), oxidation to $CO_2$ and $H_2O$ is probably prevalent.

Metal oxalates have attracted a great deal of attention as a result of the relatively straightforward reactions that result in the elimination of the supporting ligands and produce the corresponding metal oxide. The reaction of calcium oxalate according to the equations

$$Ca(C_2O_4) - \Delta \longrightarrow CaCO_3 + CO, \quad (8.66)$$

$$CaCO_3 - \Delta \longrightarrow CaO + CO_2, \quad (8.67)$$

is an example. It has also been shown that mixed metal oxalates are appropriate sources of mixed metal oxides via thermal decomposition processes.[74–77] Electronic ceramics have been most commonly prepared by thermal decomposition of mixed metal oxalates, including $BaZrO_3$, $PbTiO_3$, and $BaTiO_3$. As with inorganic precursors, the formation of oxo–oxalate species is common. The so-called oxalate process is used industrially to obtain $BaTiO_3$ powders in large quantities with precise stoichiometry.[78] Mixing of $BaCl_2$, $TiCl_4$, water, and oxalic acid yields a precipitate of $BaTiO(C_2O_4)_2 \cdot 4H_2O$. Crystalline *pseudocubic* $BaTiO_3$ is formed via thermal decomposition, and a number of different reaction pathways have been proposed. In one study, the following pathway was suggested:[79]

$$100\text{–}140°\text{C: } BaTiO(C_2O_4)_2 \cdot 4H_2O \longrightarrow BaTiO(C_2O_4)_2 + 4H_2O, \quad (8.68)$$

$$300\text{–}350°\text{C: } BaTiO(C_2O_4)_2 \longrightarrow \tfrac{1}{2}BaTi_2O_5 + \tfrac{1}{2}BaCO_3 + 2CO + \tfrac{3}{2}CO_2, \quad (8.69)$$

$$600\text{–}700°\text{C: } \tfrac{1}{2}BaTi_2O_5 + \tfrac{1}{2}BaCO_3 \longrightarrow BaTiO_3 + \tfrac{1}{2}CO_2. \quad (8.70)$$

According to this study, the bis(oxalate) species does not convert directly to $BaTiO_3$. In fact, the mechanism is similar to the conventional process described earlier, in which a mixed metal oxide of the wrong stoichiometry is formed as an intermediate. As a result, questions arise about the "molecular mixing" with this technique; however, the sizes of the $BaCO_3$ and $BaTi_2O_5$ regions are small due to the low reaction temperature and result in a decrease of the sintering temperatures for $BaTiO_3$. The resulting *pseudocubic* barium titanate phase is converted to the *tetragonal* phase on additional heat treatment.

In other studies, the thermal decomposition of barium titanyl oxalate was proposed to proceed via different intermediates, according to one or other of the following reactions:

$$Ba_2Ti_2(O)_2(C_2O_4)_4 \longrightarrow Ba_2Ti_2(O)_2(C_2O_4)_3(CO_3) + CO, \quad (8.71)$$

$$Ba_2Ti_2(O)_2(C_2O_4)_3(CO_3) \longrightarrow Ba_2Ti_2(O)_5(CO_2)(CO_3) + 2CO_2 + 3CO, \quad (8.72)$$

$$Ba_2Ti_2(O)_5(CO_2)(CO_3) \longrightarrow Ba_2Ti_2(O)_5(CO_3) + CO_2, \quad (8.73)$$

$$Ba_2Ti_2(O)_5(CO_3) \longrightarrow 2BaTiO_3 + CO_2. \quad (8.74)$$

The thermal decomposition of barium titanyl oxalate has been the subject of a large number of studies in the literature, and the phase purity and crystallinity of the $BaTiO_3$ product has been shown to depend subtly on the reaction conditions. The controversy concerning the exact nature of the barium titanyl oxalate intermediate was recently put to rest as a result of a single crystal X-ray diffraction study of $BaTi(O)(C_2O_4)_2 \cdot 5H_2O$ by the reaction between barium nitrate and ammonium titanyl oxalate, $(NH_4)_2Ti(O)(C_2O_4)_2 \cdot H_2O$.[80]

A similar approach that also relies on the formation of carboxylate precursors, the Pechini or citrate process, patented in 1967,[81] involves the thermal decomposition of a 1:1 mixed crystalline Ba · Ti tris(citrate) · water adduct ($BaTi(C_6H_6O_7)_3 \cdot xH_2O$) to give tetragonal $BaTiO_3$ at 800°C.[82] This process has been used to make metal oxide (e.g., $BaTiO_3$ and $\gamma$-$Fe_2O_3$) powders by spray pyrolysis of citrate precursors. The barium titanium citrate is obtained by mixing titanium and barium citrate solutions and adjusting the pH to less than 2.6 in order to precipitate the 1:1 Ba:Ti adduct. At levels greater than pH 3.6, a 2:1 Ba:Ti salt [$Ba_2Ti(C_6H_5O_7)_2(C_6H_6O_7) \cdot 7H_2O$] was obtained. These results were independent of the initial barium-to-titanium ratio (1:1 to 3:2). The thermal decomposition of $BaTi(C_6H_6O_7)_3 \cdot xH_2O$ via the following reactions has been studied:

$$\text{20–200°C: } BaTi(C_6H_6O_7)_3 \cdot 6H_2O \longrightarrow BaTi(C_6H_6O_7)_3 + 6H_2O, \qquad (8.75)$$

$$\text{200–210°C: } BaTi(C_6H_6O_7)_3 \longrightarrow BaTi(C_6H_4O_6)_3 + 3H_2O, \qquad (8.76)$$

$$\text{210–250°C: } BaTi(C_6H_4O_6)_3 \longrightarrow BaTi(C_5H_4O_4)_3 + 3CO_2, \qquad (8.77)$$

$$\text{250–360°C: } BaTi(C_5H_4O_4)_3 \longrightarrow TiO_2 \cdot Ba(C_5H_4O_4) + 2C_5H_4O_3, \qquad (8.78)$$

$$\text{360°C: } C_5H_4O_3 + O_2 \longrightarrow 5CO_2 + 2H_2O, \qquad (8.79)$$

$$\text{360–600°C: } BaTi(C_5H_4O_4)_3 \longrightarrow BaCO_3 + CO_2 + H_2O + TiO_2, \qquad (8.80)$$

$$> \text{500°C: } (BaCO_3)_x \cdot (TiO_2)_y \longrightarrow BaTiO_3 + CO_2. \qquad (8.81)$$

Above 360°C, the presence of small $BaCO_3$ crystallites (150 Å) was observed by X-ray powder diffraction.

The small size of the intermediate barium carbonate and titanium dioxide regions [Equations. (8.80) and (8.81)] resulted in a sintering temperature lower than that of traditional routes (typically, less than 800°C), but the process does not offer the necessary homogeneity until the reaction is complete. This process has been applied to a variety of mixed metal oxides (Mg, Ca, Sr, Ba, Pb, Zn, Cd, La, rare earths, Ga, Ti, Zr, Sn, Nb, Ta, Mo, W, Bi, and Sb) and gave good control over the stoichiometry of the final materials.[82] However, the method has disadvantages, including its high weight loss (low ceramic yield) and agglomeration during calcination. On the other hand, the agglomeration associated with solid-state processing can be avoided by *spray pyrolysis methods*. A variety of other organic reagents, including tartaric acid and catechol, have been used in similar processes, often to form analogous intermediates such as $BaTi(catecholate)_3 \cdot 3H_2O$.[83]

Similar reactions for the preparation of lead-containing mixed metal oxides are more difficult to achieve, due to the volatility of PbO. Very often, Pb is introduced in excess, or the reaction is carried out in an atmosphere of lead oxide.[34]

As with other substituents in the case of metal-organic compounds, the presence of fluorine ligands, which are frequently added in an attempt to improve volatility, can result in the formation of the corresponding metal fluoride rather than the desired metal oxide.

***8.4.2.2 Metal Alkoxides*** The thermal decomposition of metal alkoxides in the solid state has been shown to occur by a number of methods, ranging from intramolecular rearrangement according to the equations:[28, 84]

$$M(OR) \longrightarrow MOH + \text{alkene}, \qquad (8.82)$$

$$2MOH \longrightarrow M_2O + H_2O, \qquad (8.83)$$

to the hydrolysis and condensation of metal alkoxides, which form the basis of hydrolytic sol-gel routes to metal-oxide materials, via the following equations:

$$M{-}OR + H_2O \longrightarrow M{-}OH + ROH \quad \text{(hydrolysis)}, \qquad (8.84)$$

$$M{-}OH + M{-}O{-}R \longrightarrow M{-}O{-}M + ROH \quad \text{(alkoxolation)}, \qquad (8.85)$$

$$M{-}OH + M{-}O{-}H \longrightarrow M{-}O{-}M + H_2O \quad \text{(oxolation)}, \qquad (8.86)$$

$$M{-}OH_2 + M{-}OR \longrightarrow M{-}(OH){-}M + ROH \quad \text{(olation)}. \qquad (8.87)$$

Certain metal alkoxides are volatile, while others are involatile, so metal alkoxides have been used in aerosol routes that either involve or avoid gas-phase reactions. For example, the gas-phase thermal decomposition of $Ti(O{-}i{-}Pr)_4$ yields $TiO_2$, but the reaction is often conducted in the presence of added water vapor to ensure that the organic ligands are completely removed as alcohol, to avoid contaminating the product with carbon. This requires some care in controlling the method of introduction of the reagents, because premature hydrolysis and reaction of $Ti(O{-}i{-}Pr)_4$ with water is likely to result in the formation of extended oxopolymers, which are involatile. Alternatively, water may be intentionally added to a solution containing metal alkoxides to ensure that all the metal alkoxides have been oligomerized, to *avoid* gas-phase reactions. This is the case in the formation of

$BaTiO_3$ by spray pyrolysis of solution of the metal alkoxides. However, if large inorganic oxopolymers are formed, they may become insoluble and precipitate.

Volatile metal alkoxides have also been used as precursors for the CVD of metal-oxide films. Metal alkoxides can react with the surface hydroxyl groups on glass reactor walls according to the equation

$$\text{MOR} + \text{HOSi} = \longrightarrow \text{MOSi} = +\text{HOR} \tag{8.88}$$

*8.4.2.3* ***Metal Diketonates*** Metal $\beta$-diketonates—especially fluorinated derivatives—are among the most volatile metal-organic compounds available, thus, they have been used extensively for CVD.[85] Metal $\beta$-diketonates have been utilized for the deposition of a wide variety of materials, including metals, metal oxides, and metal sulfides. As with metal halides, additional reagents are often required to produce the desired material with reasonable purity. For example, the reduction of metal $\beta$-diketonates with $H_2$ produces metals (M) via the reaction

$$\text{M}(\beta\text{-diketonate})_2 + \text{H}_2 \longrightarrow \text{M} + 2\,\text{H}(\beta\text{-diketonate}), \tag{8.89}$$

reaction with water produces metal oxides (MO) in accordance with the equation

$$\text{M}(\beta\text{-diketonate})_2 + \text{H}_2\text{O} \longrightarrow \text{MO} + 2\,\text{H}(\beta\text{-diketonate}), \tag{8.90}$$

and reaction with $H_2S$ via the equation

$$\text{M}(\beta\text{-diketonate})_2 + \text{H}_2\text{S} \longrightarrow \text{MS} + 2\,\text{H}(\beta\text{-diketonate}) \tag{8.91}$$

gives metal sulfides (MS).

Problems associated with the incorporation of fluoride into metal-oxide films deposited from fluorinated $\beta$-diketonates have been avoided to some extent by the addition of water vapor during deposition. As a result, it is generally possible to deposit high-quality films using metal $\beta$-diketonates as precursors. These precursors, however, have not been used extensively for particle formation due to their relatively high cost. The chemistry of metal $\beta$-diketonates as sources for the CVD of metals has been discussed elsewhere.[4]

*8.4.2.4* ***Metal Amides*** Metal amides[66, 67] have been used primarily as precursors to metal-nitride powders and films. Volatile precursors are generally sought to make nanophase metal-nitride particles by gas-to-particle conversion or metal nitride films by CVD. Metal amides generally decompose thermally to give products that are highly contaminated with carbon. To solve this problem, an additional nitrogen-containing reagent, such as $NH_3$ is generally added to remove the amide ligands more cleanly as an amine. However, significant amounts of hydrogen may be incorporated into these materials.

The reaction proceeds as follows:

$$3\,Zr(NR_2)_4 + 4\,NH_3 \longrightarrow Zr_3N_4 + 4HNR_2 + 4H_2. \qquad (8.92)$$

(Note that the analogous reaction between $NH_3$ and $TiCl_4$, Equation (8.41), results in the reduction of Ti(IV) to Ti(III), whereas Zr(IV) is not reduced.)

*8.4.2.5 Other Derivatives* Other derivatives of metal complexes, such as metal dithiocarbamates, metal thiocarboxylates, and polyoxoanions, have been used in the aerosol synthesis of materials. Although their use is less extensive, descriptions can be found in recent reviews as well as other places in this text (Section 8.9),[1, 4, 16] and the chemistry of these compounds is not described here.

### 8.4.3 Thermal Decomposition of Organometallic Compounds

Organometallic compounds have been used extensively in the gas-phase synthesis of materials, particularly CVD and gas-to-particle conversion, because these compounds are often sufficiently volatile. Organometallic species have been used to form metals, metal carbides, metal nitrides, metal sulfides, and metal oxides, among other products. They also have been used as precursors to metal powders and films, including metal alkyls, aryls, alkene and alkyne adducts, allyls, cyclopentadienes, arenes, and carbonyls. The thermal chemistry of these compounds in the CVD of metal films has been described elsewhere.[4, 86] The formation of metals often requires the addition of a reducing agent such as $H_2$, and the reaction proceeds analogously, to Equation (8.89). Thermal decomposition of metal alkyls to form some metals can occur without the aid of a reducing agent by $\beta$-hydride migration [c.g., CVD of Al from $Al(-i-Bu)_3$], $\alpha$–C–H abstraction, or M–C bond homolysis.[37] In many cases, $\beta$-hydrogens are not present in a molecule, which precludes $\beta$-hydride elimination. Metals that are lower in a group in the periodic table tend to have weaker M–C bonds and so favor bond homolysis mechanisms. The reaction mechanisms that are prevalent in the formation of metal-containing films using organometallic compounds have been described elsewhere,[4] and surfaces processes are discussed in Section 8.9.

## 8.5 PHOTO-INDUCED REACTIONS

In this text, photo-induced reactions are relevant primarily to laser-driven processes and can be divided into two types: those that result in absorption of the photons to induce photochemical reactions and those that result in heating of the molecule or carrier gas molecules to produce photothermal reactions.[5, 87–89] Photothermal reactions have been used most extensively in aerosol synthesis of materials, whereby laser reactors employing $CO_2$ or Yttrium aluminium garnet (YAG) lasers result in powder formation by gas-to-particle conversion. Photothermal reactions have characteristics similar to those of thermally induced reactions in which the transfer of thermal energy has been affected by other means and so will not be discussed further.

Photochemical reactions result from the absorption of photons of a particular energy that coincides with an absorption band of the molecule, creating an electronic excited state (Fig. 8.14). The electronic state of the molecule into which the photonic energy is absorbed is not necessarily the photoreactive state. As the figure shows, the photonic energy can be much larger than the dissociation energy of the molecule. However, the useful photonic energy is only $h\nu_{\text{eff}}$. The remaining energy, $h\nu$—$h\nu_{\text{eff}}$, is generally lost to the system as thermal energy. As a result, the quantum efficiency of the energy transfer into a photoreactive state is often low, and this method is not widely used for aerosol processing of materials. The main advantage of

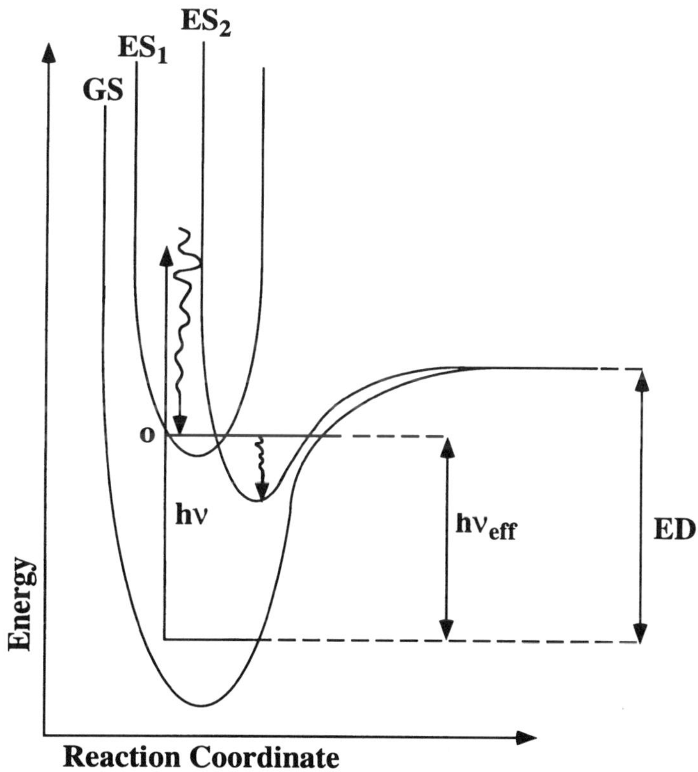


**Fig. 8.14.** Potential-energy diagram for absorption of photons.

photo-induced processes is derived from the spatial resolution of the photons in, for example, laser CVD, which can result in deposition to selective areas. Examples of these processes can be found in various papers.[5, 87, 90–92]

## 8.6 PLASMA-INDUCED REACTIONS

Plasmas can be classified into a number of different types, the most important of which, in the context of this text, are thermal plasmas, coronas, laser-produced plasmas, and glow discharge plasmas.[93] Thermal plasmas (or plasma jets) are formed by the passage of an electric current through an ionized gas medium. Typical parameters for high-temperature plasmas are atmospheric pressure, temperatures of 10,000 to 40,000°C, and discharge current densities of 10 to $10^5$ A/cm$^2$. One of the first examples of the use of a thermal plasma was the reaction between elemental carbon and hydrogen yielding acetylene, a process that was patented by DuPont in 1969 and replaced the production of acetylene from $CaC_2$. The process produced reaction temperatures of around 4200°C, as well as carbon particles, which in this case were detrimental from an aerosol point of view. At these high temperatures, all the feedstock introduced into a plasma can be dissociated to atomic species and ionized in addition. This implies that the dominant powder formation process is gas-to-particle conversion during cooling, which has been used to considerable advantage in the synthesis of inorganic materials, where the high degree of mixing of the elements involved at the atomic level is likely to result in the formation of homogeneous products with a complex composition, such as ceramic oxides.[94]

The precursors for thermal plasma-induced reactions can be delivered in numerous forms. For powder formation, the best method of delivering the species required in the final material to the plasma is by gas feed, but this requires that volatile reagents containing the desired elements be available. (See Section 8.3.2.) The alternatives that have been explored are liquid and solid (particle) injection into the plasma. Particle delivery has some obvious limitations as a result of the short residence times, including the potentially incomplete vaporization of the particles in the plasma. Liquid delivery to thermal plasmas has some advantages over the delivery of gaseous and solid precursors, for reasons similar to those that make liquid delivery advantageous in CVD processes: (1) The time required for the evaporation of liquid droplets is relatively short; (2) compositional control is possible, and the products can be homogeneous (assuming complete vaporization); (3) seeding to produce particles with a particular morphology and crystallinity is possible; (4) an extensive range of soluble precursors is available; (5) volatile precursors are no longer a requirement, because thermal decomposition of the precursors at high temperatures leads to the formation of gas-phase elemental species at the high temperatures of the thermal plasma; and (6) the residence time can be varied to some extent (e.g., by using counterflow liquid delivery), which can ensure or control the crystallinity and homogeneity of the product. For example, yttria-stabilized zirconia solid solutions have been prepared using liquid-injection plasma synthesis.[95]

Coatings have also been deposited using thermal plasmas in which the output of a plasma spray reactor is directed at a surface. Film deposition is more commonly carried out using radio frequency plasmas.

Coronas are gas discharges in which the geometry of the corona (generated by a small tip at high potential) confines the gas ionization process to high field ionization around the electrodes. The high-energy electrons produced in the ionization region around the tip interact with a cold gas, whereupon they dissociate and ionize the gaseous species. The processes that can occur in coronas are similar to those described shortly for glow-discharge plasmas.[93]

Laser-produced plasmas are a surprising phenomenon in view of the relatively low energy of the laser photons, typically less than 2 eV, compared to the excitation and ionization potentials of gaseous species, about 15 eV. This discrepancy is explained by multiphoton absorption, resulting in stepwise ionization though various virtual and allowed optical transitions.[93]

Finally, glow-discharge plasmas, which are often referred to as low-temperature plasmas, are a method used to facilitate chemical reactions at relatively low ambient temperatures (in contrast to thermal, or high-temperature, plasmas), through the formation of highly energetic species in the gas phase. Also in contrast to high-temperature thermal plasmas, which generally involve equilibrium processes, low-temperature glow-discharge plasmas involve nonequilibrium processes. Glow-discharge plasmas contain the largest variety of chemical species and have been used to form powders (although to a lesser extent than high-temperature plasmas) and to deposit thin films through plasma-enhanced or plasma-assisted CVD. Also, the chemistry that occurs in high-temperature plasmas is similar to that which occurs in high-temperature flames. (See Section 8.7.)

As a result of the presence of highly reactive species, processes that normally require temperatures greater than 1000°C can be carried out at approximately 100 to 400°C in a glow-discharge plasma. Although the chemistry and physics of a glow discharge are extraordinarily complex, the plasma performs two basic functions. First, reactive chemical species are generated through collisions of electrons; these collisions overcome kinetic limitations that may exist in thermally activated processes and are the principal sources for transferring the energy to the reagents. Second, the discharge supplies "energetic radiation" (generally in the form of positive ions, neutral species, metastable species, electrons, and photons) that bombard species or surfaces immersed in the plasma. The combination of these physical processes with the strictly chemical reactions due primarily to atoms, radicals, or molecules yields material properties unattainable with each process individually.

Glow discharges, or low-temperature plasmas, are, by definition, plasmas that exhibit no net charge; that is, the numbers of negatively charged and positively charged species are equal. The plasma, a partially ionized gas, is composed of ions, electrons, and a host of neutral species in both ground and excited vibrational states. However, the ratio of the neutral-species density to the electron density is usually greater than $10^3$, so that these plasmas are only weakly ionized. Radicals and molecules are not inherently more reactive than ions, but they are present in significantly higher concentrations. As a result, free radicals and molecules in the discharge are primarily responsible for decomposition reactions.

Free radicals are atoms or fragments of molecules that contain one or more unpaired electrons. Free radicals are generally highly reactive with respect to internal rearrangement, radical–radical recombination, or reactions with other species. A large number of free radicals are known to exist in glow-discharge plasmas. In common discharges, a free radical such as $Cl^{\bullet}$ is most often formed from gas molecules by the abstraction of an atom, although in some discharges low-molecular-weight radical species such as $CF_3^{\bullet}$ may be prevalent.

The energies required for the formation of free radicals lie well within the eV ranges normally encountered in glow discharges. For simple radicals, the collision efficiency is close to unity; that is, nearly every collision will result in recombination.

The conditions typically employed in glow-discharge plasmas are given in Table 8.10. Further understanding and characterization of the plasma processes can be gained by categorizing the reactions into two general types: homogeneous gas-phase collisions and heterogeneous surface interactions.

Homogenous gas-phase collisions generate reactive free radicals, metastable species, and ions. Chemical dissociation and ionization are therefore independent of the thermodynamic temperature. Impacts with electrons can result in a number of different reactions, depending upon the electron energies. The following reactions are listed in order of increasing energy requirements, where the notation "*" denotes an electronic excited state:[93]

**TABLE 8.10. Conditions typically employed in a glow-discharge plasma**

| Parameter | Value |
|---|---|
| Electron temperature | 1–10 eV |
| Electron density | $10^{-9}$–$10^{-12}$ $cm^{-3}$ |
| Power output | 3–1000 W |
| Operating radio frequency | 13.56 MHz |
| Pressure | 0.1–10 torr |
| Gas flow | ~ 10 $cm^3$/min |

*Excitation (rotational, vibrational, or electronic)*

$$e^- + X_2 \longrightarrow X_2^* + e^-. \quad (8.93)$$

*Heterolytic dissociation*

$$e^- + X_2 \longrightarrow X^- + X^+ + e^-. \quad (8.94)$$

*Homolytic dissociation*

$$e^- + X_2 \longrightarrow 2X^{\bullet *} + e^-. \quad (8.95)$$

*Ionization*

$$e^- + X_2 \longrightarrow X_2{}^+ + 2e^-. \quad (8.96)$$

*Dissociative ionization*

$$e^- + X_2 \longrightarrow X^+ + X^* + 2e^-. \quad (8.97)$$

Excitation and dissociation processes can occur with mean electron energies below a few electron volts. Thus, the discharge is extremely effective in producing large quantities of free radicals.

The following are the heterogeneous processes of interest that can occur at solid surfaces exposed to a glow discharge. These interactions result from the bombardment of surfaces by particles.

**Ion–surface interactions**

1. Neutralization and secondary electron emission
2. Sputtering
3. Ion-induced chemistry

**Electron–surface interactions**

1. Secondary electron emission
2. Electron-induced chemistry

**Radical- or atom–surface interactions**

1. Surface etching
2. Film deposition

The formation of the coating is likely to take place on the substrate and on the film surface by the adsorption of radicals, reaction with neighboring atoms on the surface, and chemical reaction, depositing a film with desorption of volatile by-products. The alteration and enhancement of film properties due to particle bombardment can be understood by referring to Equations (8.95) through (8.97). These are heterogeneous processes whose kinetics are dependent on temperature. Raising the temperature, however, is only one method of increasing the surface energy; particle bombardment can break surface bonds that not only can create adsorption sites, but also can assist in the desorption of products. Further chemical reactions at the surface can be promoted by the bombardment as well.

Plasma processes have been used extensively to form both films and powders. Powder formation is most relevant when metal-containing precursors are delivered to a plasma reactor, where gas-phase reactions lead to particle formation. This aspect is described in the technology of gas-to-particle conversion and in liquid-to-solid or solid-to-solid conversion.

## 8.7 FLAME-INDUCED REACTIONS

Flame reactors have been used extensively for the formation of powders, and, in some cases, chemical aspects of reactions in flames have been studied in detail. Flames are exothermic systems in which the reactions are strongly coupled to a number of powder properties.[96] Chemical reactions in flames are dominated by radical–molecule and radical–radical reactions. Radicals are generated by rapid bimolecular reactions but are consumed by slower, termolecular reactions or recombinations. As a result, at normal pressures this produces a nonequilibrium maximum in reactive radicals. Therefore, the key to understanding flame chemistry is to understand two of the methods by which radicals are formed: homolytic dissociation and exchange reactions, given, respectively, by the following equations:

$$X_2 + M^* \longrightarrow 2X^{\bullet} + M, \tag{8.98}$$

$$O_2 + H^{\bullet} \longrightarrow OH + O^{\bullet}. \tag{8.99}$$

Flames contain a fuel and an oxidizing agent. The fuel is a chemical (e.g., $H_2$, a hydrocarbon, ammonia, or $X_2$, where $X$ = F, Cl, or Br) that is oxidized by the oxidant (e.g., $O_2$, NO) after suitable initiation of the combustion reaction. Some fuels are radicals themselves—most notably, $O_2$—but the $O_2$ radical is stable at ambient temperature and pressure with respect to reaction towards most fuels (in the absence of an initiation process such as a spark). This is extremely convenient because the fuel and oxidant can be premixed and delivered to the flame, where the high temperatures and reactive radicals cause the fuel and oxidant to react, or else they are ignited by an external source of energy (a spark). Some of the reagents and principle reactions that occur in flames are presented in Table 8.11.

The temperature of the flame will be determined by the conditions under which the flame is operated (the ratio of fuel to oxidant and the flame

**TABLE 8.11. Radical reactions in flame systems**

| Flame Stoichiometry | Radicals | Generation Steps |
|---|---|---|
| $O_3 = \frac{3}{2}O_2$ | O | $O_3 + M^{*a} = O_2 + O$<br>$O + O_3 = 2O_2$ |
| $H_2 + X_2 = 2HX$ ($X$ = F, Cl, Br)[b] | H, $X$ | $X_2 + M^* = X + X + M$<br>$H + X_2 = HX + X$<br>$X + H_2 = HX + H$ |
| $2H_2 + O_2 = 2H_2O$<br>($2CO + O_2$ (trH)[c] $= 2CO_2$) | H, O, OH, $(HO_2)^e$ | $H + O_2 = O + OH$<br>$O + H_2 = OH + H$<br>$OH + H_2 = H_2O + H$ |
| $CH_4 + 2O_2 = 2H_2O + CO_2$ | H, O, OH, $(HO_2)$, $CH_3$, HCO | (See $H_2 + O_2$ flame)<br>$CH_4 + R^c = CH_3 + RH$<br>$CH_3 + O = CH_2O + H$<br>$CH_2O + R = HCO + RH$ |
| $2NH_3 + \frac{3}{2}O_2 = N_2 + 3H_2O$ | H, O, OH, $(HO_2)$, $NH_2$, NH, (N) | (See $H_2 + O_2$ flame) |
| $H_2 + N_2O = N_2 + H_2$ | H, O, OH, $(HO_2)$ | $O + N_2O = O_2 + N_2$<br>(See $H_2 + O_2$ flame) |
| $H_2 + 2NO = H_2O + N_2$ | H, O, OH, N, $O_2$ | (See NO flame)<br>(See $H_2 + O_2$ flame) |
| $2NO = N_2 + O_2$ | O, N | $O + NO = O_2 + N$ |
| $2H_2 + NO_2 = 2H_2O + N_2$ | $NO_2$, H, O, OH, $(HO_2)$ | $(NO_2 + H_2 = NO + H_2O)^f$<br>(See NO flame) |

[a] $M^*$ represents any thermally excited molecule.
[b] Iodine does not form a flame.
[c] The carbon monoxide–oxygen flame does not propagate in the absence of traces of hydrogen or any hydrogen-containing compound such as water. This occurs, because in the absence of H, radicals must be produced by high-activation dissociation of oxygen.
[e] $R$ represents any radical (H, O, OH).
[f] $NO_2$ is a reactive radical that forms a hypergolic flame.

configuration) and the energy available from the combustion of the fuel (the chemical nature of the oxidant and the fuel). Flames can be initiated at temperatures as low at 90 K and as high as 1000 K, to give final temperatures as low as 500 K and as high as 5000 K. Some examples of the chemical structure and temperatures of flames are given in Figures 8.15[97] and 8.16.[96]

*The ability to control the temperature of the flame is useful because it provides some control over the temperature at which reactions in a flame can be carried out.* Aerosol processes occur in flames where either droplets containing reagents are delivered to the flame or gaseous reagents are introduced into the flame. At high flame temperatures using gaseous reagents, it is likely that most of the species delivered to the flame are atomized, and therefore, gas-to-particle conversion occurs, resulting in the formation of nanostructured materials. However, aerosols containing reagents may not undergo

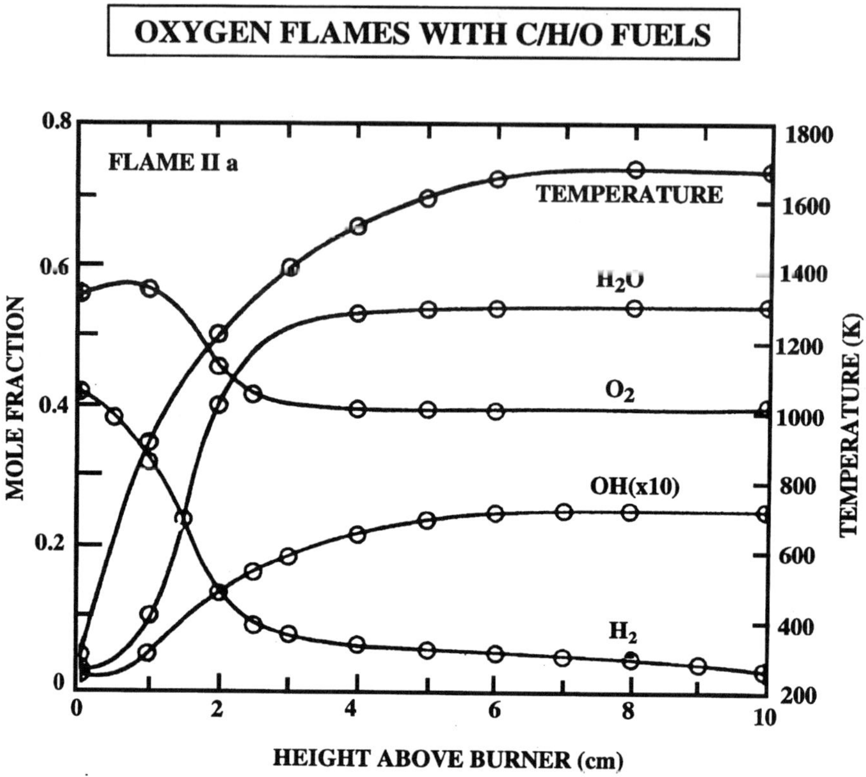


**Fig. 8.15.** Experimental structure of a fuel-lean, low-pressure hydrogen–oxygen flame ($O_2$ = 0.52). (Adapted from Brown et al.)[97]

gas-to-particle conversion even at high flame temperatures, as a result of the short residence time characteristic of most flames. At low flame temperatures it is more likely that larger particles are formed if aerosol droplets containing precursors are delivered to the flames, because insufficient energy or residence time is available to atomize either the reactants or products.

Flame reactors are currently used to produce large amounts of materials such as $SiO_2$ (e.g., Cab-O-Sil) by burning $SiCl_4$ in a hydrogen–air flame. This produces silica particles by gas-to-particle conversion. The particles agglomerate to form a high-surface-area silica ($\sim 325\ m^2/g$), which has various applications on nanostructured materials as described earlier. As a result, flame reactors have also been studied at the laboratory scale to unravel fundamental aspects of flame-induced reactions.[98–100] For example, the growth of silica particles in flames from silanes and organosilanes was studied using in situ light-scattering dissymmetry measurements of particle size and number density.[101] In addition, temperature profiles for the flames were obtained by using fine-wire thermocouples. The experiments have shown that

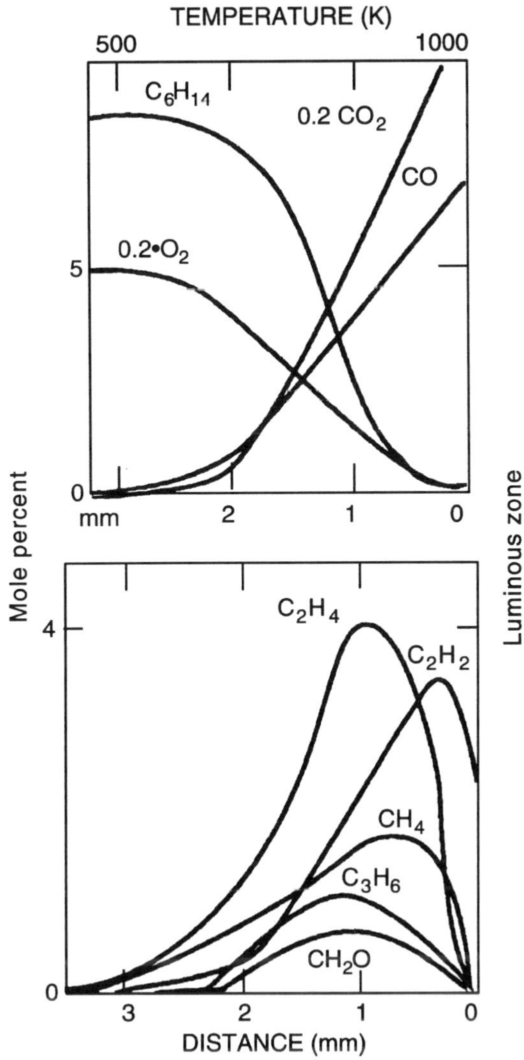


**Fig. 8.16.** Structure of a fuel-rich hexane flame. (Adapted from Fristrom et al. From *Flame Structure and Processes* by R.M. Fristrom. © 1995 by Oxford University Press, Inc. Used by permission of Oxford University Press, Inc.)[96]

the silicon source concentration and the temperature are the two dominant variables controlling particle growth and morphology. The type of silicon-containing precursor affected the mechanism of growth and, presumably, the morphology of the particles produced. Two particle growth models have been applied that account for the consumption of the gas precursor, nucleation, and subsequent growth by surface condensation and coalescence. Both models agree with the experiment. Finally, numerical simulations of the gas-phase chemistry of silane have shown the presence of multicomponent nucleation under some conditions, evidence for which has been observed by laser attenuation measurements.

## 8.8 SINGLE-SOURCE PRECURSORS

Thus far, we have focused upon the reaction of individual species, regardless of the desired characteristics of the final product. However, many of the materials prepared by aerosol methods are complex, so the formation of phase-pure, homogeneous, crystalline products, rather than inhomogeneous mixtures, is not necessarily simple to achieve. For example, if a ternary material such as $ABO_x$ is required by spray pyrolysis from precursors $AL_n$ and $BL_n$, then individual reactions to form $AO_m$ and $BO_m$ are likely to be detrimental because high temperatures are generally required to interdiffuse and convert $AO_m$ and $BO_m$ to homogeneous, crystalline $ABO_x$, regardless of the thermal decomposition temperatures of $AL_n$ and $BL_n$ (Fig. 8.17). It is noteworthy that this example is already simplified by the facts that (1) the same ligands, L, have been chosen for both metals A and B to avoid the possibility of reactions between the precursors, (2) the stoichiometries chosen are simple, (3) the reagents and the intermediates formed are assumed to be nonvolatile, (4) the reagents and intermediates are assumed to be immiscible, and (5) the only stable ternary phase that is accessible under these conditions is presumed to be $ABO_x$. The formation of a volatile intermediate results in gas-to-particle conversion, which is detrimental if spray pyrolysis is desired.

In other cases, chemically inhomogeneous products such as composites or layered materials may be desired. This requires that the metal-containing precursors and intermediates not react under the processing conditions and

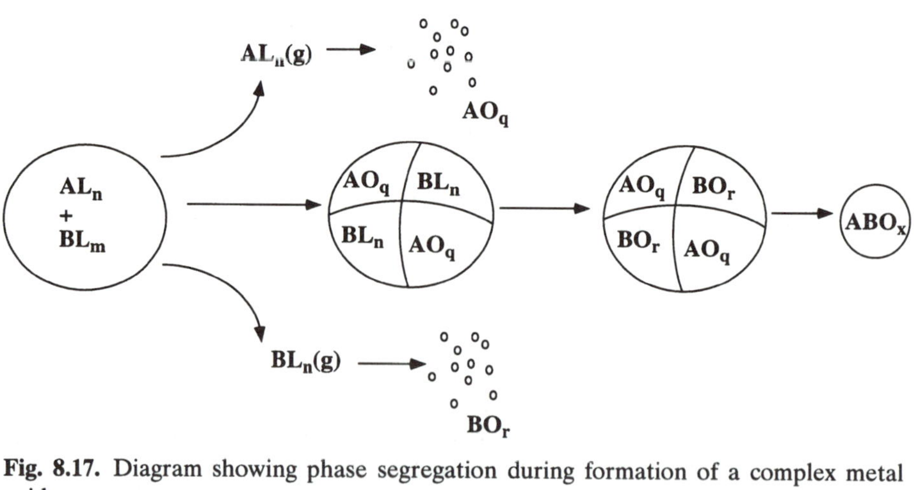


**Fig. 8.17.** Diagram showing phase segregation during formation of a complex metal oxide.

that the homogeneity be introduced and preserved. Aerosol synthesis of these materials will clearly require conditions and/or reagents different from those required to produce compositionally homogeneous materials.

Chemically homogeneous, controlled stoichiometry and phase-pure material is most often sought after, which warrants a description of the chemical factors that influence the formation of this material. The situation described in Figure 8.17 for spray pyrolysis where chemically inhomogeneous products result occurs quite frequently. A number of approaches have been developed to solve this problem and can be categorized into two types: methods that increase the likelihood of reaction of the mixture of oxides and methods that avoid the individual formation of the oxides in the first place. The first and potentially simplest method is to increase the temperature and/or time during the deposition process in an attempt to complete the reaction between the mixture of metal oxides. Unfortunately, this is not always feasible, and a sufficient increase in temperature or residence time to form a homogeneous product may not be possible or desired.

The alternative approach is to tailor the chemistry of the system to avoid phase segregation in the first place, which can be achieved in a number of ways by ensuring (1) matched reactivities of the precursors, (2) the use of a single precursor that contains both metals in the correct ratio, a so-called *single-source precursor*, to achieve a high level of uniformity, or (3) the use of precursors with complementary ligands.[1] This approach can be used for both gas-to-particle conversion and spray pyrolysis.

Matching the reactivities of precursors is an important criterion for achieving homogeneous materials in spray pyrolysis routes to powders. With unmatched reactivities, the individual precursors react independently at different temperatures to form phase-segregated products, as shown in Figure 8.17. The resulting phase segregation has been observed in a number of different systems, including yttrium–aluminum oxide, gadolinium–iron-oxide, yttrium–iron-oxide, metal ferrites, and metal ruthenates.[12, 36, 102–104] For example, in the yttrium–iron-oxide, yttrium–aluminum-oxide, and gadolinium–iron-oxide systems, it has been shown that the desired product, generally the corresponding garnet, $A_3B_5O_{12}$, is not formed directly by spray pyrolysis, because the individual metal nitrates used as precursors have markedly different thermal decomposition temperatures. For example, $Y(NO_3)_3$ thermally decomposes to $Y_2O_3$ at about 600°C, while $Al(NO_3)_3$ and $Fe(NO_3)_3$ thermally decompose to $Al_2O_3$ and $Fe_2O_3$, respectively at approximately 200°C, as shown by thermogravimetric analysis in Figure 8.18. The many inflection points in the curve for $Y(NO_3)_3 \cdot 9H_2O$ have been explained in terms of the formation of a number of oxonitrate intermediates such as $YO(NO_3)$.[12] In cases where the thermal decomposition temperatures are comparable, better control over phase purity tends to be achieved.

Similar problems have also been encountered in the CVD of films from more than one precursor where a particular composition is required in the final material.[37] In a simple case, the deposition of binary late-transition

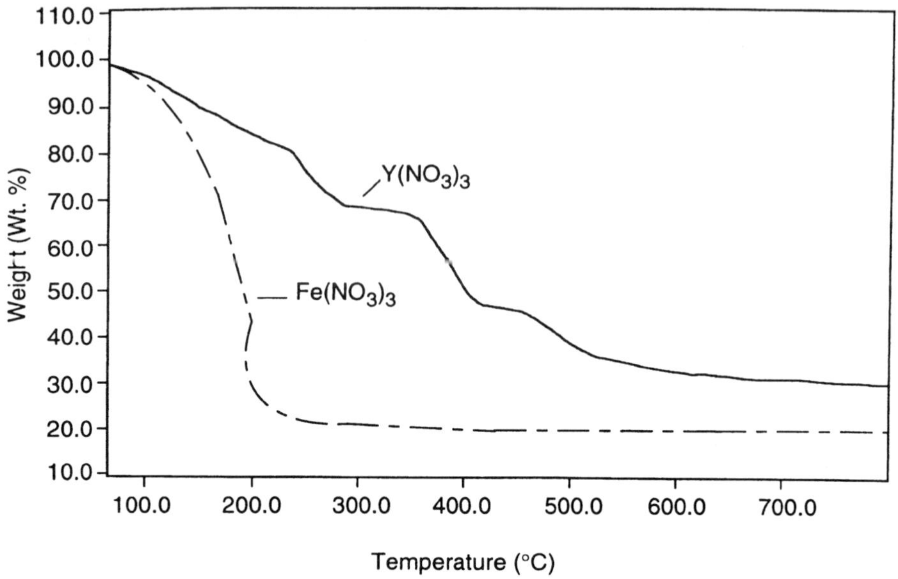


**Fig. 8.18.** Data from thermogravimetric analysis of $Y(NO_3)_3$ and $Fe(NO_3)_3$, showing a marked difference in thermal decomposition temperatures. (Adapted from Nyman et al.)[11]

metal alloy films, such as Ag/Pd alloy by AACVD, has been investigated in a model system in an attempt to understand how to control the composition in a system in which the two elements in the final product can form a solid solution over the entire composition range so that the stoichiometry of the final product is not strongly influenced by the thermodynamic stability of a particular crystalline phase.[105] It was demonstrated that when the deposition rate was limited by steps that depended directly on a chemical property of the precursor, such as evaporation from solid precursor particles or reactions on the substrate surface, the film composition obtained was reproducible, but could not be controlled. However, if the deposition was limited by a step that did not rely on a chemical property of the precursor, such as the feed rate or diffusion rate, the film stoichiometry could then be controlled to be the same as that of the solution from which the precursors were derived.

Single-source precursors have been used in attempts to control phase purity, composition, and stoichiometry in materials prepared by aerosol techniques. Examples can be found in spray pyrolysis (involatile single-source precursors),[12] gas-to-particle formation of powders,[1] and the formation of films (volatile single-source precursors)[37] by using aerosol delivery methods. The basic premise of this technique is that if a single compound contains the elements (particularly a metal and a nonmetal) required in the final material, preferably in the correct ratio, then it is likely that thermal decomposition will result in retention of that stoichiometry, thereby avoiding phase segregation because all the elements required in the final material exist in the same compound. In the case of film formation, volatile, single-source precursors such as $[Zn(S_2CNEt_2)_2]_2$[106] and $M(SOCR)_x \cdot L_y$ have been used as precursors for the formation of metal sulfides (M = Zn, Cd, Ba, Sr, Ca, Pb, Sn, Ga, In) and have been shown to deposit films with high phase and compositional purity.[107–110] An example of a reaction designed to eliminate the unwanted supporting ligands and allow deposition of a metal sulfide is

$$Zn\,(SOCMe)_2(TMEDA) \longrightarrow ZnS + (MeCO)_2S + TMEDA. \quad (8.100)$$

A similar precursor, $[Zn(S_2CNEt_2)_2(TMEDA)]$, was used for the formation of nanometer-sized ZnS particles by gas-to-particle conversion.[111]

Segregation within a particle during spray pyrolysis or aerosol decomposition can result in the formation of volatile intermediates (for example; $MoO_3$, $V_2O_5$, and PbO). Loss of these volatile components can produce powders with incorrect stoichiometry or undesired phases. For example, Pb losses of 10, 40, and 80% were observed at temperatures of 700, 800, and 900°C, respectively, during the generation of Pb–Bi–Sr–Ca–Cu–O using metal nitrates as precursors.[112] This is also the case for the formation of perovskite-phase $PbTiO_3$, in which the loss of lead generally leads to crystallization of the pyrochlore phase.[29] In a study of the formation of lead-containing perovskite-phase metal oxides by spray pyrolysis, single-source precursors were used to address the problem of controlling compositional homogeneity and dealing with the selective loss of one reagent due to its preferentially higher volatility. The preferential loss of lead as lead oxide is a common problem, especially in the formation of perovskite-phase materials containing lead, due to the relatively high volatility of PbO. Single-source, mixed metal-organic precursors $A(O_2CCMe_2O)_2Ti(O\text{-}i\text{-}Pr)_2$ (A = Pb or Ba) at 600 to 900°C were employed to prepare crystalline, phase-pure, submicron-sized $PbTiO_3$ and $BaTiO_3$ via aerosol decomposition.[34, 113, 114] The particles were both phase pure and compositionally pure by elemental analysis under conditions where greater than 10% of the lead is normally lost due to evaporation when multiple-source precursors (i.e., individual compounds as sources of each element required in the final film) are used.

Precursors with complementary ligands are another class of reagents whose use can lead to better control over composition, homogeneity, and crystallinity, including in systems where multiple-source precursors that have identical ligands can lead to the formation of solid solutions of precursors. This situation then becomes somewhat analogous to that in which single-source precursors are used for spray pyrolysis in which the dried particle is compositionally homogeneous at the molecular level. However, one drawback of the approach compared to using single-source precursors is that, although molecularly dispersed, the solid solution of precursors can still contain precursors with unmatched chemical reactivities. This has been observed in the case of mixtures of $Y(NO_3)_3$ and $Al(NO_3)_3$.[12]

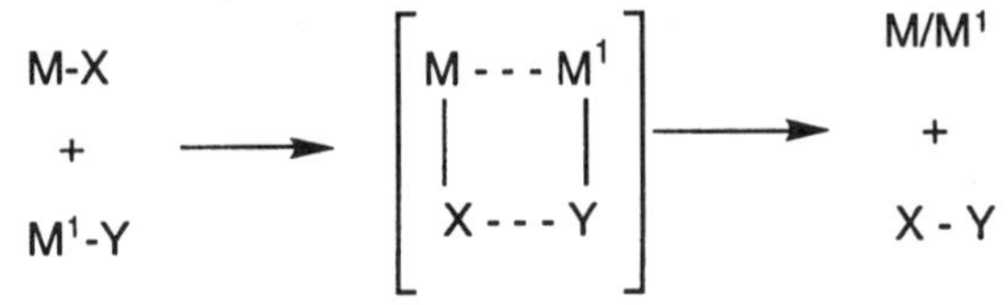


**Fig. 8.19.** Schematic representation of matched ligand reactivities.

Another example of the use of complementary ligands is illustrated schematically in Figure 8.19, in which the ligands are seen to undergo an intermolecular elimination reaction that results in both removal of the organic supporting ligands and control over stoichiometry. In the case of the simple example of a binary-metal system, in which a different ligand for each reagent reacts intermolecularly with a different ligand contained on an adjacent metal center, a better level of compositional homogeneity than independent thermal decomposition is achieved.

## 8.9 SURFACE PROCESSES

Surface processes are relevant to aerosol synthesis of materials in three specific ways: (1) the deposition of thin films on heated substrates, (2) the deposition of materials on a particle surface during aerosol synthesis (such as the synthesis of coated particles), and (3) the deposition or reaction on the walls of the vessel used for the aerosol synthesis. This section focuses on the reactions of molecules on surfaces; other physical aspects of the interaction of particles with surfaces are found in other chapters.

A schematic overview of the reaction of a metal-containing molecule with a heated surface is shown in Figure 8.20 for the simplest case of the deposition of an elemental film (i.e., a metal). The figure shows how the overall process can be considered as a number of individual steps in series, many of which are reversible, so that whether a reaction occurs (the equilibrium lies on the right in the figure) depends on the conditions of the process. A number of important steps previous to those shown can also have a dominating effect on the deposition of the films. These steps, which include the feed rate and transport rate of the molecular species to the surface and other important processes, such as gas-phase reactions above the surface, are treated in Chapter 14.

The overall deposition process proceeds as follows.[6] The molecular species is transported to the surface, where it can physisorb. Physisorption corresponds to the formation of a weak acid–base type of interaction between the surface and the molecule. Typical enthalpies of this interaction are in the range of up to 5 Kcal/mol. Physisorption can then be followed by chemisorption, in which bonds in the molecular species are broken and new

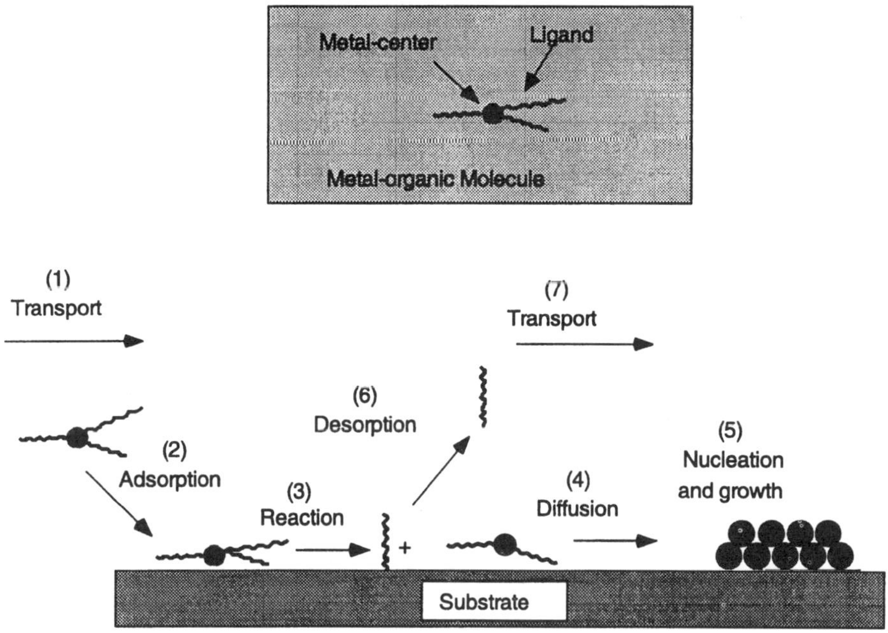


**Fig. 8.20.** Schematic overview of the reaction of a metal-containing molecule with a heated surface. (Adapted from Kodas and Hampden-Smith.)[4]

bonds are formed between fragments of the molecular species and the surface. Typical energies for these processes are much greater than 5 Kcal/mol. The species formed on the surface are then generally very reactive and can undergo further chemical reactions, which result in the formation of reactive species, such as metal atoms in the case of deposition of a metal, and organic by-products. The metal nuclei generally have high diffusion coefficients at the deposition temperature and are free to migrate across the substrate surface until they encounter a site of greater thermodynamic stability, which is often another atom or part of a growing film. This migration ultimately results in additional film growth. A variety of mechanisms of film nucleation and growth are treated in detail elsewhere.[2]

At the same time as physisorption and chemisorption take place, the organic by-products can undergo rearrangements on the surface (often they are surface catalyzed), which can result in cleavage of their bonds until a species of greater kinetic or thermodynamic stability is formed, such as an organic fragment that exhibits multiple bonding to the film surface. In this case, it is unlikely that the species will be desorbed from the surface; instead, it is likely to become entrained in the growing film. This is generally an undesirable situation, since it can lead to contamination of the film through the formation of metal carbides, or it can result in the incorporation of

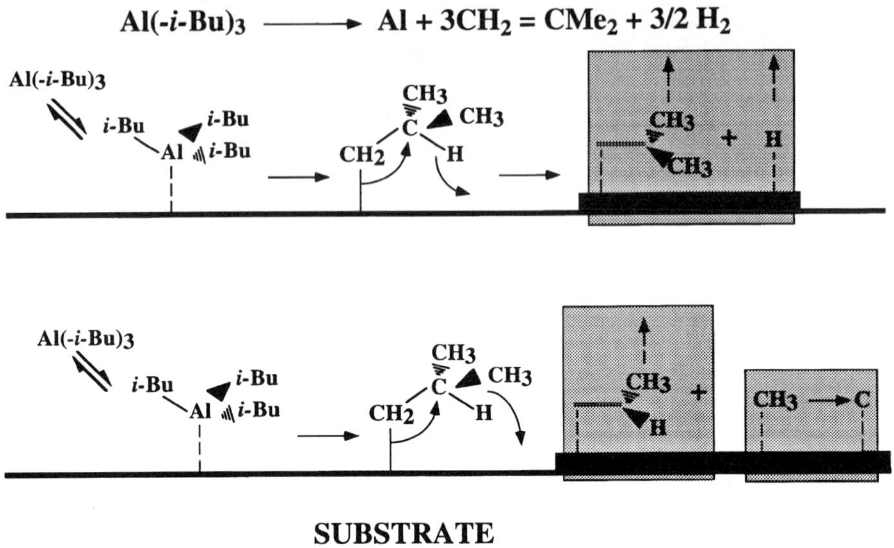


**Fig. 8.21.** Deposition of Al films from Al(-*i*-Bu)$_3$. (Adapted from Kodas and Hampden-Smith.)[4]

carbon into the film, which may be detrimental to its properties. Alternatively, thermally induced rearrangement of the organic ligands on the substrate surface can lead to the formation of strong bonds within the organic by-product and weaker bonds to the surface, reducing the strength of the interaction with the organic by-product from chemisorption to physisorption and, ultimately, desorption. The incorporation of other impurities from the organic supporting ligands or solvents, such as H, F, O, P, and Si, can occur by similar processes. A simple example in which this possibility has been studied in detail[115] is the deposition of Al films from Al(-*i*-Bu)$_3$, which is shown in Figure 8.21.

The *i*-Bu radicals liberated on the surface after chemisorption of Al(-*i*-Bu)$_3$ can undergo either $\beta$-hydride elimination or methyl group migration. $\beta$-hydride elimination results in the formation of strong C = C double bonds in the organic by-product isobutene, $H_2C = C(CH_3)_2$, which is weakly adsorbed on an Al surface and subsequently dissociated intact. The hydrogen radicals, $H^{\bullet}$, that are liberated combine and are eliminated from the surface as molecular $H_2$. In contrast methyl group migration results ultimately in contamination of the final film with carbon as a result of the lack of an efficient mechanism to remove the methyl species.

In many situations, it is ideal for all the organic ligands to be removed completely from the film surface. In some cases, the organic ligands are simply present as a vehicle for the central metal, and the design of the metal-containing complex for CVD relies upon the ligands providing a means to transport the metal to the surface, after which they should be removed intact. In other cases, such as that of $[Zn(S_2CNEt_2)_2]_2$, described earlier,[106] the organic ligands also provide a source of one of the elements in the final film, or an additional reagent (e.g., N from $NH_3$, S from $H_2S$ or thiourea, and P from $PH_3$) is added as a source of another element in the film.

Most of the information on surface chemistry that is relevant to this discussion is derived from the desire to understand conventional CVD processes and heterogeneous catalysis. Most systems that have been the subject of a thorough mechanistic analysis are relatively simple in terms of the number and type of species present, which is generally in contrast to the conditions used in aerosol synthesis of materials. In addition, most mechanistic studies have been carried out under ultra-high vacuum (UHV) conditions, which are also far removed from the typical conditions used for aerosol synthesis of materials. While the results of UHV experiments provide a useful guide to likely surface reactions in aerosol processes, the details are probably not applicable due to the vastly different conditions. For example, key steps in the surface deposition reaction being investigated change from bimolecular in a deposition experiment to unimolecular under UHV conditions, where surface concentrations can be sufficiently low.

It is important to note that surface reactions are often catalytic, as surfaces can donate or accept electron density from an adsorbed species, which can significantly lower the energy required for cleavage of bonds and chemical reactions. Homogeneous reactions between gaseous species generally have higher activation energies and so require higher temperatures than do heterogeneous reactions of the same species on a surface such as particles entrained in the gas stream or the walls of the reactor. It is also noteworthy that, on a continuous coating of the new material (which can be as small as a monolayer, depending on the film growth mechanism) deposited on the surface, the remaining molecular species react with the film surface and not the original substrate surface. As a result, the chemistry and the outcome of the surface reaction can be quite different.

## 8.10 SOLID-STATE PROCESSES: CRYSTALLIZATION, DENSIFICATION, POROSITY, AND SURFACE PROPERTIES

Some aspects of solid-state processes are predominantly chemical in nature. For example, the preparation of powders frequently involves the initial formation of amorphous materials after completion of the reactions of the metal-containing reagents. The amorphous materials then often undergo crystallization by processes that involve the cleavage and formation of bonds. In addition, many other processes, such as densification and sintering, require the cleavage and formation of bonds.[9, 116] In this section we consider just the chemical aspects of these processes; the more physical aspects, such

**TABLE 8.12. Cohesive energy**

| Type of Bond | Examples | kcal/mol | eV/ molecule | Striking Characteristics |
|---|---|---|---|---|
| Ionic | LiCl | 199 | 8.63 | Nondirected bonding giving structures with high coordination; no electrical conductivity at low temperatures; good ionic conductivity at high temperatures. |
| | NaCl | 183 | 7.94 | |
| | KCl | 166 | 7.20 | |
| | RbCl | 159 | 6.90 | |
| Covalent | Diamond | 170 | 7.37 | Spatially directed bonds giving structures with low coordination; low conductivity at low temperatures for pure crystals. |
| | Si | 108 | 4.68 | |
| | Ge | 89 | 3.87 | |
| | Sn | 72 | 3.14 | |
| Metallic | Li | 37.7 | 1.63 | Nondirected bonding giving structures with very high coordination and density; high electrical conductivity; ductile. |
| | Na | 25.7 | 1.11 | |
| | K | 21.5 | 0.934 | |
| | Rb | 19.6 | 0.852 | |
| Molecular | Ne | 0.46 | 0.020 | Low melting and boiling points; very compressible; the properties of the molecules are retained. |
| | Ar | 1.79 | 0.078 | |
| | Kr | 2.67 | 0.116 | |
| | Xe | 3.92 | 0.170 | |
| Hydrogen bonded | $H_2O$ (ice) | 12 | 0.52 | Increase in bonding energy over similar molecules without hydrogen bonds. |
| | HF | 7 | 0.30 | |

as the magnitudes of diffusion coefficients and their influence on the overall process, are described elsewhere in the text. These processes are relevant to powder (spray pyrolysis and gas-to-particle conversion) and film (AACVD) formation because the properties of the final material are often influenced by their microstructure and crystallinity.

The nature of bonding in a solid depends primarily on the distribution of electrons between the atoms in the solid. The five types of bonding found in solids are presented in Table 8.12.

The types of bonds relevant to the current discussion are ionic, covalent, and metallic bonds, which occur in amorphous as well as crystalline solids. Amorphous materials are generally a higher energy form of a material, compared to the corresponding crystalline form. The driving force for crystallization is the minimization of free energy, as indicated in the equation

$$\Delta G = \Delta H - T\Delta S, \tag{8.101}$$

where $\Delta G$ denotes the change in free energy of the system (a negative value

**TABLE 8.13. Typical values for $\Delta G$**

| Phenomenon | $\Delta G$/J mol$^{-1}$ |
|---|---|
| Chemical reaction with change in energy of large bonds | $10^5$ to $10^6$ |
| "Weaker" chemical reaction | $10^3$ to $10^5$ |
| Diffusion down concentration gradients | $10^2$ to $10^4$ |
| Solidification with a few K of undercooling | 10 to $10^2$ |
| Crystallographic phase changes | 1 to 10 |
| Sintering, grain growth, precipitate coarsening | $10^{-2}$ to 1 |

is required for a spontaneous change); $\Delta H$ is the change in enthalpy of the system, which is related to the energy gained or lost by breaking and/or reforming bonds; and $\Delta S$ is the change in entropy in the system, a measure of the amount of order (a positive value corresponds to an increase in disorder.) A spontaneous change in a system is thermodynamically possible as long as the change in free energy is negative (i.e., $\Delta G < 0$), although this does not describe the rate at which the change occurs. Typical values for $\Delta G$ are given in Table 8.13. The factors shown need to be taken into account in attempting to prepare materials by aerosol routes. For example, it is important to realize that a great deal of energy may be required to convert an amorphous powder to a crystalline powder, so that it may be more economical to prepare the crystalline phase directly.

Solid-state processes are important in all methods of aerosol synthesis of powders. In many cases, solid particles that contain precursors or mixtures of different reagents are formed. In the case of intraparticle reaction, these particles are likely to react and form an amorphous product that can subsequently crystallize. The amount of amorphous material, the extent of crystallization (i.e., the size of the crystallites), and the density of the particle will depend on the nature of the chemical system, the diffusion coefficients of the species involved, and the temperature–time history of the particle. The thermodynamic driving force in the formation of a crystalline material is based on the additional energy gained by achieving long-range order (corresponding to the lattice energy) at the expense of a decrease in disorder (which is thermodynamically unfavorable). Grain growth occurs because the energy of the system is lowered by reducing the surface energy, thereby causing a reduction in the surface area and an increase in the crystallite size.

In most aerosol processes the temperatures can be relatively high (for example, typically greater than 1000°C), but the residence times are relatively short—on the order of seconds—compared to solid-state materials synthesis techniques. In addition, with the exception of metals, most materials prepared by aerosol processes, such as oxide and nonoxide ceramics, have relatively low diffusion coefficients. As a result, if crystalline particles are required, then a high level of mixing of the metal-containing precursors is important to decrease diffusional distances (because the time available for diffusion is short), and crystallization of individual components of the final material should be avoided because greater energy (e.g., higher temperatures in a thermally induced process) is required to restructure the material over that required for an amorphous material (by an amount that corresponds to the lattice energy). Even in the case of an amorphous material, which still contains chemical bonds, but without long-range order of a crystalline material, the energy required to break the bonds may be prohibitively high. This situation has frequently been encountered in the formation of oxide phases containing aluminum, where phase segregation resulting in the formation of amorphous alumina can be difficult to overcome due to the high temperatures required to break Al–O bonds.[12]

Phase diagrams provide valuable information pertaining to the thermodynamic stability of crystalline materials within a given system.[116] However, the experimental conditions from which data for such diagrams are derived are based on solid-state reactions of large-grained reagents with long reaction times (in order to obtain thermodynamic products). Therefore, information gleaned from phase diagrams should be applied with caution to aerosol-derived materials, for the following two reasons. First, aerosol-derived materials (powders and films) are often formed under nonequilibrium conditions. As a result, unexpected phases (based on data from the phase diagram) may be formed because the system is under kinetic control (i.e., under given conditions, insufficient time was available to convert the material to a thermodynamic product). Second, the crystallite size of the material formed is often small (e.g., $< 50$ nm), and the high surface energy of small-grained materials frequently results in the stabilization of different crystalline phases relative to the same material composed of larger grains.[29]

The sintering and mechanical properties of solids can be strongly affected by the size of the grains or crystallites within each particle of the solid. Powders that are made up of nanometer-sized crystallites exhibit properties unlike those of bulk materials (i.e., materials composed of large grains), in that they exhibit size-dependent properties such as their melting point, conductivity, optical properties, and, in this context, sintering behavior. As a result, there is a strong interest in the synthesis of powders composed of nanometer-sized crystallites, since these materials are likely to have unusual, but beneficial, properties. For example, traditional ceramics have been shown to undergo plastic deformation. In addition, powders made up of nanometer-sized particles also contain nanometer-sized pores or voids. This leads to optical transparency and with low density, an unusual combination because normally, low-density porous materials have pores with sizes on the order of the wavelength of light, which results in significant scattering and therefore opacity. Aerosol processes are excellent methods for making particles that contain nanometer-sized crystallites and also powders that are composed of nanometer-sized particles. A more detailed discussion of the characteristics of nanometer-sized particles is given in Chapter 9.

## 8.11 REFERENCES


1. Roger, C., Corbitt, T., Xu, C., Zeng, D., Powell, Q., Chandler, C.D., Nyman, M., Hampden-Smith, M.J., & Kodas, T.T. *J. Nanostr. Mater.*, *4*(5), 1994, 529–535.
2. Hitchman, M.L. & Jensen, K.F. *Chemical Vapor Deposition: Principles and Applications*. London and San Diego: Academic Press, 1993.
3. Jensen, K.F. & Kern, W. Thermal Chemical Vapor Deposition. In Vossen, J.L. & Kern, W. (Eds.), *Thin Film Processes II*. San Diego: Academic Press, Inc., 1991, 283.
4. Kodas, T.T. & Hampden-Smith, M.J. *The Chemistry of Metal CVD*. Weinheim, Germany: VCH, 1994.
5. Kuech, T.F. & Jensen, K.F. Photochemical Vapor Deposition in Vossen, J.L. & Kern, W. (Eds.), *Thin Film Processes II*. San Diego: Academic Press, Inc., 1991, 369.
6. Kodas, T.T. & Hampden-Smith, M.J. The Chemistry of Metal CVD. Chapter 9 in Kodas, T.T. & Hampden-Smith, M.J. (Eds.), *The Chemistry of Metal CVD*. Weinheim, Germany: VCH, 1994,
7. Cotton, F.A. & Wilkinson, G., *Advanced Inorganic Chemistry*. New York: Wiley, 1988.
8. Huheey, J.E. *Inorganic Chemistry: Principles of Structure and Reactivity*. New York: Harper and Row, 1983.
9. Wells, A.F. *Structural Inorganic Chemistry*. Oxford, U.K.: Clarendon Press, 1986.
10. Drozdov, A. & Troyanov, S. *J. Chem. Soc. Chem. Commun.* 21, 1993, 1619–1621.
11. Nyman, M., Hampden-Smith, M.J., & Duesler, E.N. Unpublished results, 1997.
12. Nyman, M., Caruso, J., Hampden-Smith, M.J., & Kodas, T.T. *J. Am. Cer. Soc.* *80*(5), 1997, 1231–1238.
13. Norman, J.A.T. & Pez, G.P. *J. Chem. Soc., Chem. Commun 14*, 1991, 971–972.
14. Sluis, P.v.d. & Spek, A.L. *Acta Cryst. C46*, 1990, 1741.
15. Drake, S.R., Hursthouse, M.B., Malik, K.M.A., & Otway, D.J. *J Chem Soc Dalton Trans* (19), 1993, 2883–2890.
16. Wojtczak, W.A., Fleig, P., & Hampden-Smith, M.J. *Adv. Organomet. Chem. 40*, 1996, 215–340.
17. Schulz, D.L., Hinds, B.J., Neumayer, D.A., Stern, C.L., & Marks, T.J. *Chem. Mater. 5*, 1993, 1605.
18. Schulz, D.L., Hinds, B.J., Stern, C.L., & Marks, T.J. *Inorg Chem. 32*(3), 1993, 249–250.
19. Auger, M.V. & Robin, I., *Compt. Rendu 178*, 1924, 1546.
20. Gordon, R.M. & Silver, H.B., *Can. J. Chem. 61*, 1983, 1218.
21. Hiltunen, L. *Acta. Chem. Scand., A.*, 1987, 548.
22. Messing, G.L., Zhang, S.-C., & Jayanthi, G.V. *J. Am. Ceram. Soc. 76*(11), 1993, 2707–2726.
23. Zhang, S.-C. & Messing, G.L. Solid particle formation during spray pyrolysis. In Messing, G.L., Hausner, H., & Hirano, S.-C. (Eds.), *Ceramic Powder Science III*, Columbus, OH: American Ceramic Society, 1990, 49–57.

24. Zhang, S.-C., Messing, G.L., & Borden, M. *J. Am. Ceram. Soc. 73*(1), 1990, 61–67.
25. Cabanas, M.V., Vallet-Regi, M., Labeau, M., & Gonzalez-Calbet, J.M. *J. Mater. Res. 8*(10), 1993, 2694.
26. Shriver, D.F., Atkins, P.W., & Langford, C.H. *Inorganic Chemistry*. New York: W.H. Freeman and Company, 1990.
27. *CRC Handbook of Chemistry and Physics*, Boca Raton, FL: CRC Press, 1987.
28. Brinker, C.J. & Scherer, G.W. *Sol-Gel Science: The Physics and Chemistry of Sol-Gel Processing*. San Diego: Academic Press, 1990.
29. Chandler, C.D., Roger, C., & Hampden-Smith, M.J. *Chem. Rev. 93*, 1993, 1205.
30. Marks, T.J. *Pure Appl. Chem. 67*, 1995, 313–318.
31. Rees, W.S. & Barron, A.R. *Mat. Sci. Forum 137–139*, 1993, 473–494.
32. Schulz, D.L. & Marks, T.J. *Adv. Mater. 6*(10), 1994, 719–730.
33. Simmonds & Gladfelter, W. The Chemistry of Metal CVD. Chapter 2 in Kodas, T.T. & Hampden-Smith, M.J. (Eds.), *The Chemistry of Metal CVD*. Weinham, Germany: VCH, 1994.
34. Chandler, C.D., Powell, Q., Hampden-Smith, M.J., & Kodas, T.T. *J. Mater. Chem. 3*, 1993, 7.
35. Kodas, T.T. *Angew. Chem. Adv. Mater. 6*, 1989, 92.
36. Senzaki, Y., Hampden-Smith, M., Kodas, T.T., & Hussler, J.W. *J. Amer. Cer. Soc. 78*, 1995, 2977–2983.
37. Hampden-Smith, M.J. & Kodas, T.T. The Chemistry of Metal CVD. Chapter 8 in Kodas, T.T. & Hampden-Smith, M.J. (Eds.), *The Chemistry of Metal CVD*, Weinheim, Germany: VCH, 1994,
38. Doumeng, M. *Rev. Chim. Miner. 7*(5), 1970, 897–925.
39. Criado, J.M., Ortega, A., & Real, C. *React. Solids 4*(1–2), 1987, 93–103.
40. Malecki, A., Gajerski, R., Labus, S., Prochowska-Klisch, B., & Wojciechowski, K. *J. Therm. Anal. 39*(5), 1993, 545–550.
41. Gadalla, A.M. & Yu, H.-F. *J. Mater. Res. 5*(12), 1990, 2923–2927.
42. Brock, L.R. & DeVore, T.C. *High Temp. Sci. Volume Date 1991, 32*(2–3), 1994, 101–112.
43. Auffredic, J.P., Rodriguez Roldan, A.M., Garcia Martinez, O., & Louer, D. *Journ. Calorim. Anal. Therm 1*, 1980, 3.
44. Senzaki, Y., Caruso, J., Hampden-Smith, M.J., Kodas, T.T., & Wang, L.-M. *J. Amer. Cer. Soc. 78*, 1995, 2973–2976.
45. Packter, A. *Silic. Ind. 53*(11–12), 1988, 169–177.
46. Sawada, Y., Kanou, N., & Mizutani, N. *Thermochim. Acta 183*, 1991, 279–287.
47. Sawada, Y. & Manabe, K. *J. Therm. Anal. 37*(8), 1991, 1657–1663.
48. Sawada, Y., Mizutani, N., & Kato, M. *Thermochim. Acta 143*, 1989, 319–324.
49. Sawada, Y., Watanabe, N., Henmi, H., Mizutani, N., & Kato, M. *Thermochim. Acta 138*(2), 1989, 257–265.
50. Wydeven, T. *J. Catal. 16*(1), 1970, 82–89.
51. Zemtsova, Z.N., Prodan, E.A., & Pavlyuchenko, M.M. *Zh. Fiz. Khim. 52*(10), 1978, 2708.

52. Criado, J.M. *Mater. Sci. Monogr. 6* (Energy Ceram.), 1980, 1096–1105.
53. Criado, J.M., Rouquerol, F., & Rouquerol, J. *Thermochim. Acta 38*(1), 1980, 117–123.
54. Terry, B.S. & McGurk, G. *Trans. Inst. Min. Metall., Sect. C 103* (Jan.–April), 1994, C62–C68.
55. Takamori, T. & Boland, J.J. *J. Appl. Phys. 64*(4), 1988, 2130–2133.
56. McIntosh, R.M., Sharp, J.H., & Wilburn, F.W. *Thermochim. Acta 165*(2), 1990, 281–296.
57. Ball, M.C. *Proc. Soc. Anal. Chem.* 9(8), 1972, 167–168.
58. Avramov, L.K. & Yanachkova, I.M., *Z. Phys. Chem. 241*(3–4), 1969, 244–247.
59. Okada, K., Tanaka, A., & Hayashi, S. *J. Mater. Res. 9*(7), 1994, 1709–1713.
60. Cronin, J.T. & Brill, T.B. *Appl. Spectrosc. 43*(4), 1989, 650–653.
61. Jullien, J., Pechine, J.M., Perez, F., & Sadek, M.A. *Propellants, Explos., Pyrotech. 8*(4), 1983, 99–101.
62. Knauth, H.D. *Ber. Bunsenges. Phys. Chem. 82*(2), 1978, 212–216.
63. Lurie, B., Valishina, Z., & Malchevski, V. *Int. Annu. Conf. ICT 24th* (Energetic Materials: Insensitivity and Environmental Awareness), 1993, 57–1/57–19.
64. Saibova, M.T. *Uzb. Khim. Zh.* (6), 1979, 24–27.
65. Dubois, L.H. *Polyhedron 13*(8), 1994, 1329–1336.
66. Fix, R., Gordon, R.G., & Hoffman, D.M. *Chem. Mater. 5*, 1993, 614.
67. Gordon, R.G. *Mater. Res. Soc. Symp. Proc 335*, 1994, 9–19.
68. Zinn, A., CVD of Tungsten. Chapter 3 in Kodas, T.T. & Hampden-Smith, M.J. (Eds.), *The Chemistry of Metal CVD*. Weinheim, Germany: VCH, 1994.
69. French, W.G., Pace, L.J., & Foertmeyer, V.A. *J. Phys. Chem. 82*(20), 1978, 2191–2194.
70. Pratsinis, S.E., Bai, H., & Biswas, P. *Journal of the American Ceramic Society 73*(7), 1990, 2158–2162.
71. Powers, D.R. *J. Amer. Ceram. Soc. 61*(7–8), 1978, 295–297.
72. Mehrotra, R.C. & Bohra, R, *Metal Carboxylates*. New York: Academic Press, 1983.
73. Adrianova, T.N., Kholkin, A.I., Polskii, A.I., Chereshkevich, A.V., & Gulyaeva, L.V. *J Appl Chem–Engl Tr 65*(11, Part 1), 1992, 2001–2004.
74. Borker, V.M. & Rane, K.S. *Proc. Natl. Symp. Therm. Anal., 8th* , 1991, 525–535.
75. Insausti, M., Cortes, R., Arriortua, M.I., Rojo, T., & Bocanegra, E.H. *Solid State Ionics 65*(1–4), 1993, 351–357.
76. Pfaff, G., Schmidt, F., Ludwig, W., & Feltz, A. *J. Therm. Anal 33*(3), 1988, 771–779.
77. Sekar, M.A., Dhanaraj, G., Bhat, H.L., & Patil, K.C. *J. Mater. Sci. Mater. Electron 3*(4), 1992, 237–239.
78. Emmono, Y. & Yamaji, A. *Am. Ceram. Soc. Bull. 60*, 1981, 566.
79. Bind, J.M., Dupin, T., Schafer, J., & Titeux, M. *J. Metals 39*, 1987, 60.
80. Rhine, W.E., Hallock, R.B., Davis, W.M., & Wong-Ng, W. *Chem. Mater* (4), 1992, 1208.

81. Pechini, M.P., United States Patent, 3 330 697.
82. Mudler, B. *Amer. Cer. Soc. Bull. 49*, 1970, 990.
83. Davis, J.A. & Dutremez, S. *J. Amer. Cer. Soc. 73*, 1990, 1429.
84. Bradley, D.C., Mehrotra, R.C., & Gaur, D.P. *Metal Alkoxides*. London: Academic Press, 1978.
85. Mehrotra, R.C., Bohra, R., & Gaur, D.P. *Metal β-Diketonates and Allied Derivatives. New York: Academic Press*, 1978.
86. Rees, W.S. (Ed.), *CVD of Nonmetals*, Weinheim, Germany: VCH, 1996,
87. Eden, J.G., Photochemical Vapor Deposition. In Vossen, J.L. & Kern, W. (Eds.), *Thin Film Processes II*, San Diego: Academic Press, Inc., 1991, 443.
88. Mazumder, J. & Kar, A. *Theory and Application of Laser Chemical Vapor Deposition*, New York: Plenum, 1995.
89. Mogyorosi, P. & Carlsson, J.O. *J Vac Sci Technol A 10*(5), 1992, 3131–3135.
90. Auvert, G. *Appl Surf Sci 86*(1–4), 1995, 466–474.
91. Auvert, G., Pauleau, Y., & Tonneau, D. *Jpn. J. Applied Physics I: Regular Papers & Short Notes 31*(1), 1992, 100–104.
92. Davidson, J.L., John, P., Roberts, P.G., Jubber, M.G., & Wilson, J.I.B. *Chem. Mater. 6*, 1994, 1712–1718.
93. Droes, S.R., Kodas, T.T., & Hampden-Smith, M.J. Plasma-Enhanced Chemical Vapor Deposition (PECVD). Chapter 23 in Weimer, A.W. (Ed.), *Carbide, Nitride and Boride Materials Synthesis and Processing*, London: Chapman & Hall, 1997, 579–604.
94. Ichinose, N., Ozaki, Y., & Kashu, S. *Superfine Particle Technology*, London: Springer-Verlag, 1991.
95. Kong, P.C. & Pfender, E. Synthesis of Ceramic Powders in a Thermal d.c. Plasma Jet by Injection of Liquid Precursors. *Internl. Conf. Ceram. Powder Process. Sci.*, Westerville, OH: Am. Ceram. Soc., 1988, 237–243.
96. Fristrom, R.M. *Flame Structure and Processing*. Oxford, U.K.: Oxford University Press, 1995.
97. Brown, N.J., Erbius, K.H., Fristrom, R.M., Hoyermann, K.H., & Wagner, H.G., *C & F 33*, 1978, 117.
98. Chung, S.L., Sheu, Y.C., & Tsai, M.S. *Ceram. Trans. Ceram. Powder Sci. 3*(12), 1990, 275–282.
99. Vemury, S. & Pratsinis, S.E. *J. Am. Ceram. Soc. 78*(11), 1995, 2984–2992.
100. Zhu, W. & Pratsinis, S.E. *ACS Symp. Ser. 622* (Nanotechnology), 1996, 64–78.
101. Zachariah, M.R. *Chem. Eng. Sci. 45*(8), 1990, 2551–2558.
102. Hay, R.S. & Matson, L.E. *Acta Metall. Mater. 39*(8), 1991, 1981–1994.
103. Kagawa, M., Suzuki, M., Mizoguchi, Y., Hirai, T., & Syono, Y. *J. Aerosol Sci. 24*, 1991, 349–355.
104. Matsumoto, K., Yamanobe, Y., Sasaki, S., Fujii, T., Honda, K., & Miyamoto, T. *J. Applied Phys. 70*(10), 1991, 5912–5914.
105. Xu, C., Kodas, T.T., & Hampden-Smith, M.J. *Chem. Mater. 7*, 1995, 1539–1546.
106. Pike, R.D., Cui, H., Kershaw, R., Dwight, K., Wold, A., Blanton, T.N., Wernberg, A.A., & Gysling, H.J. *Thin Solid Films 224*(2), 1993, 221–226.

107. Kunze, K., Bihry, L., Atanasova, P., Hampden-Smith, M.J., & Duesler, E.N. *Adv. Mat. Chem. Vapor Deposition 2*, 1996, 105–108.

108. Nyman, M., Pennino, S., Hampden-Smith, M.J., Zeng, D., & Kodas, T.T. Aerosol-assisted Chemical Vapor Deposition of ZnS from $[Zn(SEt)(Et)]_{10}$. *MRS Fall Meeting Abstract Book*, Boston: Materials Research Society 1994, 179.

109. Shang, G., Duesler, E.N., & Hampden-Smith, M.J. *J. Chem. Soc., Chem. Commun.*, 1996, 1733.

110. Shang, G., Duesler, E.N., & Hampden-Smith, M.J. *Adv. Mater. CVD 2*, 1996, 242–246.

111. Powell, Q., Derricotti, D., Duesler, E.N., Hampden-Smith, M.J., & Kodas, T.T. Unpublished results, 1997.

112. Gurav, A.S., Kodas, T.T., Kauppinen, E.I., Joutsensaari, J., & Zilliacus, R. *J. Mater. Res. 10*(7), 1995, 1644–1652.

113. Chandler, C.D. & Hampden-Smith, M.J. *Chem. Mater. 4*(6), 1992, 1137–1139.

114. Chandler, C.D., Hampden-Smith, M.J., & Duesler, E.N. *Inorg. Chem. 31*, 1992, 4891.

115. Bent, B.E., Dubois, L.H., & Nuzzo, R.G. *Mat. Res. Soc. Proc. 131*, 1989, 327.

116. West, A.R. *Solid State Chemistry and Its Applications*, New York: Wiley, 1989

# 9 Characteristics of Nanostructured Materials

## 9.1 INTRODUCTION

This chapter is written to provide the reader with an appreciation for the attributes of nanostructured materials. Aerosol methods are among the best synthetic routes to prepare nanostructured materials, either in film or powder form, with a high level of control over their structure.

The term *nanostructured material* is a general one that applies to any material made up of structural components having dimensions on the scale of nanometers. A relatively broad class of materials fits under this classification and includes powders, macroscopic objects, and films that are composed of nanometer-sized particles, nanometer-sized crystallites, nanometer-sized pores, or any phases combined with a spatial distribution that involves the nanometer length scale. These materials have also been called *ultrafine grained*, *submicron grained*, *ultradispersed*, *highly dispersed*, *nanocrystalline*, *microcrystalline*, and *nanophase*, among other terms used. The scientific interest and the potential or, in some cases, realized technological or commercial value of nanostructured materials are derived from the unique properties associated with nanometer-sized structures. In order to understand the parameters that influence properties in the nanometer-length region, and to help define the numerical range of the nanometer length scale (1 nm, 1000 nm, etc.), it is worth considering an example.

The simplest case is that of an elemental material, such as a metal. We consider how properties change as the size of a single piece of this material varies from small (i.e., approximately 1 Å, the size of an atom) to "large" (i.e., on the order of micrometers). (See Fig. 9.1.)

The same principles apply to any material, regardless of its composition. A piece of metal that has dimensions greater than several microns exhibits all the properties characteristic of the "bulk" metal, including physical properties such as electrical (e.g., conductivity), optical (e.g., UV-visible absorption), magnetic, density, and refractive index properties, as well as chemical properties such as the material's melting point, boiling point, vapor pressure, reactivity, and crystallinity, which have been measured for larger pieces of the metal and the data for which are available in handbooks. However, as the piece of metal decreases in size, there is a gradual change in properties

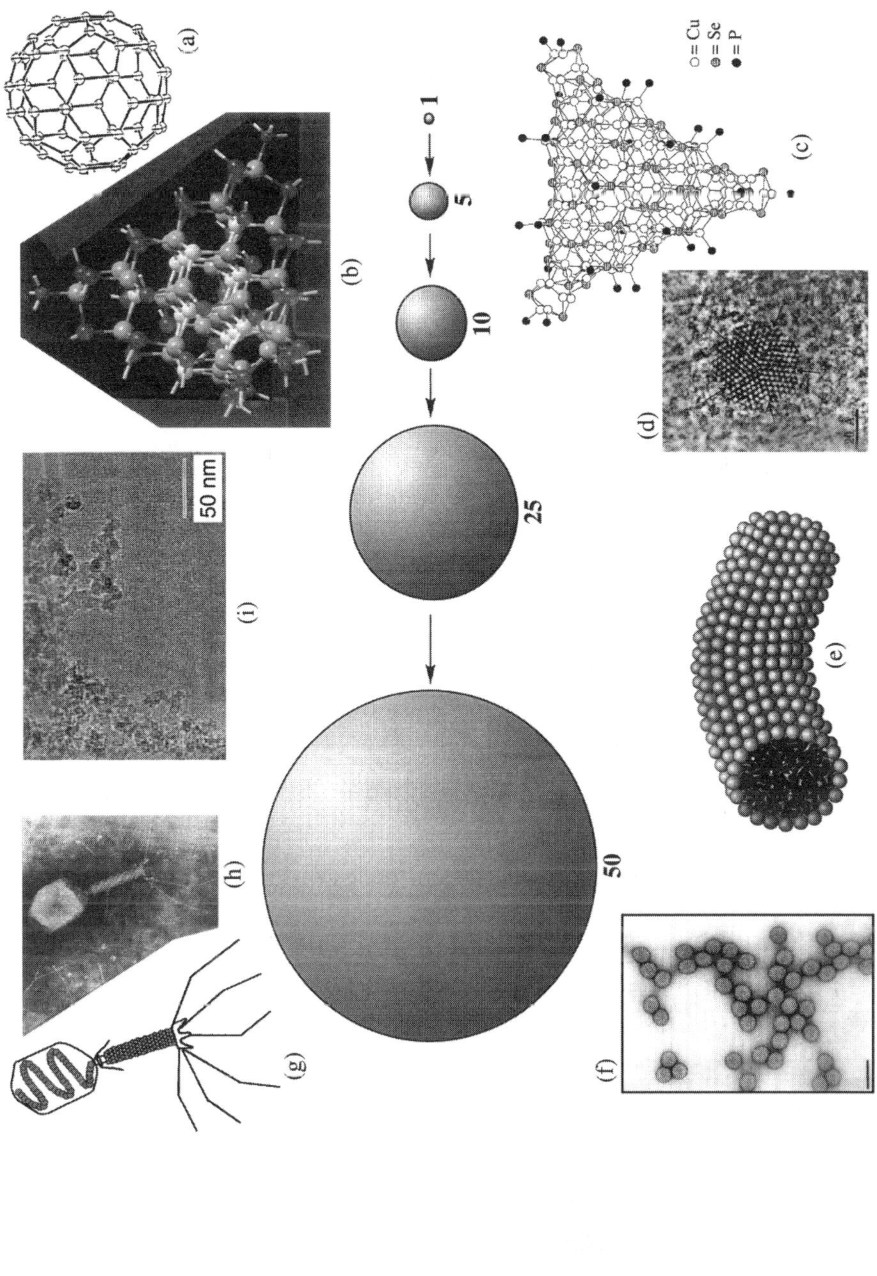
(a)
(b)
(c)
(d)
(e)
(f)
(g)
(h)
(i)
50 nm
1
5
10
25
50
○ = Cu
◍ = Se
● = P

whose onset occurs in a size range such that there is no longer a sufficient number of atoms in a single assembly for the material to exhibit the "bulk" properties. The reason for this change is that each "bulk" property is derived from a combination of the properties of a certain minimum number of atoms in a single assembly. Therefore, the unique properties of nanostructured materials are likely to depend on the nature of the material, the size of the primary particles, and the crystallites in the material, as well as external conditions such as temperature and pressure. Size effects are likely to be significant in the range from approximately 1 to 50 nm, although this depends upon the particular material and property under consideration and will be discussed later in the chapter. Below this size, the material exhibits molecular properties.

---

**Fig. 9.1.** Illustration showing the relative sizes of particles 1 to 50 nm in diameter, together with examples of various species within this size range. Species that exist in the range from just less than 1 nm to just more than 50 nm are (according to increasing size) atoms, molecules, molecular clusters, small particles, aggregates of particles, and living organisms. Clockwise: (a) $C_{60}$ (1 nm). Reprinted with permission from H.W. Kroto, A.W. Allaf, S.P. Balm, *Chem. Rev.*, $C_{60}$: *Buckminsterfullerene 91* (6), 1213–1235, Copyright 1991 American Chemical Society. (b) $Cd_{22}S_{14}(SPh)_{36} \cdot$ 4DMF (2 nm). Reprinted with permission from Herron, N., Calabrese, J.G., Farneth, W.E., Wang, Y. *Science 259*, 1426–1428. Copyright 1993 American Association for the Advancement of Science. (c) $Cu_{146}Se_{73}(PPh_3)_{30}$ (4 nm). Reprinted with permission from Wiley-VCH, Germany, Schmidt, G., Longoni, G., and Fenske, D. "Clusters in Ligand Shells," *Clusters and Colloids from Theory to Applications*. Germany: Wiley-VCH, 1994. (d) Ag particle (4 nm).

(e) Cylindrical micelles, where the ends of the cylinder are capped by hemispheres covered by surfactant head groups ($R_{micelle} = 2$ nm). Cylindrical micelles are usually polydisperse, because the rod can grow to varying lengths simply by incorporating more surfactants into the cylindrical portion of the micelle. Evans, D.F. and Wennerstrom, H., *The Colloidal Domain: Where Physics, Chemistry, Biology, and Technology Meet*, © VCH Publishers, Inc., New York, 1994. Reprinted with permission of John Wiley & Sons, Inc.

(f) Inner capsid (single-shelled particles) of rotavirus. Reprinted with permission from Palmer, E.L. and Martin, M.L., *An Atlas of Mammalian Viruses*, 1982, Copyright CRC Press, Boca Raton, FL. (g) A diagramatic representation of a T4 virion. Reprinted with permission from Birge, E.A., *Bacterial and Bacteriophage Genetics, An Introduction*, 1981, Copyright Springer-Verlag, New York Inc. (h) Transmission electron micrograph of a T2 phage. The contractile tail (the tubular part) is 20 nm in diameter. Adapted from Fraekel-Conrat, H. and Kimball, P.C., *Virology*, 1982, Prentice-Hall, Englewood Cliffs, NJ.

(i) Transmission electron micrograph of fumed silica (Cab-O-Sil, produced by the Cabot Corporation), showing the small primary particle size and the aggregation of particles to form chains.

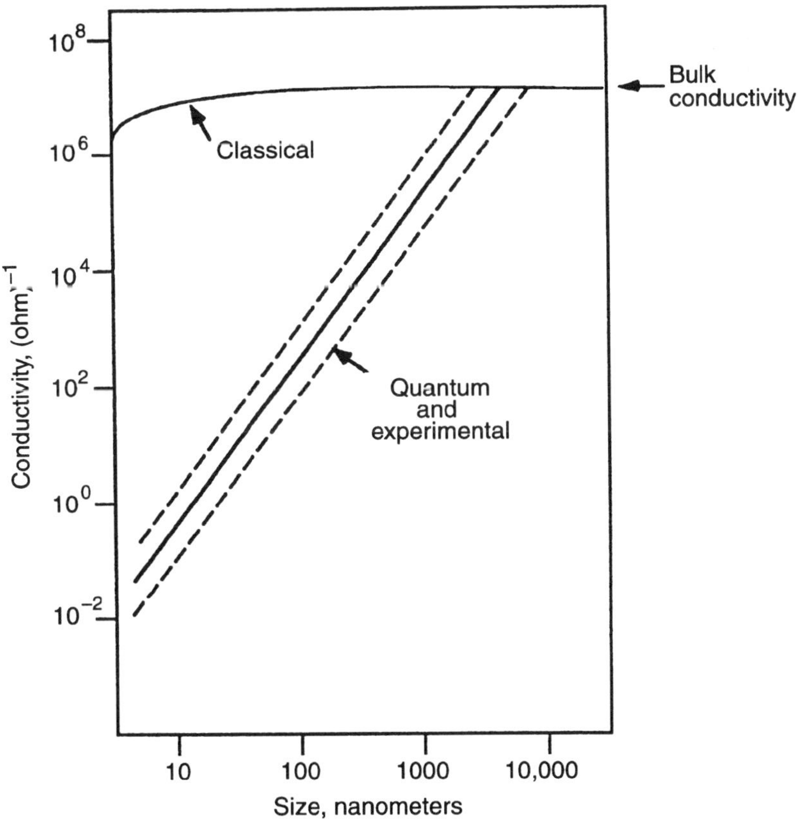


**Fig. 9.2.** Plot of conductivity of indium particles vs. particle size. Data taken at 4.2 K.

As an example of size-dependent properties, a plot of conductivity as a function of particle size for indium metal is shown in Figure 9.2. It is apparent that, for indium, the conductivity deviates from the bulk value at a particle size of approximately 1000 nm at 4.2 K. The relatively large size at which the conductivity deviates from bulk conductivity is a consequence of the low temperature. In general, the size at which this deviation begins is larger when the temperature is lower. The plot also shows the conductivity that is predicted from classical and quantum descriptions of matter. The quantum description (in which it is assumed that energy is quantized) displays a closer correlation with the experimental data than does the classical description (according to which energy is not assumed to be quantized). For this reason, materials whose components have properties (especially electrical and optical properties) in the size-dependent range are often termed "quantum confined," because the components have dimensions that are smaller than the equilibrium separation (the Bohr radius) of the electron–hole pair. (See later.) The electron–hole pair is "confined" by the dimensions of the grain (or the assembly of atoms in question) and is unlikely to result in properties similar to those of the bulk in which equilibrium separation is achieved.[1-3]

The preceding description provides a general framework for understanding the deviation of the *physical* properties of materials from their bulk values in the nanometer size regime. Deviations from bulk *chemical* and *materials* properties can be understood by considering the ratio of atoms on the surface of the assembly to those in the internal part of the assembly. Atoms on the surface are both electronically and coordinatively unsaturated and, as a result, are highly reactive—to the point that they generally react with species in the external environment to form a chemical bond that culminates in surface passivation. When the surface is passivated with such species, the clusters are considered "capped." In nanometer-sized particles, the ratio of the number of surface to internal atoms is, obviously, significantly higher than in a large piece of the material, which exhibits bulk properties.

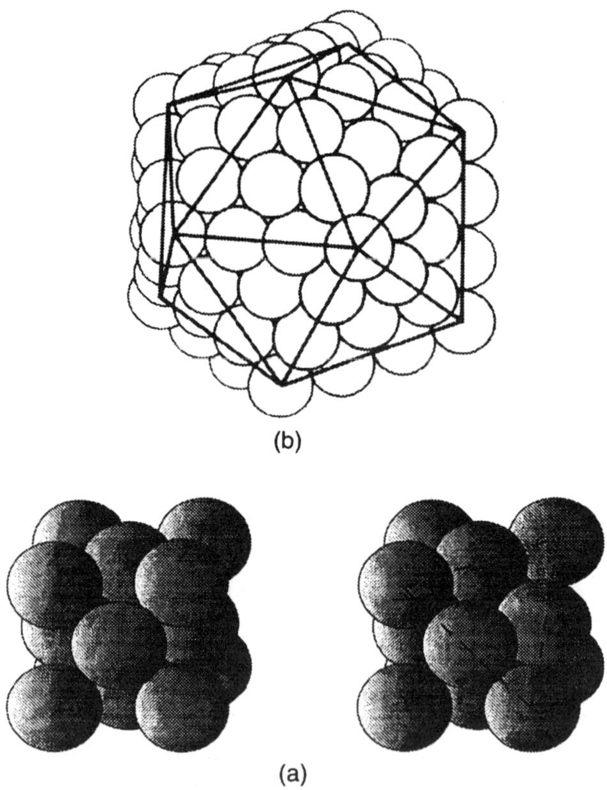


**Fig. 9.3.** Drawing of (a) 1-nm- and (b) 2-nm-diameter metal particles, showing the large proportion of atoms on the surface. Pending.

The "surface energy" of nanoparticles is therefore much higher than that of bulk materials. For example, an approximately 1-nm-diameter metal particle composed of metal atoms with radius 1.5 Å (for example, the atomic radius of In is 1.676 Å) contains a small number of atoms, most of which reside on the surface of the particle [Fig. 9.3(a)]. Only the relatively few internal atoms of the particle have a coordination environment that resembles that of the majority of the atoms in a "large" particle. [Figure 9.3(b) shows a particle approximately 2 nm in diameter in which the ratio of internal to surface atoms is greater.] As a result, the conventional definitions of thermodynamic functions must be modified to take into account the large contribution of surface energy, which is comparable to the nanoparticle volume energy, to the total energy of the system. This modification is usually achieved through the surface tension or energy, as is discussed later.

As the size of the piece of metal is reduced further, below approximately 1 nm, the properties become analogous to those of molecular species that have typical dimensions in this size regime and whose physical and chemical properties are relatively well understood. Finally, at the smallest length scale (Å), the properties of atoms are also well understood.

A number of other general aspects of nanostructured materials should be described before we undertake a more detailed discussion of size-dependent properties. Nanometer-sized features such as nanoparticles or nanocrystals are metastable objects with respect to the formation of large single crystals, and therefore it is impossible to define an "ideal" or "stable" nanocrystal. However, nanocrystals can reach a steady state when reconstruction processes have led to the removal of defects and dislocations from the bulk and distortions from the interfaces. At that point, the nanocrystal is described as "relaxed" and can be relatively stable with respect to grain growth compared to "nonrelaxed" nanocrystals. As a result, the properties of nanocrystals with equal grain sizes and chemical and phase compositions can differ noticeably, because their grain boundaries, topography, and pore structure differ, depending on the evolution of the crystals (their method of preparation, temperature, time, etc.). Therefore, nanostructured materials should be considered "condensed matter in a highly excited state."

The word *cluster* is used frequently by a variety of scientists and engineers, often with quite different meanings.[4–8] It is worth clarifying these different definitions. Synthetic chemists often refer to *molecular clusters* to describe a molecular species, neutral or charged, which generally contains two or more metal atoms with supporting ligands on the surface of the cluster that help to passivate the surface and control reactivity. (See Fig. 9.1.) *Nanoclusters* are clusters of atoms larger than molecular clusters, generally 1 to 10 nm, which generally have surfaces that are passivated (capped) by donor or acceptor ligands (molecular species that donate or accept electrons, respectively, to satisfy the electronic requirements of the surface).[9–12] Both molecular clusters and nanoclusters are generally stable, isolable species when the surface has been passivated. Physical chemists, aerosol scientists, engineers, and physicists frequently define clusters as assemblies of atoms that are often formed or transported in the gas phase.[13, 14] These clusters are generally 1 to 10 nm, are highly reactive, and are generally stable only in the gas phase at low concentrations and partial pressures of reactive species (to avoid collision with other clusters or species). Aerosol scientists also refer to *sub-* and *supercritical clusters*, often in the context of homogeneous nucleation theory. In this case, *subcritical* refers to clusters smaller than the thermodynamically stable size defined by the Kelvin equation.

The words *grain size* and *particle size* are often used interchangeably in the literature, but their distinction is important. The grain size of a material is synonymous with the crystallite size of the material; that is, it describes the length scale over which the material exists as a single crystal. The particle size, in contrast, refers to the size of separate particles. A particle may be a single crystal, in which case the particle size equals the grain size. Alternatively, the particle may be composed of a number of smaller crystallites, so that the grain size is smaller than the particle size.

Nanostructured materials have a variety of structural characteristics, from isolated particles, through three-dimensional materials made up of connected nanometer-sized particles, to thin films. There are a number of ways to describe the general structural features of nanostructured materials. One way is to characterize the dimensionality of the material, or the number of directions in which it is spatially confined. The dimensionality of quantum-confined materials can be zero, one, or two dimensional. A zero-dimensional quantum-confined material is a material in which size-dependent properties are realized in all three dimensions. An example of such a material is a spherical particle 5 nm in diameter. A one-dimensional quantum-confined material is a material in which size-dependent properties are realized in two dimensions. (In other words, the material is not spatially confined in one dimension.) An example of such a structure is a "quantum wire," which has size-dependent properties in two dimensions. Finally a two-dimensional quantum-confined material is a material in which size-dependent properties are realized in one dimension. A thin film [Fig. 9.4(a)] exemplifies this type of material. [A three-dimensional nanostructured material is a bulk solid composed of grains or primary particles (and, in some cases, pores) with sizes on the nanometer length scale, as shown in Fig. 9.4(b), part 3.]

The dimensionality of a nanostructured material will determine how it can be used. For example, a zero-dimensional quantum-confined material will have to be connected to the outside world with care, to avoid converting it into a one- or two-dimensional quantum-confined structure. Furthermore, quantum-confined structures can be placed in close proximity to each other (that is, close with respect to their own dimensions), resulting in properties that cannot be achieved by isolated, individual structures (e.g., quantum superlattices, multiple layers of materials with thicknesses less than 10 nm).

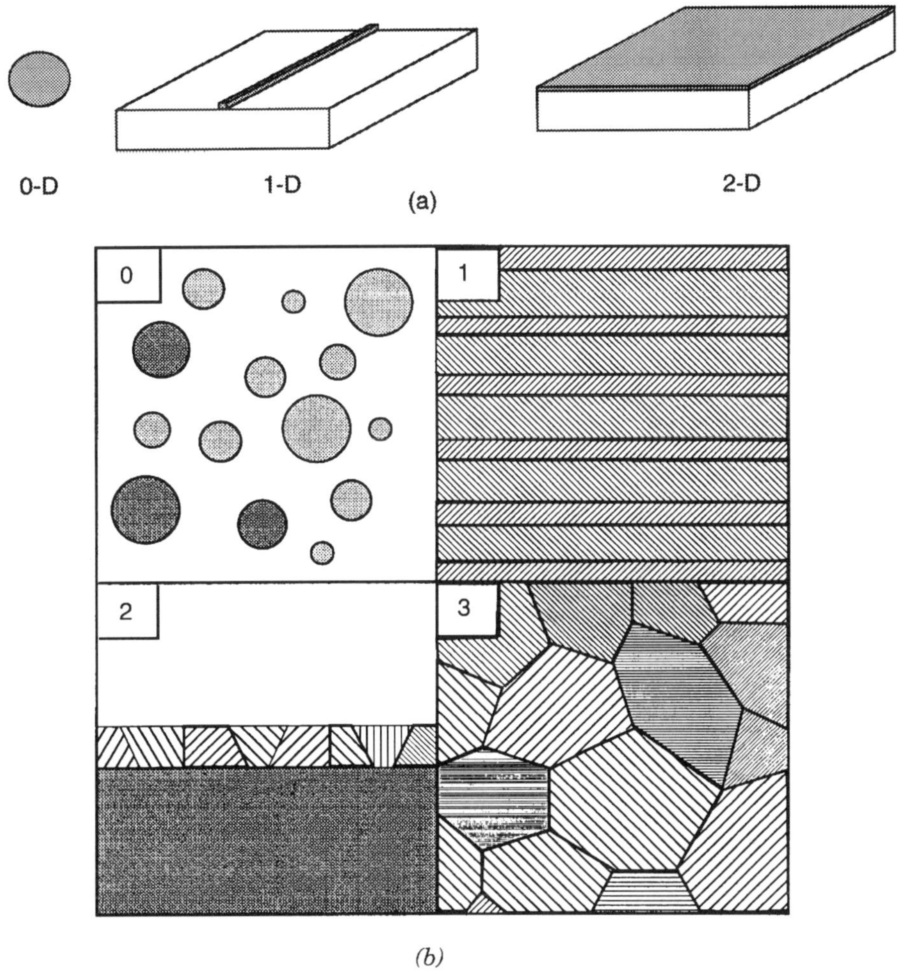


**Fig. 9.4.** Schematic diagrams of quantum-confined structures. (Adapted from Siegel.[13] Permission granted by ASM International.)

It is worth mentioning that the characterization of the properties of nanostructured materials is often difficult. The difficulty is easily illustrated by the thought, "How does one measure the conductivity of a 5-nm particle (or even a collection of 5-nm particles)?" Many problems associated with the measurement of properties of quantum-confined materials have been overcome, and measurement is an important aspect of the characterization of these materials.

Nanoparticles and nanophase materials have a variety of applications (see Section 9.3), including the possibility of forming ceramic articles from nanoparticles by mechanically deforming the material in a manner similar to that done with metals. This deformation can result in structures with unusual properties, such as high strength and good fracture toughness.

From the foregoing discussion, it should be clear that it is possible to "tailor" a particular property or a particular set of properties for a particular application to the particle size, rather than trying to find a suitable material with the appropriate intrinsic bulk properties. For example, the electrical conductivity of a film or the optical transparency of a coating may be varied as a function of its dimensions. Similarly, the hardness of a material can be controlled by varying the internal structure of the material. Aerosol synthetic methods are able to produce materials that contain or that are composed of particles or grains on the nanometer length scale. Therefore, aerosol routes hold significant promise in the synthesis of nanostructured materials.

Nanostructured materials have been synthesized by a variety of methods, including solid-state, liquid-phase, and gas-phase methods. Solid-state and liquid-phase methods are not described here; the reader is referred to many other recent reviews on these subjects.[1, 2, 7, 13, 15–25] Gas-phase routes to nanophase materials—particularly, gas-to-particle conversion—are relevant to this text, and examples using this method are provided wherever possible. Most methods require an extensive effort to control the size of the particles in the nanometer size regime and to prevent agglomeration due to the highly reactive nature of the particle surfaces.

In what follows, we will review the important size-dependent properties of materials and describe, where it is known, how these properties vary as a function of size. The discussion is not meant to be either extensive or intensive, but rather, serves to point out which properties can be varied and the relationships between properties, size, and other variables, where they are known. All properties of nanostructured materials are different from those of a bulk material, and those properties judged to be most important in the applications of nanostructured materials are described here. References to review articles are provided for more detailed information.[1, 2, 7, 13, 15–25]

## 9.2 PROPERTIES OF NANOSTRUCTURED MATERIALS

In this section, we briefly review the physical characteristics of nanostructured materials. Such materials, in the form of either films or powders, that exhibit these size-dependent properties have all been prepared by aerosol routes. In particular, gas-to-particle conversion is an excellent method of preparing nanometer-sized particles of materials with compositions that can be controlled to enhance control over the properties described in the rest of this section.

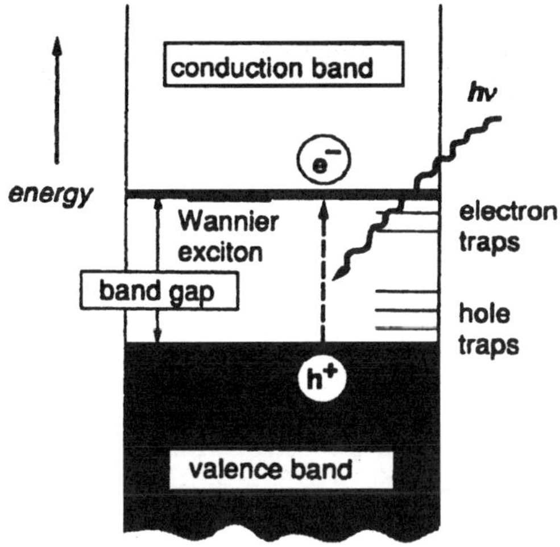


**Fig. 9.5.** Electronic structure diagram for a semiconductor. (Adapted from Photochemistry and photoelectrochemistry of quantized matter: properties of semiconductor nanoparticles in solution and thin-film electrodes, Weller, H. and Eychmueller, A., *Adv. Photochem. 20*, 165–216, Copyright 1995. Reprinted by permission of John Wiley & Sons, Inc.)[2]

### 9.2.1 Electronic Properties

The electronic properties of bulk materials are derived from the overlap of many atomic orbitals, forming a continuum of energy levels called *bands*. In the case of an undoped semiconductor (see Figure 9.5), the valence band is completely filled at 0 K, whereas the conduction band is completely empty. The energy difference between the valence band and the conduction band is called the *band gap*. An electron can be excited from the valence band to the conduction band by absorption of a quantum of light. This leaves a hole in the valence band, and the electron and hole can move independently in the crystal, resulting in electrical conductivity. The electron and hole experience coulombic interactions and can form an exciton. Recombination of the electron and hole is hindered by the thermal energy in the system, which tends to keep the electrons and holes in motion. The exciton has a low effective mass and is effectively screened (protected from the positive charges of the nuclei) by the electrons in the bulk crystal. Therefore, excitons have a relatively large radius and small binding energy (e.g., the binding energy of CdS excitons with radius 2.5 nm is 0.05 eV).[2] Because $kT$, the energy associated with thermal motion, is roughly 2 kJ/mol at room temperature, the excitons dissociate into free charge carriers (holes and electrons) at room temperature.

Quantum confinement effects arise in these systems because, in nanometer-sized particles or grains, the grain size or particle size may be on the same order as, or even smaller than, the size of an exciton in a macroscopically crystalline material. Therefore, the electron–hole pair can fit into such a particle or grain only if the charge carriers assume a higher energy state (because they are forced closer together than their equilibrium separation in bulk materials). This results in the bands splitting into discrete, quantized levels such that the band gap increases as the particle size decreases, to the point where below a certain size range (usually considered to be 1 to 2 nm), the electronic properties of the material are better described by molecular characterizations such as the highest occupied molecular orbital (HOMO) and the lowest unoccupied molecular orbital (LUMO). This feature explains how the electrical properties of nanostructured materials lie between those of bulk solids and molecules.

A number of studies of conductivity (or resistivity) as a function of particle size have been reported; data for indium, for example, are illustrated in Figure 9.2, where the conductivity was measured by microwave techniques. Other examples are given in Figures 9.6 and 9.7. From these figures and

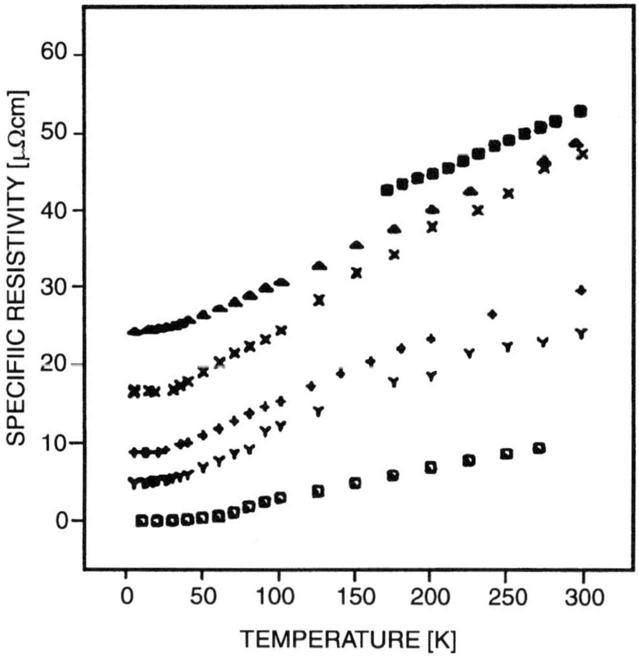


**Fig. 9.6.** Specific electronic dc resistivity of nanocrystalline Pd as a function of temperature and crystal size. (Full square = 10 nm, triangle = 12 nm, × = 13 nm, + = 22 nm, and $\gamma$ = 25 nm, in comparison to coarse-grained pure Pd, open square.) Reprinted with permission granted by Springer-Verlag. Birringer, R. and Gleiter, H., *Springer Series Chemical Physics*, *56*, 384–404, Figure 14, 1994.

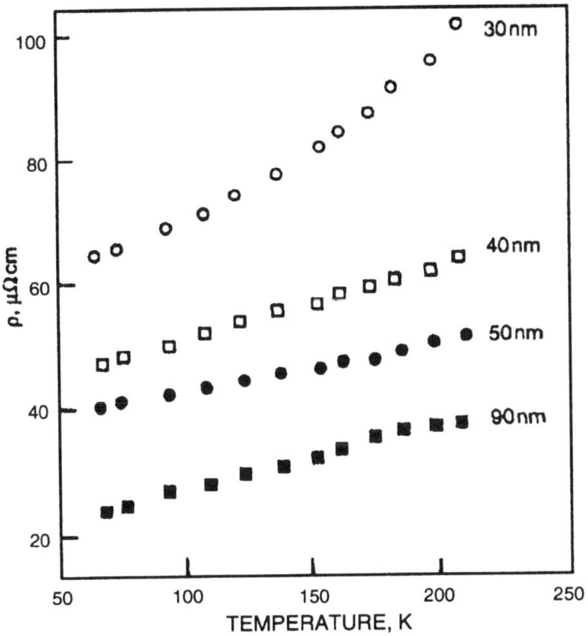


**Fig. 9.7.** Variation of electrical resistivity with temperature and grain size for nanocrystalline Fe–Cu–Si–B alloys. (Adapted from Suryanarayana.[23] Permission granted by ASM International.)

various other studies, several conclusions can be drawn:

- Resistivity (conductivity) is proportional to temperature for different grain sizes over a wide temperature range.
- At constant temperature, resistivity increases (or conductivity decreases) with decreasing grain size.
- The temperature coefficient of electrical resistivity decreases with decreasing grain size.
- A large number of grain boundaries (or an increase in grain boundary density) causes an increase in resistivity.

The last point is important when the grain size of a material is smaller than the electron mean free path, because in that case the resistivity is primarily governed by grain boundary scattering. In contrast, when the grain size is larger than the electron mean free path, the resistivity is governed by scattering inside the crystallites.

Giant magnetoresistance, a phenomenon that results in a decrease in the electrical resistance of materials exposed to a magnetic field, has been reported in a number of multilayer systems and in nanostructured materials in which, for example, nanocomposites composed of nanometer-sized cobalt particles are embedded in a continuous copper film.[26-29] The drop in electrical resistance is small (1 to 2%) in coarse-grained film structures, but is large (50 to 80%) in nanostructured materials.[30]

### 9.2.2 Optical Properties

It is natural to discuss optical properties after electrical properties because of their close relationship. The intrinsic optical properties of a material, such as its UV–visible absorption spectrum, vary in a way that is related to the band

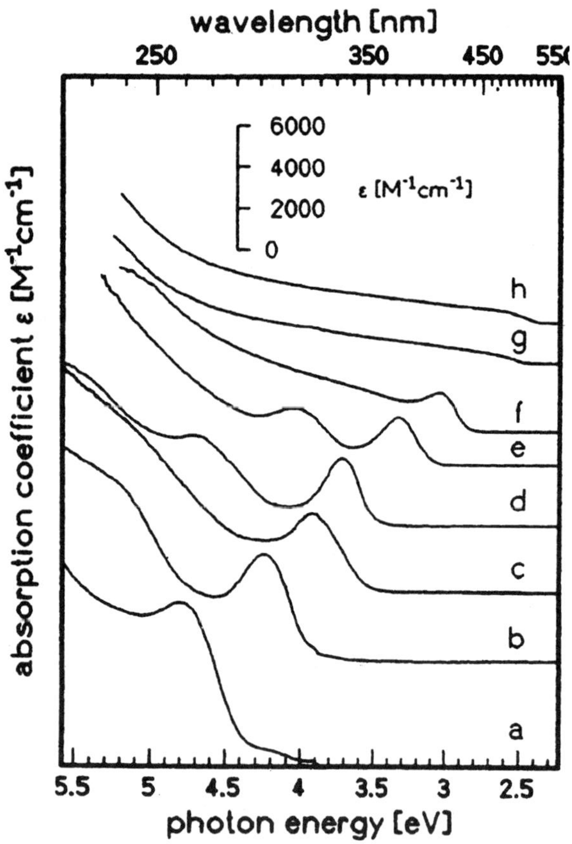


**Fig. 9.8.** Absorption spectra of CdS nanoparticles. The mean diameter of the particles increases from 1.3 (a) to 9 nm (h) in diameter (as determined by X-ray scattering). The 9-nm-diameter particle exhibits an absorption spectrum similar to that of bulk CdS. (Adapted from Photochemistry and photoelectrochemistry of quantized matter: properties of semiconductor nanoparticles in solution and thin-film electrodes, Weller, H. and Eychmueller, A., *Adv. Photochem. 20*, 165–216, Copyright 1995. Reprinted by permission of John Wiley & Sons, Inc.)

gap of the material. As the particle size decreases, the electrical conductivity decreases, so the UV–visible absorption maximum of a material is blue-shifted relative to the bulk of the material (because the band gap increases, so that the energy required by the material to promote an electron from the valence band to the conduction band increases). Figure 9.8 shows the changes in the UV–visible absorption spectrum as a function of size for particles 1.3 to 9 nm in diameter.[2, 31] For example, CdSe nanoclusters of diameter 1.2 to 1.5 nm have a band gap of 3.0 eV (~ 300 kJ/mol, or 72 kcal/mol), whereas those with diameters in the range 3.0 to 3.5 nm have a band gap of 2.3 eV (~ 230 kJ/mol, or 55 kcal/mol). By contrast, bulk CdSe has a band gap of 1.8 eV (~ 180 kJ/mol, or 43 kcal/mol).

Another optical property of a nanostructured material is optical transparency, which can be controlled by manipulating the spatial arrangement of pores in a nanoporous material. This is illustrated by comparing two disks of $Y_2O_3$, both synthesized by gas-to-particle conversion. The two samples have approximately the same pore and grain size (8 nm), but one is optically transparent while the other is not. The key parameter that results in this difference is the distribution of the pores throughout the samples. The optically transparent specimen has pores with a very low degree of aggregation; that is, the pores are isolated. At the same time, the nontransparent sample has pores with a higher degree of aggregation (i.e., the "primary" pores are connected and give a larger average pore size). The result is that light is not scattered by the "isolated" pores, due to the much smaller dimensions of the pores compared to the wavelength of the light, while light is scattered by the aggregated pores, which have larger dimensions on a length scale similar to that of the wavelength.[22, 23] An example of a transparent nanostructured zeolite monolith is presented in Figure 9.9.[32]

Recently, it was reported that doped nanocrystals of semiconductors can simultaneously yield both high luminescent efficiencies and a shortening of the lifetime of the semiconductor. Nanocrystals of Mn-doped ZnS with sizes varying from 3.5 to 7.5 nm were prepared at room temperature, yielding an external photoluminescent quantum efficiency of 18% and a luminescent decay at least five orders of magnitude faster than that of the corresponding $Mn^{2+}$ transition in the bulk crystals. The quantum efficiency increases with decreasing size of the particles. This combination of properties has not been reported previously and can be attributed to the size-dependent nature of these properties.[33]

Porous silicon, which is composed of primary particles on the order of 2 to 5 nm, has recently been shown to give rise to photoluminescence, which has stimulated interest in this material for applications in optoelectronic devices. The property has been connected to the small particle size in this form of silicon, and theoretical studies predict that an increase in the size of the silicon nanoparticles should result in a shift of the photoluminescence peak to a lower energy, with a concomitant decrease in energy until the phenomenon vanishes, as in the case of bulk silicon.[34]

**Fig. 9.9.** Slab of nanocrystalline zeolite, showing that optically transparency can be obtained by tailoring the the particle size and pore structure of the material. (Reprinted with permission from Jung, K.T., and Shul, Y.G., *Chem. Mater.* *9*, 420, copyright 1997, American Chemical Society.)

Coated semiconductor nanoparticles are also attracting a great deal of interest as a result of the modification of the properties of the nanoparticles by the coating. These so-called heterostructure quantum dots, in which a semiconductor nanoparticle is coated by a different semiconductor, can give rise to a variety of interesting optical and electrical phenomena (e.g., the quantum Hall effect and new fluorescence phenomena), in addition to added chemical stability. For example, CdS has been coated with $Cd(OH)_2$, CdSe with ZnS, and HgS with CdS. All three materials exhibited a new fluorescence near their bandgaps that was not observed in the original nanoparticles.[25] In another example, gold sulfide nanoparticles, $Au_2S$, approximately 50 nm in diameter, were coated with a thin layer ($\sim$ 2 nm) of crystalline gold by chemical reduction of the outer layer of the $Au_2S$ particles. The Au-coated particles exhibited extraordinary optical absorption spectra, compared to non-coated samples. Interest in metal-coated particles stems from the expectation of a large nonlinear optical response due to surface plasmon resonance.[35]

### 9.2.3 Magnetic Properties

Ferromagnetic properties of materials (the behavior of the magnetic moments in the materials) depend on interatomic distances. Therefore, it should be expected that in nanostructured materials, where interatomic distances are distorted, ferromagnetic properties will be changed. As a result, the

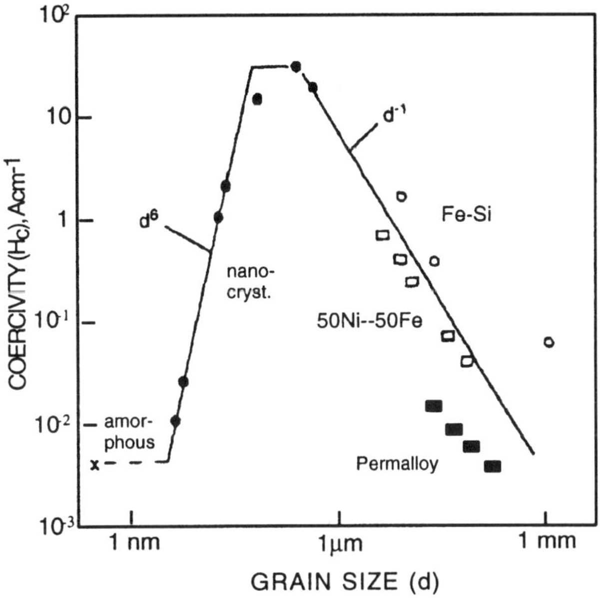


**Fig. 9.10.** Variation in coercivity as a function of grain size for various soft magnetic materials. (Adapted from Herzer.[36, 37] Permission granted by ASM International.)

saturation magnetization $M_s'$ (the maximum magnetization due to the application of a magnetic field) and the ferromagnetic transition temperature of nanocrystalline materials should be reduced, compared to these properties in bulk materials. Examples of such a reduction have been reported in various systems, including Fe, in which 6-nm Fe particles exhibit an $M_s'$ of 130 emu $g^{-1}$, compared to the bulk $\alpha$-Fe value of 220 emu $g^{-1}$.[17] The ferromagnetic Curie temperature $T_c$ was reduced by about 40 K in 70-nm Ni particles, compared to the value in bulk Ni.[17]

Herzer studied the coercivity (a measure of the difficulty of reversing a material's magnetization) of Fe–Cu–Nb–Si–B alloys with grains in the range 10 to 300 nm and found that the values varied over several orders of magnitude. In particular, the coercivity increased steeply as a function of the sixth power of the grain size, $d_{gr}^6$, up to 50 nm.[36, 37] Above 150 nm, the coercivity followed the more familiar $d_{gr}^{-1}$ dependency. This behavior is illustrated in Figure 9.10.

Magnetic materials are a case where the history of the sample is extremely important.[38–40] Materials made by different methods, but that appear to have identical structures or grain sizes, can have markedly divergent properties as a result of different domain structures, which may result from disparate thermal treatments and stress. These domains are regions over which all the magnetic moments are aligned and are not necessarily the same size as grains or particles.

Magnetic nanocomposites can exhibit a magnetocaloric effect, which occurs when small magnetic particles, or particulates, are contained in a weakly magnetic or nonmagnetic matrix. When the particulates are placed in a magnetic field, their spins align with the field, lowering the magnetic entropy of the system. If this process is carried out adiabatically, the reduction in spin entropy is counteracted by an increase in lattice entropy, resulting in a rise in temperature. The process is reversible, so that, conversely, the specimen cools upon removal of the magnetic field (causing an increase in magnetic entropy). Shull et al.[41] have studied this phenomenon extensively and have shown that nanocomposites composed of various solid solutions, such as $Gd_3Ga_{3.25}Fe_{1.75}O_{12}$, exhibit magnetocaloric effects that exceed those of the best previously known paramagnetic refrigerant ($Gd_3Ga_5O_{12}$) under a given set of conditions.

### 9.2.4 Compositional and Phase Behavior

The origin of the variation in many properties of nanostructured materials can be attributed to the fundamental changes caused by the physical reduction in size compared to macroscopic materials. The effects of particle size on these properties can be described using several approaches. In the most common one, the *capillarity approximation*, it is assumed that the free energy of a nanoparticle can be written as the sum of the free energy of the bulk material and the free energy associated with surfaces and interfaces. In this approximation, the free energy per unit mole of the material in a nanoparticle is assumed to be independent of the particle size and the same as that of the bulk material. Similarly, the free energy per unit surface area is usually assumed to be independent of particle size. Both of these approximations break down as the particle size decreases, and there is argument over the size at which the approximations fail. (See Section 9.2.4.6.) Furthermore, the approach assumes a liquid particle and is not valid for solid particles, except Curie–Wulff solids. Nonetheless, the relative simplicity of this approach makes it attractive. Simple equations can be derived in terms of the surface tension and energy, particle size, and other relevant parameters. The capillarity approximation is the approach taken in most models of aerosol reactor behavior.

The other approach to calculating the effects of particle size on properties is essentially an atomic approach in which the free energy of a cluster is calculated on the basis of potentials describing the interaction of atoms in nanoparticles. This approach is far more complex than the capillarity approach and is therefore used infrequently.

***9.2.4.1 Crystal Lattice Distortions*** Crystal lattice distortions, which are common in nanometer-sized particles, are the result of the changes in bond lengths due to the large fraction of atoms on the surface of the particles. The large number of surface atoms alters the equilibrium interatomic forces and tensile deformations. The average distortion of interatomic distances $a'$ in nanoparticles caused by surface tension is given by

$$\Delta a'/a' = \theta_2 \cdot K' \cdot \gamma/d_p, \tag{9.1}$$

where $\theta_2$ is a numerical coefficient (on the order of unity or, in the case of isotropic spherical particles, 2/3), $K'$ is the volume compressibility coefficient, $d_p$ is the average diameter of the particles, and $\gamma$ is the average surface energy of the particles.[1] This equation predicts values of lattice contractions of several tenths of a percent at typical values of these parameters for 10-nm-diameter nanoparticles. The contractions contribute to the high reactivity of the surfaces of small particles because a small change in the equilibrium bond distances results in a large change in bond energy.

***9.2.4.2 Compositional Inhomogeneities*** Like crystal lattice distortions, compositional or concentrational inhomogeneities vary as a function of particle size. Calculations concerning the segregation of one crystalline phase (with dimensions on the nanometer length scale) to the surface of a crystalline nanoparticle show that the compositional inhomogeneity in nanoparticles can increase to as much as 10% relative to that of larger particles.[1] One way to examine this phenomenon is in terms of the solubility of one phase in another. Solubilities are determined by chemical potentials, but because the atomic structure is changed in nanoparticles compared to that of the bulk material, the chemical potential, and therefore the solubility, also should change. Such effects have now been experimentally observed. For example, hydrogen is significantly more soluble in nanophase Pd than in macroscopic Pd, and the solubility of Bi in Cu increases from the value of $10^{-4}\%$ in bulk to approximately 4% in nanocrystalline Cu.

Compositional or phase inhomogeneities may be formed or removed, depending on the free energies $F$ of the individual phases (and the interfaces between them), compared to those of the homogeneous material. The problem can be examined with the inequality

$$F_{V_1, f_1} + F_{V_2, f_2} + \frac{\overline{\gamma}_1 S_1}{V_1} + \frac{\overline{\gamma}_{1,2} S_2}{V} < F_{V, f_o} + \frac{\overline{\gamma}_i S_1}{V}. \tag{9.2}$$

This inequality is written in terms of energy per unit volume. The left side is the free energy associated with the coated particle, and the right side is the energy associated with the homogeneous particles. The first term on the left, $F_{V_1, f_1}$ (erg/cm$^3$), is the free energy of the core; the second term, $F_{V_2, f_2}$ (erg/cm$^3$), is the free energy of the coating; the third term is the interfacial energy associated with the coating; and the fourth term is the energy associated with the interface. The quantities $S_1$ (cm$^2$) and $S_2$ (cm$^2$) are the surface areas of the particle and the core, respectively, and the quantities $V_1$ (cm$^3$) and $V_2$ (cm$^3$) are the volumes of the shell and the core, respectively. The quantity $\overline{\gamma}_1$ (erg/cm$^2$) is the surface free energy associated with the outer surface, $\overline{\gamma}_{1,2}$ (erg/cm$^2$) is the energy associated with the coating–core

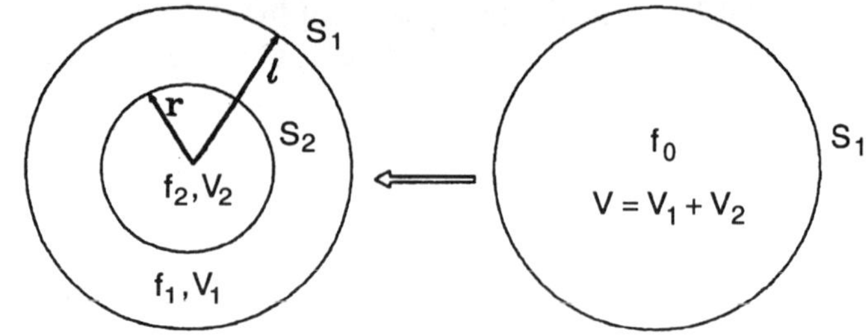


**Fig. 9.11.** Schematic transformation of two-element nanoparticle with initially homogeneous structure to inhomogeneous structure. (Reprinted from *Progress in Materials Science*, *37*, Gryaznov and Trusov, Size effects in micromechanics of nanocrystals, pp. 289–401, copyright 1993, with kind permission from Elsevier Science Ltd., The Boulevard, Langford Lane, Kidlington OX5 1GB, UK.)[1]

interface, and $\overline{\gamma}_i$ ($erg/cm^2$) is the surface free energy associated with the homogeneous material.

For example, if the free energy of a two-component system made up of *A* and *B* is considered to have the fractions $f_0$ and $1 - f_0$ of its constituent atoms at equilibrium at a given temperature, then the formation of a new phase in the shell region (the coating) and the formation of another interface $S_2$ between the two phases requires an energy loss relative to the energy of the homogeneous material. The formation of a single homogeneous phase is energetically favorable when the gain in surface energy caused by formation of the coating with surface area $S_1$ exceeds the free energy of the homogeneous particle [Equation (9.2) and Fig. 9.11].[1]

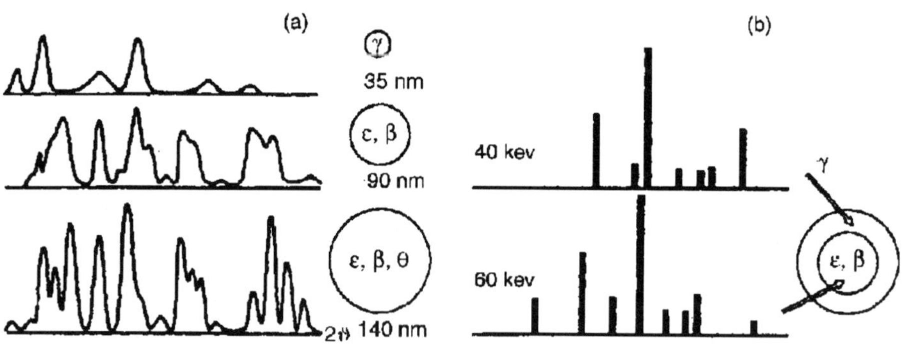


**Fig. 9.12.** X-ray diffraction data showing phases in alumina particles as a function of size and homogeneity. (a) Data for different-size alumina particles; (b) data at different voltages, showing compositional inhomogeneity. (Reprinted from *Progress in Materials Science*, *37*, Gryaznov and Trusov, Size effects in micromechanics of nanocrystals, pp. 289–401, copyright 1993, with kind permission from Elsevier Science Ltd., The Boulevard, Langford Lane, Kidlington OX5 1GB, UK.)[1]

The preceding effects have been observed in the formation of coated alumina particles.[1] A related problem is the formation of different phases of alumina as a function of particle size. This has also been observed experimentally (Fig. 9.12). The reasons that one particular crystalline phase is observed rather than another is discussed in the next section.

*9.2.4.3* ***Phase Transitions and Temperatures*** The characteristic length scale at which macroscopic thermodynamic functions are modified compared to their bulk values is approximately 10 nm. The discussion thus far has addressed how the lattice dimensions and segregation of materials can vary with particle size. Here we discuss the difference between the *phase transition temperatures* and the thermodynamically stable phase of homogeneous bulk and nanostructured materials. The starting point for this analysis is the *Clausius–Clapeyron* relationship,

$$\Delta T_t = \frac{T_t}{\lambda_p}\left(\frac{\gamma_2 S_2}{V_2} - \frac{\gamma_1 S_1}{V_1}\right), \tag{9.3}$$

where $\lambda_p$ (erg/cm$^3$) is the heat of the phase transition per unit volume, $S_1$ and $S_2$ are the surfaces areas and $V_1$ and $V_2$ are the volumes for states 1 and 2, respectively, and $\gamma_1$ and $\gamma_2$ are the respective surface energies for states 1 and 2. The change in temperature of a phase transition, $\Delta T_t$, can be roughly estimated as $T_t\gamma/\lambda l$ where l is the difference in the nanoparticle sizes in states 1 and 2.

These formulas imply that lower surface energy (more tightly packed) crystalline phases become more energetically favorable with decreasing nanoparticle size. Thus, some predictions can be made for simple structures. Face-centered cubic (FCC) crystal lattices have a lower specific volume and surface energy, compared to body-centered cubic (BCC) lattices. Therefore, a material that exhibits an FCC structure in the macroscopic state will likely remain FCC as the particle size decreases into the nanometer size regime. However, materials that exhibit BCC structures in the macroscopic state are likely to undergo a phase transition to FCC as the particle size falls into the nanometer size range. Furthermore, new phases may be observed for nanometer-sized particles or grains which have not previously been observed for that particular material in bulk. This has been borne out experimentally, where, for example, 5-to-10-nm-sized Nb, Mo, W, and Ta particles have been shown to exhibit FCC or hexagonal close-packed (HCP) structures instead of the BCC structures they exhibit in the bulk material.[1] Similar observations have been made in metal-oxide materials. For example, $BaTiO_3$, an important ferroelectric material with a perovskite type of structure, which exhibits a tetragonal unit cell in bulk at room temperature, undergoes a phase change to a cubic unit cell as the particle size decreases.[42] The higher cubic symmetry destroys the polarity of the unit cells, with a resultant loss of ferroelectric properties.

*9.2.4.4 Vapor Pressure* Another feature of the relationships described thus far is that nanoparticles should be expected to have surface atoms that have higher mobilities than surface atoms in bulk materials. This leads to changes in a series of properties, such as lower melting points, higher diffusion coefficients, differences in thermal expansion coefficients, higher solubilities, and higher equilibrium vapor pressures, compared with these same properties in the bulk material. These changes are expected as a result of the lower activation energy required to displace atoms at a surface and the high proportion of surface atoms in nanostructured materials. Here we describe several such effects.

The increase in vapor pressure and solubility at a point of high surface curvature, which is found extensively in nanometer-sized particles, has been well documented. For example, the increase in vapor pressure for a non-spherical particle is given by

$$RT\ln\left(\frac{p_d}{p_e}\right) = V_{\text{mol}}\gamma(1/r_1 + 1/r_2), \tag{9.4}$$

where $p_d$ is the vapor pressure over a curved surface, $p_e$ is the vapor pressure over a flat surface, $R$ is the gas constant, $T$ (K) is the temperature, $V_{\text{mol}}$ ($\text{cm}^3/\text{mole}$) is the molar volume, $\gamma$ ($\text{erg/cm}^2$) is the surface tension, and $r_1$ (cm) and $r_2$ (cm) are the radii of curvature of the particle. For a spherical particle,

$$RT\ln\left(\frac{p_d}{p_e}\right) = 2V_{\text{mol}}\gamma/R_p, \tag{9.5}$$

where $R_p$ is the particle radius. This relationship can be derived by considering the work expended in transferring one mole of a material from a flat surface through the vapor phase to a spherical surface. The work must be equal to the product of the surface energy $\gamma$ and the differential change in surface area, which is given by

$$\gamma dA_p = \gamma 8\pi R_p dR_p. \tag{9.6}$$

The change in volume is given by

$$dv = 4\pi R_p^2 dR_p, \tag{9.7}$$

and the change in radius for a 1-mole transfer is given by

$$dR_p = dv/4\pi R_p^2. \tag{9.8}$$

**TABLE 9.1. Effect of surface curvature on pressure difference and relative vapor pressure as a function of particle diameter**

| Material | Surface Diameter (microns) | Pressure Difference (psi) | Relative Vapor Pressure $(p/p_0)$ |
|---|---|---|---|
| Silica glass | 0.1 | 1750 | 1.02 |
| (1700°C) | 1.0 | 175 | 1.002 |
| $\gamma = 300\ \mathrm{erg/cm^2}$ | 10.0 | 17.5 | 1.0002 |
| Liquid cobalt | 0.1 | 9750 | 1.02 |
| (1450°C) | 1.0 | 975 | 1.002 |
| $\gamma = 1700\ \mathrm{erg/cm^2}$ | 10.0 | 97.5 | 1.0002 |
| Liquid water | 0.1 | 418 | 1.02 |
| (25°C) | 1.0 | 41.8 | 1.002 |
| $\gamma = 72\ \mathrm{erg/cm^2}$ | 10.0 | 4.18 | 1.0002 |
| Solid $Al_2O_3$ | 0.1 | 5250 | 1.02 |
| (1850°C) | 1.0 | 525 | 1.002 |
| $\gamma = 905\ \mathrm{erg/cm^2}$ | 10.0 | 52.5 | 1.0002 |

Equation (9.6), with auxilliary equations, can be manipulated to give the same relationship as found in Equation (9.5). Some examples of relative vapor pressures of various materials as a function of the diameter of a spherical particle are given in Table 9.1.

Equation (9.5) is equivalent to the Kelvin equation (described in Chapter 4) that describes the partial pressure of a vapor required for a particle to remain in thermodynamic equilibrium with its surroundings. That is to say, a nanoparticle in a gas will evaporate unless its vapor pressure is so low that its evaporation rate is negligible or the vapor is supersaturated with the evaporating species to the point where the partial pressure of that species is equal to the value of $p_d$ defined by Equation (9.5). For small particles, this value can be many times larger than the vapor pressure over a flat surface.

As a result of the preceding effects, nanoparticles can be considered "hot," or in a thermally excited state, compared to bulk materials, even though the ambient temperature they experience may be low. This explains why high temperature phases are often adopted by nanoparticles.

***9.2.4.5 Internal Pressure*** In a manner similar to their increased vapor pressures, nanoparticles can be considered to have a high internal pressure compared to their bulk analogs when the ambient pressure is not extreme. At equilibrium, the work of expansion, $\Delta p\, dv$ (particle growth), where $p$ is the pressure, must equal the change in surface energy, $\gamma\, dA_p$ (where $\gamma$ is the surface tension). For a spherical geometry, this equality is shown in the equation

$$\Delta p = \frac{2\gamma}{R_p}, \tag{9.9}$$

where $R_p$ is the radius of the particle. For a nonspherical geometry, this relation is given by[1]

$$\Delta p = \gamma(1/r_1 + 1/r_2), \tag{9.10}$$

where $r_1$ and $r_2$ are the radii of the principal axes of the solid. The inverse relationship between the pressure and size of the particle results in high pressures at small particles sizes. This explains why high pressure phases are commonly adopted by nanoparticles. For example, the high pressure phase of alumina, $\gamma$-alumina, was adopted when the size of alumina particles was approximately 35 nm, and the high-pressure cubic phase was observed in 5-nm CdS particles.[1] The pressure difference inside a particle relative to that outside the particle is shown in Table 9.1.

***9.2.4.6 Melting Point*** The melting point of nanoparticles can be considerably lower than that of bulk materials. The change in melting point relative to that of the bulk is given by[43]

$$T_b - T_m = \frac{2T_m}{L_f \rho_{\text{sol}} R_p}\left[\gamma_{\text{sol}} - \gamma_{\text{liq}}\left(\frac{\rho_{\text{sol}}}{\rho_{\text{liq}}}\right)^{2/3}\right], \tag{9.11}$$

where $T_b$ is the melting point of the bulk material, $T_m$ is the melting point of the nanoparticle, $R_p$ is the radius of the particle, $L_f$ is the molar latent heat of fusion, $\rho_{\text{sol}}$ is the density of the solid form of the material, $\rho_{\text{liq}}$ is the density of the liquid form of the material, $\gamma_{\text{sol}}$ is the surface tension of the solid material, and $\gamma_{\text{liq}}$ is the surface tension of the liquid material. Equation (9.11) has been used to correlate the melting point with the particle size for a variety of materials, including metals and semiconductor nanocrystals,[44, 45] and to determine the values of the surface tensions of solid or liquid materials in nanoparticle form. A fit of the melting point to the particle size gives the difference in surface tensions of the solid and liquid. If onc is known independently, both are known.

Equation (9.1), describing the lattice contraction as a function of particle size, can be used to obtain the surface tension of the solid and allows one to determine the surface tension of liquid nanoparticles. This approach has been used to show that the surface tensions of nanoparticles are much higher than those of the bulk materials.[44] The surface tension of bulk CdS is 0.750 N/m; in contrast, the surface tension of CdS nanoparticles is 2.5 N/m. *This result has serious implications for calculations using surface tensions* (*calculations involving nucleation theory, melting-point depression, solubility, internal pressure, vapor pressure, phase transitions, and compositional inhomogeneities*) *when the surface tension of the bulk material is used*; *the correct surface tension may be much different from the calculated value*. In fact, because the surface tension often appears in exponentials in these equations, the calculated quantity may be in error by more than an order of magnitude.

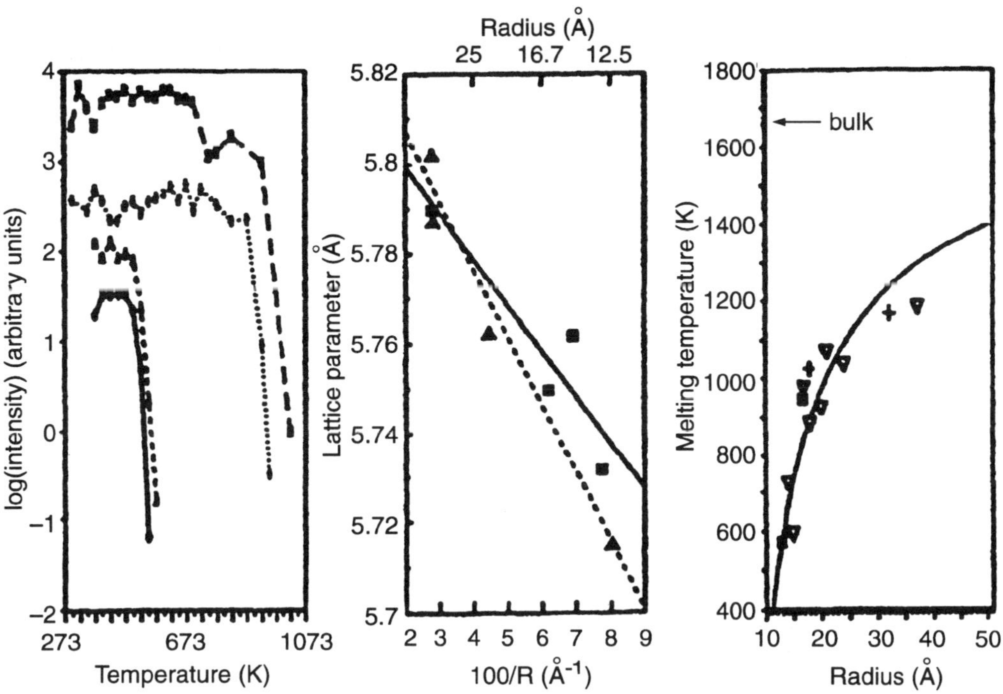


**Fig. 9.13.** (left) Intensity of electron diffraction from the (111) peak as a function of temperature for CdS nanocrystals of several sizes (12, 13, 16, and 18 angstroms, from left to right). (center) Lattice parameters for CdS nanocrystals as a function of inverse nanocrystal size. Triangles denote data for bare nanocrystals, and squares denote data for mercaptoacetic acid-capped nanocrystals. The solid line fits a surface tension of 1.74 N/m, while the dashed line fits a surface tension of 2.5 N/m. (right) Dependence of melting point on particle size. (Reprinted with permission from Goldsten, A.N., Echer, C.M., and Alivisatos, A.P., *Science*, 1425, Copyright 1992. American Association for the Advancement of Science.)[44]

Figure 9.13 (left) shows how the intensity of the (111) peak for electron diffraction of CdS nanocrystals diminishes when the nanocrystals melt. Figure 9.13 (center) shows how the lattice parameters for CdS nanocrystals vary with size, and Figure 9.13 (right) shows how the melting temperature of the particles varies with their size.

***9.2.4.7 Solubility*** The solubility of nanoparticles can be considerably higher than that of bulk materials. The analysis of this problem is analogous to that of vapor pressure. An increase in solubility because of surface curvature is given by

$$\ln \frac{C}{C_o} = \frac{4\gamma_{s1}\upsilon_a}{d_p k_B T}, \qquad (9.12)$$

where $C$ is the solubility over the nanoparticle or curved surface, $C_o$ is the solubility over a flat surface, $\gamma_{s1}$ is the solid–liquid surface energy, $v_a$ is the atomic volume in cm$^3$, $d_p$ is the particle diameter, $k_B$ is Boltzmann's constant, and $T$ is temperature. This equation shows that small particles will dissolve faster than larger particles and suggests that nanoparticles of drugs, for example, would be useful if higher delivery rates into liquids were desired.[46] The equation also suggests that sintering through mechanisms involving liquid phases may proceed more rapidly when the particles are in the nanometer size range.

***9.2.5 Mechanical Properties of Consolidated Nanoparticles*** The mechanical properties of materials are strongly affected when the bulk material is composed of grains in the nanometer size regime. Therefore, the following discussion applies more to an assembly or collection of connected nanometer-sized particles rather than to isolated nanometer-sized particles. The origin of the deviation of mechanical properties in nanostructured materials from the properties of bulk (microcrystalline) materials with larger primary particles is the large volume fraction of grain boundaries. This feature is illustrated schematically in Figure 9.14, and, assuming that the grains are spherically symmetrical, we can calculate the percentage of atoms in the grain boundary volume according to the equation[47]

$$C_b = \frac{3L_g}{d_{gr}}, \tag{9.13}$$

where $C_b$ is the percentage of atoms in the grain boundary volume, $L_g$ is the

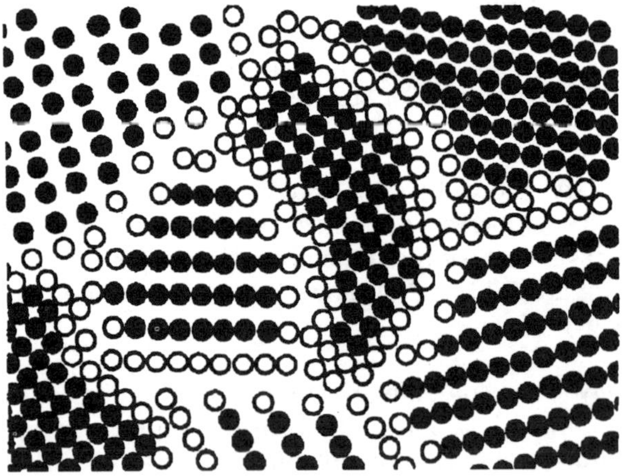

**Fig. 9.14.** Schematic representation of equiaxed nanocrystalline elemental material, distinguishing between atoms associated with the individual grains (•) and those contained in the grain boundaries (o). (Adapted from Gleiter.[47] Permission granted by ASM International.)

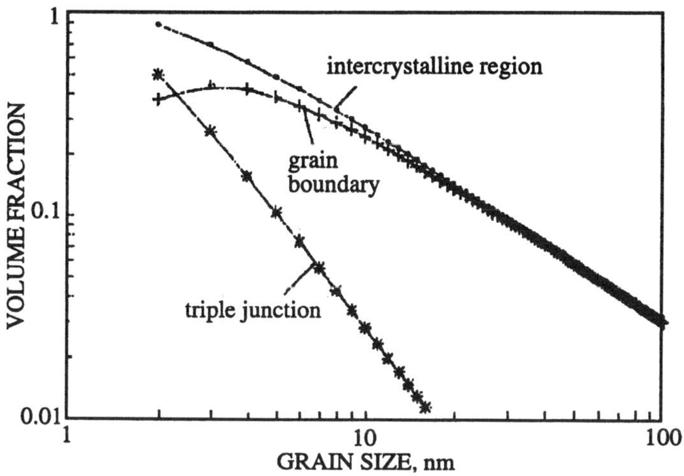


**Fig. 9.15.** Effect of grain size on calculated volume fractions for intercrystalline regions, grain boundaries, and triple junctions. (Adapted from Suryanarayana et al.[48] Permission granted by ASM International.)

average grain boundary thickness, and $d_{gr}$ is the average grain diameter. This equation gives grain boundary volumes of 50% for 6-nm grains, 30% for 10-nm grains, and 3% for 100-nm grains, assuming a constant grain boundary thickness of 1 nm.[23] However, the model is not realistic, because, in practice, the grains are unlikely to be spherical and the surfaces of the crystallites are likely to be significantly distorted. Calculations using 14-sided (tetradecahedral) grains and taking into account 1-nm grain boundaries, intercrystalline regions with a thickness of 0.5 nm (the outer layer of the grain), and triple junctions between grains show that these regions can account for up to 70 to 80% of the volume fraction of the material when it is composed of 2- to 3-nm grains (Fig. 9.15).[48]

The consequence of a large volume fraction of matter associated with the grain boundaries is that all mechanical properties related to the diffusion of atoms are drastically altered, compared to their values in bulk crystalline materials.[49-53] This alteration is caused by the huge difference in the rates of grain boundary diffusion vis-à-vis lattice diffusion, for which it has been demonstrated that in some cases the coefficient for grain boundary diffusion is 16 orders of magnitude larger than the coefficient for lattice diffusion in the bulk crystalline material. For example, self-diffusion in nanocrystalline copper composed of 8-nm-sized particles ($T$ = 293 to 393 K) showed an enhancement of approximately $10^{19}$ compared to lattice diffusion! Furthermore, a hundred fold enhancement in grain boundary diffusion of nanocrystalline copper over microcrystalline copper was observed and ascribed to the different grain boundary structures between the two materials.[54]

As a result of the small size of the grains, the highly reactive nature of the grain surfaces, the large volume fraction of grain boundaries, and the high grain boundary diffusion coefficients, nanostructured materials exhibit an exciting variety of mechanical and structural properties that are different from, and often advantageous to, those of bulk macrocrystalline materials. These properties include lower sintering temperatures, superplastic behavior, increased strength and hardness, increased ductility and toughness, increased specific-heat capacity, lower thermal conductivity, and a higher thermal expansion coefficient. However, it should be noted that many factors are involved in determining the macroscopic properties of a solid made up of nanometer-sized particles, and the trends described in the previous sentence are only generally true. For example, in many cases, the rate of growth of nanocrystal grains is very much slower than might be anticipated on the basis of the preceding discussion because of the unexpected stability of the grain boundary regions due to the presence of stabilizing impurities or pores in the boundaries. The effect of pores and impurity doping on grain growth in $TiO_2$ was recently demonstrated in a sample of the material in which the porosity was about 25% and the crystallite size was 14 nm; the grains grew to 30 nm in

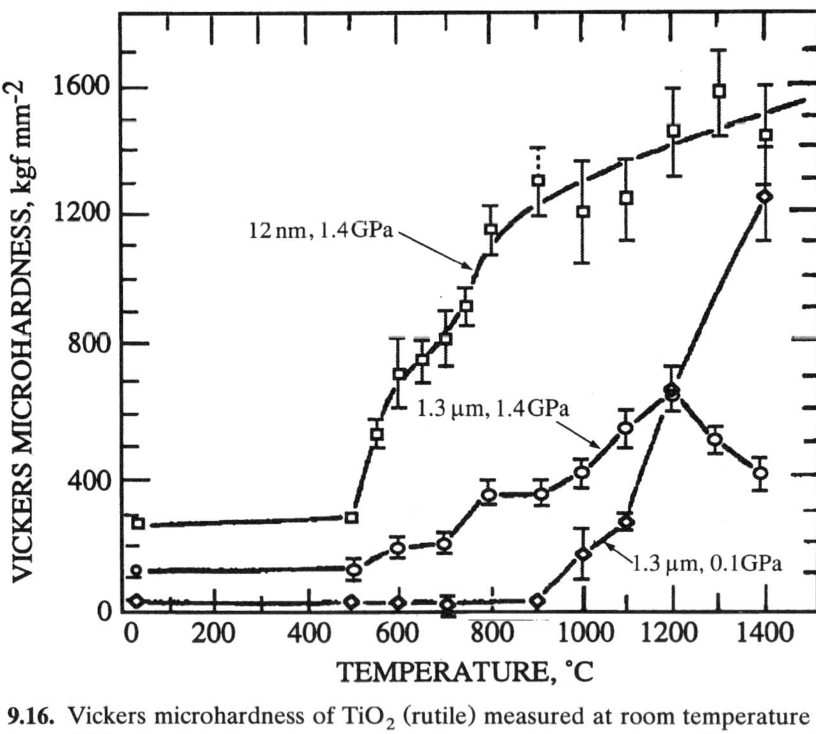


**Fig. 9.16.** Vickers microhardness of $TiO_2$ (rutile) measured at room temperature as a function of 30-min sintering at successively increased temperatures. (Adapted from Siegel et al.[57] Permission granted by ASM International.)

diameter when the material was heated to 700 °C for 20 h. In contrast, when the porosity was reduced to 10%, the grain size changed from 14 to 500 nm with the same treatment. The retardation of grain growth also was demonstrated by the presence of the appropriate impurities (e.g., Y) and the application of pressure (1 GPa).[55–57]

Studies of the sinterability of nanocrystalline $TiO_2$ have shown a drastic enhancement in the sintering temperature of the compound. For example, $TiO_2$ composed of 12-nm grains sinters at a temperature 400 to 600 K lower than the sintering temperature of microcrystalline $TiO_2$ derived from ball-milled samples with 1.3-$\mu$m-sized grains, as illustrated in Figure 9.16.[58] The plot also illustrates another trend that has generally been observed in nanostructured materials: There is often a four- to fivefold increase in the strength and hardness of such materials, compared to coarse-grained materials. However, in some cases, the opposite trend has been observed. The relationship between grain size and yield strength or hardness in conventional coarse-grained materials (the *Hall–Petch* relationship) shows that the yield strength should increase with decreasing size according to the equation

$$\sigma_{ys} = \sigma_{ys,o} + k_H d_{gr}^n, \tag{9.14}$$

where $\sigma_{ys}$ is the 0.2% yield strength (hardness), $\sigma_{ys,o}$ is the hardness of a single crystal specimen, $d_{gr}$ is the grain size, and $k_H$ is the Hall–Petch intensity parameter. This equation has been shown to hold for nanocrystalline (14 nm) Pd samples whose 0.2% yield strength was reported to be 250 MN m$^{-2}$, whereas for a 50-$\mu$m grain-size sample, the value was 52 MN m$^{-2}$ (Fig. 9.17).[59] However, there could also be a change in mechanism that accounts for increasing hardness as the grain size of the material is reduced below certain values; therefore, this relationship may not hold in all cases (which would account for the discrepancy in the trends reported in the literature).

The other properties discussed might also be expected to be significantly different for nanostructured vs. coarse-grained materials. Elastic constants are reduced by up to 30% and may be the result of the different interatomic spacings or porosity in the intergranular regions of the material. The specific heat capacity is likely to change, because this property is directly related to atomic structure. Thermal expansion coefficients normally are increased, because they are strongly influenced by the larger value for the intergrain region. As a result, it should be possible to tailor these coefficients in a predictable way by varying the grain size in a nanostructured material.

Finally, in this context, it is worth mentioning the recent interest in nanocomposite materials as a result of further possibilities to tailor their structure, and therefore properties, through combinations of different nanostructured materials.[60–66]

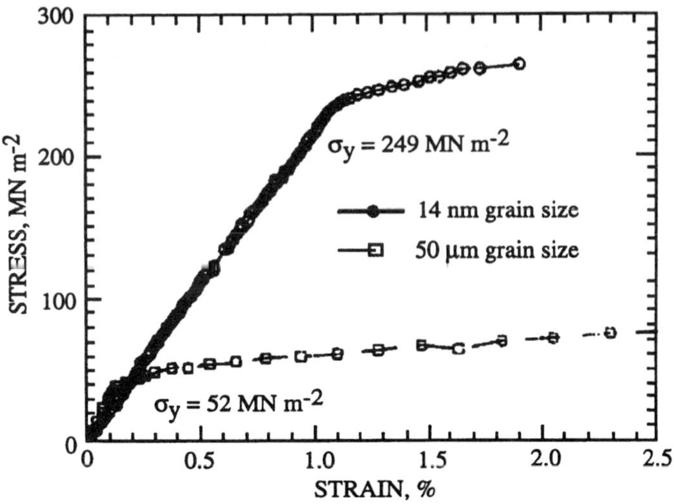


**Fig. 9.17.** Stress–strain curve for nanocrystalline and coarse-grained Pd samples. (Adapted from Nieman et al.[59] Permission granted by ASM International.)

***9.2.6 Chemical Reactivity*** Another property of nanostructured materials peculiar to their size is their chemical reactivity. The size-dependent reactivity of materials was recognized many decades ago and is the basis of heterogeneous catalysis used in the petrochemical industry. For example, nanoparticles of Rh on $Al_2O_3$ and Pt on C are used extensively in the production of petrochemicals, automobile emission control, and fine-chemical synthesis. In heterogeneous catalysis, the chemical reactivity being considered is the reactivity of the particles towards other chemicals, rather than towards each other or different materials, (as was discussed earlier). Therefore, the discussion that follows is relevant to the heterogeneous catalysis of organic transformations using nanometer-sized particle surfaces as the catalytic medium; particularly relevant are materials composed of nanometer-sized particles with accessible surfaces. The surfaces of these particles have atoms that are highly unsaturated, both coordinatively and electronically, and, as a result, are expected to be highly reactive. The reactivity per unit mass of material will be significantly higher than that of an analogous material made up of large particles because the surface area of the nanostructured material is larger (i.e., the number of reactive sites is much greater). Another aspect of the reactivity of nanometer-sized particles, which is harder to anticipate, is the specificity of their reactions. The reactivity and selectivity of a catalyst with respect to a certain organic transformation are two properties that are usually opposed to each other. The higher the reactivity of a catalyst, the less selective it generally is for a given transformation and vice versa. The selectivity of catalysts made up of nanometer-sized particles for a particular organic transformation can be quite different from that of large particles composed of the same material, because the active sites in each case are likely to have different structures.

One aspect of catalysts that is generally in contrast to other materials is the necessity for a porous, open structure or, put another way, the absence of a requirement to exhibit a fully dense structure.[67] Many of the problems that remain to be overcome in the applications of nanostructured materials are associated with their conversion to fully dense solids to take advantage of their hardness, strength, superplasticity, and other mechanical properties in structural or wear-resistant applications. This is generally not an issue in heterogeneous catalysts, whose open structures with accessible nanometer-sized catalyst particles on the surface are desirable.

There are numerous examples of enhanced reactivity and differences in selectivity of chemical reactions catalyzed by nanometer-sized particles compared to macroscopic materials. Some reference texts offer detailed discussions of the topic,[5, 6, 68–70] which are quite complex in that the catalyst support can play a direct role in the catalytic reaction, as well as influencing the reactivity of an organic transformation on the catalyst particle surface. Gas-to particle conversion has been used to prepare nanocrystalline $TiO_2$ from the evaporation followed by condensation of Ti and postoxidation with oxygen. The dissociative adsorption of $H_2S$ in a hydrogen environment (a process relevant to the catalytic hydrodesulfurization of coal) has been studied, and it was shown that the nanocrystalline $TiO_2$ material (with particle diameter 12 nm and surface area 76 $m^2/g$) prepared by this method had higher activity than did commercially available materials. The higher activity was attributed to the smaller particle size of the $TiO_2$ and also its higher oxygen deficiency compared to that of existing powders.

The coevaporation of Li and Mg, followed by condensation and postoxidation with 20% $O_2$ in He for 12 h at room temperature, produced nanocrystalline 5-wt% Li–MgO. This material exhibits crystallites with an average diameter of 5.2 nm and surface area of 750 $m^2/g$. Comparative studies of the catalytic oxidative conversion of methane to higher hydrocarbons between nanocrystalline 5-wt% Li–MgO and conventionally prepared Li-impregnated MgO showed that the nanocrystalline 5-wt% Li–MgO exhibited enhanced activity (by a factor of 3.3) at the same temperatures (above 300°C) and was still catalytically active at 200°C-lower temperatures.

Photosensitive materials such as $TiO_2$, which exhibit photoconduction, photocatalysis, and other properties of a similar nature, can be further sensitized by the presence of nanometer-sized particles of semiconductors on their surfaces (Fig. 9.18). For example, a photoelectrochemical cell was recently described in which quantum-confined particles of CdS or PbS were embedded in the pores of a porous $TiO_2$ electrode. On absorption of light, the electrons produced are injected into the $TiO_2$ substrate, and the holes oxidize the electrolyte. The electrons are transferred by an external circuit to a counterelectrode, where they rereduce the electrolyte; hence, the system is reversible. The photocurrent efficiency of the cell described in Figure 9.18

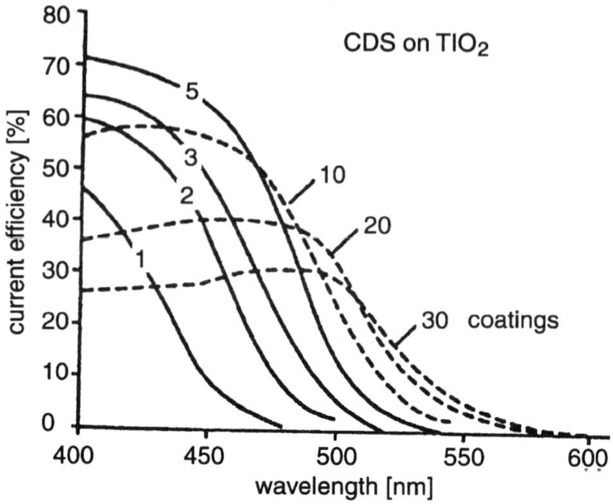


**Fig. 9.18.** Photocurrent spectra of a $TiO_2$ electrode after various numbers of coatings with CdS. (Adapted from Weller.[25] Reproduced with permission from Wiley-VCH, Germany.)

was measured after different numbers of coatings with solutions of colloidal (3-nm) CdS particles on the $TiO_2$ electrode.[25] Two trends can be observed. As the number of coatings increases, the onset of absorption is shifted to longer wavelengths, which is attributed to an increase in the particle size (due to agglomeration, borne out by TEM studies) with increasing numbers of coatings. This trend is analogous to that just described regarding the photocurrent, whose yield increases and then decreases with increasing numbers of coatings. The higher yield is accounted for by the increase in light absorption as a result of an increase in the amount of CdS present. Above an optimum coverage, energy loss occurs by processes other than electron transfer to $TiO_2$, which accounts for the decrease in efficiency. This type of system shows great promise for photovoltaic applications and has the potential to be more efficient than using dye sensitizers to increase the efficiency of light harvesting as a result of the better match between the absorption spectrum of quantum-confined particles and the solar spectrum. However, quantum particle-sensitized photovoltaics currently have insufficient long-term photostability (over weeks or months) to be useful in applications.

## 9.3 APPLICATIONS OF NANOSTRUCTURED MATERIALS

In this section, we briefly review some of the existing and potential applications of nanostructured materials. Currently, powder batches composed of nanometer-sized particles are not in widespread use, because many industries do not require or cannot handle powders with these sizes. For example, industries that rely on thick-film paste technology, such as the microelectronics industry (which uses thick-film conductors and resistors and multilayer ceramic capacitors) and the display industry (which utilizes phosphor powder screens), are currently considering reducing the particle size of the powders they use to a size on the order of a micron. However, it will be some time, if ever, before they are able to use nanopowders.

Some use is made of nanometer-sized powders or nanostructured materials in paints and pigments, organic polymers, composites, ferrofluids, and catalysis. Potential applications that are currently in the advanced stages of research include quantum devices and structural ceramics composed of nanostructured materials. There are also a number of natural nanostructures in biology and mineralogy.[71, 72] A brief description of some current and future applications of nanostructed materials is given in the rest of the section.

### 9.3.1 Current Applications

Nanostructured materials have been employed extensively in the area of heterogeneous catalysis for many years, one of the first examples of the commercial and industrial application of nanomaterials. Zeolites are nanostructured inorganic aluminosilicates that have an organized arrangement of pores in the size range of 5 to 20 Å. The extensive use of these materials is a result of the high acidity of the protons on their surfaces and their unique transport and molecular sieving properties. Shape-selective catalysis is possible with zeolites.[73]

Nanometer-sized particles are the constituents of many paints and pigments, especially those manufactured by a gas-to-particle conversion process, such as $TiO_2$ or $SiO_2$. (This application is discussed in some detail in Chapters 10 and 11.) Silica powders prepared by flame combustion have a number of applications that involve improving the flow of other powders, such as those in toners. Silica powders also are used as components of plastics. Carbon black is a large-commodity material (e.g., in toner cartridges) and is made by a combustion process. Coated titania particles are used in sunscreens, where good UV protection together with good optical transparency, is important.

Nanoparticles are employed in biomedical applications for both diagnosis and treatment.[74] The use of polymer spheres for biological investigation dates back to 1926, when carbon-black spheres were used to study surface movements in amoebas.[75] These particles have a range of morphologies and compositions and possess functions that range from the controlled release of drugs to radiation treatment of target organs in cancer chemotherapy.[46] Typical biodegradable nanoparticles have sizes from 30 to 200 nm and are used to transport drug formulations to target areas via different delivery methods (e.g., oral, intravenous, and ocular). An example of such a nanoparticle composition is a polyalkylcyanoacrylate particle with drugs (e.g., doxorubicin) either incorporated into, or adsorbed on, the surface.

Examples of nanostructured materials with structures at the nanometer length scale are the many organic polymers for which nanorheology is important:[76] polymer composites (including polymer coatings on polypropylene car bumpers),[77] concrete, and polyurethane paints, to name but a few. In many cases, these materials were in use for a good number of years and were found to have a nanostructure that was critical to the successful use of the material after the structure was investigated sometime later.

### 9.3.2 Future Applications

There are a large variety of potential applications of nanostructured materials, as one can imagine from the variety of properties that are "tunable" as a function of particle size. However, this is a case where the adage, "It is easier to carry out fundamental research than it is to develop materials that perform well in a technological application," is relevant. Many issues are important, such as long life, reliability, and uniformity (how to make all the particles exactly the same size and, in many cases, how to connect the nanostructured material to the outside world). These are the challenges being faced by scientists working in the field of quantum devices,[78-81] where the formation of a regular array of nanostructures required for the construction of nanocomputers is the Holy Grail.[82-84]

## 9.4 REFERENCES


1. Gryaznov, V.G. & Trusov, L.I. *Prog. Mater. Sci. 37*(4), 1993, 289–401.
2. Weller, H. & Eychmueller, A. *Adv. Photochem. 20*, 1995, 165–216.
3. Yoffe, A.D. *Advances in Physics 42*(2), 1993, 173–266.
4. Melinon, P., Paillard, V., Dupuis, V., Perez, A., Jensen, P., Hoareau, A., Broyer, M., Vialle, J.L., Pellarin, M., Baguenard, B., & Lerme, J. *International Journal of Modern Physics 9*(4–5), 1995, 339–397.
5. Schoen, G. & Simon, U. *Colloid Polym. Sci. 273*(2), 1995, 101–117.
6. Schoen, G. & Simon, U. *Colloid Polym. Sci. 273*(3), 1995, 202–218.
7. Siegel, R.W. *Materials Science and Engineering B—Solid State Materials for Advanced Technology 19*(1–2), 1993, 37–43.
8. Whetten, R.L. *Acc. Chem. Res. 26*(2), 1993, 49–56.
9. Herron, N. & Wang, Y. *Mater. Res. Soc. Symp. Proc. 346* (Better Ceramics through Chemistry VI), 1994, 887–899.
10. Wang, Y. & Herron, N. *Research on Chemical Intermediates 15*(1), 1991, 17–29.
11. Wang, Y. & Herron, N. *J. Phys. Chem. 95*(2), 1991, 525–532.
12. Wang, Y. & Herron, N. *Res. Chem. Intermed. 15*(1), 1991, 17–29.

13. Siegel, R.W. *Materials Science and Engineering A—Structural Materials Properties Microstructure and Processing 168*(2), 1993, 189–197.
14. Ying, J.Y. *Journal of Aerosol Science 24*(3), 1993, 315–338.
15. Adrievskii, R.A. *Powder Metallurgy and Metal Ceramics 32*(11–12), 1993, 935–941.
16. Andrievski, R.A. *Journal of Materials Science 29*(3), 1994, 614–631.
17. Birringer, R. & Gleiter, H. *Springer Ser. Chem. Phys. 56* (Clusters of Atoms and Molecules II), 1994, 384–404.
18. Gleiter, H. *Nanostruct. Mater. 6*(1–4), 1995, 3–14.
19. Gleiter, H. *Mater. Sci. Forum 189–190* (Interfaces II), 1995, 67–80.
20. Gleiter, H. *Adv. Mater. 4*(7–8), 1992, 474–481.
21. Hahn, H. *Nanostruct. Mater. 2*(3), 1993, 251–265.
22. Suryanarayana, C. *Bull. Mater. Sci. 17*(4), 1994, 307–346.
23. Suryanarayana, C. *International Materials Rev. 40*(2), 1995, 41–64.
24. Suryanarayana, C. & Froes, F.H. *Metal. Trans. A—Physical Metallurgy and Mater. Sci. 23*(4), 1992, 1071–1081.
25. Weller, H. *Adv. Mater. 5*(2), 1993, 88–95.
26. Barthelemy, A., Cros, V., Duvail, J.L., Fert, A., Morel, R., Parent, F., Petroff, F., & Steren, L.B. *Nanostruct. Mater. 6*(1–4), 1995, 217–226.
27. Chien, C.L. *Annu. Rev. Mater. Sci. 25*, 1995, 129–160.
28. Lachowicz, H.K. *Elektronika 36*(1), 1995, 12–17.
29. Li, G. *Gongneng Cailiao 26*(5), 1995, 385–388.
30. Parkin, S.S.P. *Annual Review of Materials Science 25*, 1995, 357–388.
31. Weller, H., Vossmeyer, T., Eychmueller, A., Mews, A., Katsikas, L., & Reck, G. *Mater. Res. Soc. Symp. Proc. 358* (Microcrystalline and Nanocrystalline Semiconductors), 1995, 213–218.
32. Jung, K.T. & Shul, Y.G. *Chem. Mater. 9*, 1997, 420.
33. Bhargava, R.N., Gallagher, D., Hong, X., & Nurmikko, A. *Physical Review Letters 72*(3), 1994, 416–419.
34. Fu, H.X., Ling, Y., & Xide, X. *Physical Review B—Condensed Matter 48*(15), 1993, 10978–10982.
35. Zhou, H.S., Honma, I., Komiyama, H., & Haus, J.W. *Physical Review B—Condensed Matter 50*(16), 1994, 12052–12056.
36. Herzer, G. *IEEE Trans. Magn. MAG-26*, 1990, 1397.
37. Herzer, G. *Mat. Sci. Eng. A133*, 1991, 1.
38. Dormann, J.L. & Fiorani, D. Nanophase Magnetic-Materials—Size and Interaction Effects on Static and Dynamical Properties of Fine Particles. *Journal of Magnetism and Magnetic Materials*, PO Box 211, 1000 AE Amsterdam, Netherlands, 1995, 415–418.
39. Dormann, J.L. *Materials Science and Engineering A—Structural Materials Properties Microstructure and Processing 168*(2), 1993, 217–224.
40. Zhou, G.F. & Bakker, H. *Materials Transactions Jim 36*(2), 1995, 329–340.
41. Shull, R.D., McMichael, R.D., & Ritter, J.J. *Nanostruct. Mater. 2*, 1993, 205.
42. Chandler, C.D., Roger, C., & Hampden-Smith, M.J. *Chem. Rev. 93*, 1993, 1205.

43. Buffat, P. & Borel, J.-P. *Phys. Rev. A A 13*, 1976, 2287.
44. Goldstein, A.N., Echer, C.M., & Alivisatos, A.P. *Science 256*, 1992, 1425–1427.
45. Shi, F.G. *J. Mater. Res.* (9), 1994, 1307–1313.
46. Couvreur, P. & Vauthier, C. *J. Controlled Release 17*(2), 1991, 187–198.
47. Gleiter, H. *Prog. Mater. Sci. 33*, 1989, 223.
48. Suryanarayana, C., Mukhopadhay, D., Patankar, S.N., & Froes, F.H. *J. Mater. Res. 7*, 1992, 2114.
49. Langdon, T.G. *Materials Science and Engineering A—Structural Materials Properties Microstructure and Processing 166*(1–2), 1993, 67–79.
50. Lewis, T.J. *IEEE Trans. Dielectr. Electr. Insul. 1*(5), 1994, 812–825.
51. Mon, K.K. *Physical Review B—Condensed Matter 50*(22), 1994, 16718–16721.
52. Siegel, R.W. & Thomas, G.J. *Ultramicroscopy 40*(3), 1992, 376–384.
53. Wang, N., Wang, Z.R., Aust, K.T., & Erb, U. *Acta Metallurgica et Materialia 43*(2), 1995, 519–528.
54. Schumacher, S., Birringer, R., Strauss, R., & Gleiter, H. *Acta. Metall. 37*, 1989, 2485.
55. Siegel, R.W. & Fougere, G.E. *NATO ASI Ser., Ser. E 260* (Nanophase Materials), 1994, 233–261.
56. Siegel, R.W. & Fougere, G.E. *Mater. Res. Soc. Symp. Proc. 362* (Grain Size and Mechanical Properties—Fundamentals and Applications), 1995, 219–229.
57. Siegel, R.W. & Fougere, G.E. *Nanostruct. Mater. 6*(1–4), 1995, 205–216.
58. Siegel, R.W., Ramasamy, S., Hahn, H., Li, Z., & Lu, T. *J. Mater. Res. 3*, 1988, 1367.
59. Nieman, G.W., Weertman, J.R., & Siegel, R.W. *Scr. Metall. Mater. 24*, 1990, 1367.
60. Burdon, J. & Calvert, P. *Trans. Mater. Res. Soc. Jpn. 16B* (Composites, Grain Boundaries and Nanophase Materials), 1994, 1537–1541.
61. Chakravorty, D. *Bulletin of Materials Science 15*(5), 1992, 411–420.
62. Gaffet, E., Abdellaoui, M., & Malhourouxgaffet, N. *Materials Transactions Jim 36*(2), 1995, 198–209.
63. Greer, A.L. Crystallization of Metallic Glasses. *Materials Science and Engineering A—Structural Materials Properties Microstructure and Processing*, PO Box 564, 1001 Lausanne 1, Switzerland, 1994, 41–45.
64. Greer, A.L. *Science 267*(5206), 1995, 1947–1953.
65. Komarneni, S. *Journal of Materials Chemistry 2*(12), 1992, 1219–1230.
66. Niihara, K. *Nippon Seramikkusu Kyokai Gakujutssu Ronbunshi—Journal of the Ceramic Society of Japan 99*(10), 1991, 974–982.
67. Trudeau, M.L. & Ying, J.Y. *Nanostruct. Mater. 7*(1/2), 1996, 245–258.
68. Belloni, J. *Curr. Opin. Colloid Interface Sci. 1*(2), 1996, 184–196.
69. Henglein, A. *Ber. Bunsen-Ges. 99*(7), 1995, 903–913.
70. Toshima, N. & Wang, Y. *Adv. Mater. 6*(3), 1994, 245–247.
71. Bond, G.M., Richman, R.H., & McNaughton, W.P. *Journal of Materials Engineering and Performance 4*(3), 1995, 334–345.

72. Mow, V.C., Ratcliffe, A., & Poole, A.R. *Biomaterials 13*(2), 1992, 67–97.
73. Gates, B.C. Catalytic Materials. In Interrante, L.V. (Ed.), *Materials Chemistry: An Emerging Discipline*. Washington, American Chemical Society, 1995.
74. Arshady, R. *Biomaterials 14*(1), 1993, 5–15.
75. Mast, S.O. *J. Morphol. Physiol. 41*, 1926, 347.
76. Cai, L.L., Peanasky, J., & Granick, S. *Trends Polym. Sci. 4*(2), 1996, 47–51.
77. Reihs, K. *Thin Solid Films 264*(2), 1995, 135–140.
78. Gossard, A.C. & Fafard, S. *Solid State Communications 92*(1–2), 1994, 63–70.
79. Gourley, P.L. & Warren, M.E. *Journal of Nonlinear Optical Physics and Materials 4*(1), 1995, 27–81.
80. Notzel, R. & Ploog, K.H. *International Journal of Modern Physics 7*(15), 1993, 2743–2778.
81. Sakaki, H. *Solid State Communications 92*(1–2), 1994, 119–127.
82. Corcoran, E. *Sci. Amer., November* 1990, 122.
83. Reed, M.A. *Sci. Amer., January* 1993, 118.
84. Wang, Y. *Acc. Chem. Res. 24*, 1991, 133.

# 10 Overall Qualitative Behavior of Gas-to-Particle Conversion Processes

## 10.1 INTRODUCTION

This chapter and the next one discuss gas-to-particle conversion. Here, we present the parameters that are often used to quantify the characteristics of particles and describe the different modes of overall system behavior that can be obtained. Chapter 11 covers the technology and describes the implementation of these approaches at both the laboratory and an industrial scale.

Together, the two chapters discuss the synthesis and processing of particles by approaches that exploit the conversion of gaseous species into solid form. For particle formation, gas-to-particle conversion relies on producing a supersaturated vapor of one or more species. This vapor can be created by chemical reaction of gaseous precursors, causing the formation of condensable products (Fig. 10.1), or by physical processes such as the cooling of a hot vapor. In the figure, the steps are drawn in series for convenience, but several of them may occur in parallel.

The quality of particles produced by gas-to-particle conversion and spray processes can be evaluated through the use of a number of parameters. The relative importance of these parameters depends on the specifics of the application. The parameters are listed here to remind the reader which quantities are most important in gauging the success of a process for powder generation.

**Average Size:** The average particle size is evaluated by various methods; number and mass averages are typically the most important and most commonly used parameters, often in the form of geometric mean particle diameters. Nanophase materials exhibit certain desirable properties, but only for particle sizes smaller than about 10 nm. (See Chapter 9.) Precious-metal particles for thick-film pastes must be in the size range from 0.1 to 5 $\mu$m for most applications.[1] Perhaps the most demanding application is $TiO_2$ pigments, for which the optimum size is near 200 nm. Deviations of only 10 nm are easily detected in terms of properties of the pigment, and size distributions are approximately lognormal.[2]

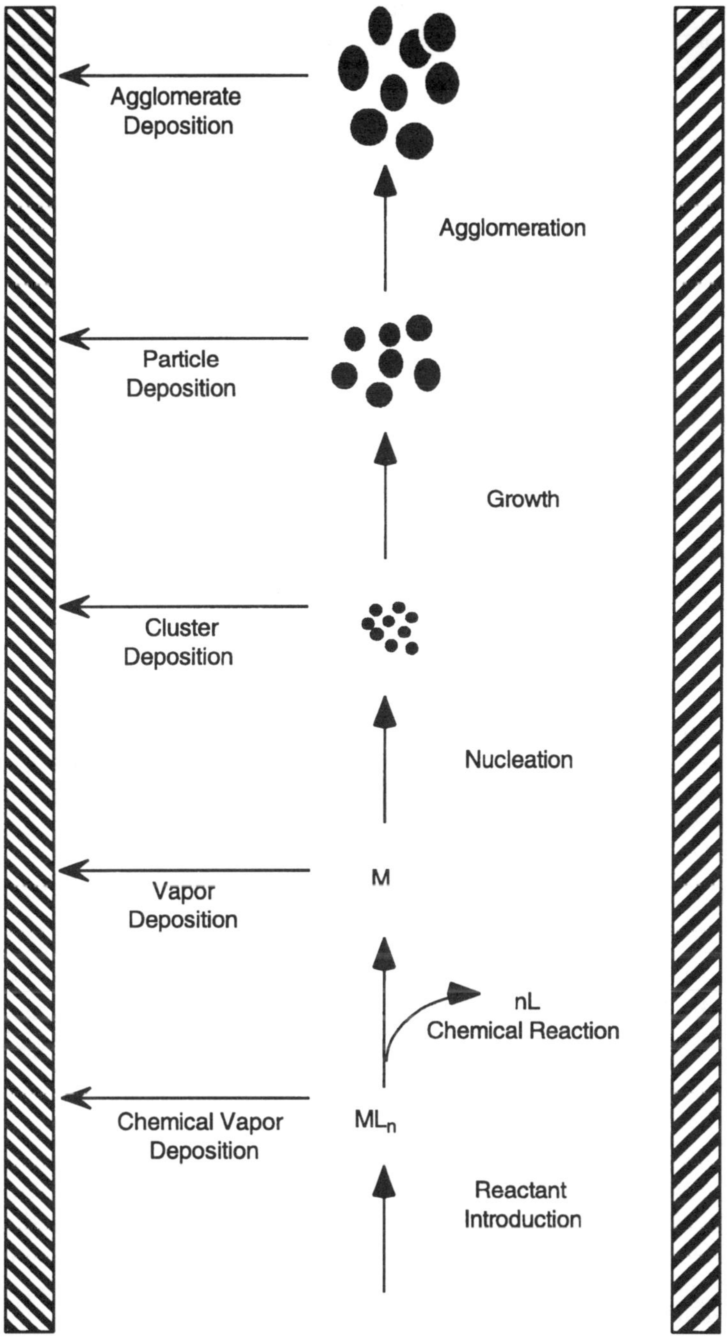


**Fig. 10.1.** Schematic representation of processes that occur in chemical routes to gas-to-particle conversion.

**Spread of Particle Size Distribution:** The spread of the particle size distribution is one of the most important parameters determining the properties of a powder product. The desired spread of the distribution depends on the particular application and can range from monodisperse to broad or multimodal. For most practical purposes, this spread is often quantified in terms of the geometric standard deviation $\sigma_g$ where $\sigma_g < 1.2$ is a sufficiently narrow distribution for the majority of applications. In the case of pigments such as $TiO_2$, it is desirable to have a particle size distribution that is as narrow as possible, centered around the size corresponding to the maximum in light-scattering efficiency (~ 200 nm). For precious-metal particles used for thick-film pastes, the particles should be as close to monodisperse ($\sigma_g = 1.0$) as possible. With ceramic powders used as precursors for the manufacture of bulk ceramics, it is desirable to have a size distribution that is not necessarily monodisperse, but is not very broad. A very narrow particle size distribution can lead to packing of the particles in domains. During sintering, these domains can result in differential shrinkage, which may cause large flaws in the ceramic.

**Yield:** The yield is the fraction of the precursor collected as product. In most cases, increases in yields can result in improved economics and minimize waste; therefore, it is desirable to have yields as high as possible. Yields of less than 100% can be caused by a number of processes, including (1) incomplete conversion of the reactant to a condensable species in gas-to-particle conversion, (2) losses of particles onto reactor walls and associated tubing (for example, thermophoretic losses of particles to cool surfaces in a variety of high-temperature processes), and (3) inefficient deposition of particles onto surfaces (e.g., in the production of optical fibers, particles are formed in the system; particles that do not deposit on the surface of the preform flow out of the system and decrease yields).[3, 4]

**Chemical Composition:** The chemical composition of particles and films strongly influences the behavior and performance of materials produced by aerosol processes. Deviations in composition from the desired ratio may be due to the loss of volatile components and incomplete conversion of a reactant in multicomponent systems. In general, a product in which all particles have the same composition is preferred.

**Phase Composition:** Various phases—that is, regions of material with different composition and/or crystalline structure—can be produced. In general, it is important to be able to produce single-phase materials of the correct phase. For example, pigment particles of $TiO_2$ need to be rutile, but $TiO_2$ used for photocatalysis must be anatase because the anatase phase has a higher photocatalytic activity than the rutile phase. The successful processing of ceramic powders relies strongly on availability of homogeneous powders. Small amounts of undesirable secondary phases often can influence sintering and grain growth processes. In some cases, such as $Ag/Y_1Ba_2Cu_3O_{7-x}$ and other composites, it may be desirable to maintain phase segregation.[5] Other cases may require particles of one material to be coated with another material.

**Morphology:** Various morphologies, including porous or solid, spherical or non-spherical, and other morphologies, are encountered. Particle morphology plays an important role in determining the behavior of the particles after production. For example, the fabrication of high-quality ceramics requires particles that are spherical, solid, and nonporous; on the other hand, the manufacture of catalysts demands porous particles with high specific surface areas. Other applications, such as piezoelectric materials, may require single-crystal particles, and in many cases, filler materials in plastics require hollow particles.

**Agglomeration:** The state of agglomeration of a powder can determine whether the powder is useful for a specific application. For ceramics processing, agglomerates—especially hard agglomerates, which are not as easily broken up as soft agglomerates—are highly undesirable, because they can lead to nonuniform grain-size distributions in product ceramics. In some cases, it may be desirable to form flocs in the gas phase in order to increase the particle size and allow easier collection of the particles. The other extreme is in the production of reinforcing agents such as carbon black, where hard agglomerates are required to give the necessary mechanical strength.

## 10.2 MODES OF OPERATION FOR SINGLE-COMPONENT SYSTEMS

In this and the following sections, we bring together the information from Chapters 1 through 9 and describe the overall qualitative behavior of systems for gas–to–particle conversion. We begin with an introduction to the modes of system operation for the simplest class, single-component materials. These materials, found in large industrial applications, are pigments ($TiO_2$), fumed materials ($SiO_2$), and reinforcing agents (carbon black). Multicomponent materials include pigments (Al-doped $TiO_2$), particles that are intermediates in optical fiber production (doped $SiO_2$), and coated materials (pigments coated with $Al_2O_3$ and $SiO_2$).

Models have been developed (see Chapter 7) to describe aerosol dynamics in a variety of systems. The types of systems modeled include virtually all of those listed in Chapter 11: tubular flow, plasma, laser, and flame systems. Although the models and systems differ in many ways, the qualitative features of the aerosol dynamics are the same for all.

Four of the most important characteristics of an aerosol are the average particle diameter ($d_g$ for a lognormal distribution), total particle concentration ($N_\infty$), spread of the size distribution ($\sigma_g$ for a lognormal distribution), and particle morphology (fractal dimension or other measure, such as surface area). These quantities can evolve along an aerosol reactor in two qualitatively different ways, depending on the mode of operation of the reactor. In one mode of operation, *nucleation–condensation controlled*, the monomer (a molecule or atom) saturation ratio increases because of monomer formation by a chemical reaction or a decrease in temperature, until nucleation occurs. Growth proceeds by condensation of the monomer onto the particles. As there are no collisions among particles, they can remain nearly spherical throughout the residence time in the reactor. Growth may also occur by surface reaction on the particles, as well as by volume reaction inside each particle. This mode is often encountered in laboratory-scale systems, where particle concentrations are relatively low; it is not found in industrial systems based on gas-to-particle conversion.

The other mode of reactor operation is *nucleation–coagulation*, in which particles are formed and their concentration builds up to high levels such that collisions and coalescence of particles occur, resulting in particle growth. In this mode, agglomerates are formed with varying degrees of hardness and different fractal dimensions, depending on the details of the reactor operation. Nucleation–coagulation is found in both industry and the laboratory.

### 10.2.1 Nucleation–Coagulation

For the case of particles formed by reaction or cooling of a vapor followed by collisions, the particles must coalesce by sintering after collision in order to become spherical. The importance of collision and coalescence in determining the particle morphology depends largely on the relative rates of collision and coalescence, quantified in terms of the characteristic times for collision and sintering, as discussed in Chapter 5. Figure 10.2 shows two modes of reactor behavior for particle growth by collision and coalescence.[6, 7] For *collision-limited growth*, with $\alpha_c = \infty$, the sintering rate is rapid relative to the collision rate, and the particles exist as spherical particles during most of the period between collisions. Here, particle size is mainly a function of collision rate and is not a strong function of the properties of the materials. For *sintering-limited growth*, with $\alpha_c = 0$, the particles exist as aggregates at all times. In this case, the primary particle size is mainly a function of the parameters that control the sintering rate, namely, the temperature and the properties of the materials. In real situations, intermediate conditions (e.g., $\alpha_c \approx 1$) may prevail, as is illustrated in the figure.[8]

In many cases, sintering is not sufficiently rapid for particles to fuse completely before colliding with other particles, as is shown in Figure 10.3. At the beginning of the process, the particles are very small and temperatures are usually high. This favors rapid sintering of particles after collisions. Once the particles grow to larger sizes, if the temperature drops, the sintering rate decreases, resulting in the formation of hard agglomerates and representing a shift from collision-limited to sintering-limited behavior. Numerous models have been developed to describe this process, as discussed

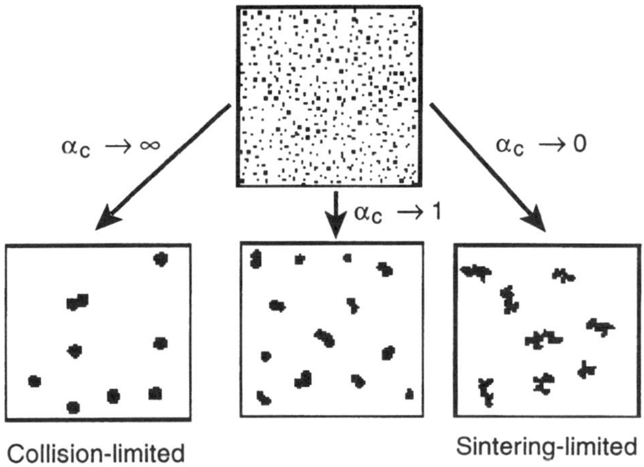


**Fig. 10.2.** Modes of aerosol reactor behavior. The quantity $\alpha_c$ is the ratio of characteristic times for collision and fusion. (Reprinted with permission of Elsevier Science Inc. from "Controlled synthesis of nanosized particles by aerosol processes," by Wu, M.K., Windeler, R.S., Steiner, C.K.R., Börs, T. and Friedlander, S.K., *Aerosol Science and Technology*, *19*, pp. 527–548, Copyright 1993 by American Association for Aerosol Research.)[8]

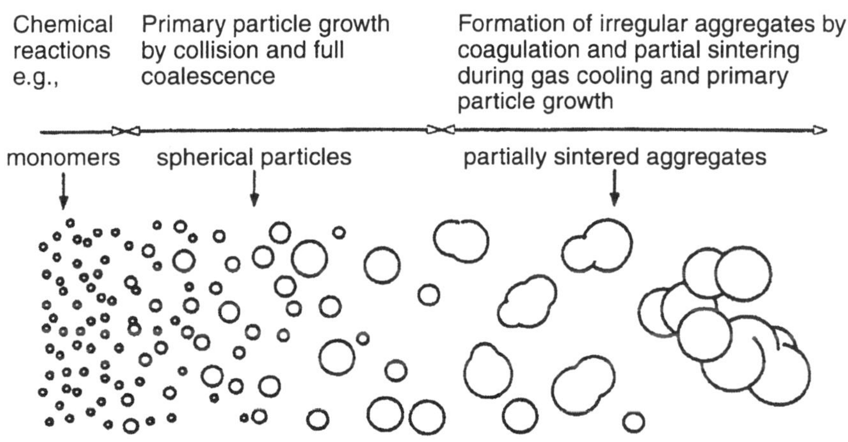


**Fig. 10.3.** Evolution of particle morphology with reactor residence time. (Reprinted with permission of Elsevier Science Inc., from "A simple model for the evolution of the characteristics of aggregate particles undergoing coagulation and sintering," by Kruis, F.E., Kusters, K., Pratsinis, S.E., and Scarlett, B., *Aerosol Science and Technology 19*, 514–526, 1993. Elsevier Science Inc., New York, NY.)[9]

in Chapter 7. Here we present results obtained using a model based on monosized particles[9] and accounting for particle sintering, as described in Chapter 7. Figure 10.4 shows how the number of particles per aggregate ($n_p$) and the average primary particle size ($d_p$) evolve with residence time in an aerosol reactor, producing silicon particles at 1073 K for different sintering rates.[9] The number of primary particles per aggregate increases initially, as new particles are formed and collide. Higher sintering rates give fewer primary particles in each aggregate, because the primary particles fuse together to create larger particles. Similarly, higher sintering rates give larger primary particles. Because the sintering rate is a function mainly of tempera-

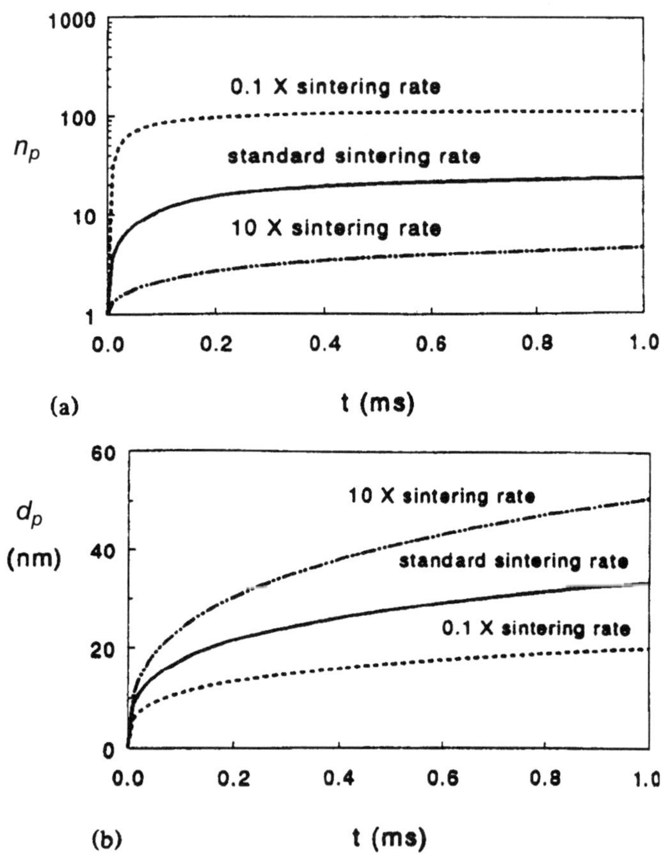


**Fig. 10.4.** Number of primary particles in an agglomerate ($n_p$) and diameter of primary particles as a function of reactor residence time for different sintering rates. (Reprinted with permission from *Aerosol Science and Technology 19*, Kruis, F.E., Kusters, K., Pratsinis, S.E., and Scarlett, B., "A simple model for the evolution of the characteristics of aggregate particles undergoing coagulation and sintering," 514–526, 1993. Elsevier Science Inc., New York, NY.)[9]

ture and the properties of materials, the primary particle size is chiefly a function of those two parameters as well. The primary particle size is also weakly related to the reactant or inlet vapor concentration.

The preceding discussion focused on primary particle size. The dependence of the overall particle size and concentration on the reactor operating conditions is also of interest and is most easily discussed for systems with instantaneous coalescence (collision-limited operation), in which spherical particles are formed. Figure 10.5 shows the geometric mean particle diameter and the number concentration of particles in an isothermal flow reactor driven by chemical reaction of a precursor for collision-limited operation. The model used for the calculation was a sectional model,[10, 11] as outlined in Chapter 7. The particle concentration initially rises at very short times ($X < 5$), but then drops for the remainder of the time in the reactor. This behavior occurs because, initially, there are no particles in the reactor. Then, as chemical reaction results in the formation of particles, the particle concentration builds up until the reactant is mostly consumed and the collision frequency is high enough for the particle concentration to be influenced and to drop. At the same time, the average particle diameter increases monotonically because reactant is being converted to product and coagulation results in an increase in the average size of the particles. Results are shown for three different rates of chemical reaction, quantified in terms of $\theta_1$, the characteristic time for reaction, which is inversely related to the rate constant of the

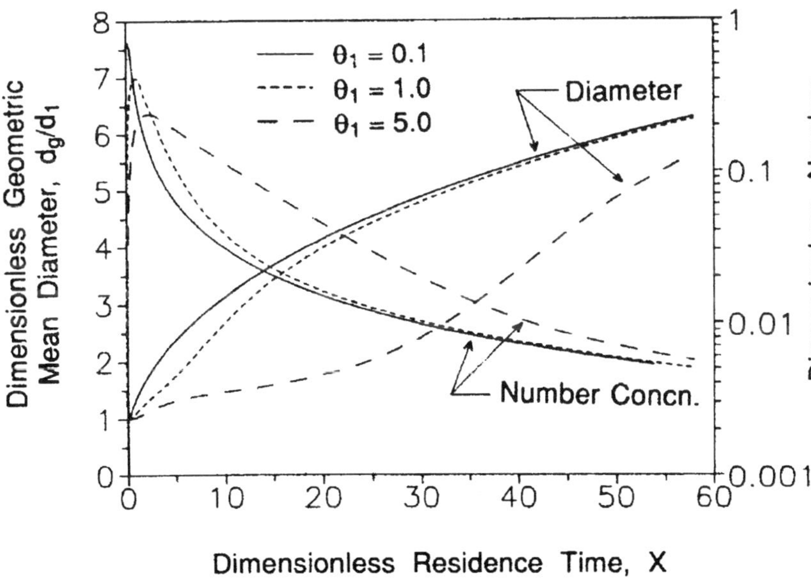


**Fig. 10.5.** Evolution of particle size and concentration with time for different dimensionless reaction rates (proportional to $1/\theta_1$). (Reprinted with permission from Landgrebe and Pratsinis.[10] *Ind. Eng. Chem. Res.* *28*(10), 1989, 1474–1481. Copyright 1989, American Chemical Society.)

reaction. Note that at the reactor outlet the difference in diameters for the highest and lowest reaction rates is small, because a fixed amount of reactant is consumed near the reactor inlet. Subsequent collision and coalescence of the particles in approximately the same amount of time results in similar particle sizes. This type of behavior is very common in industrial-scale aerosol reactors used to produce materials with high melting points where the particle concentrations are also high. Laboratory-scale experiments are not always conducted in the nucleation–coagulation mode and may be carried out instead in the mode of nucleation–condensation, in which qualitatively different behavior is observed.

The distribution of particle sizes is often an important measure of the quality of a powder, which aids in determining the usefulness of the powder in subsequent processing steps. Figure 10.6 shows how the spread of the size distribution evolves along the reactor for collision-limited operation for the same conditions as are illustrated in Figure 10.5.[10] Three reaction rates are shown, the fastest rate being the solid line. In all cases, the geometric standard deviation evolves toward an asymptotic value of $\sigma_g = 1.46$. This value corresponds to the *self-preserving size distribution* for the free-molecular regime and is the narrowest distribution that can be obtained by coagulation, with the particle diameter smaller than the gas mean free path. The value of $\sigma_g$ first rises and then falls, because when new particles are being formed in

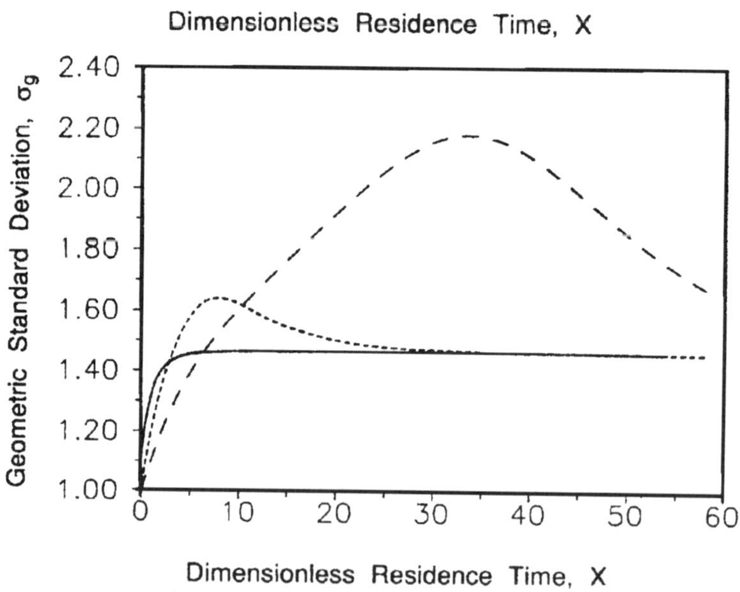


**Fig. 10.6.** Evolution of spread of particle size distribution. Curves are defined in Figure 10.5. (Reprinted with permission from Landgrebe and Pratsinis. *Ind. Eng. Chem. Res. 28*(10), 1989, 1474–1481. Copyright 1989, American Chemical Society.)[10]

the presence of particles that have already been formed and have grown to larger sizes, the spread of the size distribution is broad. As new particles stop being formed, because the reactant is consumed, coagulation removes the smaller particles and narrows the distribution toward the asymptotic self-preserving value. The time required to reach this value is governed by the reaction rate. In Figure 10.6, the value is not reached for the slowest reaction rate, because the residence time is not sufficient. This type of behavior is extremely common in systems in which particles grow by coagulation.

### 10.2.2 Nucleation–Condensation

The aerosol dynamics for particle formation by nucleation in the absence of coagulation are qualitatively different from those with coagulation. Figures 10.7 through 10.11 show the aerosol diameter, number concentration, surface area, spread of the particle size distribution, and yield as a function of reactor residence time in a reactor in which the reaction to form the condensing species occurs at a constant rate. The results presented are from a moment model with calculations shown for several different types of

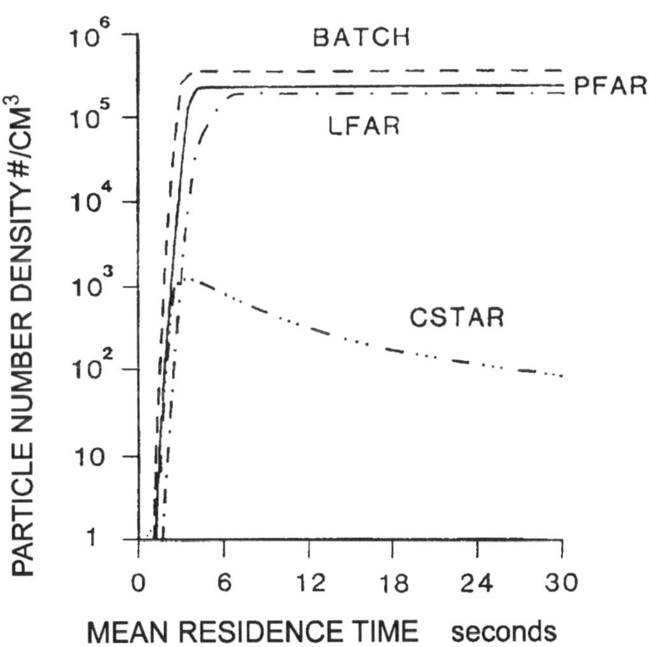


**Fig. 10.7.** Particle number concentration as a function of residence time for batch, laminar-flow (LFAR), plug-flow (PFAR) and continuous stirred-tank (CSTAR) aerosol reactors. (Reprinted with permission from Pratsinis, S.E., Kodas, T.T., Dudukovic, M.P., and Friedlander, S.K., *Ind. Eng. Chem. Process Des. Dev.* 25(3), 634–642, 1986. Copyright 1986, American Chemical Society.)[12]

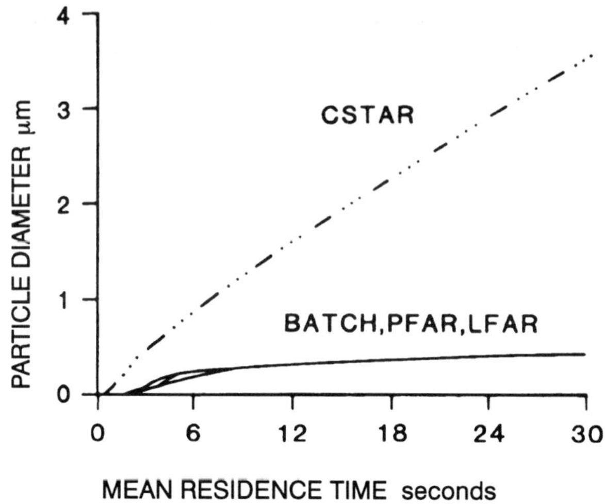


**Fig. 10.8.** Particle diameter, in $\mu$m, as a function of residence time for batch, laminar-flow (LFAR), plug-flow (PFAR), and continuous stirred-tank (CSTAR) aerosol reactors. (Reprinted with permission from Pratsinis, S.E., Kodas, T.T., Dudukovic, M.P., and Friedlander, S.K., *Ind. Eng. Chem. Process Des. Dev.* 25(3), 634–642, 1986. Copyright 1986, American Chemical Society.)[12]

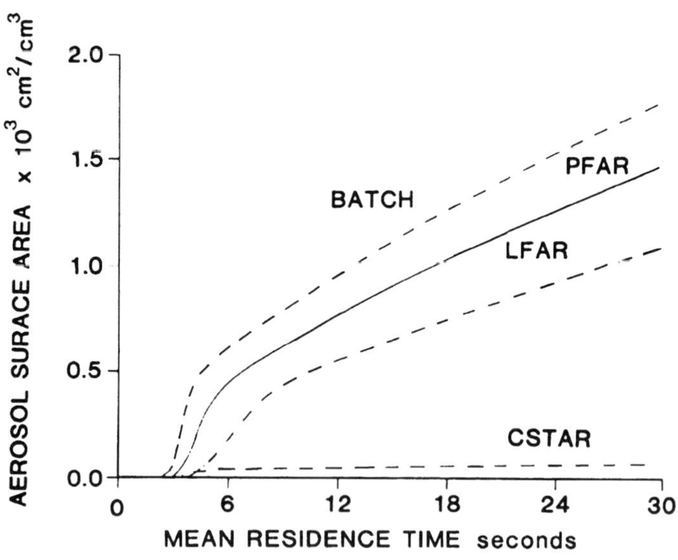


**Fig. 10.9.** Aerosol surface area ($\times 10^3$ cm$^2$/cm$^3$) as a function of residence time for batch, laminar-flow (LFAR), plug-flow (PFAR), and continuous stirred-tank (CSTAR) aerosol reactors. (Reprinted with permission from Pratsinis, S.E., Kodas, T.T., Dudukovic, M.P., and Friedlander, S.K., *Ind. Eng. Chem. Process Des. Dev.* 25(3), 634–642, 1986. Copyright 1986, American Chemical Society.)[12]

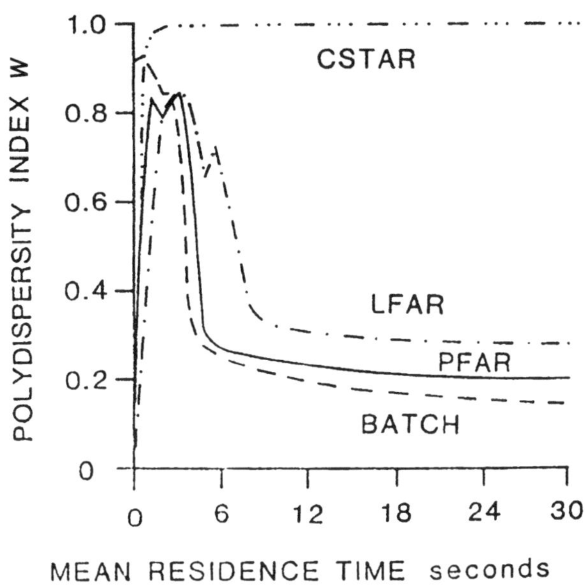


**Fig. 10.10.** Polydispersity index *W* as a function of residence time for batch, laminar-flow (LFAR), plug-flow (PFAR), and continuous stirred-tank (CSTAR) aerosol reactors. (Reprinted with permission from Pratsinis, S.E., Kodas, T.T., Dudukovic, M.P., and Friedlander, S.K., *Ind. Eng. Chem. Process Des. Dev.* *25*(3), 634–642, 1986. Copyright 1986, American Chemical Society.)[12]

reactors:[12, 13] batch (uniform conditions with no flow); a plug-flow aerosol reactor (PFAR) with a uniform velocity profile and tubular flow; a laminar-flow aerosol reactor (LFAR) with a parabolic velocity profile and tubular flow; and a continuous stirred-tank aerosol reactor (CSTAR) with well-mixed conditions and continuous flow into and out of the system. It is useful to describe some of these reactors in a bit more detail. The LFAR has a range of residence times for the gas exiting the reactor, because of the slow velocities near the walls and the higher velocities near the center of the reactor tube. The PFAR model assumes that all the gas spends the same time in the reactor, and in this sense, the flow is closer to turbulent flow, which has a flatter velocity profile than laminar flow in a tube. The PFAR and LFAR models assume that there is no mixing of the gas in the reactor. The batch reactor, the LFAR, and the PFAR exhibit similar behavior because the only difference between these systems is the residence time distribution. The CSTAR is very different from the other reactors. Assuming a system in which the gas is well mixed, the composition in the reactor is spatially uniform, and the gas exiting the system has a broad distribution of residence times relative to those of the other reactors.

Examining the curves for the LFAR, PFAR, and batch systems, we see that the particle concentration rises and then levels off at a constant value.

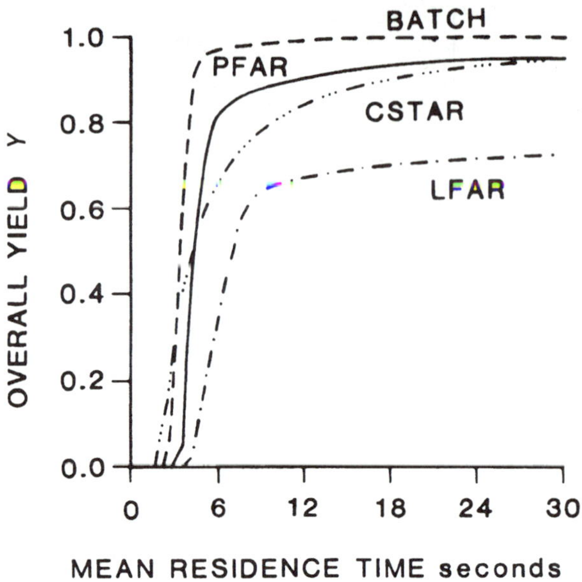


**Fig. 10.11.** Overall yield $Y$ as a function of residence time for batch, laminar-flow (LFAR), plug-flow (PFAR), and continuous stirred-tank (CSTAR) aerosol reactors. (Reprinted with permission from Pratsinis, S.E., Kodas, T.T., Dudukovic, M.P., and Friedlander, S.K., *Ind. Eng. Chem. Process Des. Dev.* *25*(3), 634–642, 1986. Copyright 1986, American Chemical Society.)[12]

This type of behavior occurs because nucleation ceases to form new particles when the saturation ratio of the nucleating species drops below the critical value for new-particle formation. The saturation ratio drops because any new monomer that is formed is consumed by condensation on the particles. Once nucleation ceases in the absence of mechanisms leading to the loss of particles, the particle concentration becomes constant. The particle diameter, surface area, and yield increase monotonically because the condensable species is formed continuously along the reactor and condenses on the particles, resulting in an increase in their size. The spread of the particle size distribution ($W$) first increases and then decreases. This trend occurs because new particles are being formed while the existing particles grow to larger sizes. Once new-particle formation ceases, the smaller particles grow to larger sizes, reducing the difference in size between the largest and smallest particles and thereby narrowing the size distribution.

The CSTAR's very different behavior takes place because the conditions are uniform everywhere in the reactor and the inlet gas stream comes into immediate contact with fluid elements that have spent long times in the system. The major differences in the behavior of the CSTAR from that of the other reactors are the lower particle concentration, larger size, lower surface area, and broader size distribution. These differences derive from the uniform, well-mixed conditions in the CSTAR and suggest that CSTARs are not attractive if narrow particle size distributions are desired.

The type of behavior represented in Figures 10.7 through 10.11 is common in systems in which particles grow by nucleation and condensation. *This type of behavior is observed only at low particle concentrations, where coagulation is not important.* As a result, it is observed primarily at the laboratory scale and not at the industrial scale. Increasing the reaction rate can move the mode of operation to nucleation–coagulation (Section 10.2.1).

## 10.3 MODES OF OPERATION FOR MULTICOMPONENT SYSTEMS

In addition to systems with one condensable species (a single type of monomer), there are a number of more complicated situations that utilize materials containing several elements. These materials are produced in a variety of forms for industrial applications requiring a homogeneous composition, composite, or coated particles. In a multicomponent particle with a homogeneous composition, the relative amounts of the various elements are the same at all positions in the particle; that is, the composition is uniform throughout each particle. As an example, in AgPd alloy particles, every point in the particle has the same Ag:Pd ratio, and no separate regions of Ag or Pd exist.[14] Such a configuration is possible for this combination of elements because they are mutually soluble. For doped particles, a small amount of another element is added to the material being considered. For example, a small amount of Ca ($< 1\%$) added to Pd metal results in improved oxidation resistance for the Pd. A homogeneously doped particle would have the same ratio of Pd to Ca everywhere in the particle. By contrast, composite particles contain two or more materials with different compositions. For example, when larger amounts of Ca are added to Pd metal, not all of the Ca can dissolve in the Pd, because the solubility limit of Ca in Pd is exceeded and regions of CaO are formed.[15] Thus, the particle is a composite of CaO and Pd containing small amounts of Ca. Finally, in coated particles, the second material is confined to the surface of the particle, rather than being interspersed through the bulk of the particle.[15] This may be achieved either through the mutual insolubility of the materials involved or by forming the material under nonequilibrium conditions.

### 10.3.1 Generation of Multicomponent Materials Using Two or More Volatile Species

The gas-to-particle conversion of multiple species can be driven by gas-phase reactions or the cooling of a vapor. In this section, we discuss, in a qualitative manner, the modes of operation of systems producing multicomponent materials. We examine (1) condensation in multicomponent systems resulting from chemical reaction or the cooling of a vapor, (2) strategies employed to avoid segregation of particles, (3) approaches to composite materials, (4) methods for generating coated particles, and (5) fluidized beds. Due to their complexity, these systems have not been modeled; therefore, only an outline of their important features will be presented.

Although multicomponent gas-to-particle conversion processes can sometimes provide particles with a uniform composition, some physical and chemical routes lead to segregation[16] (Figs. 10.12 and 10.13). The formation of multicomponent materials by condensation of the vapors of two or more elements or molecular species suffers from the difficulty that very rapid cooling rates are required for simultaneous nucleation and condensation of all the vapor species (Fig. 10.12). The gas must be cooled to well below the temperature that yields a supersaturated state for all the species present in a time much shorter than that required for significant particle formation and growth for the material with the lowest vapor pressure. This cannot always be accomplished, due to the short time scales associated with particle formation and growth and the difficulties in rapidly removing energy from flowing gas streams. The problem also becomes more difficult as systems are scaled up and heat transfer rates decrease. One way of avoiding the difficulty is to carry out nozzle expansion processes in which heat transfer is not required to lower the temperature. Without sufficient, rapid cooling, the species with the lowest vapor pressure (A in Fig. 10.12) forms particles when a sufficiently high supersaturation is reached. As the temperature decreases, species *B*, with the higher vapor pressure, can condense onto the particles of the first species or may form new particles and result in coated particles or particles of each individual species. This behavior can be succeeded by collisions of the different particles, followed by interdiffusion to give the final product, *C*, which contains the desired ratio of A to B. Depending on the details of the reactor operation, a variety of materials can be formed, ranging from separate particles of A and B to particles of C consisting of material with a uniform ratio of A to B that is constant from particle to particle.

A similar problem occurs for multicomponent particle formation using two or more *reactive* precursors[16] (Fig. 10.13). The reaction rates of the two precursors are often too sensitive to temperature and the partial pressures of the precursors to allow matching. Unequal reaction rates at a given temperature can result in the formation of particles of one component ($M$ in the figure), followed by subsequent particle formation and/or condensation of the other component $M'$. This sequence leads to chemically inhomogeneous particles with a nonuniform composition from particle to particle. If the particle concentration is high enough, the particles with different compositions collide, and the species in each resulting particle interdiffuse. If the temperature is high enough, interdiffusion results in particles with a homogeneous elemental ratio and phase composition. On the other hand, if the temperature is not high enough, the particles are too large, and the particle concentration is too low, then particles with a variety of compositions are formed.

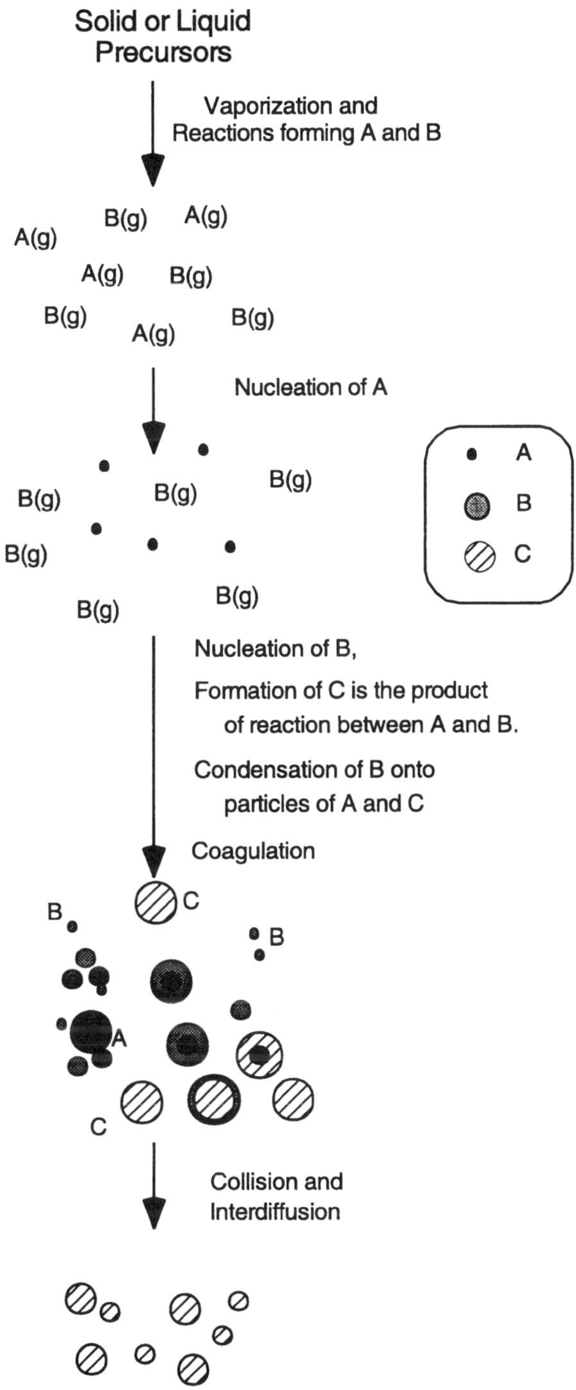


**Fig. 10.12.** Gas-to-particle conversion with two condensable vapors *A* and *B*. (Courtesy of Gurav et al.)[16]

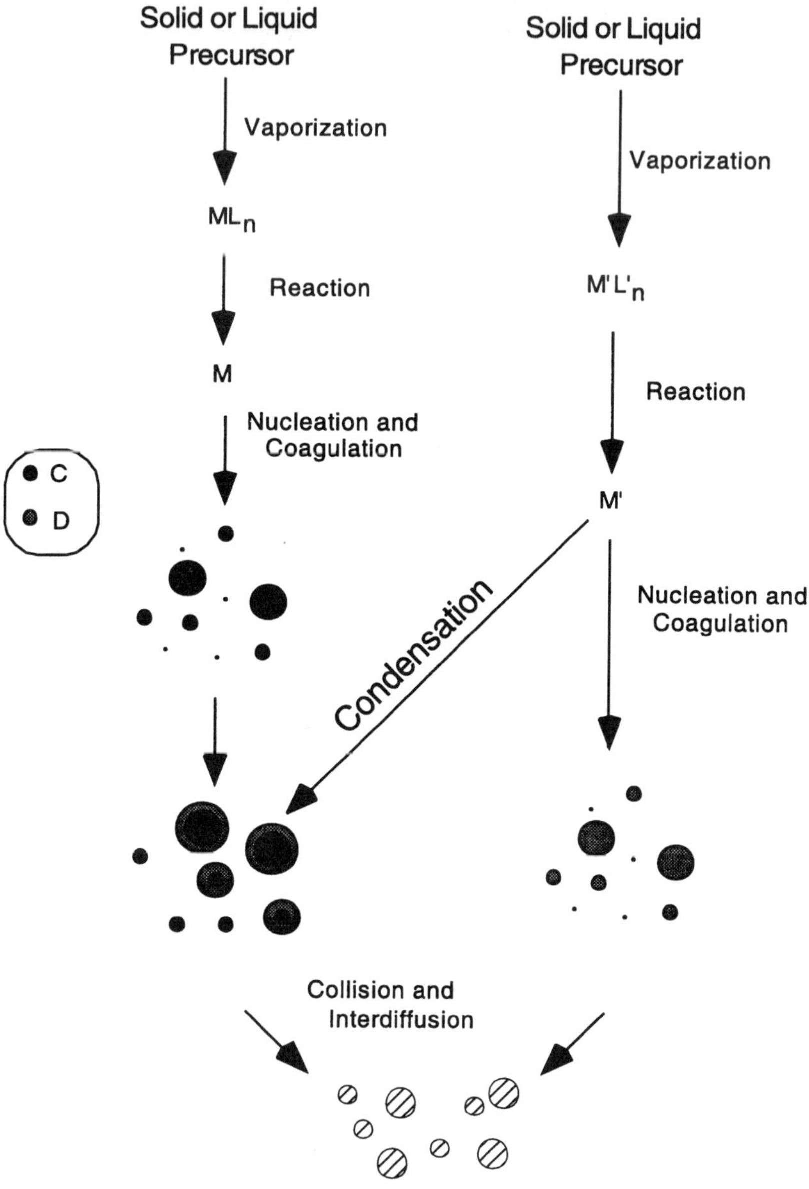


**Fig. 10.13.** Gas-to-particle conversion by chemical reaction of two precursors $ML_n$ and $M'L'_n$ to form products $C$ and $D$. (Courtesy of Gurav et al.)[16]

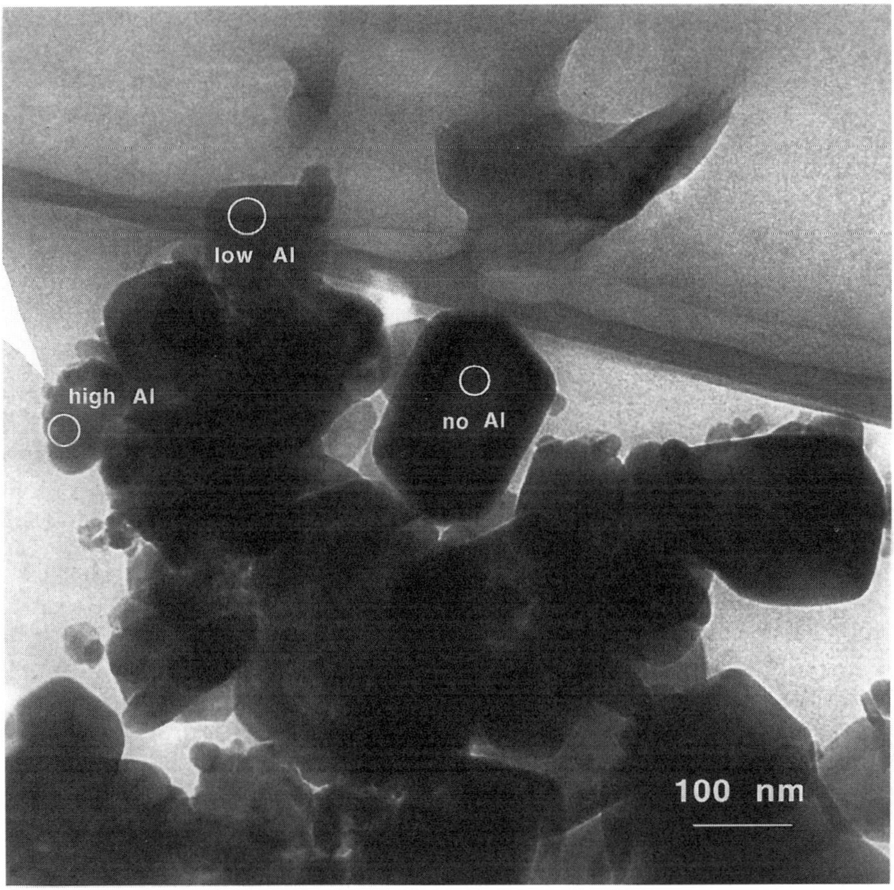


**Fig. 10.14.** Composite $TiO_2$–$Al_2O_3$ particles formed by gas-phase reaction, showing segregation. Composition was determined by energy dispersive spectroscopy (EDS). (Courtesy of Powell and Kodas, personal communication.)

If the reactivities of the precursors are matched, the formation of particles with a homogeneous composition within a particle and from particle to particle can be achieved, provided that the vapors are formed simultaneously at sufficiently low temperatures or that separate particles form, but collisions and interdiffusion of the elements erase the inhomogeneities.[17] The former process is favored if the reaction temperature is far below the temperature giving a highly supersaturated vapor ($S \gg 1$) for both species. In this case, all the reaction products are highly supersaturated, and atomically homogeneous particles can result. Operation at higher temperatures can lead to lower saturation ratios and sequential nucleation of the reaction products. The latter process is favored by high particle concentrations and high temperatures, under which the particles collide frequently and solid-state diffusion rates are rapid.

Figure 10.14 shows particles formed by the reaction of premixed $AlCl_3$ and $TiCl_4$ at 1500°C in a hot-wall flow reactor. Individual particles of $Al_2O_3$ and $TiO_2$ were found, along with agglomerates containing both Al and Ti oxides. This distribution was attributed to the reaction of $AlCl_3$ and $TiCl_4$ at different regions in the reactor to form individual particles of $Al_2O_3$ and $TiO_2$, which then collided and partially interdiffused, thus demonstrating that caution must be used when trying to obtain particles with a homogeneous composition via the chemical reaction of two or more precursors.

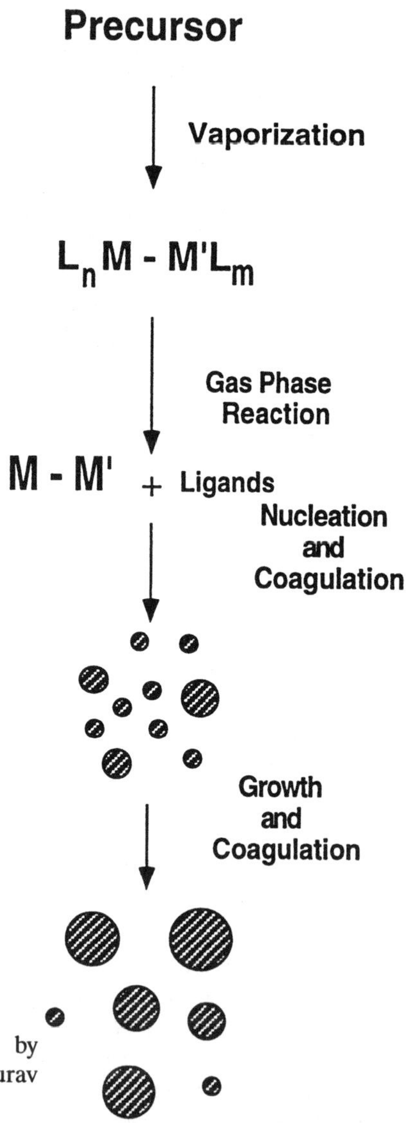


**Fig. 10.15.** Generation of binary materials by using single-source precursors. (Courtesy of Gurav et al.)[16]

### 10.3.2 Strategies for Avoiding Segregation in Multicomponent Materials

Besides relying on collision and interdiffusion, at least three other strategies have been developed for avoiding the problem of segregation (the formation of compositionally nonuniform particles) during the generation of binary and ternary materials: using single-source precursors, matching reactivities of the precursors, and using precursors that can react with each other only in a fixed ratio. Single-source precursors are volatile molecular species that combine all of the necessary elements in the correct ratio into a single molecule.[16, 18] Figure 10.15 shows two metals, *M* and *M'*, bonded to each other, each attached to several ligands *L*. Two classes of single-source precursors exist: those that can produce materials containing a metal and a nonmetal (e.g., TiN, ZnS, MgO, etc.) and those that can produce materials containing two metals (e.g., $BaTiO_3$, AgPd, $MoSi_2$, etc.). If the bonds holding the metals together are not broken during gas-phase reaction, the stoichiometry of the precursor is duplicated in the product, so that a chemically homogeneous material is produced. This approach has been used extensively for CVD of multicomponent materials,[18] but not as much for particle formation by gas-phase reaction. One example is the formation of ZnS particles from a single-source precursor containing Zn–S bonds, rather than using separate sources of Zn and S, such as $Zn(CH_3)_2$ and $H_2S$, which are dangerous to handle because they are poisonous and pyrophoric. The latter

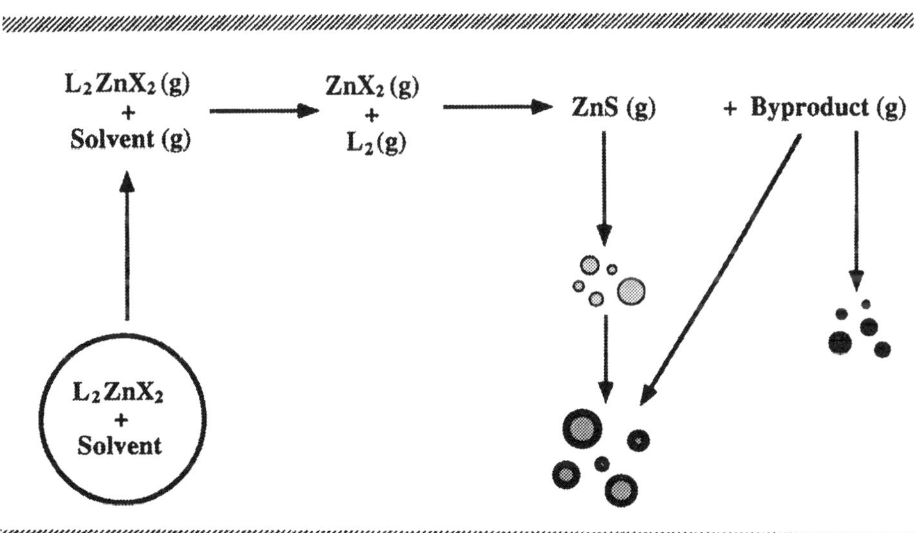


$L_2$ = TMEDA
$X = S_2CNEt_2$

**Fig. 10.16.** Aerosol dynamics with single-source precursor for ZnS. (Courtesy of Powell, Kodas and Hampden-Smith, personal communication.)

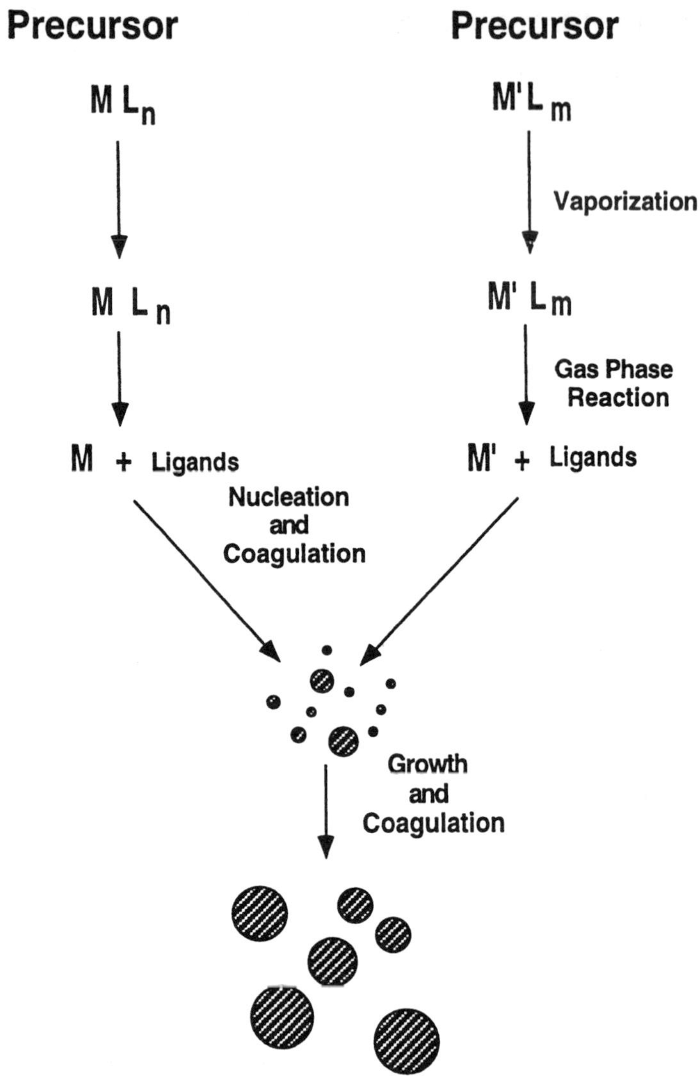


**Fig. 10.17.** Generation of binary and ternary materials by matching of precursors' reactivities. (Courtesy of Gurav et al.)[16]

situation is shown in Figure 10.16: The precursor $L_2Zn(S_2CNEt_2)_2$ (where L = tetramethylethylenediamine and Et = ethyl) reacts in the gas phase to form ZnS particles.

The strategy of matching the reactivities of two or more precursors is outlined in Figure 10.17. Two or more precursors are chosen to give similar reaction rates at the same temperature. As a result, both precursors react at the same position in the reactor, yielding condensable species that combine to form particles with a uniform composition. The reaction temperature must

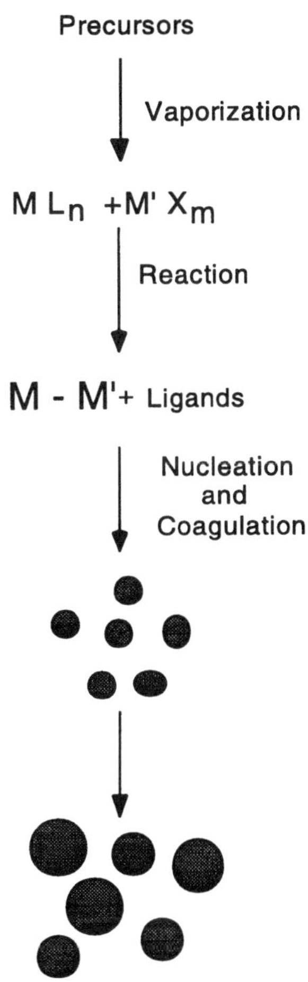


**Fig. 10.18.** Formation of binary materials by reaction between precursors. (Courtesy of Gurav et al.)[16]

be low enough for the reaction products to be supersaturated in order to drive condensation of all the reaction products at the same region in the reactor. As an example, $Pd(hfac)_2$ and $Cu(hfac)_2$ can be reacted at temperatures such that both reactants form the metal at similar rates. This allows the formation of nanometer-sized PdCu particles without segregation.[19]

Figure 10.18 shows the strategy for choosing two precursors ($ML_n$ and $M'X_m$) that cannot react by themselves below a certain temperature, but can react with each other.[16] For example, elimination reactions can be exploited in which ligands from both precursors are required to form a reaction product ($LX$, for instance). An example of this type of reaction is the elimination of an ester ($R'CO_2R$) between metal alkoxides and metal carboxylates, which results in the formation of metal oxides according to the reaction

$$M(OR)_4 + M'(O_2CR')_4 \longrightarrow MM'O_4 + 4R'CO_2R. \qquad (10.1)$$

Although this approach has not often been intentionally used for aerosol generation of materials, numerous examples exist in which complexes form between two species in the gas phase, as has been proposed for $AlCl_3$ and $TiCl_4$.

A general approach that is used in industry for the generation of multicomponent powders (mainly $TiO_2$ containing Al) with uniform composition is the formation of particles at high temperatures and concentrations, where collisions between particles and interdiffusion of elements after the collisions permit the mixing of the elements required to form homogeneous materials. This approach is shown in Figures 10.13 and 10.18. The pigment industry uses it for the formation of $TiO_2$ particles containing a few percent of Al. The compounds $TiCl_4$ and $AlCl_3$ are reacted with oxygen at high temperatures and extremely high aerosol loadings, where particle collisions are extremely frequent and solid-state diffusion rates are very high. (The chemical reactions that occur in the formation of $TiO_2$ from $TiCl_4$ are described in Chapter 8.)

### 10.3.3 Strategies for Producing Composite Particles

Composite particles can be produced by two main approaches: kinetic and thermodynamic control. In the case of kinetic control, two or more precursors can react in the gas phase to form products. If separate particles form for each species, as described in Section 10.3.2, then collisions without significant interdiffusion can result in composite materials. An example of this technique is the formation of $TiO_2$ and $Al_2O_3$ at low temperatures from chlorides of the metals. (See Figure 10.14.) At temperatures much below 1100°C, the metal chlorides can react to form particles of $TiO_2$ and $Al_2O_3$, which collide. Because of the relatively low temperature, the elements cannot interdiffuse, thereby producing a composite material made up of particles with a nonuniform composition.

In the case of thermodynamic control, the phase diagram for the materials under consideration must show that two or more phases exist at thermodynamic equilibrium. Again, an example is the formation of $TiO_2$ containing Al from chlorides of the metal and oxygen. At high temperatures (roughly 1500°C) and low levels of Al, the Al can be accommodated in the $TiO_2$ lattice, and a single-phase material is produced. Above a critical value, however, the solubility of Al in $TiO_2$ is exceeded, and the Al forms separate $Al_2O_3$ or $Al_2TiO_5$ phases.

### 10.3.4 Strategies for Producing Coated Particles

Coated particles can be formed under both thermodynamic and kinetic control, starting with two or more species.[2, 15, 20] In the case of thermodynamic control, one of the phases segregates to the particle surface, where it wets the material in the core, thereby forming a coating.[15] In the case of kinetic control, vapor condensation routes can be used to exploit differences

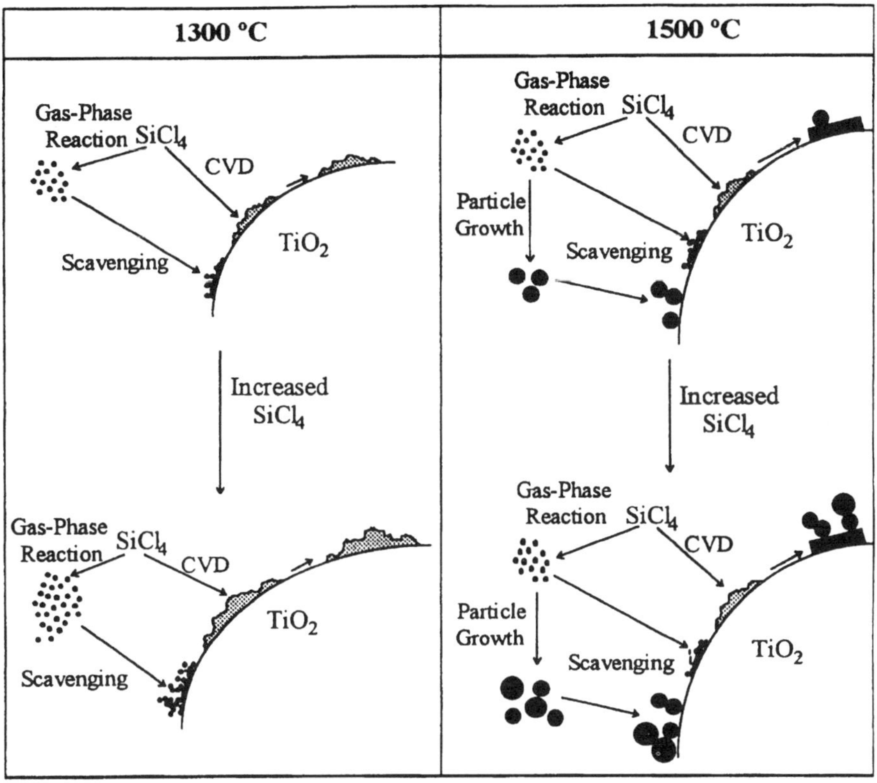


**Fig. 10.19.** Physical and chemical phenomena involved in coating of particles by gas-phase and surface reactions. (Adapted from Powell et al.)[25]

in the vapor pressures of the two materials. Here, particles of one material with a low vapor pressure are formed when the temperature is reduced. As the temperature drops further, the second species becomes supersaturated and condenses on the existing particle surface area to give a coating. Although this route has been demonstrated for some combinations of materials, it is not common because it relies on the physical properties of the two product phases.

A more general approach based on kinetic control is to use the chemical reaction of a precursor to drive the coating.[21-24] This approach is outlined in Figure 10.19 for the case of $SiO_2$ coatings on $TiO_2$.[25] First, the $TiO_2$ particles are generated in a flow system by reaction of $TiCl_4$. Further downstream, the particles are mixed with $SiCl_4$ vapor. The $SiCl_4$ reacts in the gas phase to produce small $SiO_2$ particles, which collide with the $TiO_2$ particles. After collision, if the temperature is high enough and the $SiO_2$ particles are small enough, the $SiO_2$ particles fuse onto the surface of the $TiO_2$ particles and form a coating. The coating can also be formed by CVD of $SiCl_4$ on the particle surface. As shown in Figure 10.20, higher tempera-

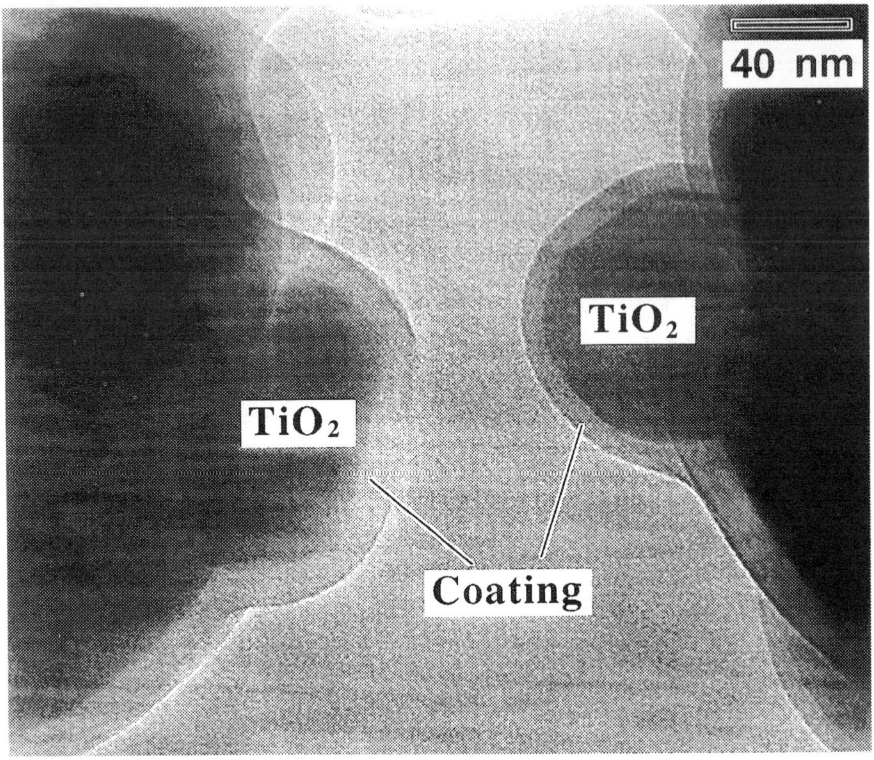


**Fig. 10.20.** Coated $TiO_2$ particles with $SiO_2$ layer formed by process outlined in Figure 10.19. (Courtesy of Powell and Kodas, unpublished results.)

tures favor smoother coatings because of the higher sintering rate of $SiO_2$ under those conditions. This approach has been demonstrated for a variety of coatings. Models based on a lognormal distribution and moments of the distribution have been used to describe the process.[20] An advantage of the method is that coating can be carried out in virtually any type of system—for example, agglomerates formed spontaneously in fluidized beds, laser reactors, furnace reactors, flame reactors, and, indeed, most other systems.

## 10.4 FLUIDIZED BEDS

Aerosol particles may be present when fluidized beds are operated in several different modes involving gas-to-particle conversion.[26–29] The formation of aerosol particles is encountered during the coating of particles in fluidized beds. Typically, a gaseous reactant is introduced into the bed, which is heated to a sufficient temperature for CVD to form a coating. If the reaction temperature or reactant concentration is too high, gas-phase particle formation can occur. This is undesirable in most cases, and strategies have been developed to avoid new-particle formation. The situation is closely related to the coating of aerosol particles as described in Section 10.3.4.

Fluidized beds can be used to coat agglomerates of fine particles. Agglomerate formation is undesirable for powder generation in most aerosol systems, but is desirable for micron-sized particles in fluidized beds. The primary problem with working with fine particles in fluidized beds is that the small sizes of the particles result in low settling velocities, which lead to entrainment of particles in the gas flow. It has been found experimentally that many materials in the form of fine particles spontaneously agglomerate in fluidized beds.[30-32] The agglomerates can be of sufficient size (100 $\mu$m) to allow smooth fluidization. This has led to the use of fluidized beds for the coating of agglomerates to form composite particles.

In another variation of the fluidized bed approach, the agglomerated fine particles are *allowed* to be entrained in the gas flow, forming a lean, entrained bed of particles.[33] The particles in this stream can then be coated by combining the particle-laden stream with a reactant gas stream and passing the mixture through a furnace reactor. Aluminum particles have been converted into AlN using this approach.[33]

## 10.5 REFERENCES


1. Majumdar, D., Kodas, T.T., & Glicksman, H. *Adv. Mater. 8*(12), 1996, 1020–1022.
2. Jain, S., Kodas, T.T., Wu, M.K., & Preston, P. *J. Aerosol. Sci. 28*(1), 1997, 133–146.
3. Kim, K.-S. & Pratsinis, S.E. *AIChE J. 34*(6), 1988, 912–921.
4. Kim, K.-S. & Pratsinis, S.E. *Chem. Eng. Sci. 44*(11), 1989, 2475–2482.
5. Ward, T.L., Kodas, T.T., Carim, A.H., Kroeger, D.M., & Hsu, H. *J. Mater. Res. 7*(4), 1992, 827–836.
6. Wu, M.K. & Friedlander, S.K. *J. Aerosol Sci. 24*(3), 1993, 273–282.
7. Wu, M.K. & Friedlander, S.K. *J. Colloid Interface Sci. 159*(1), 1993, 246–248.
8. Wu, M.K., Windeler, R.S., Steiner, C.K.R., Börs, T., & Friedlander, S.K. *Aerosol Sci. Technol. 19*, 1993, 527–548.
9. Kruis, F.E., Kusters, K., Pratsinis, S.E., & Scarlett, B. *Aerosol Sci. Technol. 19*, 1993, 514–526.
10. Landgrebe, J.D. & Pratsinis, S.E. *Ind. Eng. Chem. Res. 28*(10), 1989, 1474–1481.
11. Landgrebe, J.D. & Pratsinis, S.E. *J. Colloid Interface Sci. 139*(1), 1990, 63–86.
12. Pratsinis, S.E., Kodas, T.T., Dudukovic, M.P., & Friedlander, S.K. *Ind. Eng. Chem. Process Des. Dev. 25*(3), 1986, 634–642.
13. Pratsinis, S.E., Kodas, T.T., Dudukovic, M.P., & Friedlander, S.K. *Chem. Eng. Sci. 41*, 1986, 693–700.
14. Pluym, T., Kodas, T.T., Wang, L.-M., & Glicksman, H. *J. Mater. Res. 10*(7), 1995, 1661–1673.
15. Jain, S., Atanasova, P., Kodas, T.T., & Hampden-Smith, M.J. *J. Electrochem. Soc. 143*(11), 1996, 3762–3770.

16. Gurav, A., Kodas, T., Pluym, T., & Xiong, Y. *Aerosol. Sci. Tech. 19*, 1993, 411–452.
17. Dericotte, D. *Formation of Nanoscale Single and Multicomponent Particles via Two Gas-Phase Routes: Matched Reactivities and Single-Source Precursors*. Master's thesis, University of New Mexico, 1997.
18. Kodas, T.T. & Hampden-Smith, M.J. (Eds.). *The Chemistry of Metal CVD*. Weinheim, West Germany: VCH, Verlagsgesellschaft, 1994.
19. Dericotte, D., Powell, Q.H., Gurav, A., Kodas, T.T., Hampden-Smith, M.J., Zeng, D., & Wang, L. *NanoStr. Mater.*, forthcoming.
20. Jain, S., Fotou, G.P., & Kodas, T.T. *J. Colloid Interface Sci. 185*, 1997, 26–38.
21. Fotou, G.P. & Kodas, T.T. *Adv. Mater. 9*(5), 1997, 420–423.
22. Fotou, G.P. & Kodas, T.T., unpublished.
23. Powell, Q.H., Fotou, G.P., & Kodas, T.T. *Chem. Mater. 9*(3), 1996, 685–693.
24. Powell, Q.H., Fotou, G.P., Kodas, T.T., Anderson, B.M., & Guo, Y. *J. Mater. Res. 12*(2), 1997, 552–559.
25. Powell, Q.H., Kodas, T.T., & Anderson, B.M. *J. Chem. Vapor Deposition 2*, 1996, 179–181.
26. Kimura, T. & Kojima, T. *J. Chem. Eng. Jpn. 17*(1), 1991, 81–87.
27. Kimura, T. & Kojima, T. *J. de Physique IV 1* (Colloque C2), 1991, 103–110.
28. Kojima, T. & Morisawa, O. *J. de Physique IV 1* (Colloque C2), 1991, 475–482.
29. Matsukata, M., Odagiri, T., & Kojima, T. *J. de Physique IV 1* (Colloque C2), 1991, C2-483–C2-490.
30. Kusakabe, K., Kuriyama, T., & Morooka, S. *Powder Technol. 58*, 1989, 125–130.
31. Morooka, S., Okubo, T., & Kusakabe, K. *Powder Technology 63*(2), 1990, 105–112.
32. Morooka, S., Kusakabe, K., Kobata, A., & Kato, Y. *J. Chem. Eng. Jpn. 21*(1), 1988, 41–46.
33. Kimura, I., Ichiya, K., Ishii, M., Hotta, N., & Kitamura, T. *J. Mater. Sci. Lett. 8*, 1989, 303–304.

# 11 Technology of Gas-to-Particle Conversion

## 11.1 INTRODUCTION

Particles can be formed by intraparticle reaction of precursors with negligible vapor pressures or by nucleation and growth of low-vapor-pressure species, otherwise known as gas-to-particle conversion. The technology of gas-to-particle conversion is the subject of this chapter and has also been discussed in numerous review articles.[1-6] The technology of intraparticle reaction routes to particle generation is examined in Chapter 13.

Particle formation, as discussed in Chapter 10, can be driven by chemical reaction of gaseous precursors (chemical routes), which results in the formation of condensable products, or by cooling of a vapor (physical routes), as shown in Figure 11.1. Examples of chemical routes are processes driven by hot-wall tubular furnaces, flames, lasers, and plasmas. Physical processes often rely on the cooling of a hot vapor produced by heating a precursor to a temperature at which it has a significant vapor pressure. Other methods for vaporization include laser ablation and sputtering with ions or electrons. Physical processes can also be classified as confined and free processes. Confined processes include condensation in a nozzle; free processes include a free convective plume of a hot vapor and laser ablation.

In some cases, gas-phase routes can be combined with spray processes or intraparticle reaction to give hybrid approaches, as shown in Figure 11.2. An example of gas-to-particle conversion incorporating intraparticle reaction is the nucleation of a metal-organic precursor vapor to form droplets or particles, which are then reacted with a gas to form the product. An example of combining a spray route with gas-to-particle conversion is the spraying of solutions containing fuels (which burn in the gas phase) around the particles, resulting in volatilization (in some cases, only partial volatilization) of the solids in the particles, followed by gas-to-particle conversion. Particles can be formed in the gas phase during cooling of a gas containing volatile species with high vapor pressures that have evaporated during intraparticle reaction routes. The delivery of volatile precursors as droplets also combines these different approaches.

The chemical and physical routes are identical in terms of the aerosol dynamics that can occur once the condensable species is generated. The

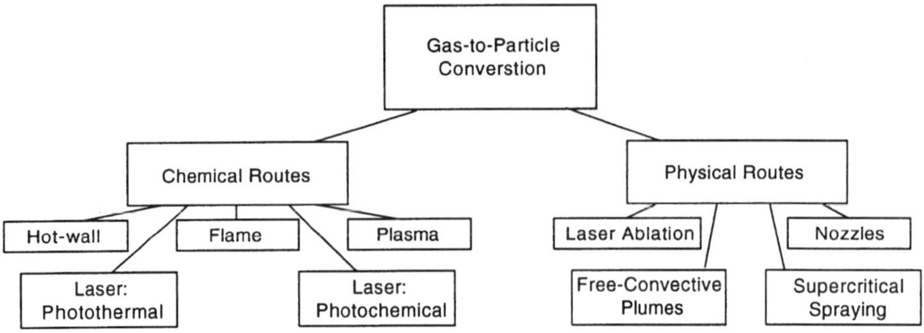


**Fig. 11.1.** Types of gas-to-particle conversion systems.

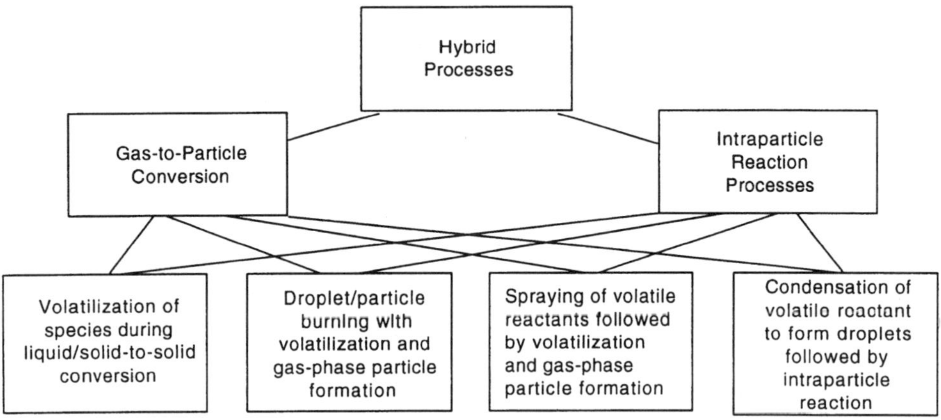


**Fig. 11.2.** Hybrids of intraparticle reaction and gas-to-particle conversion systems.

saturation ratio (the ratio of partial pressure to vapor pressure) of the vapor builds up as the vapor cools or more product is formed by chemical reaction. When the vapor becomes sufficiently supersaturated, nucleation occurs to form thermodynamically stable clusters. Some materials may have such low vapor pressures at the temperatures at which they are formed, that there is no barrier to cluster formation, and particles are generated as atoms or molecules of the species collide with themselves. Once clusters are formed, they can grow by vapor condensation, by collision and coalescence, and by surface reaction, in which the precursor reacts on the cluster surfaces. The clusters can grow to a size at which they are considered particles. Particle collisions at high temperatures and for small particle sizes result in sintering, to form spherical particles. However, for larger particles and lower temperatures, the particles do not sinter rapidly, and hard agglomerates are formed. At even lower temperatures and for much larger particle sizes, agglomeration may result in the formation of flocs, which are collection of particles weakly held together by van der Waals forces. Other physical processes, such as thermophoresis, sedimentation, and diffusion, may result in particle deposition and, along with resuspension (also known as reentrainment), can modify the particle size distribution.

The gas-to-particle conversion route is used industrially to generate a variety of materials. Titanium dioxide pigment, fumed silica, and carbon black are produced on massive scales and are the largest volume applications of aerosol manufacture of powders. The majority of methods for the generation of nanophase materials are also gas-to-particle conversion approaches. As a result, the gas-to-particle conversion route is the most studied of all aerosol systems for powder generation. The route is versatile and has been used to produce metals, semiconductors, and oxide and nonoxide ceramics. Gas-to-particle conversion routes are most useful for the generation of elemental or binary materials (simple oxides, nitrides, borides, carbides), but are used by industry for ternary and more complex materials also.

The primary advantages of most gas-to-solid routes relative to solid-state and liquid-phase routes are:

- Smaller particle size (nm to 10 $\mu$m) than in solid-state routes
- Narrow particle size distribution (nearly monodisperse for nucleation–condensation and broader for nucleation–coagulation) relative to that of solid-state routes
- Solid particles produced at high temperatures relative to those of intraparticle reaction routes
- High purity relative to that in solid-state routes

Disadvantages, in some cases, include:

- Difficulty in producing multicomponent materials relative to some solid-state routes
- Complexity relative to solid-state routes
- Expensive gaseous reactants relative to solid-state routes
- Hard agglomerates at high particle concentrations

Depending on the type of energy source and whether or not chemical reactions take place, there are a number of variations of gas-to-particle routes for powder generation, including flame, plasma, hot-wall, laser, evaporation–condensation in natural plumes and nozzles, and evaporation–condensation–reaction. In spite of their variety, these routes have in common the basic processes of particle nucleation, condensation, coagulation, and the other fundamental processes.

This chapter divides gas-to-particle conversion processes into chemical routes and physical routes and then into groups of processes that use the same type of energy source to drive the processes (Fig. 11.1). Each type of reactor is covered in a section that is divided into an overall description of the route, the precursors and products, experimental details, and theory that is common to each particular type of reactor and has not already been covered in the sections on overall reactor behavior.

## 11.2 COMPARISON OF DIFFERENT GAS-TO-PARTICLE CONVERSION ROUTES AND THEIR COMPARISON WITH LIQUID-TO-SOLID AND SOLID-TO-SOLID CONVERSION ROUTES

The two primary types of aerosol processes for materials generation are intraparticle reaction (liquid-to-solid conversion, solid-to-solid conversion, or spray pyrolysis) and gas-to-particle conversion. These two classes of aerosol routes have distinct features that make them suitable for different applications.

Intraparticle reactions with sprayed precursors have the major advantages that they can handle multicomponent materials containing a large number of elements and that involatile precursors can be used. This makes spray routes the natural choice for generating materials containing two or more metals. An examination of the classes of materials of interest shows that they include mixed-metal titanates, metal ferrites, superconductors, metal alloys, and a variety of composites. Spray routes also are more attractive for materials for which volatile precursors either are not available or are too expensive. Examples of elements for which sufficiently volatile, inexpensive precursors are not available are elements used in mixed-metal titanates (Pb, La, Ba, Ca, and Sr); precious-metal alloys (Au, Pd, Pt, and Ag); ceramic superconductors (Y, Ba, Cu, Bi, Sr, Ca, and Cu); and ferrites (Fe, Sr).[7] Spray routes with intraparticle reaction are not useful for generating large quantities of unagglomerated nanometer-sized particles, because generators capable of producing large concentrations of small droplets are not available. Furthermore, if high throughputs were available, coagulation would still result in agglomerates. This class of routes is, in general, most attractive for the production of particles with diameters of at least 0.1 $\mu$m. Spray routes also can have poor control over particle morphology, leading to the formation of hollow or highly porous particles in certain cases.

In contrast, gas-to-particle conversion processes are the best choice for the generation of solid particles of single-component and binary systems such as metals, semiconductors, simple oxides, nitrides, borides, etc., with particle sizes smaller than 1 $\mu$m; however, these processes require the use of volatile precursors. High production rates are possible when highly volatile reactants are available, and because the gaseous precursors can be purified more easily than solid or liquid precursors, these processes allow the generation of materials with ultrahigh purities. This combination of features favors the use of this class of aerosol routes for generating simple materials like $TiO_2$, fumed $SiO_2$, optical fibers, and carbon black. Unfortunately, a disadvantage of the route is that the synthesis of high-purity volatile precursors is complicated and expensive. In industry, for example, a large portion of a typical plant producing $TiO_2$ pigment is devoted to synthesizing and purifying $TiCl_4$. For this reason, these precursors tend to be more expensive than involatile precursors, which are often mined and are commercially available in a form in which they can be immediately used. The major advantage that gas-to-particle conversion routes have over spray routes is that they can produce nanometer-sized particles.

In general, gas-to particle conversion processes are more complex to design and operate than systems with intraparticle reaction with liquid-to-solid or solid-to-solid conversion. A typical gas-to-particle system requires a gas-handling subsystem for the controlled introduction of gaseous reactants. In contrast, spray systems require only an atomizer. The reactor portion of the gas-to-particle system varies in complexity, depending on the method used to heat the gas. All aerosol processes share the same types of powder collection units.

All of the gas-to-particle conversion processes provide particles with similar characteristics. (See Table 11.1, at end of chapter.) All are capable of producing particles with sizes down to the nanometer scale because the particles are built up from atomic or molecular species in the gas phase. Most of the processes usually produce agglomerated particles, unless care is taken to keep particle concentrations below the coagulation limit. However, this strategy is not industrially viable because it severely limits the production rate. In industrial processes, agglomerates are formed at particle concentrations above the coagulation limit because particle collisions and vapor condensation often occur simultaneously. If rates of particle sintering are low, particles collide, and then vapor condensation, surface reaction, and sintering fuse the particles to form hard agglomerates. The maximum temperatures for gas-to-particle conversion processes vary greatly, with the highest being many tens of thousands of degrees centigrade for plasma routes and the lowest a few hundred degrees centigrade for evaporation and condensation of organometallic reactants followed by intraparticle reaction. Most of the processes allow the generation of metals, metal oxides, and nonoxides. The complexity of the systems varies widely: Plasma reactors are complex (and expensive), while hot-wall reactors require only a tube furnace.

In industry, the choice of gas-to-particle conversion process is made on the basis of the complexity and economics of the process. The most successful commercial processes, used by companies like DuPont, SCM, Tioxide, Kemira, Kronos, Cabot, and Degussa, are those that are the simplest and have the lowest cost. Many of these routes are carried out at massive scales (hundreds of tons per day per system) and exploit chemical reactions in tubular flow reactors, chemical reactions in flame reactors, and the physical route of evaporation and condensation in a tubular flow system. The first of these routes is used for the generation of $TiO_2$ pigment, the second for fumed silica and carbon black, and the last for nanophase materials. Commercial applications of the other types of chemical systems (laser and plasma reactors) and physical systems (laser ablation, sputtering, nozzle expansion, and supercritical expansion) have been limited by their complexity and high cost. The behavior of specific systems has been of interest primarily for scientific studies.

## 11.3 DESIGN

Before discussing the design of aerosol reactors, it is important to stress that all gas-to-particle conversion systems involve similar phenomena, including nucleation, coagulation, condensation, surface reaction, coalescence, and deposition. The primary differences among the systems are the temperature profiles, species concentration profiles, and fluid velocity profiles. Because of the amazingly wide variety of aerosol systems and the lack of suitable models for many types of aerosol processes, it is difficult to give generalized design approaches. This section aims instead to provide an outline of the approaches.

The design of a system for powder production by gas-to-particle conversion in an aerosol reactor may, *in principle*, be carried out rigorously by collecting data on precursors and products, modeling the system, and determining the operating conditions required to produce the desired product. In practice, however, these systems are highly complex, and not enough data are available to predict product properties. As a result, taking into account certain general guidelines, the usual procedure is to design the system and then to optimize its behavior empirically with guidance from detailed modeling.

The primary variables that can be controlled include the reactor temperature profile, reactant concentration, and reactor residence time. Parameters that are usually fixed include the reactor geometry, properties of the reactants and products, and the total operating pressure. In practice, the reactant concentration and residence time are usually confined to fairly narrow ranges in order to maximize yields and keep the size of the reactor as small as possible. Many methods are used to drive the reactions or physical processes leading to particle formation. However, the vast majority of these methods tend to control the temperature profiles in the system. Thus, in most cases, the main experimental control parameter is the temperature, which controls the rate of the processes that determine the saturation ratio of the condensable product. In this sense, the rate of reaction to form the gaseous product species also can be viewed as a control parameter for systems in which chemical reactions occur.

The typical design procedure begins with experiments in which powders are produced in a pilot or laboratory-scale system and the product particle size distribution is measured. At this point, guidance from modeling can be used to determine how to modify the operating conditions. If the particle size is too small and the reactor is operated in the nucleation–condensation mode with no coagulation, the temperature can be decreased, resulting in a lower reaction rate and the formation of fewer particles. If the same amount of reactant is used in both cases, the result is more condensable species for each particle and a larger particle size. Care must be taken to make sure that all of the reactant is consumed before it can reach the reactor exit. If the particle size is too large, the temperature can be increased, yielding smaller particles. In flame reactors, increasing the reactor temperature changes the residence time.

Similar approaches can be taken if the reactor is operated in the nucleation–coagulation mode. If the particle size has reached an unacceptably broad distribution as a result of not having attained a self-preserving particle size distribution, the temperature can be increased, resulting in more rapid formation of particles, which in turn increases the coagulation rate and reduces the time to reach the self-preserving distribution. Among other considerations in the design of these reactor systems, multistage units can be used to provide a region where particles are formed, followed by a region where particles are coated.

Another major consideration is control of agglomeration. Hard agglomerates in particular are highly undesirable for most applications (except, for example, the generation of carbon black). A more rapid temperature reduction following the particle generation stage can minimize this problem. Control over the crystallinity of the product may require control over the temperature or cooling rate in the reactor. This in turn may demand either relatively slow cooling or very rapid cooling, depending on the desired properties.

Plasma reactors are the most complicated of all types of reactors and, as a result, are the most poorly understood. The primary control parameters are the plasma temperature, cooling rate after leaving the plasma, reactant concentration, and residence time distribution.

## 11.4 CHEMICAL ROUTES: TUBULAR FLOW REACTORS

### 11.4.1 Introduction

Hot-wall reactors involve driving chemical reactions using a furnace-heated reactor. In other cases, preheated reactant streams of exothermically reacting species are used, and the walls of the reactor are cooled. If the reactant streams are not premixed, two or more reactants are mixed inside the reactor to allow the reaction to form low-vapor-pressure species, which then nucleate to form particles (Fig. 11.3). The particles can grow by a variety of mechanisms, depending on conditions. In laboratory systems, gas-phase reaction competes with CVD of the reactant on the walls. This is less of a problem in industrial systems, in which transport distances are longer. Hot-wall tubular

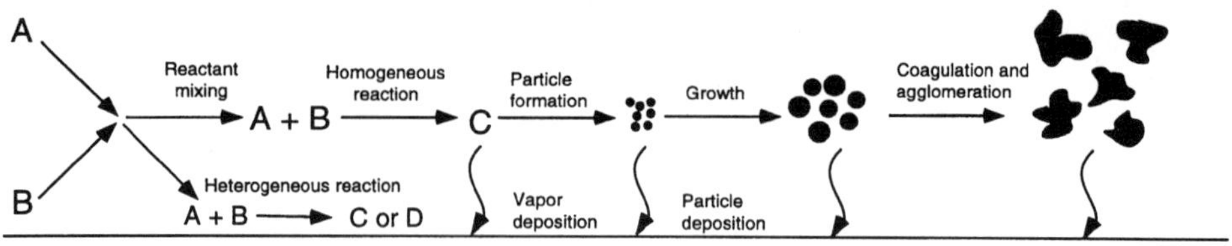


**Fig. 11.3.** Typical gas-to-particle conversion process for powder generation.

flow reactors are more common in lab-scale studies rather than industry, because of their high energy requirements. They not only provide better control of particle characteristics, but often are ideal for studying the effect of individual parameters on powder properties. Tubular flow reactors have been the subject of several review articles.[1–3, 5, 6, 8–10]

In industry, the most important example of a tubular flow reactor is in the production of $TiO_2$ pigment, where pure $TiCl_4$ and $AlCl_3$ vapors are often premixed and introduced into a tubular flow system along with a separate stream of preheated oxygen. The two streams mix and react exothermically to form $TiO_2$ containing Al.

There have been few systematic studies of the influence of mixing reactants on the aerosol dynamics. However, results from the literature suggest that, at the laboratory scale in reactors having low Reynolds numbers, mixing reactants can have a dramatic effect on the aerosol dynamics.[11] A variety of mixing arrangements have been used at the laboratory and industrial scales, as shown in Figure 11.4.[12] Premixed streams are often employed in laboratory-scale experiments in which a dilution gas is used and reaction outside the reactor can be avoided. On the other hand, in industry, reactants often cannot be premixed because the two or more reactant streams would react spontaneously before entering the reactor. In these cases, reactant mixing also can influence reactor behavior. In general, the information available about reactant-mixing problems in industry has come through the patent literature: Numerous patents discuss mixing arrangements for introducing the $TiCl_4$–$AlCl_3$ stream and the oxygen stream into tubular flow systems for pigment production.[13–27] Fortunately, industrial systems frequently operate with very high Reynolds numbers (often $10^5$ to $10^6$), for which mixing can be rapid.

In contrast to other types of gas-to-particle conversion systems, tubular flow reactors are perhaps the simplest in construction. Temperatures are usually less than 1700°C, and the systems are generally operated at atmospheric pressure, although some industrial processes operate at higher pressures. The reactors operate at lower temperatures than plasma and flame processes, but at higher temperatures than evaporation–condensation–reaction processes.

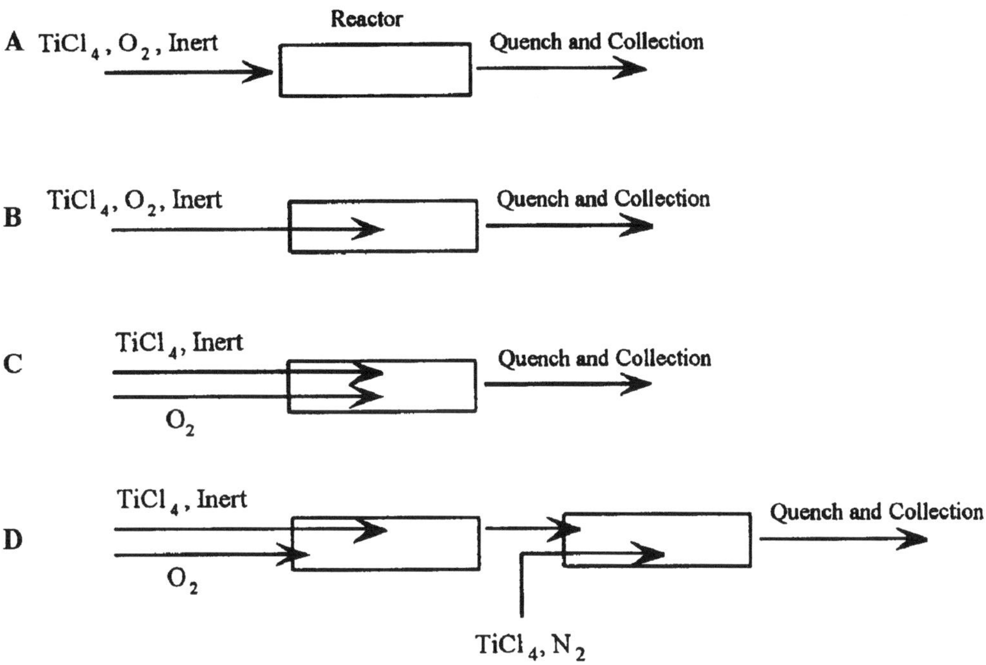


**Fig. 11.4.** Reactant-mixing configurations used for generation of $TiO_2$ by gas-to-particle conversion. (Adapted from Powell.)[12]

Tubular flow reactors have several attractive features for powder generation:

- Simplicity
- Scale-up to roughly 100-ton-per-day levels demonstrated for materials such as $TiO_2$
- Production of oxides, nonoxides, semiconductors, and metals
- Particle size from atomic to 10 microns
- Mechanism relatively well understood

Disadvantages of this method include:

- Requirement of precursors with significant vapor pressure and stability below reaction temperature
- Expensive precursors
- Hard agglomerates formed at high aerosol loadings
- High energy requirements

### 11.4.2 Precursors and Products

Hot-wall systems rely on the use of volatile reactants that are usually introduced into the reactor as gases. (In some cases, they are introduced as an aerosol of a solution: (see Fig. 11.5). Ideally, both the solution and the precursors in the droplet evaporate in the reactor before gas-phase reaction occurs. Thus, the droplets serve only as a convenient method for introducing the reactants into the reactor. This approach is useful for thermally sensitive precursors, which cannot be heated for long periods of time, to increase their vapor pressures. It also provides more reproducible delivery of solid precursors because the rate of delivery depends only on the liquid feed rate. In addition, liquid delivery allows convenient feeding of several reactants simultaneously in a reproducible manner. The major drawback, however, is that intraparticle reaction of the volatile precursor before vaporization can result in the formation of remnant particles, leading to a bimodal particle size distribution.

A wide variety of reactants have been used in hot-wall reactors. (See Table 11.2 at end of chapter.) However, due to cost considerations and the requirement of high vapor pressure and purity of the product, the most commonly employed reactants are metal chlorides such as $TiCl_4$ and $SiCl_4$. Less work has been performed using organometallic precursors, because of their higher cost and the possibility of contamination by carbon and oxygen (for metals).

Most of the volatile precursors that have been used for particle formation by gas-phase reaction can be classified as metal halides, oxyhalides, hydrides, carbonyls, alkyls, alkoxides, and $\beta$-diketonates. In general, the requirements for precursors for particle formation are the same as those for precursors for CVD: high volatility and a decomposition pathway leading to the formation of high-purity material. In a CVD process, gaseous reactants are transported to a hot surface, where heterogeneous reaction takes place and forms an

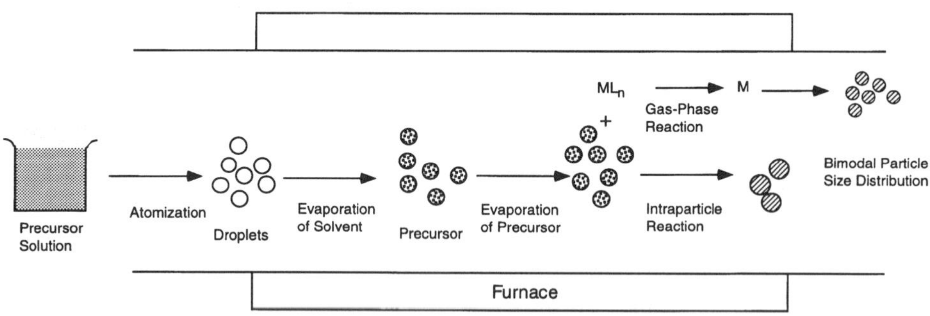


**Fig. 11.5.** Hot-wall reactor with single volatile precursor introduced into reactor by spraying of solution.

involatile product (ceramic, metal, semiconductor), along with volatile byproducts that desorb from the surface.[7] The main differences between powder formation and CVD are that the production of powders requires precursors with low costs and that the precursors do not have to react at low temperatures.

Table 11.2 shows precursors and products for tubular flow systems. Most studies of such systems have focused on the generation of $TiO_2$ because of its applications as a pigment, and on Si, because it is a convenient model element; there are also potential applications for the conversion of precursors into elemental Si used in microelectronics. Various other materials have been formed, including many simple metal oxides, carbides, and nitrides. Most of these materials have applications as structural ceramics. Very little work has been accomplished to produce elemental metals, sulfides, and other materials with microelectronics applications.

By far the most extensively studied system is Al-doped $TiO_2$ because of its applications as a pigment. This material is produced industrially from the reaction of $TiCl_4$, $O_2$, and a few mass percent of $AlCl_3$. The oxygen is preheated and is usually introduced through a separate inlet. In some systems, however, the $AlCl_3$ and $TiCl_4$ are premixed and heated to temperatures near 1000°C before being introduced along with the oxygen into the reactor. The two streams must mix, react, form particles, and cool in less than one second. There is a vast literature on all aspects of this system, including the chemistry, reactor design (patents), theory, and laboratory-scale experiments. Although the majority of the studies have examined simple metal oxides and elemental materials, coated and composite particles have also been produced. As an example, $TiO_2$ particles have been coated with $SiO_2$, $Al_2O_3$, $ZrO_2$, and various mixtures.[28-31]

Little is known about the precursor chemistry in most systems. Very few studies have analyzed volatile reaction products. Even fewer studies have examined the kinetics of the gas-phase reactions of the precursors. However, some generalizations can be made: As already mentioned, the classes of precursors that have been used most often for particle formation are metal halides, oxyhalides, hydrides, carbonyls, alkyls, alkoxides, and $\beta$-diketonates. These precursors are discussed in more detail in Chapter 8.

### 11.4.3 Experimental Considerations

Tubular flow reactors are relatively simple at the laboratory scale, as shown in Figure 11.6, which depicts a system for the generation of $TiO_2$ particles.[32] The precursor $TiCl_4$ is held in the bubbler (a) where a carrier gas (usually Ar or $N_2$) is passed through to give the vapor. The precursor vapor is premixed with oxygen and sent into the reactor (b), where conversion to $TiO_2$ occurs. The exiting gas stream is diluted with nitrogen gas to minimize losses due to thermophoresis and is then collected at (h). The exiting gas stream is scrubbed to remove undesirable gases and is then vented.

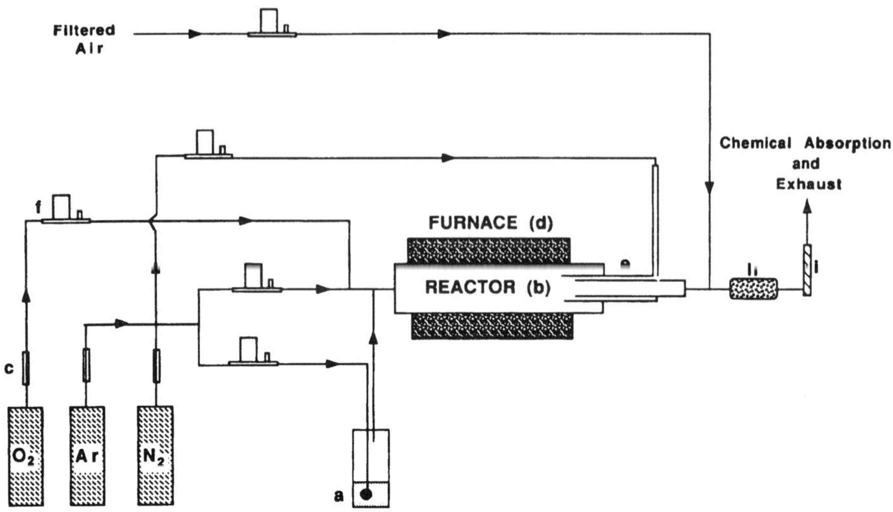


**Fig. 11.6.** Schematic of laboratory-scale system for $TiO_2$ powder generation. (Adapted from Akhtar M.K., Xiong, Y., and Pratsinis, S.E., "Vapor synthesis of titania powder by titanium tetrachloride oxidation," *AIChE J. 37* (10), 1561–1570, 1991. Reproduced with permission of the American Institute of Chemical Engineers. Copyright © 1991 AIChE. All rights reserved.)[32]

Hot-wall tubular flow reactors have also been used at the laboratory scale to produce coated particles. Figure 11.7 shows a system for the generation of $TiO_2$ particles coated with metal oxides.[31] The $TiO_2$ particles are formed by bubbling a carrier gas through $TiCl_4$ liquid to carry the vapor into the reactor, where it is converted to $TiO_2$. The coating precursor, such as $SiCl_4$ is vaporized and then introduced into the reactor near the exit, where it reacts to give the coating.

Tubular flow reactors also have utilized other methods of precursor delivery at the laboratory scale. Figure 11.8 shows a system for delivering volatile precursors into a tubular flow system using droplet delivery. Such a system is much simpler than one utilizing conventional precursor delivery.[2] In it, the precursor is dissolved in a suitable solvent (water is the easiest to use), which is atomized to form droplets. The droplets are carried in a carrier gas into the reactor, where the solvent and precursor evaporate. This approach allows convenient delivery of multiple precursors into a hot-wall system.

Tubular flow reactors operated at the laboratory scale are often operated in the coagulation-limited mode with particle number concentrations on the order of $10^8/cm^3$ and can have higher concentrations for short periods of time in the reactor while chemical reaction is taking place. In addition, pressures in laboratory reactors are usually atmospheric. Although temperatures can be as high as 1700°C in laboratory-scale systems, they usually are

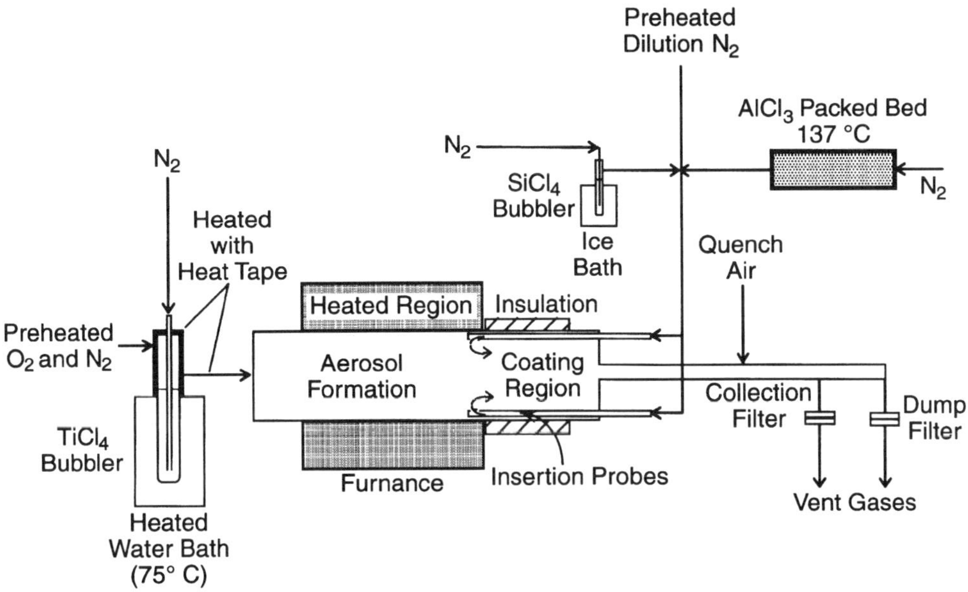


**Fig. 11.7.** Schematic of laboratory-scale system for generation of coated $TiO_2$ particles. (Adapted from Powell et al.)[31]

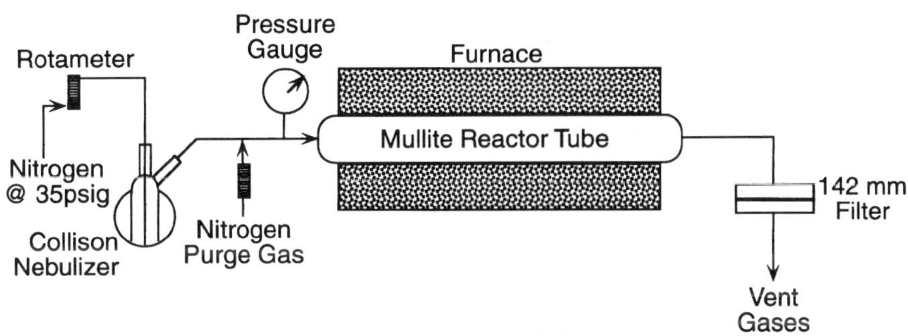


**Fig. 11.8.** Schematic of laboratory-scale system for delivery of volatile reactants. (Adapted from Powell et al., unpublished material.)

lower. Gas flow rates are often on the order of 10 liters per minute (lpm) at the laboratory scale, with residence times of 1 to 10 seconds. Reynolds numbers are usually less than 1000 corresponding to laminar flow.

Industrial processes often use very high loadings of the reactants and, in extreme cases, such as the formation of $TiO_2$ pigment, operate with no carrier gas. Pressures in industrial systems are frequently several atmospheres to provide higher throughputs. Reynolds numbers are often 1 million, corresponding to highly turbulent flow. Particle concentrations are much higher, by many factors of 10, than in laboratory-scale systems. Residence times are usually from 0.01 to 1 second. Particle size falls in the range of 50 to 500 nm.

There is a vast patent literature on industrial reactor configurations for the generation of $TiO_2$ pigment. A variety of patents discuss different methods and locations for the introduction of reactants, including the use of additives such as sand to scour the walls of the reactor tubes and addition of species like KCl or other ionic additives for control of particle size. Powell summarizes many of these results up to the year 1968.[33]

### 11.4.4 Theory

Because of the simplicity of tubular flow reactors and because they have been extensively modeled, the behavior of these reactors is well understood. The primary processes are gas-phase reaction, nucleation, collisions of particles, condensation, surface reaction, solid-state diffusion, and sintering of particles after collision. The types of reactors that have been examined are plug-flow, turbulent-flow, laminar-flow, and well-mixed or continuously stirred tank reactors. This class of reactors is archetypical of gas-to-particle conversion systems, and there are no complicating features such as plasmas, photons, severe temperature gradients, combustion processes, etc. The theory of these systems is covered in Chapter 10 as a representative example of the basic behavior of all gas-to-particle conversion systems.

## 11.5 CHEMICAL ROUTES: FLAME REACTORS

### 11.5.1 Introduction

Flame reactors, one of the most common reactor designs, have been discussed in several reviews.[34–36] The energy from a flame is used to drive chemical reactions that result in the formation of species with low vapor pressures, which then nucleate to form particles (Fig. 11.9). A typical sequence of events occurs over a short distance as products and then clusters form. The cluster concentration is very high ($> 10^{10}/cm^3$), leading to collisions among the clusters to form particles. At sufficiently high temperatures and for sufficiently small particle sizes, the particles coalesce to form larger units that become the primary particles. Further collisions among these primary particles while the gas is cooling leads to the formation of hard aggregates—collections of primary particles that do not have sufficient time to fuse together. The aggregates can range from a few up to many thousands of primary particles. The size range for the primary particles is from a few nm to 0.5 micron. Maximum flame temperatures usually reach 2500 K or may be as low as 1200 K.

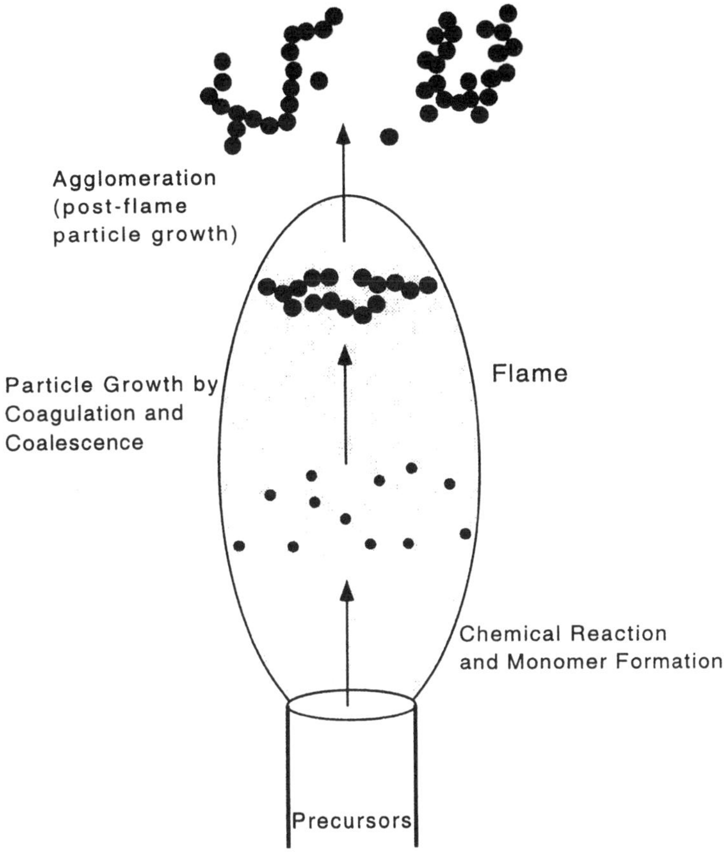


**Fig. 11.9.** Particle formation with single reactant in flame reactor.

The temperatures in flame reactors are lower than in plasma reactors, but usually higher than those found in hot-wall, evaporation–condensation, and laser reactors. In addition, flame reactors are simple in construction relative to most other systems. Such factors have combined to make the flame reactor an attractive method for particle generation. However, a major drawback is that this type of reactor often provides broad distributions of temperatures and residence times for particles. Complex temperature and flow profiles make laboratory investigations under controlled conditions difficult.

There are two types of flames: diffusion and premixed. In diffusion flames, the reactants do not contact each other until they leave the burner, after which they diffuse together. This diffusion results in a flame front at the region where the fuel and oxygen or air contact each other. In premixed flames, the fuel and oxygen source are mixed beforehand. After exiting the burner, they combust. This type of flame is quite dangerous to work with, because the fuel and oxygen source are combined in lines leading to the burner.

The advantages of flame reactors include:

- Simple oxides easily produced in short residence times (a few milliseconds)
- Scale up demonstrated for $SiO_2$, $TiO_2$, $Al_2O_3$, and other oxides
- Simple and relatively inexpensive system to build and operate
- Use of volatile or involatile precursors
- High purity
- Large range of diameters for particles in aggregates

Disadvantages of flame reactors include:

- Formation of hard agglomerates under most conditions
- Broad particle size distribution (coagulation controlled)
- Nonuniform temperature and velocity profiles
- Difficulty producing many (easily oxidized) materials, including nitrides, borides, etc.
- Inability to produce large primary particles ($> 1\,\mu m$)
- Particle properties significantly affected by mixing of precursors

### 11.5.2 Precursors and Products

Generation of powders using flame processes relies on introducing species into the flame reactor, followed by reaction to form gaseous species with low vapor pressures. The reactants that have been used most often are metal chlorides, because of their low cost and high volatility. Some processes introduce precursors in an aqueous solution in the form of droplets that are sent into the flame, where evaporation takes place to provide gaseous reactants. Table 11.3 at the end of the chapter lists some of the reactants and products that have been used in flame reactors. Virtually every simple metal oxide has been formed with these reactors. Typical temperatures range from 1500 to 2500°C.

The major industrial applications of flame reactors are generation of carbon black and fumed silica. These materials are made at massive scales by companies such as Cabot and Degussa. Carbon black is used primarily as a reinforcing agent, while fumed silica is used as a free-flowing agent in numerous applications.

Flame reactors also have been used to form coated and composite particles. For example, titania particles have been coated with silica, and alumina particles have been coated with $V_2O_5$.[37]

The most commonly used reactants by far are the metal halides. The chemistry of this class of precursors is discussed in Chapter 8.

### 11.5.3 Experimental Considerations

Figure 11.10 shows a diffusion flame system designed to generate fibers coated with $SiO_2$ particles themselves generated in the flame from $SiCl_4$.[38, 39] The $SiCl_4$ is vaporized using a bubbler and is introduced into the annular region surrounding the core, which contains the fibers suspended as an aerosol in air. The outermost annular region carries the methane fuel. This type of flame is called a diffusion flame because the reactants are not premixed and must diffuse together for combustion to take place. The powder is collected by using a pump to draw the airstream through a filter.

Figure 11.11 shows an oil furnace reactor used to produce carbon black.[40] The air, gas, and feedstock are introduced separately into a flow system in which combustion takes place under conditions favoring soot formation. This is an example of a flame reactor in which the fuel is also the reactant that is converted to the aerosol product.

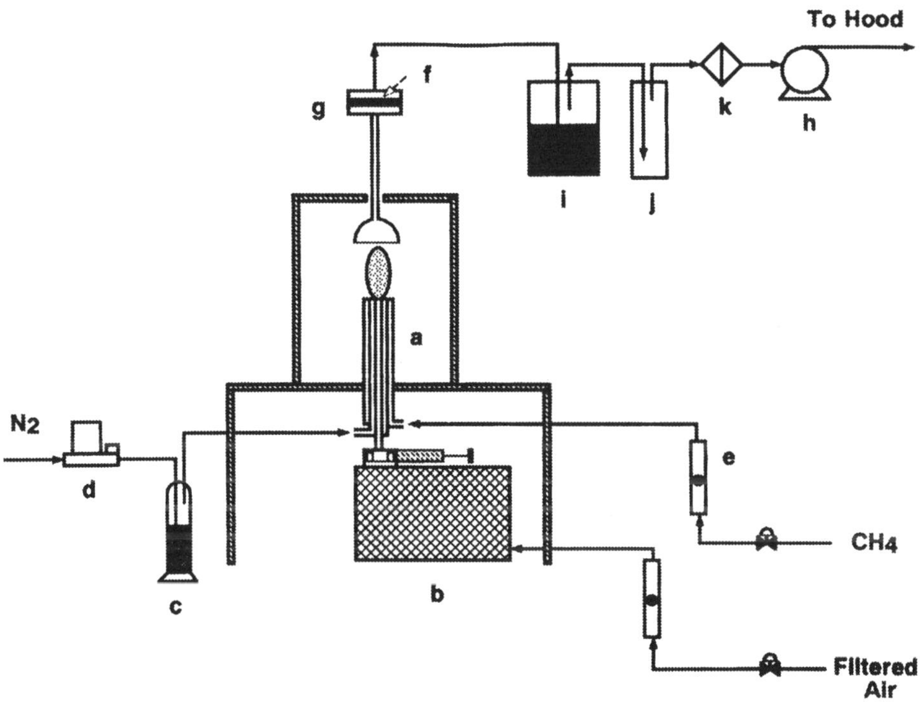


**Fig. 11.10.** Schematic diagram of experimental apparatus for powder generation in a flame reactor: (a) flame reactor, (b) fiber generator, (c) bubbler, (d) mass flow controller, (e) rotameter, (f) metallic screen, (g) filter holder, (h) vacuum pump, (i) chemical trap, (j) cold trap, (k) exhaust filter. (Adapted from Fotou and Pratsinis. Reprinted with permission of The American Ceramic Society, Westerville, Ohio. Copyright 1994 by The American Ceramic Society. All rights reserved.)[38]

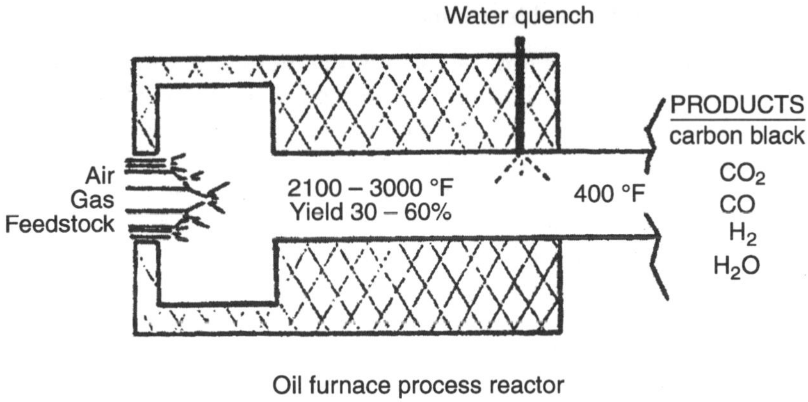


**Fig. 11.11.** Schematic diagram of flame reactor for carbon black generation. (Adapted from Dannenberg.)[40]

Typical operating conditions for a laboratory-scale flame reactor are as follows. Gas flow rates are on the order of 10 lpm. High-temperature thermocouples are usually used to measure the temperature, but this is not a simple measurement, due to the effects of radiative heat transfer. Temperatures can be as high as 2500°C or as low as roughly 1500°C and are varied by controlling the type of reactants. The fuel–air ratio ($H_2$–$O_2$) produces the hottest temperatures, and hydrocarbons and air give lower temperatures. The residence time in the highest temperature region is often on the order of 10 to 100 msec. Surface areas of powders can be greater than 100 $m^2/g$, with primary particle sizes as small as 10 nm. Because residence times are shorter than in tubular flow systems, the particle concentrations tend to be higher in flame reactors, with values greater than $10^{10}$ $cm^3$ being common. Industrial-scale systems are often similar, but with higher gas flow rates.

### 11.5.4 Theory

The theory of flame reactors is qualitatively similar to that discussed in Chapter 10. However, there are some peculiarities of flames, such as the distribution of residence times and temperatures, that complicate the simple picture. Residence times in flame reactors are usually shorter by at least a factor of 10 than those in tubular flow reactors. The temperatures are generally much higher in flame reactors and can be as high as 3000°C, while the maximum temperature in tubular flow reactors is usually 1500 to 1700°C. The temperature and species concentrations are usually more uniform in tubular flow reactors, both in the axial and the radial directions. However, given these differences, the theory of flame reactors is still very similar to that of tubular flow reactors operated in the nucleation–collision–coalescence mode. For example, it is commonly found that the size distribution of particles formed in flame reactors is self-preserving. The same size distributions can be formed in tubular flow reactors. Because of the qualitative similarity of these systems, the reader is referred to Chapter 10 for an introduction to the theory.

The primary parameters available to control the particle size distribution include the flame temperature, residence time, and reactant concentration. The flame temperature controls the rate of formation of the condensable species and the sintering rate of particles after collisions. Shorter residence times favor smaller particles, but time can be varied only over a limited range.

## 11.6 CHEMICAL ROUTES: LASER-DRIVEN PHOTOTHERMAL REACTORS

### 11.6.1 Introduction

Laser light can be used to drive chemical reactions that result in particle formation. Two types of laser reactors exist. In the photothermal process, the laser light is absorbed by the reactant or other gaseous species, and then energy is transferred to the carrier gas, resulting in heating of the gas. After that, the reactions are thermally driven, and the process is said to be photothermal. In the photochemical process, shorter wavelengths are used, and the reactant molecules absorb the laser light at specific wavelengths, resulting in dissociation to form the condensable species. Thus, temperatures can be near ambient, in contrast to the photothermal case, where temperatures are near 1000°C or higher. The photochemical process is discussed in Section 11.7.

The photothermal process uses laser light to drive thermally induced chemical reactions that result in the formation of species with low vapor pressures, which then nucleate to form particles (Fig. 11.12). The aerosol dynamics are similar to that found in flame reactors and hot-wall flow reactors. Typically, a $CO_2$ laser is used to heat gas molecules directly by light absorption. The laser acts only as a heat source, because the molecules are in thermal equilibrium with the gas. The absence of walls can lead to a highly pure product. The advantage of using a laser as opposed to conventional light sources is its narrow spectral width, which allows efficient coupling between the light source and the molecular precursor. The coupling requires near coincidence between laser light wavelength and the wavelength of absorption. Efficiencies for use of the laser power are about 15%.

Laser reactors operate at lower temperatures than plasma and flame reactors. Temperatures are on the order of those present in evaporation–condensation and hot-wall reactors, but are higher than in evaporation–condensation–reaction processes. The systems require expensive and complex lasers.

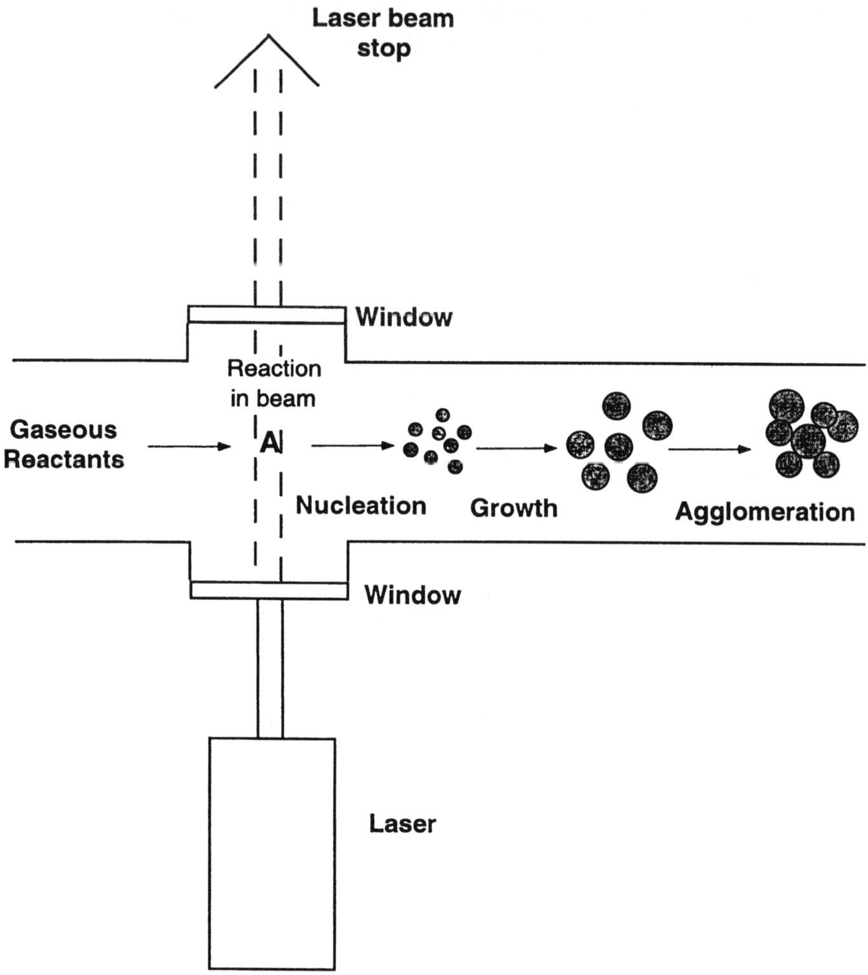


**Fig. 11.12.** Schematic diagram of particle formation in photothermal and photochemical laser reactors.

The primary advantages of the laser-driven process are:

- Production of simple oxides and nonoxide ceramics
- Scale-up demonstrated to kg level
- Production of high-purity powders

The major disadvantages include:

- Requirement for expensive reactants
- Formation of hard agglomerates under most conditions
- Complex and expensive system

### 11.6.2 Precursors and Products

The precursor requirements for photothermal laser reactors are similar to those for flame and other gas-to-particle conversion systems. Volatile reactants that do not contain carbon are required to provide a high-purity product, unless metal carbides are being formed. Table 11.4 (end of chapter) gives examples of reactants and products produced using photothermal laser reactors.

There are very few, if any, commercially operated units for powder production by laser reactors. This may be attributable to the combined problems of complexity, high capital cost, and high operating cost due to the need for relatively expensive precursors ($SiH_4$, for example).

Most studies with laser reactors have examined the formation of silicon-based ceramics, including $Si_3N_4$ and SiC; the most commonly used reactant has been $SiH_4$. Some Ti-based materials also have been formed, including $TiB_2$ and $TiSi_2$. The chemistry of this class of reactions is discussed in Chapter 8.

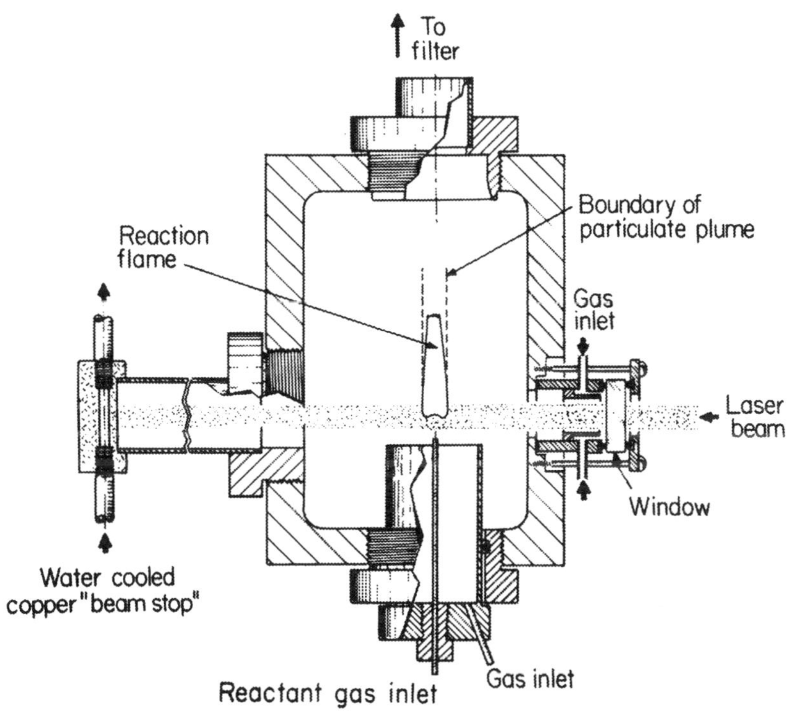


**Fig. 11.13.** Experimental system for powder generation by laser reaction. (Adapted from Cannon et al.[41] Reprinted with permission of The American Ceramic Society, PO Box 5136, Westerville, OH 43086-6136. Copyright 1982 by The American Ceramic Society. All rights reserved.)

### 11.6.3 Experimental Details

Figure 11.13 shows an experimental system for particle generation by photothermal laser reaction.[41] This process is typically carried out in a vacuum chamber into which the reactant is introduced as a vapor. A laser beam is brought perpendicularly up to the reactant vapor stream. The intersection of the laser beam and precursor stream gives a reaction flame that cools rapidly. The particles are then collected on a filter.

A typical reactor uses a cross-flow geometry with laminar flow. The laser power is on the order of 100 W at a wavelength of 10.6 microns. The reactor pressure can range from 0.01 to 1 atm, with maximum temperatures of about

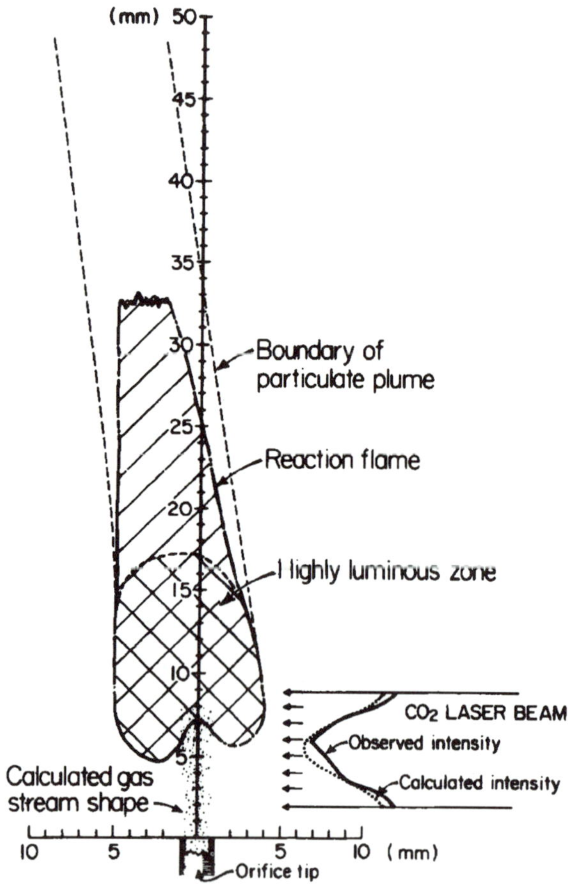


Fig. 11.14. Plume formed during laser reaction showing regions of importance. (Adapted from Cannon et al.[41] Reprinted with permission of The American Ceramic Society, PO Box 5136, Westerville, OH 43086-6136. Copyright 1982 by The American Ceramic Society. All rights reserved.)

800 to 1500°C. Although the temperatures are not very high compared to those in plasma and flame reactors, gas heating rates of 1 million°C/sec can be achieved. A conversion of 80% of the reactant to product is typical, and particle concentrations are often greater than $10^{10}/cm^3$. Typical results for $Si_3N_4$ are surface area of 100 $m^2/g$, Brauner–Emmett–Teller (BET); equivalent particle size, 10–20 nm; mean particle size, from transmission electron microscope (TEM), 10–20 nm; and size range, 10–30 nm.

### 11.6.4 Theory

The qualitative behavior of photothermal laser reactors is similar to that of gas-to-particle conversion systems. The aerosol dynamics in laser reactors has the greatest similarity to that in flame reactors. These two approaches share the feature that they have relatively broad distributions of residence times and temperatures for particles and that their temperature gradients are steep. Figure 11.14 shows the size and shape of the plume in which particle formation and growth occur.[41] The primary parameters available to control the particle size distribution include the laser power and reactant concentration.

## 11.7 CHEMICAL ROUTES: LASER-DRIVEN PHOTOCHEMICAL REACTORS

### 11.7.1 Introduction

The photothermal laser process discussed in the previous section relies on the thermal decomposition of gaseous molecules. Another type of laser-driven process is photochemical, in which light with shorter wavelengths (ultraviolet) is used to dissociate molecules photochemically. The design can be the same as that used for photothermal laser reactors (Fig. 11.12), although irradiation along the axis of the flow has also been used. Photodissociation results in the formation of a product with a low vapor pressure and supersaturation, which drives particle formation. The process has not been used extensively for particle formation because expensive reactants and lasers are required. Also, the efficiency of the process is low due to the small cross sections for the absorption of UV light by most reactant species. The sole advantage of this method is that it can produce nanophase materials as a result of a low gas temperature. The primary disadvantages are that it is a complex and expensive system, that it is limited to volatile precursors which absorb at the laser wavelength, and that no scale-up has ever been demonstrated.

### 11.7.2 Precursors and Products

Table 11.5 (at the end of the chapter) shows examples of materials produced by photochemical laser reactors. All of the examples are rather specialized cases that are of no great interest for materials applications. This route may have applications for nanophase particle generation in very specific cases where suitable reactants are available.

### 11.7.3 Experimental Considerations

Photochemical laser reactors have been operated in various configurations. In one example, a pulsed UV laser beam is passed down the core of a tubular flow reactor.[42] This arrangement gives a longer path length of the laser light through the gaseous medium. The details of the laser characteristics, reactant concentrations, gas flow rates, etc., are not discussed here because of the limited application of the approach.

### 11.7.4 Theory

The theory of photochemical reactors is the same as that of the other aerosol reactors (see Chapter 10), except for the chemical reaction kinetics, which are discussed in Chapter 8. Perhaps the most unique feature is that the reaction rate expression requires knowledge of the laser light intensity and absorption cross section for the reactant. Because these processes operate at such low temperatures relative to the melting points of the materials of interest, they are able to produce very fine-grained materials. Some models for particle formation in axially irradiated photochemical systems have been developed.[42]

## 11.8 CHEMICAL ROUTES: THERMAL PLASMA REACTORS

### 11.8.1 Introduction

Thermal plasma reactors use high-temperature plasmas to provide the energy required to drive the physical and chemical processes that lead to particle formation and growth.[43–56] A plasma is a system with a high energy content in which a significant fraction of the species are ionized and are conductors of electricity. Two types of plasma systems are of interest: high-temperature equilibrium thermal plasmas and low-temperature nonequilibrium plasmas. In the thermal plasmas, the temperatures of the electrons and ions are equal, and the pressures are high. In low-temperature plasmas such as glow discharges, the temperature of the electrons is much greater than that of the ions, and pressures are on the order of mtorr. The most common system for powder generation is the high temperature thermal plasma, although particle formation in glow discharges is also common. Applications of thermal plasma processing of materials include plasma synthesis, spraying, and consolidation of materials. Glow discharges have not been used extensively for powder generation or processing, but particle formation in glow discharges can provide very fine-grained materials.

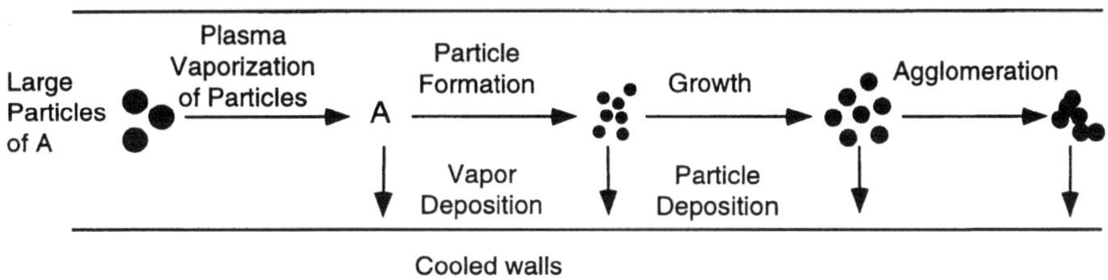


**Fig. 11.15.** Plasma reactor with vaporization of solid feed. In other variations, the condensable species A is formed from gaseous reactants or from species dissolved in droplets.

Numerous variations of thermal plasma reactors exist that depend on how the reactants are introduced into the system. Reactants can be introduced as powders that are vaporized (Fig. 11.15), as premixed gaseous reactants, or as gaseous reactants downstream of the plasma. The last two processes follow similar schemes, but the source of the condensable species is different. All of the systems have several features in common. Temperatures (5000 to 25,000 K) are higher than in all other types of aerosol reactors, and complete vaporization and dissociation of the reactants into their atomic form is common. This allows the use of involatile molecular and solid feed streams, so almost any material can be processed. A common process is the formation of gaseous species that nucleate to form particles during cooling while exiting the plasma.

Advantages of plasma processes for materials processing include:

- Use of materials with a high melting point
- Use of inexpensive powder feed
- Rapid cooling rates, which can provide unique phases
- Demonstrated scale-up
- Production of nonoxide powders

The primary disadvantages are:

- Difficulty of completely vaporizing solid reactants
- Complicated and poorly understood process
- Broad particle size distributions
- Expensive reactants if powder feed is not used
- Agglomerated particles (often hard agglomerates)
- High energy requirements

Plasmas are unique relative to other gas-to-particle conversion systems in that the extremely high temperatures provide the alternative of using liquid, gaseous, or solid reactants. The spray inductively coupled plasma method, for example, relies on the introduction of a solution containing precursors into the plasma. In some plasma systems, the reactants are fed into the system in the form of powders; in others, gaseous reactants are used. The difficulty with using solid particles and liquids is that it is difficult to control the rate of introduction of reactants.

### 11.8.2 Precursors and Products

Tables 11.6 and 11.7 (end of the chapter) give examples of the materials produced in plasma reactors. Commercial applications of this approach are limited for powder generation. The major emphasis is the generation of nanometer-sized particles. However, plasma systems are used extensively for spheroidization of particles and for plasma spraying to form coatings. Virtually all classes of materials have been produced, including elemental materials, nitrides, carbides, borides, and oxides. Coated and composite particles have also been produced.

The chemistry of these processes begins with the complete destruction of molecular species (unless the reactants are injected sufficiently far down the tail flame of the plasma) to form ionic species. As the temperature drops, molecular species with low vapor pressures are generated and eventually nucleate to form particles.

### 11.8.3 Experimental Considerations

Two broad classes of thermal plasma reactors are in widespread use: dc arc jets and high-frequency induction systems. In the case of the dc arc jet, current is supplied to the plasma by physical contact with a metallic electrode surface. This type of system is relatively simple and inexpensive. However, the electrodes are consumed and end up contaminating the product. An advantage is that the temperature and flow fields are less complex than those in radio frequency systems. Typical temperatures are 15,000 K in an argon plasma. Figure 11.16 shows a system used to generate a variety of nonoxide materials.[57] This schematic gives a feeling for the complexity of these processes, which involve high-frequency and high-power electrical components.

In the case of a high-frequency induction plasma, there is no contact between the plasma and its power source. Energy transfer takes place through the electromagnetic field of the induction coil. A drawback is that the efficiency is lower than for dc jet reactors. However, this is usually offset by the advantage that, since the electrodes are not consumed, there is no contamination of the product. Variables that can be controlled are the plasma composition and frequency. The most common case is an argon plasma operated at 200 kHz to 20 MHz.

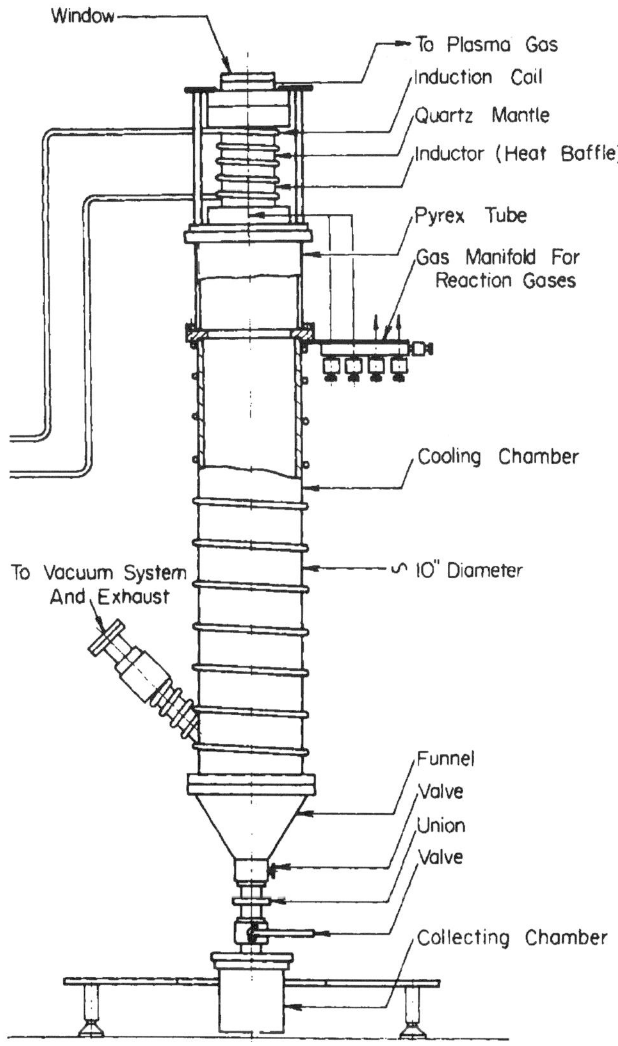


**Fig. 11.16.** Plasma reactor system, showing complexity of process. Reprinted from *Journal of Material Science 18*, (11), 1983, "R.F.-plasma system for the production of ultrafine, ultrapure silicon carbide powder," Hollabaugh, C.M., Hull, D.E., Newkirk, L.R., and Petrovic, J.J., Chapman & Hall.

A popular system for powder production is what has been called the spray inductively coupled plasma (ICP) process, which is shown in Figure 11.17.[58] A solution containing precursors is atomized and passed through an inductively coupled plasma, where the solution is vaporized and the reactants are vaporized and dissociated into atomic form. As the gas cools and the vapor pressures of the various species drop, particle formation takes place. This

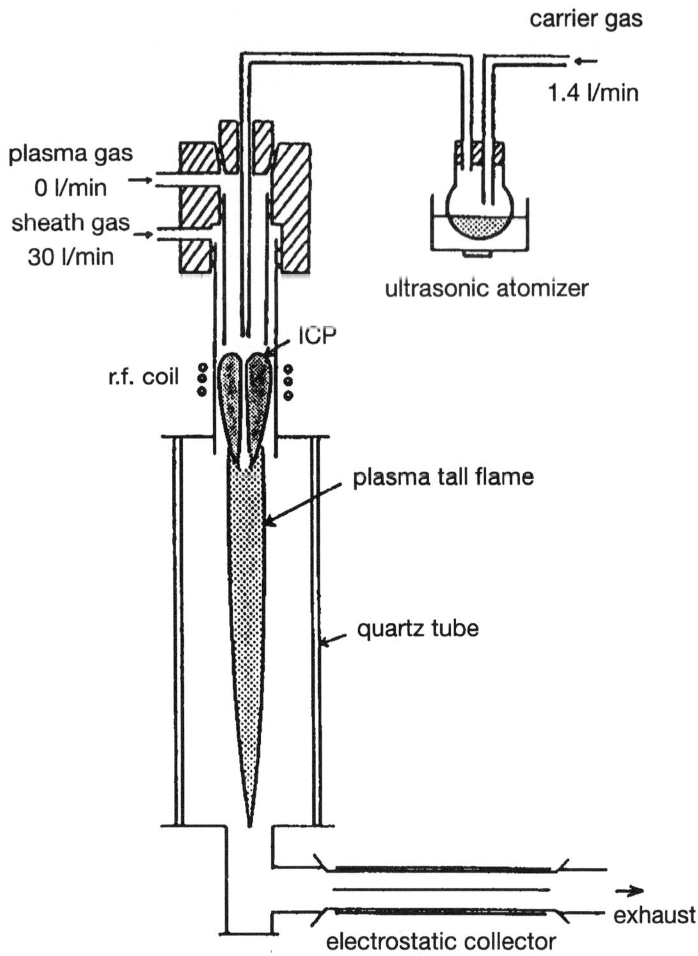


**Fig. 11.17.** Spray inductively coupled plasma (ICP) system, showing ultrasonic generation of precursor droplets. (Adapted from Suzuki, M., Kagawa, M., Syono, Y., and Hirai, T. *Ceramic Powder Sci IV*, *Ceramic Trans. 22*, 1991, 147–152. Reprinted with permission of The American Ceramic Society, Westerville, OH. Copyright 1991 by The American Ceramic Society. All rights reserved.)[58]

technique allows the use of a wide variety of molecular and colloidal precursors without the difficulties of feeding powder or finding inexpensive volatile reactants.

Typical operating parameters for a radio frequency plasma are 13.56 MHz, 20 kW power input, atmospheric pressure, production rates of 100 g/hr, gas flow rates of 10 lpm, and residence times of less than 1 second. Particle sizes of less than 10 to 500 nm are routinely obtained.

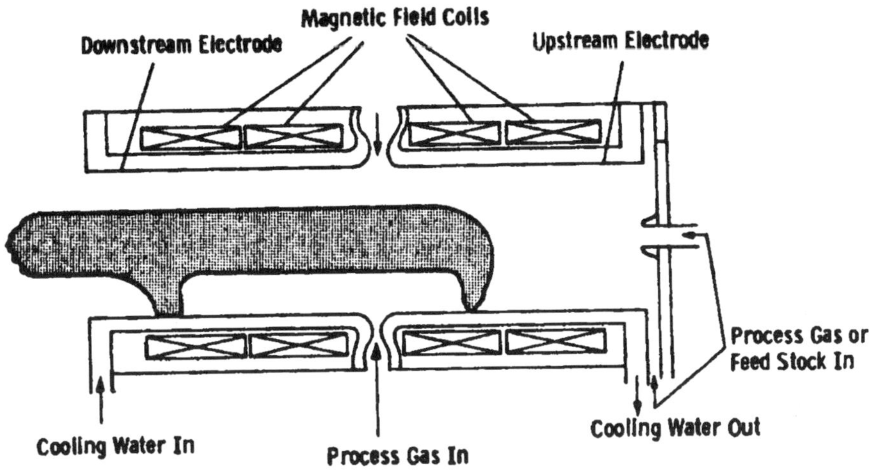


**Fig. 11.18.** Typical plasma reactor with dark area showing region of plasma. (Adapted from Heberlein.)[47]

### 11.8.4 Theory

A full theoretical description of the aerosol dynamics of a plasma process is very complex due to the presence of ions, large temperature gradients, chemical reactions, etc.[59–67] As an example, Figure 11.18 shows a typical plasma configuration. A general solution for a system such as this would require solving Maxwell's equations, the Navier–Stokes equations, and the equations describing the aerosol dynamics. Fortunately, the theoretical analysis can be dramatically simplified by noting that when the plasma is present, the temperature is so high that all the species are dissociated and ionized. It is only downstream of the plasma that the aerosol dynamics begin, in regions where the temperature is a few thousand K or less. This is illustrated in Figure 11.19, which shows the various regimes present in a plasma reactor.

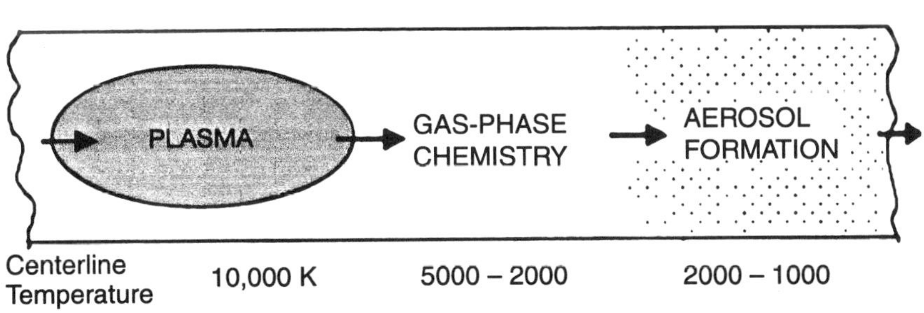


**Fig. 11.19.** Sequence of events during plasma generation of particles. Particle formation and growth occur downstream of the plasma simplifying theoretical analyses of the problem. (Adapted from Girshick and Chiu.[61] Reproduced with permission from Steven Girshick.)

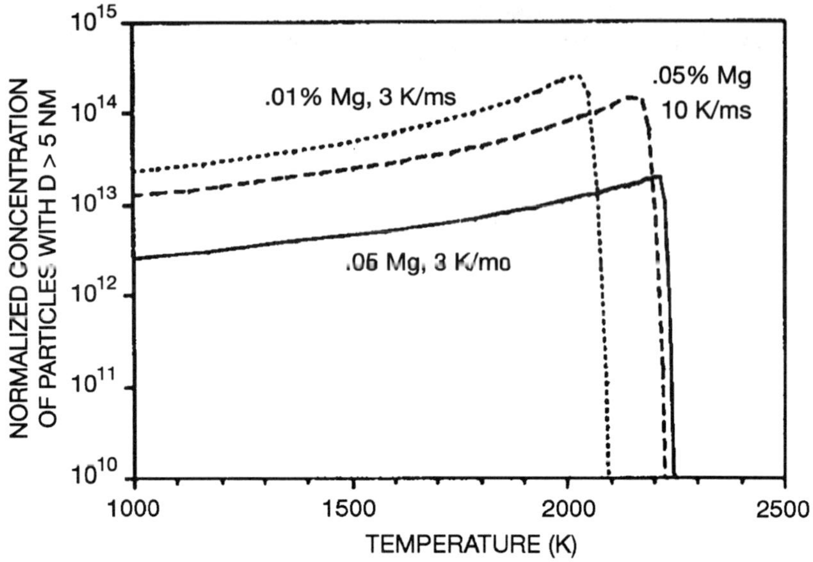


**Fig. 11.20.** Effect of cooling rate and metal loading on particle concentration. (Reprinted from *Journal of Aerosol Science 21*, Girshick, S.L. and Chiu, C.P., 641–650, Copyright 1990, with kind permission from Elsevier Science Ltd., The Boulevard, Langford Lane, Kidlington OX5 1GB, UK.)[63]

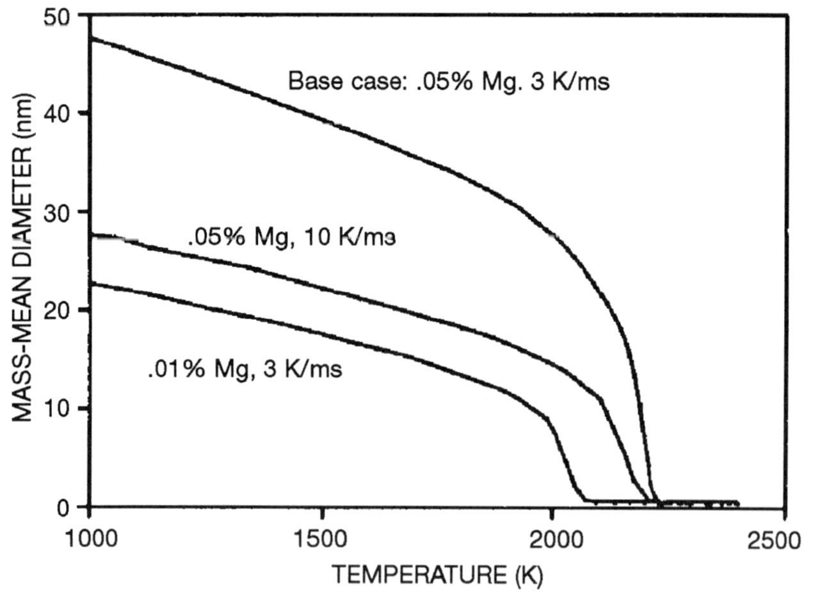


**Fig. 11.21.** Effect of cooling rate and metal loading on particle size. Temperature axis corresponds to axial position. (Reprinted from *Journal of Aerosol Science 21*, Girshick, S.L. and Chiu, C.P., 641–650, Copyright 1990, with kind permission from Elsevier Science Ltd., The Boulevard, Langford Lane, Kidlington OX5 1GB, UK.)[63]

The aerosol dynamics take place where the temperature has dropped sufficiently to reduce the vapor pressures of the various species to levels at which they become supersaturated.

The same theory that applies to hot wall reactors applies to plasma reactors. The high cooling rates make the plasma process similar to that of laser and flame reactors. It is common for the size distribution to be controlled by coagulation, and the particle size distribution is often self-preserving.

As an example specific to plasma reactors, calculations have been performed assuming plug flow with the rapid cooling rates associated with plasmas. Figures 11.20 and 11.21 show calculated particle concentrations and mass mean diameters for MgO particle formation in a plasma reactor.[63] Particle formation begins near 2200 K, at which point the vapor pressure of MgO is sufficiently low to allow particle formation. The rapid cooling rates of up to 10 K/ms result in the formation of very high particle concentrations, on the order of $10^{12}$ to $10^{15}/cm^3$. At the same time, small particles are formed with diameters on the order of 20 to 50 nm. These results are typical for plasma reactors.

## 11.9 CHEMICAL ROUTES: LOW-TEMPERATURE PLASMA REACTORS

### 11.9.1 Introduction

Low-temperature plasmas such as glow discharges and microwave plasmas contain un-ionized atoms and molecules, ionized species, and electrons. The temperature of the electrons is much greater than that of the ions and neutral species, which have temperatures near ambient. The pressures in these plasmas are on the order of mtorr. Although the most common type of plasma system for powder generation is a thermal plasma, glow discharges (another name for low-temperature plasmas) are discussed here because they can provide fine-grained materials.

Particle formation is a major problem during etching and vapor deposition processes used to form and process coatings. For this reason, a number of researchers have investigated the formation of particles during etching and CVD.[68–98] This body of literature provides numerous insights into the operation of low-temperature plasma reactors for particle formation.

Advantages of low-temperature plasma processes for powder generation include:

- Small crystallite size
- Production of nonoxide powders

Primary disadvantages are:

- Poorly understood, very complicated process
- Requirement of expensive volatile reactants
- No demonstration of scale-up

### 11.9.2 Precursors and Products

Only a few materials have been formed by low-temperature plasma processes. Indeed, few, if any, materials are produced at commercial levels by low-pressure plasma systems. With the growing interest in nanophase materials, however, the approach is being investigated in more detail.

Hydrogenated silicon particles have been formed from silane gas. The particle size distribution was bimodal, with sizes of 20 to 50 nm for the smaller particles and 80 to 150 nm for the larger particles. Silicon-nitride precursor powders have also been formed, using silane and ammonia. The particle size distribution appeared bimodal, with the smallest particles being 10 nm and the largest particles being 200 nm. The particles produced were amorphous, but could be converted to $Si_3N_4$ after prolonged heating at 1600°C. Metallic iron nanoparticles have been formed by introducing ferrocene into the afterglow region of a low-pressure, low-power microwave plasma. In all cases, the particles contained significant amounts of graphitic carbon around the metal particles.[99] Particle sizes varied from 10 to 100 nm, with most of the particles being in the range of 50 nm. The materials $SiO_2$, TiN, $Si_3N_4$, and Si also have been produced.[99–103]

Vollath and colleagues have produced a variety of materials using a microwave plasma. The materials include $ZrO_2$, $Y_2O_3$, $Al_2O_3$, $Fe_2O_3$, and coated particles.[104–111]

### 11.9.3 Experimental Considerations

Low-temperature plasma reactors used for materials synthesis have typically been modified etching or vapor deposition reactors. The characteristics of these reactors are discussed in detail elsewhere.[68–98] Few, if any, systems have been designed specifically for particle formation, and the reader is referred to the sources previously cited for more detailed information.

### 11.9.4 Theory

A full theoretical description of a plasma process has not been achieved and would be very complex due to the presence of neutrals, ions, radicals, electrons, electric fields, gas-phase and surface chemical reactions, charged particles, and numerous other phenomena. The reader is referred to sources already cited.[68–98]

## 11.10 CHEMICAL PROCESSES: FLUIDIZED BEDS

### 11.10.1 Introduction

A fluidized bed is formed by passing a stream of gas through a bed of particles with sizes corresponding to settling velocities higher than the gas velocity.[112] The gas stream holds the particles in suspension, forming a bed that has many useful properties, including high heat and mass transport rates (uniform temperature and rapid mixing). Fluidized beds have a wide variety of applications. Some involve the synthesis and processing of materials, including converting gaseous materials to particulate form and coating larger particles ($> 10\ \mu m$) or agglomerates of smaller particles by CVD. A gaseous reactant is introduced into the bed of fluidized particles, where it reacts on the surfaces of the particles to deposit material. If the material deposited is the same as the particles, this approach serves to convert the gaseous reactant into a solid product in powder form. The primary example of such an application is the conversion of chlorosilanes into elemental silicon particles used to form silicon wafers for microelectronics. The use of a fluidized bed has several advantages over the competing technology of conversion in a batch-type bell jar reactor.[113] If the particles consist of a different material from that of the elements in the reactants, coated or composite particles are formed. Here the applications primarily involve the synthesis of powders used for ceramic or metallurgical applications. In both cases, gas-phase particle formation may occur to form fine aerosol particles that can be entrained in the gas flow and removed from the system, resulting in reduced yields.[113–116] In these approaches, aerosol particle formation is a problem, and strategies for reduced particle formation are desirable.

Although submicron-sized particles cannot be fluidized, they can in some cases spontaneously agglomerate in the bed to form agglomerates with diameters on the order of $100\ \mu m$, large enough to be fluidized in a conventional manner.[117–130] In this case, composite particles can be formed by introducing a gaseous reactant into the system where it reacts on the surfaces of the particles in the agglomerates.

Fluidized beds can also be operated in an entrained mode (sometimes called a *lean bed* or *floating bed*) wherein fine particles, too small to be fluidized in a conventional sense, are carried at high loadings in a gas along the inside of a tube.[131] During their transit time in the gas, a gaseous reactant can be introduced into the aerosol, where it can react on the surfaces of the particles to form a coating. Gas–solid reactions can also be carried out to convert metals, for example, to metal nitrides. This mode of operation is similar to that of a conventional aerosol reactor with high mass loadings.

Potential advantages of fluidized bed processes for coating of particles include:

- The entrained bed has a large throughput and can handle large particle sizes relative to those in conventional aerosol processes, which have lower mass loadings and smaller particle sizes.
- Coatings and composites can be obtained with any material that can be deposited by CVD.

The primary disadvantages are:

- Fine particles agglomerate in fluidized beds, but the process is understood only empirically.
- Without agglomeration, the product has particles with dimensions larger than 10 $\mu$m
- The process cannot produce unagglomerated materials with primary particles of submicron size.

### 11.10.2 Precursors and Products

Many materials have been formed and coated by fluidized-bed technology. Here, we restrict our attention to fine-particle formation during fluidized-bed gas-to-particle conversion, materials produced using micron-sized or smaller particles in floating-bed systems, and fine-particle agglomeration in fluidized beds. Systems in which large particles ($> 10$ $\mu$m) are coated are not reviewed. Examples of these approaches are given in Tables 11.8 and 11.9 (at the end of the chapter).

### 11.10.3 Experimental Considerations

Floating-bed systems have the closest resemblance to conventional aerosol systems. Figure 11.22 shows a typical experimental system.[131] In the simplest form, they are just tubes heated by a tubular furnace through which particles and one or more gaseous reactants are carried in a gas. The gaseous precursor reacts on the surfaces of the particles to form a coating. As with aerosol processes, gas-phase reaction can also occur, leading to the formation of fines. In this sense, these systems are aerosol reactors with high loadings of particles. A unique experimental requirement is a powder feeder capable of entraining a dry powder into a flow system.

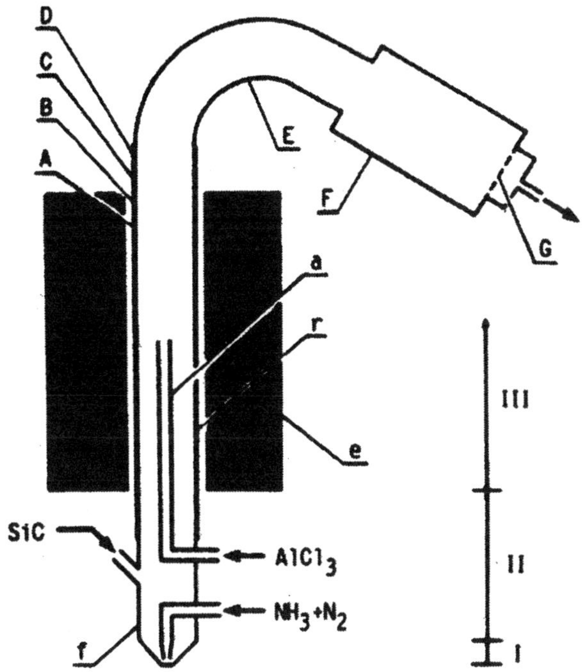


**Fig. 11.22.** Floating-bed system for coating of aerosol particles. (Reprinted from *Powder Technology 68*, Kimura, I., Hotta, N., Niwano, J.-I., and Tanaka, M. "Synthesis of AlN/SiC composite powders by floating-type fluidized-bed CVD," pp 153–158, Copyright 1991, with kind permission from Elsevier Science, The Boulevard, Langford Lane, Kidlington OX5 1GB, UK.)[131]

### 11.10.4 Theory

Several theoretical descriptions have been developed for CVD onto particles in fluidized beds. These models rely on the basic principles of aerosols presented in Chapter 10, along with models of fluidized beds. The theory is discussed elsewhere.[113–116]

## 11.11 PHYSICAL PROCESSES: FREE CONVECTIVE PLUMES AND TUBULAR FLOW SYSTEMS

### 11.11.1 Introduction

A convenient method for the formation of nanometer-sized particles is evaporation followed by condensation of a hot vapor of a species such as a metal.[132–143] This has most often been carried out in a semibatch method (Fig. 11.23) at the laboratory scale in a variety of studies that have concen-

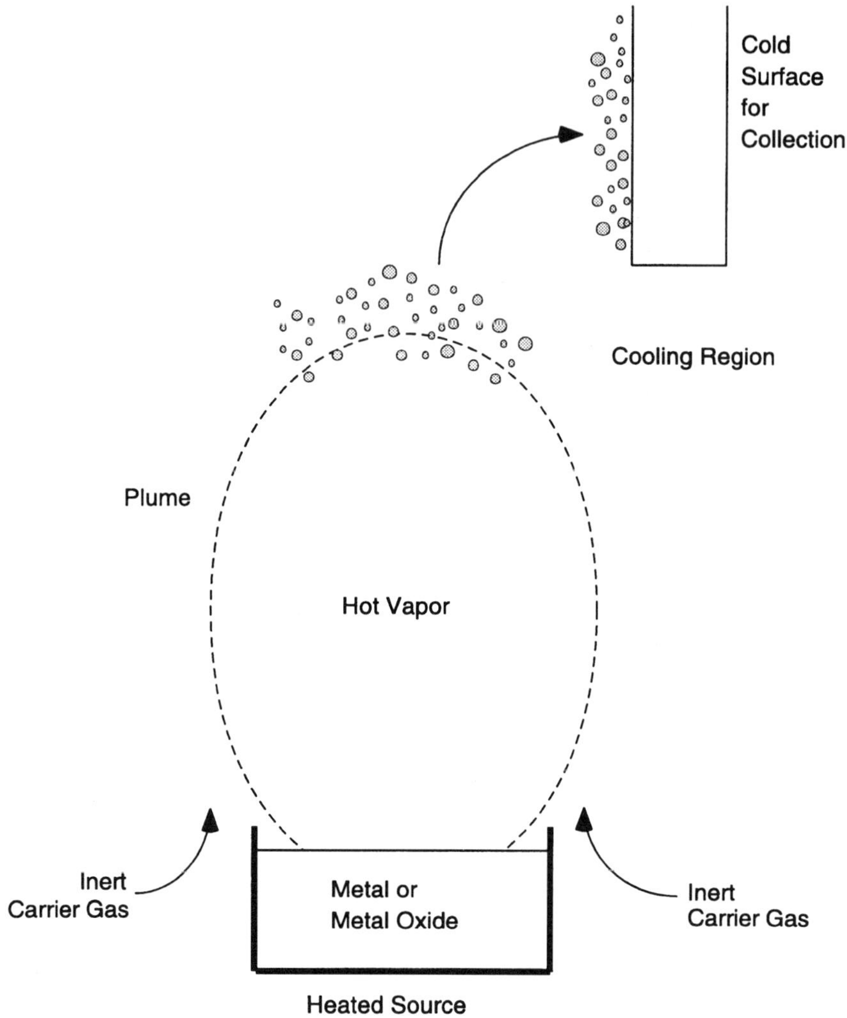


**Fig. 11.23.** Free convective plume process for generation of nanoparticles by evaporation and nucleation.

trated mainly on producing materials with novel properties. A metal is vaporized into a host gas in which the vapor is cooled, resulting in particle formation by nucleation.

The process of evaporation–condensation is often carried out near or below atmospheric pressure in inert gases. A typical system involves a vacuum chamber that can be baked out to base pressures of $10^{-6}$ to $10^{-10}$ torr in order to reduce contamination levels. Often, a boat is used to hold a metal, which is heated to vaporize it into the gas phase. A natural plume is established by free convection (hotter vapor rising) from a heated source. In

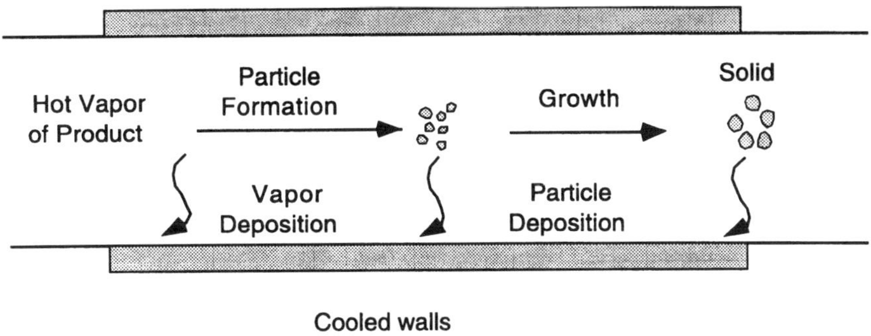


**Fig. 11.24.** Forced-convection tubular flow system for generation of nanoparticles by evaporation and nucleation.

this system, temperatures are usually much lower than those in plasma and flame reactors. Gaseous reactants are not required, as they are for hot-wall reactors and evaporation–condensation–reaction processes. In addition, evaporation–condensation systems are simple in construction and do not involve chemical reactions in the gas phase.

An alternative approach to the semibatch method is a continuous process in which a gas is flowed over a source of the vapor, which is then cooled (Fig. 11.24). Because of its continuous nature, this process is used industrially for the manufacture of nanophase powders. As the vapor cools, it becomes supersaturated, resulting in the formation of particles.

The advantages of this continuous method are:

- Demonstrated scale-up to at least the kg/hour level
- Simple system
- Possible narrow size distribution
- Spherical particles
- Possible production of nanophase-sized particles

The primary disadvantages are:

- Complex, nonisothermal system
- Need for temperatures high enough to vaporize material
- Few examples of nonmetallic products

### 11.11.2 Precursors and Products

The evaporation–condensation process has been used to produce powders of most metals. Numerous reviews have examined this process. The characteristics of the particles have been studied in detail. Relatively few experiments, however, have investigated the generation of materials other than metals and simple metal oxides (primarily $TiO_2$). (See Table 11.10, at end of chapter.) Nanophase $TiO_2$ and several other materials produced by the process are available commercially. Commercial applications of the process include the generation of 10- to 15-nm-sized $TiO_2$ for use in sunscreens, as well as other materials for net shape forming (fabrication of a ceramic part with the final dimensions of the part being determined during forming) of ceramics.

### 11.11.3 Experimental Considerations

Figures 11.25 and 11.26 show typical experimental systems exploiting evaporation–condensation processes. Figure 11.25 sketches a laboratory-scale system for particle generation and in situ particle consolidation.[136] A metal or

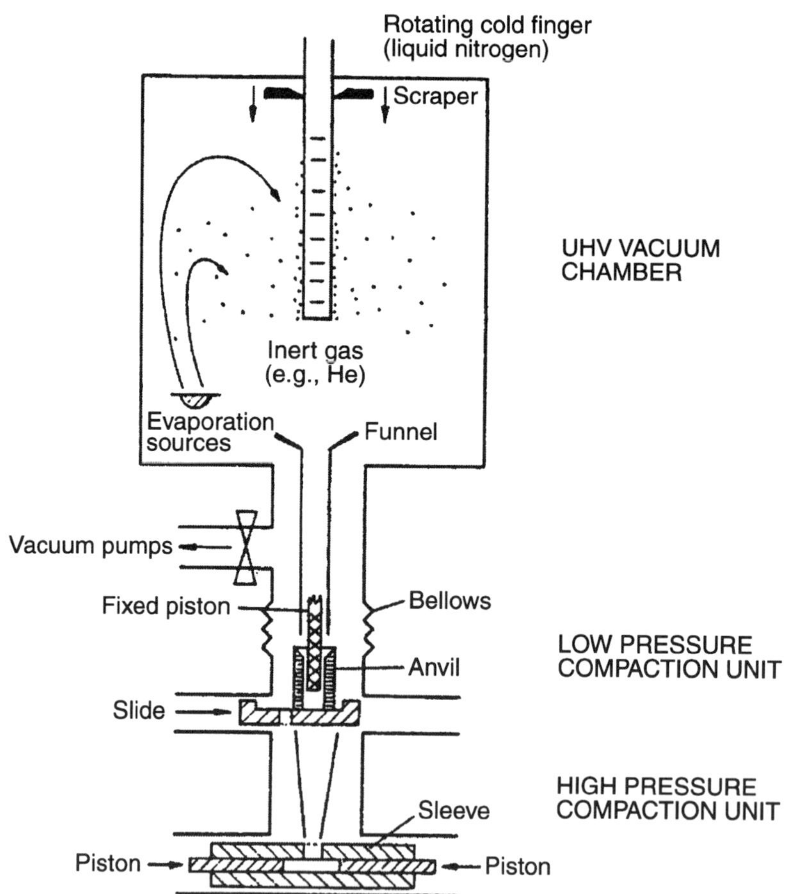


**Fig. 11.25.** Schematic of experimental system for powder generation by evaporation–condensation. (Reprinted from *Progress in Materials Science*, *33*, Gleiter, "Nanocrystalline Materials," pp 223–315, Copyright 1989, with kind permission from Elsevier Science, The Boulevard, Langford Lane, Kidlington OX5 1GB, UK.)[136]

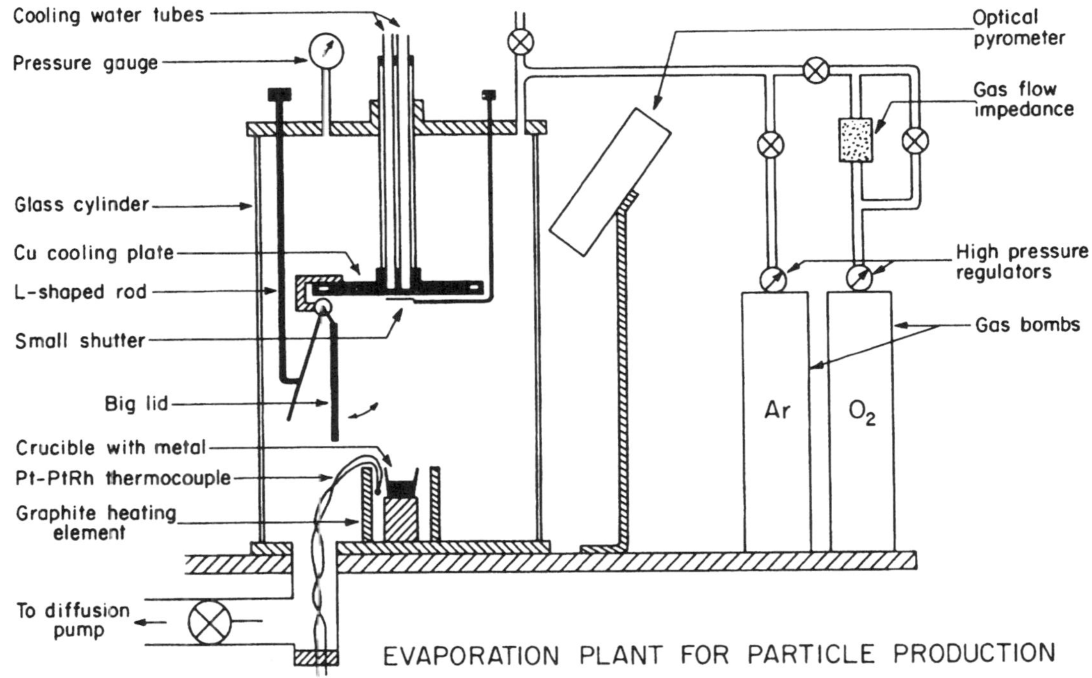


**Fig. 11.26.** Small-scale production facility for nanoparticles exploiting evaporation–condensation approach. (Reprinted with permission from Granqvist, C.G. and Buhrman, R.A. *J. Appl. Phys.* *47* (5), 1976, 2200–2219. Copyright 1976 American Institute of Physics.)[144]

other material is vaporized to give a free convective plume in which particles are formed. The particles are collected by thermophoresis onto a liquid-nitrogen-cooled cold finger. After the required amount of powder is produced, the material can be processed further by oxidation with controlled amounts of oxygen, for example, to form metal oxides such as $TiO_2$ from a metal such as Ti. The powder is then scraped off the cold finger and falls into a compaction unit, where it is formed into a pellet. The pellet can be removed and processed further.

Figure 11.26 shows a larger scale system for powder production using evaporation–condensation.[144] Again, the powder is produced by heating a metal or other material to a high temperature at which it has a sufficient vapor pressure. The vapor cools as it rises in the plume, where it nucleates to form particles. The particles are collected on a cooled surface.

Other methods of vaporizing the material may be used. Spark ablation has been utilized to introduce Si into the gas phase, from which it nucleated to form nanometer-sized particles. The particles are collected thermophoretically onto a cooled surface, as in many of the other approaches with ultrafine particles.

A typical experimental system for nanophase powder production involves semibatch operation with an ultrahigh vacuum chamber (base pressure $<10^{-9}$ torr). The chamber is pumped out and backfilled with high-purity inert gas at the desired pressure, usually a fraction of an atmosphere. The material of interest is vaporized by heating in a crucible, using an electron beam or by induction. Particle formation occurs as described before, and the particles are collected on a cold surface by thermophoresis. In the case of metals such as Ti, if the metal oxide is desired, oxygen is introduced into the system after the particles are formed and collected. In this semibatch operation, there is no continuous flow of gas. The vapor pressure of the evaporating species is often 1 to 25 torr, with lower values giving smaller particle sizes. Production rates are low—less than 1 gram per hour in many cases. Particle size is often in the range from 5 to 50 nm. The spread of the particle size distribution in terms of a geometric standard deviation is often 1.3 to 1.6.[144] Continuous-flow systems provide similar powder properties, but with much higher production rates.

### 11.11.4 Theory

The theory of particle formation by vapor condensation in flow systems is qualitatively similar to that of other gas-to-particle conversion processes. The situation for free convective plumes is comparable, except that the flow is not confined and occurs because the hot vapor rises. Both processes are in widespread use.

Some theoretical work has been done to determine the processes controlling evaporation–condensation in a natural plume,[144, 145] and a physical picture has emerged from this and other work. Metal vapor evaporates from the source, which can be approximated as a spherical source of vapor. At low source temperatures, the supersaturation is not high enough to cause particle formation. However, increasing the source temperature results in the formation of a smoke envelope around the source. The smoke-free inner region is the result of thermophoresis, which drives particles away from the source. Diffusion and convection of vapor away from the source lead to a concentration of metal vapor that decays monotonically with distance from the source. Heat conduction into the gas results in a temperature profile that also decreases radially away from the source. This leads to a maximum in supersaturation some distance away from the source, where particle formation occurs. Coagulation may or may not be important, depending on the operating conditions.

Theoretical studies have shown the formation of particles in forced-convection evaporation–condensation systems in which reactive precursors are vaporized, cooled, and nucleated to form particles and reacted with gaseous species to form solid products.[146] However, the qualitative behavior of the condensation portion of the process is similar to that of other gas-to-particle conversion approaches.

Two mechanisms of particle formation have been reported for condensation from the vapor: the vapor-to-liquid and the vapor-to-solid mechanisms.[147–151] These mechanisms partially determine the morphology and crystal habit of the particles produced. Vapor–solid nucleation can lead to well-formed prisms (dendritic structures), while vapor–liquid nucleation can lead to spherical particles. Metals such as Bi and Pb follow the gas-to-liquid mechanism; Zn, Cd, and Mg follow either the gas-to-liquid or gas-to-solid mechanism, or a combination of both, depending on the temperature.

Only a few studies have examined the effect of the addition of oxygen.[152] Once produced, metal particles can be converted to oxides by exposing the particles to oxygen at elevated temperatures, such as in the formation of $TiO_2$.[152] These results suggest that oxide particles can be formed directly or that in some cases oxide particles coated with metal can be formed.

The design of evaporation–condensation systems for particle formation depends strongly on the details of the powder generation system. Small changes in geometry can change the characteristics of the product powder. Once the system geometry is fixed, the primary control parameters include the source temperature, inert-gas flow rate in the case of a forced convection system, pressure, and nature of the source. The key parameter, however, is the temperature of the evaporator, since the vapor pressure of the precursor depends exponentially on temperature. Typical free convective plume systems do not allow control over the residence time at a given temperature. However, forced-convection tubular flow systems do allow control over residence times and cooling rates.

## 11.12 PHYSICAL PROCESSES: NOZZLE EXPANSION

### 11.12.1 Introduction

An alternative method for generating particles by evaporation–condensation is vaporization of a material followed by cooling using expansion in a nozzle. Multiple expansions are often used. Historically, this process has been used to produce clusters for studies of their basic behavior. Figure 11.27 shows a schematic of particle formation in a free-jet expansion. A vapor in a carrier gas is passed through a nozzle into an expansion region. The adiabatic expansion of the gas results in rapid cooling, which drives particle formation.

The advantages of this method are:

- Possible narrow size distribution
- Spherical particles
- Possible production of nanophase materials

The primary disadvantages are:

- Low production rates
- Difficulty to produce anything other than simple oxides and metals
- Complex, nonisothermal system
- Need for temperatures high enough to vaporize material
- No demonstration of scale-up

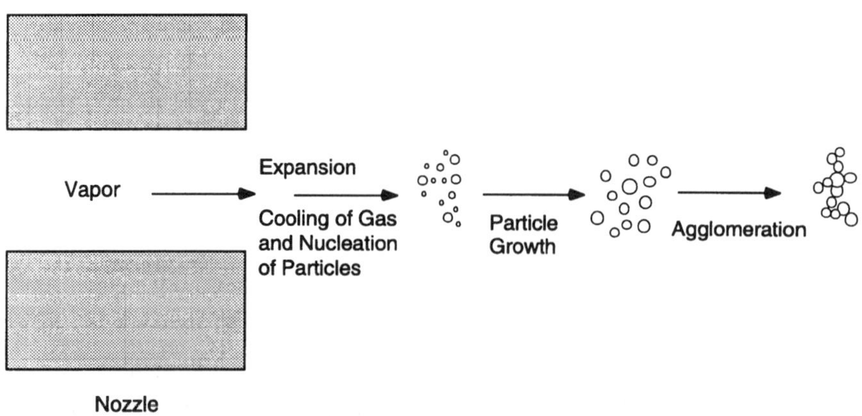


**Fig. 11.27.** Aerosol phenomena occurring during particle formation driven by gas cooling during expansion from a nozzle.

### 11.12.2 Precursors and Products

Table 11.11 (at end of chapter) shows some of the materials produced by nozzle expansion. Overall, only very simple materials have been produced by this route, but often in the form of small particles with dimensions less than 10 nm. Most studies of the subject examined fundamental aspects of cluster formation and behavior and were not directed at forming bulk quantities of powder; the route is not used commercially for the production of powders.

### 11.12.3 Experimental Considerations

The implementation of a nozzle expansion system for particle generation is rather complicated. Figures 11.28 and 11.29 show a typical system.[153] The generator consists of a heated boat containing the material to be vaporized. A carrier gas is passed over the boat to give a gas stream containing the vapor, which is then expanded through a sonic orifice into a chamber at low pressure (1 torr). The expansion and subsequent cooling of the gas drives the formation of the particles, which are then collected in a chamber, and the system is opened to obtain the powder. Often, only 30 to 40% of the

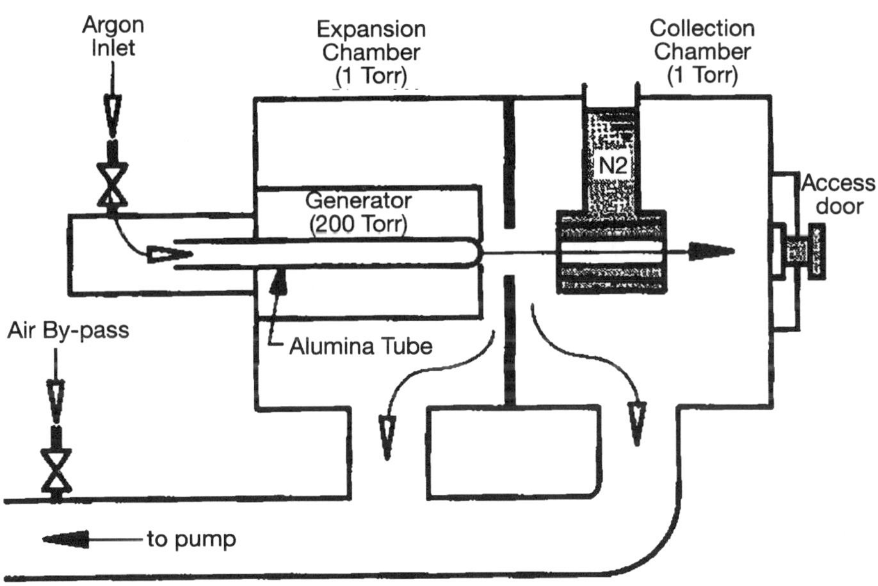


**Fig. 11.28.** Schematic diagram of experimental system for particle generation by nozzle expansion. (Reprinted with permission of Elsevier Science Inc. from "Design and operation of a nanocluster generation and collection system," by Recknagle, K., Leung, A., Brown, D., Martian, P., Chung, J.N., Xia, Q., Hamilton, H., and Crowe, C.T. *Aerosol Science and Technology*, *22*, pp. 3–10, Copyright 1995 by American Association for Aerosol Research.)[153]

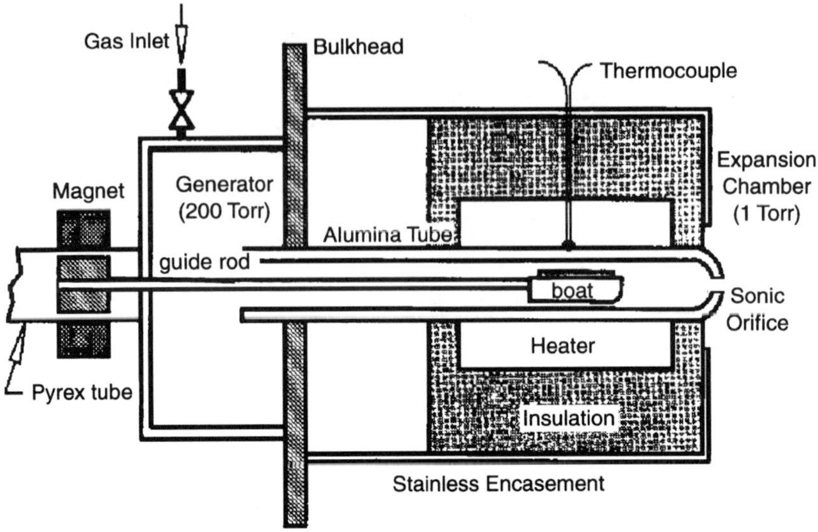


**Fig. 11.29.** Schematic diagram of nozzle used for particle generation by nozzle expansion. (Reprinted with permission of Elsevier Science Inc. from "Design and operation of a nanocluster generation and collection system," by Recknagle, K., Leung, A., Brown, D., Martian, Pl, Chung, J.N., Xia, Q., Hamilton, H., and Crowe, C.T. *Aerosol Science and Technology*, *22*, pp. 3–10, Copyright 1995 by American Association for Aerosol Research.)[153]

vaporized material is collected. The nozzle expansion system requires significant gas-pumping capacity.

### 11.12.4 Theory

The theory of nozzle expansion processes is relatively straightforward. The features of gas-to-particle conversion systems are relevant. The major problem, however, is describing the temperature and pressure of the gas as it expands. Most models treat the system as one-dimensional flow and calculate the evolution of temperature and pressure using standard equations. The equations describing the aerosol dynamics are solved with these temperature and pressure profiles to give the evolution of the aerosol characteristics along the axis of the jet. A number of groups of researchers have studied this problem theoretically.[154–163]

## 11.13 PHYSICAL PROCESSES: LASER ABLATION

### 11.13.1 Introduction

Laser ablation is a process that can produce small amounts of nanometer-sized particles (Fig. 11.30). The mechanism is similar to spark ablation and some plasma processes, in that material is introduced into the gas phase as atomic

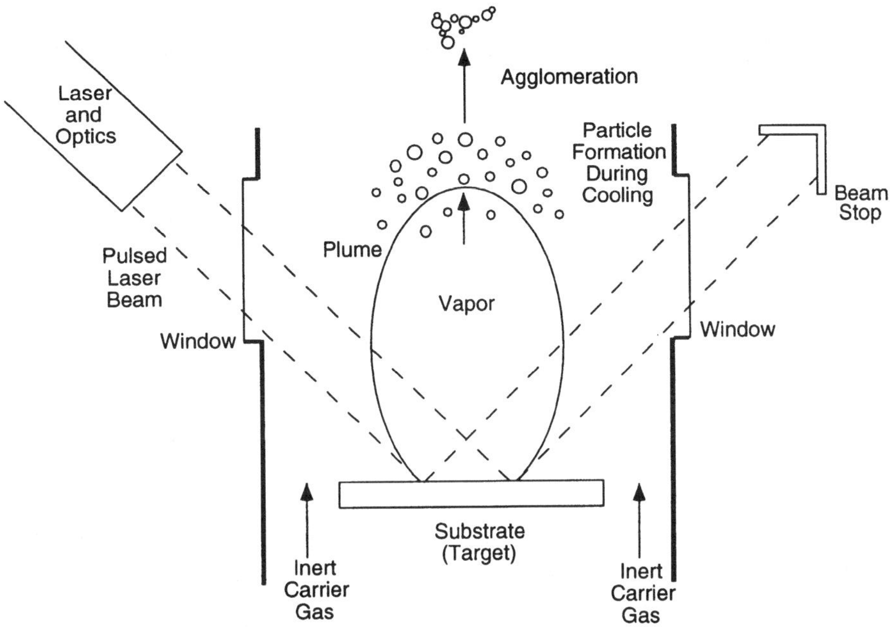


**Fig. 11.30.** Particle formation by laser ablation.

and ionic species at very high temperatures and then is cooled to supersaturate the vapor and drive particle formation. In the case of laser ablation, pulsed lasers with very high energy densities are used. The laser beam is brought into a chamber through a window, where the beam strikes the surface of a target composed of the material to be produced in particulate form. In reactive laser ablation, the material that is vaporized reacts in the gas phase with a gaseous reactant containing oxygen or nitrogen, for example, to give the metal oxide or nitride. When the laser beam strikes the surface, an extremely large amount of energy is deposited into a very small volume of material, causing the material to be both vaporized and ionized. The material then forms a plume, which moves away from the surface. As the plume expands, it cools, and the species in the plume interact to generate the vapors that eventually condense to form the particles. Laser ablation is also used to form films when particle formation is undesirable. The chief advantage of the method is that it can produce nanophase-sized materials. The primary disadvantages are that laser ablation is a complex, nonisothermal system, that scale-up has never been demonstrated, that the process possesses very low production rates ($<1$ gm/day), and that expensive equipment is required.

### 11.13.2 Precursors and Products

Numerous materials have been produced in nanophase form using laser ablation. Table 11.12 (at the end of the chapter) gives examples of these materials. Only simple materials such as metals and simple metal oxides and nitrides have been formed, usually in very small quantities. This route is not used commercially for the production of powders.

### 11.13.3 Experimental Considerations

The experimental apparatus required for laser ablation is quite sophisticated: The major components include a laser (often a ytrium aluminum garnet (YAG) or Excimer) and a vacuum chamber in which the target is placed. Typical experimental parameters are a laser power of 200 mJ/pulse with a pulse rate of 30 Hz. Production rates corresponding to these conditions are roughly 10 to 20 mg/hour. A problem is that only a fraction of the laser energy is absorbed by the target, because of the reflectivity of the substrate. Little can be done about this problem once a material is chosen. Chamber pressures are usually small fractions of an atmosphere (on the order of 10 torr). Collection efficiencies for the powder are often 50%. Small particles are usually obtained, with diameters less than 10 nm. Only limited information is available on powder characteristics because of the small amount of powder produced during a run.

### 11.13.4 Theory

Very little work has been done on the theory of particle formation by laser ablation. Some insight can be gained by examining the results of models for plasma systems, due to the similarities of the two approaches. Particle formation occurs in the manner described in Chapter 10.

## 11.14 PHYSICAL PROCESSES: SUPERCRITICAL SPRAYING

### 11.14.1 Introduction

Supercritical spray processes can be used for powder generation as well as film formation. A precursor or the actual product material is dissolved in a supercritical fluid such as $H_2O$ or $CO_2$. The supercritical fluid is formed by applying high temperatures and pressures to the liquid. This approach allows many materials to be dissolved into supercritical liquids that cannot be dissolved into other liquids. An example is $SiO_2$ in supercritical water. The fluid is passed through a nozzle, after which the pressure and temperature decrease and the fluid converts to the normal state (gas or liquid). The conversion is accompanied by a drastic decrease in the solubility of the species in solution, leading to particle formation. If the solvent is a gas or a liquid with a high vapor pressure, particle formation will take place from the gas or vapor.

The advantages of this method are:

- Capability of processing a wide variety of materials
- Ability to produce nanophase materials

The primary disadvantages are:

- Complex, nonisothermal system
- Difficulty controlling system due to poor understanding of its dynamics
- No demonstration of scale-up
- Few studies of method

### 11.14.2 Precursors and Products

Only a limited number of inorganic materials ($SiO_2$, $GeO_2$, $Al_2O_3$, polymers) have been processed by supercritical spraying. (See Table 11.13 at end of chapter.) Organic species have been processed, but most studies have examined the formation of particles for applications in drug delivery.

The system $SiO_2$–$H_2O$ has been examined extensively because of the well-characterized solubility of $SiO_2$ in subcritical and supercritical water.[164–166] An experimental difficulty is that the relatively low solubility of $SiO_2$ in neutral water solutions requires the use of an autoclave in which the silica is dissolved at elevated temperatures. For silica, the autoclave must operate near 400°C and 600 bar. In experiments, the temperature of the preexpansion fluid from which the powder was generated was 600°C. Particle sizes were on the order of 0.1 $\mu$m, and the powders were unagglomerated. At lower concentrations (< 3000 ppm), particle size was roughly related to solute concentration in the solution prior to expansions, while at higher concentrations, the particles produced tended to be smaller and agglomerated. Particles of $GeO_2$ were also made under similar conditions. Homogeneously mixed $SiO_2$–$GeO_2$ powders were produced using two autoclaves maintained at different temperatures. The solutions were thoroughly mixed prior to expansion through the nozzle. X-ray fluorescence analysis of the product showed a uniform distribution of Si and Ge throughout the particles, demonstrating the potential for the production of homogeneously mixed multicomponent materials

Supercritical routes also allow preparation of powders from ceramic precursor salts, which usually have much higher solubilities than the corresponding oxides. Soluble organometallic ceramic precursors offer similar advantages. With these precursors, the particles often consist of the precursor material, and the process is similar to spray drying in that it allows removal of a solvent and homogeneous mixing of two or more species.

A variety of pharmaceutical materials have been produced using supercritical spraying.[167–170]

### 11.14.3 Experimental Considerations

Figure 11.31 shows an experimental setup used to produce $SiO_2$ particles.[166] The main experimental parameters that can be varied are the type of solvent, the solubility of the solute, and the pressure and temperature upstream of the expansion. The basic components of the system include a pump capable of reaching pressures several times the critical pressure for the fluid of interest and a region in which the fluid is heated above its critical temperature. A nozzle is used to expand the solution from the critical pressure to near atmospheric pressure. For solute–solvent combinations with low solubilities, such as $SiO_2$ in $H_2O$, the solute and solvent must be combined in an autoclave before passing into the region in which supercritical temperatures are attained.

The nozzle typically consists of capillary tubing at the end of the heated region in which supercritical conditions are maintained. Capillary materials that have been used include stainless steel, platinum–iridium alloy, and fused silica. Production rates tend to be low, and most studies have collected the particles onto surfaces rather than separating them from the gas flow onto filters.

### 11.14.4 Theory

The physical basis for supercritical spraying is the rapid reduction in fluid density as a supercritical fluid solution expands through a nozzle into a low-pressure environment. The solubility and the supercritical fluid density are related, and the latter can be controlled by varying the fluid pressure. Supercritical fluids can have densities that are a significant fraction of the corresponding liquid density at pressures of 2 to 10 times the critical pressure used in the process. The supercritical fluid can be expanded to a single, low-density, gaseous solvent phase. Expansion forces nonvolatile solute species to nucleate and condense rapidly as the fluid solvating capacity drops, and particles are formed. The detailed mechanisms for these nucleation and condensation processes have been modeled by Debenedetti and others.[171–176] The expansion takes place in a time of roughly $10^{-8}$ to $10^{-5}$ sec. The short time and relatively low temperatures result in powders that are usually amorphous. The short time frame for particle formation and growth also allows the production of homogeneous powders from solutions containing several solutes.

The expansion nozzles typically are short lengths ($<5$ mm) of capillary tubing ($<100$ $\mu$m, inside diameter). There are several expansion stages, including the expansion within the nozzle; an isentropic expansion after the jet exits the nozzle, but before interaction with background gases; expansion in the region of interaction with the background gas, in which a series of shock fronts are formed; and the bulk of the expansion. Modeling of the supercritical particle formation process requires coupling the description of the fluid behavior to the aerosol dynamics as described in Chapter 10.

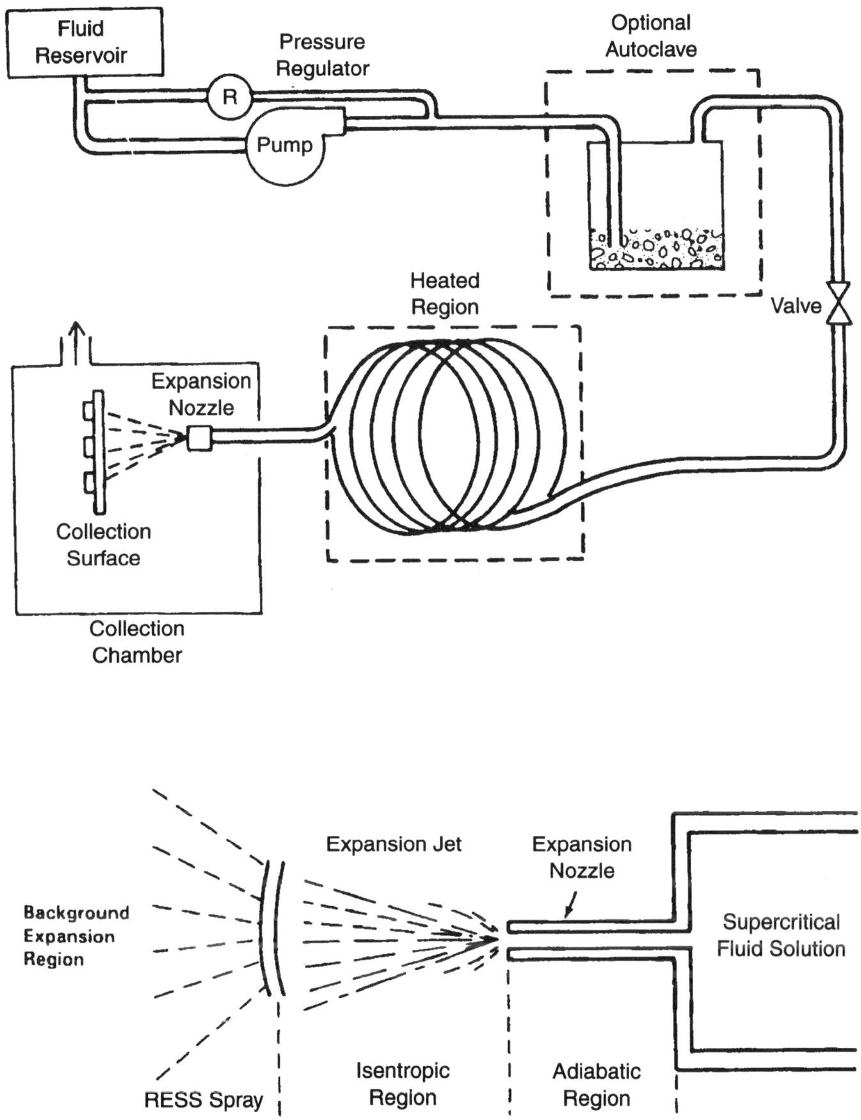


**Fig. 11.31.** Schematic diagram of experimental system for powder generation by supercritical fluid spraying. (Adapted from Matson D.W., Fulton, J.L., Petersen, R.C., and Smith, R.D. Reprinted with permission of The American Ceramic Society, Westerville, OH. Copyright 1987 by The American Ceramic Society. All rights reserved.)[166]

## 11.15 HYBRID PROCESSES: EVAPORATION–CONDENSATION–REACTION

### 11.15.1 Introduction

In the hybrid process of evaporation–condensation–reaction, an organometallic compound is vaporized into a carrier gas, which is then cooled, resulting in droplet formation by nucleation (Fig. 11.32). The droplets are then mixed with a gaseous reactant such as water, which results in chemical reaction in the droplets to form solid particles. The particles are collected and heated to form the desired product. Vaporization and condensation temperatures are usually less than 500°C, and the systems are typically operated at atmospheric pressure. Temperatures are usually much lower than in laser, plasma, hot-wall, and evaporation–condensation systems. In contrast to systems such as hot-wall, laser, and plasma reactors, powder production requires several stages—one for vaporization of the reactant, one for condensation, and one for reaction with a gaseous reactant such as water.

The advantages of this process are:

- Production of simple oxides
- Production of coated particles possible
- Demonstrated scale-up to kg/day level
- Production of spherical particles

The primary disadvantages are:

- Complex system
- Requirement of volatile precursors
- Difficulty controlling process
- Expensive precursors
- Formation of porous particles in some cases

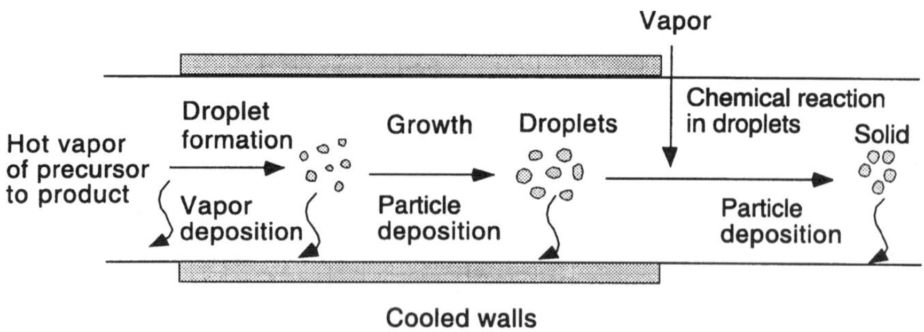


**Fig. 11.32.** Hybrid process consisting of vapor nucleation and droplet growth followed by intraparticle reaction.

### 11.15.2 Precursors and Products

The evaporation–condensation–reaction process relies on the availability of volatile precursors that can be condensed and then reacted with a gaseous species to form a nonvolatile product. For this reason, metal alkoxides have been the primary form of precursor that has been used. These materials are volatile and can be hydrolyzed to form solid products that can be reacted further, once they are collected, to form the product ceramic.

This process is versatile and can form homogeneous, coated, and composite particles, as described in Table 11.14 (at end of chapter). The most commonly studied materials have been $TiO_2$ and $Al_2O_3$, partly because of the availability of volatile metal alkoxide precursors for these materials. Product particles are formed by gas–solid reaction, which can lead to the formation of porous particles, unless high temperatures are used to densify the particles. Polymer particles and composite polymer–metal-oxide particles have also been formed.

The characteristics of alumina powder produced by this route have been examined.[177] The powder was 99.995% pure, with no other elements such as Fe or Na detected. X-ray diffraction showed boehmite and amorphous phases. The powder had a surface area of 227 $m^2/g$, demonstrating that the micron-sized particles were porous as a result of the gas–solid hydrolysis reaction. The spread of the particle size distribution showed that the ratio of the diameters below which 5 and 95% of the particles, by mass, existed was 3 to 3.5.

### 11.15.3 Experimental Considerations

Figure 11.33 shows an experimental system used to generate $SrTiO_3$ particles at the laboratory scale.[178] In this system, seed particles are formed in a furnace, and then a vapor is introduced by passing the gas through a system in which the gas is exposed to the hot liquid of the metal-alkoxide precursor. The vapor and seed particles in the gas are then passed through a condenser in which the vapor condenses onto the seed particles to form droplets. The droplets are mixed with water, and the reaction forms a solid product, which is collected. This approach has also been implemented at larger scales capable of producing kg quantities of $Al_2O_3$ from Al(sec-butoxide)$_3$.[177]

In the experimental system for the production of 100-gram quantities, nitrogen was used as a carrier gas; the vaporizer consisted of a bubbler holding 22 liters of the metal alkoxide; the condenser was a coiled, 4-m-long tube of 1 cm inside diameter, immersed in a chilled-water bath. Hydrolysis of the alkoxide droplets formed in the condenser was carried out by mixing water vapor with the alkoxide droplet stream. A typical gas flow was 50 l/min, corresponding to a Reynold's number of 5000. The average residence time was on the order of 1 second in the hydrolyzer, where reaction to form the product occurred. The powder was collected on high-efficiency filters.

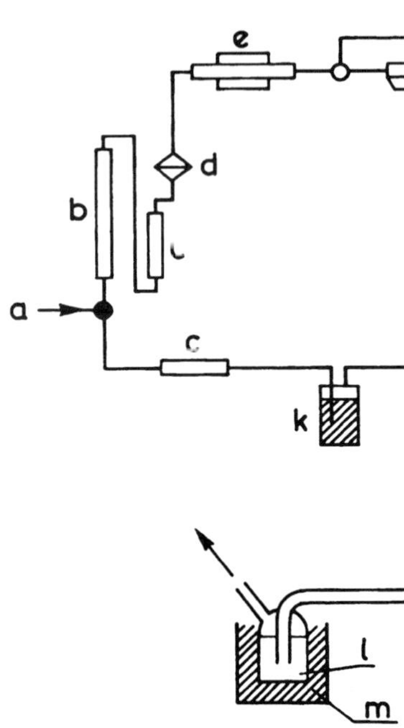


**Fig. 11.33.** Schematic of experimental system for powder generation by hybrid process. a: Gas tank. b: Drying column. c: Flowmeter. d: Filter. e: Nuclei generator. f: Reheater. g: Vapor evaporation chamber. m: Constant-temperature bath. (Reprinted from *Journal of Material Science Letters 9* (7), 1990, Continuous production of spherical strontium titantate at low temperature, Ocana M., and Serna, C.J., Chapman & Hall.)[178]

For unseeded conditions, the majority of the alkoxide vapor was deposited on the condenser walls. Vapor deposition on the walls was reduced to 10% by seeding with particles generated by the hydrolysis of a small fraction of the alkoxide vapor.[177]

### 11.15.4 Theory

Several theoretical studies of evaporation–condensation–reaction have been carried out that have investigated the effect of vaporizer temperature, carrier gas properties, precursor properties, length of condensor, flow rate, and other parameters on the distribution of droplet sizes formed during the condensation step[179–182].

Davis and Liao examined condensation onto seed nuclei and were the first to identify the controlling parameters.[179] Their model did not include the possibility of new-particle formation. The major finding from their study was that the Peclet numbers for heat and vapor transport were the critical parameters. Also, anything that enhanced heat transfer relative to vapor transport resulted in narrower droplet size distributions. A broad size distribution resulted from the tendency of particles near the wall to act as sinks for condensing vapor that would diffuse to the droplets from the center of the condensor. Thus, rapid cooling before appreciable vapor transport could occur resulted in the narrowest size distributions. As a result, switching from $N_2$ to He as the carrier gas produced narrower size distributions.

Pesthy et al.[182] examined the same problem, but included the possibility of new-particle formation by homogeneous nucleation. The most important parameter in their study was not the Peclet number, but the type of nucleation theory used in their model. Classical nucleation theory resulted in much larger wall losses of vapor (75%) than did Lothe–Pound theory. Pesthy et al.'s findings suggested that the importance of homogeneous nucleation in systems with seed particles depends strongly on the type of precursor used. Similarly, the concentration and size of the seed particles strongly influence the product droplet size distribution.

## 11.16 REFERENCES

1. Akhtar, M.K. & Pratsinis, S.E. Thermal processes for aerosol synthesis of non-oxide ceramic powders, in Weimer, A.W. (Ed.), *Carbide, Nitride and Boride Materials Synthesis and Processing*. London: Chapman & Hall, 1997, 671.
2. Gurav, A., Kodas, T., Pluym, T., & Xiong, Y. *Aerosol. Sci. Tech. 19*, 1993, 411–452.
3. Kodas, T.T. & Sood, A. Alumina powder production by aerosol processes. in Hart, L. (Ed.), *Handbook of Science and Technology of Alumina Chemicals*. Westerville, OH, American Ceramics Society, 1990, 375–383.
4. Kusters, K.A. & Pratsinis, S.E. *Powder Technol. 82*(1), 1995, 79–91.
5. Marijnissen, J. & Pratsinis, S.E. (Ed.), *Synthesis and Measurement of Ultrafine Particles*. Delft, The Netherlands: Delft University Press, 1993.
6. Pratsinis, S. & Kodas, T.T. Manufacturing of materials by aerosol processes. In Willeke, K. & Baron, P.A. (Eds.), *Handbook of Aerosol Measurement: Principles, Techniques and Applications*. New York: Van Nostrand Reinhold, 1993, 721–746.
7. Kodas, T.T. & Hampden-Smith, M.J. (Eds.), *The Chemistry of Metal CVD*. Weinheim, West Germany: VCH, Verlagsgesellschaft, 1994.
8. Okuyama, K. *J. Aerosol Sci. 22*(1), 1991, S7–S10.
9. Okuyama, K., Ohshima, K., & Tsuto, K. *Kona 9*, 1991, 79–93.
10. Schulz, O. & Hausner, H. *Ceramic Powder Processing Science*, 1988, 275–282.
11. Kodas, T.T. & Friedlander, S.K. *Ind. Eng. Chem. Res. 26*, 1987, 1999–2007.
12. Powell, Q.H. *Gas-phase production and coating of titania in tubular, hot-wall reactors*. Master's thesis, University of New Mexico, 1995.
13. Arkless, K. & Cleaver, D. United States Patent 3,501,262, 3/17/70.
14. Herriman, J.D. & Hare, A.L. United States Patent 3,464,792, 9/2/69.
15. Hoekje, H.H., Strain, F., & Wilson, W.L. United States Patent 3,642,442, 2/15/72.

16. Inoue, H. & Komiyama, H. United States Patent 4,842,832, 6/27/89.
17. Kilgren, A.W. United States Patent 3,632,313, 1/4/72.
18. Lane, F.W. United States Patent 2,653,078, 9/22/53.
19. Magyar, J.C., McDill, R.G., & Adams, E.R. International Patent Patent WO 89/11450, 11/30/89.
20. Monroe, E.S. United States Patent 3,449,076, 6/10/69.
21. Morris, A.J. & Coe, M.D. United States Patent 4,803,056, 2/7/89.
22. Nutting, R.D. United States Patent 2,670,272, 2/23/54.
23. Olson, C.M. & Tully, J.N. United States Patent 2,915,367, 12/1/59.
24. Pfender, E. United States Patent 3,695,840, 10/3/72.
25. Rick, C.E. United States Patent 3,639,100, 2/1/72.
26. Salter, D. & Scott, D. United States Patent 4,455,288, 6/19/84.
27. Stern, D.R., Gundzik, R.M., Jones, P.M., & Lynskey, P.J. United States Patent 3,615,202, 5/19/70.
28. Jain, S., Fotou, G.P., & Kodas, T.T. *J. Colloid Interface Sci. 185*, 1997, 26–38.
29. Jain, S., Kodas, T.T., Wu, M.K., & Preston, P. *J. Aerosol. Sci. 28*(1), 1997, 133–146.
30. Powell, Q.H., Kodas, T.T., & Anderson, H. *J. Chem. Vapor Deposition 2*, 1996, 179–181.
31. Powell, Q.H., Fotou, G.P., & Kodas, T.T. *Chem. Mater. 9*(3), 1997, 685–693.
32. Akhtar, M.K., Xiong, Y., & Pratsinis, S.E. *AIChE 37*(10), 1991, 1561–1570.
33. Powell, R. *Chemical Process Review No. 18, Titanium Dioxide and Titanium Tetrachloride*. Parkridge, New Jersey: Noyes Development Corporation, 1968.
34. Frenklach, M. *Chem. Phys. Processes Combust.*, 1989, C1–C8.
35. Kriechbaum, G.W. & Kleinschmit, P. *Ang. Chemie, Advanced Materials 10*(10), 1989, 330–337.
36. Ulrich, G.D. *Chem. Eng. News*, Aug. 6, 1984, 22–29.
37. Hung, C.-J. & Katz, J.L. *J. Mater. Res.* 7(7), 1992, 1861–1875.
38. Fotou, G.P. & Pratsinis, S.E. Particle coating of silica fibers by aerosol and sol-gel processes. In Bansal, N. (Ed.), *Ceramic Transactions: Advances in Ceramic-Matrix Composites*. Westerville, OH, American Ceramic Society, 1993, 737–747.
39. Fotou, G.P. & Pratsinis, S.E. *Chem. Eng. Sci. 49*(10), 1994, 1651–1662.
40. Dannenberg, E.M. *J. IRI*, October, 1971, 190–195.
41. Cannon, W.R., Danforth, S.C., Flint, J.H., Haggerty, J.S., & Marra, R.A. *J. Am. Ceram. Soc. 65*(7), 1982, 324–330.
42. Raes, F., Kodas, T.T., & Friedlander, S.K. *Aerosol Sci. Technol. 12*, 1990, 856.
43. Aivazov, M.I., Volod'ko, V.V., Evseev, B.A., & Nikulin, Y.N. *Sov. Powder Met. & Met. Ceramics 20*, 1981, 1–3.
44. Carson, D.R. & Holden, C.B. United States Patent 4,022,872, 5/10/77.
45. Chang, Y. & Pfender, E. *Plasma Chem. and Plasma Processing* 7(3), 1987, 275–297.

46. Fauchais, P., Boudrin, E., Coudert, J.F., & McPherson, R. High pressure plasmas and their application to ceramic technology. In *Top. Curr. Chem. 107* 1983, 59–183.

47. Heberlein, J.V.R. *Mat. Res. Soc. Symp. Proc. 30*, 1984, 101–115.

48. Holmgren, J.D., Gibson, J.O., & Sheer, C. *J. Electrochem. Soc. 111*(3), 1964, 362–369.

49. Miller, T.N., Izvestiya Akad. Nauk SSSR, *Neorg. Mater.* (*Trans.*) *15*(4), 1979, 557–562.

50. Phillips, D.S. & Vogt, G.J. *MRS Bulletin*, Oct/Nov, 1987, 54–58.

51. Pickles, C.A. & McLean, A. *Ceram. Bull. 62*(9), 1983, 1004–1009.

52. Rykalin, N.N. *Pure & Appl. Chem. 48*, 1976, 179–194.

53. Szekely, J. *Mat. Res. Soc. Symp. Proc. 30*, 1984, 1–11.

54. Veale, C.R. *Fine Powders Preparation, Properties and Uses*. London: Applied Science Publishers, Ltd., 1972.

55. Waldie, B. *The Chemical Engineer*, March 1972, 92–96.

56. Young, R.M. & Pfender, E. *Plasma Chem. and Plasma Processing 5*(1), 1985, 1–37.

57. Hollabaugh, C.M., Hull, D.E., Newkirk, L.R., & Petrovic, J.J. *J. Mater. Sci. 18*, 1983, 3190–3194.

58. Suzuki, M., Kagawa, M., Syono, Y., & Hirai, T. *Ceramic Powder Science IV, Ceramic Transactions 22*, 1991, 147–152.

59. Bilodeau, J.-F. & Proulx, P. *Aerosol Sci. Tech. 24*, 1996, 175–189.

60. Boulos, M.I. *Mat. Res. Soc. Symp. Proc. 30*, 1984, 53–60.

61. Girshick, S.L. & Chiu, C.-P. *Plasma Chem. Plasma Process 9*(3), 1989, 355–369.

62. Girshick, S.L. & Chiu, C.-P. *J. Chem. Phys. 93*(2), 1990, 1273–1277.

63. Girshick, S.L. & Chiu, C.-P. *J. Aerosol Sci. 21*(5), 1990, 641–650.

64. Girshick, S.L., Chiu, C.-P., & McMurry, P.H. *Aerosol Sci. and Tech. 13*, 1990, 465–477.

65. Girshick, S.L. *J. Chem. Phys. 94*(1), 1991, 826–827.

66. Vardelle, M., Trassy, C., Vardelle, A., & Fauchais, P. *Plasma Chem. and Plasma Processing 11*(2), 1991, 185–201.

67. Zyatkevich, D.P., Pokrovskii, D.D., Makarenko, G.N., & Kosolapova, T.Y. *Izvestiya Akad. Nauk SSSR, Neorg. Mater. 15*(4), 1979, 599–601.

68. Anderson, H.M., Radovanov, S., Mock, J.L., & Resnick, P.L. *Plasma Sources Science and Technology 3*(3), 1994, 302–309.

69. Bertran, E., Costa, J., Sardin, G., Campmany, J., Andujar, J.L., & Canillas, A. *Plasma Sources Science and Technology 3*(3), 1994, 348–354.

70. Blain, M.G., Tipton, G.D., Holber, W.M., Selwyn, G.S., Westerfield, P.L., & Maxwell, K.L. *Plasma Sources Science and Technology 3*(3), 1994, 325–333.

71. Boeuf, J.P., Belenguer, P., & Hbid, T. *Plasma Sources Science and Technology 3*(3), 1994, 407–417.

72. Bouchoule, A. & Boufendi, L. *Plasma Sources Science and Technology 3*(3), 1994, 292–301.

73. Boufendi, L. & Bouchoule, A. *Plasma Sources Science and Technology 3*(3), 1994, 262–267.
74. Buss, R.J. & Hareland, W.A. *Plasma Sources Science and Technology 3*(3), 1994, 268–272.
75. Carlile, R.N., O'Hanlon, J.F., Hong, L.M., Garrity, M.P., & Collins, S.M. *Plasma Sources Science and Technology 3*(3), 1994, 334–339.
76. Choi, S.J., Ventzek, P.L.G., Hoekstra, R.J., & Kushner, M.J. *Plasma Sources Science and Technology 3*(3), 1994, 418–425.
77. Dalvie, M., Surendra, M., Selwyn, G.S., & Guarnieri, C.R. *Plasma Sources Science and Technology 3*(3), 1994, 442–447.
78. Garscadden, A., Ganguly, B.N., Haaland, P.D., & Williams, J. *Plasma Sources Science and Technology 3*(3), 1994, 239–245.
79. Girshick, S.L. *Plasma Sources Science and Technology 3*(3), 1994, 388–394.
80. Goree, J. *Plasma Sources Science and Technology 3*(3), 1994, 400–406.
81. Graves, D.B., Daugherty, J.E., Kilgore, M.D., & Porteous, R.K. *Plasma Sources Science and Technology 3*(3), 1994, 433–441.
82. Haaland, P.D., Garscadden, A., Ganguly, B., Ibrani, S., & Williams, J. *Plasma Sources Science and Technology 3*(3), 1994, 381–387.
83. Havnes, O., Nitter, T., Tsytovich, V., Morfill, G.E., & Hartquist, T. *Plasma Sources Science and Technology 3*(3), 1994, 448–451.
84. Hollenstein, C., Dorier, J.-L., Dutta, J., Sansonnens, L., & Howling, A.A. *Plasma Sources Science and Technology 3*(3), 1994, 278–285.
85. Kamata, T., Kakuta, S., Yamaguchi, Y., & Makabe, T., *Plasma Sources Science and Technology 3*(3), 1994, 310–313.
86. Kroesen, G.M.W., Stoffels, W.W., Stoffels, E., Haverlag, M., denBoer, J.H.W.G., & deHoog, F.J. *Plasma Sources Science and Technology 3*(3), 1994, 246–251.
87. Perrin, J., Bohm, C., Etemadi, R., & Lloret, A. *Plasma Sources Science and Technology 3*(3), 1994, 252–261.
88. Rader, D.J. & Geller, A.S. *Plasma Sources Science and Technology 3*(3), 1994, 426.
89. Reents, W.D., Jr., Downey, S.W., Emerson, A.B., Mujsce, A.M., Muller, A.J., Siconolfi, D.J., Sinclair, J.D., & Swanson, A.G., *Plasma Sources Science and Technology 3*(3), 1994, 369–372.
90. Reents, W.D.J. & Mandich, M.L. *Plasma Sources Science and Technology 3*(3), 1994, 373–380.
91. Sato, N. *Plasma Sources Science and Technology 3*(3), 1994, 395–399.
92. Seebock, R.J., Bohme, W., Kohler, W.E., Romheld, M., & Veprek, S. *Plasma Sources Science and Technology 3*(3), 1994, 359–368.
93. Selwyn, G.S. *Plasma Sources Science and Technology 3*(3), 1994, 340–347.
94. Steinbruchel, C. & Yoo, W.J. *Plasma Sources Science and Technology 3*(3), 1994, 273–277.
95. Stoffels, W.W., Stoffels, E.C., Kroesen, G.M.W., Haverlag, M., denBoer, J.H.W.G., & deHoog, F.J. *Plasma Sources Sci. Technol. 3*(3), 1994, 320–324.

96. Tachibana, K., Hayashi, Y., Okuno, T., & Tatsuta, T. *Plasma Sources Science and Technology 3*(3), 1994, 314–319.

97. Watanabe, Y., Shiratani, M., Fukuzawa, T., & Kawasaki, H. *Plasma Sources Science and Technology 3*(3), 1994, 355–358.

98. Watanabe, Y. & Shiratani, M. *Plasma Sources Science and Technology 3*(3), 1994, 286–291.

99. Chou, C.H. & Phillips, J. *J. Mater. Res. 7*(8), 1992, 2107–2118.

100. Ho, P., Buss, R.J., & Loehman, R.E. *J. Mater. Res. 4*(4), 1989, 873–879.

101. Selwyn, G.S., Singh, J., & Bennett, R.S. *J. Vac. Sci. Technol. A 7*(4), 1989, 2758–2765.

102. Singh, R.P. & Doherty, R.D. *Mat. Lett. 9*(2,3), 1990, 87–89.

103. Yoo, W.J. & Steinbruchel, C. *J. Vac. Sci. Technol. B 12*(4), 1994, 2758–2762.

104. Vollath, D., Varma, R., & Sickafus, K.E. *Mat. Res. Soc. Symp. Proc. 269*, 1992, 379–384.

105. Vollath, D. & Sickafus, K.E. *NanoStruct. Mater. 1*, 1992, 427–437.

106. Vollath, D. *Ceram. Trans. 36*, 1993, 147–156.

107. Vollath, D. & Sickafus, K.E. *J. Mater. Sci. 28*(21), 1993, 5943–5948.

108. Vollath, D. & Sickafus, K.E. *NanoStruct. Mater. 2*(5), 1993, 451–456.

109. Vollath, D. & Sickafus, K.E. *J. Mater. Res. 8*(11), 1993, 2978–2984.

110. Vollath, D. *Mat. Res. Soc. Symp. Proc. 347*, 1994, 629–634.

111. Vollath, D. & Szabo, D.V. *NanoStruct. Mater. 4*(8), 1994, 927–938.

112. Froment, G.F. & Bischoff, K.B. *Chemical Reactor Analysis*. New York: Wiley, 1990.

113. Matsukata, M., Odagiri, T., & Kojima, T. *J. de Physique IV 1* (Colloque C2), 1991, 483–490.

114. Kimura, T. & Kojima, T. *J. Chem. Eng. Jpn. 17*(1), 1991, 81–87.

115. Kimura, T. & Kojima, T. *J. de Physique IV 1* (Colloque C2), 1991, 103–110.

116. Kojima, T. & Morisawa, O. *J. de Physique IV 1* (Colloque C2), 1991, 475–482.

117. Chaouki, J., Chavarie, C., Klvana, D., & Pajonk, G. *Powder Technol. 43*, 1985, 117–125.

118. Geldart, D. *Powder Technol. 7*, 1973, 285–292.

119. Gibilaro, L.G., Di Felice, R., & Foscola, P.U. *Powder Technol. 56*, 1988, 21–29.

120. Huang, C.-C. & Kono, H.O. *Powder Technol. 55*, 1988, 35–49.

121. Iyer, S.R. & Drzal, L.T. *Powder Technol. 57*, 1989, 127–133.

122. Kato, K., Takarada, T., Koshinuma, A., Kanazawa, I., & Sugihara, T. Decarbonization of silicon carbide–carbon fine particles mixture in a fluidized bed. In Grace, G.R., Shemilt, L.W., & Bergougnou, M.A. (Eds.), *Fluidization VI*. New York: AIChE, 1989, 351–358.

123. Kono, H.O., Huang, C.C., Morimoto, E., Nakayama, T., & Hikosaka, T. *Powder Technol. 53*, 1987, 163–168.

124. Kusakabe, K., Kuriyama, T., & Morooka, S. *Powder Technol. 58*, 1989, 125–130.

125. Morooka, S., Kusakabe, K., Kobata, A., & Kato, Y. *J. Chem. Eng. Jpn. 21*(1), 1988, 41–46.

126. Morooka, S., Okubo, T., & Kusakabe, K. *Powder Technology 63*(2), 1990, 105–112.

127. Pacek, A.W. & Nienow, A.W. *Powder Technol. 60*, 1990, 145–158.

128. Tardos, G., Mazzone, D., & Pfeffer, R. *Canadian J. Chem. Eng. 63*, 1985, 377–383.

129. Visser, J. *Powder Technol. 58*, 1989, 1–10.

130. Wierenga, C.R. & Morin, T.J. *AIChE J. 35*(9), 1989, 1555–1558.

131. Kimura, I., Hotta, N., Niwano, J.-I., & Tanaka, M. *Powder Technology 68*(2), 1991, 153–158.

132. Andres, R.P., Averback, R.S., Brown, W.L., Brus, L.E., Goddard, W.A., Kaldor, A., Louie, S.G., Moscovits, M., Peercy, P.S., Riley, S.J., Siegel, R.W., Spaepen, F., & Wang, Y. *J. Mater. Res. 4*(3), 1989, 704–736.

133. Castleman, A.W., Jr. & Keesee, R.G. *Acc. Chem. Res. 19*, 1986, 413–419.

134. Eastman, J. & Siegel, R.W. *Research & Development*, January 1989, 56–60.

135. Fissan, H., Karow, H.V., & Kauffeldt, T. (Eds.), *Aerosol Methods and Advanced Techniques for Nanoparticle Science and Nanophase Technology*. Strasbourg, France: European Science Found., 1994.

136. Gleiter, H. *Prog. Materials Science. 33*, 1989, 233–315.

137. Gleiter, H. *Adv. Mater. 4*(7/8), 1992, 474–481.

138. Hayashi, C. *J. Vac. Sci. Technology 5*(4), 1987, 1375–1384.

139. Ichinose, N., Ozaki, Y., & Kashu, S. *Superfine Particle Technology*. London: Springer-Verlag, 1992.

140. Kai, H.Y. *Nanocrystalline meterials: A study on their preparation and characterization*. Ph.D. dissertation, Amsterdam University, 1993.

141. Pool, R. *Science 248*, 1990, 1186–1188.

142. Siegel, R.W. *MRS Bull.*, October, 1990, 60–67.

143. Uyeda, R. *Prog. Materials Science, 35*, 1991, 1–95.

144. Granqvist, C.G. & Buhrman, R.A. *J. Appl. Phys. 47*(5), 1976, 2200–2219.

145. Kim, S.G. & Brock, J.R. *J. Appl. Phys. 60*(2), 1986, 508–513.

146. Panda, S. & Pratsinis, S.E. *NanoStruct. Mater. 5*(7–8), 1995, 755–767.

147. Buckle, E.R. & Mawella, K.J.A. *J. Aerosol Sci. 18*(2), 1987, 139–146.

148. Hayashi, T., Ohno, T., Yatsuya, S., & Uyeda, R. *Jpn. J. Appl. Phys. 16*(5), 1977, 705–717.

149. Kimoto, K., Kamiya, Y., Nonoyama, M., & Uyeda, R. *Jpn. J. Appl. Phys. 2*(11), 1963, 702–713.

150. Thölén, A.R. *Arts Metallurgica 27*, 1979, 1765–1778.

151. Uyeda, R. *J. Crystal Growth 24*(25), 1974, 69–75.

152. Parker, J.C. & Siegel, R.W. *J. Mater. Res. 5*(6), 1990, 1246–1252.

153. Recknagle, K., Leung, A., Brown, D., Martian, P., Chung, J.N., Xia, Q., Hamilton, H., & Crowe, C.T. *Aerosol Sci. Technol. 22*, 1995, 3–10.

154. Buckle, E.R. & Pouring, A.A. *Nature 208*(5008), 1965, 367–369.

155. Be, S.H., Yano, K., & Kawai, S. *Jpn. J. Appl. Phys. 21*(12), 1982, 1755–1758.

156. Friedlander, S.K., Windeler, R.S., & Weber, A.P. *NanoStructured Mater.* *4*(5), 1994, 521–528.

157. Hagena, O.F. & Obert, W. *J. Chem. Phys.* *56*(5), 1972, 1793–1802.

158. Hill, P.G. *J. Fluid Mech.* *25*(3), 1966, 593–620.

159. McBride, D.D., Sherman, P.M., & Pierce, T.H. *Appl. Sci. Res.* *25*, November 1971, 83–96.

160. Pouring, A.A. *Phys. Fluids* *8*(10), 1965, 1802–1810.

161. Turner, J.R., Kodas, T.T., & Friedlander, S.K. *J. Chem. Phys.* *88*(1), 1988, 457–465.

162. Wegener, P.P. *J. Appl. Phys.* *25*(12), 1954, 1485–1491.

163. Wegener, P.P. & Pouring, A.A. *Phys. Fluids* *7*(3), 1964, 352–361.

164. Matson, D.W., Petersen, R.C., & Smith, R.D. *Advances in Ceramics* *211*, 1987, 109–120.

165. Matson, D.W., Petersen, R.C., & Smith, R.D. *J. Mater. Sci.* *22*, 1987, 1919–1928.

166. Matson, D.W., Fulton, J.L., Petersen, R.C., & Smith, R.D. *Ind. Eng. Chem. Res.* *26*, 1987, 2298–2306.

167. Debenedetti, P.G., Tom, J.W., Yeo, S.-D., & Lim, G.-B. *J. Controlled Release* *24*, 1993, 27–44.

168. Kohn, J., Yu, C., Yeo, S.D., & Debenedetti, P.G. 21*st Intern. Symp. Control. Rel. Bioact. Mater.* *21*, 1994, 132–133.

169. Tom, J.W. & Debenedetti, P.G. *Biotechnol. Prog.* *7*, 1991, 403–411.

170. Tom, J.W., Debenedetti, P.G., & Jerome, R. *J. Supercritical Fluids* *7*(1), 1994, 9–29.

171. Debenedetti, P.G., Petsche, I.B., & Mohamed, R.S. *Fluid Phase Equilibria* *52*, 1989, 347–356.

172. Debenedetti, P.G., Tom, J.W., Kwauk, X., & Yeo, S.-D. *Fluid Phase Equilibria* *82*, 1993, 311–321.

173. Debenedetti, P.G. (Ed.), *Supercritical fluids as particle formation media.* Norwell, MA: Kluwer Academic Pub., 1994.

174. Kwauk, X. & Debenedetti, P.G. *J. Aerosol Sci.* *24*(4), 1993, 445–469.

175. Mohamed, R.S. *ACS Symp. Ser.* *406*, 1989, 355–378.

176. Tom, J.W. & Debenedetti, P.G. *J. Aerosol Sci.* *22*(5), 1991, 555–584.

177. Kodas, T.T., Sood, A., & Pratsinis, S.E. *Powder Technology* *50*, 1987, 47–53.

178. Ocaña, M. & Serna, C.J. *J. Mater. Sci. Lett.* *9*, 1990, 772–773.

179. Davis, E.J. & Liao, S.C. *J. Colloid Interface Sci.* *50*(3), 1975, 488–502.

180. Nadkarni, A.R. & Mahalingam, R. *AIChE J.* *31*(4), 1985, 603–614.

181. Pesthy, A.J., Flagan, R.C., & Seinfeld, J.H. *J. Colloid Interface Sci.* *82*(2), 1981, 465–479.

182. Pesthy, A.J., Flagan, R.C., & Seinfeld, J.H. *J. Colloid Interface Sci.* *91*(2), 1983, 525–545.

**TABLE 11.1. Comparison of gas-to-particle conversion systems for particle generation. Processes are grouped according to sections in this chapter. "Maximum size" refers to the largest size imposed by practical constraints; larger sizes have been obtained under specialized conditions. "Spread of the size distribution" and "morphology" refer to high production rates at which coagulation occurs. Maximum temperatures are those most commonly found in practice. Materials are those for which majority of results are available**

| | Processes | Maximum Size | Morphology | Maximum Temperature, °C | Material | Complexity |
|---|---|---|---|---|---|---|
| Chemical | Hot-wall | 1 μm | Agglomerated | 1,700 for most conventional furnaces | Nonoxides, oxides, metals, semiconductors | Low |
| | Plasma | 1 μm | Agglomerated | 25,000 | Nonoxides, oxides | High |
| | Laser | 1 μm | Agglomerated | 1,000 | Nonoxides, oxides, metals, semiconductors | Medium |
| | Flame | 1 μm | Agglomerated | 2,500 | Oxides, nonoxides in Na flame | Low |
| Physical | Evaporation condensation | 0.1 μm | Agglomerated | Varies, depending on heat source | Metals, oxides, nonoxides, semiconductors | Low |
| Hybrid | Evaporation condensation intraparticle reaction | 1 μm | Spherical, unagglomerated | 300 (limited by stability of reactant) | Oxides | High |

**TABLE 11.2. Precursors and products in tubular flow reactors**

| Product | Reactant | Comments | References |
|---|---|---|---|
| $TiO_2$, $SiO_2$, $Al_2O_3$, 0.1–1 μm | $TiCl_4$, $SiCl_4$, $AlCl_3$ | $T_r$ = 1300–1500°C, $TiO_2$ particles coated with $SiO_2$ and $Al_2O_3$ | [1, 2] |
| $TiO_2$, 55–65 nm | $TiCl_4$ | $T_r$ = 1123–1473 K | [3] |
| $TiO_2$ | $TiCl_4$ | $T_r$ = 900–1300°C | [4] |
| $TiO_2$, 0.3–0.7 μm | $TiCl_4$ | $T_r$ = 850–1115°C | [5] |
| $TiO_2$ doped with P, Si, B, Al, Ge | $TiCl_4$, $POCl_3$, $BCl_3$, $SiCl_4$, $AlCl_3$ | $T_r$ = 1000–1500°C | [6–10] |
| $TiO_2$ | $TiCl_4$ | $T_r$ = 800–900°C, | [11] |
| $TiO_2$ | $TiCl_4$ | Au precipitated after collection for photocatalysis $T_r$ = 830°C | [12] |
| $TiO_2$ | $TiCl_4$ | $T_r$ = 700–1000°C, studied kinetics of reaction | [13] |
| $TiO_2$ doped with Zr, Si, Fe, Al | $TiCl_4$, $SiCl_4$, $ZrCl_4$, $FeCl_3$, $AlCl_3$ | $T_r$ = 1000–1100°C, studied effect of dopants | [14] |
| $TiO_2$ | $TiCl_4$ | $T_r$ = 600–1100°C, measured kinetics of reaction | [15] |
| $TiO_2$ | $TiCl_4$ | $T_r$ = 850–1200°C, studied anatase–rutile transformation | [16] |
| $TiO_2$ | $TiCl_4$ | $T_r$ = 850–1200C, two reactors in series | [3, 17–19] |
| $TiO_2$, < 40 nm | $TiCl_4$ | $T_r$ = 250–600°C | [20] |
| $TiO_2$, 30–70 nm | $TiCl_4$ | $T_r$ = 900°C | [21] |
| $TiO_2$, 0.05–0.95 μm | $TiCl_4$, $H_2$, $CO_2$, and $H_2O$, $H_2$ | $T_r$ = 700–1200°C | [6, 22] |
| $TiO_2$, 0.11–0.66 μm | $TiCl_4$, $H_2$, $CO_2$ | $T_r$ = 900–1200°C | [23] |
| $TiO_2$, $Al_2O_3$, $ZrO_2$, $Fe_2O_3$, 0.2–0.7 μm | $TiCl_4$, $AlBr_3$, $ZrCl_4$, $FeCl_2$ | $T_r$ = 700–1200°C | [24, 25] |
| $TiO_2$, 0.02–1 μm | Ti(O–*i*–Pr)$_4$ | $T_r$ = 443–973 K | [26, 27] |
| $TiO_2$ | Ti(O–*i*–Pr)$_4$ | $T_r$ = 400–600°C | [28] |
| $TiO_2$, 1.5–14 nm $SiO_2$, 10–25 nm $Al_2O_3$, 14–18 nm | $Ti(O_3H_7)_4$ $Si(OEt)_4$ $Al(OC_4H_9)_3$ | $T_r$ = 400–1000°C | [29] |
| $Al_2O_3$, 5 150 nm | $Al(CH_3)_3$ | $T_r$ = 550–950°C | [30] |
| Mullite, 0.05 μm | $AlCl_3$, $SiCl_4$, $O_2$ | $T_r$ = 900–1200°C | [31] |
| MgO, 0.01–0.2 μm | Mg vapor, $O_2$ | $T_r$ = unknown because exothermic | [32, 33] |
| ZnO | Zn vapor, $O_2$ | Metal vapor reacted with oxygen | [34] |

**TABLE 11.2.** (*Continued*)

| Product | Reactant | Comments | References |
|---|---|---|---|
| $In_2O_3$, $Sb_2O_3$, $Bi_2O_3$, $CuO$, $Cu_2O$, $SnO_2$, $WO_3$, PbO, SnO | Metals | Burning metal in Ar–$O_2$ | [35] |
| $MoO_3$, $WO_3$ | Mo, W | Burning metal in Ar–$O_2$ | [36] |
| ZnO | Zn vapor + $O_2$ | $T_r$ = 750–1300°C | [34] |
| $ZrO_2$, $Y_2O_3$–$ZrO_2$ | Y, Zr metals | Y and Zr particles mixed and reacted with oxygen | [37] |
| $Y_2O_3$, $CeO_2$, $La_2O_3$ $d_p \sim 0.05$ μm | $Y(tmhd)_3$ $Ce(tmhd)_3$ $La(tmhd)_3$ | $T_r$ = 1100°C | [38] |
| $ZrO_2$ | $ZrCl_4$ | $T_r$ = 900–1250°C | [39] |
| NiO–$SiO_2$ composites, 4–17 nm for NiO | $NiCl_2$, $SiCl_4$, $O_2$ | $T_r$ = 400–600°C | [40] |
| $Fe_2O_3$, 0.01–0.1 μm | $FeCl_3$ | $T_r$ = 550–600°C | [41] |
| $Fe_2O_3$, 0.01–0.1 μm | $Fe(acac)_3$ | $T_r$ = 500–900°C | [42] |
| $SiO_2$, 0.05 μm | $SiCl_4$, $O_2$ | $T_r$ = 1300°C | [43] |
| ZnO, 0.3–0.8 μm | Zn vapor, $O_2$ | $T_r$ = 850–1070°C | [14] |
| $Al_2O_3$, 50–300 nm | $AlBr_3$, $O_2$ | $T_r$ = 950–1200°C | [44] |
| MgO, 0.004–0.4 μm | Mg, $O_2$ | $T_r$ = 890–1130°C | [45, 46] |
| $ZrO_2$ $ZrO_2$:Ag composites, 0.035–0.15 μm | Zr(*t*–butoxide)$_4$ | $T_r$ = 600–700°C | [47] |
| $Fe_2O_3$, 0.3 – 1.6 μm | $FeCl_2$ | $T_r$ = 700–1000°C | [48] |
| Bi–Sr–Ca–Cu–O, 0.05–0.06 μm | $Cu(dpm)_2$ $Sr(dpm)_2$ $Bi(C_6H_6)_2$ $Ca(dpm)_2$ | $T_r$ = 750–900°C | [49] |
| α–WC, β–$W_2C$, 0.07–0.3 μm | $WCl_6$ | $T_r$ = 1000–1200°C | [50] |
| $B_4C$, $B_4C$–$TiB_2$ | $B_2O_3$ + carbon | $T_r$ = 1950°C | [51] |
| PdC, 5 nm | Pd particles + acetone | $T_r$ = unknown | [52] |
| β–SiC | $SiCl_xH_y$ + hydrocarbon | $T_r$ = 1700°C | [53] |
| SiC, $d_p \leq 0.5$ μm | $SiH_4$ + $CH_4$ | $T_r$ = 800–3650 K | [54] |
| SiC–$W_2C$ nanocomposite | $SiH_4$, $CH_4$, $WF_6$, $H_2$ | SiC shell, $W_2C$ core, tubular flow system | [55] |
| $Mo_2C$, 0.02–0.4 μm | $MoCl_4$ + $CH_4$ | $T_r$ = 1200–1400°C | [56] |
| SiC, 0.01–0.2 μm | $(CH_3)_4Si$ | 900–1400°C | |
| $Si_3N_4$, 0.01–0.1 μm | $SiCl_4$ + $NH_3$ | 1000–1500°C | |
| TiC, 0.01–0.15 μm | $TiCl_4$ + $CH_4$ | 1200–1400°C | |
| TiN, 0.01–0.4 μm | $TiCl_4$ + $NH_3$ | 700–1500°C | |
| VN, 0.01–0.1 μm | $VCl_4$ + $CH_4$ | 700–1200°C | |
| WC, 0.02–0.3 μm | $WCl_6$ + $CH_4$ | 1300–1400°C | |
| $Zr_3N_4$, < 0.1 μm | $ZrCl_4$ + $NH_3$ | 1000–1500°C | |

**TABLE 11.2.** (*Continued*)

| Product | Reactant | Comments | References |
|---|---|---|---|
| TiN | $TiCl_4$, $NH_3$ | $T_r$ = 900–1173°C | [57] |
| AlN, 0.01–1 μm | Al vapor, $NH_3$ | $T_r$ = 1050°C | [58] |
| AlN, 0.07–0.2 μm | $Al(C_2H_5)_3$, $NH_3$ | $T_r$ = 700–1100°C | [59] |
| $Si_3N_4$, $d_p$–0.2 μm | $SiH_4$, $NH_3$ | $T_r$ = 1000°C | [60] |
| $Si_3N_4$, 35–120 nm | $SiCl_4$, $NH_3$ | $T_r$ = 900–1150°C | [61] |
| $Si_3N_4$, 0.03–0.2 μm | $SiH_4$, $NH_3$ | $T_r$ = 550–920°C | [62] |
| $Si_3N_4$, 0.1–1 μm | $(Me_2N)_3$ SiH | $T_r$ = 700–1400°C | [63] |
| Cu | $Cu(hfac)_2$, $H_2$ | $T_r$ = 350–450 °C | [64] |
| W, Mo, 0.01–0.1 μm | $WCl_6$, $MoCl_5$, $H_2$ | $T_r$ = 430–905°C | [65] |
| Mo, Pd, Ta, Nb | $MoCl_5$, $Pd(acac)_2$ $TaCl_5$, $NbCl_5$, $H_2$ | $T_r$ = 500–900°C | [66] |
| Si, 0.3–0.9 μm | $SiH_4$ | $T_r$ = 700–950°C | [67] |
| Si | $SiH_4$ | $T_r$ = 700–100°C | [68] |
| Si, 0.001–10 μm | $SiH_4$ | $T_r$ = 700–1600°C | [69–76] |
| Si, 0.01–0.03 μm | $SiH_4$ | $T_r$ = 520–620°C | [77] |
| GaAs, 10–20 nm | $AsH_3$, $Ga(CH_3)_3$ | $T_r$ = 600–800°C | [78–80] |
| α-Si:H | $SiH_4$ | $T_r$ = 520–730°C | [77] |
| MnGe | $Me_2CC(CH_2)_2Ge[Mn(CO)_5]_2$ | | [81] |
| ZnS | (TMEDA)-$Zn(S_2CNMe_2)_2$ | | [81] |
| $ZrF_4$–$BaF_2$–$LaF_3$–$AlF_3$–NaF–based glass, 70–100 nm | $Zr(hfac)_4$, $Ba(ppm)_2$, $La(ppm)_3$, $Al(fod)_3$, Na(ppm) ppm = Pentafluoro-pentanedione | $T_r$ = not reported | [82] |
| Diamond | Dichloromethane, trichloroethylene, oxygen | Microwave plasma | [83, 84] |

## References to Table 11.2


1. Powell, Q.H., Kodas, T.T., & Anderson, B.M. *J. Chem. Vapor Deposition 2*, 1996, 179–181.
2. Powell, Q.H., Fotou, G.P., & Kodas, T.T. *Chem. Mater. 9*(3), 1997, 685–693.
3. Morooka, S., Kobata, A., & Kusakabe, K. *Ceramic Powder Science IV*, 1997, 109–114.
4. Suyama, Y., Ito, K., & Kato, A. *J. Inorg. Nucl. Chem. 37*, 1975, 1883–1888.
5. Suyama, Y. & Kato, A. *J. Am. Ceram. Soc. 59*(3–4), 1976, 146–149.
6. Akhtar, M.K., Pratsinis, S.E., & Mastrangelo, S.V.R. *J. Am. Ceram. Soc. 75*(12), 1992, 3408–3416.

7. Akhtar, M.K., Pratsinis, S.E., & Mastrangelo, S.V.R. *J. Mater. Res.* *9*(5), 1994, 1241–1249.
8. Akhtar, M.K., Vemury, S., & Pratsinis, S.E. *AIChE J.* *40*(7), 1994, 1183–1192.
9. Akhtar, M.K., Vemury, S., & Pratsinis, S.E. *NanoStruct. Mater.* *4*(5), 1994, 537–544.
10. Akhtar, M.K., Xiong, Y., & Pratsinis, S.E. *AIChE 37*(10), 1991, 1561–1570.
11. Toyama, S., Nakamura, N., Mori, H., Kanai, K., & Nachi, T. The production of ultrafine particles by vapor phase oxidation from chloride. *Proc. Second World Congress Particle Technology, September 19–22*, 1990. Kyoto, Japan: 360–367. Society of Powder Technology in association with VDI-Gesellschaft Verfahrenstechnik, Germany, Fine Particle Society, U.S.A., [1990]
12. Gao, Y.-M., Lee, W., Trehan, R., Kershaw, R., Dwight, K., & Wold, A. *Mat. Res. Bull.* *26*, 1991, 1247–1254.
13. Pratsinis, S.E., Bai, H., Biswas, P., Frenklach, M., & Mastrangelo, S.V.R. *J. Am. Ceram. Soc.* *73*(7), 1990, 2158–2162.
14. Suyama, Y. & Kato, A. *J. Am. Ceram. Soc.* *68*(6), 1985, C154–C 156.
15. Antipov, I.V., Koshunov, B.G., & Gofman, L.M. *Transl. Zhurnal Prikladnoi Khimii 40*(1), 1967, 11–15.
16. Kobata, A., Kusakabe, K., & Morooka, S. *AIChE 37*(3), 1991, 347–359.
17. Morooka, S., Kobata, A., Umeda, T., & Kusakabe, K. *J. Chem. Eng. Japan 22*(1), 1989, 94–95.
18. Morooka, S., Yasutake, T., Kobata, A., Ikemizu, K., & Kato, Y. *Intrnl. Chemical Engineering 29*(1), 1989, 119–126.
19. Morooka, S., Yasutake, T., Kobata, A., Ikemizu, K., & Kato, Y. *Transl. Kagaku Kogaku Ronbunshu 13*(2), 1987, 159–164.
20. Inoue, H. & Komiyama, H. United States Patent 4,842,832, 6/27/89.
21. Jang, H.D. & Jeong, J. *Aerosol Sci. Technol.* *23*, 1995, 553–560.
22. Suyama, Y., Ohmura, K., & Kato, A. *The Chemical Society of Japan 4*, 1976, 584–590.
23. Kato, A. & Suyama, Y. *Chemistry Letters*, 1974, 961–962.
24. Kato, A. *Adv. in Ceramics 21*, 1987, 181–192.
25. Kato, A. (Ed.), *Recent production methods of ultrafine ceramics powders*. Amsterdam: Elsevier, 1987.
26. Kirkbir, F. & Komiyama, H. *Canadian J. Chem. Eng.* *65*, 1987, 759–766.
27. Kirkbir, F. & Komiyama, H. *Adv. Ceram. Mater.* *3*(5), 1988, 511–515.
28. Seto, T., Shimada, M., & Okuyama, K. *Aerosol Sci. Technol.* *23*, 1995, 183–200.
29. Okuyama, K., Kousaka, Y., Tohge, N., Yamamoto, S., Wu, J.J., Flagan, R.C., & Seinfeld, J.H. *AIChE J.* *32*(12), 1986, 2010–2019.
30. Kim, K.H., Ho, C.H., Suh, T.-G., Prakash, S., & Bunshah, R.F. *J. Mater. Eng.* *13*, 1991, 199–205.
31. Itatani, K., Kubozono, T., Howell, F.S., Kishioka, A., & Kinoshita, M. *J. Mater. Sci.* *30*, 1995, 1158–1165.

32. Nishida, A., Yoshida, K., Igarashi, H., & Kobayashi, W. *Adv. in Ceramics 21*, 1987, 271–277.

33. Nishida, A., Ueki, A., & Yoshida, K. *Adv. in Ceramics 21*, 1987, 265–269.

34. Bolsaitis, P.P., McCarthy, J.F., Mohiuddin, G., & Elliott, J.F. *Aerosol Sci. Technol. 6*, 1987, 225–246.

35. Kaito, C., Fujita, K., Shibahara, H., & Shiojiri, M. *Jpn. J. Appl. Phys. 16*(5), 1977, 697–704.

36. Kaito, C. *Jpn. J. Appl. Phys. 17*(4), 1978, 601–609.

37. Foster, C.M., Bai, G.R., Parker, J.C., & Ali, M.N. *J. Mater. Res 8*(8), 1993, 1977–1982.

38. Kanapilly, G.M., Tu, K.W., Larsen, T.B., Fogel, G.R., & Luna, R.J. *J. Colloid Interface Sci. 65*(3), 1978, 533–547.

39. Suyama, Y., Mizobe, T., & Kato, A. *Ceramurgia Intern. 3*(4), 1977, 141–146.

40. Nishikawa, T., Ootsuki, A., Ooi, H., & Yano, M. *J. Ceram. Soc. Jpn. 101*(11), 1993, 1229–1233.

41. Funaki, K., Saeki, Y., & Nakai, T. *Denki Kagaku 40*(1), 1972, 73–77.

42. Okuyama, K., Kousaka, Y., Warren, D.R., Flagan, R.C., & Seinfeld, J.H. *Aerosol Sci. Technol. 6*(1), 1987, 15–27.

43. Biswas, P., Li, X., & Pratsinis, S.E. *J. Appl. Phys. 65*(6), 1989, 2445–2450.

44. Takeuchi, H., Nagano, M., Suyama, Y., & Kato, A. *Yogyo-Kyokai-Shi 83*(1), 1975, 1975.

45. Watari, T., Nakayoshi, K., & Kato, A. *The Chemical Society of Japan 6*, 1984, 1075–1076.

46. Watari, T., Nakayoshi, K., & Kato, A. *The Chemical Society of Japan 4*, 1985, 790–791.

47. Adachi, M., Okuyama, K., Moon, S., Tohge, N., & Kousaka, Y. *J. Mater. Sci. 24*(6), 1989, 2275–2280.

48. Kato, A. & Matuya, S. *Denki Kagaku 45*(5), 1977, 285–290.

49. Okuyama, K., Shimada, M., Adachi, M., & Tohge, N. *J. Aerosol Sci. 24*(3), 1993, 357–366.

50. Won, C.-W., Chun, B.-S., & Sohn, H.Y. *J. Mater. Res. 8*(10), 1993, 2702–2708.

51. Weimer, A.W., Moore, W.G., Roach, R.P., Haney, C.N., & Rafaniello, W. *AIChE J. 37*, 1991, 759–768.

52. Yamamoto, T., Adachi, M., Kawabata, K., Kimura, K., & Hahn, H.W. *Appl. Phys. Lett 63*(22), 1993, 3020–3022.

53. Murakawa, N., Nakajima, M., Maruyama, K., Isogaya, K., & Yoshida, K. Synthesis of an easily sinterable $\beta$-SiC powder. In Vincenzini, P. (Ed.), *High Tech Ceramics*. Amsterdam: Elsevier Science Publishers, 1987, 3080.

54. Carmer, C.S. & Frenklach, M. *Appl. Phys. Lett. 54*(15), 1989, 1430–1432.

55. Chen, L., Goto, T., & Hirai, T. *J. Mater. Sci. 28*(20), 1993, 5543–5547.

56. Kato, A., Hojo, J., & Okabe, Y. *Memoirs of Faculty of Engineering, Kyushu University 41*(4), 1981, 319–334.

57. Dekker, J.P., van der Put, P.J., Veringa, H.J., & Schoonman, J. *J. Mater. Chem. 4*, 1994, 689–694.

58. Itatani, K., Sano, K., Howell, F.S., Kishioka, A., & Kinoshita, M. *J. Mater. Sci 28*, 1993, 1631–1638.
59. Adjaottor, A.A. & Griffin, G.L. Aerosol reactor design for aluminum nitride powder synthesis. In Messing, G.L., Hausner, H., & Hirano, S.-C. (Eds.), *Ceramic Powder Science III*. Columbus, OH: American Ceramic Society, 1990, 299–304.
60. Gomez-Aleixandre, C., Albella, J.M., & Martinez-Duart, J.M. *J. Mater. Res.* *7*(10), 1992, 2864–2868.
61. Jacquemijns, E.J., van der Put, P.J., & Schoonman, J. *High Temp.–High Pressures* *20*(1), 1988, 31–34.
62. Prochazka, S. & Greskovich, C. *Ceramic Bulletin 57*(6), 1990, 579–586.
63. Shen, W.M. & Chang, C.F. *Adv. in Ceramics 21*, 1987, 192–201.
64. Dericotte, D., Powell, Q.H., Gurav, A., Kodas, T.T., Hampden-Smith, M.J., Zeng, D., & Wang, L. personal communication.
65. Lamprey, H. & Ripley, R.L. *J. Electrochem. Soc. 109*(8), 1962, 713–715.
66. Pluym, T.C., Kodas, T.T., Wang, L.M., Wiedensohler, A., Hansson, H.C., Maximov, I., & Samuelson, L. Gas-phase generation of nanometer–sized metal particles for fabrication of zero–dimensional quantum structures. *Mat. Res. Soc. Symp. Proc.*, 1993, 300, 465.
67. Lay, J.R. & Iya, S.K. Silane pyrolysis in a free–space reactor. IEEE Photovoltaic Specialists Conf., Kissimmee, FL, 1981.
68. Slootman, F. & Parent, J.-C. *J. Aerosol Sci. 25*(1), 1994, 15–21.
69. Alam, M.K. & Flagan, R.C. *Aerosol Sci. Technol. 5*, 1986, 237–248.
70. Gregory, O.J., Lee, S.-B., & Flagan, R.C. *Comm. Am. Ceram. Soc. 70*(3), 1987, C52–C55.
71. Nguyen, H.V. & Flagan, R.C. *Langmuir 7*(8), 1991, 1807–1814.
72. Wu, J.J. & Flagan, R.C. *J. Appl. Phys. 61*(4), 1987, 1365–1371.
73. Wu, J.J., Flagan, R.C., & Gregory, O.J. *Appl. Phys. Lett. 49*(2), 1986, 82–84.
74. Wu, J.J., Nguyen, H.V., & Flagan, R.C. *Langmuir 3*(2), 1987, 266–271.
75. Wu, J.J., Nguyen, H.V., & Flagan, R.C. Synthesis of fine powders in aerosol reactors. *2nd Int. Aerosol Conf.* Berlin: 1986, 910–913.
76. Wu, J.J., Nguyen, H.V., Flagan, R.C., Okuyama, K., & Kousaka, Y. *AIChE J.* *34*(8), 1988, 1249–1256.
77. Onischuk, A.A., Strunin, V.P., Ushakov, M.A., & Panifilov, V.N. *Chem. Phys. Reports 13*(5), 1994, 931–940.
78. Flagan, R.C., Atwater, H.A., & Vahala, K.J. *J. Aerosol Sci. 22*(1), 1991, S31–S33.
79. Saunders, W.A., Sercel, P.C., Flagan, R.C., Atwater, H.A., & Vahala, K.J. *Z. Phys. D 26* (*Suppl.*), 1993, 219–221.
80. Sercel, P.C., Saunders, W.A., Atwater, H.A., Vahala, K.J., & Flagan, R.C. *Appl. Phys. Lett. 61*(6), 1992, 696–698.
81. Powell, Q.H., Garvey, J., Gurav, A., Hampden-Smith, M.J., & Kodas, T.T. *J. Aerosol Sci. 23*, 1992, S249.
82. Nishida, Y., Fujiura, K., Sato, H., Sugawara, S., Kobayashi, K., & Takahashi, S. *J. Appl. Phys. 31*, 1992, L1692–L1694.
83. Frenklach, M., Kematick, R., Huang, D., Howard, W., Spear, K.E., Phelps, A.W., & Koba, R. *J. Appl. Phys. 66*(1), 1989, 395–399.
84. Frenklach, M. *Chem. Phys. Processes Combust.*, 1989, C1–C8.

**TABLE 11.3. Precursors and products in flame reactors**

| Product | | Reactant | Notes | References |
|---|---|---|---|---|
| $Al_2O_3$, | 30–1000 Å | $Al(acac)_3$ | 2100°C | [1] |
| $SiO_2$, | 50–200 Å | $SiCl_4$ | 1800–2200 K | [2, 3] |
| Carbon black, | 0.01 μm | Alkanes | 2100–3000 K | [4] |
| $Al_2O_3$ | 6–200 nm | Chlorides | 2500–3000 K | [5] |
| $Cr_2O_3$ | " | " | " | |
| $Fe_2O_3$ | " | " | " | |
| $GeO_2$ | " | " | " | |
| $SiO_2$ | " | " | " | |
| $SnO_2$ | " | " | " | |
| $TiO_2$ | " | " | " | |
| $V_2O_5$ | " | " | " | |
| $ZrO_2$ | " | " | " | |
| $Al_2O_3$ | 100–200 nm | Chlorides | 1700–3000 K | [6] |
| $Fe_2O_3$ | " | " | " | |
| $GeO_2$ | " | " | " | |
| $SiO_2$ | " | " | " | |
| $TiO_2$ | " | " | " | |
| $V_2O_5$ | " | " | " | |
| $ZrO_2$ | " | " | " | |
| $Al_2O_3$, | 20 nm | $Al(O\text{-}i\text{-}Pr)_3$ | 3100°C | [7] |
| $SiO_2$, primary | 0.01–0.05 μm | hexamethyl disiloxane $(CH_3)_6Si_2O$ | — | [8] |
| $ZrO_2$ | | $ZrCl_4$ | 1600–1800°C | [9] |
| $ZrO_2$, $ZrO_2$–$SiO_2$ $ZrO_2$–$TiO_2$ | 8–10 nm | $ZrCl_4$, $SiCl_4$, $TiCl_4$ | Homogeneous multicomponent materials | [10] |
| $ZrO_2$–$Al_2O_3$, | 50 nm | $AlCl_3 + ZrCl_4$ | 1600–1800°C | [11] |
| $SiO_2$ $TiO_2$ | | $SiH_4$ $TiCl_4$ | 1500–1600 K | [12] |
| $SiO_2$–$Fe_2O_3$, | 5–50 nm | $Fe(Cp)_2$ $SiCl_4$ | — | [13] |
| V–P–O, | 5–50 nm | $VOCl_3$, $PCl_3$ | Counterflow diffusion flame, 2780–2290 K maximum | [14] |
| $Al_2O_3$, $GeO_2$, $P_2O_5$, $SiO_2$, $TiO_2$ $V_2O_5$ | | $Al(CH_3)_3$, $GeCl_4$ $PCl_3$, $SiCl_4$, $TiCl_4$, $VOCl_3$ | Wide variety of materials formed | [14] |
| $TiO_2$, $GeO_2$, $SiO_2$ < 0.5 μm | | $TiCl_4$, $SiH_4$, $SiCl_4$, $GeCl_4$ | Applying electric field increased particle size 3 × – 10 × | [15] |
| $Al_2O_3$–$TiO_2$, $SiO_2$–$GeO_2$, $TiO_2$–$SiO_2$ | 10–100 nm | Metal halides | 2300–2700 K | [16] |

**TABLE 11.3.** (*Continued*)

| Product | Particle Size | Reactant | Notes | References |
|---|---|---|---|---|
| $V_2O_5$–$TiO_2$, $V_2O_5$–$Al_2O_3$ | 40–70 nm | $VOCl_3$, $TiCl_4$ $VOCl_3$, $Al(CH_3)_3$ | 2300–2700 K | [16] |
| $SiO_2$ | | $SiCl_4$ | 1100–1700 K | [17] |
| $TiO_2$ | | $TiCl_4$, $Ti(O\text{-}i\text{-}Pr)_2(acac)_2$ | Ultrasonic droplet formation, droplet delivery into flame | [18] |
| $TiO_2$ | | $TiCl_4$ | 1178–1440 K | [19] |
| $TiO_2$ | | $TiCl_4$ | 100–2200 K | [20] |
| $MoO_3$–$TiO_2$ | | $Mo_2(acac)_4$, $Ti(O\text{-}i\text{-}Pr)_2(acac)_2$ | Ultrasonic droplet formation, droplet delivery into flame | [21] |
| $WO_3$–$TiO_2$ | | $W(OEt)_5$ $Ti(OEt)_4$ | Ultrasonic droplet formation, droplet delivery into flame | [22] |
| $SiO_2$ | | $Me_3SiCl$ | Na and K additives reduced particle size | [23] |
| MgO, ZnO | | $Mg(CH_3CO_2)_2$ $Zn(NO_3)_2$ | Droplet delivery into flame | [24] |
| $SiO_2$ | | $SiH_4$, $Si(CH_3)_4$, $(Me_3Si)_2O$ | Mechanism and particle characteristics influenced by precursor properties | [25] |
| $SiO_2$ | 0.01–0.1 $\mu$m | $SiCl_4$, $SiH_4$ | Studied overall characteristics | [26] |
| $TiB_2$ | 15 nm | $TiCl_4$, $BCl_3$, Na vapor | NaCl combustion product removed by washing | [27] |

## References to Table 11.3


1. Sokolowski, M., Sokolowska, A., Michalski, A., & Gokieli, B. *J. Aerosol Sci. 8*, 1977, 219–230.
2. Ulrich, G.D. *Combustion Sci. Technol 4*, 1971, 47–57.
3. Ulrich, G.D., Milnes, B.A., & Subramanian, N.S. *Combust. Sci. & Tech. 14*, 1976, 243–249.
4. Dannenberg, E.M. *J. IRI*, October 1971, 190–195.
5. Juillet, F., Lecomte, F., Mozzanega, H., Teichner, S.J., Thevenet, A., & Vergnon, P. Inorganic oxide aerosols of controlled submicronic dimensions. 1973, 57–62.
6. Formenti, M., Juillet, F., Meriaudeau, P., Teichner, S.J., & Vergnon, P. Preparation in a hydrogen–oxygen flame of ultrafine metal oxide particles. In Hidy, G.M. (Ed.), *Aerosols and Atmospheric Chemistry*. New York: Academic Press, 1972, 45–55.

7. Buchanan, R.K., Grose, P.J., & Gani, M.S.J. Ultrafine mixed oxide powders prepared in an oxy–acetylene flame. *Proc. 9th Aust. Ceram. Conf.* Sydney, Australia: Dept. Ceramic Engineering, Univ. New South Wales, 1980, 102–104.
8. Hurd, A.J. & Flower, W.L. *J. Colloid Interface Sci. 122*(1), 1988, 178–192.
9. Kriechbaum, G.W. & Kleinschmit, P. *Ang. Chemie, Advanced Materials 10*(10), 1989, 330–337.
10. Liu, A.T. & Kleinschmit, P. Production of fumed oxides by flame hydrolysis. *British Ceramic Proceedings*, 1986, 38, 1–10.
11. Hori, S., Yoshimura, M., Somiya, S., Kurita, R., & Kaji, H. *J. Mater. Sci. Letters 4*, 1985, 413–416.
12. Lindackers, D., Strecker, M.G.D., & Roth, P. *J. Aerosol Sci. 25*(1), 1994, S293–S294.
13. Fotou, G.P., Scott, S.J., & Pratsinis, S.E. *J. Aerosol Sci. 25*(1), 1994, S65–S66.
14. Katz, J.L. & Miquel, P.F. *NanoStruct. Mater. 4*(5), 1994, 551–557.
15. Katz, J.L. & Hung, C.-H. Initial studies of electric field effects on ceramic powder formation in flames, *Twenty–third Symposium (Intern.) on Combustion*, Pittsburgh, PA: The Combustion Institute, 1990, 1733–1738.
16. Hung, C.–J. & Katz, J.L. *J. Mater. Res. 7*(7), 1992, 1861–175. particle formation using the counter-flow diffusion flame burner. *Combust. Plasma Synt. High Temp. Mater.*, 1990, 373–390.
18. Lee, W., Gao, Y.-M., Dwight, K., & Wold, A. *Mat. Res. Bull. 27*, 1992, 685–692.
19. Fotou, G.P., Vemury, S., & Pratsinis, S.E. *Chem. Eng. Sci. 49*(24B), 1994, 4939–4948.
20. Rulison, A.J., Miquel, P.F., & Katz, J.L. *J. Mater. Res. 11*(12), 1996, 3083–3089.
21. Lee, W., Do, Y.R., Dwight, K., & Wold, A. *Mat. Res. Bull. 28*, 1993, 1127–1134.
22. Do, Y.R., Lee, W., Dwight, K., & Wold, A. *J. Solid State Chemistry 108*, 1994, 198–201.
23. Wu, M.K., Windeler, R.S., Steiner, C.K.R., Börs, T., & Friedlander, S.K. *Aerosol Sci. Technol. 19*, 1993, 527–548.
24. Matsoukas, T. & Friedlander, S.K. *J. Colloid Interface Sci. 146*(2), 1991, 495–506.
25. Zachariah, M.R., Chin, D., Katz, J.L., & Semerjian, H.G. *Chem. Phys. Processes Combust. 38*, 1987, 38-1–38-4.
26. Chung, S.L., Lin, H.D., Kang, C.F., & Tsai, M.S. Synthesis of silica powder using a counterflow diffusion flame. *Ceramic Powder Processing Science*. Berchtesgaden, Germany: 1988, 289–304. Ceramic powder processing science: Proceedings of the second international conference, Berchtesgaden (Bavaria) FRG, October 12–14, 1988 edited by H. Hausner, G.L. Messing, S. Hirano.
27. Axelbaum, R.L., DeFaux, D.P., Frey, C.A., Kelton, K.F., Lawton, S.A., Rosen, L.J., & Sastry, S.M.L. *J. Mater. Res. 11*(4), 1996, 948–954.

**TABLE 11.4. Precursors and products in photothermal laser reactors**

| Product | Particle Size | Reactant | Notes | References |
|---|---|---|---|---|
| $Si_3N_4$, | 17 nm | $SiH_4$, $NH_3$ | $CO_2$ laser, 150 W | [1, 2] |
| SiC, | | $SiH_4$, $C_2H_4$ | | |
| Si | 30–100 nm | $SiH_4$ | | |
| Si, | 22–65 nm | $SiH_4$ | $CO_2$ laser | [3] |
| SiC | 20–35 nm | $SiH_4$, $C_2H_4$ | | |
| Si–C–N | | $SiH_4$, $Me_2NH$, $NH_3$ | $CO_2$ laser, 50 W | [4] |
| Si | 70–300 nm | $SiH_4$ | $T_m$ < 1455°C | [5] |
| Si, | 19–250 nm | $SiH_4$ | $T_m$ < 1955°C | [6] |
| $Si_3N_4$, | 7.5–50 nm | $SiH_4$, $NH_3$ | | |
| SiC | 20–70 nm | $SiH_4$, $C_2H_4$ | | |
| $Si_3N_4$ | 10–20 nm | $SiH_4$, $NH_3$ | — | [7] |
| Si–C–N | 100 nm | $(Me_3Si)_2NH$, $(Me_3Si)_2O$ | $CO_2$ laser | [8] |
| $Si_3N_4$ | 15–55 nm | $SiH_4$, $NH_3$ | $CO_2$ laser, 1500–1700°C | [9] |
| Si(C, N) | 20–200 nm | $[CH_3SiHNH]_n$ | $CO_2$ laser | [10] |
| $Si_3N_4$ | 20–90 nm | $SiH_4$, $NH_3$ | Studied crystallization of $Si_3N_4$ during sintering | [11] |
| Si | 10–80 nm | $SiH_4$ | Studied crystal structure of particles | [12] |
| SiC, B–doped SiC | 20–50 nm | $SiH_4$, $C_2H_4$, $B_2H_6$ | $T_r$ = 1100–1550°C | [13] |
| $Si_3N_4$, | 20–40 nm | $SiH_2Cl_2$, $NH_3$, | Dichlorosilane needed sensitizer due to low absorption at 10.6 $\mu$m | [14] |
| Si | 50 nm | $SiCl_4$ | | |
| $Si_3N_4$ | 15–110 nm | $SiH_2Cl_2$, $NH_3$ | Reactants mixed at high temperatures to minimize low-temperature reactions | [15] |
| SiC | | $SiH_4$, hydrocarbons | Examined sintering behavior | [16] |
| $Fe_3C$ | 3–13 nm | $Fe(CO)_5$, $C_2H_4$ | Examined catalytic activity | [17] |
| SiC, Si | | $SiH_4$, $C_2H_4$ | Measured maximum temperature of 2000°C | [18] |
| SiC:$Si_3N_4$ composites | | $SiH_4$, $C_2H_4$, $NH_3$ | Characterized by solid–state nuclear magnetic resonance | [19] |
| SiC | | $SiH_4$, $C_2H_2$ | $CO_2$, 50 W | [20] |
| $TiSi_2$ | 50 nm | $SiH_4$, $TiCl_4$ | $CO_2$ laser | [21] |

**TABLE 11.4.** (*Continued*)

| Product | Particle Size | Reactant | Notes | References |
|---|---|---|---|---|
| B, $TiB_2$ | 35–60 nm<br>8–100 nm | $BCl_3$, $H_2$<br>$TiCl_4$, $B_2H_6$ | $CO_2$ laser, 150 W | [22] |
| $Ti_{1-x}V_xO_2$ | 5–20 nm | $Ti(OC_3H_7)_4$, $VO(OC_3H_7)_3$ | $CO_2$ laser, 600 W | [23] |
| $B_4C$ | 34 nm | $BCl_3$, $H_2$, $CH_4$ | $CO_2$ laser $T_r = 1500°C$ | [24] |
| B | | $BCl_3$, $H_2$ | Studied mechanism of B formation | [25] |
| $Al_2O_3$ | 15–20 nm | $Al(CH_3)_3$ | $CO_2$ laser | [26] |
| $Fe_2O_3$ | 6–9 nm | $Fe(CO)_5$ | $CO_2$ laser | [27] |
| $TiO_2$ | | Ti alkoxides | Studied anatase–rutile transformation | [28] |

## References to Table 11.4


1. Cannon, W.R., Danforth, S.C., Flint, J.H., Haggerty, J.S., & Marra, R.A. *J. Am. Ceram. Soc. 65*(7), 1982, 324–330.
2. Cannon, W.R., Danforth, S.C., Haggerty, J.S., & Marra, R.A. *J. Am. Ceram. Soc. 65*(7), 1982, 330–335.
3. Cauchetier, M., Croix, O., Luce, M., Michon, M., Paris, J., & Tistchenko, S. Laser synthesis of ultrafine powders. In Vincenzini, P. (Ed.) *High Tech Ceramics*. Amsterdam: Elsevier, 1987, 545–553.
4. Borsella, E., Botti, S., Fantoni, R., Alexandrescu, R., Morjan, I., Popescu, C., Dikonimos-Makris, T., Giorgi, R., & Enzo, S. *J. Mater. Res.* 7(8), 1992, 2257–2268.
5. Flint, J.H. & Haggerty, J.S. *Aerosol Sci. Technol. 13*, 1990, 72–84.
6. Flint, J.H. & Haggerty, J.S. Ceramic powders from laser driven reactions. *Proc. SPIE Applications of Lasers to Industrial Chemistry, 458*, 1984, 108–113.
7. Zheng, Z.J., Zuo, D.L., & Li, D.H. *J. Vac. Sci. Technol. B 12*(3), 1994, 1832–1834.
8. Rice, G.W. & Woodin, R.L. Laser synthesis of powders from large molecules. *Proc. SPIE, Applications of Lasers to Industrial Chemistry, 458*, 1984, 98–107.
9. Symons, W. & Danforth, S.C. *Advances in Ceramics 21*, 1987, 249–256.
10. Gonsalves, K.E., Strutt, P.R., Xiao, T.D., & Klemens, P.G. *J. Mater. Science 27*, 1992, 3231–3238.
11. Sawhill, H.T. & Haggerty, J.S. *Comm. Am. Ceram. Soc. 65*(8), 1982, C131–C132.
12. Marra, R.A. & Haggerty, J.S. *J. Am. Ceram. Soc. 70*(1), 1987, C15–C17.
13. Suyama, Y., Marra, R.M., Haggerty, J.S., & Bowen, H.K. *Am. Ceram. Soc. Bull. 64*(10), 1985, 1356–1359.
14. Bauer, R.A., Smulders, R., Becht, J.G.M., van der Put, P.J., & Schoonman, J. *J. Am. Ceram. Soc. 72*(7), 1989, 1301–1304.
15. Bauer, R. *Laser Chemical Vapor Precipitation of Ceramic Powders*, Ph.D. dissertation, Delft University, 1991.

16. Luce, M., Croix, O., Gounot, M., Bergez, P., & Cauchetier, M. Synthesis and processing of laser formed silicon carbide powders. *Ceramic Powder Processing Science*. Berchtesgaden, Germany: 1988, 305–311. Ceramic powder processing science: Proceedings of the second international conference, Berchtesgaden (Bavaria) FRG, October 12–14, 1988 edited by H. Hausner, G.L. Messing, S. Hirano.
17. Eklund, P.C., Stencel, J.M., Bi, X.-X., Keogh, R.A., & Derbyshire, F.J. *Am. Chem. Soc.* *36*(2), 1991, 551–559.
18. Flint, J.H., Marra, R.A., & Haggert, J.S. *Aerosol Sci. Technol.* *5*, 1986, 249–260.
19. Suzuki, M., Hasegawa, Y., Aizawa, M., Nakata, Y., Okutani, T., & Uosaki, T. *J. Am. Ceram. Soc.* *78*(1), 1995, 83–89.
20. Fantoni, R., Borsella, E., Piccirillo, S., Ceccato, R., & Enzo, S. *J. Mater. Res.* *5*(1), 1989, 143–150.
21. Gupta, A., Beeson, K.W., Donlan, J.P., & West, G.A. *J. Appl. Phys.* *61*(3), 1987, 1162–1167.
22. Casey, J.D. & Haggerty, J.S. *J. Mater. Sci.* *22*, 1987, 737–744.
23. Musci, M., Notaro, M., Curcio, F., & Casale, C. *J. Mater. Res.* *7*(10), 1992, 2846–2852.
24. Knudsen, A.K. *Advances in Ceramics*, *21*, 1987, 237–247.
25. Sunouchi, K., Komiyama, H., & Takeuchi, K. *Reza Kenkyu* *18*(9), 1990, 761–763.
26. Borsella, E., Botti, S., Giorgi, R., Martelli, S., Turtu, S., & Zappa, G. *Appl. Phys. Lett.* *63*(10), 1993, 1345–1347.
27. Majima, T., Miyahara, T., Haneda, K., & Takami, M. *Jpn. J. Appl. Phys. Lett.* *33*, 1994, L223–226.
28. Depero, L.E., Bonzi, P., & Zocchi, M. *J. Mater. Res.* *8*(10), 1993, 2709–2715.

**TABLE 11.5. Precursors and products in photochemical laser reactors**

| Product | Reactant | Comments | References |
|---|---|---|---|
| $UF_5$ | $UF_6$ | UV dye laser | [1] |
| CsD | $Cs + D_2$ | CW dye laser | [2] |
| NaH | $Na + H_2$ | $Ar^+$ laser | [3] |
| $SiO_2$ | $SiCl_4$, $O_2$ | ArF pulsed laser | [4] |

**References Table 11.5**

1. Borsella, E., Catoni, F., & Freddi, G. *J. Chem. Phys.* *73*(1), 1980, 316–321.
2. Tam, A.C. & Happer, W. *Chem. Physics Lett.* *49*(2), 1977, 320–323.
3. Yabuzaki, T., Sato, T., & Ogawa, T. *J. Chem. Phys.* *73*(6), 1980, 2780–2783.
4. Tezaki, A., Morita, K., & Matsui, H. *J. Phys. Chem.* *98*(41), 1994, 10529–10534.

**TABLE 11.6. Precursors and products in thermal plasma reactors**

| Product | Particle Size | Reactant | Comments | References |
|---|---|---|---|---|
| Fe | 10 nm | Fe powder | Powder feed | [1] |
| Fe, Ag | < 100 nm | $Fe_2O_3$, $AgNO_3$ | Solution feed, suspension feed | [2] |
| Fe, Co, Cu, Ag | 10 nm | Metal vapor source | Vapor feed | [3] |
| Fe, $Fe_2O_3$ | 10 nm | $Fe(Cp)_2$ | Gaseous feed, both before and after flame | [4] |
| Fe | 10–60 nm | Fe powder | — | [5] |
| Fe | 20–70 nm | Fe powder | Compared to model | [6] |
| B | 50–150 nm | $BCl_3$ | Gaseous feed | [7] |
| Si + polymers + gases | | $SiCl_4$ | Gaseous feed | [8] |
| Si | 30 nm | Si blocks | Tungsten inert-gas welding machine | [9] |
| Ag, Cr, SiC | 10–40 nm | Metal powders | Metals vaporized from crucible and then sent into plasma | [10] |
| TiN | 5–150 nm | Ti powder, $N_2$ | Powder feed | [11] |
| $Si_3N_4$ | 10–30 nm | $SiCl_4$, $NH_3$ | Gaseous feed | [12] |
| $Si_3N_4$ | 5–15 nm | $SiCl_4$, $NH_3$ | Gaseous feed | [12] |
| AlN | 60 nm | Al powder, $N_2$ | Good sinterability | [13] |
| AlN | 50 nm | Al powder, $N_2$ | Powder feed | [14] |
| AlN | 30 nm | Al anode, $N_2$ | Anode was source of Al | [15] |
| AlN | 60 nm | Al powder, $N_2$ | Powder feed | [16] |
| AlN | 5–25 nm | $Al(CH_3)_3$, $NH_3$ | Gaseous feed | [17] |
| $Si_3N_4$, AlN, SiC, WC, TiC | 20–40 nm | Metals, $NH_3$ or $CH_4$ | Powder feed | [18] |
| SiC, $Si_3N_4$ | 0.01–0.02 $\mu$m | $SiCl_4$, $NH_3$ $Si(OC_2H_5)_4$, $(CH_3)_2Si(OC_2H_5)_2$, $(CH_3)_2Si(OCH_3)_2$ | Downstream injection of gaseous reactants | [19] |
| SiC | 10–20 nm | Rice hulls with $SiO_2$ | Rice hulls: inexpensive source of carbon; XPS and FTIR analysis | [20] |
| SiC | 0.1 $\mu$m | $SiCl_4$, $CH_4$, $H_2$ | 200g/hour, studied effect of process parameters on powder characteristics | [21] |
| SiC | 10–30 nm | $SiH_4$, $CH_4$ | Gaseous feed | [22] |
| TaC | 5–100 nm | TaC powder, $CH_4$ | Powder feed | [23] |
| SiC | 10–20 nm | $SiCl_4$, $CH_4$, $H_2$ | Gaseous feed | [24] |

**TABLE 11.6.** *Continued)*

| Product | Particle Size | Reactant | Comments | References |
|---|---|---|---|---|
| $Al_2O_3$–$TiO_2$ | 30–200 nm | $Al_2Br_6$, $TiCl_4$, $O_2$ | Vapor feeds from bubblers, homogeneous product | [25] |
| $TiO_2$ | | $TiCl_4$, $O_2$ | Gaseous feed | [26] |
| $TiO_2$ | | $TiCl_4$, $O_2$ | Gaseous feed, measured kinetics | [27] |
| Oxides of Si, Al, Fe, Th, Mn, Cb, Ni, Y, U, Mo, Zr, Mg, W. Metals: C, Al, Li, Ni, Fe, W, Mo. Carbides: ThC, TiC, $B_4C$, UC, TaC, SiC. Several multi-component materials | | Elements | Powder feed or anode consumption | [28] |
| NiO, $CeO_2$, $Al_2O_3$, bauxite | $< 1\ \mu m$ | Powdered oxides | Powder feed | [29] |
| Ag–Pd, Ag–Cu, Fe–Ni | | Metals | Molten metal | [30] |
| $SiO_2$ | | $Si(OC_2H_5)_4$ | AlN substrate particles 40 $\mu m$, plasma CVD reactor | [31] |

## References to Table 11.6


1. Yoshida, T., Kawasaki, A., & Akashi, K. *Commun.–Symp. Int. Chim. Plasmas 3*, 1977, 1–8.
2. Tarroni, G., Prodi, V., Melandri, C., Bompane, G.F., DeZaiacomo, T., & Formignani, M. *J. Aerosol Sci. 6*, 1975, 305–310.
3. Girardin, D. & Maurer, M. *Mat. Res. Bull. 25*(1), 1990, 119–127.
4. Chou, C.H. & Phillips, J. *J. Mater. Res. 7*(8), 1992, 2107–2118.
5. Yoshida, T. & Akashi, K. *Trans. Japan Inst. Metals 22*(6), 1981, 371–378.
6. Girshick, S.L., Chiu, C.–P., Muno, R., Wu, C.Y., Yang, L., Singh, S.K., & McMurry, P.H. *J. Aerosol Sci. 24*(3), 1993, 367–382.
7. Diana, M., Luponio, C., & Russo, G. *Revue de Physique Applique 12*(9), 1977, 1237–1242.
8. Coudert, J.F., Bourdin, E., & Fauchais, P. *Plasma Chem. & Plasma Processing 2*(3), 1982, 399–419.
9. Tanaka, K., Ishizaki, K., Yumoto, S., Egashira, T., & Uda, M. *J. Mater. Sci. 22*, 1987, 2192–2198.

10. Oya, H., Ichihashi, T., & Wada, N. *Japanese J. Appl. Phys. Part 1 21*(3), 1982, 554–555.
11. Yoshida, T., Kawasaki, A., Nakagawa, K., & Akashi, K. *J. Mater. Sci. 14*, 1979, 1624–1630.
12. Perugini, G. Silicon nitride production: II. On the formation of $Si_3N_4$ submicron powders from $SiCl_4$ and $NH_3$ in an arc-plasma furnace. *Proc. Int. Symp. Plasma Chem.*, Novara, Italy, *2*, 1979, 779–785.
13. Baba, K., Shohata, N., & Yonezawa, M. Preparation and properties of ultrafine AlN powder by rf plasma. *Ceramic Powder Processing Science*. Berchtesgaden, Germany: 1988, 283–288. Ceramic powder processing science: Proceedings of the second international conference, Berchtesgaden (Bavaria) FRG, October 12–14, 1988 edited by H. Hausner, G.L. Messing, S. Hirano.
14. Vissokov, G.P. & Brakalov, L.B. *J. Mater. Sci. 18*, 1983, 2011–2016.
15. Etemadi, K. *Plasma Chem. and Plasma Processing 11*(1), 1991, 41–56.
16. Baba, K., Shohata, N., & Yonezawa, M. *Appl. Phys. Lett. 54*(23), 1989, 2309–2311.
17. Kim, K.-H., Ho, C.-H., Doerr, H., Deshpandey, C., & Bunshah, R.F. *J. Mater. Sci. 27*, 1992, 2580–2588.
18. Ishizaki, K., Egashira, T., Tanaka, K., & Celis, P.B. *J. Mater. Sci. 24*, 1989, 3553–3559.
19. Kuibira, A., Mitsui, A., Hojo, J., & Kato, A. *J. Ceram. Soc. Jpn. 95*, 1987, 89–91.
20. Stachowicz, L., Singh, S.K., Pfender, E., & Girshick, S.L. *Plasma Chem. Plasma Process 13*(3), 1993, 447–461.
21. Allaire, F., Parent, L., & Dallaire, S. *J. Mater. Sci. 26*(15), 1991, 4160–4165.
22. Vogt, G.J., Phillips, D.S., & Taylor, T.N. *Advances in Ceramics 21*, 1987, 203–215.
23. Canteloup, J., Tueta, R., & Braguier, M. Synthesis in an R.F. plasma reactor of silica powders doped with alumina, *4th Int. Symp. Plasma Chem.*, Evry, France, 1979, *2*, 438–443.
24. Hollabaugh, C.M., Hull, D.E., Newkirk, L.R., & Petrovic, J.J. *J. Mater. Sci. 18*, 1983, 3190–3194.
25. Gani, M.S.J. & McPherson, R. *J. Mater. Sci. 15*, 1980, 1915–1925.
26. Musa, G., Ionescu, G., Lungu, C.P., Mustata, I., Popa, S., Chichirez, I., & Seicaru, O. *Contrib. Plasma Phys. 27*(3), 1987, 181–186.
27. Mahawili, I. & Weinberg, F.J. *Plasma Cheical Processing 75*(186), 1979, 11–24.
28. Holmgren, J.D., Gibson, J.O., & Sheer, C. *J. Electrochem. Soc. 111*(3), 1964, 362–369.
29. Pickles, C.A. & McLean, A. *Ceram. Bull. 62*(9), 1983, 1004–1009.
30. Uda, M. *NanoStruct. Mater. 1*, 1992, 101–106.
31. Tsugeki, K., Yan, S., Maeda, H., Kusakabe, K., & Morooka, S. *J. Mater. Sci. Lett. 13*(1), 1994, 43–45.

**TABLE 11.7. Precursors and products in spray inductively coupled plasma reactors**

| Product | Particle Size | Reactant | Comments | References |
|---|---|---|---|---|
| MgO | 0.016–0.044 μm | $Mg(NO_3)_2$ | Ultrasonic atomizer | [1] |
| Oxides of Y, Sm, Zr, Al, Ti, Bi, Cu, Cr, La, Nd, Pr, Ce, Ca, Ni, Co, Mn, Sn, Zn, Pb, Mo, and various mixed metal oxides | 15–35 nm | Metal nitrates, $(NH_4)_6Mo_7O_{21}$, $(NH_4)_2TiO(C_2O_4)_2$ | Ultrasonic atomizer | [2] |
| $ZrO_2$–$Al_2O_3$ | 0.01–0.04 μm | Nitrates | $Al_2O_3$ on surfaces $ZrO_2$ | [3] |
| $ZrO_2$–$SiO_2$ | 10–20 nm | $ZrO(NO_3)_2 \cdot 2H_2O$, silicone oil | $SiO_2$ coating on $ZrO_2$ | [4] |
| $BaPb_{1-x}Bi_xO_3$ | 10–40 nm | Metal nitrates | Multicomponent | [5] |
| MgO, $Al_2O_3$, $Fe_2O_3$, $ZrO_2$, $Al_2O_3$–$ZrO_2$, $SiO_2$–$ZrO_2$ | 2–20 nm | Metal nitrates | Al–Mg solid solutions | [6] |
| M–$PO_4$, M = Al, Mg, Ca, Zr | not available | $Al(NO_3)_3$, $Mg(NO_3)_2$, $Ca(NO_3)_2$ $ZrO(NO_3)_2$, $H_3PO_4$ | Ultrasonic atomizer, powders amorphous as prepared | [7] |
| Y–Ba–Cu–O | not available | Metal nitrates | Counterflow injection of solution into plasma | [8] |

## References to Table 11.7


1. Kagawa, M., Kikuchi, M., Ohino, R., & Nagae, T., Preparation of ultrafine MgO by the spray–ICP technique. *Comm. Am. Ceram. Soc.*, 1981, C7–8.
2. Suzuki, M., Kagawa, M., Syono, Y., & Hirai, T. *Ceramic Powder Science IV*, *Ceramic Transactions 22*, 1991, 147–152.
3. Kagawa, M., Imamura, Y., Usui, S., & Syono, Y. *J. Mater. Sci. Lett. 3*, 1984, 699–702.
4. Ono, T., Kagawa, M., & Syono, Y. *J. Mater. Sci. 20*, 1985, 2483–2487.
5. Ono, T., Kagawa, M., Syono, Y., Ikebe, M., & Muto, Y. *Plasma Chemistry and Plasma Processing* 7(2), 1987, 201–209.
6. Pollinger, J.P. & Messing, G.L. *Advances in Ceramics 21*, 1987, 217–227.
7. Kagawa, M., Ohomori, M., Suzuki, M., Syono, Y., Yamada, A., & Watanabe, T. *J. Mater. Sci. Lett. 8*, 1989, 1163–1164.
8. Kong, P.C. & Pfender, E. Synthesis of ceramic powders in a thermal DC plasma jet by injection of liquid precursors. *Ceramic Powder Processing Science*. Berchtesgaden, Germany: 1988, 237–243. Ceramic powder processing science: Proceedings of the second international conference, Berchtesgaden (Bavaria) FRG, October 12–14, 1988 edited by H. Hausner, G.L. Messing, S. Hirano.

**TABLE 11.8. Precursors and products in fluidized-bed systems for coating of particles**

| Product | Reactant | Comments | References |
|---|---|---|---|
| Carbon | $CH_4$ | $Al_2O_3$, Si substrate particles, 150 μm, microwave plasma fluidized bed | [1] |
| Pyrolytic carbon, SiC | $CH_4$, $C_3H_8$, $C_3H_6$, $C_2H_2$, $CH_3SiCl_3$ | Substrate particles not discussed, fluidized bed | [2] |
| Surface-modified carbon | $O_2$ | Carbon fibers, 13 μm diameter by 370 μm long, plasma fluidized bed | [3] |
| Diamond | $CH_4$ | Si, $SiO_2$ substrate particles, 250–700 μm, plasma fluidized bed | [4] |
| Diamond | $CH_4$ | Si, $SiO_2$, SiC substrate particles, 150–4000 μm, microwave plasma fluidized bed | [5] |
| Si | $SiH_4$ | $Si_3N_4$ substrate particles, 5–20 μm, fluidized bed | [6] |
| TiC, TiN, 15–200 nm | — | $Al_2O_3$ substrate particles, 30–60 μm, fluidized bed | [7] |
| TiN ($\alpha$-$Al_2O_3$–TiN composite) | $TiCl_4$, $NH_3$ | $\alpha$-$Al_2O_3$ substrate particles, agglomerates of 200 μm, fluidized bed | [8] |
| TiN | $N_2$ for nitriding Ti | Ti substrate particles, 46–74 μm, plasma fluidized bed | [9] |
| TiN | $TiCl_4$, $NH_3$ | $Si_3N_4$ substrate particles, 20–400 nm, fluidized bed, agglomerated particles | [10] |
| AlN | $NH_3$ | Al substrate particles, 15 μm, transported fluidized bed | [11] |

**TABLE 11.8.** (*Continued*)

| | | | |
|---|---|---|---|
| AlN | $AlCl_3$, $NH_3$ | SiC substrate particles, 0.15 μm, floating fluidized bed | [12] |
| Al, $Al_2O_3$, Ti, TiN | TIBA, $TiCl_4$ | Mica, Ni substrate particles, 1–5 μm, fluidized bed | [13] |
| TiC | $TiCl_4$ | $Al_2O_3$–FeTi substrate particles, 175 μm, fluidized bed | [14] |
| $SiO_2$ | $Si(OC_2H_5)_4$ | MgO substrate particles, 16–90 nm, fluidized bed | [15] |

## References to Table 11.8


1. Matsukata, M., Suzuki, K., Ueyama, K., & Kojima, T. *Int. J. Multiphase Flow 20*(4), 1994, 763–773.
2. Kaae, J.L. *Mat. Res. Soc. Symp. Proc. 372*, 1995, 139–150.
3. Kawamura, H., Okubo, T., Kusakabe, K., & Morooka, S. *J. Mater. Sci. Lett. 9*, 1990, 1033–1035.
4. Shin, H.S. & Goodwin, D.G. *Mater. Lett. 19*(3–4), 1994, 119–122.
5. Takarada, T., Tamura, K., Takezawa, H., Nakagawa, N., & Kato, K. *J. Mater. Sci. 28*, 1993, 1545–1550.
6. Kojima, T., Wakatsuki, R., Matsukata, M., Ozawa, E., & Friedt, J.M. *Kagaku Kogaku Ronbunshu 18*(3), 1992, 274–280.
7. Jouanny–Tresy, C., Vardavoulias, M., & Jeandin, M. *Materials and Manufacturing Processes 9*(5), 1994, 901–920.
8. Tsugeki, K., Kato, T., Koyanagi, Y., Kusakabe, K., & Morooka, S. *J. Mater. Sci. 28*(12), 1993, 3168–3172.
9. Okubo, T., Kawamura, H., Kusakabe, K., & Morooka, S. *J. Am. Ceram. Soc. 73*(5), 1990, 1150–1152.
10. Morooka, S., Kusakabe, K., & Kobata, A. Modification of submicron particles by chemical vapor deposition in fluidized bed. *Fluidization VI*, 1989, 359–366.
11. Kimura, I., Ichiya, K., Ishii, M., Hotta, N., & Kitamura, T. *J. Mater. Sci. Lett. 8*, 1989, 303–304.
12. Kimura, I., Hotta, N., Niwano, J.–I., & Tanaka, M. *Powder Technology 68*(2), 1991, 153–158.
13. Wood, B.J., Sanjurjo, A., Tong, G.T., & Swider, S.E. *Surface and Coatings Technology 49*(1–3), 1991, 228–2332.
14. Kinkel, S., Angelopoulos, G.N., & Dahl, W. *Surface Coatings Technol. 64*, 1994, 119–125.
15. Toda, Y. & Kato, A. *Ceramics International 15*, 1989, 161–166.

**TABLE 11.9. Precursors and products in fluidized-bed systems using agglomerated fine particles**

| Product | Reactant | Comments | References |
|---|---|---|---|
| Si | $SiH_4$ | 200–500 μm seed Si particles, studied formation of fines | [1] |
| Si | $SiH_4$ | 350 μm seed Si particles, grew to 550 μm, observed formation of fines | [2] |
| SiC, $TiO_2$, Ni agglomerates, 80–150 μm | SiC, $TiO_2$, Ni particles | 0.02–0.14 μm particles, agglomerated and fluidized | [3] |
| WC–Co agglomerates | WC–Co | 0.5–2 μm particles, agglomerated and fluidized | [4] |
| SiC | SiC with carbon | 5 μm SiC particles decarbonized in fluidized bed | [5] |
| Ni, $Si_3N_4$, SiC, $Al_2O_3$, $TiO_2$ agglomerates, 70–700 μm | Ni, $Si_3N_4$, SiC, $Al_2O_3$, $TiO_2$ particles | 20–400 nm particles, agglomerated and fluidized | [6] |
| CuO–$Al_2O_3$ aerogel agglomerates | CuO–$Al_2O_3$ aerogels | Particles, agglomerated and fluidized | [7] |

## References to Table 11.9


1. Kojima, T. & Morisawa, O. *J. de Physique IV 1* (Colloque C2), 1991, 475–482.
2. Matsukata, M., Odagiri, T., & Kojima, T. *J. de Physique IV 1* (Colloque C2), 1991, C2-483–C2-490.
3. Kusakabe, K., Kuriyama, T., & Morooka, S. *Powder Technol. 58*, 1989, 125–130.
4. Pacek, A.W. & Nienow, A.W. *Powder Technol. 60*, 1990, 145–158.
5. Kato, K., Takarada, T., Koshinuma, A., Kanazawa, I., & Sugihara, T. Decarbonization of silicon carbide–carbon fine particles mixture in a fluidized bed. In Grace, G.R., Shemilt, L.W., and Bergougnou, M.A. (Eds.), *Fluidization VI*. New York: AIChE, 1989, 351–358.
6. Morooka, S., Kusakabe, K., Kobata, A., & Kato, Y. *J. Chem. Eng. Jpn. 21*(1), 1988, 41–46.
7. Chaouki, J., Chavarie, C., Klvana, D., & Pajonk, G. *Powder Technol. 43*, 1985, 117–125.

**TABLE 11.10. Precursors and products in evaporation–condensation processes**

| Product | Reactant | Notes | References |
|---|---|---|---|
| Ag, Al, Au, Co, Cr, Cu, Fe, Mg, Ni, Zn | Metals | 10–100 nm | [1] |
| Au, Pt, Pd | Metals | 4–25 nm | [2] |
| Au, Pb, Bi, Mg, Al, Cr, Mn, Fe, Co, Ni, Cu, Zn, Ag, Cd, Sn | Metals | 10–100 nm | [3] |
| Al, In, Pb, Fe, Cu, Co, Pd, Ag, Au, Ni | Metals | 20–200 nm | [4] |
| Zn | Metal | Crystal structure study | [5] |
| Cd | Metal | Crystal structure study | [6] |
| Zn | Metal | 5–5000 nm | [7] |
| Pb | Metal | For inhalation experiments | [8] |
| Zn, Bi, Mg, Pb, Cd | Metals | Vapor sources compared | [9] |
| Mg, Pb | Metals | Nucleation study | [10] |
| Ag | Metal | Optical property study | [11] |
| Mg, Zn, Sn, Al | Metals | Size distribution measured 1–10 nm | [12] |
| Many elements | Metals | Habit study | [13] |
| Pd, Cu | Metals | Mechanical behavior study 3–50 nm | [14] |
| Fe, Co, Ni | Metals | Magnetism study | [15] |
| MgO, CaO, $Al_2O_3$, $SiO_2$, $ZrO_2$, $CeO_2$, $U_3O_8$ | Bulk oxides | Focused electron beam | [16] |
| $UO_2$ | $UO_2$ | Resistance heating of pellet | [17] |
| ZnO | ZnO | Air annealing required for oxidation | [18] |
| CdTe | CdTe | Stacking faults suggested collisions in gas phase | [19] |
| TiAl | Ti, Al | Magnetron sputtering | [20] |
| $TiO_2$ | Ti, $O_2$ | Evaporation to form particles, which are collected and oxidized | [21–28] |
| $Al_2O_3$ | $Al_2O_3$ | Radio frequency sputtering | [29] |
| Ag, Cu, Bi, Te | Metals | Continuous flow | [30] |
| Ag | Ag | Coated particles with surfactant to prevent agglomeration, continuous flow | [30] |
| Si | Si | Spark discharge with continuous flow | [31, 32] |

**References to Table 11.10**

1. Thölén, A.R. *Arts Metallurgica 27*, 1979, 1765–1778.
2. Renou, A. & Gillet, M. *Surface Science 196*, 1981, 27–34.
3. Kimoto, K., Kamiya, Y., Nonoyama, M., & Uyeda, R. *Jpn. J. Appl. Phys. 2*(11), 1963, 702–713.
4. Hayashi, T., Ohno, T., Yatsuya, S., & Uyeda, R. *Jpn. J. Appl. Phys. 16*(5), 1977, 705–717.
5. Buckle, E.R. & Pointon, K.C. *J. Mater. Sci. 12*, 1977, 75–89.
6. Buckle, E.R. & Pointon, K.C. *J. Mater. Sci. 10*, 1975, 365–378.
7. Eversole, J.D. & Broida, H.P. *J. Appl. Phys. 45*(2), 1974, 596–602.
8. Homma, K., Kawai, K., & Nozaki, K. (Eds.), *Metal-fume generation and its application to inhalation experiments*. Ann Arbor Science Pub., Ann Arbor, Michigan: 1980.
9. Buckle, E.R., Mawella, K.J.A., & Tsakiropoulos, P. *J. Colloid and Interface Sci. 112*(1), 1986, 42–53.
10. Buckle, E.R. & Mawella, K.J.A. *J. Aerosol Sci. 18*(2), 1987, 139–146.
11. Abe, H., Schulze, W., & Tesche, B. *Chemical Physics 47*, 1980, 95–104.
12. Granqvist, C.G. & Buhrman, R.A. *J. Appl. Phys. 47*(5), 1976, 2200–2219.
13. Uyeda, R. *J. Crystal Growth 24*(25), 1974, 69–75.
14. Nieman, G.W., Weertman, J.R., & Siegel, R.W. *J. Mater. Res. 6*(5), 1991, 1012–1027.
15. Gangopadhyay, S., Hadjipanayis, G.C., Dale, B., Sorensen, C.M., & Klabunde, K.J. *NanoStruct. Mater. 1*, 1992, 77–81.
16. Ramsay, J.D.F. & Avery, R.G. *J. Mater. Sci. 9*, 1974, 1681–1688.
17. Narayana, D.S.S., Sundararajan, A.R., & Harvey, J. *J. Aerosol Sci. 25*(5), 1994, 909–922.
18. Mayo, M.J., Siegel, R.W., Liao, Y.X., & Nix, W.D. *J. Mater. Res.* 7(4), 1992, 973–979.
19. Fujii, S. *Proceedings Institute of Environmental Sciences*. Las Vegas, NV: 1985, 70–73.
20. Chang, H., Altstetter, C.J., & Averback, R.S. *J. Mater. Res. 7*(11), 1992, 2962–2970.
21. Beck, D.D. & Siegel, R.W. *J. Mater. Res.* 7(10), 1992, 2840–4845.
22. Hahn, H., Logas, J., & Averback, R.S. *J. Mater. Res. 5*(3),1990, 609–614.
23. Li, Z., Ramasamy, S., Hahn, H., & Siegel, R.W. *Materials Lett. 6*(7), 1988, 195–201.
24. Mayo, M.J., Siegel, R.W., Narayanasamy, A., & Nix, W.D. *J. Mater. Res. 5*(5), 1990, 1073–1082.
25. Melendres, C.A., Narayanasamy, A., Maroni, V.A., & Siegel, R.W. *J. Mater. Res. 4*(5), 1989, 1246–1250.
26. Parker, J.C. & Siegel, R.W. *J. Mater. Res. 5*(6), 1990, 1246–1252.
27. Parker, J.C. & Siegel, R.W. *NanoStruct. Mater. 1*, 1992, 53–57.
28. Siegel, R.W., Ramasamy, S., Hahn, H., Zongquan, L., Ting, L., & Gronsky, R. *Mater. Res. 3*(6), 1988, 1367–1372.
29. Ying, J.Y. *Ceram. Trans. 44*, 1994, 67–76.
30. Tohno, S., Itoh, M., & Takano, H. *J. Aerosol Sci. 25*(1), 1994, S39–S40.

31. Saunders, W., Sercel, P.C., Lee, R.B., Atwater, H.A., Vahala, K.J., Flagan, R.C., & Escorcia-Aparcio, E.J. *Appl. Phys. Lett.* *63*(11), 1993, 1549–1551.
32. Saunders, W.A., Sercel, P.C., Flagan, R.C., Atwater, H.A., & Vahala, K.J. *Z. Phys. D 26* (*Suppl.*), 1993, 219–221.

**TABLE 11.11. Precursors and products in nozzle expansion processes**

| Product | Particle Size | Reactant | Notes | References |
|---|---|---|---|---|
| Zn | 10–20 nm | Zn | Thermophoretic collection | [1] |
| Zn | size not reported | Zn | Studied condensation in supersonic jet | [2] |
| Zn | 5–50 nm | Zn | Subsonic turbulent jet | [3] |
| Cu, Ag, Au, Ni | 2–25 Å | Cu, Ag, Au, Ni | Collection by impaction | [4] |
| Ag | 40–100 Å | Ag | Use of inert carrier gas increased production of clusters | [5] |
| Al | < 10 Å | Al | Clusters formed on surface and desorbed | [6] |
| Na, K, Zn, $Na_2O$, ZnO | 10–100 nm | Na, K, Zn, $O_2$ | Oxidized vapor once it formed | [7] |
| Sb, Bi, Pb, up to 650 atoms | | Sb, Bi, Pb | Studied how to vary cluster size | [8] |
| Fe | < 45 atoms | Fe | Pulsed-laser vaporization | [9, 10] |
| In, Pb, Bi | 40–95 Å | In, Pb, Bi | Studied structure of particles | [11] |

## References to Table 11.11

1. Recknagle, K., Leung, A., Brown, D., Martian, P., Chung, J.N., Xia, Q., Hamilton, H., & Crowe, C.T. *Aerosol Sci. Technol.* *22*, 1995, 3–10.
2. McBride, D.D., Sherman, P.M., & Pierce, T.H. *Appl. Sci. Res.* *25* November 1971, 83–96.
3. Hastings, K.A. & Friedlander, S.K. *J. Aerosol Sci.* *22*(1), 1991, S39–S41.
4. Bowles, R.S., Kolstad, J.J., Calo, J.M., & Andres, R.P. *Surface Sci.* *106*, 1981, 117–124.
5. DeBoer, B.G. & Stein, G.D. *Surface Sci.* *106*, 1981, 84–94.
6. Hawley, J.H. & Ficalora, P.J. *J. Appl. Phys.* *63*(8), 1988, 2884–2885.
7. Hecht, J., West, W.P., & Norton, M.A. *Surface Sci.* *106*, 1980, 131–135.
8. Mühlbach, J., Recknagel, E., & Sattler, K. *Surface Sci.* *106*, 1981, 188–194.
9. Whetten, R.L., Cox, D.M., Trevor, D.J., & Kaldor, A. *Phys. Rev. Lett.* *54*(14), 1985, 1494–1497.
10. Whetten, R.L., Cox, D.M., Trevor, D.J., & Kaldor, A. *J. Phys. Chem.* *89*, 1985, 566–569.
11. Yokezeki, A. & Stein, G.D. *J. Appl. Phys.* *49*(4), 1978, 2224–2232.

**TABLE 11.12. Precursors and products in laser ablation processes**

| Product | Particle Size | Reactant | Comments | References |
|---|---|---|---|---|
| $Al_2O_3$ | 3–7 nm | Al, $O_2$ | $O_2$ atmosphere | [1] |
| Cu | | Cu | Size of agglomerates measured | [2] |
| $SiO_2$, MgO, $Al_2O_3$, $Fe_3O_4$, $Mg_2SiO_4$, $CaTiO_3$, $MgAl_2O_4$ | | Metals and oxides | $CO_2$ laser heating, 10-mg/min output | [3] |
| ZnO, $SiO_2$, $Fe_2O_3$, $Bi_2O_3$, PdO, NiO, $Ag_2O$, TeO, $Sb_2O_3$, $TiO_2$, $ZrO_2$, $Al_2O_3$, $In_2O_3$, SnO, $V_2O_5$, MgO | 10 Å | Metals, $O_2$ | Nd: yttrium aluminum garnet laser in diffusion cloud chamber | [4, 5] |
| $Y_2O_3$, $SiO_2$, $Al_2O_3$ | 4–40 nm | Oxides | $CO_2$ and excimer lasers | [6] |
| $ZrO_2$, MgO | | Zr wire, Mg wire | Studied powder sintering | [7–16] |
| Cu, Ni, Mo, W, Al | | Metals | Excimer and yttrium aluminum, garnet lasers, studied agglomerates formed | [17] |
| $Cu_2$ and larger | | Cu | Target within supersonic nozzle | [18] |
| Au, Ag, Cu small clusters | | Au, Ag, Cu | Vaporized in mass spectrometer | [19] |
| Ni–Cr, Ni–Al small clusters | | Binary alloys | Clusters sent into nozzle | [20] |
| $Ga_xAs_y$ | ~ 5–25 atoms | GaAs | Supersonic nozzle after vaporization | [21] |
| Bi–Cr, Mn–Ag | 2–20 atoms | Metal rods | Target within nozzle | [22, 23] |
| $SiO_2$ particles | | $SiO_2$ | W particles also formed by evaporation and co-deposited | [24] |
| Carbon | | Polyimide | Particulate debris from laser processing | [25] |
| Y–Ba–Cu–O unknown composition | | $YBa_2Cu_3O_{7-x}$ | Undesired particles | [26] |

**TABLE 11.12.** (*Continued*)

| Product | Particle Size | Reactant | Comments | References |
|---|---|---|---|---|
| Y–Ba–Cu–O, unknown composition | | $YBa_2Cu_3O_{7-x}$ | Undesired particles | [27] |
| Ti, $TiO_2$, TiN | < 100 nm | Ti | Studied beam–plume interaction | [28] |
| AlN | 11 nm | Al | Nitrogen atmosphere | [29] |
| Fe, Al, Ti, Zr, Cr, Mo, Ta, $Fe_2O_3$, NiO, $Al_2O_3$, $TiO_2$, $ZrO_2$, $Cr_2O_3$, $Ta_2O_5$, $WO_3$, $MoO_3$, TiN, $Zr_3N_4$, $Cr_2N$, $Ta_2N$ | 10–30 nm | Metals, $O_2$, $N_2$ | Controlled size by varying chamber pressure | [30] |

## References to Table 11.12


1. Johnston, G.P., Muenchausen, R., Smith, D.M., Farenholtz, W., & Foltyn, S. *J. Am. Ceramic Soc. 75*(12), 1992, 3298–3298.
2. Boulaud, D., Chouard, J.C., Briand, A., & Chartier, F. *J. Aerosol Sci. 23*(1), 1992, S225–228.
3. Kato, A., Hojo, J., & Okabe, Y. *Memoirs of Faculty of Engineering, Kyushu University 41*(4), 1981, 319–334.
4. El-Shall, M.S., Bahta, A., Rabeony, H., & Reiss, H. *J. Ceram. Phys. 87*(2), 1987, 1329–1345.
5. El-Shall, M.S., Slack, W., Vann, W., Kane, D., & Hanley, D. *J. Phys. Chem. 98*(12), 1994, 3067–3070.
6. Gartner, G., Janiel, P., Lydtin, H., & Rehder, L. Generation of very small particles by laser ablation in a carrier gas deposition & characterization. In Marijnissen, J.C.M. & Pratsinis, S.E. (Eds.), *Synthesis and Measurement of Ultrafine Particles*. Delft, The Netherlands: Delft Univ. Press, 1993, p.41.
7. Lee, H., Villamil, S., & Burton, L.C. "Grain-boundary impedance in ferroelectric ceramic," *IEEE Trans. Ultrasonics, Ferroelect., Freq. Control 33*(6), pp. 808–808, 1986.
8. Lee, Y.H., Haun, M.J., Safari, A., & Newnham, R.E. *IEEE*, 1986, 318–322.
9. Lee, H.D., Pober, R.L., Calvert, P.D., & Bowen, H.K. *J. Mater. Sci. Lett. 5*, 1986, 81–83.
10. Lee, T.J., Fey, G.T.K., Lin, W.J., Yao, P.C., & Hsu, S.E. *Mater. Res. Soc. Symp. Proc., 135*, 1989, 597–602.
11. Lee, K.W., Curtis, L.A., & Chen, H. *Aerosol Sci. Technol. 12*, 1990, 457–462.

12. Lee, W., Shen, H.-S., Dwight, K., & Wold, A. *J. Solid State Chem. 106*, 1993, 288–294.
13. Lee, W., Do, Y.R., Dwight, K., & Wold, A. *Mat. Res. Bull. 28*, 1993, 1127–1134.
14. Lee, J.-K., Rubow, K.L., Pui, D.Y.H., & Liu, B.Y.H. *Aerosol Sci. Technol. 23*, 1995, 481–490.
15. Lee, D.H., Vuong, K.D., Williams, J.A.A., Fagan, J., Condrate, R.A., Sr., & Wang, X.W. *J. Mater. Res. 11*(4), 1996, 895–903.
16. Lee, K.N., Jacobson, N.S., & Miller, R.A. *MRS Bulletin*, October 1994, 35–38.
17. Boulaud, D. Aerosol production by laser ablation. In Marijnissen, J.C.M. & Pratsinis, S.E. (Eds.), *Synthesis and Measurement of Ultrafine Particles*. Delft, The Netherlands: Delft Univ. Press, 1993, p. 31.
18. Powers, D.E., Hansen, S.G., Geusic, M.E., Pulu, A.C., Hopkins, J.B., Dietz, T.G., Duncan, M.A., Langridge-Smith, P.R.R., & Smalley, R.E. *J. Phys. Chem. 88*(14), 1982, 2556–2560.
19. Moini, M. & Eyler, J.R. *J. Chem. Phys. 88*(9), 1988, 5512–5515.
20. Rohlfing, E.A., Cox, D.M., Petkovic–Luton, R., & Kaldor, A. *J. Phys. Chem. 88*(25), 1988, 6227–6231.
21. O'Brien, S.C., Liu, Y., Zhang, Q., Heath, J.R., Tittel, F.K., Curl, R.F., & Smalley, R.E. *J. Chem. Phys. 84*(7), 1986, 4074–4079.
22. LaiHing, K., Cheng, P.Y., & Duncan, M.A. *J. Phys. Chem. 91*(26), 1987, 6521–6525.
23. LaiHing, K., Wheeler, R.G., Wilson, W.L., & Duncan, M.A. *J. Chem. Phys. 87*(6), 1987, 3401–3409.
24. Chow, G.-M., Klemens, P.G., & Strutt, P.R. *J. Appl. Phys. 66*(7), 1989, 3304–3308.
25. Singleton, D.L., Paraskevopoulos, G., & Irwin, R.S. *J. Appl. Phys. 66*(7), 1989, 3324–3328.
26. Becker, C.H. & Pallix, J.B. *J. Appl. Phys. 64*(10), 1988, 5152–5156.
27. Chiba, H., Murakami, K., Eryu, O., Shihoyama, K., Mochizuki, T., & Masuda, K. *Jpn. J. Appl. Phys. 30*(4B), 1991, L732–L734.
28. Matsunawa, A., Yoshida, H., & Katayama, S. Beam–plume interaction in pulsed YAG laser processing. Boston: Laser Inst. America Intern. Congress Applications of Lasers and Electro–optics, 1984,
29. Johnston, G.P., Meunchausen, R.E., Smith, D.M., & Foltyn, S.R. *J. Am. Ceram. Soc. 75*, 1992, 3465–3468.
30. Matsunawa, A. & Katayama, S. Laser production of ultra-fine metallic and ceramic particles. San Francisco: Intern. Congress Applications of Lasers and Electro–optics, 1985, 205–211.

**TABLE 11.13. Precursors and products in supercritical fluid spray processes**

| Product | Supercritical Solvent | Comments | References |
|---|---|---|---|
| $SiO_2$ | Water | Amorphous product | [1–3] |
| $GeO_2$ | Water | | [4] |
| Organics–pharmaceuticals | $CO_2$, ethylene | Numerous materials | [5, 6] |

**References to Table 11.13**


1. Matson, D.W., Petersen, R.C., & Smith, R.D. *Advances in Ceramics 211*, 1987, 109–120.
2. Matson, D.W., Petersen, R.C., & Smith, R.D. *J. Mater. Sci. 22*, 1987, 1919–1928.
3. Matson, D.W., Fulton, J.L., Petersen, R.C., & Smith, R.D. *Ind. Eng. Chem. Res. 26*, 1987, 2298–2306.
4. Petersen, R.C., Matson, D.W., & Smith, R.D. *J. Amer. Chem. Soc. 108*(8), 1986, 2100–2102.
5. Tom, J.W. & Debenedetti, P.G. *J. Aerosol Sci. 22*(5), 1991, 555–584.
6. Tom, J.W. & Debenedetti, P.G. *Biotechnol. Prog. 7*, 1991, 403–411.


**TABLE 11.14. Precursors and products in hybrid evaporation–condensation–intraparticle reaction processes**

| Product | Reactant | Comments | References |
|---|---|---|---|
| $Al_2O_3$ | Al(O–Sec–Bu)$_3$ | Amorphous product | [1] |
| $TiO_2$ | Ti(OEt)$_3$ | 0.06–0.6 nm | [2] |
| $TiO_2$–$Al_2O_3$ | Al(O–Sec–Bu)$_3$, Ti(OEt)$_4$ | Ti rich at surface | [3] |
| $Al_2O_3$ | Al(O–Sec–Bu)$_3$ | Mixing chamber for droplet formation | [4] |
| Silazane polymer ($Si_3N_4$ after calcination) | $Cl_2SiMe_2$ | — | [5] |
| $SrTiO_3$ | Ti(OR)$_4$ $Sr^{2+}$ | $TiO_2$ formed and then reacted in liquid with Sr source | [6] |
| $Al_2O_3$ $TiO_2$ | Al(O–Sec–Bu)$_3$, Ti(OEt)$_4$ | Kinetics of hydrolysis studied | [7] |
| Styrene polymer divinyl–benzene/ ethylvinylbenzyne copolymer | styrene, divinylbenzene | 10–20 $\mu$m reacted droplets with trifluoro-methanesulfonic acid | [8] |
| Poly(*p*-tertiary) butylstyrene | *p*-tertiary butyl styrene, trifluoromethane-sulfonic acid | Ratio of monomer to initiator critical | [9] |
| Polyurea mixed polyurea–metal-oxide particles | toluene diisocyanate, hexamethyl diisocyanate, ethylenediamine, Ti(OEt)$_4$ | Metal oxides incorporated into polymer beads | [10] |

## References to Table 11.14


1. Ingebrethsen, B.J. & Matijevic, E. *J. Aerosol Sci. 11*, 1980, 271–280.
2. Visca, M. & Matijevic, E. *J. Colloid Interface Sci. 68*(2), 1979, 308–319.
3. Ingebrethsen, B.J., Matijevic, E., & Partch, R.E. *J. Colloid Interface Sci. 95*(1), 1983, 228–239.
4. Marini, L., Balducci, L., Montino, F., & Spoto, G. Preparation via chemical routes of ceramic powders having controlled characteristics. In Vincenzini, P. (Ed.), *High Tech Ceramics*. Amsterdam: Elsevier Science Publications, 1987, 471–479.
5. Zimmerman, C.J., Partch, R.E., & Matjevic, E. *J. Aerosol Sci. 22*(7), 1991, 881–886.
6. Ocaña, M. & Serna, C.J. *J. Mater. Sci. Lett. 9*, 1990, 772–773.
7. Ingebrethsen, B.J. & Matijevic, E. *J. Colloid Interface Sci. 100*(1), 1984, 1–16.
8. Nakamura, K., Partch, R.E., & Matijevic, E. *J. Colloid Interface Sci. 99*(1), 1984, 118–127.
9. Partch, R., Matijevic, E., Hodgson, A.W., & Aiken, B.E. *J. Polymer Science 21*, 1987, 961–967.
10. Partch, R.E., Nakamura, K., Wolfe, K.J., & Matijevic, E. *J. Colloid Interface Sci. 105*(2), 1985, 560–569.

# 12 Overall Qualitative Behavior of Liquid-to-Solid and Solid-to-Solid Conversion Processes

## 12.1 INTRODUCTION

This chapter describes the roles of the phenomena discussed in Chapters 1 through 8 in liquid-to-solid and solid-to-solid conversion processes. The conversion of droplets and solid reactant particles to product particles by intraparticle reactions is one of the most complicated of aerosol processes due to the occurrence of a wide variety of physical and chemical phenomena. All of the processes that occur for gas-to-particle conversion systems can also occur for spray pyrolysis when volatile reactants, intermediates, or products are present. In addition, liquid-to-solid and solid-to-solid conversion processes have the added complexity that a number of intraparticle phenomena occur in both droplets and solid particles. Furthermore, fuels can be added to the droplets or solid particles, resulting in combustion processes inside the particle or in the gas phase. Examples of the different types of processes are conventional spray pyrolysis (conversion of solution droplets to solid products); intraparticle reaction in either liquids or solids with a gaseous co-reactant; spray pyrolysis with liquid fuels such as ethanol, where combustion processes take place in the gas phase; spray pyrolysis with solid fuels such as sugar, where the release of heat occurs because of the oxidation–reduction processes within a particle; and atomization of molten materials. Most of these processes are used industrially for the manufacture of powder.

The development of a single, comprehensive model for this class of aerosol processes that accounts for all of the physical and chemical phenomena has been difficult due to the wide variety of phenomena. Even restricting efforts to an individual technology does not simplify the problem sufficiently to allow a complete theoretical description. Fortunately, although the separate technologies have not been described with models, the individual steps involved are well understood and are common to many of the technologies. For this reason, we begin by discussing the individual physical and chemical phenomena that occur at the particle scale, which include droplet generation, heat transfer, solvent transport, solute transport, solute precipitation, transport of gaseous species in solid particles, and intraparticle reaction. We follow with a discussion of the different types of overall reactor behavior, including examinations of droplet evaporation in flow systems and the influence of solvent evaporation, precursor characteristics, and intraparticle reaction on particle morphology. We conclude with an analysis of the different modes of reactor behavior, including intraparticle reaction with no volatile species and with volatile species. Specific examples of these modes of behavior are discussed in Chapter 13.

## 12.2 OVERVIEW OF BASIC PHYSICOCHEMICAL PHENOMENA

In this section, we describe the individual steps involved in the formation of powders by liquid-to-solid and solid-to-solid routes using the formation of solid particles by spray pyrolysis as an example, because this process involves the widest range of phenomena. The process consists of the following basic steps: formation of droplets, transport of these droplets into a heated reactor, evaporation of the solvent, and thermal conversion of the solid precursors to the final solid particles (Fig. 12.1). These steps involve a wide variety of physicochemical phenomena that occur at different length scales ranging from the diameter of a single-phase droplet, to the diameter of the liquid core in a two-phase particle, the thickness of the porous shell in a two-phase particle, the thickness of the gaseous boundary layer surrounding a droplet or solid particle, the crystallite sizes of the reactant, intermediate, and product species, and on up to the reactor radius and length. The phenomena occurring at the particle scale are shown in Figures 12.2 through 12.5 and have been thoroughly investigated.[1–13] Various aspects of these phenomena are identified next, together with the major factors that influence thcm.

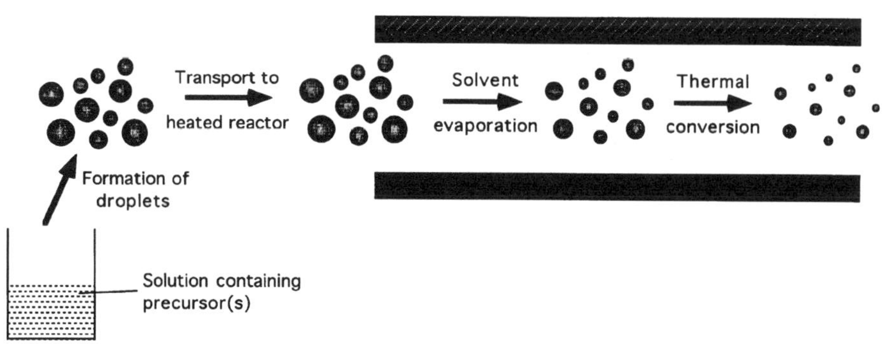


**Fig. 12.1.** Overview of key steps in spray pyrolysis.

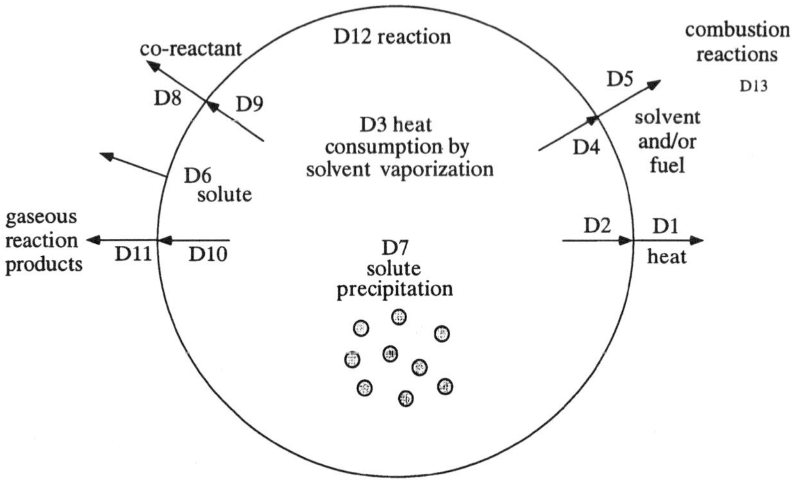


**Fig. 12.2.** Schematic representation of the processes that can occur in droplets suspended in a gas.

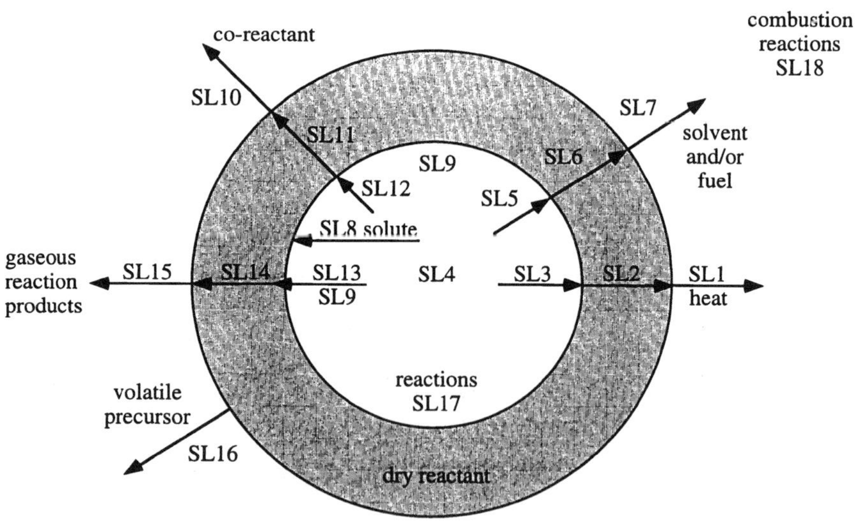


**Fig. 12.3.** Schematic representation of the processes that can occur in a droplet containing both liquids and solids.

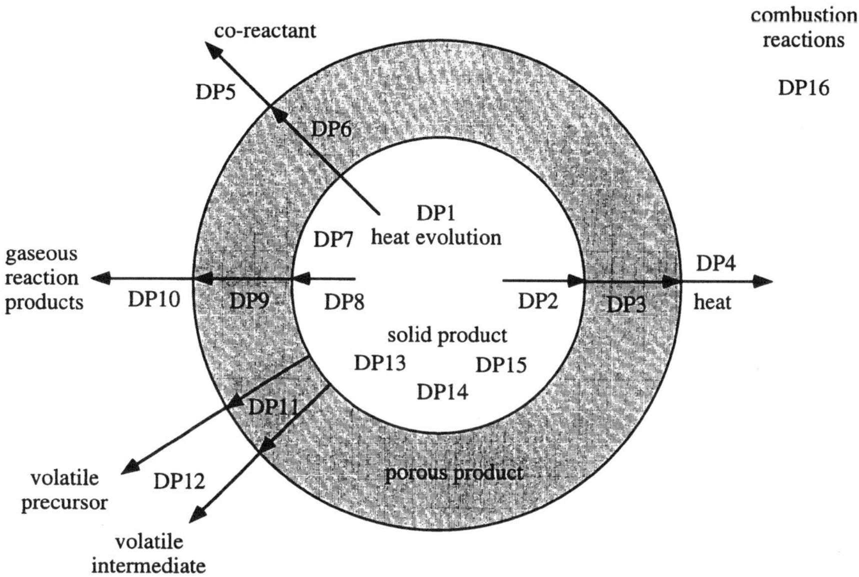


**Fig. 12.4.** Schematic representation of the processes that can occur in a dry particle.

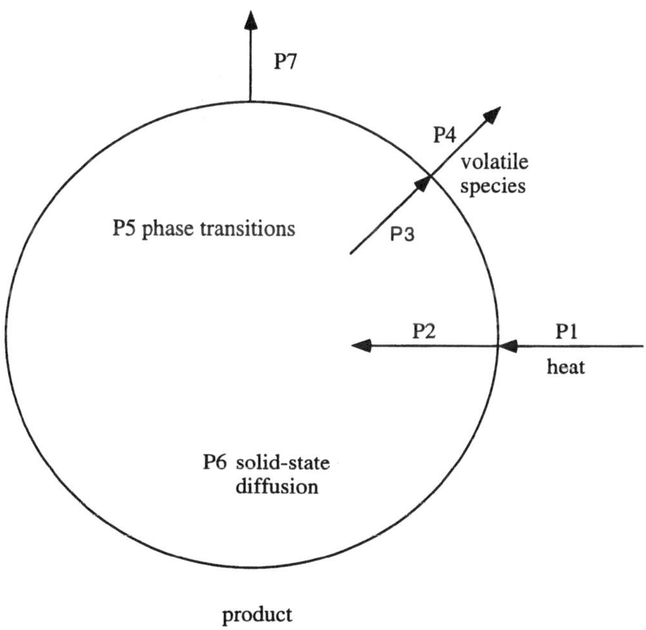


**Fig. 12.5.** Schematic representation of the processes that can occur in a product particle.

### 12.2.1 Droplet Formation

The aerosol formation process determines the magnitude and spread of the size distribution of the droplets. The most important factors that influence the formation of droplets are the method of droplet formation and the physical nature of the solution or suspension (surface tension, viscosity, conductivity, and density).[14–17]

### 12.2.2 Processes within Droplets

The processes that can occur inside droplets include heat transfer, the transport of solvent, solute, co-reactant, and reaction product in and around the droplets, and chemical reactions inside and outside the particles. Also, the liquid inside particles can move because of the possible gas flow relative to the particle induced by the relative motion between the particle and the surrounding gas and because of temperature nonuniformities created by heat transfer processes.[18] The critical steps are shown in Figure 12.2 and include:

- (D1) Heat transfer in the gas surrounding the droplet
- (D2) Heat transfer in the droplet
- (D3) Heat consumption by solvent vaporization
- (D4) Solvent transport in the droplet
- (D5) Solvent transport in the gas phase
- (D6) Solute transport in the droplet
- (D7) Solute precipitation
- (D8) Co-reactant transport in the gas phase
- (D9) Co-reactant transport in the liquid
- (D10) Gaseous reaction product transport in the liquid
- (D11) Gaseous reaction product transport in the gas phase
- (D12) Reactions in the liquid (endothermic or exothermic)
- (D13) Combustion process in the gas surrounding the droplet when the solvent is fuel or when fuel is added

### 12.2.3 Particles Containing Solids and Liquids

The physical and chemical phenomena occurring in particles containing both solids and liquids are complicated by the presence of the solid. Figure 12.3 shows many of these phenomena for the case of a solid shell around a liquid core:

- (SL1) Heat transfer in the gas surrounding the particle
- (SL2) Heat transfer in the solid
- (SL3) Heat transfer in the liquid

- (SL4) Heat consumption by solvent vaporization
- (SL5) Solvent transport in the liquid
- (SL6) Solvent transport through the solid
- (SL7) Solvent transport through the gas surrounding the particle
- (SL8) Solute transport in the liquid (concentration profiles)
- (SL9) Solute precipitation
- (SL10) Co-reactant transport in the gas phase
- (SL11) Co-reactant transport in the solid
- (SL12) Co-reactant transport in the liquid
- (SL13) Gaseous reaction product transport in the liquid
- (SL14) Gaseous reaction product transport in the solid
- (SL15) Gaseous reaction product transport in the gas phase
- (SL16) Transport of volatile precursors in the gas phase
- (SL17) Reactions in the liquid
- (SL18) Combustion processes when fuels are added
- Shrinkage, bursting, swelling, and other morphological changes

### 12.2.4 Dried Particles

Numerous phenomena can occur in dry particles. In the example discussed here, we assume that the dried reactant has been partially converted to a porous reaction product and that there is a well-defined boundary between the two materials. (This model is often called the shrinking core model.)[19] Figure 12.4 shows the many processes that are possible:

- (DP1) Heat generation (or consumption) by the chemical reaction
- (DP2) Heat transfer in the dry reactant
- (DP3) Heat transfer in the solid product
- (DP4) Heat transfer in the gas phase
- (DP5) Co-reactant transport through the gas phase
- (DP6) Co-reactant transport through the porous reacted solid
- (DP7) Co-reactant transport in the unreacted core
- (DP8) Gaseous reaction product transport in the unreacted core
- (DP9) Gaseous reaction product transport in the porous reacted solid
- (DP10) Gaseous reaction product transport in the gas phase
- (DP11) Transport of volatile precursors and reaction intermediates in the porous reacted solid
- (DP12) Transport of volatile precursors and reaction intermediates in the gas phase

- (DP13) Reaction of precursors with co-reactant to form solid and gaseous products
- (DP14) Phase transitions (melting of one or more precursors, crystallization)
- (DP15) Solid-state diffusion (sintering, grain growth)
- (DP16) Combustion processes when solid fuels are added

### 12.2.5 Product Particles

The phenomena that can occur in product particles are more limited than in the other types of particles. Figure 12.5 shows the following processes for a particle in which the reactant has been completely converted to solid products:

- (P1) Heat transfer in the gas phase
- (P2) Heat transfer in the solid state
- (P3) Transport of volatile products in the particle
- (P4) Transport of volatile products in the gas phase
- (P5) Phase transitions (melting, crystallization)
- (P6) Solid-state diffusion (sintering, grain growth)
- (P7) Evaporation of the product

### 12.2.6 Aerosol Dynamics

The individual physical and chemical phenomena just described can lead to various changes in aerosol properties. The various phenomena can occur at the scale of a single particle or at the reactor scale and include the following processes:

- Formation of new particles if volatile species are present
- Coagulation–agglomeration–reaction of new particles with each other or with existing particles
- Particle transport by sedimentation, inertia, diffusion, and thermophoresis
- Dynamics of particle size distribution

### 12.2.7 Reactor-scale Phenomena

All of the preceding phenomena are influenced by, and coupled to, the conditions in the reactor. This relationship requires a consideration of the following reactor-scale phenomena:

- Axial and radial temperature profiles
- Axial and radial species concentration profiles (solvent vapor, volatile reactants, intermediates, products)

- Velocity profiles
- Condensation and reaction of volatile species on reactor walls
- Reactor residence time distribution (distribution of times droplets or particles spend in reactor)

Only a few studies have investigated these phenomena.[20, 21]

The sections that follow examine the most important aspects of the phenomena outlined in this section. The information is grouped by phenomenon (heat transfer, mass transfer, etc.) and not by the type of particle (liquid, solid, liquid–solid).

## 12.3 INDIVIDUAL PHYSICAL AND CHEMICAL PHENOMENA

### 12.3.1 Droplet Generation

Droplet generation can be an important process, depending on the type of liquid-to-solid or solid-to-solid conversion to powder. Three different processes may be encountered:

- Processes with no particle–droplet disruption and no postprocessing to change the particle size (e.g., spray pyrolysis[22, 23] and molten material atomization),[24 - 26] where the final particle size distribution is directly related to the droplet size distribution.
- Processes with particle–droplet disruption (e.g., combustion synthesis), where the product particle size distribution may be unrelated to the droplet size distribution.[27]
- Processes with no particle–droplet disruption and with postprocessing (e.g., spray- or freeze-drying), where the final product size distribution is unrelated to the droplet size distribution.[28 - 33]

The importance of these three process variations depends on where the value lies in the product. In some applications (e.g., thick-film pastes), the presence of a small fraction of particles above a critical size may be extremely detrimental. In this case, large particles must not be allowed to form.

Droplet generation is a critical step in spray pyrolysis because it can control the particle size and size distribution of the product. The size distribution is less important in combustion processes, where subsequent processes reduce the particle size, and in freeze-drying and spray-drying, where subsequent processing steps such as milling reduce the particle size. A variety of aerosol generators are available, including pressure atomizers and rotary, air-assist, air-blast, ultrasonic, and electrostatic generators; these are discussed in more detail in Chapter 16. In general, these droplet generators produce particle size distributions with Sauter mean diameter (SMD) values

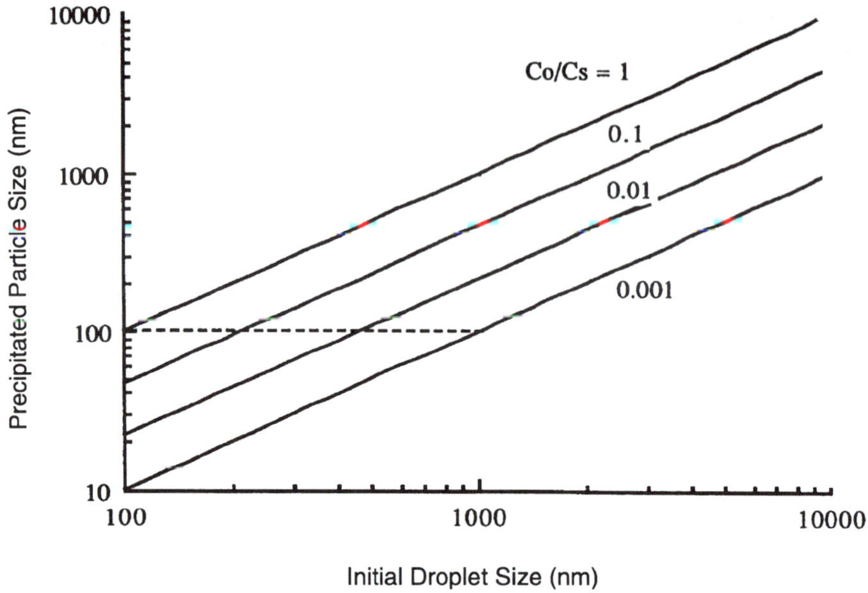


**Fig. 12.6.** Plot of particle size vs. initial-droplet size. (Reproduced with permission from Messing.)[34]

of 0.1 micron to 100 microns, roughly lognormal shapes, and geometric standard deviations greater than 1.4. The spread of the droplet size distribution obtained from droplet generators is often more than adequate for applications.

Several aspects of the droplet size distribution are relevant. The average droplet size determines the average size of the particle that is precipitated, as shown in Figure 12.6.[34] This plot demonstrates that, for average droplet sizes of 1 $\mu$m and 5 $\mu$m, the dry particle can have an average diameter of 0.1 to 0.9 $\mu$m and 0.5 to 4 $\mu$m, respectively, depending on the solution concentration. The spread of the droplet size distribution plays a role in determining the dynamics of droplet evaporation in the reactor and the evolution of particle morphology. Smaller droplets have shorter evaporation times, leading to complete solute precipitation (if it occurs) under different reactor conditions (lower temperature) than for larger droplets. This difference in precipitation conditions can lead to differences in morphology of the dry particles. Particle morphology also depends more directly on particle size, with smaller droplets or particles often being solid because the size of the crystallites can be similar to the size of the overall precursor particles. In contrast, the small crystallite size relative to the particle size for larger particles favors the formation of hollow or porous particles. For this reason, it is important to minimize the number of larger droplets.

In molten material atomization, the droplet size directly determines the final particle size.[35-38]

### 12.3.2 Heat Transfer

The heating of droplets, two-phase liquid–solid particles consisting of a precipitate and the original solution, and solid particles consisting of dried reactant, intermediates, and products drives transport processes, chemical reactions, phase changes, and other phenomena that influence particle characteristics. In the case of a highly exothermic reaction inside a particle (metal nitrates and solid fuel—e.g., sugar), the heat released can be enough to explode the particle into smaller fragments if the particle size is large enough. When a liquid fuel (ethanol, methanol) is used in droplets, the fuel can vaporize into the gas phase, whereupon it can react with a gaseous reactant (oxygen) to release heat.[39] This release of heat can cause the particle to heat up and blow apart. Because these approaches are the bases for several industrial spray pyrolysis routes, it is important to know the temperature of the particle itself (as opposed to the gas temperature, which can be very different) during processing. The discussion that follows is focused on this problem. Several studies have examined the coupled problem of heat and mass transfer for solution droplets;[1-13] more exotic approaches exploiting laser heating have also been used.[40]

In general, an aerosol of the precursor solution droplets is typically sent into a flow system in which the temperature increases from room temperature to the maximum temperature and back to near room temperature over a time determined by the details of the reactor design and operating conditions. This time is often less than 10 seconds and, in many cases, less than 1 second. To calculate the temperature of a particle, it must be determined whether the particle is in thermal equilibrium with the gas as the gas temperature changes. If equilibrium is maintained, the temperature of the particle tracks that of the gas, and heat transfer to and from particles does not have to be considered. In this case, the problem of determining the temperature and time histories of the particles is reduced to determining the temperature and velocity profiles of the gas in the reactor. If the particles are not in thermal equilibrium with the gas, the particle temperature must be calculated using more involved approaches.

Several possibilities arise when one considers this problem. For liquid droplets, the latent heat required for evaporation of the solvent must be balanced by heat transport from the hotter gas to the cooler droplet and into the droplet (Steps D1, D2, and D3 in Fig. 12.2). In this case, the particle temperature is determined by a balance between the rate of heat consumption and the rates of heat transfer inside the droplet and from the gas to the surface of the droplet. If a solid is present around the liquid, heat transport through the shell (Steps SL1, SL2, SL3, and SL4 in Fig. 12.3) must be taken into account.[41] For solid particles, if exothermic chemical reactions occur, the temperature in the particle is determined by a balance between the rate of heat released and the rate of heat transfer through the reactant core, reacted layer, and gas phase (Steps DP1, DP2, DP3, and DP4 in Figure 12.4). The simplest case is that of a completely reacted solid particle in a gas that is changing temperature as it flows through a reactor in which the temperature is changing (Fig. 12.5). In this case, the temperature of the particle is determined by the rate at which heat can be transported into the particle (Steps P1 and P2 in the figure) and the heat capacity and density of the particle. This situation is different from the others in that there is no heat source or sink; instead, the heat transfer is driven by the difference between the gas and particle temperatures.

The information on characteristic times presented in Chapters 2 through 8 can be used to gain an insight into the problem of determining the temperature of a solid particle in a gas whose temperature is changing as a function of time (Fig. 12.5). Two characteristic times for heating of the particles correspond to the heat transfer in the gas phase (P1) and in the solid (P2). It was shown in Chapter 4 that the characteristic time for heat transfer in the gas (P1) is roughly $d_p^2/\alpha_g$, where $\alpha_g$ is the thermal diffusivity of the gas ($cm^2/sec$) and $d_p$ is the particle diameter. Similarly, the characteristic time for heat transfer inside the particle is $d_p^2/\alpha_p$, where $\alpha_p$ is the thermal diffusivity of material inside the particle. The analysis presented in Chapter 4 showed that, for most practical situations, solid particles in which no reaction occurs will be in thermal equilibrium with the gas, and this problem does not have to be considered in describing the overall process.

Determining the temperature for evaporating solution droplets requires a comparison of the rates of heat consumption (the latent heat of evaporation of the solvent, Step D3) and heat transfer into the droplet (Steps D1 and D2). Because the thermal conductivities of liquids are higher than those of gases, the characteristic time for heating the interior of the droplet, $d_p^2/\alpha_p$, is much shorter than the time for heat transfer in the gas phase. As a result, the temperature of the liquid in the droplet is nearly uniform, and only gas-phase heat transfer (D1) must be considered. A pseudosteady state is reached in which, although the droplet diameter is continuously changing, the temperature difference between the droplet and the gas surrounding it can be calculated from an equation that assumes a steady state, namely[41]

$$T_\infty - T_d = \frac{D_g M L_v}{Rk}\left(\frac{p_d}{T_d} - \frac{p_\infty}{T_\infty}\right), \tag{12.1}$$

where $T_\infty$ is the gas temperature, $T_d$ is the droplet temperature, $D_g$ is the diffusion coefficient of solvent in the gas, $R$ is the gas constant, $M$ is the molecular weight of the evaporating species, $L_v$ is the latent heat of evaporation of the liquid, $k$ is the thermal conductivity of the gas, $p_d$ is the vapor pressure at the droplet surface, and $p_\infty$ is the solvent partial pressure far from the droplet. This equation predicts temperature depressions of up to 25°C at low relative humidities and at ambient temperatures near 50°C. The temperature depression approaches zero as the relative humidity approaches 100%, because the evaporation rate drops as the relative humidity increases. Hinds gives a more detailed discussion of this problem and provides representative numbers.[41]

If a crust is formed around the liquid due to solvent evaporation from the droplet surface, the rate of heat transfer can change. The heat must flow through the gas (SL1), through the crust (SL2), and into the liquid (SL3). Because the thermal diffusivity of the solid is much higher than the thermal diffusivity of the gas, the crust does not introduce a resistance to the flow of heat,[9] and the equations given earlier for a pure liquid droplet can be used to describe the heat transfer. As a result, the temperature difference between the droplet and gas can be estimated using Equation 12.1.

The situation for an exothermic chemical reaction in a dry particle is handled similarly. Balancing the heat released in the particle by chemical reaction (DP1) with the heat flow out of the particle (DP2, DP3, DP4) gives an expression for the increase in temperature of the particle. As in the case of a liquid, the heat transfer steps in the solid (DP2 and DP3) are rapid relative to heat transfer in the gas phase (DP4) and can be neglected, thereby simplifying the analysis. If the time constant for heat release in the particle (DP1) is much shorter than that for heat transfer from the particle (DP4), the temperature of the particle will increase. In this case, the temperature is determined by a balance between the rate of heat release in the particle and the rate of heat transfer away from the particle into the gas phase, according to the equation

$$\frac{\pi}{6} d_p^3 H_r r_v = \pi d_p^2 h_g (T_p - T_\infty), \tag{12.2}$$

where $d_p$ is the particle diameter (cm), $H_r$ is the heat of reaction (ergs/mol), $r_v$ is the rate of the reaction inside the particle (moles/cm$^3$ sec), $h_g$ is the heat transfer coefficient in the gas surrounding the particles (erg/cm$^2$ sec K), $T_p$ is the temperature of the particle, and $T_\infty$ is the temperature far from the particle. This equation can be solved for the temperature of the particle:

$$T_p = \frac{d_n H_r r_v}{6 h_g} + T_\infty. \tag{12.3}$$

Elevated temperatures in the particle are favored by larger particle sizes, higher exothermicity, higher rates of reaction, and slower rates of heat transfer into the gas phase. Heating of the particle with subsequent explosion is the basis of several industrial processes for powder manufacture. The analysis is identical for an endothermic chemical reaction.

Calculating the temperature of a particle containing an evaporating fuel and the temperature of the surrounding gas is more complicated but still relies on the basic principles just presented. This problem is discussed in detail elsewhere.[17]

### 12.3.3 Solvent Transport

Solvent evaporation is important in both spray-drying and spray pyrolysis processes because it can modify the particle size distribution and influence particle morphology. For situations without fuels, solvent evaporation leads to one of the greatest disadvantages of these processes: the formation of hollow or porous particles. The morphology of particles produced by solvent evaporation strongly influences the properties of ceramics produced by sintering the particles. Porous or hollow particles can lead to difficulties in sintering to near-theoretical density and can result in degraded mechanical, electrical, and magnetic properties. The evaporation rate of a fuel in a particle also determines the rate of heat release in the gas phase. High rates of heat release can result in changes in particle morphology. Thus, for most applications of ceramic and metal powders, it is critical to be able to control the evaporation or subsequent steps to allow the formation of solid particles.

The solvent evaporation process begins with evaporation of the solvent from a pure solution droplet or droplet containing a colloidal precursor (Steps D4 and D5 in Fig. 12.2). Let us consider the case of a molecular solute. Once enough solvent has evaporated, the solution becomes supersaturated, and one or more of the solutes eventually precipitates to form crystallites, often near the surface of the droplet. As further evaporation takes place, the crystallites grow. If a solid shell is formed at the droplet surface by solute precipitation, the solvent must evaporate through a shell composed of the solute (Steps SL5, SL6, and SL7 in Fig. 12.3). The shell may retain its shape, burst, shrivel, collapse, or undergo numerous other fates. In some cases, a shell will not be formed, and evaporation will take place from a particle consisting of crystallites in a solvent. The presence or absence of a shell surrounding the liquid and the permeability of the shell play an important role in determining the evaporation time. Some of the possibilities are shown in Figure 12.7.[42]

In order to calculate the residence time required for the conversion of solution droplets to the solid product, it is important to calculate drying times for droplets. Equations for droplet evaporation times were given in Chapter 4. The problem can be complicated by the formation of a solid phase in the droplet during drying, but in many cases the bulk of the solvent evaporates before a crust is formed, and estimates of drying times for droplets composed of pure solvents provide an adequate approximation.

Solution droplet evaporation (steps D4 and D5) is more complicated than the evaporation of a pure solvent droplet, because the diffusion of solvent must occur in both the liquid (D4) and the gas (D5) phases and these steps

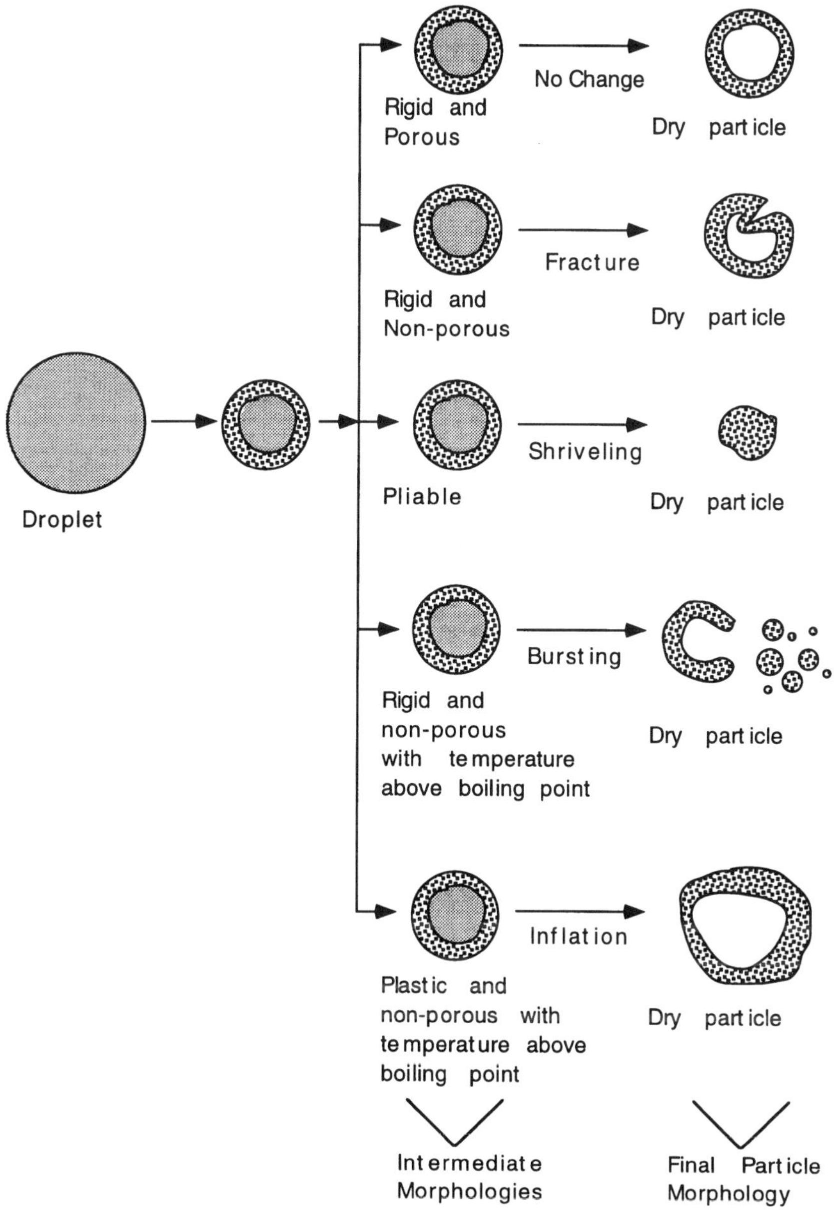


**Fig. 12.7.** Schematic representation of the influence of particle drying phenomena on particle morphology. (Adapted from Charlesworth and Marshall.)[42]

are coupled. As will be shown, the transport of solvent away from the surface of the droplet into the gas phase results in a lower concentration of the solvent in the liquid at the droplet surface, the value of which can be calculated only by considering the coupled transport processes. This lower concentration corresponds to a lower effective vapor pressure and a lower evaporation rate. However, the analysis can be simplified by noting that the evaporation time can be calculated by using an assumed value of the solvent concentration at the surface that corresponds to an assumed value of the vapor pressure.

The theory given in Chapter 4 described the evaporation time for pure solution droplets much larger than the gas mean free path ($\mathrm{Kn} = \lambda/d_p \ll 1$; $\lambda$ is roughly 0.06 $\mu$m for air at standard temperature and pressure). Here, an equation is presented that includes the effects of reduced droplet temperature relative to the temperature of the gas,[41] namely,

$$\tau_e = \frac{R\rho_1 d_{po}{}^2}{8D_g M\left(\dfrac{p_d}{T_d} - \dfrac{p_\infty}{T_\infty}\right)}, \tag{12.4}$$

where $\tau_e$ is the evaporation time, $R$ is the gas constant, $d_{po}$ is the initial droplet diameter, $D_g$ is the gas-phase diffusion coefficient of the solvent, $M$ is the molecular weight of the solvent, $\rho_1$ is the droplet density, $T_\infty$ is the temperature far from the droplet, $T_d$ is the temperature at the surface of the droplet, $p_\infty$ is the partial pressure of the solvent far from the droplet, and $p_d$ is the vapor pressure of the solvent at the surface of the droplet. This equation assumes evaporation into a stagnant gas and is therefore valid only for droplet diameters less than about 50 $\mu$m, for which droplet settling velocities are small. (See Chapter 3.) The evaporation time for a 10 $\mu$m water droplet at 20°C and 50% relative humidity is 0.15 sec. Evaporation times for smaller droplets can be much shorter. For example, the evaporation time for a 0.1-$\mu$m droplet under the same conditions is $5 \times 10^{-5}$ sec.

The solvent mass transport rates inside and outside the droplet are coupled through the relationship between the vapor pressure and concentration. The effective vapor pressure of the solvent is reduced by the presence of a solute. This lowers the evaporation rate of droplets. For ideal gas-phase behavior, the vapor pressure of a solution droplet is

$$\frac{p_d}{p_{\text{sat}}} = \gamma_A\, x_A \tag{12.5}$$

where $p_{\text{sat}}$ is the vapor pressure of the pure solvent, $\gamma_A$ is the activity coefficient (a thermodynamic property that is given in the literature for only a few systems of interest), and $x_A$ is the mole fraction of solvent in the liquid. For ideal, dilute solutions, the activity coefficient is unity. Equation (12.5) must be used for accurate modeling of these systems.

The preceding discussion examined the case where there was no convective transport (gas flow) of water vapor in the gas phase surrounding the droplet. This means that the gas surrounding the droplet was assumed stagnant. Droplets with diameters on the order of 50 $\mu$m and larger have appreciable settling velocities. In this case, the droplet moves relative to the gas, and convective heat and mass transport occur in the gas phase, leading to the need for a more involved method for calculating evaporation rates of droplets. Estimates of heat and mass transfer rates accounting for convective transport are made using chemical engineering correlations for heat and mass transfer coefficients that are obtained empirically and represented as a function of dimensionless groups.[43] In a typical process, the rate of heat transfer to the droplet controls the evaporation rate and is equal to the energy required for evaporation (the latent heat of evaporation). Evaporation times are shorter than would be predicted ignoring convection, but are still long: 1 to 10 sec for 500-$\mu$m water droplets in dry air and 0.1 to 1 sec for 100-$\mu$m droplets.

Evaporation of solvent in the presence of a crust (Step SL6) formed by solute precipitation requires transport of the solvent through a porous layer. The equation for this rate, accounting for diffusion through the crust and the gas surrounding the particle in series, is[9]

$$\frac{dm_s}{dt} = \frac{4\pi R_p D_g}{1 + \dfrac{D_g \delta_c}{D_{\text{cr}}(R_p - \delta_c)}}(m_d - m_g), \tag{12.6}$$

where $m_s$ is the mass of solvent in the droplet, $R_p$ is the outer radius of the particle, $\delta_c$ is the thickness of the crust, $D_{\text{cr}}$ is the effective diffusivity of the vapor in the crust, $D_g$ is the diffusion coefficient of solvent in the gas phase, $m_d$ is the mass concentration of the solvent at the droplet surface, and $m_g$ is the mass concentration of the solvent far from the particle. The effective diffusivity (the diffusion coefficient) is a function of the permeability of the crust, which depends in a complicated manner on the properties and operating parameters of the solute. When the diffusivity in the crust is much smaller than the diffusivity in the gas phase (a common situation) and the crust thickness is a significant fraction of the particle radius, the evaporation rate can be much lower than that for a single-phase droplet, as illustrated by the following equation:[9]

$$\frac{dm_s}{dt} = \frac{4\pi R_p (R_p - \delta_c) D_{\text{cr}}}{\delta_c}(m_d - m_g). \tag{12.7}$$

Because the effective diffusivity for the crust ($D_{\text{cr}}$ ) can be orders of magnitude lower than a typical gas-phase diffusivity ($D_g$), crust formation can increase evaporation times by many orders of magnitude.

Overall, the foregoing discussion indicates that typical droplet evaporation times of a few milliseconds for single-phase 10-$\mu$m droplets are much less than typical reactor residence times of 10 sec, but crust formation can dramatically slow down droplet evaporation. Evaporation from two-phase particles several hundred micrometers in diameter consisting of a precipitate and a saturated solution of the precursors has been investigated in the context of spray-drying. Charlesworth and Marshall showed that evaporation rates from water droplets and solution droplets were nearly identical until the completion of crust formation for droplet diameters of several hundred millimeters.[42] After that point, the solvent evaporation rate dropped off sharply. Similar results have been obtained in experiments by Nesic and Vodnik.[9] The evaporation rate after complete crust formation depended on the nature of the solute and the surrounding air temperature and humidity, which determined the particle morphology. A porous or structurally open crust resulted in a faster evaporation rate than did a solid crust. A common occurrence was that a solid crust would form that was impervious to the vapor and resulted in increases in the internal pressure of the droplet, followed by fracturing or rupturing of the crust. The evaporation rate temporarily increased while the crust re-formed. In general, the decrease in the evaporation rate after a crust is formed cannot be predicted, but must be examined experimentally. This is because of the wide variety of crust permeabilities that result from different types of solutes and drying conditions.

For large droplets on the order of 100 $\mu$m, with very high loadings of water where the relative humidity can be near 100% and when an impermeable crust is formed on the particles, the droplet evaporation time usually must be taken into account in determining the required reactor residence time. Because impermeable crusts, if formed, result in explosions of particles, releasing the liquid, the droplet evaporation time does not usually have to be considered in determining the reactor residence time required for spray pyrolysis. Droplet evaporation can in some cases occur even before the droplets have entered the reactor if the inlet gas is not at 100% relative humidity. Thus, the reactor design problem reduces to providing a temperature and residence time that will result in complete conversion of the precursors to the product by intraparticle reactions and transport processes.

### 12.3.4 Solute Transport in Droplets

Spray pyrolysis processes utilize a wide variety of precursors, including simple metal salts, colloidal precursors, and more complicated oligomeric species. As a result, a solid phase can be formed by a variety of routes—precipitation, oligomerization, gelling, etc. However, the most common route involves metal salts that precipitate from solution as crystalline materials.

Solute transport in solution (D6, SL8) is coupled to solvent transport (D4, SL5) through a diffusion equation of the form presented in Chapter 4. Comparison of the characteristic times for solute diffusion in the liquid phases, $d_p^2/D_1$, and in the gas phase, $d_p^2/D_g$, suggests that diffusion in the liquid is slower than droplet evaporation times, leading to solute and solvent concentration gradients in the liquid. In physical terms, the solvent leaves the droplet surface at a rate given by the evaporation rate. This tends to reduce the concentration of the solvent in the liquid at the droplet surface and increase the solute concentration at the surface. Solvent diffuses inside the droplet towards the particle surface, and solute diffuses towards the interior of the particle, but the rates are not rapid enough to keep up with the loss of solvent from the surface of the droplet into the gas phase, leading to a concentration gradient inside the droplet.

Evaporation of solvent proceeds from a single-phase solution droplet until a sufficiently high solute supersaturation develops in the droplet. Because the solute concentration is highest at the perimeter of the droplet, the solute nucleates and begins to precipitate out of the solution to form a crust at the surface of the particle. This can result in reduced evaporation rates and the formation of hollow particles as described earlier, both of which may be undesirable. Dealing with the problem requires an expression for the solute concentration profile in the droplet.

Several studies have developed expressions for the solute concentration profiles in droplets during evaporation.[2, 5, 19, 21, 44] The solution is obtained by solving the diffusion equation (see Chapter 6) inside the droplet with a decreasing droplet radius as a function of time. A solution for the case without convective transport (bulk motion of the liquid) in the droplet has been obtained for the special circumstance in which $-2R_p(dR_p/dt) = G_c$, a constant, where $R_p$ is the droplet radius.[5] This assumption is often reasonable and allows the analytical solution

$$\frac{C_s}{C_o} = \exp\left(\frac{-G_c}{4D_L}\right) = \exp\left(\frac{D_g M\left(\frac{p_\infty}{T_\infty} - \frac{p_d}{T_d}\right)}{2R\rho_1 D_L}\right), \tag{12.8}$$

where $R$ is the gas constant, $C_s$ is the concentration of solute at the droplet surface, $C_0$ is the concentration of solute at the center of the droplet, and $D_L$ is the liquid-phase diffusion coefficient of the solute. The assumption of a constant $G_c$ is valid because, considering the expression for the evaporation rate (see Chapter 4) of a droplet,

$$G_c = \frac{2\,D_g M}{R\rho_1}\left(\frac{p_\infty}{T_\infty} - \frac{p_d}{T_d}\right) \tag{12.9}$$

is constant under realistic conditions.

A key result of this analysis is that the solute concentration at the droplet surface ($C_s$) can be 25% greater than at the center of the droplet ($C_0$). This situation is shown in Figure 12.8. A second important result is that there is

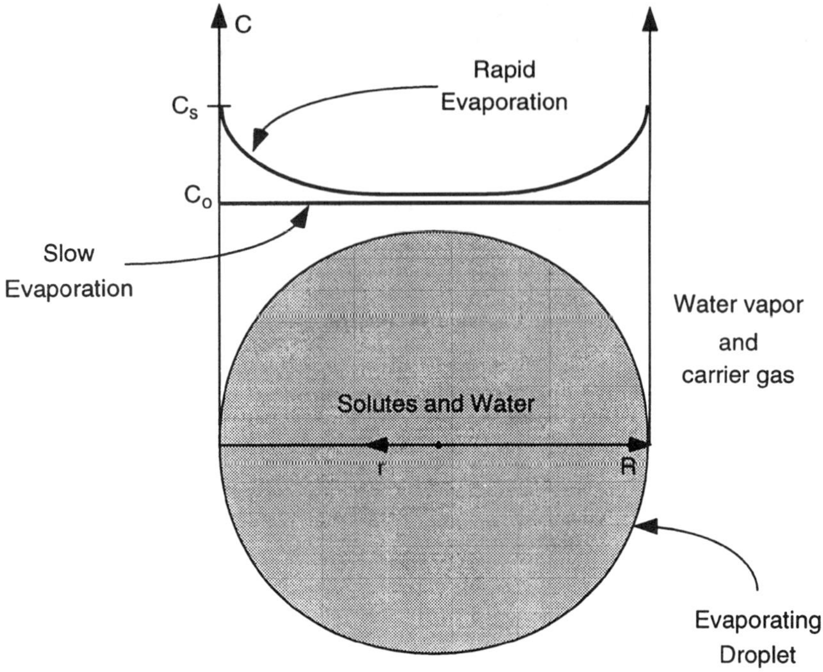


**Fig. 12.8.** Schematic representation showing how the solute concentration varies as a function of droplet radius during drying of droplets.

no dependence of the solute concentration on droplet diameter, which suggests that reducing the droplet size will not mitigate the problem of solute precipitation at the surface of the droplet.

The evolution of the solute concentration in the droplet with time is influenced by several factors, such as relative humidity ($p_\infty$), the initial solute concentration, and the ambient temperature ($T_\infty$). In some cases, these factors can be manipulated to favor the formation of solid particles. Messing (see Jayanthi et al.)[46] has modeled the solute concentration profiles in evaporating droplets using a version of Fick's second law (as outlined in Chapter 4). The system examined was the precursor zirconium hydroxy chloride in water, for which all the important physical and chemical characteristics are known. Solute precipitation was assumed to occur when the concentration reached the critical supersaturation for precipitation, a value that is significantly higher than the equilibrium concentration (the solubility limit). Figure 12.9 shows the predicted solute concentration as a function of radial position in the droplet for different initial solute concentrations. For low initial concentrations, the solute concentration reached the critical supersaturation at the edge of the droplet for conditions under which the concentration in most of the droplet was below the solubility limit. Under these

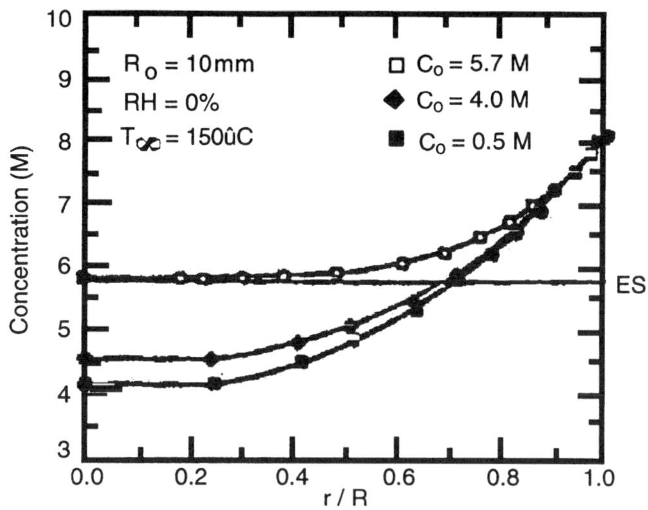


**Fig. 12.9.** Effect of initial concentration on solute concentration profile inside droplet at time of precipitation. Ambient temperature of 150°C. (Adapted from Jayanthi, Zhang, and Messing with permission of Elsevier Science, Inc., *Aerosol Science and Technology*, *19*, No. 4, pp. 478–490, Copyright 1993 by American Association for Aerosol Research.[46]

conditions, the solute should precipitate only at the perimeter of the droplet, and not at the interior, forming a shell around a liquid. At the highest initial solute concentration, the precipitation occurred under conditions where the solution in the entire droplet was above the solubility limit and precipitation occurred throughout the droplet. Because particle morphology is influenced by where precipitation occurs, this plot demonstrates that relatively small changes in operating parameters can have a large influence on particle properties.

### 12.3.5 Solute Precipitation in Droplets

Solute precipitation in droplets has not been described by any theory in a quantitative manner.[47] For a given solute, the supersaturation required for precipitation must be measured, and even then, it is a function of the exact composition of the solution (i.e., whether impurities are present to act as heterogeneous nucleation sites). Although a quantitative theory is not available, the qualitative features of the problem are reasonably well understood. As shown by Equation 12.8, the solute concentration profile in a droplet is controlled primarily by the relative humidity, which is related to $p_\infty$ and $T_\infty$. According to the theory of nucleation in liquids,[47] the rates of nucleation and crystal growth of the solute inside the droplet depend on the solute concentration through the saturation ratio $C/C_{sat}$ (where $C_{sat}$ is the saturation concentration) and properties of the salt such as surface energy. Because the rates of nucleation and growth determine particle morphology, it is $p_\infty$ and $T_\infty$ that ultimately control that morphology. This also suggests that the rate of crystallization can be limited by the drying rate.

Another problem associated with crystallization of solutes in droplets is sequential precipitation of the solutes in multicomponent systems, which can lead to segregation. As solvent evaporates, the solute concentrations increase and are highest at the surface of the droplet. Because the solubilities of the precursors are likely to be different, the solute with the lowest solubility will precipitate from solution first (in the simplest case). This will be followed by growth of crystals. After sufficient solvent has evaporated, the solute with the next-lowest solubility will begin to precipitate from solution. The extent of segregation will depend strongly on the nature of the solutes, but, in any case, segregation is limited to the particle diameter. Any segregation that does occur must be eliminated by holding the particle in the gas phase at a sufficiently high temperature for a sufficiently long residence time in order to allow intraparticle transport and reaction processes to proceed to completion. A solution to this problem is to use single-source precursors, which, for purposes of this discussion, are molecular, oligomeric, polymeric, or colloidal precursors, which contain all of the required elements uniformly dispersed in one structure. Such an approach, discussed in Chapter 8, has been used to form metal titanates by spray droplet-to-solid conversion.

### 12.3.6 Transport of Gaseous Species in Solid Particles

Several other mass transport processes occur during liquid-to-solid and solid-to-solid conversion. When a liquid precursor reacts with a gaseous co-reactant, co-reactants must diffuse into single-phase droplets (D8, D9) in order to react (D12). Gaseous products must be transported out of the droplet (D10, D11). For two-phase particles, with a solid layer surrounding a liquid, the same processes can occur and require diffusion through a solid layer (SL10, SL11, SL12 for the co-reactant; SL13, SL14, SL15 for the products) and reaction SL17. An example is the formation of metal oxides and metal-oxide-coated oils by reaction of metal alkoxide droplets with water vapor.[48, 49] An example in which this process is undesirable is the reaction of CVD precursors in solutes with a gaseous co-reactant, such as the reaction of hydrogen during AACVD.[50] For the case of solid-state reactions, however, gaseous co-reactants must diffuse into particles (DP5, DP6, DP7) in order for reaction to occur (DP13), and the reaction products must diffuse out (DP8, DP9, DP10). This class of processes occurs during almost all liquid-to-solid and solid-to-solid conversion processes, in which reaction of a solid is required to form the desired product.

Many materials undergo reactions without the need for a co-reactant. Very often, the precursor may contain waters of hydration. As the particles are heated, the water can be liberated and must diffuse to the surface of the particle and into the gas phase. An example is the reduction of PdO to Pd at temperatures above roughly 900°C where solid-state diffusion rates are high.[51] Once formed, the palladium particles are thermodynamically stable. As the particles cool to below roughly 900°C while exiting the reactor, PdO can re-form. However, because the diffusion coefficient is low, the kinetics of oxidation are slow enough that nearly pure Pd particles emerge from the reactor. This is an example of a process in which intraparticle diffusion of oxygen is not rapid enough to convert the Pd to PdO.

The loss of volatile reaction intermediates and products from solid particles involves diffusion through solid particles (DP11, DP12). This loss is usually undesirable as witness the loss of volatile species such as PbO from particles during the generation of such materials are $PbTiO_3$, Pb–Bi–Sr–Ca–Cu–O, and related compounds.[52]

The theory describing these transport processes is well established and is presented in Chapters 4 and 6. The most commonly used model is the shrinking-core model,[19] in which it is assumed that there exists a well-defined interface between the reacted and unreacted materials. Chapters 1 through 7 presented equations giving the time required for the conversion of particles for different limiting steps. For example, if the gas-phase transport steps (SL10, SL15, DP5, and DP10) are rapid and the reaction (SL17, DP13) is also rapid, so that the co-reactant and products do not penetrate the unreacted material (SL13, DP8) and reaction occurs at the interface, diffusion of the co-reactant (DP5, SL11) or reaction products (DP9, SL14) through the reacted layer controls the rate of the conversion. Commonly, the diffusion coefficient of the co-reactant is lower than that of the reaction products, and diffusion of the co-reactant through the reacted layer is the rate-limiting step. In this case, the time for complete reaction of the particle is[19]

$$\tau_{as} = \frac{\rho_b R_p^{\,2}}{6bC_{Ag} D_e}, \tag{12.10}$$

where $D_e$ is the effective diffusion coefficient in the film, $\rho_b$ is the molar density of the solid reactant, $R_p$ is the particle radius, $b$ is the number of moles of solid required for reaction with one mole of the gaseous reactant, $C_{Ag}$ is the concentration of gaseous co-reactant in the gas. The effective diffusion coefficient depends on the permeability of the layer, which cannot be predicted in most cases. For a porous material, the effective diffusion coefficient can be many orders of magnitude lower than the diffusion coefficient in a gas. A typical calculation is $\rho_b = 0.1$ mole/cm$^3$, $C_{Ag} = 10^{-5}$ mole/cm$^3$, $R_p = 10^{-4}$ cm, $D_e = 10^{-3}$ cm$^2$/sec, and $b = 1$. In this case, the time for complete reaction is roughly 0.01 second, a period short enough to allow conversion in a typical flow system. If the reacted layer is dense and diffusion must occur through a nonporous solid, the diffusion coefficient can be $10^{-10}$ $cm^2/sec$, and the time for complete reaction is roughly $10^5$ sec. This time is much longer than typical reactor residence times and indicates that conversion would not have enough time to occur. The example also illustrates the importance of the particle morphology in determining the time for conversion of solid reactants to products.

As a second example, the time for a gaseous reactant to diffuse into a solid particle can be estimated from the equations in Chapters 1 through 7. By assuming that the particle is spherical and that the diffusion coefficient is constant in the particle, we can calculate the time for a species such as oxygen atoms to diffuse into the center of a particle such as Pd, roughly $d_p^2/D$, where $d_p$ is the particle diameter and $D$ is the effective diffusion coefficient for the diffusing species in the solid.[51] For particles 0.1 $\mu$m in diameter and a diffusion coefficient of $10^{-10}$ $cm^2/sec$, the time taken is 1 sec. This suggests that if the diffusion coefficient is significantly higher than $10^{-10}$ $cm^2/sec$, intraparticle transport will be rapid for a residence time of 1 sec in the region of the reactor of interest. However, solid-state diffusion coefficients can be lower than $10^{-10}$ $cm^2/sec$,[53] and then the time available for diffusion to the center of the particle is not sufficient.

### 12.3.7 Intraparticle Reaction

Intraparticle reactions occur during spray pyrolysis both before and after solvent evaporation. Intraparticle reactions are responsible for converting the reactants to the final product and can also influence particle morphology. Rates of these reactions in the solid state are difficult to predict and are usually studied by thermogravimetric analysis, differential thermal analysis, and differential scanning calorimetry, as described in Chapter 8. These methods give information about temperatures of decomposition of the various reactants, as well as about the phases that are formed.

Gardner and Messing examined the thermal decomposition of magnesium acetate, nitrate, chloride, and sulfate.[54] Thermal analysis (both thermogravimetric and differential thermal analysis) results showed that the powder characteristics from spray pyrolysis were a direct result of the decomposition path of the precursor salt. The acetate required a temperature of 400 to 700°C, the nitrate 500 to 600°C, the chloride 800 to 1100°C, and the sulfate greater than 1100°C for full decomposition at heating rates of 10 to 100°C/min in air and nitrogen. The acetate and nitrate salts were acceptable precursors for spray pyrolysis, but the sulfate and the chloride required unacceptably high reactor temperatures for full conversion to the product. The results of differential scanning calorimetry (DSC) and spray pyrolysis studies showed that the acetate salt decomposed exothermically and led to the fragmenting of hollow, dried salt aggregates formed as the droplets dried. From these, individual MgO particles formed that were free from strong interparticle contacts. In contrast, the nitrate precursor led to the formation of hard agglomerates.

Several other types of intraparticle reaction processes have been examined by many researchers. Salata et al. generated CdS nanoparticles encapsulated in a polymer via intraparticle reaction,[55, 56] Jang et al. formed $YBa_2Cu_3O_7$ particles by intraparticle reaction of precursors driven by a laser,[57] and several groups formed AlN particles by reacting Al particles with $N_2$.[58–61] Also, gallium particles have been reacted with an As source to form GaAs particles,[62] and Adcock reacted hydrocarbon droplets with $F_2$ to produce fluorinated hydrocarbons.[63]

## 12.4 OVERALL QUALITATIVE BEHAVIOR OF LIQUID-TO-SOLID AND SOLID-TO-SOLID CONVERSION PROCESSES

Sections 12.2 and 12.3 discussed the individual physical and chemical phenomena that together make up the spray pyrolysis process. In this section, we discuss the overall behavior of these systems. We begin with a discussion of droplet evaporation in flow systems. This is followed by an analysis of the different factors that influence the morphology of particles. Finally, the different modes of reactor behavior are discussed.

### 12.4.1 Droplet Evaporation in Flow Systems

The theory describing the evaporation of isolated solution droplets was discussed in Chapter 4. In actual spray systems, evaporation of single droplets is influenced by evaporation of the other droplets and the nature of the surrounding gas. The characteristics of evaporation vary with position in the system because of reactor-scale phenomena such as heat and mass transfer. Droplet evaporation takes place in a nonisothermal system with a large number of droplets containing both a solute with a temperature-dependent solubility and a solvent whose vapor pressure depends on temperature. Furthermore, the relative humidity of the gas surrounding the droplets varies with position in the system. Because of these considerations, the qualitative behavior of this kind of system cannot be predicted by considering equations for single droplets. For this reason, a description of the process requires a full model that accounts for a variety of coupled phenomena.

The droplet evaporation phase of spray processes carried out in tubular flow systems has been modeled using an approach that describes the changes in droplet size, solvent vapor concentration in the carrier gas, and both droplet and gas temperatures along the axis of a reactor with radially uniform conditions.[21] Figure 12.10 shows the system that was modeled.[21] The model also accounted for solute concentration profiles in each droplet and could predict the onset of solute precipitation in the solution droplets. The model was used to describe the evaporation of sodium-chloride aqueous solution droplets in isothermal drying systems with adsorbents at the walls and in hot-wall flow systems with temperature gradients along the walls as a function of residence time, droplet size, solution molality (up to 2 M),

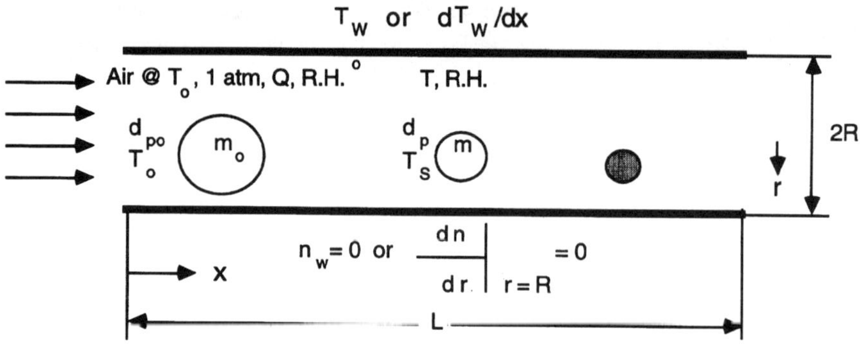


**Fig. 12.10.** Parameters used in modeling the droplet evaporation phase of spray pyrolysis. (Adapted from Xiong and Kodas.)[21]

droplet concentration ($10^6$ to $10^7$ number/cm$^3$), relative humidity of the carrier gas (0 to 50%), and wall temperature profiles. Sodium chloride was chosen because it is a model system for which the values of all the required parameters are known.

The equations describing this system are presented here as examples of the approach required to model such systems in general and to show how the equations in Chapters 1 through 7 must be coupled together to describe the overall behavior of the system.[21] More complicated, spatially two-dimensional approaches with non-monodisperse droplets can, in principle, be developed, but have not yet been used to describe the behavior of this class of systems.

The residence time is related to position along the reactor axis by

$$\frac{dt}{dx} = \frac{0.06\pi R_t^2}{Q}\left(\frac{T_o}{T}\right)\frac{(1 - y_w)}{1 - y_w{}^o}, \tag{12.11}$$

where $R_t$ is the tube radius, $t$ (sec) is the residence time, $x$ (cm) is the reactor axial coordinate, $Q$ (lpm) is the inlet air flow rate, $T$ (K) is the air temperature, $y_w$ is the mole fraction of water vapor in the air, and the super- or subscript "o" denotes properties at the reactor inlet ($x = 0$). The factor $(1 - y_w)/(1 - y_w{}^o)$ accounts for the increase in the volume of the gas associated with solvent evaporation from the droplets.

The change in droplet diameter $d_p$ with time along the reactor axis is

$$\frac{dd_p}{dx} = \left(\frac{dd_p}{dt}\right)\left(\frac{dt}{dx}\right) = \frac{4D_v m_1}{\rho_p d_p}(n - n_s)\left(\frac{dt}{dx}\right), \tag{12.12}$$

where $D_v$ is the vapor diffusion coefficient in air at temperature $T$, $m_1$ is the mass of a water molecule, $n$ and $n_s$ are the vapor concentration (molecules/cm$^3$) in air at $T$ and at the droplet surface at $T_S$, respectively, and $\rho_p$ is the droplet density, which is a function of the particle size.

A balance on the water vapor concentration gives

$$\frac{dn}{dx} = \left(\frac{dn}{dt}\right)\left(\frac{dt}{dx}\right) = \left[-2\pi d_p D_v N_\infty (n - n_s) - \frac{2k_m(n - n_w)}{R_t}\right]\left(\frac{dt}{dx}\right), \tag{12.13}$$

where $N_\infty$ is the droplet number concentration (number/cm$^3$), $n_w$ is the vapor concentration at the reactor wall, and $k_m$ is the vapor mass transfer coefficient for laminar tube flow. The first term on the right side accounts for evaporation of water vapor into the air from droplets; the second term accounts for transport of solvent vapor to the reactor walls.

The droplet temperature $T_S$ is influenced by the latent heat of water evaporation and heat transfer from the surrounding air to the droplet surface according to the equation

$$\frac{dT_S}{dx} = \left(\frac{dT_S}{dt}\right)\left(\frac{dt}{dx}\right) = \frac{1}{C_p d_p}\left[3H_L \frac{dd_p}{dt} + \frac{6h_s}{\rho_p}(T - T_S)\right]\left(\frac{dt}{dx}\right), \tag{12.14}$$

where $C_p$ is the heat capacity of the droplet, $H_L$ is the latent heat of water evaporation, and $h_s$ is the heat transfer coefficient around the droplet.

The air temperature changes because of loss of heat by transfer to the droplets and increases because of transport from the reactor walls according to the equation

$$\frac{dT}{dx} = \frac{1}{F_a C_{\text{pa}}}\left[-\pi^2 R_t^2 d_p^2 N_\infty n_s (T - T_S) + 2\pi R_t h_w (T_w - T)\right], \tag{12.15}$$

where $F_a$ is the air molar flow rate, $C_{\text{pa}}$ is the heat capacity of the wet air, and $h_w$ is the heat transfer coefficient at the reactor wall.

These coupled equations, along with several other auxiliary equations, can be solved, using the appropriate boundary conditions, by standard numerical techniques.

The initial vapor concentration is related to the inlet relative humidity R.H.$^{\text{o}}$ by the equation

$$n_{\text{o}} = \frac{\text{R.H.}^{\text{o}} p_{\text{sat}}{}^{\text{o}}}{k_B T_{\text{o}}} \tag{12.16}$$

where $p_{\text{sat}}^{\text{o}}$ is the equilibrium water vapor pressure at $T_{\text{o}}$ and $k_B$ is Boltzmann's constant. Similarly, the vapor concentration at the droplet surface is

$$n_s = \frac{p_{\text{sat}}}{k_B T_S}, \tag{12.17}$$

where $p_{\text{sat}}$ is the vapor pressure at the droplet surface at $T_S$. The presence of the solute lowers the equilibrium vapor pressure over the solution according to

$$p_{\text{sat}}(T_S) = a p_{\text{sat}}{}^{\text{o}}(T_S), \tag{12.18}$$

where $a_a$ is the water activity in the solution droplet and is related to the solution molality $m$ and the water osmotic coefficient $\phi$ by

$$a_a = \exp(-0.036 m \phi). \tag{12.19}$$

The coefficient $\phi$ is in turn a function of $m$ and can be evaluated using standard approaches.

For ideal mixing of the solute and water, the droplet density is

$$\rho_p = \rho_L + \left(\frac{d_{\text{ps}}}{d_{\text{p}}}\right)^3 (\rho_s - \rho_L), \tag{12.20}$$

where $\rho_L$ is the density of pure water, $\rho_s$ is the density of the solute, and $d_{\text{ps}}$ is the equivalent sphere diameter of a completely dried droplet. The average molality of the droplet is

$$m_m = \frac{1000 \rho_s}{M_s \rho_L \left[(d_p/d_{\text{ps}})^3 - 1\right]}, \tag{12.21}$$

where $M_S$ is the molecular weight of the solute.

As shown earlier, a concentration profile develops within the droplet, with the highest solute concentration found at the surface. The solution concentration at a radial distance $r$ from the center of an evaporating droplet before precipitation takes place can be related to the corresponding saturation ratio

$$S = \frac{a_{\text{sol}}}{a_c} = \frac{m_s \gamma_s}{m_c \gamma_c}, \tag{12.22}$$

where $a_{\text{sol}}$ and $a_c$ are the solute activities at $m_s$ and at the equilibrium saturation concentration $m_c$, respectively, and $\gamma_{\text{sol}}$ and $\gamma_{\text{c}}$ are the corresponding solute activity coefficients at $m_s$ and $m_c$. This value of the saturation ratio was compared with the critical supersaturation ratio $S^*$ to determine the onset of solute nucleation at the droplet surface.

Classical homogeneous nucleation theory gives the critical supersaturation ratio for nondissociating solutes[44, 47] as

$$S^* = \exp\left[\frac{16\pi\sigma^3 M_s^2}{3k_B T_S^3 \rho_s^2 R_g^2 \ln(A_r/I)}\right], \tag{12.23}$$

where $\sigma$ is the surface energy of the solute, $R_g$ is the gas law constant, $A_r$ is a rate constant depending on the nature of the solute, and $I$ is the nucleation rate, usually taken to be 1 nucleus/s · cm$^3$ for spontaneous nucleation to occur. This equation can be used to evaluate the critical supersaturation ratio and to determine the onset of solute nucleation by comparing the critical ratio with the actual saturation ratio $S$.

The model can provide information about the influence of various process parameters on the droplet diameter along the flow system. The gas temperature in a typical laboratory-scale system increases rapidly at the entrance, as shown in Figure 12.11.[21] The time associated with the temperature increase can be on the order of 1 second. The corresponding heating rates are several hundreds of degrees Celsius per second. In industrial systems, the droplets are often introduced into a gas that is heated by a combustion process. In that case, the heating rates are similar or even higher.

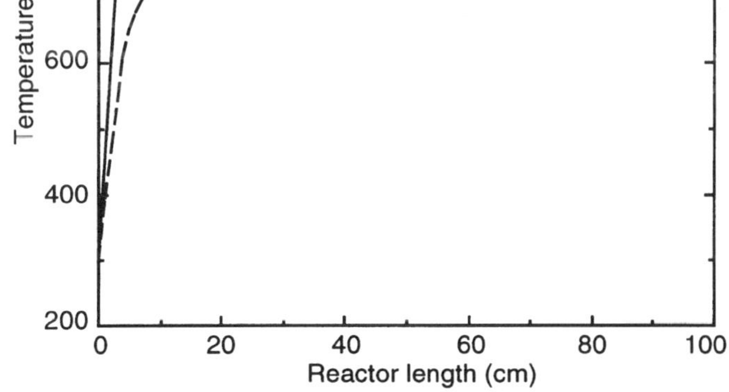


**Fig. 12.11.** Carrier gas temperature profile along the reactor length for a number concentration of $10^6$ number/cm$^3$ at two reactor wall temperatures. (Adapted from Xiong and Kodas.)[21]

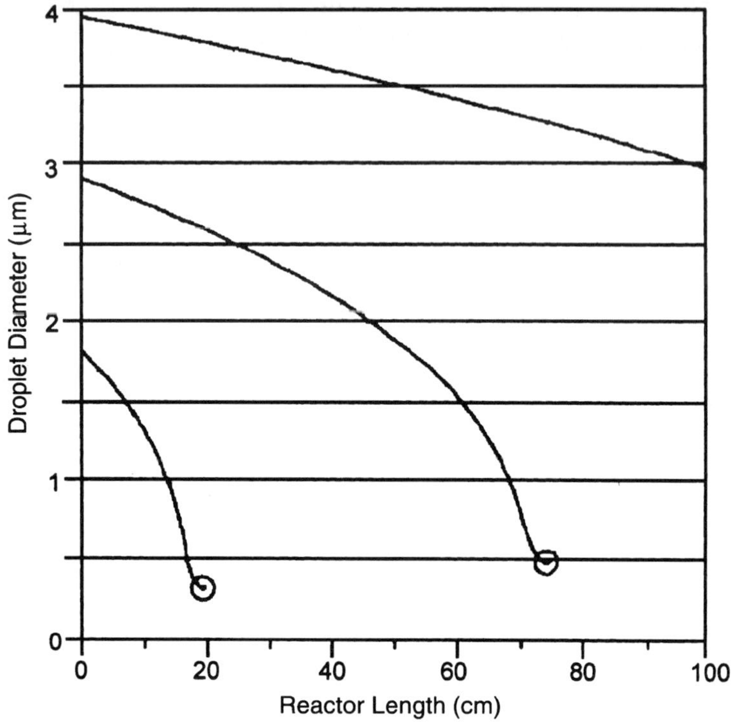


**Fig. 12.12.** Evolution of NaCl droplet diameter along the axis of a diffusion dryer for various initial droplet sizes ($d_{po}$) at R.H. = 0. $N_o = 5 \times 10^6$ number/cm$^3$. $T_w = 50$°C and $m_o = 0.036$ mol kg$^{-1}$. Circles denote the onset of solute nucleation. (Adapted from Xiong and Kodas.)[21]

Figure 12.12 shows that decreasing the initial droplet size accelerates droplet evaporation.[21] Changing from 4 to 2-$\mu$m droplets results in a dramatic decrease in the length of the section required to dry the particles. Larger droplets contain much more water, which evaporates and increases the relative humidity in the gas stream, slowing down droplet evaporation. This illustrates that simply changing the type of aerosol generator and thereby changing the droplet size can result in large changes in the length of the system required to dry the particles.

Figure 12.13 shows that increasing the droplet number concentration from $1 \times 10^6$ to $5 \times 10^6$ to $1 \times 10^7$ number/cm$^3$ increases the time required to evaporate the droplets.[21] Again, this occurs because the water vapor entering the gas from the particles increases the relative humidity, thereby slowing down droplet evaporation. This effect is demonstrated more directly in Figure 12.14, which shows how the droplet diameter depends on the inlet

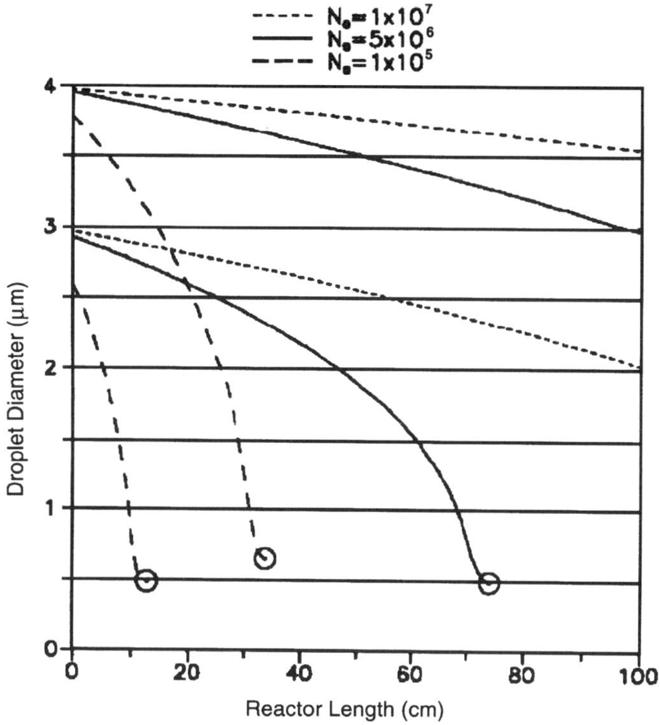


**Fig. 12.13.** Evolution of NaCl droplet diameter along the axis of a diffusion dryer for various droplet concentrations ($N_o$) at different initial droplet sizes ($d_{po}$) R.H. = 0, $T_w$ = 50°C, and $m_o$ = 0.036 mol $kg^{-1}$. (Adapted from Xiong and Kodas.)[21]

relative humidity.[21] The influence of the wall temperature is greater than of the inlet relative humidity because the vapor pressure of water increases exponentially as the gas flows down the tube and the temperature increases. This shows that, by changing droplet generators, the droplet concentration is changed, which can lengthen the time required for solvent evaporation, although the effect can be overcome by high wall temperatures.

Figure 12.15 shows that increasing the wall temperature gradient (up to 100°C/cm) along the reactor axis increases the droplet evaporation rate but does not appreciably change the droplet size at the point of precipitation for a given droplet size and solute concentration.[21] Droplets evaporate at less than 10 cm versus at 80 cm for gradients of 100 and 1°C/cm, respectively. Thus, control of the droplet size at the onset of solute nucleation by varying process parameters other than the solution concentration and initial droplet size is limited.

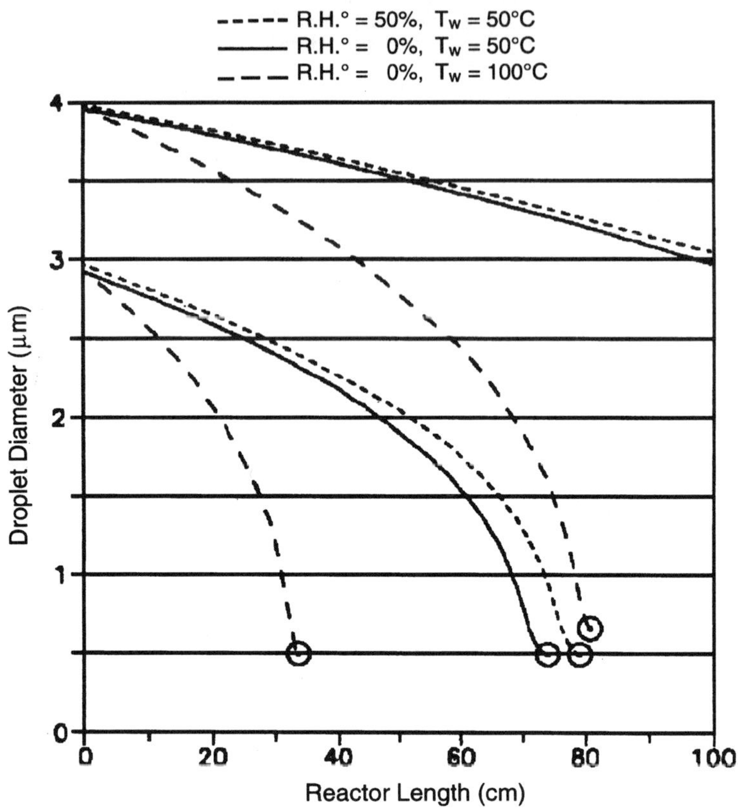


**Fig. 12.14.** Evolution of NaCl droplet diameter along the axis of a diffusion dryer for various inlet gas relative humidities (R.H.) and wall temperatures ($T_w$) at different initial droplet sizes ($d_{po}$), $N_o = 5 \times 10^6$ number/cm$^3$ and $m_o = 0.036$ mol kg$^{-1}$. (Adapted from Xiong and Kodas.)[21]

### 12.4.2 Influence of Solvent Evaporation on Particle Morphology

The previous model described spray systems up to the point of solute precipitation. The details of solvent evaporation during solute precipitation are what determine the morphology of the dried particles. Unfortunately, the complexity of the solute precipitation process precludes quantitative modeling of this phase of spray pyrolysis. Here we describe, at a qualitative level, the influence of precursor characteristics on particle morphology. Several studies have examined this problem.[44, 64–66]

Solvent evaporation can result in precipitation of the solute in the particle at different radial positions, depending on the process conditions. In some cases, a solid shell can be formed at the droplet surface, and the solvent must evaporate through this shell composed of the solute. (See Fig. 12.7.) During evaporation, the shell may retain its shape, crack, burst open, shrivel,

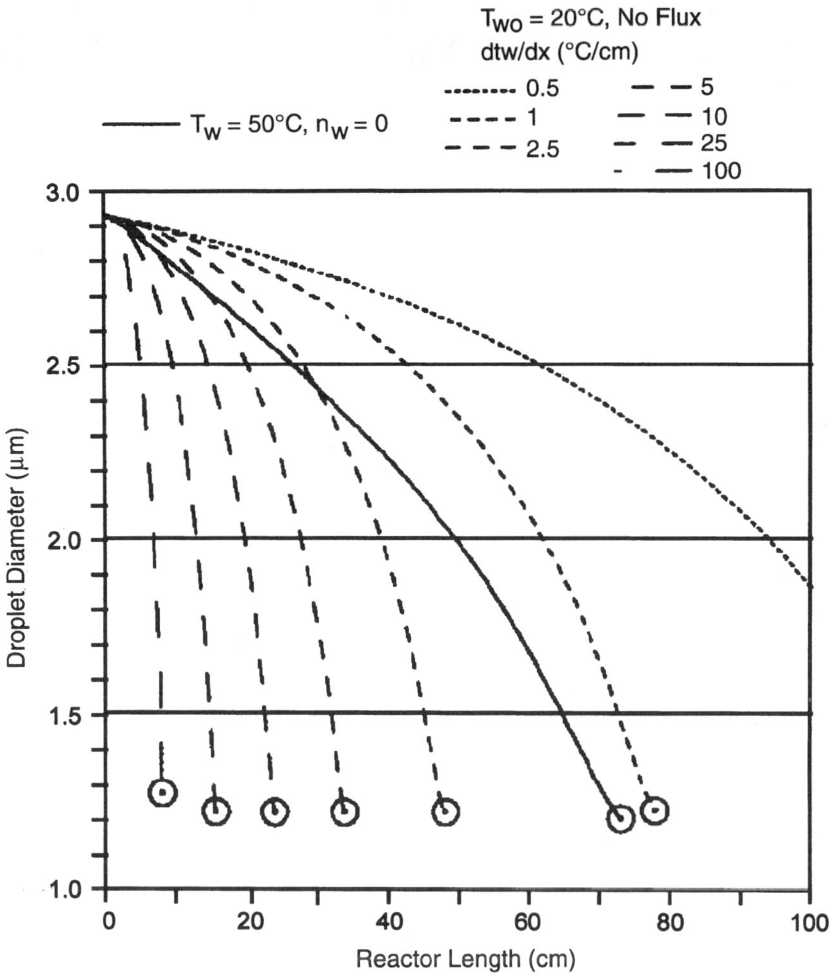


**Fig. 12.15.** Evolution of NaCl droplet diameter for $d_{po} = 3\ \mu m$, R.H. = 0, $N_o = 5 \times 10^6$ number/cm$^3$, and $m_o = 0.036$ mol kg$^{-1}$ in a diffusion dryer at $T_w = 50°C$ (solid line) and in a hot reactor tube with various wall temperature gradients ($dT_w/dx$) at $T_{wo} = 20°C$ and no-flux mass boundary condition at the wall (dashed lines). (Adapted from Xiong and Kodas.)[21]

collapse, or undergo numerous other changes. Bursting of the shell can occur due to slow transport of solvent through the shell. At the other extreme, a shell may not be formed at all, and evaporation will take place from a particle consisting of crystallites in a solvent. An example of hollow particles resulting from solvent evaporation with a dissolved metal salt is shown in Figure 12.16. Figure 12.17 shows hollow $SrTiO_3$ particles formed by spray

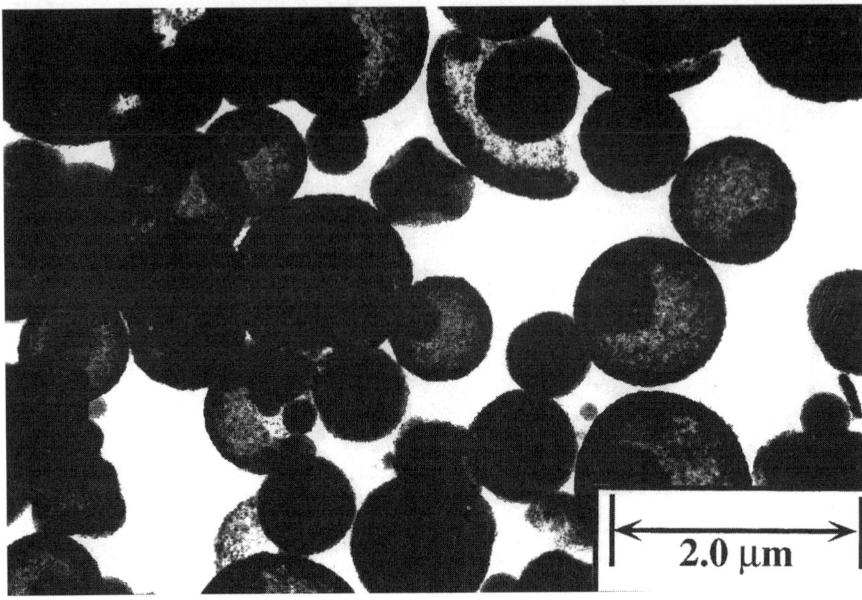


**Fig. 12.16.** Hollow MgO particles made from $Mg(NO_3)_2 \cdot 6H_2O$ at 800°C in ultrasonic aerosol generator using *dry* carrier air. (Courtesy of Ortega and Kodas.)

pyrolysis using strontium nitrate and titanium citrate solutions.[67] The latter example shows how droplet evaporation with precursors that form oligomeric species during drying can lead to hollow and exploded particles. The low reactant solubilities, on the order of 0.01M, led to solute precipitation before most of the water had evaporated from the particles. The internal pressure produced by the evaporating solvent resulted in bursting of the particles.

There have been few studies of the morphology of dry *precursor* particles formed by solvent evaporation for particles smaller than 10 $\mu$m and precursors of relevance to materials applications. In one of these studies, Leong examined the influence of droplet drying conditions on particle morphology for 10- to 100-$\mu$ droplets.[64] Very low particle concentrations were used, less than 1000/cm$^3$, many orders of magnitude lower than in spray processes. The characteristics of the dried salt particles were examined by scanning electron microscopy. Limited control of particle morphology was achieved through control of the temperature and humidity of the carrier gas used in generating the aerosol. However, precursor characteristics also had a very strong influence on particle morphology. The relative humidity and temperature, along with the crystal habit of the solute and its solubility, surface energy, and mass, determined the rates of formation and growth of crystals. The number and size of crystals formed in a particle and their habit and arrangement in turn controlled particle morphology. Relative humidity, temperature, solute solubility, and mass are well-defined quantities. However,

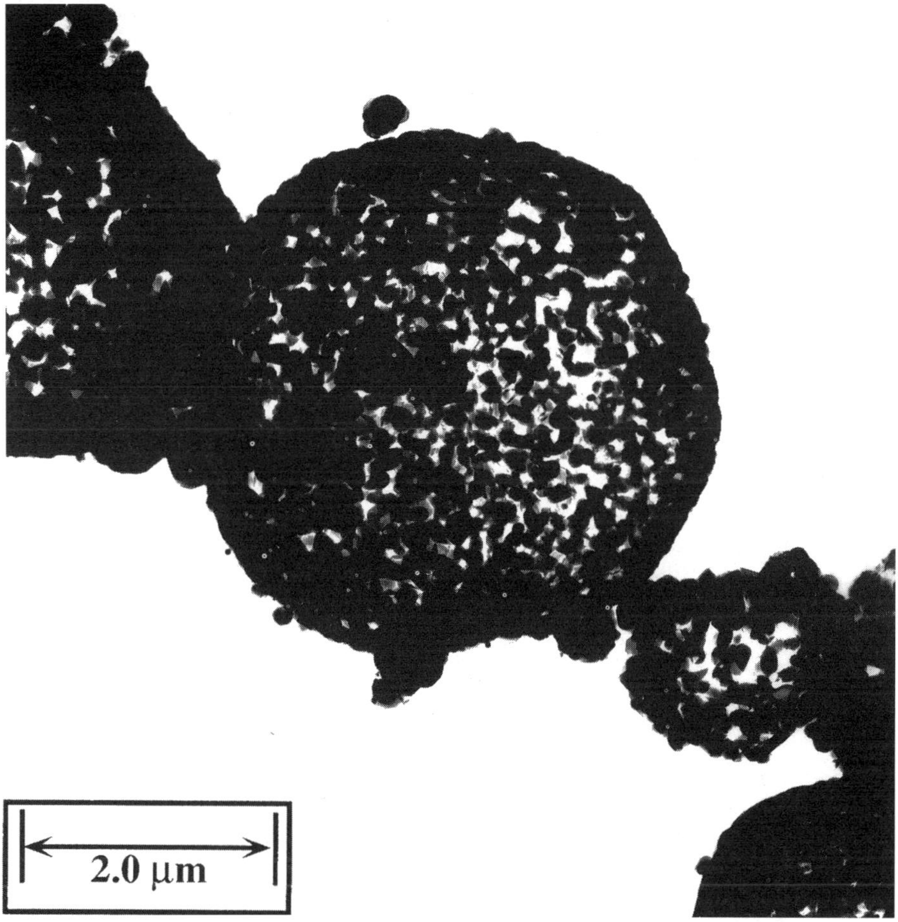


**Fig. 12.17.** $SrTiO_3$ produced by spray pyrolysis from a solution of strontium nitrate and titanium citrate at 1560°C. (Courtesy of Hikida.)[67]

because the crystal habit and surface energy of a compound are usually not known or well characterized, Leong[64] concluded that the final particle morphology must be determined empirically. In one example of how drying conditions influenced particle morphology, solution droplets of NaCl in $H_2O$ were dried and transformed from shells of many crystals to more solid polycrystalline particles and, finally, to solid single crystals by increasing the evaporation rate. This change was attributed to a higher relative reduction in the time for solute diffusion relative to the time for activation of a second nucleus and resulted in a decrease in the number of nuclei that were activated as the drying rate was increased. Nuclei that were formed could grow rapidly, reduce the saturation ratio, and halt the formation of new crystallites.

The influence of drying conditions on precursor and product particle morphology has also been examined. In one case, the particles were converted to solid products by chemical reaction, and the role of solvent evaporation in determining morphology was inferred from the final morphology.[68] Zhang and Messing have shown that, for typical spray pyrolysis conditions with 10-micron droplets and high heating rates, solid particles of $ZrO_2$ are formed when a droplet's initial relative solution saturation is $10^{-2}$ and the precipitated salt is sufficiently permeable to permit evolution of the remaining solvent after precipitation.[68] The formation of solid particles was

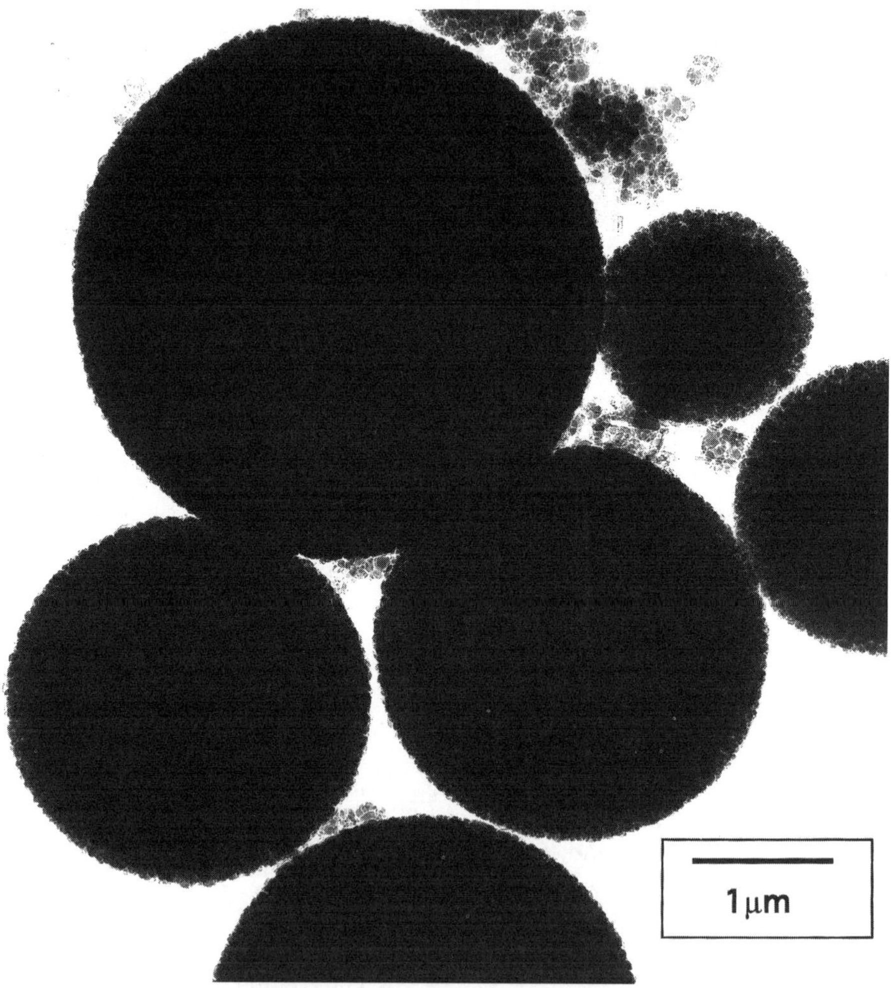


**Fig. 12.18.** $SiO_2$ particles produced by spray pyrolysis using a solution of colloidal $SiO_2$ as a precursor. (Courtesy of Majumdar and Kodas.)

attributed to a longer drying time before precipitation and a shorter diffusion distance in which to maintain chemical homogeneity before precipitation.

Spray pyrolysis can also be carried out using solid precursors in suspension in a liquid. In this case, evaporation can proceed in a qualitatively different manner from evaporation with dissolved molecular precursors. Figure 12.18 shows $SiO_2$ particles produced using colloidal $SiO_2$ as the precursor. With two or more colloidal precursors and no molecular precursors, there is no solute to precipitate, and it is possible to form solid particles during the drying phase. The drying of droplets containing colloidal material and slurries has been studied by Cheong et al.[1] and by Sano and Keey.[13]

With one or more colloidal species and one or more molecular species, solute precipitation can occur onto the colloidal particles in the droplet.[69, 70] Nucleation of the solute can still occur at the surface of the droplet when the loading of the solute is high. Solid mullite particles have been formed from colloidal silicon dioxide and aluminum nitrate.[69] For low loadings of solute, nucleation can take place on the suspended particles, and crust formation can be avoided. This example demonstrates how the use of colloidal precursors can lead directly to the formation of solid particles at relatively low reactor temperatures. Solvent evaporation resulted in precipitation of the aluminum nitrate on the surface of the silicon dioxide instead of at the surface of the droplet, thereby avoiding the formation of a crust. The conditions favoring the formation of solid particles during evaporation are discussed in Section 12.4.

### 12.4.3 Influence of Precursor Characteristics on Particle Morphology

Precursor characteristics vary widely and in many cases determine whether or not hollow or porous particles are formed. A key distinction is whether a precursor melts before reacting. Precursors that melt before reacting can form hollow particles after solvent evaporation, but can still eventually generate solid particles because the hollow precursor particle melts, giving a droplet of the precursor before reaction occurs. However, melting of the precursor does not ensure the formation of solid particles because the gases that have evolved during reaction can cause the particles to expand. Also, if the volumetric yield of the precursor is low, the reaction of molten precursor particles can give highly porous or hollow particles. Shriveled $ZnO:Al_2O_3$ particles formed from $Zn(NO_3)_2$ and $Al(NO_3)_3$ are shown in Figure 12.19.[71]

One of the most important conditions required for the formation of solid particles from precursors which melt is that the molten precursor and the solid reacted layer be sufficiently permeable to allow the gaseous reaction products to escape from the particle. This approach to solid-particle formation is highly dependent on the characteristics of the precursors, and general conclusions cannot be easily drawn. The other important condition is that, upon reaction, the metal or ceramic yield is high enough to exceed the percolation limit.[72]

**Fig. 12.19.** $ZnO:Al_2O_3$ particles formed by spray pyrolysis from a solution of $Zn(NO_3)_2$ and $Al(NO_3)_3$. (Courtesy of Ortega.)[71]

### 12.4.4 Influence of Intraparticle Reaction on Particle Morphology

A further consideration regarding particle morphology is the change in volume associated with converting the precursor to the product by chemical reaction. Jayanthi et al. have used a percolation criterion to correlate particle morphology with the volume fraction of product in a particle.[46] When the particle contains a volume fraction of more than 0.16 of the precursor or reaction product at the time of solute precipitation, the material inside the particle can form an interconnected three-dimensional structure corresponding to a porous, but not-hollow, particle. With a droplet, at the point of precipitation throughout the entire particle volume, the amount of porous precursor will not be enough to fill the space within the particle unless the volume fraction of the solid is greater than 0.16. In the case of a particle of

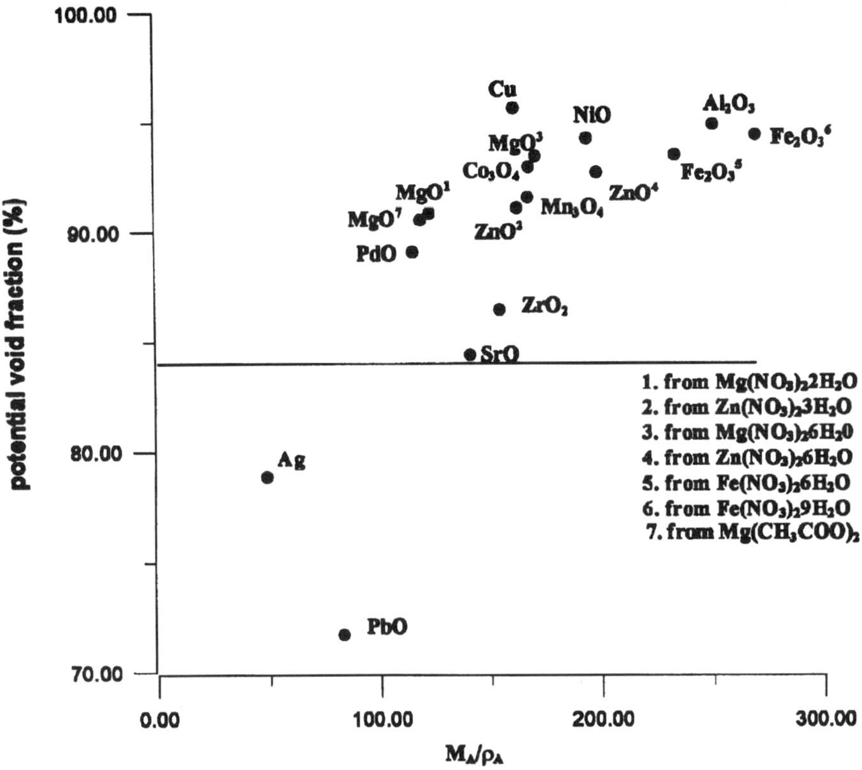


**Fig. 12.20.** Plot of potential void fraction (%) vs. $M_A/\rho_A$ for a number of metal-oxide powders. (Adapted from Jain et al.)[72]

molten precursor, however, a small change in volume with reaction can provide nearly solid particles with low porosity if the particle morphology is not influenced by other phenomena, such as melting and puffing up due to the evolution of reaction product gases. However, a large reduction in volume on reaction can result in the formation of highly porous or hollow particles if the percolation criterion is not satisfied.

Figure 12.20 shows the potential void fraction (the void fraction if a solid dry particle of precursor stayed constant in volume during conversion of precursor to product) for conversion of a variety of precursors to products.[72] This fraction ranges from roughly 70 to 95% (below and above the percolation criterion of 0.84), reflecting the low density of the reactants and the high density of the products. The plot demonstrates that molten precursors with low volumetric yields are likely to produce porous or hollow particles, regardless of the morphology of the precursor particle formed by solvent evaporation. In contrast, solid particles are favored for the cases of $AgNO_3$ and $Pb(NO_3)_2$.

## 12.5 MODES OF REACTOR BEHAVIOR

The process of forming particles by spray pyrolysis can be viewed in the simplest case as drying of droplets followed by solid- and liquid-phase chemical reactions in a microreactor (the particle) suspended in a gas to form the product. However, in many cases, the behavior is more complicated. If only nonmetallic species (O, CO, $CO_2$, NO, $NO_2$, etc.) are transported across the particle–gas interface, only intraparticle and surface reactions occur, and the product stoichiometry can be precisely controlled during mixing of the precursor solution. This case will be discussed in Section 12.5.1. When volatile species are present, gas-phase, surface, and intraparticle chemical reactions can occur when metal-containing molecules are transported across the particle–gas interface. In this case, control of the stoichiometry and phase purity becomes far more difficult than in the case of a purely intraparticle reaction. These situations are discussed after Section 12.5.1.

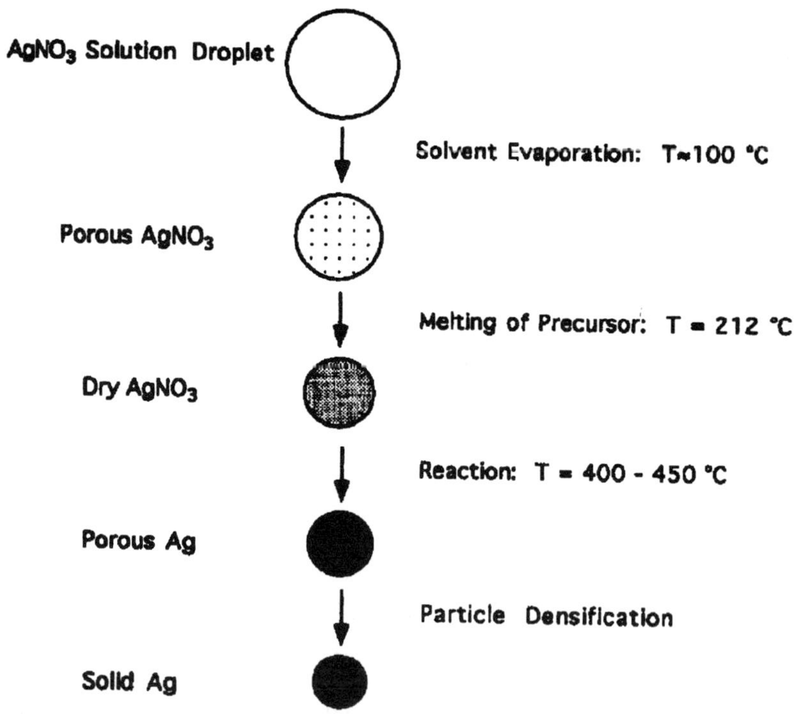


**Fig. 12.21.** Example of intraparticle reaction steps that occur in the formation of Ag particles. (Adapted from Pluym et al.)[73]

### 12.5.1 Intraparticle Reaction with No Volatile Components

In the purely intraparticle reaction mode (Fig. 12.21), the solvent evaporates from the droplets formed by the aerosol generator, and then metal salts in the particle nucleate to form one or more solid phases.[73-75] This is followed at higher temperatures by decomposition of the metal salts in the particles, forming reaction intermediates. As the reaction proceeds, these intermediates diffuse and react to form the product. Because no metal-containing species leave the particles, the spread of the particle size distribution is limited by the spread of the size distribution of the droplets formed by the aerosol generator, unless the morphology of the particles varies with particle size.

An example of the intraparticle reaction mode for a more complex material is the formation of $YBa_2Cu_3O_{7-x}$ by decomposition of metal nitrates. The melting points of the oxides of Cu, Y, and Ba (1326 to 2410°C) and other species that could be intermediates are far above typical reactor operating temperatures (900 to 1000°C); thus, evaporation of the oxides does not occur. Similarly, most precursors to these materials, such as metal nitrates, are also not volatile except under special circumstances. Because the Y, Ba, and Cu are confined to the particles, it is possible to form phase-pure $YBa_2Cu_3O_{7-x}$ by intraparticle reaction.

Although particles with a uniform chemical composition often may form, cases exist in which intraparticle reaction leads to chemically nonuniform particles due to either kinetic or thermodynamic limitations (Fig. 12.22).[76] One example of a kinetic limitation is the incomplete reaction of precursors due to diffusion limitations. Two or more precursors may partially segregate during solvent evaporation and solute precipitation or during precursor decomposition. Subsequent reaction forms two or more phases in the particle that are distributed throughout it as crystallites with sizes much smaller than the particle. Diffusion is then required to bring all the elements together in the correct ratios, but there may not be sufficient time to erase inhomogeneities.

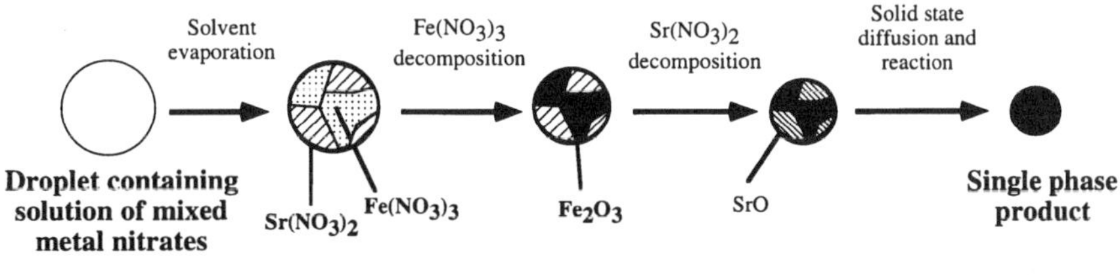


**Fig. 12.22.** Example of the formation of particles in the Sr–Fe–O system with nonuniform composition due to the different thermal decomposition temperatures of $Fe(NO_3)_3$ and $Sr(NO_3)_2$. (Adapted from Senzaki et al.)[76]

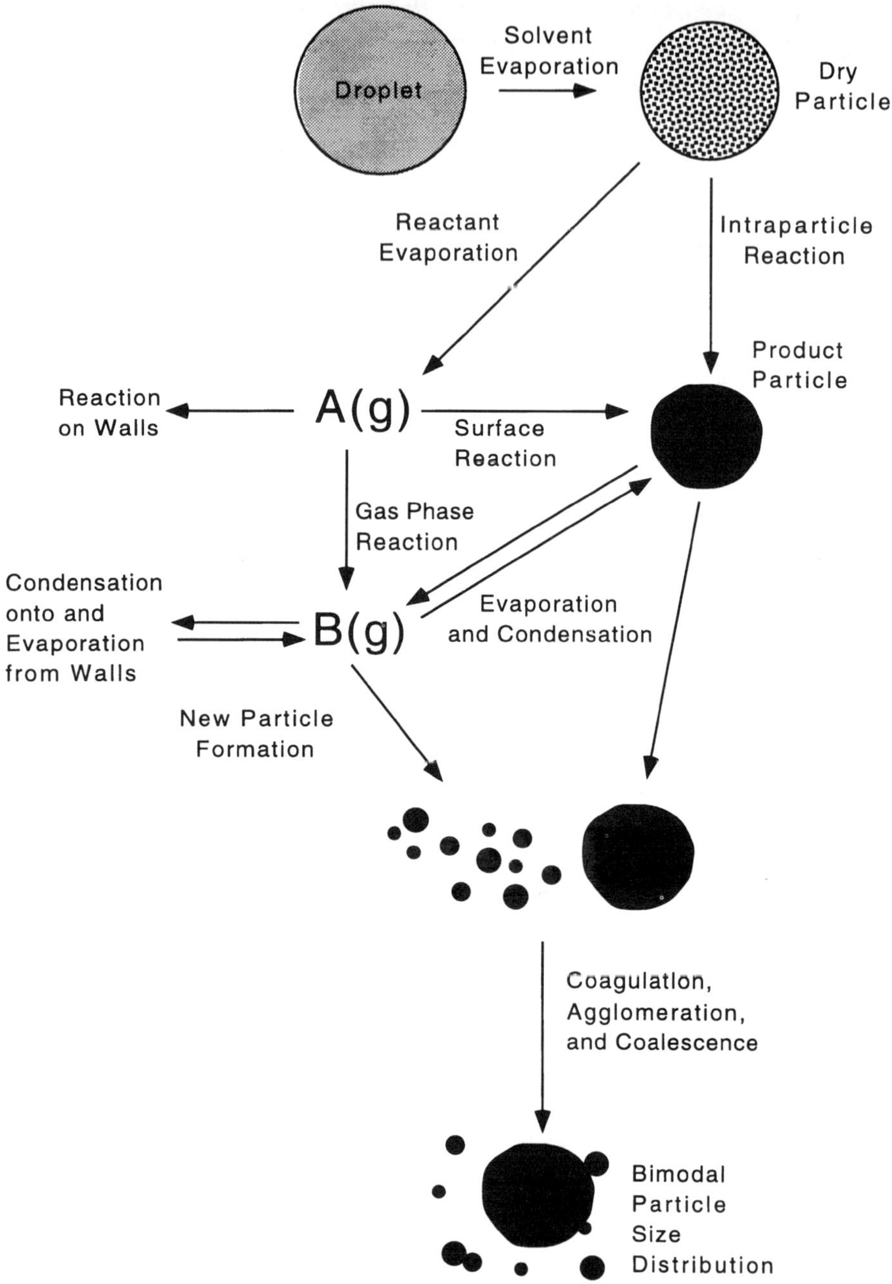


**Fig. 12.23.** Schematic representation of the steps that can occur during formation of a powder by spray pyrolysis when a volatile species is present.

### 12.5.2 Intraparticle Reaction with Evaporation of Volatile Reactants

Spray pyrolysis systems for powder generation can also be operated under conditions where both intraparticle reaction and gas-to-particle conversion of the reactant occur (Fig. 12.23). Gas-to-particle conversion can occur if the precursor (A in the figure) is volatile and is transported across the particle–gas interface. The precursor can react to form a product that may be different from the products of intraparticle reaction. The vapor of the product species, either after being formed by chemical reaction in the gas phase or during cooling while exiting the reactor, may condense on the walls of the system, condense on the particles, or nucleate homogeneously to form new particles. Such behavior has been observed during the operation of single- and multicomponent systems. Zinc acetate (or, more likely, $Zn_4O(OAc)_6$—see Chapter 8) precursor evaporation, gas-phase chemical reaction, and nucleation have been observed during ZnO particle formation by spray pyrolysis. Liu et al.[77] have also reported evaporative losses of silazane precursors from particles.

### 12.5.3 Intraparticle Reaction with Volatile Intermediates or Products

Evaporation of a volatile reaction product can result in changes in particle size and morphology in single-component systems or in the loss of a volatile component in multicomponent systems. Figure 12.23 shows the processes that occur if a single reaction product (B) is involved and is volatile. This species may evaporate and condense back onto the particles, be lost to the reactor walls, or form new particles as described in the next section. Figure 12.24 shows how the morphology of $V_2O_5$ particles is influenced by evaporation and condensation.[78] At lower temperatures spherical nanocrystalline particles are formed, while at higher temperatures platelets are formed as the result of evaporation and condensation of $V_2O_5$.

Another example of a single-component system in which evaporation of an oxide occurs is the synthesis of $PbO_x$. Figure 12.25 shows the aerosol dynamics in schematic form.[79] Evaporation and wall losses of PbO were evidenced by a rapidly declining powder collection rate for $T > 500°C$. Approximately 45% of the Pb entering the reactor was lost at 700°C, while 95% was lost at 800°C. This result is more general and has been observed in a variety of volatile metal oxides. Figure 12.26 (Xiong et al.)[80] shows how powder production rate decreased with increasing temperature because of wall losses and how these losses depend primarily on the vapor pressure of the metal oxide during spray pyrolysis.[79]

The physicochemical processes shown in Figure 12.23 can influence the properties of the product powders, and hence the properties of ceramics produced from the powders, in multicomponent systems. Figure 12.27 illustrates how the evaporation of PbO can modify the composition of Pb–Bi–Sr–Ca–Cu–O particles.[52] Figure 12.28 shows how these losses of

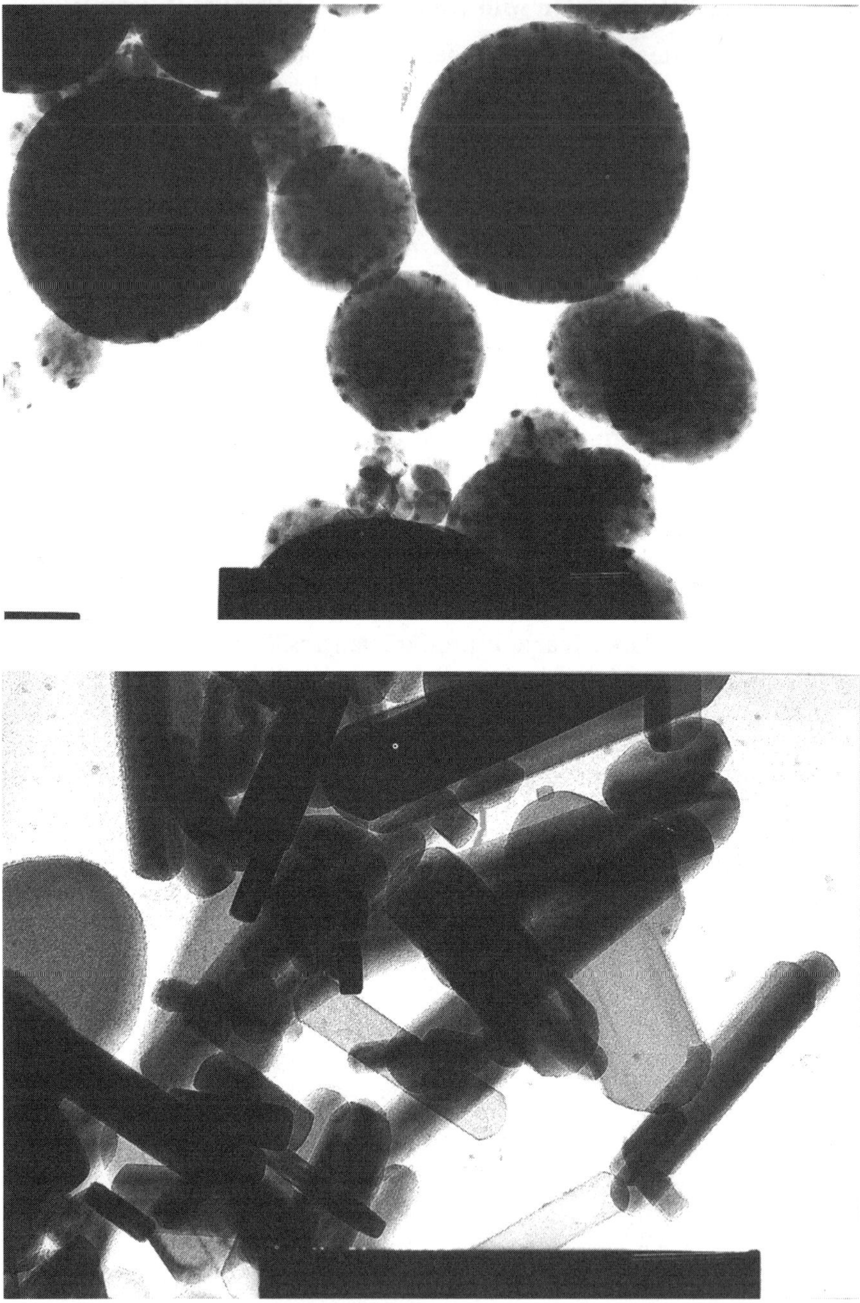

**Fig. 12.24.** Scanning electron microscope pictures showing the different morphologies of $V_2O_5$ obtained at different temperatures. (Courtesy of Lyons et al.)[78]

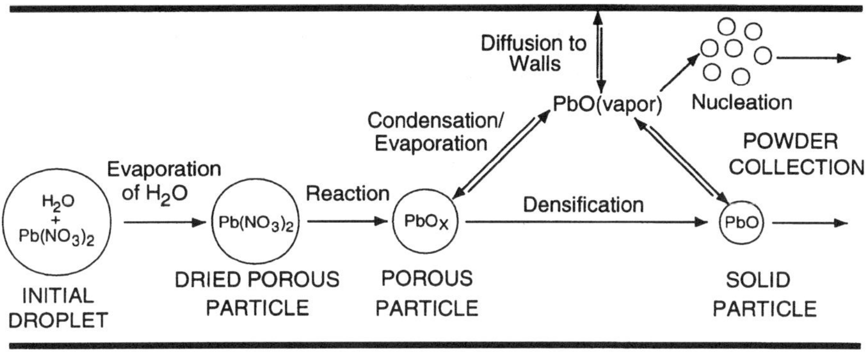


**Fig. 12.25.** Schematic representation of the processes that can occur during the formation of PbO as a result of the evaporation of that compound. (Adapted from Lyons et al.)[79]

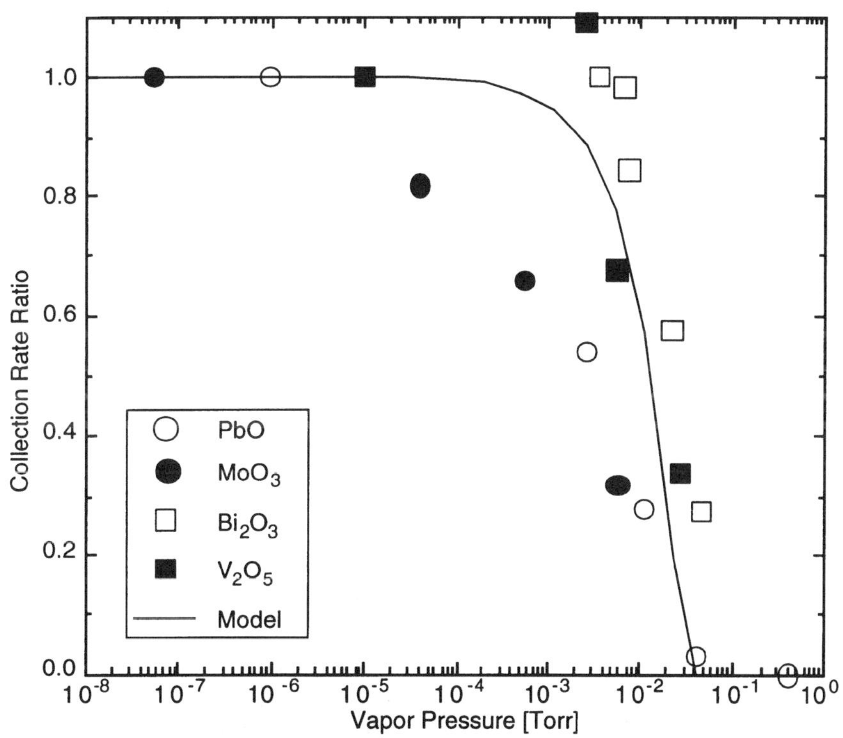


**Fig. 12.26.** Plot of collection rate vs. vapor pressure for a number of volatile metal oxides. (Adapted from Xiong et al.)[80]

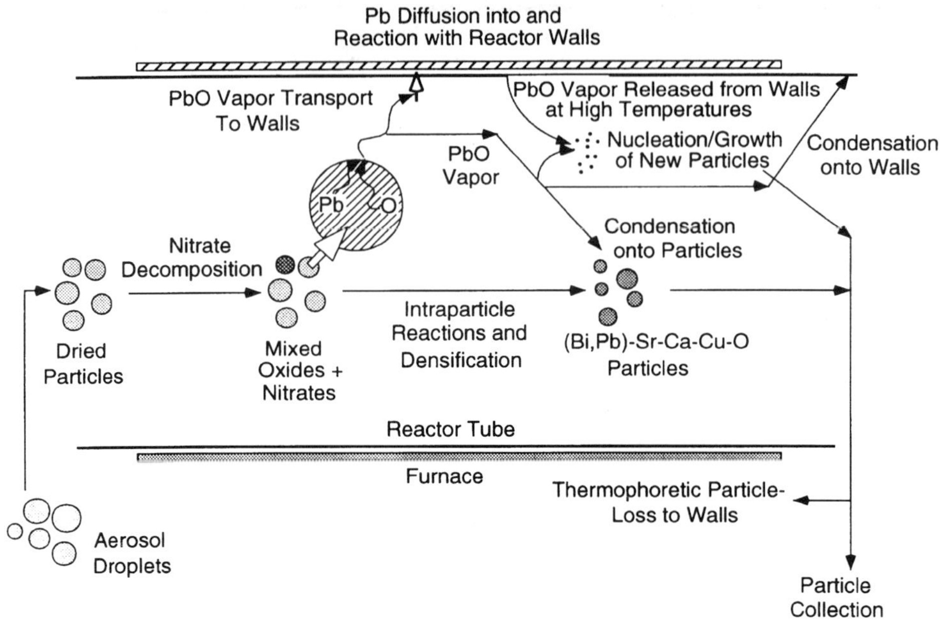


**Fig. 12.27.** Schematic representation illustrating the loss of volatile PbO during the formation of Bi–Pb–Sr–Ca–Cu–O particles. (Adapted from Gurav et al.)[52]

PbO can change the stoichiometry of the product powder.[81] Powders were produced at several reactor temperatures (700, 800, and 900°C) in nitrogen (with a residence time of 1 min), using the precursor stoichiometry $Pb_xBi_{2x}Sr_2Ca_2Cu_3O_y$, with $x = 0.6$ and $x = 0.2$. The vapor pressure of PbO, which may be formed during reaction of the metal nitrates, is significant at these reactor temperatures. Thus, PbO can evaporate from the particles, enter the gas phase, and be lost to the walls, leading to depletion of Pb in the particles. No loss was observed for Bi, Sr, Ca, or Cu at all reactor temperatures studied. However, the percentage of Pb lost varied from approximately 10% at 700°C to roughly 80% at 800°C. By operating at or slightly below 700°C, Pb losses could be eliminated to produce particles with controlled stoichiometries.

The problem can also be alleviated by the use of single-source precursors, which prevent the formation of separate species, such as PbO, as intermediates, thereby avoiding evaporative losses. (See Section 12.2.)

Another example of a chemical system in which evaporation occurs is the reaction of Tl, Ca, Ba, and Cu nitrates to form Tl–Ca–Ba–Cu–O superconductors.[82] The high volatility of $Tl_2O$ and $Tl_2O_3$ allows Tl to enter the gas phase, resulting in loss of Tl from the particles. Because of the high vapor pressures of these species, the Tl exists mainly in the gas phase until cooling while exiting the reactor, whereupon the Tl-containing species condense and

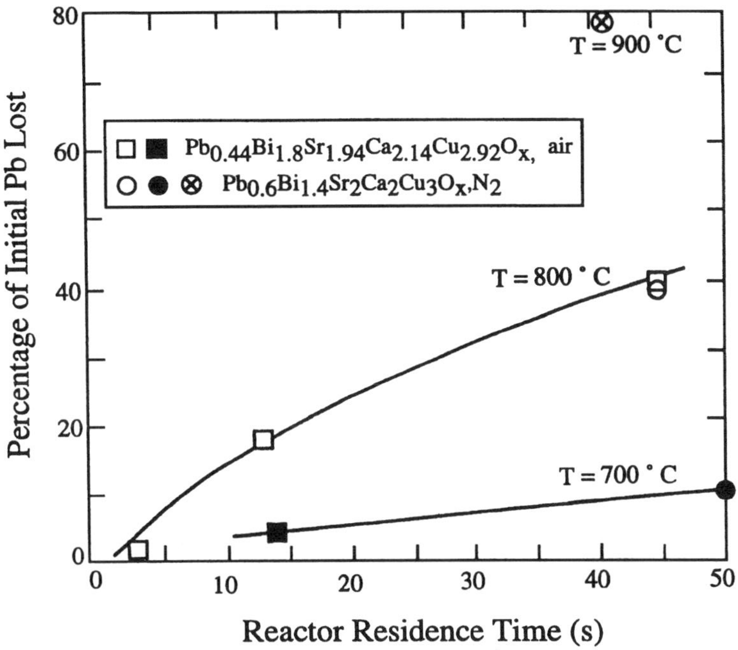


**Fig. 12.28.** Plot of percentage of initial Pb lost vs. reactor residence time, indicating the substantial loss of Pb (as PbO) at higher temperatures and longer residence times. (Adapted from Ward et al.)[81]

coat the existing particles. The loss of $Na_2O$ has been observed during the generation of powders containing Na, Mg, and Al by spray pyrolysis. Qualitatively similar results have been reported by Kato et al.[83] for the loss of $Na_2O$ and by Rosetti for the loss of $MoO_3$.[84]

In general, depending on the details of the chemistry involved, various scenarios are possible when a volatile species enters the gas phase in a multicomponent system. Total condensation of gaseous species onto the existing particles restores the initial stoichiometry, but only in a volume average sense; the particles are no longer chemically homogeneous and consist of a multicomponent core covered by a single-component layer of the condensed material. Homogeneous nucleation to form new particles without condensation on the walls again conserves the composition of the powders, averaged over the total particle volume. However, the powder consists of fine particles of one material mixed with larger, multicomponent particles now depleted of the volatile species. Condensation on the walls results in a change in the volume average composition of the powder. Since the fraction of the volatile species remaining in the particles is difficult to control, this situation is usually the least desirable of the possibilities. Fortunately, wall losses can be eliminated by appropriate design and operation of the reactor.

Studies of gas-to-particle conversion in tubular flow reactors suggest that rapid cooling rates which lead to high saturation ratios will result in new-particle formation. Slow cooling rates allow condensation on the particles and walls, since the saturation ratio can be kept below the critical value for particle formation. The relative amounts of condensation on the walls and on the particles are controlled mainly by the surface area of the preexisting particles and the reactor radius. Larger aerosol surface areas and larger reactors favor condensation onto existing particles.

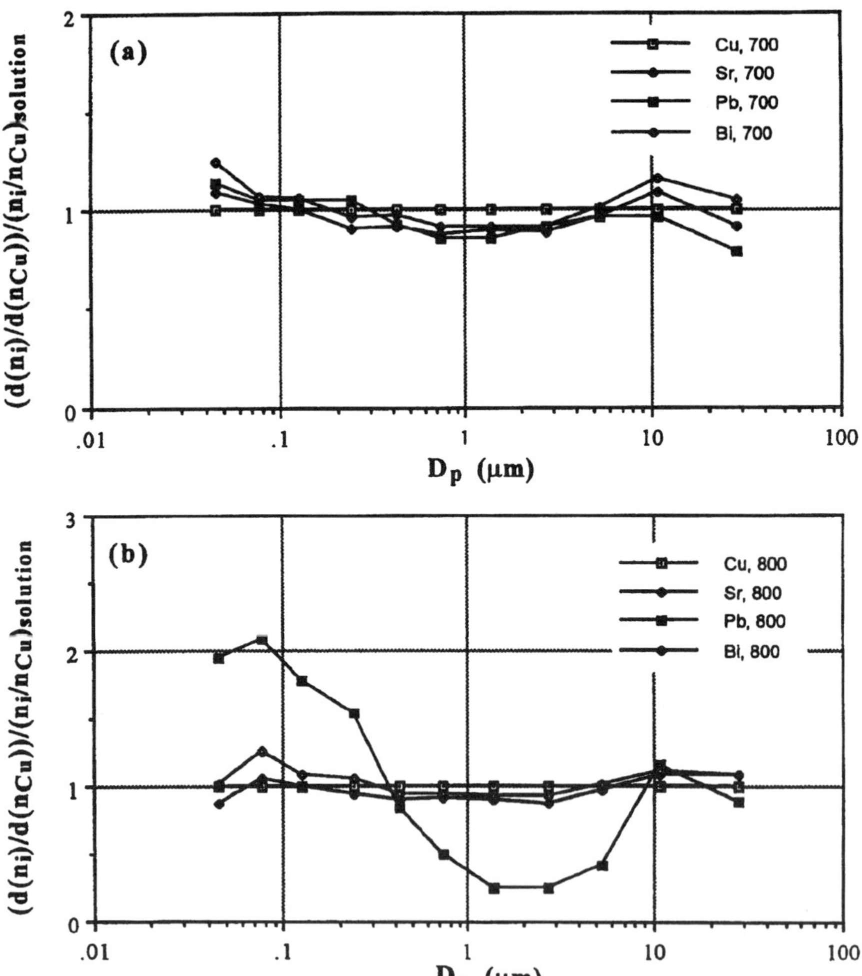


**Fig. 12.29.** Plots of mole fraction of various metals vs. particle size during formation of Pb–Bi–Sr–Ca–Cu–O at different temperatures, showing substantial formation of small Pb-containing particles at 800°C. (Adapted from Gurav et al.)[85]

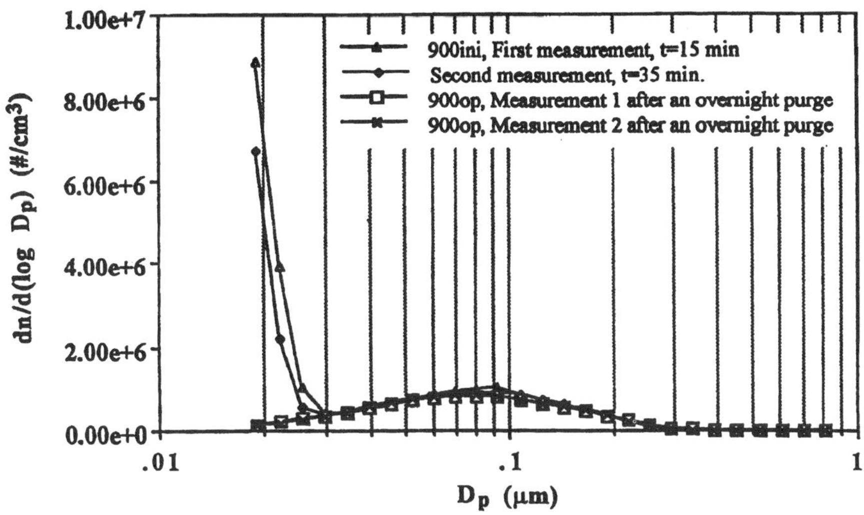


**Fig. 12.30.** Plot of $dn/d(\log d_p)$ (number/cm$^3$) vs $d_p$ ($\mu$m) for the Pb–Bi–Sr–Ca–Cu–O system. (Adapted from Gurav et al.)[52]

### 12.5.4 New-Particle Formation with a Single Volatile Component

New-particle formation can occur during spray pyrolysis when volatile components enter the gas phase and then nucleate during cooling of the gas. The formation of new particles usually leads to a bimodal particle size distribution, with the smaller mode being the result of the new particle formation.

Figures 12.29 and 12.30 show how new particle formation can occur during spray pyrolysis to form Pb–Bi–Sr–Ca–Cu–O particles.[52, 85, 86] The latter figure shows that a significant fraction of the Pb can be lost from particles in the size range 0.5 to 10 $\mu$m.[52] No measurable losses were found at 700°C, whereas extensive losses were observed at 800°C. At the same time, the formation of PbO particles was observed in the size range below 0.03 $\mu$m and resulted from nucleation of PbO vapor as the gas cooled while exiting the reactor.

### 12.5.5 Spray Combustion Synthesis

Spray processes have been combined with combustion processes to give a unique class of methods for powder generation. These methods begin with a solution containing precursors and a dissolved fuel. More gaseous fuel may be added to increase the temperature. The solution is atomized and then combusts.

In one variation of spray combustion synthesis, the exothermic combustion process results in the particles being blown apart into much smaller product

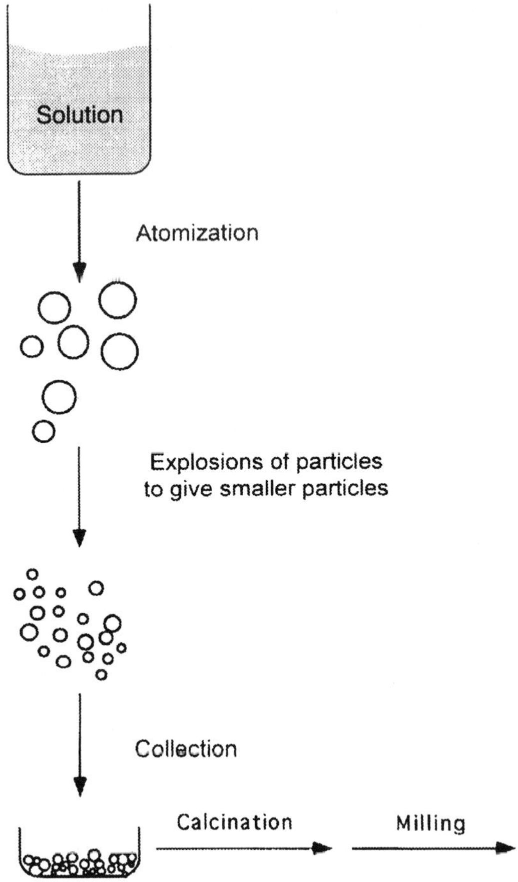


**Fig. 12.31.** Spray pyrolysis with fuels.

particles (Fig. 12.31). The heat released within each particle is sufficient to cause it to blow apart into smaller particles. The reacted material is collected, heated further to convert it into the final product, and then milled.

In a second variation, volatile fuels such as ethanol are added to the solution, resulting in a flame surrounding each droplet. The heat released can cause the droplet to blow apart into smaller droplets.[87] In a different variation of this approach, used to produce nanometer-sized particles, the high temperatures resulting from the combustion process result in the vaporization of species that then nucleate to form particles. (See Fig. 12.32.)[39] The reacted material is collected on a filter.

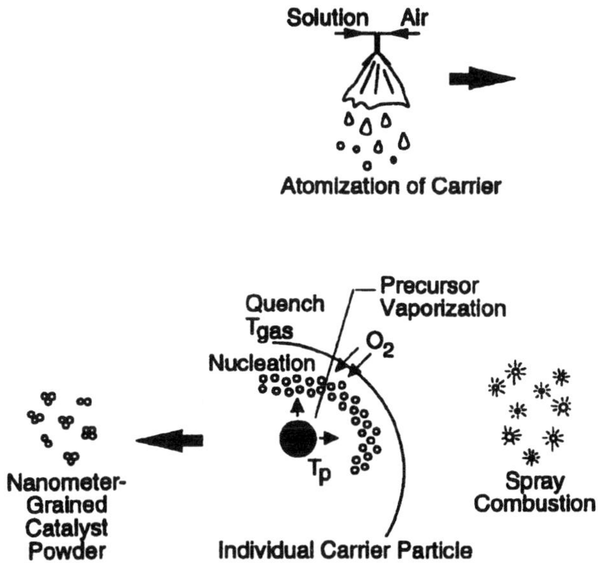


**Fig. 12.32.** Schematic representation of processes that can lead to the formation of nanometer-sized particles by combustion spray pyrolysis. (Reproduced with permission from Joseph J. Helble.)[39]

### 12.5.6 Aerosol-Phase Densification

Once particles consisting of the desired product are formed by reaction in the particles, the morphology of the particles can be modified by solid-state diffusion processes such as grain-boundary, bulk, and surface diffusion. Other sintering mechanisms—e.g., viscous sintering, evaporation–condensation, and liquid-phase-assisted sintering can also be important. Kingery et al.[88] and German[54] provide an introduction to the different types of densification mechanisms. If reactor temperatures are high enough or flames are used, particles can melt. Examples of materials for which this can occur are solid Pd, Ag–Pd alloy, and Ag particles. (See Chapter 13.)[89] The mechanism of particle densification, however, cannot always be predicted—as in the case of many multicomponent materials. When a phase diagram is available, it is possible to obtain guidelines for the temperature required for densification. Materials that melt congruently are the simplest case. Bringing the temperature near or past the melting point results in densification.

Incongruently melting materials form two or more phases when the melting point is reached: a liquid phase and one or more solid phases.[88] The specific phases and their quantities depend on the composition of the material being produced. Bringing the material to the incongruent melting point can, in many cases, provide densification, depending on the quantity of each phase produced and its properties. A more complicated situation arises when two or more solid phases are present in the material. As the temperature is increased, one of the materials may melt congruently or incongruently, to form a liquid phase that may allow densification. Even more complex situations are possible. An example of a congruently melting material that can be formed as solid particles is Ag, whose low melting point (962°C) allows solid particles to be generated.[73-75] An example of an incongruently melting material is $YBa_2Cu_3O_{7-x}$.[90]

The particle densification process during spray pyrolysis is in many ways similar to the sintering of particles during gas-to-particle conversion, as shown in Figure 12.33.[91] Studies of the morphology of powders produced by spray pyrolysis have been carried out under a range of temperatures, residence times, and droplet sizes by numerous groups. However, most of these studies have reported only qualitative observations. Few reports of quantitative measurements are available.[92] Figures 12.34 and 12.35 show how the surface area of $BaTiO_3$-based powders and the green density of pellets pressed from the powders depend on the temperature at which the particles were produced. At higher temperatures the particles were solid, resulting in low surface areas and high densities.

Measurements of gas-phase particle size distributions, surface areas, crystallite sizes, and other properties of the particles in the literature suggest the picture of the particle formation process shown in Figures 12.1 and 12.33. At

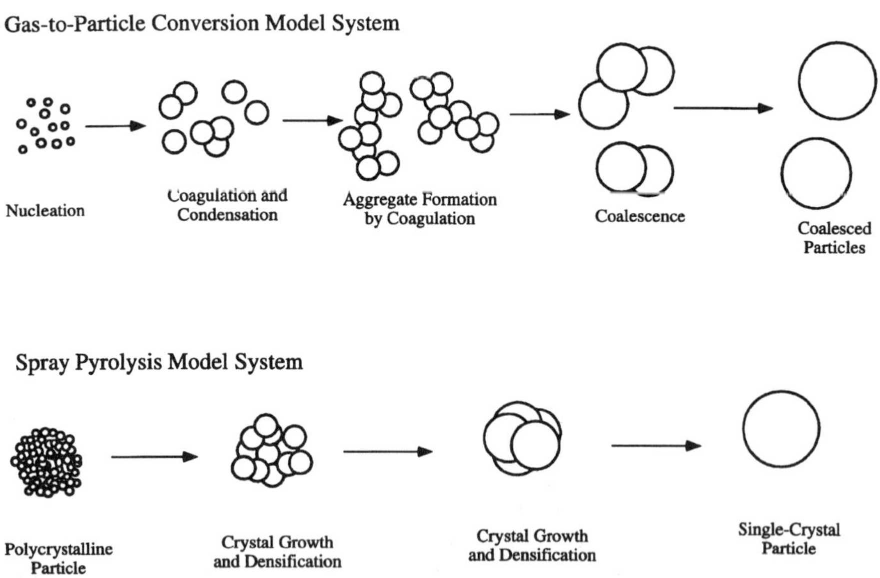


**Fig. 12.33.** Schematic representation of gas-to-particle conversion model system and spray pyrolysis model system. (Adapted from Gurav et al.)[91]

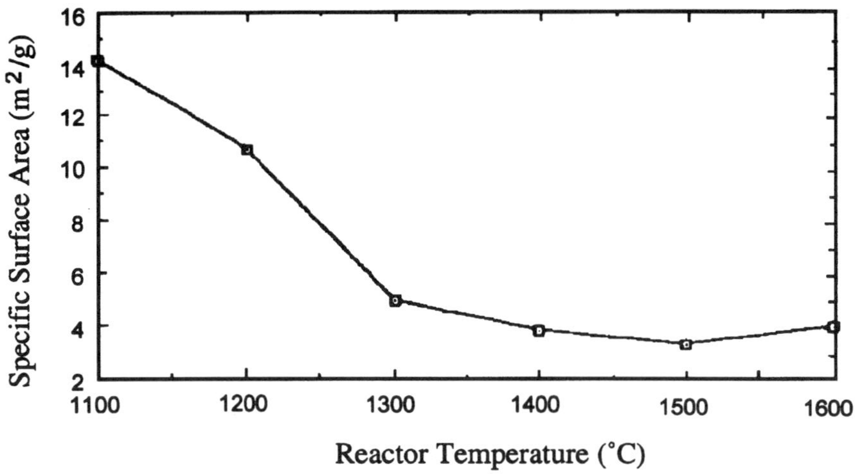


**Fig. 12.34.** Plot of surface area vs. reactor temperature, showing how low surface areas (and presumably high densities) are achieved at higher temperatures. (Adapted from Ortega et al.)[92]

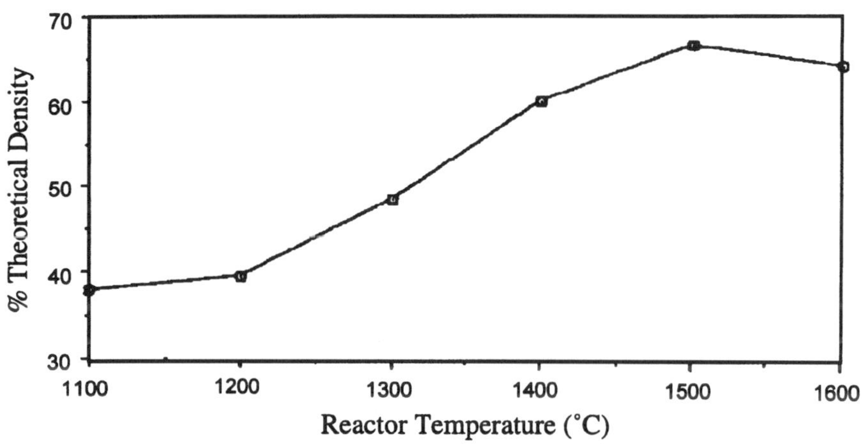


**Fig. 12.35.** Plot of theoretical density vs. reactor temperature, showing that higher densities are achieved at higher temperatures. (Adapted from Ortega et al.)[92]

temperatures high enough to evaporate the water, but below the melting point of the reactants, the particles that are formed are porous, with diameters significantly larger than that corresponding to a fully dense particle. As the temperature is increased past the melting point of the reactants, the particle melts and the average particle size decreases. Further increases in temperature do not change the particle diameter until the decomposition point of the reactant is reached, at which time the size of the particle again decreases. The decomposition process results in porous particles that may be amorphous. As the temperature is increased, the crystallites within each particle grow until the temperature approaches the melting point of the product, and densification of the particles occurs rapidly as a result of sintering and grain growth.

## 12.6 REFERENCES


1. Cheong, H.W., Jeffreys, G.V., & Mumford, C.J. *AIChE J.* *32*(8), 1986, 1334–1346.
2. Davis, E.J. *Aerosol Sci. Technol.* *2*, 1983, 121–144.
3. Dlouhy, J. & Gauvin, W.H. *AIChE J.* *6*(1), 1960, 29–34.
4. Ferron, G.A. & Soderholm, S.C. *J. Aerosol Sci.* *21*(3), 1990, 415–429.
5. Gardner, G.C. *Int. J. Heat Mass Transfer* *8*, 1965, 667–668.
6. Golli, S.E., Arnaud, G., Brichard, J., & Treiner, C. *Aerosol Sci.* *8*, 1977, 39–54.
7. Golli, S.E., Brichard, J., Turpin, P.-Y., & Treiner, C. *J. Aerosol Sci.* *5*, 1974, 273–292.
8. Liu, T.-Q., Sakurai, O., Kato, M., & Mizutani, N. *J. Ceram. Soc. Jpn. Inter. Ed.* *98*, 1990, 73–79.
9. Nesic, S. & Vodnik, J. *Chemical Engineering Science* *46*(2), 1991, 527–537.
10. Newbold, F.R. & Amundson, N.R. *AIChE J.* *19*(1), 1973, 22–30.
11. Ranz, W.E. & Marshall, W.R., Jr. *Chemical Engineering Progress* *48*, 1952, 141–146.
12. Ranz, W.E. & Marshall, W.R., Jr. *Chemical Engineering Progress* *48*, 1952, 173–180.
13. Sano, Y. & Keey, R.B. *Chemical Engineering Science* *37*(6), 1982, 881–889.
14. Bayvel, L. & Orzcchowski, Z. *Liquid Atomization.* Washington, DC: Taylor and Francis, 1993.
15. Kerker, M. *Advances in Colloid Interfacial Sci* *5*, 1975, 105–172.
16. Lampkin, C.M. *Prog. Crystal Growth Charact.* *1*, 1979, 405–416.
17. Lefebvre, A.H. *Atomization and Sprays.* New York: Hemisphere Publishing Co., 1989.
18. Bird, R., Stewart, W.E., & Lightfoot, E.N. *Transport Phenomena.* New York: John Wiley and Sons, 1960.
19. Levenspiel, O. *Chemical Reaction Engineering.* New York: John Wiley and Sons, 1972.
20. Grader, G.S., Machado, D.R., & Semiat, R. *J. Mater. Res.* *9*(10), 1994, 2490–2500.
21. Xiong, Y. & Kodas, T. *J. Aerosol Sci.* *24*, 1993, 893.
22. Kodas, T.T. *Angew. Chemie: Internat. Ed. Eng.*, *28*, 1989, 794–807.
23. Messing, G.L., Zhang, S.-C., & Jayanthi, G.V. *J. Am. Ceram. Soc.* *76*(11), 1993, 2707–2726.
24. Lee, B.-T., Chun, B.-S., & Hiraga, K. *J. Mater. Res.* *9*(10), 1994, 2519–2523.
25. Levi, C.G., Jayaram, V., Valencia, J.J., & Mehrabian, R. *J. Mater. Res.* *3*(5), 1988, 969–983.

26. Wentzell, J.M. *J. Vac. Sci. Tech. 11*(6), 1974, 1035–1037.

27. Essenhigh, R.H. & Fells, I. *Combustion of liquid and solid aerosols.* Faraday Soc. *30*, 1960, 208–221.

28. Hoste, S., Vlaeminck, H., DeRyck, P.H., Persyn, F., Mouton, R., & Van der Kelen, G.P. *Supercond. Sci. Technol. 1*, 1989, 239–241.

29. Johnson, D.W.J. *Adv. Ceram. 21*, 1987, 3–19.

30. King, C.J. *Chemical Engineering Progress*, June 1990, 33–39.

31. Lukasiewicz, S.J. *J. Amer. Ceram. Soc. 72*(4), 1989, 617–624.

32. Shaw, F.V. *Ceramic Bull. 69*(9), 1990, 1434–1489.

33. Srivatsan, T.S. & Lavernia, E.J. *J. Mater. Sci. 27*, 1992, 5965–5981.

34. Messing, G. Nanoparticulates Workshop, Monterey, CA, 1994.

35. Amano, N., Odani, Y., Takeda, Y., & Akechi, K., *MPR March*, 1989, 186–190.

36. Broer, M.M., Bautista, J.R., Chui-Sabourin, M., & Bruce, A.J. *Mater. Sci. Forum. 67–68*, 1991, 39–44.

37. Lierke, E.G. & Griesshammer, G., *Ultrasonics*, October, 1967, 224–228.

38. Yamauchi, I., Ohnaka, I., Kawamoto, S., & Fukusako, T. *Trans. Jpn. Inst. Metals 27*(3), 1986, 187–194.

39. Helble, J.J. Nanoparticulates Workshop, Monterey, CA, 1994.

40. Cauchetier, M., Herlin, N., & Luce, M. *J. Am. Ceram. Soc. 77*(4), 1994, 993–998.

41. Hinds, W.C. *Aerosol Technology*. New York: Wiley Interscience, 1982.

42. Charlesworth, D.H. & Marshall, W.R. Jr., *AIChE J. 6*, 1960, 9–23.

43. Masters, K. *Spray Drying*. New York: John Wiley and Sons, 1972.

44. Leong, K.H. *J. Aerosol Science 18*(5), 1987, 511–524.

45. Motojima, S. & Ueno, S. *Appl. Phys. Lett. 54*(11), 1989, 1001–1003.

46. Jayanthi, G.V., Zhang, S.C., & Messing, G.L. *Aerosol Sci. Technol. 19*(4), 1993, 478–490.

47. Melia, J. *J. Appl. Chem 15*, 1965, 345.

48. Amiel, O., Heintz, J.-M., Duboudin, F., & Salmon, R. *J. Aerosol Sci.* 21(6), 1990, 799–810.

49. Matijevic, E., Zhong, Q., & Partch, R.E. *Aerosol Sci. Technol. 22*, 1995, 162–171.

50. Xu, C., Hampden-Smith, M.J., & Kodas, T.T. *Chem. Mater. 7*, 1995, 1539–1546.

51. Jain, S., Kodas, T.T., & Hampden-Smith, M.J. *J. Electrochem. Soc. 14*(4), 1997, 1222–1231.

52. Gurav, A.S., Kodas, T.T., Kauppinen, E.I., Joutsensaari, J., & Zilliacus, R. *J. Mater. Res. 10*(7), 1995, 1644–1652.

53. German, R.M. *Sintering Theory and Practice*. New York: John Wiley and Sons, 1996.

54. Gardner, T.J. & Messing, G.L. *Am. Ceram. Soc. Bull. 6*(12), 1984, 1498–1504.

55. Salata, O.V., Dobson, P.J., Hull, P.J., & Hutchinson, J.L. *Thin Solid Films 251*(1), 1994, 1–3.

56. Salata, O.V., Dobson, P.J., Hull, P.J., & Hutchinson, J.L. *Adv. Mater. 6*(10), 1994, 772–775.

57. Jang, G., Mukherjee, K., & Kim, H. *Jpn. J. Appl. Phys. Lett. 33*, 1994, L215–L218.

58. Hotta, N., Kimura, I., Ichiya, K., Saito, N., Yasukawa, S., Tada, K., & Kitamura, T. *J. Ceram. Soc. Jpn. 96*, 1988, 715–719.

59. Hotta, N., Kimura, I., Tsukuno, A., Saito, N., & Matsuo, S. *J. Ceram. Soc. Jpn. 95*, 1987, 246–248.

60. Hotta, N., Kimura, I., Tsukuno, A., Saito, N., & Matsuo, S. *Yogyo-Kyokai-Shi 95*(2), 1987, 136–139.

61. Weimer, A.W., Cochran, G.A., Eisman, G.A., Henley, J.P., Hook, B.D., Mills, L.K., Guiton, T.A., Knudsen, A.K., Nicholas, N.R., Volmering, J.E., & Moore, W.G. *J. Am. Ceram. Soc. 77*(1), 1994, 3–18.

62. Saunders, W.A., Sercel, P.C., Flagan, R.C., Atwater, H.A., & Vahala, K.J. *Z. Phys. D 26* (*Suppl.*), 1993, 219–221.

63. Adcock, J.L., Horita, K., & Renk, E.B. *J. Am. Chem. Soc. 103*(23), 1981, 6937–6947.

64. Leong, K.H. *J. Aerosol Sci. 5*, 1987, 525–552.

65. Leong, K.H. *J. Aerosol Sci. 12*(5), 1981, 417–435.

66. Sproson, D.W. & Messing, G.L. *Advances in Ceramics* 21, 1987, 99–108.

67. Hikida, V. *Synthesis of $SrTiO_3$ powder by aerosol decomposition.* Master's thesis, University of New Mexico, 1992.

68. Zhang, S.-C. & Messing, G.L. Solid particle formation during spray pyrolysis. In Messing, G.L., Hausner, H., & Hirano, S.-C. (Eds.), *Ceramic Powder Science III.* Columbus, OH: American Ceramic Society, 1990, 49–57.

69. Moore, K.A., Cesarano, J.I., Smith, D.M., & Kodas, T.T. *J. Amer. Cer. Soc. 75*(1), 1992, 213–215.

70. Ohshima, K., Tsuto, K., Okuyama, K., & Tohge, N. *Aerosol Sci. Technol. 19*(4), 1993, 468–477.

71. Ortega, J. *Control of the morphology of ceramic particles produced by aerosol decomposition.* Master's thesis, University of New Mexico, 1992.

72. Jain, S., Skamser, D., & Kodas, T.T. *Aerosol Sci. Technol. 27*(5), 1997, 575–590.

73. Pluym, T.C., Powell, Q.H., Gurav, A.S., Ward, T.L., Kodas, T.T., Wang, L.M., & Glicksman, H.D. *J. Aerosol Sci. 24*(3), 1993, 383–392.

74. Pluym, T.C., Kodas, T.T., & Glicksman, H.D. *J. Aerosol Sci. 24*, 1993, S327.

75. Pluym, T.C., Kodas, T.T., Wang, L.M., Wiedensohler, A., Hansson, H.C., Maximov, I., & Samuelson, L. *J. Aerosol Sci. 24*, 1993, S569.

76. Senzaki, Y., Caruso, J., Hampden-Smith, M.J., Kodas, T.T., & Wang, L.-M. *J. Am. Ceram. Soc. 78*(11), 1995, 2973–2976.

77. Liu, T.-Q., Sakurai, O., Kieda, N., Kato, M., & Mizutani, N. *Nippon Seramikkusu Kyokai Gakujutsu Rombunshi 97*(10), 1989, 1321–1325.

78. Lyons, S.W., Wang, L.M., & Kodas, T.T. *NanoSt. Mater. 1*, 1992, 293.

79. Lyons, S., Xiong, Y., Ward, T., Kodas, T.T., & Pratsinis, S. *J. Mater. Res. 7*(12), 1992, 3333–3334.

80. Xiong, Y., Lyons, S., Kodas, T.T., & Pratsinis, S. *J. Am. Ceramic Soc. 78*(9), 1995, 2490–2496.

81. Ward, T., Lyons, S., Kodas, T.T., Brynestad, Y., Kroeger, D., & Hsu, H. *Physica C 200*, 1992, 31–42.

82. Kodas, T.T., Engler, E., & Lee, V. *Appl. Phys. Lett. 54*, 1989, 1923–1925.

83. Kato, A., Hirata, Y., Tokunaga, Y., Maeda, T., & Nishi, H. Sintering behavior of $\beta''$-alumina powders prepared by spray-pyrolysis technique. In Vincenzini, P. (Ed.), *Ceramic Powders*. Amsterdam: Elsevier Scientific, 1983, 835–841.

84. Rosetti, G.A., Jr., Burger, J.L., & Sisson, R.D., Jr. *J. Am. Ceram. Soc. 72*(10), 1989, 1811–1815.

85. Gurav, A., Kodas, T.T., Kauppinen, E., Joutsensaari, J., & Zilliacus, R. *Nano-Struct. Mater. 4*(5), 1994, 583–590.

86. Gurav, A., Kodas, T.T., Wang, L., Kauppinen, E., & Joutsensaari, J. *J. NanosSt. Mater. 4*(5), 1994, 491–496.

87. Gurav, A. & Kodas, T.T. Personal communication, 1997.

88. Kingery, W.D., Bowen, A.K., & Uhlmann, D.R. *Introduction to Ceramics*. New York: John Wiley and Sons, 1976.

89. Brewster, J. & Kodas, T.T. *AIChE J., 42*(11), 1997, 2665, Special issue on materials processing.

90. Kodas, T.T., Datye, A., Lee, V., & Engler, E. *J. Appl. Phys. 65*, 1989, 2149–2151.

91. Gurav, A., Pluym, T., Kodas, T.T., & Pratsinis, S.E. *Chem. Eng. Commun. 151*, 1996, 211–226.

92. Ortega, J., Kodas, T.T., Chadda, S., Smith, D.M., Ciftcioglu, M., & Brennan, J. *Chem. Mater. 3*, 1991, 746–751.

# 13 TECHNOLOGY: LIQUID-TO-SOLID AND SOLID-TO-SOLID CONVERSION

## 13.1 LIQUID-TO-SOLID AND SOLID-TO-SOLID CONVERSION

### 13.1.1 Introduction

*Technologies* for the generation of powders by conversion of liquid or solid aerosols to solid products (powders) include:

- Spray pyrolysis
- Flame pyrolysis
- Spray-drying
- Freeze-drying
- Vacuum atomization of molten metals
- Droplet-phase reactions to form liquid or solid products
- Gas–solid reactions to form solid products

Processes utilizing precursors that evaporate to give vapors, which then react in the gas phase, were discussed in Chapter 11; the emphasis here is on spray pyrolysis.

Liquid-to-solid and solid-to-solid routes can be classified as chemical or physical. Chemical processes rely upon reactions of precursors and include spray pyrolysis, droplet-phase reactions to form liquid or solid products, and gas–solid reactions to form solid products. Physical processes, on the other hand, rely upon drying or cooling and include spray-drying, freeze-drying, and vacuum atomization of molten materials.[1–3] The key distinction between chemical and physical processes is that in chemical processes a precursor undergoes a chemical reaction to form a product that contains some of the elements in the precursor, whereas in physical processes, chemical reactions generally do not take place, and the compositions of the reactants and products are often identical.

The common step in all of the processes using liquid reactants or solids dispersed in liquids is the formation of droplets containing the precursors by means of an aerosol generator. In spray pyrolysis, an example of a chemical process, the droplets are passed through a high-temperature environment, where the solvent evaporates and the precursors react to form the product (Figure 13.1).[1-3] The goal is often to prepare a complex material from a solution containing a variety of different reagents. A variation of this route is to entrain solid precursor particles in liquid precursor droplets and react them in a heated gas stream or with a gaseous species. In spray-drying, an example of a route in which only physical processes occur, temperatures are much lower, and the droplets evaporate to form dried particles, which are collected and converted to the final product in subsequent processing steps.[4-8] In freeze-drying, the droplets are sent into a cold liquid such as hexane or nitrogen, where they freeze and are collected.[9] Again, this is followed by subsequent processing steps that convert the precursors to the

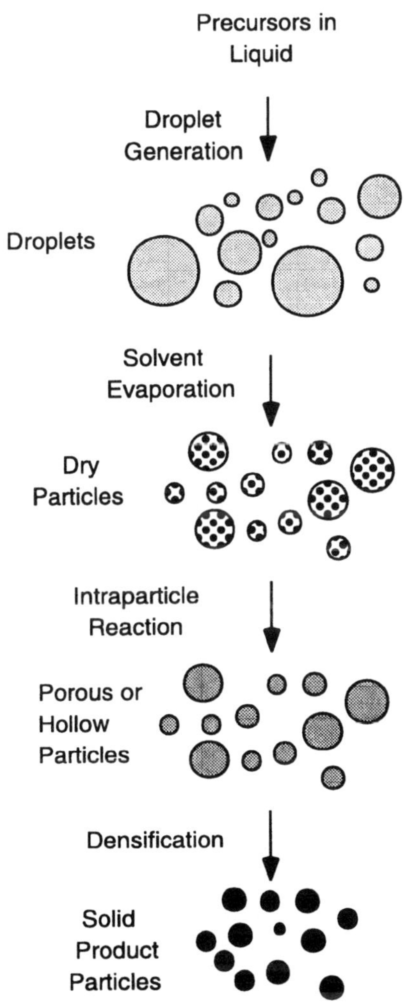


**Fig. 13.1.** Schematic representation of the important steps that can occur in a simplified spray pyrolysis process.

final product. The goal of physical processes is generally to convert a material in one form (e.g., large particles, a broad size distribution) to another (e.g., smaller particles, a narrow size distribution) that has higher commercial value. All of these methods are typically operated at atmospheric pressure and involve droplets and particles in the 0.1- to 100-$\mu$m size range. A variety of energy sources have been used to drive the reactions, but the most common are electrically heated furnaces, plasmas, and flames.

The advantages that these technologies for powder generation have over existing methods can be divided into two kinds: advantages that result from the enhanced powder properties and advantages derived from the production method. In general, advantages related to powder properties are:

- Simplicity
- Ability to process single- and multicomponent organic and inorganic materials
- Doping with controlled levels of many elements possible
- Same chemical composition from particle to particle
- Easy switch from one composition to another in multicomponent systems
- Use of a variety of commercially available precursors: metal nitrates, chlorides, fluorides, acetates, citrates, oxalates, etc., and colloidal precursors such as alumina, silica, etc.
- Direct generation of unagglomerated particles without milling in the case of spray pyrolysis
- Formation of a high-purity powder in the case of spray pyrolysis
- Control of grain size within each particle in the case of spray pyrolysis
- Control of particle size and spread of size distribution in some cases

The primary disadvantages related to powder properties are:

- Formation of porous or hollow particles under some conditions in the case of spray pyrolysis
- Limitation of spread of particle size distribution due to spread of droplet generator in the case of spray pyrolysis
- Occurrence of compositional segregation and inhomogeneities within particles under some conditions
- Evaporation of volatile intermediates, possibly leading to poor control over stoichiometry in the case of spray pyrolysis

In general, advantages related to the technology of powder production are:

- Cost-effectiveness [only a few unit (separate) operations]
- Continuous, rather than batch, processing

- Environmental benignity
- Demonstrated scale-up to ton quantities
- Relatively low production costs, leading to economical processes
- Relatively safe process, because volatile precursors are not required

The primary disadvantages related to powder production are:

- Large volumes of gas for spray pyrolysis and spray-drying
- Limited to particle sizes greater than 0.1 $\mu$m

### 13.1.2 Commercial Potential

We begin this section with a brief discussion of the factors that motivate the commercialization of powder production by liquid-to-solid and solid-to-solid conversion. We also examine some of the market potential for the powders that are produced, distinguishing materials that are mere laboratory curiosities from those that have commercial value. We consider two market extremes for the commercialization of powders: powders with low added value relative to the raw materials that must be produced in large amounts at low production costs to be economically viable and powders with high added value and for which a lower scale of production (and the ability to withstand higher production costs) is necessary. (Added value is value that is added by the processing or production method.) Examples of high-volume, low-added-value materials made by aerosol processing are $TiO_2$ and $SiO_2$ powders produced by gas-to-particle conversion for use in pigment (paint) and structural reinforcement markets. These materials are produced in large quantities (200 tons per day per reactor) at relatively low cost, and their selling price is relatively low (approximately \$1/lb). In contrast, there are niche markets for low-volume, high-added-value powders such as nanophase $TiO_2$ pigments, also produced by gas-to-particle conversion, for applications in sunscreens, and spray-dried powders such as Co/WC used for cutting-tool applications. One company, Nanophase Technologies, which prepares metal (e.g., Al, Ag, Au, Cu, and Ni) and ceramic ($TiO_2$, $Al_2O_3$, iron oxide, chromium oxide, $ZrO_2$, $Y_2O_3$, $SiO_2$, cerium oxide, and manganese oxide) and other powders by a modified inert-gas condensation method, supposedly is planning to produce at least tens of metric tons per year. The selling price for nanocrystalline $TiO_2$ was \$110/kg in 1995 and is expected to drop to \$20 to 40/kg and lower as demand increases. Another company, Nanodyne, Inc., produces WC–Co powders (see Section 13.3 and Chapter 9) by spray-drying and subsequent processing for a selling price of about \$77/kg or lower, which is higher than the selling price of existing powders (\$44/kg), but may still be economical due to the higher value of the spray-dried-derived materials. (They are expected to result in tool bits with a service life up to four times longer than that of existing bits.) Whatever the application, the powder under consideration should have certain competitive advantages, among which could be better performance in the final application (e.g., a faster, more reliable, higher efficiency product) or cost-effectiveness (e.g., a less expensive material than existing materials), and preferably both.

In this chapter, the majority of the examples presented are based on laboratory-scale experiments, with some examples of commercial-scale production. The reasons for the emphasis on laboratory-scale experiments is that laboratories are where the majority of the work has been carried out to date, and a number of papers relating these experiments address the fundamental aspects of powder production by liquid-to-solid and solid-to-solid conversion, giving some insight into the mechanisms by which powders are formed.

## 13.2 SPRAY PYROLYSIS

### 13.2.1 Introduction

Spray pyrolysis involves passing a solution or suspension of precursors through an aerosol generator to form droplets (Fig. 13.2) suspended in a gas. The droplets are then heated by passing the aerosol through a tube contained in a furnace or into a flame (or other thermal energy sources), where thermally induced chemical reactions in the particles or reaction with gaseous co-reactants convert the precursors to the product powder, which is then collected (Fig. 13.3). The schematic representation in Figure 13.2 is a simplified one, in order to show the formation of a simple powder, e.g., an elemental material. Details of the chemical and physical processes that occur were described in Chapters 1–8 and will be further examined later in this chapter. Spray pyrolysis has many variations and has been called aerosol decomposition,[2] evaporative decomposition,[10] spray roasting,[11] spray calcination,[12] liquid aerosol thermolysis,[13] and more. Its popularity is due primarily to its simplicity (Fig. 13.3) and the fact that it is one of the few aerosol methods that has demonstrated the ability to produce chemically homogeneous complex metal oxides.

Spray pyrolysis has several advantages over gas-to-particle conversion technologies for the generation of multicomponent materials. Gas-to-particle conversion or nucleation processes cannot easily produce multicomponent materials, due to difficulties in obtaining volatile precursors and controlling the nucleation processes when several species are involved. Spray pyrolysis, by contrast, allows the generation of particles with the same chemical composition from particle to particle because the precursors are present in the correct stoichiometry in each droplet produced by the aerosol generator; that is, each droplet has an identical composition. Volatile precursors are not required; thus, a wide variety of reactants can be used, while avoiding the need for carbon-containing precursors. This flexibility allows the use of inexpensive reactants such as metal nitrates, chlorides, fluorides, etc. In

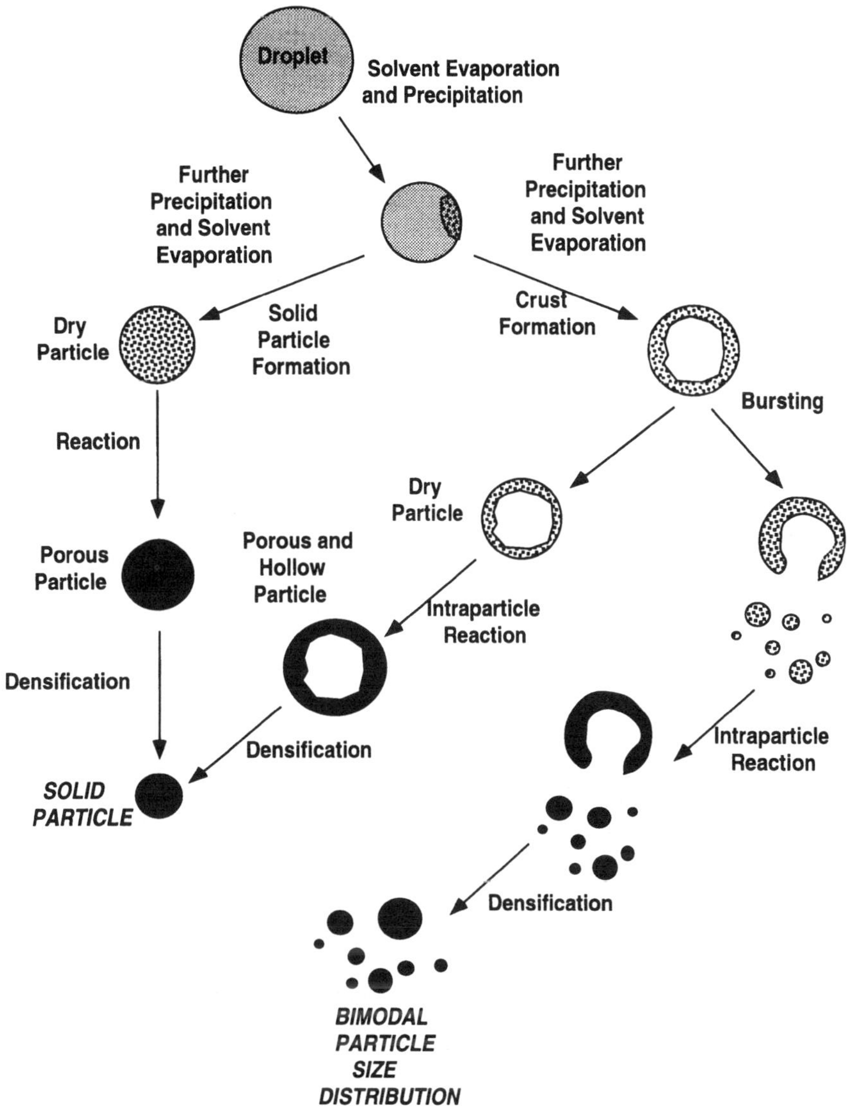


**Fig. 13.2.** Schematic representation of the important steps in a spray pyrolysis process, showing more detail than shown in Figure 13.1.

addition, polymeric and colloidal precursors can be utilized. The generation of multicomponent ceramic powders can be carried out on an industrial scale using only a few unit operations; the ability to scale up these systems for the production of quantities of conventional ceramic-oxide powders measured in tons has been demonstrated.

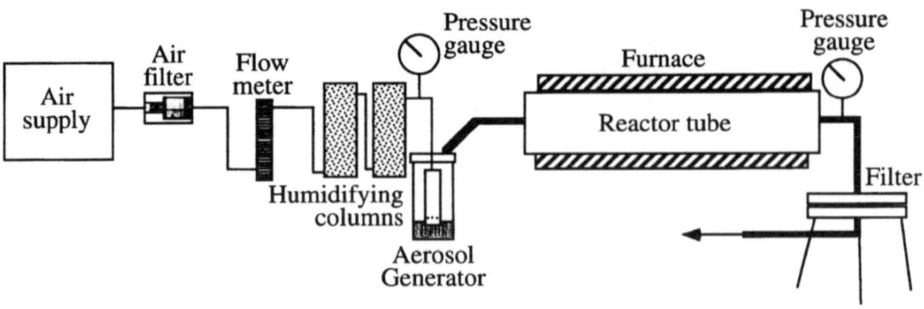


**Fig. 13.3.** Schematic drawing of a typical spray pyrolysis reactor.

Powders produced by spray pyrolysis often have superior properties relative to powders produced by conventional routes such as liquid-phase precipitation. Figure 13.4 shows silver particles produced by spray pyrolysis. For comparison, Figure 13.5 shows commercially available silver particles produced by liquid-phase precipitation. The powders produced by spray pyrolysis are more crystalline and less agglomerated than the powders produced by

**Fig. 13.4.** Scanning electron micrograph of Ag particles formed by spray pyrolysis at 800°C. (Courtesy of Kodas, unpublished results.)

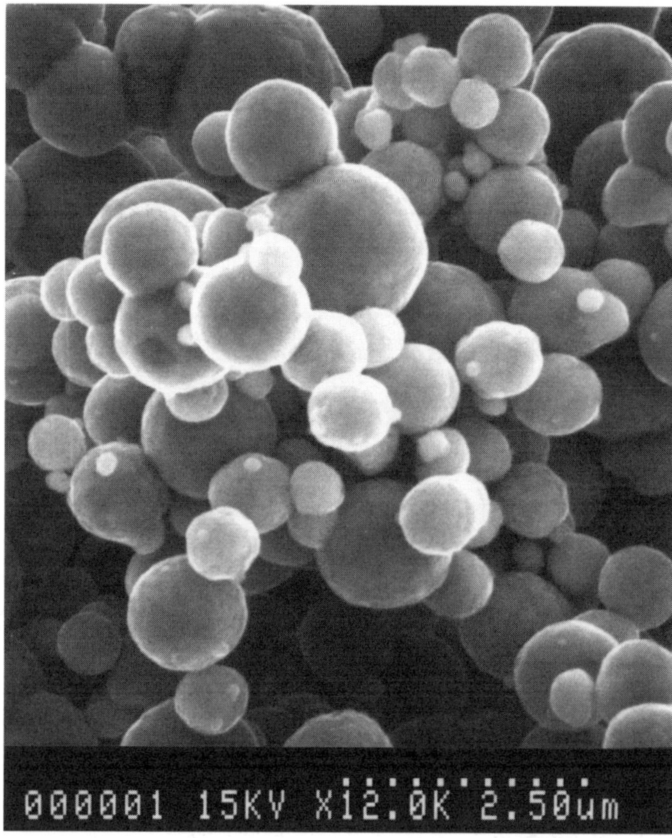


**Fig. 13.5.** Scanning electron micrograph of Ag particles formed by liquid-phase precipitation. (Courtesy of Kodas, unpublished results.)

liquid-phase precipitation. This illustrates a unique advantage of spray pyrolysis: the exposure of the particles to high temperatures while avoiding agglomeration. Liquid-phase routes are capable of producing submicron unagglomerated particles, but they generally have poor crystallinity. At the same time, solid-state routes can usually produce crystalline material, but control of the size distribution and sphericity of particles in the submicron range, as well as agglomeration, is problematic. Furthermore, most metals cannot be milled to reduce the particle size.

Another advantage of spray pyrolysis is its ability to produce composite particles containing elements that would be difficult or impossible to bring together into submicron particles using other routes. Figure 13.6 shows a sample of polished Ag–$YBa_2Cu_3O_7$ ceramic produced from composite Ag–$YBa_2Cu_3O_7$ particles; each particle contained the same ratio of silver to ceramic.[14, 15] The ability to produce these composite powders allowed the powder to be sintered to form a dense bulk material, while maintaining a small grain size for the Ag.

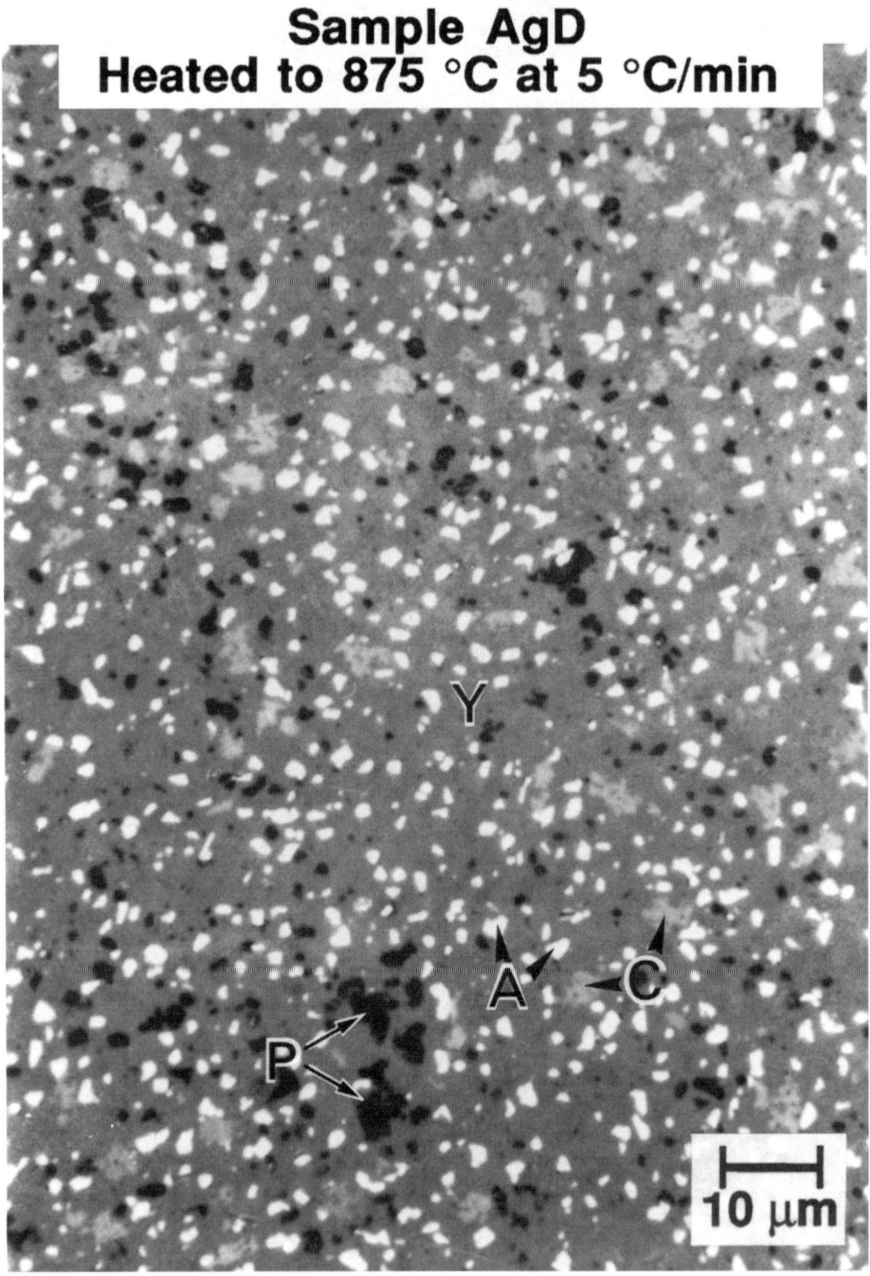


**Fig. 13.6.** Transmission electron micrograph of sintered AgYBCO formed by spray pyrolysis. (Adapted from Ward et al.)[15]

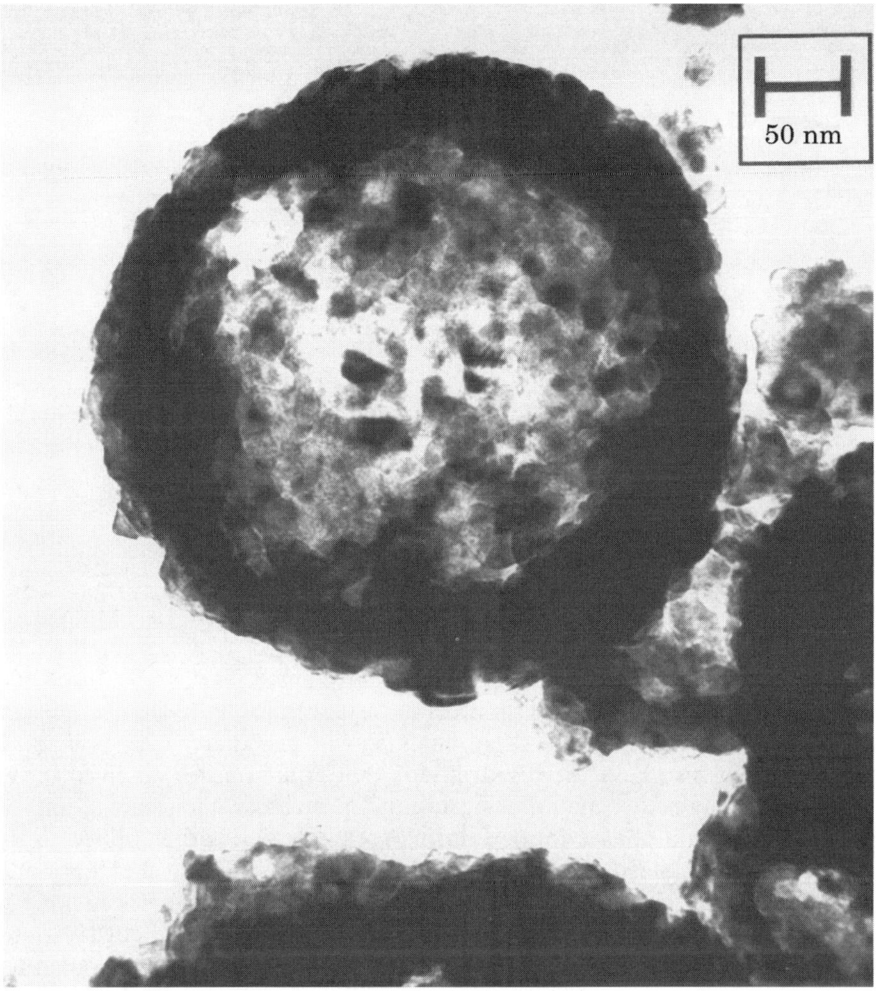


**Fig. 13.7.** Scanning electron micrograph of hollow $BaTiO_3$ particles formed by spray pyrolysis. (Courtesy of Ortega and Kodas, unpublished results.)

The primary disadvantage of spray pyrolysis is the difficulty one has, in some cases, controlling the morphology of the particles. This problem is illustrated in Figure 13.7, which shows a hollow $BaTiO_3$ particle produced by spray pyrolysis.

Spray pyrolysis has several advantages over solid-state chemical reaction routes. Solid-state routes typically begin with two or more powders, each containing one of the elements to be incorporated into the final material. The powders are heated for long periods, to drive solid-state diffusion and reaction to form the product. The product is milled to smaller particle sizes,

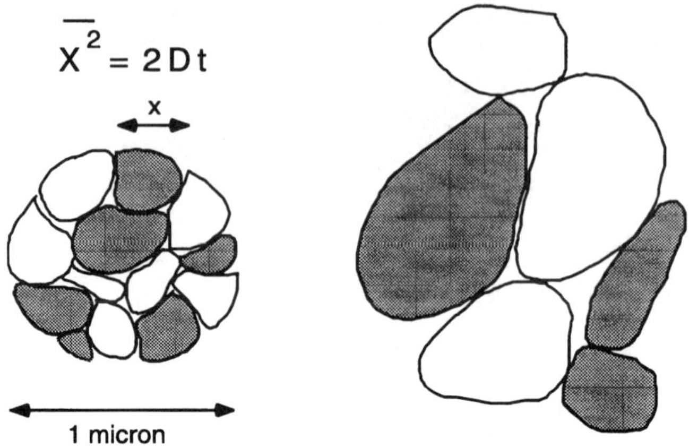


**Fig. 13.8.** Comparison of the diffusion distances in particles formed by spray pyrolysis (left) and conventional solid-state processes (right).

and the process is repeated until a homogeneous material is obtained. In spray pyrolysis processes, the scale of the segregation is no larger than the particle size (usually, < 1 μm) and is frequently much less (< 0.1 μm) than the particle size. In contrast, the diffusion distances for solid-state routes are often 1 μm or greater. This situation is shown in Figure 13.8. Solid-state reaction routes also have the disadvantage that liquid phases which form can percolate through the material, resulting in macroscopic segregation. In contrast, any liquid phases formed during spray pyrolysis are confined to the individual micron-sized or submicron-sized particles. Thus, the thermodynamically favored product can usually be formed even when prior melting of the reactants, intermediates, or products has occurred. Spray pyrolysis routes also have the advantage of directly providing submicron, unagglomerated particles without milling. Milling can result in the introduction of impurities.

Another advantage over solid-state routes is that the spread of the particle size distribution can often be narrower for powders produced by spray pyrolysis. Figure 13.9 shows size distributions obtained from liquid-phase sedimentation velocities for a powder made by solid-state reaction and milling and a powder made by spray pyrolysis (P-1600), both powders formed from the same precursor solution.[16] The powder made by spray pyrolysis had a narrower size distribution and smaller particle size.

Spray pyrolysis also has advantages over many liquid-phase routes such as sol-gel and precipitation techniques. These techniques typically begin with solutions of precursors in a solvent. Precursors for the liquid-phase routes are often expensive. In precipitation routes, the precursors are precipitated from solution by rapidly mixing the solution with one or more reactants or altering the temperature. In a typical sol-gel route, water is added to the precursor

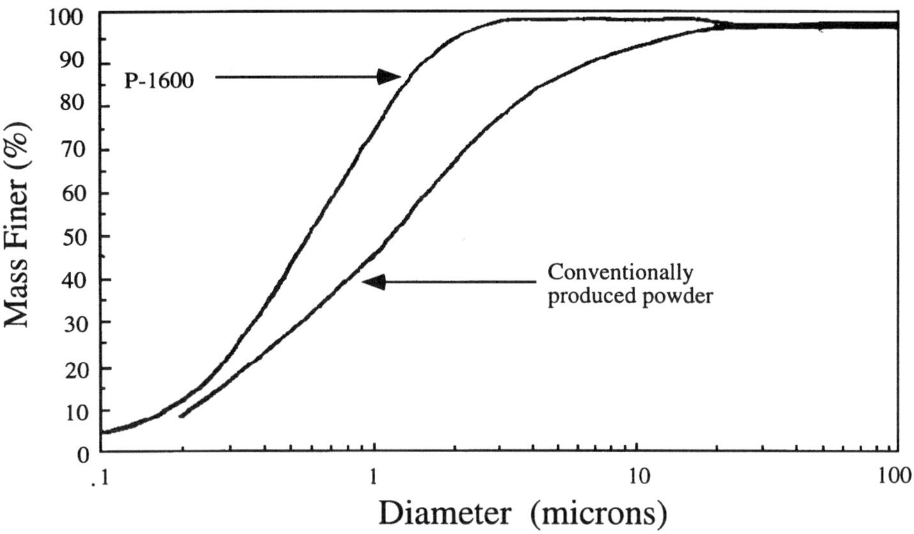


**Fig. 13.9.** Plot of the particle size distribution of $Ba_{1-x}Ca_xTiO_3$ prepared by spray pyrolysis. (Adapted from Ortega et al.)[16].

solution at a controlled rate, resulting in reactions that form a sol or a gel.[17] In both precipitation and sol-gel routes, the material must be dried, heated, milled, and classified in order to obtain a useful powder. Milling of the powders frequently results in contamination. Control over stoichiometry is frequently difficult for multicomponent materials because the precursors have to be precipitated from a solution in a required stoichiometric ratio. Liquid-phase solution processes tend to be batch processes, whereas spray pyrolysis is a continuous process.

The surface chemistry of powders prepared by spray pyrolysis is likely to be significantly different from powders prepared by solid-state and liquid-phase methods. This difference may affect a variety of the chemical and physical properties of the powder, such as its optical or magnetic properties, as well as its dispersibility in organic media.

Finally, the disadvantages of the spray pyrolysis process are the relatively low production rates and relatively high processing costs associated with processing large volumes of gas, which make this technology more suitable for the production of low-volume, high-added-value materials rather than high-volume, low-cost materials.

### 13.2.2 Precursors and Products

The goal of this section is to provide an overview of the types of powders that have been prepared by spray pyrolysis and how the characteristics of these powders compare with those of similar powders prepared by alternative methods.

In general, powders made by spray pyrolysis are spherical and unagglomerated, with a size and size distribution that can be controlled, given the method by which the aerosol was produced. The powders may be dense, hollow, porous, nonporous, single crystal, amorphous, coated, homogeneous, or inhomogeneous (composite), depending on the reactor conditions and reagents used. Many of these powder characteristics can be controlled by the experimental conditions employed, which gives spray pyrolysis much more flexibility compared with solid-state and solution powder-production methods. Perhaps the biggest asset of spray pyrolysis is the ability to make crystalline and pure micron- and submicron-sized powders, which is difficult to achieve by competing methods.

A wide variety of materials have been made by spray pyrolysis, including metal oxides, sulfides, nitrides, borides, metals, and metal alloys. An equally broad range of precursor chemistries, which have been described in Chapter 8, have been used to produce these materials.

The choice of reactants used for the spray pyrolysis process depends on a number of considerations. Economic factors may dictate the use of inexpensive precursors. Control over particle morphology may require species with high solubility or colloidal species. Requirements for precursors for powder generation by spray pyrolysis include high solubility and purity, low cost, low decomposition temperature, stability at room temperature in solution, a high-purity product, matched reactivity with other precursors, and low toxicity. Because gas-phase reactions should be avoided, metal-containing precursors should be involatile—volatile molecular species are generally undesirable.

Numerous kinds of precursors have been used for particle formation by spray pyrolysis. Representative examples of materials employed in spray pyrolysis are tabulated according to the nature of the product, which may fall into any of the following categories: simple oxides, complex oxides, ceramic superconducting metal oxides, nonoxides, and metal powders. The tables at the end of the chapter identify the products, the precursor(s) used, the approximate particle size, the morphology of the products, and the temperature at which the powders were made. The chemical implications of the choice of these precursors were discussed in Chapter 8. The sections that follow give a brief overview to illustrate the types of powders that have been produced.

### 13.2.3 Metal Oxide Powders

The chemical processes involved in the formation of metal oxide powders from various precursors were described in Chapter 8; included was a discussion of potential problems, such as phase segregation and precipitation. Specific examples of metal oxide powders and reagents used are listed in Tables 13.1, 13.2, and 13.3 at the end of the chapter.

The most commonly used precursors for metal oxides are metal nitrates. The motivation for their use is their lack of carbon (which avoids carbon contamination), their relatively high solubilities in many cases, and their low cost. Drawbacks are the potential formation of $HNO_3$ and $NO_x$, which can lead to back-reaction of metal or metal-oxide particles collected on the filter as these particles are exposed to the flowing exhaust gas stream, and the necessity to scrub the exhaust gas.

Metal halides have also been used as precursors, but their use is more restricted than that of the metal nitrates, because of the volatility of many metal halides. Examples include Mg and Sn halides. A problem with halide precursors is the possible back-reaction of the collected powder with acids and halogens during exposure on the filter.

Metal carboxylates such as acetates have been used as precursors.[18–22] These often have the advantage of reasonable solubility, and they generate gaseous by-products such as CO or $CO_2$, which are easily removed. Drawbacks include the relatively high decomposition temperatures of some metal carbonates formed as intermediates, the potential for carbon contamination, and the relatively high cost of these precursors.

Colloidal precursors have been used as sources of Al, Si, and Ti in the form of AlOOH, $SiO_2$, and $TiO_2$.[23–27] Colloidal precursors offer the advantages of low cost, high purity, and the ability to form solid particles. Drawbacks include the need for diffusion or reaction to form the final product and difficulties in forming small droplets containing the desired amount of colloidal precursor.

Single-source, water-soluble precursors containing Ba, Sr, Ca, Ti, etc., have been used.[18, 28] The use of a single precursor containing all the elements necessary in the final powder avoids segregation during precipitation of the precursors, as described in Chapter 8. The main drawback is the high cost associated with these precursors.

Spray pyrolysis has produced powders with all of the characteristics desirable for powder processing. Figure 13.10 shows a particle size distribution obtained from sedimentation velocity measurements.[15] This size distribution is sufficiently narrow, and the average particle size is sufficiently small, to allow powder processing using conventional and advanced processing techniques.

Figures 13.11[15] and 13.12[16] show comparisons of the densification of $YBa_2Cu_3O_7$ and $Ba_{1-x}Ca_xTiO_3$ powders produced by spray pyrolysis on the one hand and solid-state reaction and milling on the other. The spray pyrolysis powders densify at significantly lower temperatures, partly because of the smaller particle size.

In most cases, powder production is carried out in ambient air, which is often the source of oxidation and removal of ligands as $CO_2$ and $H_2O$. In contrast, with nonoxide materials, air must often be rigorously excluded. Conditions under which metals are formed were described in Chapter 8.

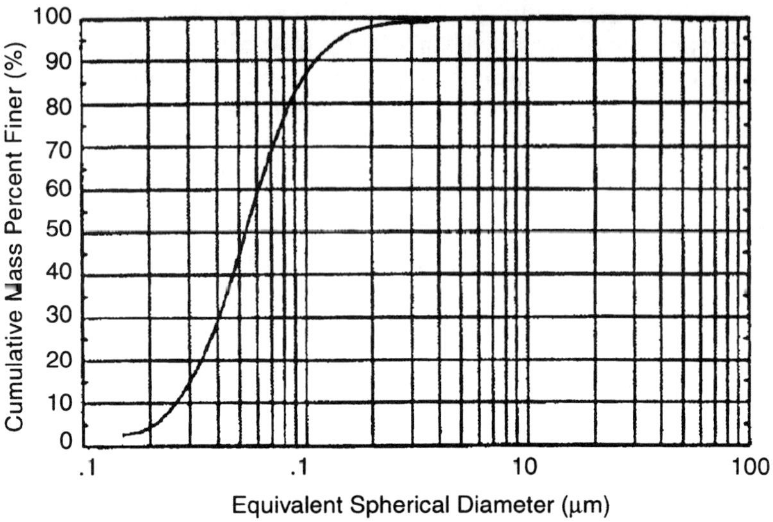


**Fig. 13.10.** Sedigraph particle size distribution of $AgYBa_2Cu_3O_x$ powder produced at 930°C. (Adapted from Ward et al.)[15]

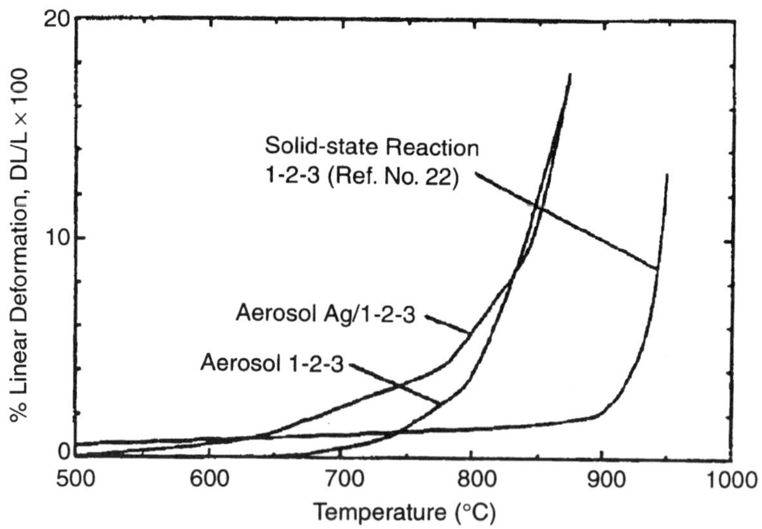


**Fig. 13.11.** Dilatometry results (air) for pellets from aerosol-derived $YBa_2Cu_3O_x$ and $AgYBa_2Cu_3O_x$ powders, compared with coarser powder prepared by solid-state methods. (Adapted from Ward et al.)[15]

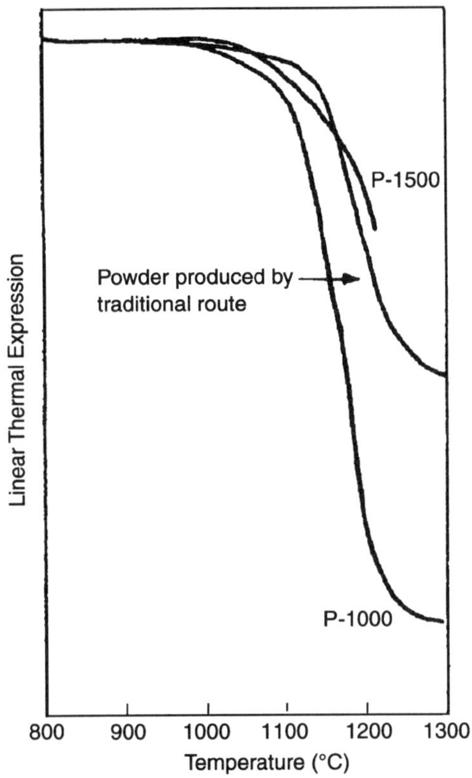


**Fig. 13.12.** Dilatometry results (air) for pellets from aerosol-derived $Ba_{1-x}Ca_xTiO_3$ powders, compared with coarser powder prepared by solid-state methods. (Adapted from Ortega et al.)[16]

### 13.2.4 Nonoxide Powders

A variety of nonoxide metal-containing powders have been prepared by spray pyrolysis, most commonly metal nitrides and metal chalcogenides (especially sulfides). In this case, the strategy is somewhat different compared with the formation of metal oxides, because a separate source of the nonmetallic element (S, N) is added, although there are some cases of "single-source" precursors to these powders (see Chapter 8).

Metal sulfides have been prepared by routes that include the reaction between a common source of a metal-containing reagent and a sulfur-containing reagent. For example, entries 5 and 8 in Table 13.4 at the end of the chapter show the reaction with $H_2S$ as a sulfur source. An example of a metal sulfide that has been prepared by single-source precursors is SrS, which was prepared from $Sr(SAc)_2$ in an aqueous solution and, surprisingly, did not result in formation of either the oxide or the sulfate.[29]

Metal-nitride powders have been prepared by the reactions of metal–nitrogen polymers such as polyborazene and polysilazane, as well as by reactions between sources of metals and ammonia.[30]

### 13.2.5 Metal Powders

Metal powders have been produced by spray pyrolysis; some representative examples are presented in Table 13.5 at the end of the chapter. Many of these materials are late transition metals, so metal complexes with oxygen-containing ligands are viable as precursors because the metal oxide, which may be formed as an intermediate, undergoes thermally induced deoxygenation at high temperatures, to form the metal. This is the case in the formation of Pd powders from $Pd(NO_3)_2$ solutions in which PdO has been identified as an intermediate.[31] Interestingly, the patent literature indicates that these materials can be produced on a commercially viable scale and that the powders have significant commercial potential.[32–34] Noble-metal powders such as Pd and Ag–Pd are used as the metallization layer in multilayer ceramic capacitors (MLCCs). Because these metals are expensive, methods to reduce the amount of powder required, while imparting the same or better properties, would lead to a significant reduction in their cost. This may be possible for metal powders produced by spray pyrolysis, in which the smaller size and better control over size distribution compared with alternative production methods could result in lower "lay-downs" (less powder used). In addition, chemical modification of the powder by even small amounts (e.g., $<$ 2 wt.%) of other elements, such as Ca, results in drastically different oxidation behavior.[35, 36] Such modification could provide added value by minimizing the extent of oxidation of the metal during sintering of green MLCCs. The oxidation of the metal layer is a significant problem encountered during the sintering of MLCCs because complete oxidation of Pd to PdO results in a 25% change in volume, which in turn results in delamination of the layers and destruction of the device.

### 13.2.6 Composite Powders

Composite particles are of great technological interest because of their potential for exhibiting properties derived from the presence of different materials in the same particle. Heterogeneous catalysts, which are commonly composed of nanometer-sized metal particles dispersed in a metal-oxide matrix, are a typical example of a composite material that takes advantage of the retention of high dispersion at high temperatures as a result of the ceramic matrix and the catalytic reactivity of the highly dispersed metal particles, often enhanced or modified by the presence of the metal-oxide matrix. This is an area in which spray pyrolysis has many advantages over other technologies (such as gas-to-particle conversion), due to the high level of control spray pyrolysis exhibits over powder composition and morphology.[37]

The formation of composite powders has been classified according to the method by which the powders form:

*Sequential Precipitation.* In this process, two or more precursors with different solubilities are dissolved in a droplet and precipitate at different times within each droplet as the solvent evaporates (according to the solubility of the precursors). This mechanism causes the initial segregation that ultimately leads to the formation of a composite powder. However, unmatched solubilities of the precursors are necessary for the formation of composite powders. Sequential precipitation can lead to the formation of single-phase materials if the temperatures at which the separate precipitates or their products can react together is exceeded. Therefore, classification of a process as sequential precipitation depends to some extent on the phase diagrams of the materials in question.

*Precipitation–Reaction.* With this technique, one component contained within the droplet is either a soluble precursor or a colloidal precipitate, and the other component or components thermally decompose to form another phase such that the (two or more) materials remain, for thermodynamic reasons, phase segregated in the final powder. An example is the formation of $Ag-YBa_2Cu_3O_x$ composites.[15] The Ag precursor typically reacts at a lower temperature than the other precursors do. But, in this case, Ag is also not soluble in $YBa_2Cu_3O_x$. As another example, highly dispersed nanocomposites composed of $C_{60}$ fullerene and Rh were prepared by the precipitation of soluble fullerene and the chemical reaction of $[(COD)RhCl]_2$ by spray pyrolysis.[38] Mixtures of soluble precursors, often to metals, and colloids, often of metal oxides, have been used to form metal–metal-oxide composites. For example, $Pd-TiO_2$ and $Pd-Al_2O_3$ composites have been prepared by these methods.[39]

*Vapor-phase-induced Reactions.* Not all of the reagents have to be contained in the droplets (Fig. 13.13). For example, chemical reactions of metal

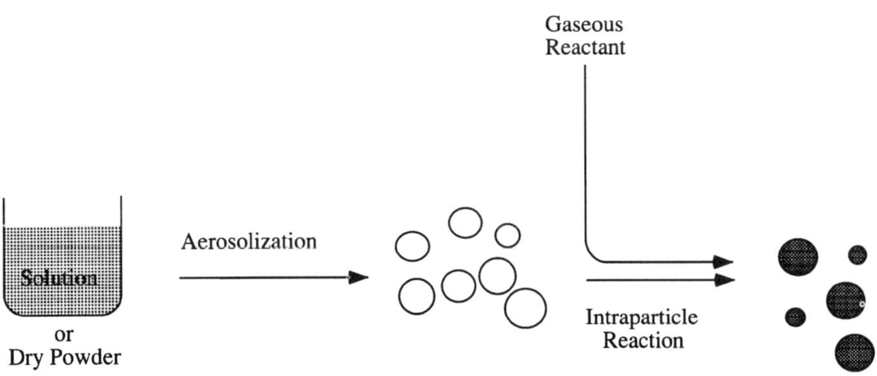


**Fig. 13.13.** Schematic representation of spray intraparticle conversion.

complexes can be initiated by the addition of a gaseous reagent (as described earlier for the preparation of metal-sulfide powders). These reactions could occur for a gaseous species that reacts at the surface of a droplet with a precursor in the droplet (a gas–liquid reaction), or the gaseous species could react at the surface of a dried particle (a gas–solid reaction). Either reaction can be carried out in a way that leads to segregation within the particle. The major point is that the reagents do not all have to be contained in the droplet. Instead, it may be desirable to add reagents such as gases directly into the system after the metal-containing precursors have been atomized, especially if the gaseous reagents are highly reactive towards the metal-containing precursors in solution, which might cause precipitation.

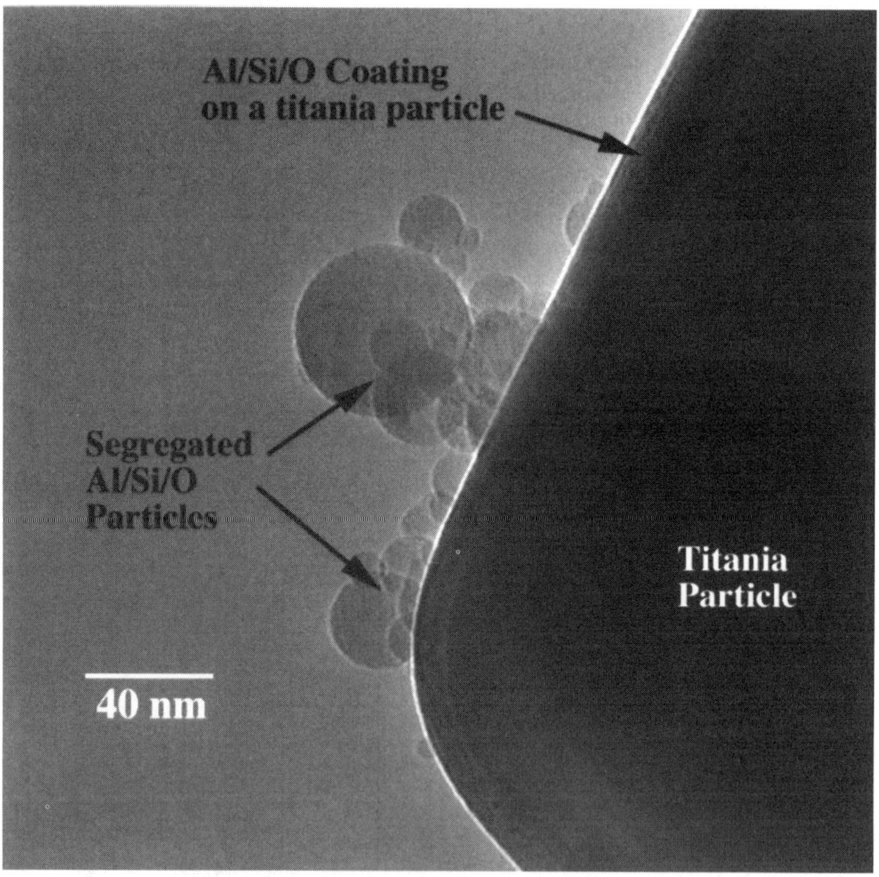


**Fig. 13.14.** Transmission electron micrograph of a $TiO_2$ particle with a 10-nm continuous $SiO_2$ coating. (Courtesy of Powell et al.)[43]

### 13.2.7 Coated Powders

Coated powders can be produced in situ during spray pyrolysis by a number of methods:

- CVD onto particles after their formation[40-43]
- Gas-to-particle conversion, followed by cluster deposition onto a substrate particle surface[40-43]
- Segregation of a component to a particle surface during spray pyrolysis of a mixture of precursors[36]

In situ CVD of precursors to form coatings on particles produced by spray pyrolysis and to form coatings by gas-to-particle conversion has already been described in Chapters 10 and 11. (See Fig. 13.14.) This method can provide uniform coatings with thicknesses on the order of nanometers.

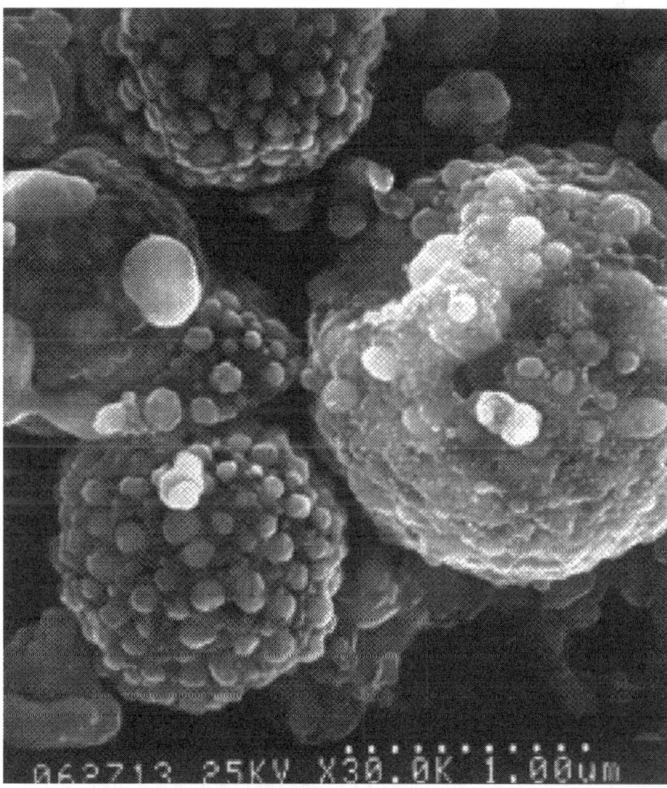


**Fig. 13.15.** Scanning electron micrograph photograph of Pd–$TiO_2$ composite particles. (Courtesy of Chandler, Kodas, and Hampden-Smith, unpublished results.)

An alternative method of producing coated particles is to dissolve a precursor for the coating together with the precursor for the powder in the same solution, which is then sprayed into the reactor to produce the powder by spray pyrolysis. During reaction and subsequent heating, one of the components diffuses to the surface. This mechanism is driven by a reduction in overall free energy. The appropriate choice of materials and precursors allows the formation of uniform, dense coatings. Clearly, there are a number of requirements for both the precursors and the products, as well as for the reactor conditions, to achieve coatings. The coated powders should be formed under conditions where the coating material and the particles do not react to form a single-phase material or interdiffuse to form a composite particle. An example of a situation in which these requirements are met is the formation of a metal-oxide coating on palladium particles;[35] numerous types of metal-oxide–metal-coated and composite particles can be formed by spray pyrolysis. (See Figs 13.15 through 13.17.)

**Fig. 13.16.** Scanning electron micrograph photograph of Pd–$Al_2O_3$ composite particles. (Courtesy of Chandler, Kodas, and Hampden-Smith, unpublished results.)

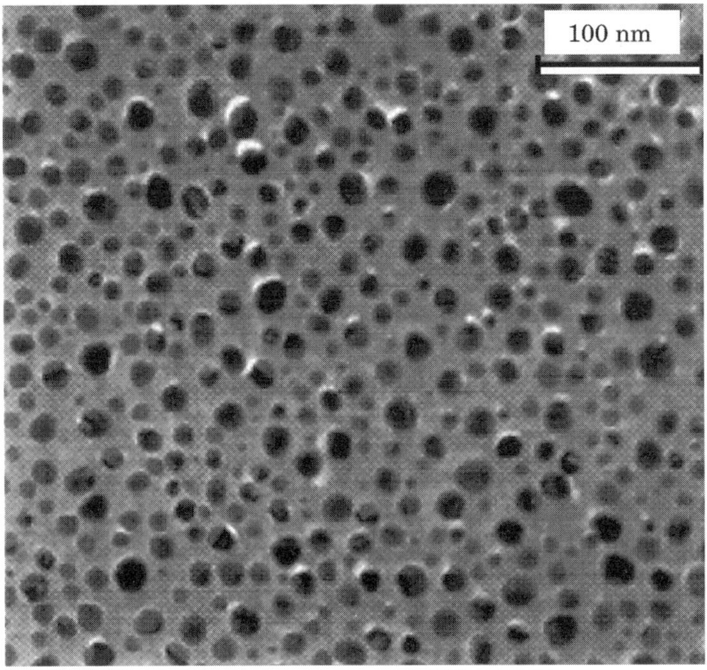


**Fig. 13.17.** Transmission electron micrograph photograph of surface of Pd–$Al_2O_3$ composite particles. (Courtesy of Chandler, Kodas, and Hampden-Smith, unpublished results.)

### 13.2.8 Influence of Precursor Characteristics on Particle Morphology

The nature of the reactants used for spray pyrolysis can strongly influence particle morphology.[2, 3] Control over particle morphology may require the use of highly soluble species or colloidal precursors. In general, requirements for molecular precursors include high solubility and purity, low cost, stability at room temperature in solution, and low toxicity. Tables 13.1 through 13.5 at the end of the chapter show the types of precursors that have been used to form metallic and ceramic particles by spray pyrolysis.

The influence of solute characteristics on particle morphology has been examined by several investigators. Zhang et al. investigated $ZrO_2$ formation from precursors including $ZrOCl_2$, ZrO(OH)Cl, $ZrO(NO_3)_2$, $Zr(CH_3COO)_4$, and $ZrO(OH)_{0.5}(CH_3COO)_{1.5}$.[44] For typical spray pyrolysis conditions with 10-$\mu$m droplets and high heating rates, solid particles of $ZrO_2$ were formed when a droplet's initial relative solution saturation was $10^{-2}$ and the precipitated salt was sufficiently permeable to permit evaporation of the remaining solvent after precipitation. The formation of solid particles was attributed to a longer drying time before precipitation and a shorter diffusion distance in which to maintain chemical homogeneity before precipitation. This solution is not always practical, however, since low solution concentrations lead to low reactor throughputs.

Leong has shown that, under identical conditions, compounds with similar solubilities could result in totally different particle morphologies.[45, 46] For instance, agglomerated crystals and hollow particles were formed from LiF and $Li_2CO_3$ under identical conditions. This was attributed to the tendency for LiF to form crystals and for $Li_2CO_3$ to form very fine-grained material.

A few studies have investigated the use of fumed and colloidal precursors as a method for generating solid particles. An example is the formation of mullite from $Al_2O_3$ and $SiO_2$ and from $Al(NO_3)_3$ and $SiO_2$.[25] In the latter case, heterogeneous nucleation of the $Al(NO_3)_3$ occurs onto the $SiO_2$, instead of homogeneous nucleation of $Al(NO_3)_3$ at the surface of the droplet, which could lead to the formation of hollow particles. In the case of two colloidal precursors with no molecular precursors, no solute precipitation is involved, again leading to the formation of solid particles. In contrast, other attempts to form mullite powder from molecular precursors have resulted in the formation of hollow particles.[47]

The generation of $SrTiO_3$ particles from titanium-citrate and strontium-nitrate solutions provides another example of how the nature of the precursor dramatically influences particle morphology. Precursor solutions made with low- (0.01 M) and high- (1.0 M) solubility titanium citrate and a mixture of colloidal $TiO_2$ and strontium nitrate led to dramatically different particle morphologies.[48] The solution with a solubility limit of 0.01 M formed hollow particles with extremely thin walls; the solutions with 1.0-M solubilities formed hollow particles with thick walls. Finally, a mixture of colloidal $TiO_2$ and strontium nitrate resulted in the formation of solid particles.[48]

### 13.2.9 Conditions Conducive to the Formation of Solid Particles

The discussions in the previous sections suggest that the following requirements are necessary for the formation of solid particles:

**High-priority Requirements**

1. Highly soluble salts (e.g., metal nitrates in water) allow no precipitation from solution until most of the solvent is evaporated.[3, 49] This results in a higher production rate.
2. The temperatures must be near the melting point to provide high mobilities for species inside particles (e.g., many metals) to densify by solid-state processes. This can be done in residence times of a few seconds or less, but requires sufficiently high temperatures.[16, 31, 50–55]
3. Allowing the product to melt[52, 54–56] is an excellent approach but requires operation above the melting point, which is possible only for some materials (some metals).
4. Colloidal precursors give solid particles after evaporation of the liquid containing the particles.[48, 57] However, colloidal precursors are not

always available, and their use may or may not be practical, depending on the material to be produced. (Some colloidal solutions containing metals and metal oxides, silica (Cab-O-Sperse), and titania are commercially available.)

5. Precursors that melt after evaporation of the solvent result in solid particles of the precursor.[49] Subsequent reaction of the precursor, however, may generate gases that cause the molten particle to expand or burst. (Many metalorganic precursors and some inorganic precursors —e.g., $Al(NO_3)_3 \cdot 9H_2O$—have low melting points.)
6. Precursors with high ceramic or metallic yields are desirable, to minimize the porosity associated with the conversion of the precursor to product. [Colloidal precursors, cluster precursors (e.g., polyoxoanions), and ligands of low molecular weight (e.g., hydroxyl groups) are desirable.][49]
7. Precursors that crystallize with well-defined crystal habits and that do not form oligomeric structures can provide porous precursor particles from which the solvent can escape without blowing the particle apart.
8. Slow reaction rates reduce the rate at which gaseous reactants are formed.

**Lower Priority Requirements**

1. A long residence time can avoid the buildup of gaseous products that can lead to bursting of the particles. However, long residence times are not practical in most cases.
2. A small droplet size provides a short diffusion distance. However, the production rate decreases rapidly as the droplet size is reduced, making this requirement impractical.
3. A slow drying rate produces uniform conditions in droplets before precipitation, but requires a long residence time, which is not practical in most cases.
4. A low concentration of salt in the solvent yields small particles once the solvent evaporates. However, the production rate decreases as the concentration is reduced, making this technique impractical.
5. An absence of nucleation sites in the droplets can keep the solute in solution for a longer time, thereby allowing more solvent to evaporate before precipitation. This method, too, is difficult to implement in practice.

The highest priority for the formation of solid particles is to choose the appropriate precursors. Precursors that have high solubilities and allow most of the solvent to be lost before precipitation occurs inside the droplet favor solid-particle formation. Densification of particles in the gas phase has been demonstrated in many cases and is an excellent approach because it allows hollow or porous particles to be formed earlier in the process. Unfortunately, this option is viable only for materials with relatively low melting points or temperatures at which significant amounts of liquid phase are formed. However, many ceramic superconductors, metals, titanates, and other materials have such low melting points or temperatures at which liquid phases are produced to allow solid particles to form at reasonable reactor temperatures. In addition, the formation of liquid phases can lead to segregation that may be difficult to eliminate at lower temperatures. Colloidal precursors do not precipitate from solution and therefore can result in the formation of solid particles. The lower priority techniques also favor the formation of solid particles, but are often difficult to carry out in practice.

## 13.3 SPRAY-DRYING

Spray-drying is similar to spray pyrolysis, but is physical (drying) as opposed to chemical (reacting).[58, 59] The major goal in using this approach is to form intermediate particles with uniform mixing of the reagents and, eventually, to produce particles with sizes, uniformities, and morphologies (generally spherical particles) that cannot be prepared by other methods. The primary difference compared to spray pyrolysis is that temperatures are much lower in the drying process (usually $< 400°C$) than in the pyrolysis process. The method begins by spraying a solution containing dissolved compounds or suspended particulates into a heated gas (usually hot air), where the liquid evaporates, leaving solid particles. For materials synthesis (as opposed to drying), the dried precursors are then collected and reacted (by heating), after which milling gives the final powder.

Advantages of spray drying for materials synthesis include:

- A simple system
- A wide variety of materials
- Cost-effectiveness (only a few unit operations)
- Many choices for liquid-phase precursors
- Doping possible
- Multicomponent materials
- Demonstrated scale-up to ton quantities

Among the disadvantages of the process are:

- For powder generation, powder collected must be heated and converted to product, resulting in agglomeration.
- Spread of particle size distribution is limited by milling process.
- Impurities can be introduced during milling.

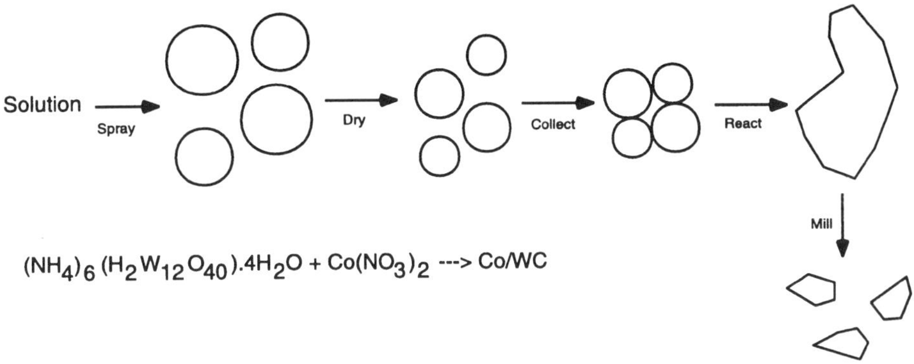


**Fig. 13.18.** Schematic representation of the Nanodyne process for the formation of Co–WC composites.

In one example of this method, also termed "spray conversion coating," an aqueous solution of sodium metatungstate $[(NH_4)_6(H_2W_{12}O_{40}) \cdot 4H_2O]$ and cobalt chloride $(Cl_2 \cdot nH_2O)$, cobalt nitrate, or cobalt acetate was spray-dried to form a powder composed of an intimate mixture of these materials.[60] (See Fig. 13.18.) This powder was then reduced with $H_2$ and reacted with CO to form a Co–WC nanocomposite powder wherein the WC is the inclusion phase in a Co matrix. The tungsten-carbide-phase crystallites are typically 20 to 40 nm in size and are entrained in hollow, porous spheres 75 microns in diameter, composed mainly of Co. This material has properties (chiefly, hardness) significantly better than those of powders prepared by alternative methods, which generally have larger WC grain sizes.

Spray-drying of precursors is particularly useful for the formation of high-purity ceramic powders. Spray-drying methods are often preferred over other methods of powder manufacture when milling and grinding are particularly undesirable (for example, in the preparation of phosphor powders, in which small contaminants destroy the luminescent properties of the material) and when control over powder morphology is desired. Powders produced by spray-drying are typically larger than powders produced by spray pyrolysis, and their features depend upon similar features of the system, such as the concentration of solution, the droplet production method, etc. If the powders are end milled, the difference in size is irrelevant. In the case of materials that are not subject to further processing, however, this constraint becomes crucial. Commercial spray-drying units generally use spray nozzles for liquid atomization, resulting in mean particle diameters of 30 to 1000 $\mu$m (Fig. 13.19). These units can have very high throughput—up to 150,000 kg/hr.

Spray-drying is used commercially to process various ceramic powders such as $(Mn,Zn)Fe_2O_4$ and $Al_2O_3$.[58] The process is commercially viable because it is economical, continuous, and capable of producing powders with

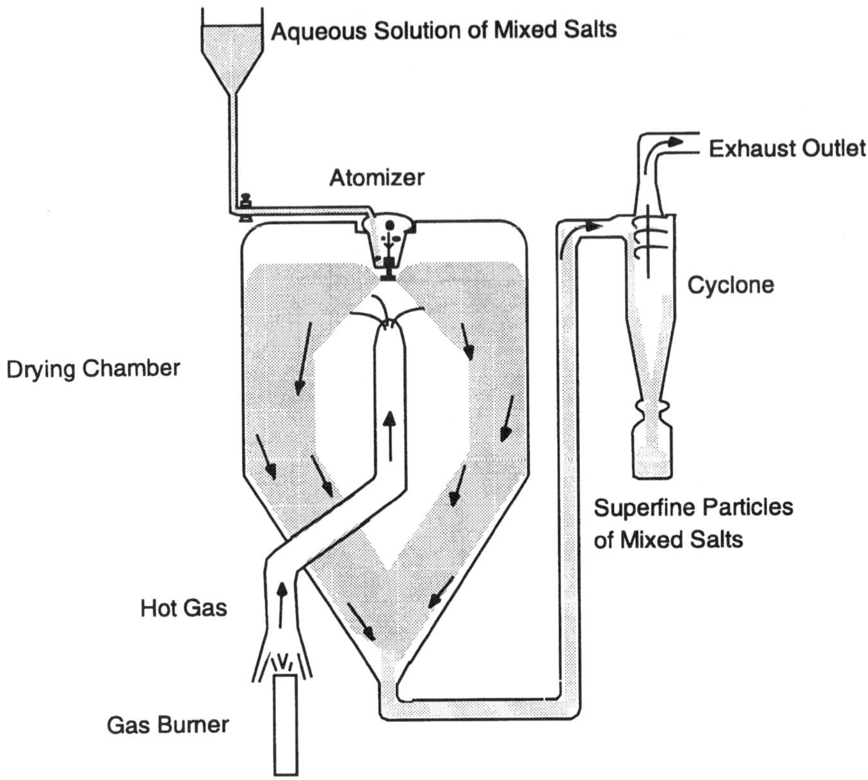


**Fig. 13.19.** Apparatus for spray drying. (Adapted from Inchinose.)[61]

a uniform, reproducible character. Control over a number of factors, including the slurry preparation, the liquid atomization, the droplet–air mixing, and the droplet drying, is crucial. The slurries (or colloids) are often water based, and the ceramic is dispersed in the slurry through the use of a dispersing agent, which is typically an organic polymer adsorbed on the particle surface through coulombic interactions, hydrogen bonding, or van der Waals forces. The higher the solids content of the slurry, the higher is the production rate of the process (other things being equal). Binders are also frequently added to the slurry, to provide green strength to the compacted solid. Binders are typically polymers, such as polyvinyl alcohol, and they are often required to form low-viscosity solutions to achieve slurries with a high solids content.

Atomization of the slurry is generally achieved by rotary atomizers, pressure nozzles, and pneumatic nozzles. The selection of the type of atomizer depends upon the nature of the slurry and the droplet or particle size required. This relationship has been discussed elsewhere.[59]

The method by which the droplets are mixed with air has a significant influence on the droplet drying time, the residence time, and the extent of removal of moisture from the droplets. The droplet–air mixing classifications commonly used in spray-dryers are concurrent, countercurrent, and mixed flow. These three methods have different effects on the evaporation rates, residence times, and temperatures assumed by the droplets or particles, so the choice of which method to use depends to large extent on the requirements of the system.[59]

As with spray pyrolysis, there is a possibility that low-density or hollow particles will be formed, which is generally deleterious to the material. Various reasons for the formation of these kinds of particles have been proposed and are analogous to the reasons described earlier with respect to spray pyrolysis.[59]

## 13.4 FREEZE DRYING

Freeze-drying is similar to spray-drying as far as the generation of droplets is concerned, but the droplets are frozen and then dried by sublimation once they are in the solid state.[61] The solution containing dissolved compounds is sprayed into liquid nitrogen or some other cold liquid, whereupon the droplets freeze. The frozen liquid is removed by sublimation, and then the dry precursor is converted to the product, usually by further heating. Smaller droplets are preferable because they freeze faster, which might avoid preferential precipitation of the precursors. If the cooling medium is a liquid, then the cooling medium and the liquid droplets must be immiscible. For example an aqueous salt solution could be sprayed into dry ice and hexane (−77°C) or liquid nitrogen (−196°C), but not dry ice and acetone. In order to evaporate the solvent from the droplet, it is important that the temperature of the system be as high as possible (to increase the vapor pressure of the solvent, and therefore its evaporation rate, under nonequilibrium conditions). Consequently, the solubility of the sol in the solvent has a strong impact on the process. For example, in using water as solvent (which is likely to be the most practical solvent), it is desirable to remove the water at approximately −10°C. However, if the solute concentration is too high, lower temperatures are necessary to avoid melting of the particle caused by depression of the freezing point of the frozen droplet.

Freeze-drying is used to prepare a range of consumable products, such as instant coffee and milk powder. It has also been used to prepare alumina, spinel, lithium ferrite, and nickel-oxide powders.[4, 5, 8, 9, 61, 62] For example, aluminum sulfate was dissolved in water (0.6 mol/l) and freeze-dried. On warming, the spherical particles retained their shape and were amorphous until approximately 300°C, at which point they crystallized as amorphous anhydrous aluminum sulfate.[61] On heating above 700°C, thermal decomposition occurred to form alumina, which crystallized on heating above 1200°C, to form agglomerated, fine-grained alumina powders.

Freeze-drying has also been used to prepare ceramic superconducting powders by freezing an aerosol in liquid nitrogen. (See Fig. 13.20.) It has been shown that the thermal-processing behavior of freeze-dried powders is fundamentally different from that of powders made by mechanical mixing and grinding. The freeze-drying method avoids impurities that are often introduced during milling.[5, 63] In some cases, the bulk freeze-dried precursor powder is heated to thermally decompose the mixed metal nitrates and form the final ceramic. This can lead to agglomeration between the particles and relative large agglomerate sizes, which may be detrimental, depending on the application. In other cases, the freeze-dried particles are suspended in a gas flow and passed through a furnace, where thermal decomposition takes place. The particles are then collected with minimum agglomeration. This process is similar to spray pyrolysis, except that the steps of particle formation and particle decomposition have been separated into different operations.[9]

Advantages of freeze-drying include:

- Simple system
- Wide variety of materials
- Cost-effectiveness (only a few unit operations)
- Many choices for liquid-phase precursors
- Doping possible
- Multicomponent materials
- Need for grinding obviated

Disadvantages of freeze-drying for materials synthesis include:

- Particles must be reacted to convert to product, resulting in agglomeration
- Milling of reaction products limits their purity and size distribution

## 13.5 SPRAY–COMBUSTION SYNTHESIS

Spray processes have been combined with combustion processes, as described in the previous chapter, to give a unique class of methods for powder generation. In one approach, the metal-containing precursors and an organic compound are dissolved in a solvent, commonly water, and sprayed into a heated gas in the presence of an oxidant such as oxygen. The organic compound acts as a fuel, and the mixture combusts under certain conditions to generate high temperatures resulting in the formation of inorganic materials, generally metal oxides that are homogenous.

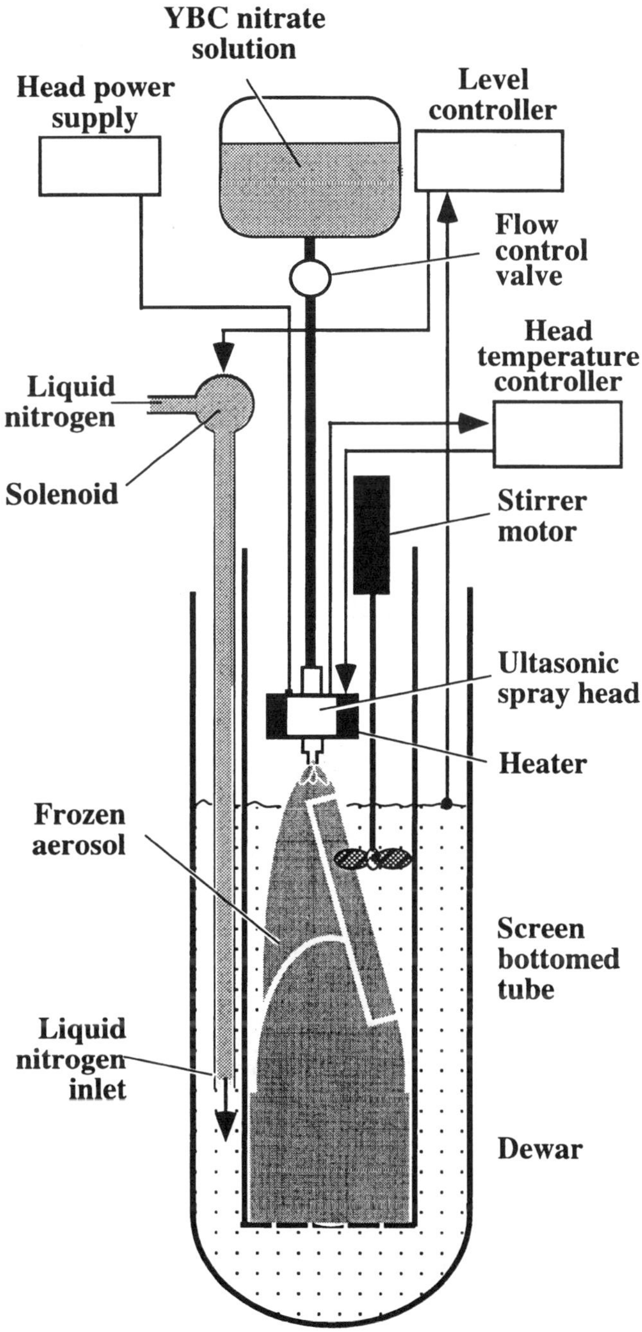


**Fig. 13.20.** Apparatus for the production and freezing of aerosols of aqueous YBCO solutions. (Adapted from Kimura et. al.)[9]

For example, amino acids[64] have been used as the fuel to produce a variety of metal-oxide powders, including lanthanum strontium chromite ($La_{0.84}Sr_{0.16}CrO_3$), manganese ferrite (MnFe), and yttrium calcium manganite ($Y_{1-x}Ca_xMnO_3$), with primary particle sizes in the range 10 to 700 nm.[64] No attempt was made to spray the material into a furnace; the mixture of fuel and reagent were ignited in the solid state.

In a combination of a combustion process together with flame spray pyrolysis, involatile reagents such as metal nitrates mixed with fuels such as sugars (e.g., sucrose) have been introduced into a flame through a piezoelectric nozzle such that an aerosol was formed. The aerosol droplets then burned as the sugar underwent combustion in the flame, resulting in the formation of spherical ceramic particles with mean sizes of approximately 37 nm. The fact that nanometer-sized particles were frequently observed by this method is probably indicative of evaporation of a reagent or, more likely, an intermediate species, resulting in gas-to-particle conversion at the high temperature of the flame.[65] Alternatively, the droplets could have exploded, producing fine particles.

## 13.6 FLAME SPRAY PYROLYSIS

Flame spray pyrolysis is discussed briefly here because there are several examples of liquid-to-solid or solid-to-solid conversion. However, care has to be taken concerning the high potential for gas-to-particle conversion as a result of the high temperatures that can occur and lead to the evaporation of reagents or products. Since there are relatively few examples of liquid-to-solid or solid-to-solid conversion in flames, the primary focus of this section will be the trends involved in the flame pyrolysis of precursors that are only volatile at high temperatures.

Flame pyrolysis involves the delivery of droplets to a flame, with the flame providing the energy source that results in intraparticle chemical reactions which cause conversion of the reagents to the final product.[66] This process is analogous to ordinary spray pyrolysis, except that the thermal energy is delivered by a flame rather than a furnace. The significant differences between these two heating methods are the temperature gradients (greater in flames than in furnaces), the residence times (shorter in flames than in furnaces), the temperatures (can be higher in flames than in furnaces), and the exposure to chemical species formed as a result of combustion in a flame that are not present in a spray pyrolysis reactor. It is likely that primarily inorganic materials will be formed by flame pyrolysis, since organic materials combust at typical flame temperatures. In order to achieve flame pyrolysis rather than gas-to-particle conversion, the inorganic precursors should be involatile at the temperature of the flame and therefore should be delivered as either liquid precursors or solids. This takes the form of aerosol delivery, with the gas stream containing particles of droplets of the precursors. One reason for producing materials by a flame pyrolysis route is to make larger particles ($\sim$ 1.0 $\mu$m), compared with gas-to-particle conversion in a flame using volatile (gaseous) reagents, which typically give particle sizes in the nanometer range. Therefore, depending on the applications for the powder product, it may be important to avoid "contamination" of a particularly desired particle size distribution with particles of a significantly smaller size. Provided that involatile precursors are fed into the flame, gas-to-particle conversion is likely to occur only as a result of the evaporation of a reactive intermediate or the product. Therefore, it should be possible to control the occurrence of gas-to-particle conversion, depending on the residence time, flame temperature, and nature of the chemical system under study.

Aside from the precursor characteristics of volatility and reactivity (see Chapter 8), two other characteristics of the flame pyrolysis system have a significant impact on the nature of the powder product: the flame temperature and the residence time. The residence times in the flames (milliseconds) are significantly shorter that residence times in spray pyrolysis reactors (seconds), and as a result, the extent of conversion of precursor to final product and crystallinity of the product can be lower than that found in conventional spray pyrolysis. This is counteracted to some extent by the potential for higher temperatures in flame pyrolysis than is common in spray pyrolysis. The shorter residence time (the time for chemical reaction and crystallization to occur) could be offset by carrying out the reaction at higher temperatures. This aspect will also depend on the reagents and products involved. For example, it is likely that materials with low melting points (such as some metals) will be more crystalline than materials with high melting points (such as some metal oxides).[67, 68]

It is also worth considering flame pyrolysis under two temperature extremes: low temperatures and high temperatures (provided that involatile reagents are delivered). At low flame temperatures, approximately 1000°C, it is likely that the temperature is too low and the residence time too short to cause significant gas-to-particle conversion, unless unusually volatile intermediates or products are formed. At high flame temperatures (e.g., $\sim$ 2,000°C), it is likely that some reagents, intermediates, or products will evaporate, resulting in the formation of particles (by gas-to-particle conversion) with a significantly smaller size than those obtained by flame spray pyrolysis. Much higher temperatures are often necessary to ensure that all the material introduced into the flame is converted to nanometer-sized particles.

Submicron $YBa_2Cu_3O_{7-\delta}$ particles have been produced by flame spray pyrolysis in which an aerosol composed of an aqueous solution of Y, Ba, and Cu nitrates was introduced into a flame.[69-71] The product showed a transition temperature of 92 K, as determined from magnetic susceptibility measurements. The particles had a wide size distribution, ranging from 10 to 1000 nm, probably as a result of both gas-to-particle conversion or bursting of particles and spray pyrolysis. Due to the effects of water vapor reactions at high temperatures, only a diffusion flame produced the correct phase. The

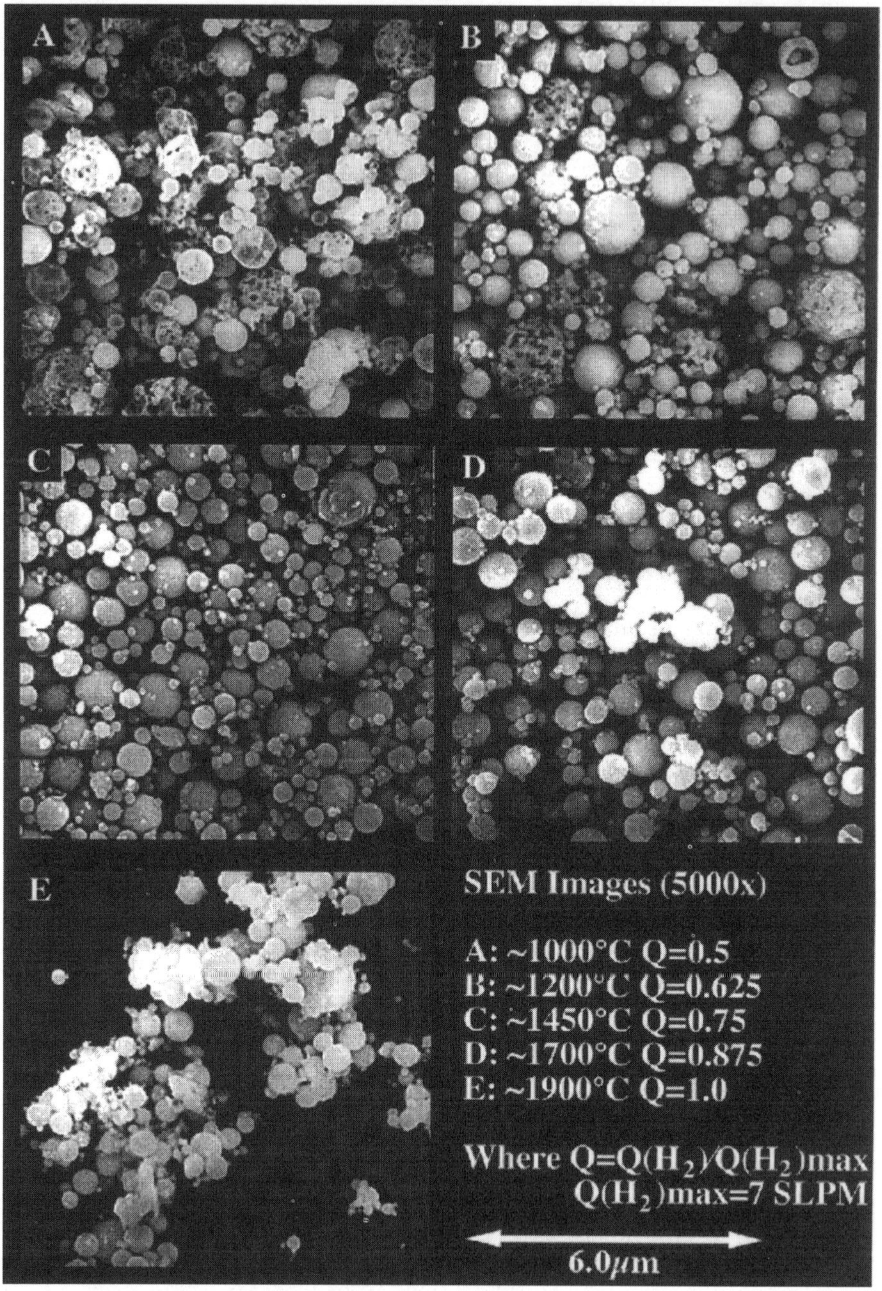


**Fig. 13.21.** Scanning electron micrograph of $BaTiO_3$ powders derived from flame spray pyrolysis, showing a transition from porous to solid particles with increasing flame temperature. (Courtesy of Brewster and Kodas.)[56]

configuration of the flame was particularly important for the successful application of the method.

As shown in Figure 13.21, fine particles of dense, high-purity, crystalline $BaTiO_3$ were produced by flame spray pyrolysis.[56] For all of the temperatures investigated (~ 1000 to 2000°C, adiabatic), the powders that were generated were chemically pure, crystalline (primarily tetragonal phase with hexagonal and cubic polymorphs), and unagglomerated. At a low flame temperature (~ 1000°C, adiabatic), the particles produced were hollow and irregularly shaped. Particles produced at higher flame temperatures (≥ 1500°C, adiabatic) were dense and homogeneous. This work demonstrated that flame spray pyrolysis provides a useful method for forming dense particles of materials with a high melting point by aerosol-phase densification.

Pure palladium particles were generated via flame spray pyrolysis using various concentrations (0.05 M to 0.2 M) of palladium-nitrate solutions.[72] The Pd powder produced contained relatively large particles (0.5 to 1.0 $\mu$m) and, at high temperatures, smaller, nanophase particles (10 to 50 nm). The large particles were formed by intraparticle reaction and were dense and spherical. The small particles were formed via gas-to-particle conversion. Thermogravimetric analysis showed that the powder (doped with 0.05% wt. Ca) produced using flame spray pyrolysis had similar oxidation resistance to Pd powder formed by conventional spray pyrolysis with an overall weight increase of approximately 3 to 5%. Similar results were obtained for silver and silver–palladium alloy.

## 13.7 MELT ATOMIZATION

Atomization of molten materials is used industrially for the generation of a wide variety of powders.[61] This approach is simple and relies on using a gas (generally under high pressure) to atomize molten material to produce droplets. The molten droplets cool in the gas phase and are then collected, often by settling. The technique is most useful for producing particles larger than 1 $\mu$m, because of the difficulty in producing submicron droplets of molten materials. Cooling rates as high as 10,000°C/sec have been reported using this technique under certain conditions. Melt atomization is employed to make powders composed of spherical (or near-spherical) particles that are highly crystalline and that can be employed in the formation of parts where high packing densities of highly crystalline materials are required. Materials with a low melting point are most suitable for this process, and the method has been used extensively for the formation of metal and metal alloy powders, as well as ceramic materials and fluoride glasses.[73–79]

Variations on this process include atomizing molten metals in an electric-arc spray gun followed by rapidly quenching the resulting droplets in a liquid cryogen to form fine, highly spherical oxide-free powders.[80] Examples of the materials produced by this method include Sn–Pb alloys with particles having a mean diameter of 15.2 $\mu$m, with 85 wt% of the particles having a diameter between 5.5 and 31 $\mu$m.[80]

## 13.8 POWDER SYNTHESIS AND DENSIFICATION IN THERMAL PLASMAS

### 13.8.1 Thermal-Plasma Densification

Densification of a variety of materials as powders has been effected by thermal-plasma processing. The method involves entraining the particles in a gas stream and using the energy of the plasma to melt the particles, followed by cooling and collecting the particles. Direct-current or inductively coupled radio frequency plasmas have been used to densify metals and metal-oxide powders and make them spheroidal. Powders that have been treated in this way tend to have large particle sizes, most having a mean particle diameter 25 to 100 $\mu$m.

### 13.8.2 Thermal-Plasma Synthesis of Powders

Thermal-plasma synthesis of powders is quite analogous to flame synthesis of powders. For example, both processes employ relatively high temperatures and short residence times, compared with electrical heating of gases. Thermal plasmas suitable for powder synthesis are generally produced by high-frequency (radio or microwave frequency) discharges or high-intensity (ac or dc) arcs. A variety of novel reactor designs have been developed to improve the properties of the powders by optimizing residence and mixing times. This method can be considered as spray pyrolysis with thermal plasma heating (as opposed to gas-to-particle conversion).

As with other higher temperature powder-synthesis methods, there is a trade-off between the high temperature and the short residence time that affects particle characteristics such as density, crystallinity, particle size, and size distribution. High temperatures often lead to the formation of volatile species and the presence of small particles of a size characteristic of homogeneous gas-phase reactions. Therefore, powders produced by spray pyrolysis may be "contaminated" by smaller particles with different compositions that are derived from gas-phase reactions.

## 13.9 REFERENCES

1. Gurav, A., Kodas, T., Pluym, T., & Xiong, Y. *Aerosol. Sci. Tech. 19*, 1993, 411–452.
2. Kodas, T.T. *Advanced Mater. 6*, 1989, 180–192.
3. Messing, G.L., Zhang, S.-C., & Jayanthi, G.V. *J. Am. Ceram. Soc. 76*(11), 1993, 2707–2726.

4. Armor, J.N., Fanelli, A.J., March, G.M., & Zambri, P.M. *J. Am. Ceram. Soc.* *71*(11), 1988, 938–942.
5. Coppa, N.V., Hults, W.L., Smith, J.L., & Brynestad, J. *J. Mater. Res.* *9*(10), 1994, 2510–2518.
6. Dayal, R., Prasad, C.D., & Lal, R. *Mat. Res. Bull.* *25*, 1990, 1339–1346.
7. Luo, P., Strutt, P.R., & Xiao, T.D. *Materials Science and Engineering B17*, 1993, 126–130.
8. Tripathi, R.B. & Johnson, D.W., Jr. *Materials Letters* *10*(3), 1990, 118–120.
9. Kimura, Y., Ito, T., Yoshikawa, H., & Hiraki, A. *Jpn. J. Appl. Phys.* *30*(5a), 1991, L798–L801.
10. Gardner, T.J., Sproson, D.A., & Messing, G.L. *Mat. Res. Soc. Symp. Proc.* *32*, 1984, 227–232.
11. Ruthner, M.J. (Ed.). *Industrial production of multicomponent ceramic powders (metal oxides) by means of the spray roasting technique*. Amsterdam: Elsevier Scientific, 1983.
12. Vollath, D. *J. Mater. Sci.* *25*, 1990, 2227–2232.
13. Jayanthi, G.V., Zhang, S.C., & Messing, G.L. *Aerosol Sci. Technol.* *19*(4), 1993, 478–490.
14. Carim, A., Doherty, P., Kodas, T.T., & Ott, K. *Materials Letters* *8*, 1989, 335–339.
15. Ward, T.L., Kodas, T.T., Carim, A.H., Kroeger, D.M., & Hsu, H. *J. Mater. Res.* *7*(4), 1992, 827–836.
16. Ortega, J., Kodas, T.T., Chadda, S., Smith, D.M., Ciftcioglu, M., & Brennan, J. *Chem. Mater.* *3*, 1991, 746–751.
17. Brinker, C.J. & Scherer, G.W. *Sol-Gel Science*. San Diego: Academic Press, 1990.
18. Chandler, C., Powell, Q., Hampden-Smith, M.J., & Kodas, T.T. *J. Mater. Chem.* *3*(7), 1993, 775–776.
19. Cho, S.-Y., Lee, J.-H., & Park, S.J. *J. Mater. Sci.* *30*, 1995, 3274–3278.
20. Gardner, T.J. & Messing, G.L. *Thermochimica Acta* *78*, 1984, 17–27.
21. Gardner, T.J. & Messing, G.L. *Am. Ceram. Soc. Bull.* *63*(12), 1984, 1498–1504.
22. Seitz, K., Ivers-Tiffee, E., Thomann, H., & Weiss, A. (Eds.). *Influence of zinc acetate and nitrate salts on the characteristics of undoped ZnO powders*. Amsterdam: Elsevier Science, 1987.
23. Jada, S.S. *J. Mater. Sci. Lett.* *9*, 1980, 565–568.
24. Matijevic, E., Zhong, Q., & Partch, R.E. *Aerosol Sci. Technol.* *22*, 1995, 162–171.
25. Moore, K.A., Cesarano, J.I., Smith, D.M., & Kodas, T.T. *J. Amer. Cer. Soc.* *75*(1), 1992, 213–215.
26. Oshima, K., Tsuto, K., Okuyama, K., & Tohge, N. *Aerosol Sci. Technol.* *19*(4), 1993, 468–477.
27. Sano, Y. & Keey, R.B. *Chemical Engineering Science* *37*(6), 1982, 881–889.
28. Roger, C., Corbitt, T., Xu, C., Zeng, D., Powell, Q., Chandler, C.D., Nyman, M., Hampden-Smith, M.J., and Kodas, T.T. *J. NanoSt. Mater.* *4*(5), 1994, 529–535.
29. Caruso, J., Kodas, T.T., & Hampden-Smith, M.J. Unpublished results.
30. Lindquist, D.A., Kodas, T.T., Smith, D.M., Xiu, X., Hietala, S.L., & Paine, R.T. *J. Am. Ceram. Soc.* *74*, 1991, 3126–3128.

31. Pluym, T.C., Lyons, S.W., Powell, Q.H., Gurav, A.S., Kodas, T.T., Wang, L.M., & Glicksman, H.D. *Mat. Res. Bull. 28*, 1993, 369–376.
32. Asada, E. & Nagashima, K. European Patent Application, Patent 0 593 167 A1, 4/20/94.
33. Asada, E., Ono, S., Matsuo, M., & Kagaku, S. Japan Patent SHO 62[1987]–2404, 1/8/87.
34. Asada, H., Akimoto, Y., Ono, S., Nagashima, K., Iwasaki, M., & Maekawa, M. Japan Patent HEI 6[1994]–172802, 6/21/94.
35. Jain, S., Atanasova, P., Kodas, T.T., & Hampden-Smith, M.J. *J. Electrochem. Soc. 143*(11), 1996, 3762–3770.
36. Jain, S., Kodas, T.T., & Hampden-Smith, M.J. *J. Electrochem. Soc. 14*(4), 1997, 1222–1231.
37. Moser, W.R. & Lennhoff, J.D. *Chem. Eng. Comm. 83*, 1989, 241–259.
38. Gurav, A., Duan, Z., Wang, L., Hampden-Smith, M.J., & Kodas, T.T. *Chem. Mater. 5*(2), 1993, 214.
39. Chandler, C.D., Hampden-Smith, M.J., & Kodas, T.T., personal communication.
40. Fotou, G.P. & Kodas, T.T. Unpublished
41. Fotou, G.P. & Kodas, T.T. *Adv. Mater. 9*(5), 1997, 420–423.
42. Jain, S., Fotou, G.P., & Kodas, T.T. *J. Colloid Interface Sci. 185*, 1997, 26–38.
43. Powell, Q.H., Fotou, G.P., Kodas, T.T., Anderson, B.M., & Guo, Y. *J. Chem. Mater. 9*(3), 1997, 685–693.
44. Zhang, S.-C., Messing, G.L., & Borden, M. *J. Amer. Ceram. Soc. 73*(1), 1990, 61–67.
45. Lcong, K.H. *J. Acrosol Sci. 5*, 1987, 525–552.
46. Leong, K.H. *J. Aerosol Sci. 12*(5), 1981, 417–435.
47. Kanzaki, S., Tabata, H., Kumazawa, T., & Ohta, S. *J. Am. Ceram. Soc. 68*(1), 1985, C6–C7.
48. Ortega, J. *Control of the morphology of ceramic particles produced by aerosol decomposition.* Master's. Thesis, University of New Mexico, 1992.
49. Jain, S., Skamser, D., & Kodas, T.T. *Aerosol Sci. Technol. 27*(5), 1997, 575–590.
50. Gurav, A., Pluym, T., Kodas, T.T., & Pratsinis, S.E. *Chem. Eng. Commun. 151*, 1996, 211–226.
51. Kodas, T.T., Datye, A., Lee, V., & Engler, E. *J. Appl. Phys. 65*, 1989, 2149–2151.
52. Majumdar, D., Kodas, T.T., & Glicksman, H. *Adv. Mater. 8*(12), 1996, 1020–1022.
53. Majumdar, D., Shefelbine, T., Kodas, T.T., & Glicksman, H. *J. Mater. Res. 11*, 1996, 128.
54. Pluym, T., Kodas, T.T., Wang, L.-M., & Glicksman, H. *J. Mater. Res. 10*(7), 1995, 1661–1673.
55. Pluym, T.C., Powell, Q.H., Gurav, A.S., Ward, T.L., Kodas, T.T., Wang, L.M., & Glicksman, H.D. *J. Aerosol Sci. 24*(3), 1993, 383–392.
56. Brewster, J. & Kodas, T.T. *AIChE J. 42*(11), 1997, 2665–2669.
57. Ohshima, K., Tsuto, K., Okuyama, K., & Tohge, N. *Aerosol Sci. Technol. 19*(4), 1993, 468–477.
58. Lukasiewicz, S.J. *J. Amer. Ceram. Soc. 72*(4), 1989, 617–624.

59. Masters, K. *Spray Drying*. New York: John Wiley and Sons, 1972.

60. Kear, B.H., Siegel, R.W., & Tsakalakos, T. (Eds.). *Nanostructured Materials*. Elmsford, NY: Pergamon Press, 1992.

61. Ichinose, N., Ozaki, Y., & Kashu, S. *Superfine Particle Technology*. London: Springer Verlag, 1992.

62. Takao, Y. & Awano, M. *J. Mater. Res.* *71*(11), 1992, 2942–2948.

63. Coppa, N.V., Myer, G.H., Salomon, R.E., Bura, A., O'Reilly, J.W., Crow, J.E., & Davies, P.K. *J. Mater. Res.* *7*(6), 1992, 2017–2026.

64. Pederson, L.R., Chick, L.A., & Exarhos, G.J. United States Patent 5 114 702, 5/19/92.

65. Helble, J.J., Moniz, G.A., & Morse, T.F. United States Patent 5 447 708, 9/5/95.

66. Brewster, J. *Aerosol Generation of* $BaTiO_3$. Master's thesis, University of New Mexico, 1997.

67. Chung, S.-L., Sheu, Y.-C., & Tsai, M.-S. Flame synthesis of $SiO_2$, $Al_2O_3$ and $3Al_2O_3$–$2SiO_2$ powders. In Messing, G.L., Hausner, H., and Hirano, S.-C. (Eds.), *Ceramic Powder Science III*, Columbus, OH: American Ceramic Society, 1990, 275–282.

68. Nagashima, K., Morimitsu, Y., & Kato, A. *Nippon Kagaku Kaishi 12*, 1987, 2293–2300.

69. Zachariah, M.R. & Huzarewicz, S. *J. Mater. Res.* *6*(2), 1991, 264–269.

70. Zachariah, M.R. & Huzarewicz, S. *Combustion and Flame 87*(1), 1991, 100–103.

71. Zachariah, M.R. & Huzarewicz, S. *Chem. Phys. Processes Combust. 96*, 1990, 4.

72. Brewster, J., Kodas, T.T., & Hampden-Smith, M.J. Unpublished results.

73. Amano, N., Odani, Y., Takeda, Y., & Akechi, K. *MPR March*, 1989, 186–190.

74. Broer, M.M., Bautista, J.R., Chui-Sabourin, M., & Bruce, A.J. *Mater. Sci. Forum. 67–68*, 1991, 39–44.

75. Carlin, B. *J. Accoustical Soc. Am. 30*(2), 1958, 153–154.

76. Hirano, T. & Fujita, T. *Light Metal Films 37*(10), 1987, 670–676.

77. Lee, B.-T., Chun, B.-S., & Hiraga, K. *J. Mater. Res.* *9*(10), 1994, 2519–2523.

78. Lierke, E.G. & Griesshammer, G. *Ultrasonics*, October, 1967, 224–228.

79. Yamauchi, I., Ohnaka, I., Kawamoto, S., & Fukusako, T. *Trans. Jpn. Inst. Metals 27*(3), 1986, 187–194.

80. Zurecki, Z., Berger, K.B., & Swan, R.B. United States Patent 5 294 242, 3/15/9.

## 13.10 TABLES

**TABLE 13.1. Simple metal oxides prepared by spray pyrolysis**

| Product | Precursors | Size, μm | Morphology | Temperature, °C | Notes | References |
|---|---|---|---|---|---|---|
| $Fe_2O_3$ | $Fe(NO_3)_3$ | 2 | Hollow | Up to 1100 | | [1] |
| $Fe_2O_3$ | $Fe_2(SO_4)_3$ | 0.2 | Not reported | 800 | | [2] |
| $Fe_2O_3$ | Metal nitrates, oxalates, and lactates | 0.2 | Not reported | | | [3–7] |
| $Fe_2O_3$, $Mn_3O_4$, NiO, ZnO, $NiMn_2O_4$, $ZnFe_2O_3$ | Metal acetates and nitrates | 1–10 | Nitrates gave hard spherical aggregates, acetates gave hard aggregate to free-fine aggregate particles | 1000 | Acetates gave aggregates | [8–11] |
| $ZrO_2$, $Y_2O_3$, $Al_2O_3$ | Metal chlorides and nitrates | 8–9 | Hollow | 500–800 | 100 g/hr production rate | [12] |
| $Y_2O_3$ stabilized $ZrO_2$ | Zirconium *n*-butoxide, yttrium alkoxides | 0.25 | Porous | 700 | Polymeric intermediate formed in droplet | [13] |
| PdO | $Pd(NO_3)_2$ | 0.1–1 | Solid | 500–900 | Studied change in surface area | [14] |
| $ZrO_2$ | $ZrOCl_2 \cdot 8H_2O$<br>$ZrO(OH)Cl$<br>$ZrO(NO_3)_2 \cdot xH_2O$<br>$Zr(CH_3COO)_4$<br>$ZrO(OH)_{0.5}(CH_3COO)_{1.5}$ | 1–5 | Solid favored by high-solubility salt at low concentration | Up to 750 | | [15] |
| $Y_2O_3$ stabilized $ZrO_2$ | $ZrOCl_2 \cdot 8H_2O$<br>$Y(NO_3)_3 \cdot 5H_2O$ | 0.2–1.8 | Porous, hollow particles eliminated using temperature gradient | 400 | Reaction burst particles | [16] |
| $Y_2O_3$ stabilized $ZrO_2$ | $ZrO(NO_3)_2 \cdot xH_2O$<br>$Y(NO_3)_3 \cdot 3H_2O$ | 0.4 | Hollow | 950 | | [17] |
| MgO, ZnO, $Co_3O_4$ | Metal nitrates | 1–20 | Porous and hollow | | Ethanol solution to fracture particles | [3–7] |

| | | | | | | |
|---|---|---|---|---|---|---|
| MgO | $MgCl_2$ | > 10 | Industrial scale by Aman process | | | [18] |
| ZnO | $Zn(NO_3)_2$ | 0.1–1 | Solid | — | Luminescence behavior studied | [19] |
| ZnO | $Zn(CH_3COO)_2 \cdot 2H_2O$ | 0.15 | Porous | 500 | Precursor evaporation, reactions, and new-particle formation observed | [20] |
| ZnO | $Zn(C_2H_3O_2)_2 \cdot 2H_2O$ $Zn(NO_3)_2 \cdot 4H_2O$ | 10 | Shells, collapsed shells, and porous particles | | Zinc-acetate sublimation altered stoichiometry | [21] |
| ZnO | Nitrate, acetate $(Co(NO_3)_2 \cdot 6H_2O$ | < 1 | Spherical hollow | 600–1,000 | Also co-doped ZnO | [22] |
| ZnO | $Zn(NO_3)_2$ | 0.5–1.0 | Not reported | 200–500 | | [23] |
| $Al_2O_3$, $CaAl_2O_4$ | $Al(NO_3)_3 \cdot 9H_2O$ $Ca(NO_3)_2$ | 0.5–1.0 | Not reported | 750–950 | | [24] |
| $Al_2O_3$ | $Al(NO_3)_3$ | 20–80 | Milling required | | Loss of $Na_2O$ to walls | [25] |
| $Al_2O_3$ | $Al(NO_3)_3$–$Al_2(SO_4)_3$ $TiO(SO_4) \cdot 4H_2O$ | ~ 10 | Spherical particles and broken shells | 900 | | [26] |
| $TiO_2$ | $TiO(SO_4) \cdot 4H_2O$ | ~ 1 | Spherical | Up to 800 | Particle size studied as a function of temperature and gas flow | [27] |
| $Al_2O_3$ | $Al(NO_3)_3$ | 10 | Hollow | 700 | | [28] |
| $PuO_2$ | $Pu(OH)_4$ | 3 | Solid | 1100 | Precipitated $Pu(OH)_4$ in droplet before drying | [29] |
| Corresponding Oxides | Chlorides of La, Tm, and Y; lactates of La, Ce, Y; cerium citrate; ethylene-diaminetetraacetate of Ce, La, Y, zirconium oxalate | 0.1–5 | Shells and solid | | | [30] |

**TABLE 13.1.** *(Continued)*

| Product | Precursors | Size, μm | Morphology | Temperature, °C | Notes | References |
|---|---|---|---|---|---|---|
| $V_2O_5$ and $V_2O_3$ | Vanadium nitrate | 4–9 | Spherical | Up to 790 | Heated above oxide melting point | [31] |
| $V_2O_5$, $Bi_2O_3$, PbO, $MoO_3$ | | 0.1–1 | Spherical and platelets | Up to 1000 | Volatile species lost to walls | [32–34] |
| Partially decomposed nitrates | $Sr(NO_3)_2$ $CaNO_3$ | 0.3 | Hollow | 1000 | | [35] |
| PbO | Nitrates | < 1 | Spherical particles | 600 | | [36] |
| PbO | $Pb(NO_3)_2$ | < 1 | Dense, spherical | Up to 800 | | [37] |
| $Y_2O_3$ | Nitrates in propanol | < 1 | Dense spheres | 550 | | [38] |
| $Fe_2O_3$–$Fe_3O_4$ | $FeC_6H_5O_7 \cdot 4H_2O$ | < 1 | Spherical | Up to 800 | | [39] |
| $SiO_2$ | $Si(OEt)_4$–$NH_4OH$ | < 1 | Spherical porous | 400–700 | | [40] |
| $SnO_2$ | $SnCl_2$–oxalic acid | < 1 | Hollow spherical particles | Up to 1100 | | [41] |
| $SnO_2$ | $SnCl_4$–$NH_3$ | < 2 | Solid spheres | Up to 400 | | [42] |
| $TiO_2$ | $Ti(O\text{-}i\text{-}Pr)_3(acac)$ | < 0.15 | Spherical | Not reported | Electrospray | [43] |
| $TiO_2$ | $TiCl_4$–$HN_4OH$ $TiO_2$–Citric acid | < 1–2 | Hollow spherical | Up to 900 | | [44] |
| MgO | Magnesium nitrate, chloride, acetate | < 1 | Large aggregates | 500–1000 | Studied morphology as a function of precursor | [10, 45, 46] |
| $ZrO_2$–CaO | Chlorides, oxalates, lactate, citrate, ethylenediaminetetraacetate | 1–5 | Solid spheres or broken shells | 1150 | | [30] |

**References to Table 13.1**


1. Gadalla, A.M. & Yu, H.-F. *J. Mater. Res. 5* (6), 1990, 1231–1236.
2. Nafis, S., Tang, Z.X., Dale, B., Sorensen, C.M., & Hadjipanayis, G.C. *J. Appl. Phys. 64* (10), 1988, 5835–5837.
3. Kato, A., Ishimatsu, H., & Suyama, Y. *J. Japan Soc. Powder & Powder Metallurgy* (3), 1979, 131–135.
4. Kato, A., Hirata, Y., Tokunaga, Y., Maeda, T., & Nishi, H. (Eds.). *Sintering behavior of β″-alumina powders prepared by spray-pyrolysis technique*. In Vincenzini, P. (Ed.), Ceramic Powders. Amsterdam: Elsevier Scientific, 1983.
5. Kato, A. & Tokunaga, F. *J. Japan Soc. Powder & Powder Metallurgy 24* (8), 1977, 219–222.
6. Kato, A. & Hirata, Y. *Memoirs of the Faculty of Engineering, Kyushu Univ. 45* (4), 1985, 251–271.
7. Kato, A., Takayama, A., & Morimitsu, Y. *Nippon Kagaku Zaishi 12*, 1985, 2342–4.
8. Sproson, D.W. & Messing, G.L. *Comm. Amer. Ceram. Soc.*, May 1984, C-92–C-93.
9. Sproson, D.W. & Messing, G.L. *Advances in Ceramics 21*, 1987, 99–108.
10. Gardner, T.J., Sproson, D.A., & Messing, G.L. *Mat. Res. Soc. Symp. Proc. 32*, 1984, 227-232.
11. Sproson, D.W., Messing, G.L., & Gardner, T.J. *Ceramics International 12*, 1986, 3–7.
12. Vollath, D. *J. Mater. Sci. 25* (4), 1990, 2227–2232.
13. Ishizawa, H., Sakurai, O., Mizutani, N., & Kato, M. *J. Am. Ceram. Soc. Bull 65* (10), 1986, 1399–1404.
14. Gurav, A., Pluym, T., Kodas, T.T., & Pratsinis, S.E. *Chem. Eng. Commun. 151*, 1996, 211–226.
15. Zhang, S.-C., Messing, G.L., & Borden, M. *J. Am. Ceram. Soc. 73* (1), 1990, 61–67.
16. Dubois, B., Ruffier, D., & Odier, P. *J. Am. Ceram. Soc. 72* (4), 1989, 713–715.
17. Pebler, A.R. *J. Mater. Res. 5* (4), 1990, 680–682.
18. Epstein, J.A. *Hydrometallurgy 2*, 1976, 1–10.
19. Vanheusden, K., Seager, C.H., Warren, W.L., Tallant, D.R., Caruso, J., Hampden-Smith, M.J., Kodas, T.T. *J. Luminescence 75*, 1997, 11–16.
20. Liu, T.-Q., Sakurai, O., Mizutani, N., & Kato, M. *J. Mater. Sci. 21*, 1986, 3698–3702.
21. Seitz, K., Ivers-Tiffee, E., Thomann, H., & Weiss, A. (Eds.). *Influence of zinc acetate and nitrate salts on the characteristics of undoped ZnO powders*. Amsterdam: Elsevier Science Publications, 1987.
22. Carreno, T.G., Mifsud, A., & Serna, C.J. *Mater. Chem. and Phys. 27*, 1991, 287–296.
23. Szabo, Z.G., Jover, B., & Juhasz, J. *Materials Science Monographs 28B*, 1985, 659–664.
24. Roy, D.M., Neurgaonkar, T.R., O'Holleran, T.P., & Roy, R. *Am. Ceram. Soc. Bull. 56* (11), 1977, 1023–1024.
25. Baukal, W., Beck, H.-P., Kuhn, W., & Sieglen, R. *Power Sources 6*, 1977, 655–671.

26. Okada, K., Tanaka, A., & Hayashi, S. *J. Mater. Res. 9* (7), 1994, 1709–1713.

27. Deguchi, S., Matsuda, H., & Hasatani, M. *Drying Technology 12* (3), 1994, 577–591.

28. Ciminelli, R.R. & Messing, G.L. *Ceramica 30* (174), 1984, 131–138.

29. LaBauve, R.J., Kanapilly, G.M., Stalnaker, N.D., & Mewhinney, J.A. *J. Aerosol Sci. 10*, 1979, 251–256.

30. Kanapilly, G.M., Raabe, O.G., & Newton, G.J. *Aerosol Sci. 1*, 1970, 313–323.

31. Sullivan, R.J., Srinivasan, T.T., & Newnham, R.E. *J. Am. Ceram. Soc. 73* (12), 1990, 3715–3717.

32. Lyons, S.W., Wang, L.M., & Kodas, T.T. *NanoSt. Mater. 1*, 1992, 293.

33. Lyons, S., Xiong, Y., Ward, T., Kodas, T.T., & Pratsinis, S. *J. Mater. Res. 7* (12), 1992, 3333–3334.

34. Xiong, Y., Lyons, S., Kodas, T.T., & Pratsinis, S. *J. Am. Ceramic Soc. 78* (9), 1995, 2490–2496.

35. Baumgartner, F. & Schauer, C. *J. Aerosol Sci. 20* (8), 1989, 879–882.

36. Yaparpalvi, R., Loyalka, S.K., & Tompson, R.V. *J. Mater. Sci. Lett. 13* (10), 1994, 749–751.

37. Kirkbir, F., Katz, D., Lysse, R., & Mackenzie, J.D. *J. Mater. Sci. 27*, 1992, 1748–1756.

38. Rulison, A.J. & Flagan, R.C. *J. Am. Ceram. Soc. 77* (12), 1994, 3244–3250.

39. Cabanas, M.V., Vallet-Regi, M., Labeau, M., & Gonzalez-Calbet, J.M. *J. Mater. Res. 8* (10), 1993, 2694.

40. Nogami, K., Sakurai, O., Mizutani, N., & Kato, M. *Yogyo-Kyokai-Shi 95* (7), 1987, 682–685.

41. Vallet-Reg, M. *Solid State Ionics 63–65*, 1993, 201–206.

42. Ocaña, M. & Matijevic, E. *J. Aerosol Sci. 21* (6), 1990, 811–820.

43. Park, D.G. & Burlitch, J.M. *Chem. Mater. 4* (3), 1992, 500–502.

44. Vallet-Reg, M., Ragel, V., Román, J., Martínez, J.L., Labeau, M., & Gonzlez-Calbert, J.M. *J. Mater. Res. 8* (1), 1993, 138–144.

45. Gardner, T.J. & Messing, G.L. *Am. Ceram. Soc. Bull. 63* (12), 1984, 1498–1504.

46. Gardner, T.J. & Messing, G.L. *Thermochimica Acta 78*, 1984, 17–27.

**TABLE 13.2. Complex metal oxides prepared by spray pyrolysis**

| Product | Precursors | Size, μm | Morphology | Temperature, °C | Notes | References |
|---|---|---|---|---|---|---|
| Mullite | $Al(NO_3)_3-SiO_2$ $Al_2O_3-SiO_2$ | 0.1–1 | Solid | | | [1] |
| Mullite | $Al(NO_3)_3 \cdot 9H_2O$ ethyl silicate. Water–methanol solvent | 1.0 | Hollow with large surface area | 1650 | | [2, 3] |
| $LaO_{0.95}Mg_{0.05}Cr_{0.5}Al_{0.5}O_3$ | Not reported | 0.5–20 | Hollow | Up to 1300 | 1 kg/hr | [4] |
| Mn–Zn ferrites | Nitrates | Size not reported | Hollow | | | [5] |
| Mn–Zn ferrites | Slurries of oxides | 50–250 | Agglomerates | Up to 1200 | Further reaction and milling required | [6] |
| $NiFe_2O_4$ | Metal nitrates | 1–5 | Hollow | 450–810 | | [7] |
| $CoFe_2O_4$ | Metal nitrates | Bimodal, 0.5–2 and < 0.02 | Flammable ethanol solution used to fragment droplets | | Flame reactor | [8] |
| $NiFe_2O_x$, $ZnFe_2O_x$ $MnFe_2O_x$ | Metal nitrates | 5–50 | Morphology not reported | | | [9] |
| $LaCoO_3$, $LaFeO_3$ $MgAl_2O_4$, and others | Metal nitrates | 1–10 | Hollow | | | [10] |
| $NiMgO_2$, MnO, $Fe_2O_3$, $MgCr_2O_4$, $NiCr_2O_4$, $CoCr_2O_4$, and others | Nitrates, chlorides | 15–400 | Hollow | ~ 760 | Industrial scale | [11–14] |
| $CoLaO_3$ | Cobalt and lanthanum nitrate | 0.7 | Porous, spherical | > 700 | Additives used to adjust surface area | [15] |
| Carnegieite | Silica and alumina sols, $NaAlO_2$, and NaOH | < 1 | Solid | Up to 900 | | [16] |
| $Co_2Mo_3O_8$, $CoMoO_4$ | $Co(NO_3)_2 \cdot 9H_20$ | 1–6 | Hollow | 700 | Evaporative loss of vapor pressure | [17] |
| $MgAl_2O_4$ | $Mg(NO_3)_2 \cdot 6H_2O$ $Al(NO_3)_3 \cdot 9H_2O$ | > 5 | Hollow | 740–1030 | | [18] |

**TABLE 13.2.** (*Continued*)

| Product | Precursors | Size, μm | Morphology | Temperature, °C | Notes | References |
|---|---|---|---|---|---|---|
| $Na_2O \cdot 0.6MgO \cdot 5Al_2O_3$ | Na, Mg, and Al nitrates | 1–15 | Exploded particles with water–ethanol–glycerine | | Hollow with water–ethanol Na loss by evaporation | [19, 20] |
| $MgAl_2O_4$ | $Mg(NO_3)_2 \cdot 6H_2O$ $Al(NO_3)_3 \cdot 9H_2O$ | 0.07–0.1 | Porous | 800 | | [2, 3] |
| $YFeO_3$ and $Fe_2O_3$ | Yttrium and iron sulfates | Not reported | Not reported | 900 | | [21] |
| $Al_2O_3$–15% $ZrO_2$ | $\alpha$-$Al_2O_3$ slurry and $ZrO(NO_3)_3$ | Not reported | Not reported | 1000 | | [22] |
| $ZnO–CoO–MnO_2–Bi_2O_3–Sb_2O_3$ | Zn, Co, Mn, Sb acetates $BiONO_3 \cdot H_2O$ | 9–12 | Hollow particles that required milling | | Zinc acetate sublimation altered stoichiometry | [23] |
| $ZrO_2–SiO_2$ | $Si(OC_2H_5)_4$ $ZrOCl_2 \cdot 8H_2O$ | 0.8–1.0 | Solid | 900 | t-$ZrO_2$ and amorphous $SiO_2$ phases were present Polymer formed in droplets | [24, 25] |
| $YZrSiO_4$ | $SiO_2$ sol and $ZrO_2$ salt | 2–8 | Hollow | ~ 1500 | | [26] |
| Hydroxyapatite | $Ca(NO_3)_2$ $(NH_4)HPO_4$ | Not reported | Hollow | | | [27] |
| M–Ru–O, where M = Bi, Y, Pb, and many rare earths | $RuNO(NO_3)_2$, nitrates of Bi, Y, Pb, and many rare earths | 0.1–1.0 | Solid, porous, or hollow, depending on material | | | [28] |
| M–Fe–O, where M = Sr, Y | $Fe(NO_3)_3$, nitrates of Sr and $Y(NO_3)_3$, $Al(NO_3)_3$ | 0.1–1.0 | Solid, porous or hollow, depending on material | | | [29] |
| $NiFe_2O_4$ | Methoxides | < 1 | Shattered particles | Up to 570 | | [30] |
| $NiFe_2O_4$ | Nitrates | | Hollow noncrystalline to dense crystalline | 450–810 | | [7] |
| $ZnO-TiO_2$ | $TiO_2$ particles in $Zn(NO_3)_2$ solution | ~ < 0.5 | Hollow or solid agglomerates | 800 | | [31] |
| $Ca_{10}(PO_4)_6(OH)_2$ | $CaCl_2$ and ammonium phosphate | < 1 | Spherical, dense, or hollow, raspberry shaped | 1050 | | [32] |
| $BaFe_{12}O_{19}$ | Nitrates, citrates | < 2 | Squares or spheres | Up to 1000 | | [33] |

| | | | | | | |
|---|---|---|---|---|---|---|
| $BaFe_{12}O_{19}$ | Nitrates | < 2 | Spherical, hollow | Up to 1000 | | [34] |
| $Pb(Mg_{0.33}Nb_{0.66})O_3$ | Nitrates, oxalates | < 1 | Spherical | 750 | | [35] |
| $LaAlO_3$ | Nitrates | ~ 1 | Spherical | Up to 1500 | | [36] |
| $CeNO_3$, $LaCoO_3$, $LaFeO_3$, $SrTiO_3$ $La_{1-x}Sr_xFeO_3$ | Salts | < 5 | Spherical, hollow | Up to 1000 | | [10] |
| $\beta''$Alumina | Na, Al, and Mg nitrates | < 1–20 | Spherical, hollow | 1100 | | [37] |
| $BaTiO_3$, $SrTiO_3$ | $Ba(NO_3)_2$, $Sr(NO_3)_2$ $Ti(OC_3H_7)_4$ | ~ 3 | Spherical, hollow | 1000 | | [38] |
| $Pb(Zr, Ti)O_3$ | Alkoxides | < 1 | Spherical | 700 | PbO evaporation | [39] |
| $BaFe_{12}O_{19}$ | Nitrates | < 1 | Spherical | 800 | | [40] |
| $Gd_3Fe_5O_{12}$ | Nitrates | < 1 | Spherical | Up to 800 | Required postheating to achieve garnet phase | [41] |
| $Zn_2FeO_4$ | Nitrates | ~ 0.5 | Spherical | > 800 | Flame spraying | [42] |
| $BaFe_{12}O_{19}$ | Barium and carbonate iron oxides | 50–100 | Spherical | Up to 1065 | Fast-reaction sintering or spray roasting | [11] |
| Yttrium aluminum garnet | Nitrates | 0.1–1.0 | Spherical, hollow | — | Phase evolution studied | [43] |
| $BaTiO_3$ | Titanium oxynitrate Barium nitrite Acac H | 1 | Spherical, dense | Up to 700 | | [44] |
| $BaTiO_3$, ZnO | Chloride | 3–4 | Hollow, spherical shells and shell fragments | 900 | | [45] |
| $Ba_{0.86}Ca_{0.14}TiO_3$ | Nitrates, acetates, lactates | 0.1–1.0 | Hollow at low temperatures | | Solid at high temperatures by densification | [46] |
| $BaTiO_3$ | Nitrates, lactates | 0.1–1.0 | Solid in flame | — | Flame used to melt product | [47] |
| $SrTiO_3$ | Isopropoxides | ~ 1.0 | Hollow particles | 1200 | Not reported | [48, 49] |
| $PbTiO_3$ | not identified | 0.1–1 | Spherical, hollow | Up to 1000 | Single-source precursor | [50] |
| $Zr_{0.8}Sn_{0.2}TiO_4$ | Chlorides–acetate | 0.2–2 | Spherical with many shells | Up to 900 | | [51] |
| $Zn_2SiO_4$:Mn | $Zn(NO_3)_2 \cdot 6H_2O$ $Mn(NO_3)_2 \cdot 6H_2O$ $Si(OEt)_4$ | 1–10 | Spherical | 1000, 1150 | Emission intensity comparable to that of commercial powders | [52] |

**References to Table 13.2**


1. Moore, K.A., Cesarano, J.I., Smith, D.M., & Kodas, T.T. *J. Amer. Cer. Soc. 75* (1), 1992, 213–215.
2. Kanzaki, S., Tabata, H., Otsuka, N., Nakagawa, Z., & Hamano, K. *Yogyo-Kyokai-Shi 91* (7), 1983, 344–346.
3. Kanzaki, S., Tabata, H., Kumazawa, T., & Ohta, S. *J. Am. Ceram. Soc. 68* (1), 1985, C6–C7.
4. Roettenbacher, R. & Schmidberger, R. (Eds.). *Preparation of special ceramics for high temperature applications*. In Vincenzini, P. (Ed.), Ceramic Powders, Amsterdam: Elsevier Scientific Publ. Co., 1983.
5. Kumar, K., Petrovich, A., Williams, C., & VanderSande, J.B. *J. Appl. Phys. 64* (10), 1988, 5665.
6. Wagner, U. *J. Magnetism and Magnetic Materials 19*, 1980, 99–104.
7. Gadalla, A.M. & Yu, H.-F. *J. Mater. Res. 5* (12), 1990, 2923–2927.
8. Yamazaki, Y. & Satou, M. *The Society of Kyoto 18*, 1972, 280–283.
9. Akashi, T., Tsuji, T., & Onoda, Y. Apparatus for the preparation of starting materials. *International Conf. Sintering and Related Phenomena*, University of Notre Dame, Notre Dame, IN, 1965, 748–754.
10. Moser, W.R. & Lennhoff, J.D. *Chem. Eng. Comm. 83*, 1989, 241–259.
11. Ruthner, M.J. *Science of Sintering 11* (3), 1979, 203–214.
12. Ruthner, M.J., Richter, H.G., & Steiner, I.L. Spray-roasted iron oxide for the production of ferrites. *Proc. Intern. Conf. Ferrites*, Japan, 1970, 75–78.
13. Ruthner, M.J. (Ed.). *Industrial production of multicomponent ceramic powders (metal oxides) by means of the spray roasting technique*. In Vincenzini, P. (Ed.), Ceramic Powders, Amsterdam: Elsevier Scientific Pub. Co., 1983.
14. Ruthner, M.J. *Science of Sintering 6* (1/2), 1974, 81–94.
15. Imai, H. & Orito, H. *Nippon Kagaku Kaishi* (6), 1984, 851–855.
16. Imai, H. & Takami, K. *J. Mater. Sci. 20* (10), 1985, 1823–1327.
17. Rossetti, G.A., Jr., Burger, J.L., & Sisson, R.D., Jr. *J. Am. Ceram. Soc. 72* (10), 1989, 1811–1815.
18. Suyama, Y. & Kato, A. *Ceramics International 8*, 1982, 17–21.
19. Kato, A. & Tokunaga, F. *J. Japan Soc. Powder & Powder Metallurgy 24* (8), 1977, 219–222.
20. Kato, A., Ishimatsu, H., & Suyama, Y. *J. Japan Soc. Powder & Powder Metallurgy* (3), 1979, 131–135.
21. Multani, M., Nandikar, N.G., Venkataramani, N., Ragupathy, V., Pansare, A.K., & Gurjar, A. *Mater. Res. Bull. 14* (10), 1979, 1251–8.
22. Sproson, D.W. & Messing, G.L. *Comm. Amer. Ceram. Soc.*, May 1984, C92–C93.
23. Ivers-Tiffee, E. & Seitz, K. *Am. Ceram. Soc. Bull. 66* (9), 1987, 1384–1388.
24. Kanno, Y. & Suzuki, T. *J. Mater. Sci. Lett. 7*, 1988, 386–388.
25. Kanno, Y. & Suzuki, T. *J. Mater. Sci. 23*, 1988, 3067–3072.
26. Jada, S.S. *J. Mater. Sci. Lett. 9*, 1990, 565–568.
27. Itatani, K., Takahashi, O., Kishioka, A., & Kinoshita, M. *Gypsum & Lime 213*, 1988, 19.
28. Senzaki, Y., Hampden-Smith, M., Kodas, T.T., & Hussler, J.W. *J. Amer. Cer. Soc. 78*, 1995, 2977–2983.

29. Senzaki, Y., Caruso, J., Hampden-Smith, M.J., Kodas, T.T., & Wang, L.-M. *J. Am. Ceram. Soc. 78* (11), 1995, 2973–2976.

30. Clearfield, A., Gadalla, A.M., Marlow, W.H., & Livingston, T.W. *J. Am. Ceram. Soc. 72* (10), 1989, 1789–1792.

31. Oshima, K., Tsuto, K., Okuyama, K., & Tohge, N. *Aerosol Sci. Technol. 19* (4), 1993, 468–477.

32. Vallet-Regí, M., Gutirrez-Ros, M.T., Alonso, M.P., deFrutos, M.I., & Nicolopoulos, S. *J. Solid State Chemistry 112* (1), 1994, 58–64.

33. Cabanas, M.V., Gonzalez-Calbet, J.M., & Vallet-Regi, M. *J. Mater. Res. 9* (3), 1994, 712–716.

34. Kaczmarek, W.A., Ninham, B.W., & Calka, A. *J. Appl. Phys. 70* (10), 1991, 5909–5911.

35. Kim, B.-H., Ueda, K., Sakurai, O., & Mizutani, N. *J. Ceramic Soc. Japan 100*, 1991, 258–261.

36. Lux, B.C., Clark, R.D., Salazar, A., Sveum, L.K., & Krebs, M.A. *J. Am. Ceram. Soc. 76* (10), 1993, 2669–2672.

37. Hirata, Y., Maeda, T., & Kato, A. *Nippon Kagaku Kaishi 10*, 1981, 1595–1603.

38. Nonaka, K., Hayashi, S., Okada, K., Otsuka, N., & Yano, T. *J. Mater. Res. 6* (8), 1991, 1750–1756.

39. Sakurai, O., Miyauchi, M., Mizutani, N., & Kato, M. *Nipon Seramikkusu Kyokai Gakujutsu Rombunshi 97* (4), 1989, 407–412.

40. Tang, Z.X., Nafis, S., Sorensen, C.M., Hadjipanayis, G.C., & Klabunde, K.J. *IEEE Transactions on Magnetics 25* (5), 1989, 4236–4238.

41. Xu, H.K., Sorensen, C.M., Klabunde, K.J., & Hadjipanayis, G.C. *J. Mater Res. 7* (3), 1992, 712–716.

42. Tokunaga, F., Suyama, Y., & Kato, A. *24* (2), 1977, 43–47.

43. Nyman, M., Caruso, J., Hampden-Smith, M.J., & Kodas, T.T. *J. Am. Cer. Soc. 80* (5), 1997, 1231–1238.

44. Wada, S., Kubota, A., Suzuki, T., & Noma, T. *J. Mater. Sci. Lett. 13*, 1994, 190–193.

45. Milosevic, O. & Uskokovic, D. *Materials Science and Engineering A 168*, 1993, 249–252.

46. Ortega, J., Kodas, T.T., Chadda, S., Smith, D.M., Ciftcioglu, M., & Brennan, J. *Chem. Mater. 3*, 1991, 746–751.

47. Brewster, J. & Kodas, T.T. *AIChE J.*, *42*(11), 1997, 2665–2669.

48. Sakurai, O., Mizutani, N., & Kato, M. *J. Ceramics Soc. Jap. 96* (6), 1988, 639–645.

49. Sakurai, O., Mizutani, N., & Kato, M. *Microstructure of $SrTiO_3$ spherical fine particles prepared by ultrasonic spray pyrolysis of metal alkoxides*. Dept. Inorganic Materials, Tokyo Institute of Technology, 12-1-2 Ookayama, Meguro-ku, Tokyo 152, 1986.

50. Chandler, C., Powell, Q., Hampden-Smith, M.J., & Kodas, T.T. *J. Mater. Chem. 3* (7), 1993, 775–776.

51. Cho, S.-Y., Lee, J.-H., & Park, S.J. *J. Mater. Sci. 30*, 1995, 3274–3278.

52. Morimo, R., Mochinaga, R., & Nakamura, K. *Mater. Res. Bull. 29* (7), 1994, 750–757.

**TABLE 13.3. Superconductors prepared by spray pyrolysis**

| Product | Precursors | Size, μm | Morphology | Temperature, °C | Notes | References |
|---|---|---|---|---|---|---|
| $La_{1-x}Sr_xCuO_4$ | Nitrates | 1.7–3.0 | Agglomerates | 550 | | [1] |
| $YBa_2Cu_3O_{7-x}$ | Nitrates | < 1 | Solid spheres with scalelike features | Up to 1050 | | [2] |
| $YBa_2Cu_3O_{7-x}$ | Nitrates, acetate, and oxalates | 1–10 | Shelled and non-hollow microparticles to nonhollow agglomerates | 700–1000 | | [3] |
| $YBa_2Cu_3O_{7-x}$ | Nitrates | 1–3 | Spherical | 550–960 | | [4] |
| $YBa_2Cu_3O_{7-x}$ | Nitrates | < 1 | Not reported | 920 | | [5, 6] |
| $YBa_2Cu_3O_{7-x}$ | Nitrates | < 1 | Single crystal | 900–1000 | | [7–10] |
| $YBa_2Cu_3O_{7-x}$ | Nitrates, acetate, citrates | < 1 | Microporous spheres | | | [11] |
| $YBa_2Cu_3O_{7-x}$ | Nitrate, ethylenediamine-tetraacetate | < 1 | Hollow | > 900 | | [12–14] |
| $YBa_2Cu_3O_{7-x}$ | Acetates | < 1 | Single crystal | 1000 | | [15] |
| $YBa_2Cu_3O_{7-x}$ | Nitrates | < 1 | Solid | 948 | | [16] |
| $YBa_2Cu_3O_{7-x}$ | Nitrates | < 1 | Solid | 900–1000 | | [17–20] |
| $YBa_2Cu_3O_{7-x}$ | Nitrates | ~ 1 | Spherical, dense | Up to 1000 | | [21] |

| | | | | | | |
|---|---|---|---|---|---|---|
| $YBa_2Cu_3O_{7-x}$ Ag–$SiO_2$ composite | Nitrates, $Ag_2O$, $SiO_2$ | Not reported | Spherical | 850 | | [22] |
| $YBa_2Cu_3O_{7-x}$–Ag | Nitrates | < 1 | Solid | | | [23, 24] |
| Ag–$YBa_2Cu_3O_{7-x}$ | Nitrates | < 1 | Solid | 950 | | [25] |
| $YBa_2Cu_3O_{7-x}$ | Nitrates | < 1 | Solid | | Flame reactor | [26] |
| $YBa_2Cu_3O_{7-x}$ | Nitrates | < 20 | Porous | | Used exothermic reaction to fragment particles | [27] |
| Bi–Sr–Ca–Cu–O | Citrates | 0.5 | Not reported | | | [13, 14] |
| $Bi_2Sr_2Ca_1Cu_2O_x$ | Citrates | < 1 | Solid | | Morphological control demonstrated | [17–20] |
| Bi–Sr–Ca–Cu–O | Nitrates | < 1 | Solid at high temperatures | | Porous and hollow at low temperatures | [28–31] |
| $Bi_2Sr_2Ca_1Cu_2O_x$ $Bi_{1.8}Pb_{0.2}Ca_2Sr_2Cu_3O_x$ | Nitrates | < 1 | Spherical | 770–870 | | [18] |
| $Bi_{1.6}Pb_{0.4}Ca_2Sr_2Cu_3O_x$ | Nitrates | 1–4 | Not reported | 855 | | [32] |
| $Bi_2CaSr_2Cu_2O_x$ | Nitrates | ~ 1 | Spherical, faceted | 800 | | [33] |

## References to Table 13.3


1. Manoharan, S.S., Prasad, V., Subramanyam, S.V., & Patil, K.C. *Physica C 190*, 1992, 225–228.
2. Hitchcock, D.C., Rusin, R.P., & Johnson, D.L. *J. Am. Ceram. Soc. 74*(9), 1991, 2165–2169.
3. Grader, G.S., Machado, D.R., & Semiat, R. *J. Mater. Res. 9*(10), 1994, 2490–2500.
4. Wang, Y.-L., Tan, Z.Q., Zhu, Y., Moodenbaugh, A.R., & Suenaga, M. *J. Mater. Res. 7*(12), 1992, 3175–3184.
5. Biswas, P., Zhou, D., Zitkovsky, I., Blue, C., & Boolchand, P. *Materials Lett. 8*(6,7), 1989, 233–237.
6. Zhou, D., Biswas, P., Oostens, J., & Boolchand, P. *J. Am. Ceram. Soc. 76*(3), 1993, 678–682.
7. Chadda, S., Ward, T., Carim, A., Kodas, T.T., Ott, K.C., & Kroeger, D. *J. Aerosol Sci. 22*, 1991, 601–616.
8. Kodas, T.T., Engler, E.M., Lee, V.Y., Jacowitz, R., Baum, T.H., Roche, K., Parkin, S.S.P., Young, W.S., Hughes, S., Kleder, J., & Auser, W. *Appl. Phys. Lett. 52*(19), 1988, 1622–1624.
9. Kodas, T.T., Datye, A., Lee, V., & Engler, E. *J. Appl. Phys. 65*, 1989, 2149–2151.
10. Kodas, T.T., Engler, E., & Lee, V. *Appl. Phys. Lett. 54*, 1989, 1923–1925.
11. Odier, P., DuBois, B., Gervais, M., & Douy, A. *Mat. Res. Bull. 24*, 1989, 11–22.
12. Pebler, A. & Hughes, C.W., *TEM examination of $YBa_2Cu_3O_{7-x}$ particles synthesized by the aerosol/pyrolysis process*. Westinghouse R & D Center, Pittsburgh, PA, 15235, 1988.
13. Pebler, A. & Charles, R.G. *Mat. Res. Bull. 23*(99), 1988, 1337–1344.
14. Pebler, A. & Charles, R.G. *Mat. Res. Bull. 24*, 1989, 1069–1076.
15. Setaka, R., Komatsu, W., Shibata, T., & Nakajima, M. *Japanese J. Applied Physics 27*(11), 1988, L2100–L2102.
16. Shimazu, S., Uematsu, T., Matsuda, S., & Kuwasawa, Y. *Modern Physics Letters B 2*(1), 1988, 501–504.
17. Tohge, N., Tatsumisago, M., Minami, T., Adachi, M., Kousaka, Y., & Okuyama, K. *J. Am. Ceram. Soc. 74*(9), 1991, 2117–2122.
18. Tohge, N., Tatsumisago, M., Minami, T., Okuyama, K., Arai, K., Inada, Y., & Kousaka, Y. *J. Mat. Sci. Materials in Electronics 1*, 1990, 46–48.
19. Tohge, N., Tatsumisago, M., Minami, T., Okuyama, K., Arai, K., & Kousaka, Y. *Japanese J. Applied Physics 28*(7), 1989, L1175–L1178.
20. Tohge, N., Tatsumisago, M., Minami, T., Okuyama, K., Adachi, M., & Kousaka, Y. *Japanese J. Applied Physics 27*(6), 1988, L292–L294.
21. Ogihara, T., Ookura, T., Yanagawa, T., Ogata, N., & Yoshida, K. *J. Mater. Chem. 1*(5), 1991, 789–794.
22. Takao, Y. & Awano, M. *J. Mater. Res. 71*(11), 1992, 2942–2948.
23. Carim, A., Doherty, P., Kodas, T.T., & Ott, K. *Materials Letters 8*, 1989, 335–339.
24. Ward, T.L., Kodas, T.T., Carim, A.H., Kroeger, D.M., & Hsu, H. *J. Mater. Res. 7*(4), 1992, 827–836.

25. Tsuchida, K., Tsudo, H., & Kato, A. *J. Mater. Sci. Lett. 8*, 1989, 222–224.

26. Zachariah, M.R. & Huzarewicz, S. *Combustion and Flame 87*, 1991, 100–103.

27. Martin, C.B., Kurosky, R.P., Maupin, G.D., Han, C., Javadpour, J., & Aksay, I.A. Spray pyrolysis of multicomponent ceramic powders. In Messing, G.L., Hausner, H., & Hirano, S.-C. (Eds.), *Ceramic Powder Science III*. Columbus, OH: American Ceramic Society, 1990, 99–106.

28. Gurav, A.S., Kodas, T.T., Kauppinen, E.I., Joutsensaari, J., & Zilliacus, R. *J. Mater. Res. 10*(7), 1995, 1644–1652.

29. Gurav, A., Kodas, T.T., Kauppinen, E., Joutsensaari, J., & Zilliacus, R. *Nano-Struct. Mater. 4*(5), 1994, 583–590.

30. Jin, S., Kammlott, G.W., Tiefel, T., Kodas, T.T., Ward, T.L., & Kroeger, D. *Physica C 181*, 1991, 57–62.

31. Ward, T., Lyons, S., Kodas, T.T., Brynestad, Y., Kroeger, D., & Hsu, H. *Physica C 200*, 1992, 31–42.

32. Tatsumisago, M., Tohge, N., Minami, T., Okuyama, K., Adachi, M., Inada, Y., & Kousaka, Y. *Japanese J. Appl. Physics 30*(5), 1991, 952–953.

33. Okuyama, K., Seto, T., & Shimada, M. *J. Mater. Sci., Mater. in Elec. 5*, 1994, 210–214.

**TABLE 13.4. Nonoxides prepared by spray pyrolysis**

| Product | Precursors | Size, $\mu m$ | Morphology | Temperature, °C | Notes | References |
|---|---|---|---|---|---|---|
| $Si_3N_4$ | Polysilazane | 0.5 | Solid | 1400 | Vaporization of silazanes, reaction and nucleation observed | [1] |
| BN | Polyborazene | 0.1–5 | Hollow and solid | | Liquid ammonia solution | [2] |
| AlN | Al and $NH_3$ | 10 | Spherical, cracked | 1250 | Particles of AlN-coated Al under some conditions | [3] |
| $CaLa_2S_4$ | $Ca(NO_3)_2$ $La(NO_3)_3 \cdot 6H_2O$ | 2.0 | Hollow and solid | 1000 | Liquid ammonia solution | [4] |
| PbS polymer | $Pb(NO_3)_2$, $H_2S$ | 5 | Spherical particles embedded in polymer matrix | Not reported | Electrospray | [5, 7] |
| ZnS | $Cd(NO_3)_2 \cdot 4H_2O$ | 0.5–1.5 | Spherical, smooth | 400–800 | $N_2$ carrier gas, ultrasonic atomizers. Powder collected by electrostatic precipitation. At 800°C hexagonal ZnS, below 800°C some amorphous material, and > 800°C ZnO present | [6] |
| CdS | $Zn(NO_3)_2 \cdot 4H_2O$ thiourea | 0.5–1.5 | Spherical, porous | 400–800 | $N_2$ carrier gas, ultrasonic atomizers. Hexagonal CdS formed at 600–750°C | [6] |

**References to Table 13.4**

1. Liu, T.-Q., Sakurai, O., Kieda, N., Kato, M., & Mizutani, N. *Nippon Seramikkusu Kyokai Gakujutsu Rombunshi 97*(10), 1989, 1321–1325.
2. Lindquist, D.A., Kodas, T.T., Smith, D.M., Xiu, X., Hietala, S.L., & Paine, R.T. *J. Am. Ceram. Soc. 74*, 1991, 3126–3128.
3. Kimura, Y., Ito, T., Yoshikawa, H., & Hiraki, A. *Japanese J. Appl. Physics 30*(5a), 1991, L798–L801.
4. Chess, D.L., Chess, C.A., & White, W.B. *J. Amer. Ceram. Soc. 66*, 1983, C205–C207.
5. Salata, O.V., Dobson, P.J., Hull, P.J., & Hutchinson, J.L. *Advanced Materials 6*(10), 1994, 772–775.
6. Tamaki, S., Tohge, N., & Okuyama, K. *J. Mat. Sci. 14*, 1995, 1388–1390.
7. Salata, O.V., Dobson, P.J., Hull, P.J., & Hutchinson, J.L. *Thin Solid Films 251*, 1994, 1–3.

**TABLE 13.5. Representative examples of metals prepared by spray pyrolysis**

| Product | Precursors | Size, μm | Morphology | Temperature, °C | Notes | References |
|---|---|---|---|---|---|---|
| Ag | $AgNO_3$ | 0.5–5.0 | Solid near melting point | | | [1] |
| Ag | $AgNO_3$ | < 1 | Spherical | > 650 | Ethanol added | [2] |
| Ag–CdO | $AgNO_3$–$Cd(NO_3)_2$ | 1–5 | Not reported | 950 | | [3] |
| Ag, Pd, Ag–Pd alloy | Metal nitrates | 0.5–5.0 | Solid above melting temperature | | High temperatures resulted in evaporation and nucleation. Flame reactor | [4–7] |
| Ag–Pd alloy | $AgNO_3$, $Pd(NO_3)_2$ | 0.5–5.0 | Solid near melting point | | | [8] |
| Pd, PdO | $Pd(NO_3)_2$ | 0.5–5.0 | Solid near melting point | | | [9] |
| Pd | $Pd(NO_3)_2$ | 1 | Spherical | 1650 | | [10] |
| Pd or Ag | $Pd(NO_3)_2$ | ~ 1–7 | | ~ 1100 | | [11] |
| Pd–alkaline earth metals | $Pd(NO_3)_2$–$Ca(NO_2)_2$ | ~ 0.6 | Spherical | 1600 | | [12] |
| Ag–Pd–Au, and composites | Nitrites and halides | < 1 | Smooth spheres or plates | Up to 300 | Reducing agents added | [13] |
| Pd | $Pd(NO_3)_2$ | 0.1–1 | Spherical | 1000 | Coated with $Al_2O_3$, ZnO, $SiO_2$, CaO | [14, 15] |
| Rh, fullerene | — | < 1 | Dense | — | Nanocomposites | [16] |
| Cu | $Cu(NO_3)_2$ | 0.5–5.0 | Solid and spherical above melting point | High temperature flames | Very high temperatures resulted in evaporation and nucleation | [4–7] |
| Ni | $Ni(NO_3)_2$ and $NiCl_2$ | 0.6 | Hollow at low T | 900–1500 | Solid at high temperatures | [4–7] |
| Cu, $Cu_2O$, CuO | $Cu(NO_3)_2$ | 0.5–5.0 | Solid near melting point | | | [17] |
| Au | Au(III) in nitric acid | 0.5–5.0 | Solid near melting point | | | [18] |

## References to Table 13.5


1. Pluym, T.C., Powell, Q.H., Gurav, A.S., Ward, T.L., Kodas, T.T., Wang, L.M., & Glicksman, H.D. *J. Aerosol Sci.* *24*(3), 1993, 383–392.
2. Kato, A., Takayama, A., & Morimitsu, Y. *Nippon Kagaku Zaishi 12*, 1985, 2342–2344.
3. Schmidberger, R. & Keil, A., *Process for making Ag powder with oxides*. Dornier Systems, Inovan-Stroebe KG, Fed. Rep. Germany, United States Patent *4,396,420*, February 16, 1982.
4. Nagashima, K., Wada, M., & Kato, A. *J. Mater. Res.* *5*(12), 1990, 2828–2834.
5. Nagashima, K., Iwaida, T., Sasaki, H., Katatae, Y., & Kato, A. *Nippon Kagaku Zasshi 1*, 1990, 17–24.
6. Nagashima, K., Morimitsu, Y., & Kato, A. *Nippon Kagaku Kaishi 12*, 1987, 2293–1300.
7. Nagashima, K., Himeda, T., & Kato, A. *J. Mater. Sci.* *26*, 1991, 2477–2482.
8. Pluym, T., Kodas, T.T., Wang, L.-M., & Glicksman, H. *J. Mater. Res.* *10*(7), 1995, 1661–1673.
9. Pluym, T.C., Lyons, S.W., Powell, Q.H., Gurav, A.S., Kodas, T.T., Wang, L.M., & Glicksman, H.D. *Mat. Res. Bull.* *28*, 1993, 369–376.
10. Asada, E. & Nagashima, K. US Patent *5,420,744*, March 30, 1995.
11. Asada, E., Ono, S., Matsuo, M., & Kagaku, S. Japanese Patent, SHO 62[1987]-2404, Jan. 1987.
12. Asada, H., Akimoto, Y., Ono, S., Nagashima, K., Iwasaki, M., & Maekawa, M.United States Patent *5,402,305*, March 28, 1995.
13. Douglas, P. & Stewart, T.I. *Method of producing powdered materials*. United States Patent *4,023,961*, May 17, 1977.
14. Jain, S., Atanasova, P., Kodas, T.T., & Hampden-Smith, M.J. *J. Electrochem. Soc.* *143*(11), 1996, 3762–3770.
15. Jain, S., Kodas, T.T., & Hampden-Smith, M.J. *J. Electrochem. Soc.* *14*(4), 1997, 1222–1231.
16. Gurav, A., Duan, Z., Wang, L., Hampden-Smith, M.J., & Kodas, T.T. *Chem. Mater.* *5*(2), 1993, 214.
17. Majumdar, D., Shefelbine, T., Kodas, T.T., & Glicksman, H. *J. Mater. Res.* *11*, 1996, 128.
18. Majumdar, D., Kodas, T.T., & Glicksman, H. *Adv. Mater.* *8*(12), 1996, 1020–1022.

# 14 Film Formation

## 14.1 INTRODUCTION

Our discussion on aerosol routes to films is divided into two parts: overall qualitative behavior and technology. In this chapter, we discuss the different modes of system operation (droplet and precursor evaporation followed by CVD, for example). In Chapter 15, we describe the actual systems that have been used to form, etch, and modify films by means of aerosol processes.

Relative to aerosol routes for particle formation, aerosol routes for the fabrication or modification of films are less well understood, are more varied in the physical and chemical phenomena taking place, and have technologies that are less well developed. Figure 14.1 shows a schematic of aerosol routes that can be used to form films and emphasizes that these processes for film generation rely on particle formation by the routes discussed in Chapters 10 and 12, gas-to-particle conversion and liquid-to-solid and solid-to-solid conversion, respectively. In general, particles of all sizes (from clusters to particles larger than 1 $\mu$m) can be formed either by gas-phase processes such as evaporation–condensation or chemical reaction or by intraparticle reaction processes, and then deposited. Particle deposition can also occur simultaneously with CVD. Etching, which involves the *removal* of material from a surface, may also occur. It is important to point out that CVD involves the transport of a volatile species, generally called a precursor, to a surface (usually heated), where it undergoes thermally induced reactions that result in the elimination of some ligands of the precursor molecule with concomitant formation of a film. The precursor is normally a metal-containing compound if an inorganic film is formed, and more than one precursor may be delivered, depending on the composition of the desired film. The molecular nature of the precursors is critical to the CVD process, because it results in attributes such as conformal coverage and selective deposition. Here we discuss CVD only in relation to problems in which particles are involved. (See Chapter 8 for further information on CVD.)[1]

The wide range of phenomena involved in film formation means that many of the concepts discussed in Chapters 2 through 9 must be brought into play. Because of this, and because of the lack of comprehensive models, this chapter cannot provide a unique qualitative description that encompasses all aerosol routes to films. Rather, several separate descriptions are required.

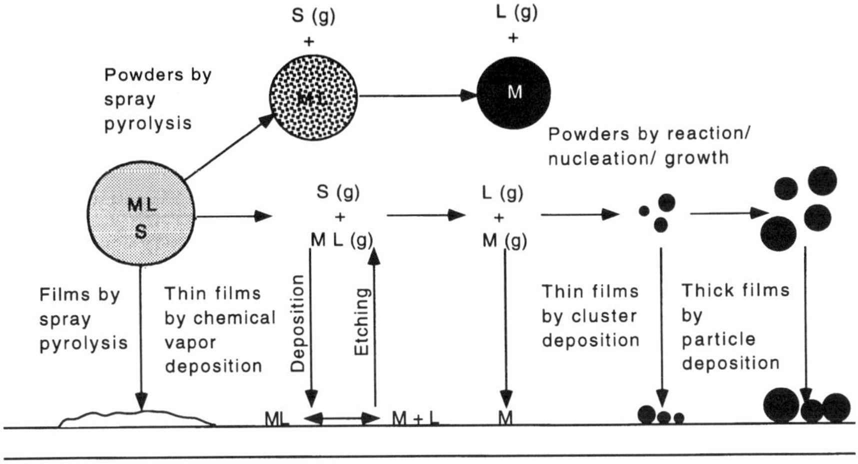


**Fig. 14.1.** Films can be formed on or removed from surfaces by a variety of aerosol approaches involving both gas-to-particle conversion and spray routes. Both particles and vapor can interact with the substrate.

For purposes of discussion, aerosol routes for film fabrication can be divided into four categories: (1) *Deposition or etching using preformed droplets or solid particles*. (Particle formation is decoupled from the primary problem of depositing an aerosol onto a surface.) The particles can be formed by any process, including spray routes, gas-to-particle conversion, and suspension of powders. (2) *In situ formation and deposition of particles, leading to film formation or etching*. (Particle formation is coupled to the problem of depositing an aerosol onto a surface.) The particles are formed by gas-to-particle conversion. (3) *Delivery of volatile precursors as aerosols, followed by evaporation and deposition by CVD or etching*. (Particle formation is decoupled from the main problem of coupled particle evaporation and transport, vapor transport, other reactor-scale phenomena, and CVD.) The particles are usually formed by spray routes. (4) *Hybrid processes relying on simultaneous particle and vapor deposition*. (The deposition of particles formed in situ or ex situ is coupled to CVD.) The particles can be formed ex situ by spray routes, gas-to-particle conversion, and suspension of powders or in situ by gas-to-particle conversion. The major considerations for all of these categories are enhancing or eliminating particle deposition onto surfaces, gas-to-particle conversion, and liquid-to-solid and solid-to-solid conversion. Particle deposition is emphasized, and the reader is referred to Chapters 10 and 12 for discussions of particle formation.

## 14.2 FILM CHARACTERISTICS

Aerosol processes for deposition of films compete with other approaches, including physical vapor deposition (sputtering, evaporation), CVD, spin coating of precursors followed by reaction, and variations on these processes. In general, the usefulness of a film and the relative advantages of using aerosol routes must be quantified using certain properties or criteria dictated by the application. There are criteria for the film and for the process used to produce the film. Criteria for the film are similar to those for powders and include purity, microstructure, morphology, stoichiometry, and uniformity, as well as a number of other, more specialized criteria such as uniformity of coverage and crystallographic orientation of the film. Criteria for the process are discussed next. The relative advantages and disadvantages of these routes are examined in Chapter 15. A critical point is that control over film characteristics is required for the deposition of films with tailor-made properties.

*Purity*. Most applications require high-purity films to optimize electrical, magnetic, mechanical, and optical properties. Impurities can drastically alter the physical properties and render devices useless. Metal films, for example, often must possess low resistivities, which cannot be attained when carbon and oxygen impurities are present. Many doped metal oxides, such as $BaTiO_3$, must have controlled levels of dopants at low concentrations to achieve their optimum characteristics, e.g., a ferroelectric transition temperature. In cases such as wear-resistant coatings, lower purities can be tolerated.

*Microstructure*. Depending on the application, amorphous, polycrystalline, oriented, or single-crystal films are desired. For instance, high-temperature superconducting ceramics must be single-crystal films to maximize their current-carrying capacity. For hard coatings, polycrystalline, fine-grained films may be desired; an example is the high hardness of nanocrystalline $TiO_2$. The manufacture of membranes or ceramic filters requires generation of porous films with controlled pore-size distributions.

*Morphology*. Films may have different morphologies, depending on the method used to fabricate them. Morphology has a profound effect on film properties. For example, columnar grains are often formed by vapor deposition processes. In other cases, even though the film may be of high purity, poorly connected grains can be formed, leading to high resistivities for metals. Porous films are often required for some membrane applications, while in other cases dense films are necessary. Surface roughness is a critical measure of film quality, especially in microelectronics applications.

*Stoichiometry*. A controlled stoichiometry is needed to ensure the formation of the correct product. An off-stoichiometry film can contain extraneous phases that degrade film properties. An example is the $Y_2BaCuO_x$ phase in $YBa_2Cu_3O_{7-x}$ ceramic superconductors, which degrades current-carrying capacity. Another example is mullite, in which extra $SiO_2$ can lead to severely degraded high-temperature mechanical properties.

*Substrate-Scale Film Uniformity.* The film generally must be uniform in thickness and composition over the area of the substrate, unless selective-area deposition is required. (See shortly.)

*Adhesion.* In almost all cases, excellent adhesion is required.

*Conformality.* In many cases, three-dimensional objects or surfaces with trenches, holes, and other complex features must be coated uniformly.

*Film-Thickness-Scale Uniformity.* Controlled gradients in composition across the thickness of a film are often desirable for functionally gradient materials. In other cases, the film must be uniform throughout.

*Homogeneity.* Composite (heterogeneous) films are often of interest, while at the other extreme, films with a homogeneous composition may be required.

Just as the film must meet certain requirements in order to be useful, the deposition process must also satisfy certain criteria.

*Precursors.* The process must use inexpensive and safe precursors and, at the same time, form by-products that can be recycled or be cleanly disposed of. The precursors must also be capable of depositing films with the characteristics described previously (e.g., high purity, a certain deposition rate, etc.).

*Operating Pressure.* Operation under atmospheric pressure is often desirable from the perspective of operating cost considerations. Although better film quality can be obtained at lower pressures, the cost of operations must be considered. An example is glass coating, which is carried out at atmospheric pressure and cannot in many cases be done economically at reduced pressures. The other extreme is the manufacture of microelectronics, in which film quality is often more important than cost, although cost of ownership is becoming an important issue.

*Complexity.* The system must be simple enough to allow controlled changes in operating parameters and resultant film properties.

*Deposition Rate.* The process generally must provide high deposition rates, usually 0.1 $\mu$m/min or better. An example of the importance of this feature is in glass-coating technology, in which deposition rates must often be greater than 1 $\mu$m/sec. In some applications in which film quality (e.g., morphology, crystallinity, and orientation) is critical, such as single-crystal films or epitaxial growth, low deposition rates may be more desirable.

*Selective-Area Deposition.* In some cases, deposition is desired only on certain areas of a substrate, usually those with a composition different from that of the other regions on the substrate, for which no coating is desired. This selective-area deposition has kinetic and thermodynamic origins and must be carefully controlled to be exploited in a useful manner.

*Composition Control.* The system must allow convenient control over the composition of the film.

*Reproducibility.* The process must provide film characteristics that are reproducible, often to within less than a few percent.

## 14.3 DEPOSITION OR ETCHING USING NONEVAPORATING PARTICLES FORMED EX SITU

Deposition or etching with nonevaporating particles formed ex situ includes processes in which particle formation is decoupled from the primary problem of depositing an aerosol onto a surface. The particles can be formed by any process, including spray routes, gas-to-particle conversion, and suspension of powders outside the system used to deposit the particles. The technologies that rely on ex situ formation of particles which do not evaporate, followed by deposition, include inertial solid-particle deposition, thermophoretic solid-particle deposition, and molten-particle deposition (including flame and plasma spraying).

The deposition of droplets and solid particles onto surfaces has been modeled for a variety of situations. Solid-particle deposition is simpler than other aerosol routes to films from the perspective that the particles do not change size during their time in the system. Several approaches have been used to model this problem.[2]

As an example of solid-particle deposition by gravitational settling, consider a stream of isothermal gas containing particles directed at a surface. Figure 14.2 shows calculated streamlines for flow directed at a solid surface at a low reactor Reynolds number of 4.[3] The flow field was axially symmet-

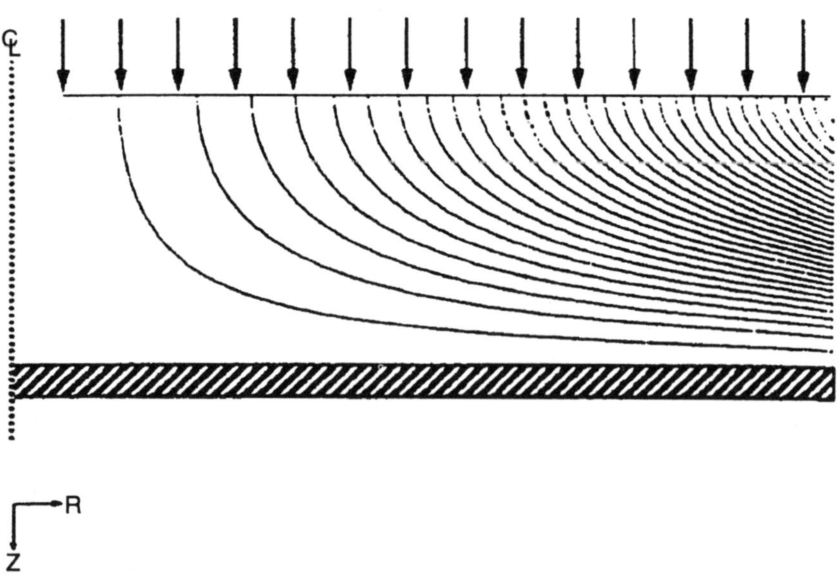


**Fig. 14.2.** Streamlines for $S = 4$ cm, $U_o = 1$ cm s$^{-1}$ and Re = 4, showing paths gas follows as it flows through the system. (Courtesy of Rader & Geller.)[3]

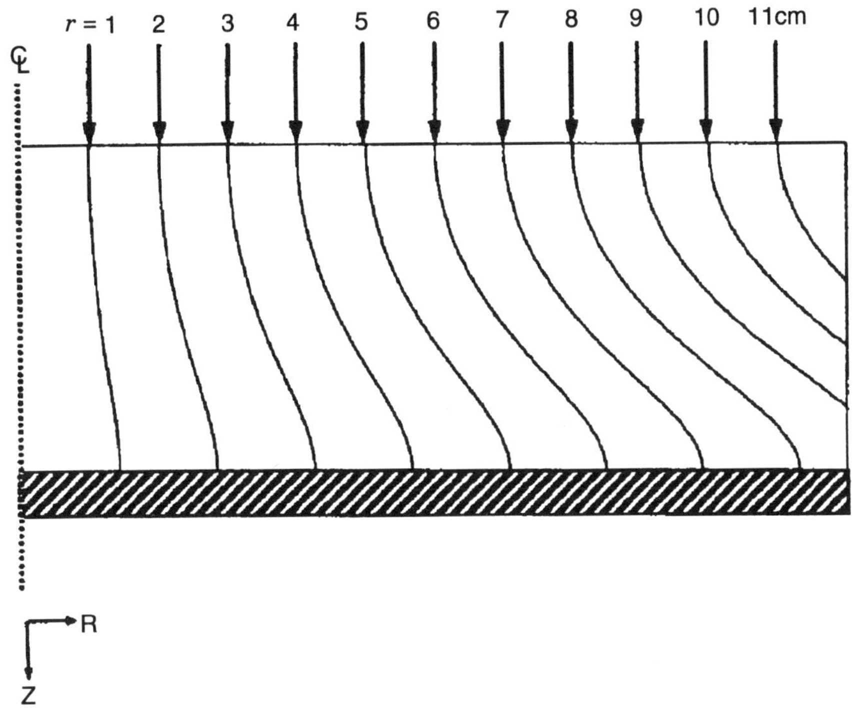


**Fig. 14.3.** Particle tracks for response time $t = 0.0025$ s with terminal velocity of 1 cm $s^{-1}$. ($S = 4$ cm, $U_o = 1$ cm $s^{-1}$, and Re = 4.) Only particles at the edge of the substrate do not deposit. (Courtesy of Rader & Geller.)[3]

ric, with the gas entering through a top plate (similar to the showerhead arrangement used in many CVD and etching reactors), and was assumed to be uniform and steady in velocity and concentration. The Reynolds number was based on the distance from the upper reactant introduction plate to the deposition surface. The specific parameters are given by Rader and Geller.[3] The particle terminal velocity was taken as constant and directed toward the plate. This is the situation found in many cases of gravitional settling of particles. By using this information in the equation describing particle motion (See Chapter 3), the two researchers calculated the trajectories of particles. Figure 14.3 shows trajectories for particles introduced at different evenly spaced positions from the center of the surface. (Note that such a configuration does not correspond to a uniform concentration of particles in the gas).[3] Particles with a very short relaxation time (0.0025 sec, the time for a particle with a given velocity to come to zero velocity with respect to a gas that is not moving) were used, corresponding to conditions under which the particles always moved at their net terminal velocity in the vertical direction. The particles were given gravitational settling velocities (1 cm/sec) that were equal to the inlet gas velocity, leading to deposition of 85% of the particles. The gas flow in the radial direction forced the particles to move in that direction, resulting in 15% of the particles—the ones nearest the outside edges of the substrate—missing the substrate. For a uniform distribution of particles in the flow, these calculations showed that 50% of the particles entering the system would deposit. The particles were assumed not to bounce because of the low settling velocities. Increasing the relaxation time increased the collection efficiency to the point where all of the particles continued to travel in the vertical direction and deposited. Physically, increased relaxation time corresponds to a decreased ability of the particles to follow thc gas as it makes the sharp turn above the substrate.

The behavior of systems with particle deposition by convection and diffusion is qualitatively different from that of systems with deposition in which inertial effects are important. Figure 14.4 shows the deposition efficiency of particles calculated using the diffusion–convection equations presented in Chapter 3.[3] The values of the parameters are given by Rader and Geller.[3] These equations show that the deposition efficiency depends on the dimensionless Peclet number (Pe), which is defined as $LU_o/D$, where $L$ is the dimension of the system (in this case, the distance from the upper to the lower plate), $U_o$ is the velocity of the gas far from the surface (in this case, the inlet velocity at the upper plate), and $D$ is the diffusion coefficient of the

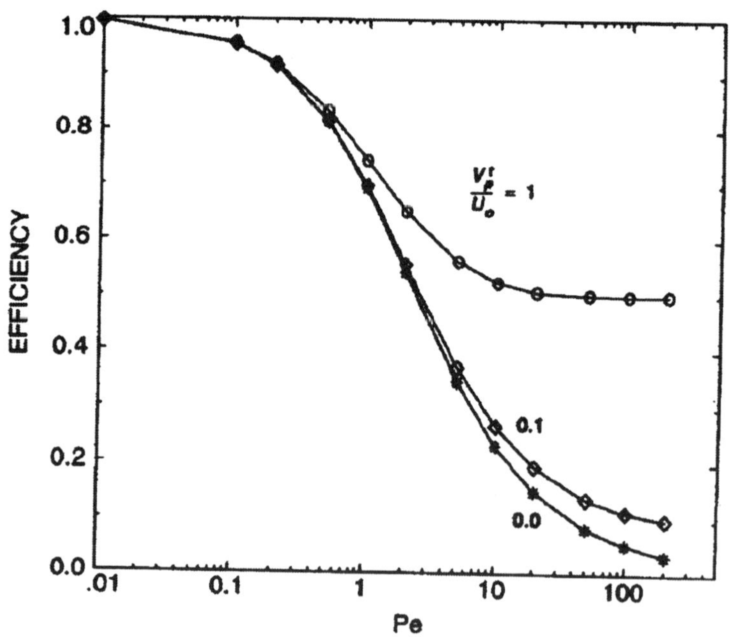


**Fig. 14.4.** Collection efficiencies as a function of the Peclet number (Pe) for three particle terminal velocities ($S = 4$ cm, $U_o = 1$ cm s$^{-1}$, and Re = 4). A small Pe corresponds to a small particle size for which particles deposit by diffusion. (Courtesy of Rader & Geller.)[3]

particles. The deposition efficiency also depends on the velocity of the particles due to gravitational effects. In the limit of small Peclet numbers, the diffusion coefficient is large, particles diffuse rapidly, and the collection efficiency approaches unity. In the limit of large Peclet numbers, the particles have small diffusion coefficients, the deposition is controlled by the settling velocity, and the collection efficiency tends to $V_p/(U_o + V_p)$. In the example presented, the gas velocity and settling velocity were equal, leading to a collection efficiency of 50% for large Peclet numbers. These results illustrate that high deposition efficiencies can be obtained for particles that are sufficiently small.

Deposition of solid particles also requires that particles striking a surface stick because of van der Waals, electrostatic, or other forces, and not bounce. The problem of particle bounce can be severe in cases where high particle velocities are encountered. This occurs mainly during inertial deposition of particles, wherein the particles can strike the surface with velocities on the order of the gas velocity. These velocities can be high, even approaching sonic velocities (the speed of sound). In contrast, particle deposition by diffusion involves particle velocities that are many orders of magnitude lower, and particle bounce is often not a concern because the van der Waals, electrostatic, and other forces are of sufficient magnitude to overcome it. The problem of particle bounce is discussed in detail elsewhere.[4]

## 14.4 DEPOSITION OR ETCHING USING EVAPORATING PARTICLES WITH INVOLATILE REACTANTS FORMED EX SITU

Describing deposition for evaporating particles (solid or liquid) is more complicated than for nonevaporating particles, because the particles can change size during their residence time in the system, even to the extent that they enter the system with sizes such that inertial effects are important, but evaporate to sizes at which diffusion becomes important (to the extent that they evaporate completely). The problem is complicated, and most approaches in the literature have used diffusional or inertial deposition as limiting cases. The technologies in which these kinds of deposition are important include solution droplet deposition (spray pyrolysis) and droplet deposition with etching (spray etching). The situation in which a volatile reactant is dissolved in the liquid is covered in Section 14.5.

The process using evaporating particles for generation of films has several variations. *Spray pyrolysis* or *droplet deposition* involves the deposition of evaporating droplets containing reactants onto surfaces, followed by further solvent evaporation and chemical reaction on the surface.[5] A note of caution must be added here. The literature contains numerous papers discussing what is called "spray pyrolysis" for film deposition. In reality, many of these processes rely on evaporation of the precursors, followed by CVD, and should more properly be grouped with vapor deposition processes. For this

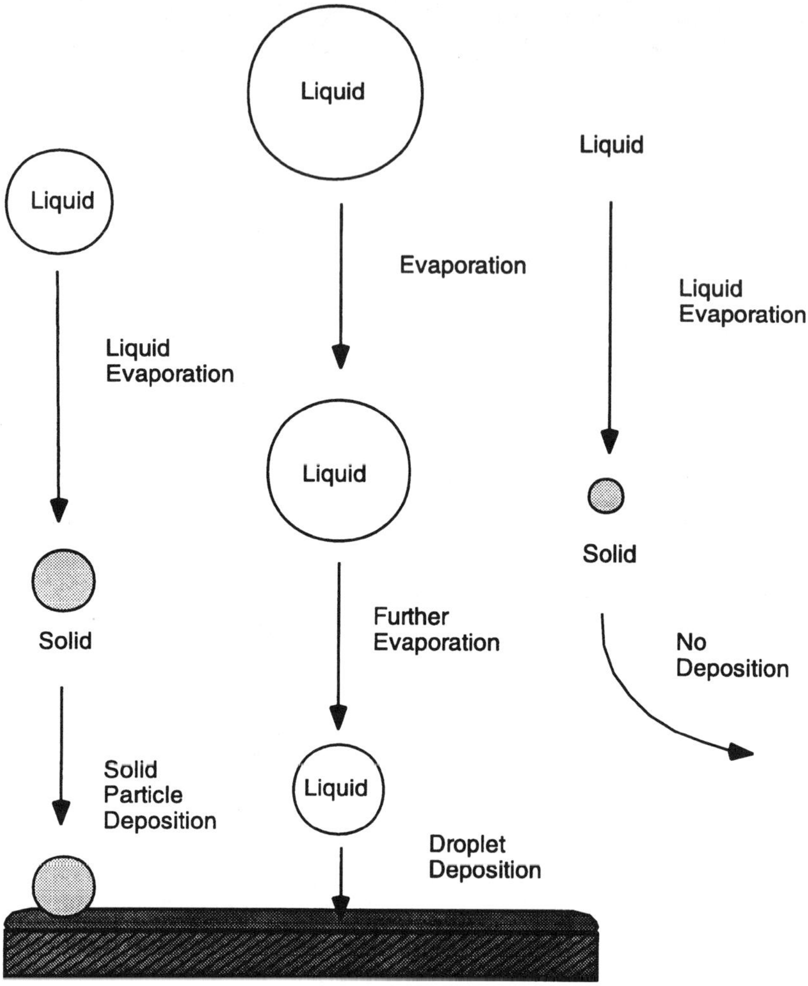


**Fig. 14.5.** Droplets of different sizes can behave differently during deposition. Larger droplets may deposit as droplets, while smaller droplets may lose all solvent and form solid particles that may or may not deposit.

reason, AACVD which uses volatile precursors, is described in a separate section. In many cases, the distinction between vapor, droplet, and particle deposition is not made, so the exact nature of those processes can only be speculated upon. In Figure 14.5, the particles are shown as being formed by a droplet-to-solid conversion route, but gas-to-particle conversion routes can also be used.

Several more complicated and undesirable situations may occur during droplet deposition. For instance, *solid precursor particle deposition* can occur even though liquid precursors are used (Fig. 14.5). As the droplets approach the heated substrate, enough solvent may evaporate to leave solid particles of the precursor(s), which can deposit to form the film. This is often undesirable because it leads to porous films. If temperatures in the gas above the substrate are sufficiently high, the precursors may decompose in the particulate phase to form particles of intermediates or the product, which can then deposit on the substrate. This, again, is often undesirable. Particles that become sufficiently small in the gas phase may not deposit if their inertia is low enough to avoid impaction and the particles follow the gas flow away from the substrate.

The theory describing the deposition of droplets is similar to that for non-evaporating particles, except that the droplets can decrease in size as they evaporate. For slow evaporation, the theory is identical to that for solid particles. For evaporation rates leading to changes in droplet diameter during the time the droplet is in the system, the description must be modified. This has been done for several situations of practical interest, although far less work has been performed on this class of problems than on solid-particle deposition.

The droplet mass and diameter vary as a function of position in the system, and auxiliary equations are used to describe the change in particle size as a function of position. This situation is illustrated qualitatively in Figure 14.5. Figure 14.6 shows trajectories of droplets with diameters of 1 and 5 $\mu$m in a typical droplet deposition geometry.[6] The trajectories were calculated using the equations given in Chapter 3 and include the effects of particle inertia, gravity, thermophoresis, and droplet evaporation. Two key points are that larger droplets [5 $\mu$m, (b) in the figure] have a greater

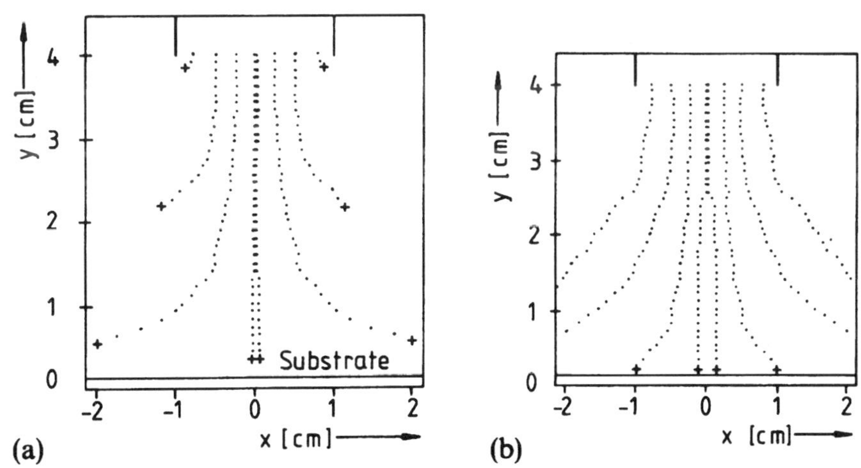


**Fig. 14.6.** Trajectories of (a) 1-$\mu$m and (b) 5-$\mu$m evaporating droplets. Deposition is favored by a larger droplet size and introduction along the center of the jet of gas directed at the substrate. (Courtesy of Siefert.)[6]

probability of striking the surface than do smaller droplets [1 $\mu$m, (a) in the figure] and that the initial trajectory of the droplets plays an important role in determining the likelihood of their impaction. Droplets introduced along the centerline of the spray have the highest probability of striking the surface, because the flow streamlines are straight until a short distance from the substrate, where the gas must turn suddenly. Droplet evaporation plays the role of reducing the size of the droplets, which leads to lower inertia and a higher probability that the droplets will follow the gas streamlines and not strike the surface.

The dependence of the deposition rate on temperature is a key operating characteristic for most deposition processes.[1] This information, along with information about the purity and morphology of the droplets as a function of temperature, provides guidance in determining the optimum deposition conditions. Furthermore, the temperature dependence is a key diagnostic that provides clues about the deposition mechanism. Deposition by CVD processes, for example, provides a rate that usually increases with temperature or, in some cases, is nearly independent of temperature. Only in rare instances will CVD rates drop with increases in temperature, such as when gas-phase particle formation occurs. Droplet deposition processes can show the opposite behavior: a rate that *decreases* with temperature, resulting from the different mechanism of deposition.

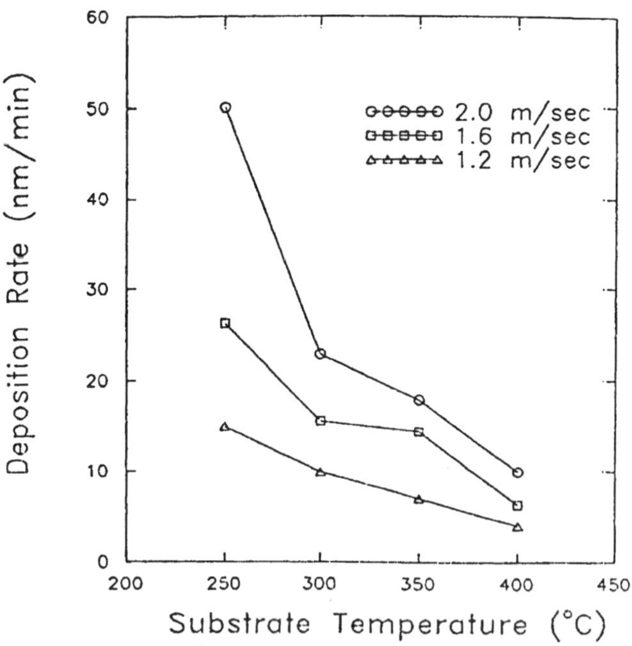


**Fig. 14.7.** Plot of droplet deposition rate as a function of temperature of substrate at three gas velocities. Higher gas velocities favor deposition by increasing the probability of inertial impaction. (Courtesy of Chung et al.[7]

Figure 14.7 shows a typical dependence of the droplet deposition rate on temperature.[7] When the temperature of the substrate is increased, the deposition rate decreases because the thermophoretic velocity driving the droplets away from the surface is higher at higher substrate temperatures and droplet evaporation is more rapid, leading to smaller droplets that are better able to flow with the gas and therefore avoid inertial and gravitational deposition. Another point is that higher droplet velocities towards the surface give higher deposition rates; higher velocity not only gives the particles greater inertia, allowing them to cross flow streamlines and strike the surface, but also shortens the time available for droplet evaporation. This form of a rate dependence on temperature is diagnostic of droplet deposition; very few conventional CVD processes, for example, exhibit rates that decrease with increasing temperature, unless gas-phase particle formation occurs.

## 14.5 AEROSOL-ASSISTED CHEMICAL VAPOR DEPOSITION

The approach of AACVD includes routes wherein particle formation is decoupled from the main problem of solvent (if used) and reactant evaporation from particles. The major distinction between this approach and droplet deposition is that in AACVD the precursor also evaporates and then adsorbs and reacts on the surface to form a film. The droplets are usually formed by spray routes in which volatile precursors are dissolved in liquids that are atomized. Alternatively, solid particles of the reactant can be entrained in a gas to form an aerosol that evaporates at least partially before reaching the substrate. This process is then followed by CVD. In many cases, solvents evaporate from droplets to form precursor particles. The general approach has been variously described as AACVD, spray pyrolysis, spray metal-organic CVD, and the pyrosol process. We use the terminology "aerosol-assisted CVD," or "AACVD" to avoid confusion with processes for powder generation (spray pyrolysis) and droplet deposition (also called spray pyrolysis in the literature).

Aerosol-assisted CVD relies on CVD to form the film. The main distinction between AACVD and "conventional" CVD is the method by which the precursors are delivered to the substrate. In conventional delivery, the precursor is generally evaporated from a bubbler or entrained in a gas stream. The basic steps are outlined in Chapter 8. The precursor must adsorb on the surface and react, and the volatile reaction products must desorb and diffuse away from the substrate. The theory of CVD processes is well developed, and numerous models of varying complexity are available.[8] However, no full models have been developed to describe AACVD.

Some of the qualitative features of AACVD are shown in Figure 14.8. We will consider multiple precursors, to emphasize that one of the largest applications for this technique is the deposition of multicomponent films. For example, three cases can be distinguished for the precursors A, B, and C. The simplest case is complete evaporation of all the precursors, followed by

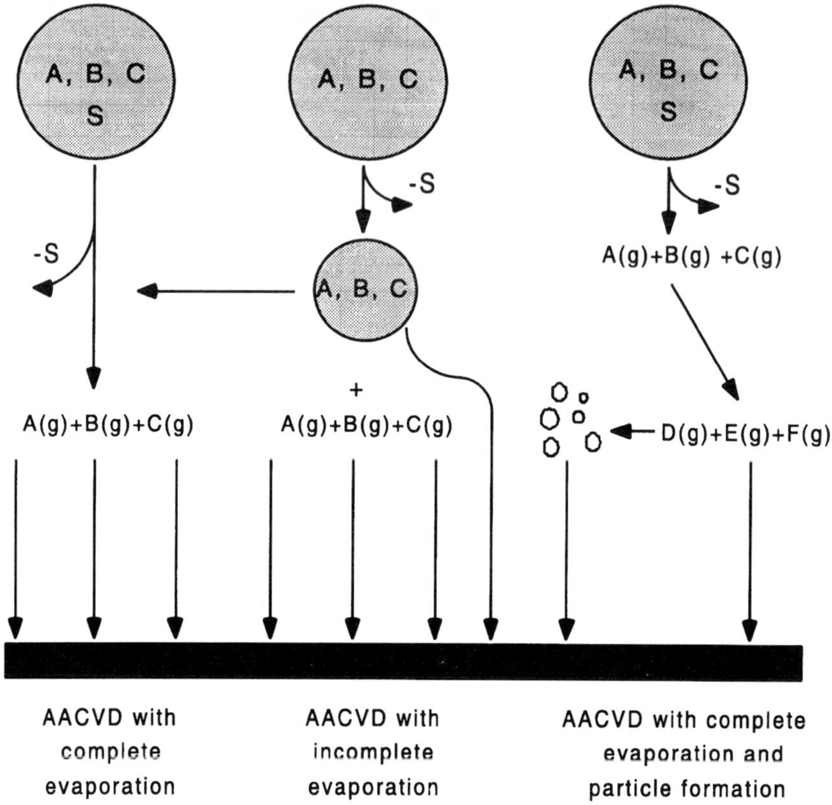


**Fig. 14.8.** Possible modes of behavior of aerosol-assisted CVD in which volatile precursors can react in the gas phase.

CVD. A more complicated case is AACVD with incomplete evaporation of the precursor particles. Finally, the precursors may evaporate either partially or completely and then react in the gas phase to give products (D, E, and F) that deposit on the surface and perhaps also form particles in the gas phase. In the latter case, CVD of the precursors A, B, and C can also lead to a complicated situation in which gas-phase and surface reactions and precursor particle evaporation are all important phenomena. Several more complicated situations are common and arise when the individual precursors have different vapor pressures, gas-phase reaction rates, and surface reaction rates and when thermally driven intraparticle decomposition occurs. Further complications are intraparticle reaction of the precursor(s) with the solvent or gaseous species.

Complete evaporation of precursors before entering the CVD reactor reduces the problem in the deposition chamber to conventional CVD. In this case, the only aerosol dynamics occur in the preheating system, where the precursor evaporates. The major concern is providing enough time for the precursor to evaporate completely before reaching the CVD reactor. This time can be estimated from the equations in Chapter 4, but the increase in reactant partial pressure in the gas phase as the reactant evaporates must be taken into account.

Partial evaporation of the precursor is a common situation and is shown in Figure 14.9 for the case of a single precursor, (hfac)CuL, where hfac is 1,1,1,5,5,5 hexafluoroacetylacetonate and L is a neutral ligand.[9–11] Partial evaporation can occur in the preheating chambers, which are designed to evaporate the precursor, when the precursor vapor pressure (and temperature) is low enough to saturate the gas phase before all of the precursor can

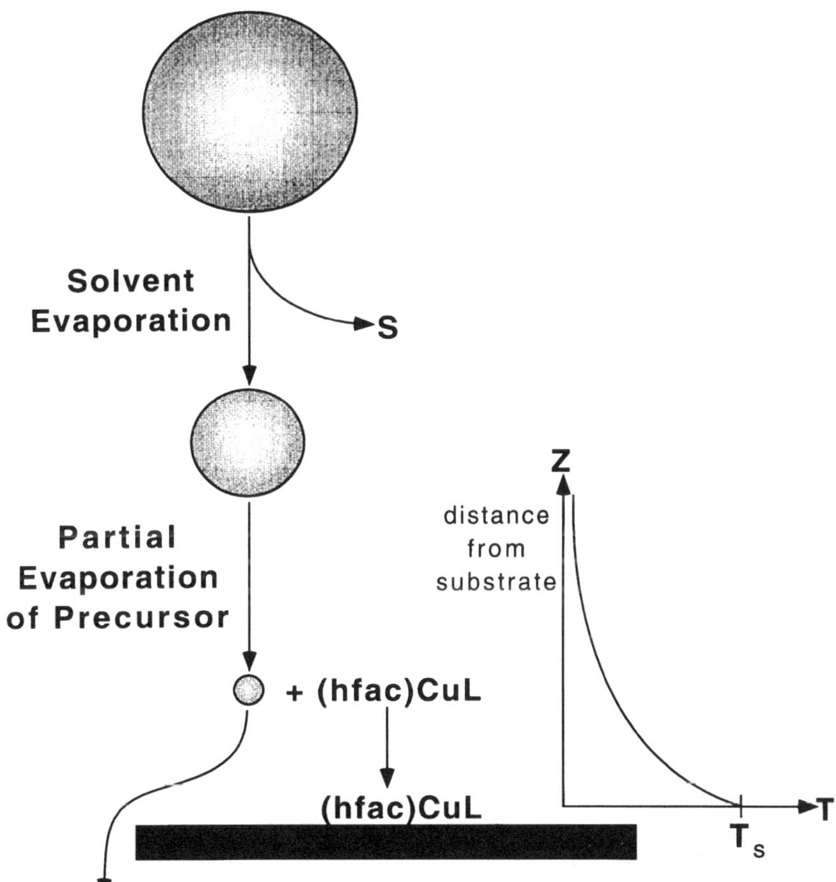


**Fig. 14.9.** Droplets containing volatile precursors may evaporate only partially during AACVD. Inset shows how temperature decreases away from substrate.

evaporate. This limitation on the evaporation temperature in the preheating zone is often imposed by precursors with low decomposition temperatures that would decompose by intraparticle reactions if heated to higher temperatures. With or without a preheating zone, the aerosol will evaporate further in the gas above the heated substrate because the gas temperature increases with decreasing distance to the substrate. As the particle approaches the substrate, further evaporation of the precursor will occur, but may not result in complete evaporation unless the vapor pressure is sufficiently high and the particles have sufficient time to evaporate. The gaseous precursor (hfac)CuL can then diffuse to the surface and undergo surface reactions, resulting in deposition. This situation is complicated by thermophoresis, which hinders the approach of the particles to the substrate and keeps the particles in regions of the gas with lower temperatures. However, thermophoresis is a valuable asset in the case where the attributes of pure CVD are required (such as conformal coverage), because this phenomenon prevents large particles from reaching the surface.

Aerosol-assisted CVD has many features in common with conventional CVD, including the dependence of the rate of *deposition* on the temperature of the substrate, but only in some cases. The major differences arise from the method of delivery of the precursor(s). Figure 14.10 shows how the logarithm of the deposition rate depends on the inverse of the substrate temperature in three idealized cases. Here, the feed rate, in terms of mass of precursor in the form of aerosol per unit time, is constant; the precursor partial pressure can vary as the gas temperature in the preheater or reactor varies. These idealized cases also assume the limiting case of rapid precursor particle evaporation relative to changes in the temperature encountered by the particle. This means that as the particles enter hotter regions of gas, the precursor evaporates, keeping the partial pressure equal to the vapor pressure.

In case A in Figure 14.10, it is important to note the variation of the rate with the substrate temperature. The precursor is completely evaporated before entering the CVD reactor, and the feed rate is high enough so that it does not limit the rate at low substrate temperatures. At these temperatures, the rate increases exponentially with the substrate temperature according to the equation

$$\text{rate} = k_s C_r v_m = k_o C_r v_m \exp(-E_A/RT) \quad \text{(surface-reaction limited)}, \tag{14.1}$$

where $k_s$ is the surface reaction rate constant, cm/sec; $T$ is the surface temperature, K; $C_r$ is the reactant concentration, molecules/cm$^3$; $v_m$ is the volume of deposited species, cm$^3$; $k_o$ is the pre-exponential constant, cm/sec; and $E_A$ is the activation energy for surface reaction (erg/mol K). The dependence of the rate on temperature is given by $E_A$ in the surface-

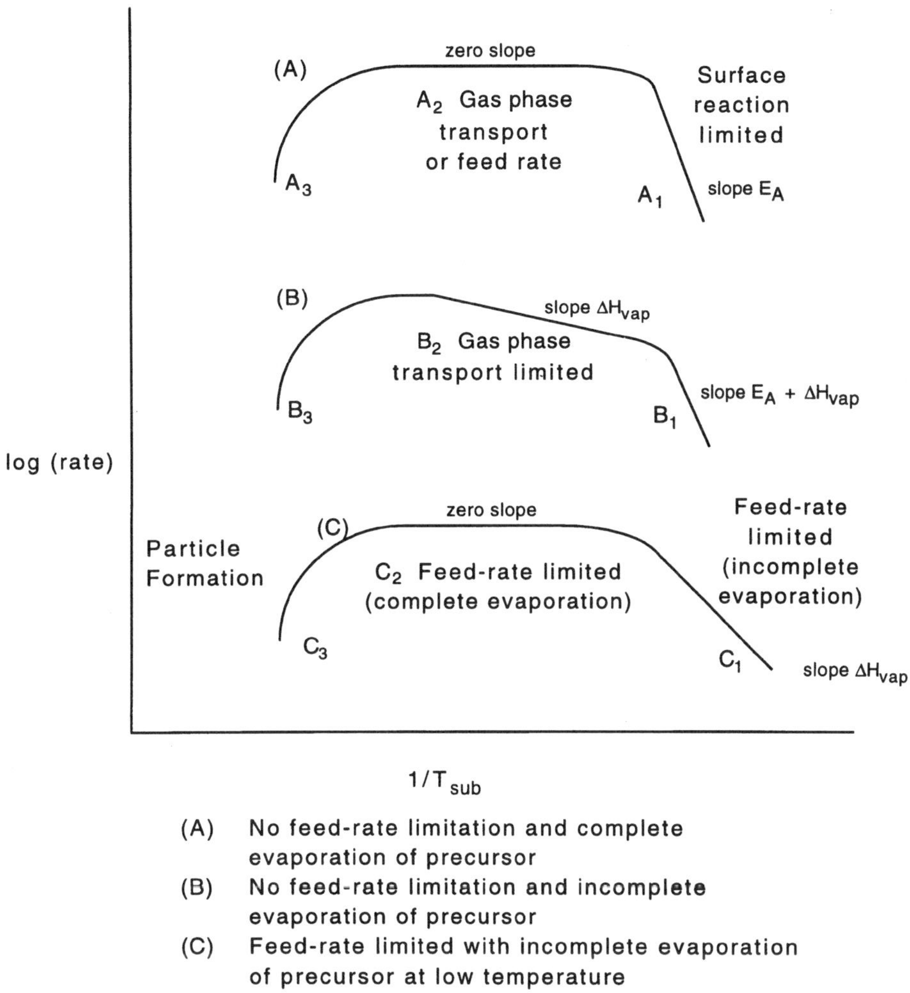


**Fig. 14.10.** Plots of log (rate) vs. $1/T$, showing possible dependencies of deposition rate on temperature in AACVD where the substrate temperature is varied.

reaction-limited regime. As the substrate temperature is increased, the transport rate cannot keep up and begins to limit the deposition rate. There is only a weak temperature dependence in the transport-limited regime. Feed rate limitations also lead to a weak temperature dependence and are often difficult to distinguish from transport limitations. This behavior occurs because changing the temperature does not increase the amount of reactant in the gas phase. At even higher temperatures, gas-phase reaction can occur, leading to a reduction in precursor concentration and therefore a reduction in deposition rate with increasing temperature. Overall, this is the situation that has the closest resemblance to conventional CVD.

In case C, surface-reaction-limited operation is not observed for any temperature range, and the behavior is different from that of conventional CVD. The vapor feed rate of the precursor is low and limits the deposition rate, and the precursor particles are not fully evaporated at any temperature. The situation shown is for a constant feed rate of the aerosol into the reactor and allows for the possibility that the partial pressure of the reactant can change with substrate temperature in the deposition chamber. At low temperatures, increasing the substrate temperature results in heating of the gas, and more precursor enters the gas phase. Thus, the slope at low temperatures reflects the dependence on temperature of the precursor vapor pressure if evaporation is rapid enough to keep the gas phase saturated with the vapor. This can be seen by examining the expression for feed-rate-limited deposition with a temperature-dependent precursor partial pressure, namely,

$$\begin{aligned}\text{rate} &= F_r M_m/(A_s \rho) \\ &= C_r Q M_m/(A_s \rho) = C_T \exp(-\Delta H_{\text{vap}}/RT)[QM_m/(A_s \rho)], \quad (14.2)\end{aligned}$$

where "rate" denotes the deposition rate, cm/sec; $F_r$ is the feed rate of the reactant, moles/sec; $C_r$ is the molar concentration of gaseous reactant, moles/cm$^3$; $Q$ is the volumetric gas flow rate, cm$^3$/sec; $M_m$ is the atomic weight of depositing species, g/mole; $A_s$ is the heated area over which deposition takes place, cm$^2$; $\rho$ is the density of the depositing film, g/cm$^3$; $C_T$ is the pre-exponential term for the temperature-dependent form of the precursor vapor pressure, moles/cm$^3$; and $\Delta H_{\text{vap}}$ is the enthalpy of evaporation of the precursor, ergs/mol K. Changing the substrate temperature causes more precursor to evaporate, leading to a higher deposition rate. [See the Clausius-Clapeyron relationship, Equation (8.5)]. The temperature dependence of the rate is given by the enthalpy of evaporation. At higher temperatures, the particles evaporate completely, and the rate no longer changes as the substrate temperature is increased, because increases in temperature no longer result in increases in precursor partial pressure. At even higher temperatures, particle formation can occur by gas phase-reaction and reduce the rate.

In case B, the feed rate is high and does not limit the rate, but the precursor particles are not completely evaporated. This provides a complex temperature dependence for the rate. In the example shown, the concentration of the reactant is lower in case B than in case A, leading to overall rates that are lower than for case A (where all of the reactant was in the gas phase and the partial pressure of the precursor in the gas phase was high), but higher than for case C. At low substrate temperatures, the rate increases because of two contributions: (1) As the substrate temperature is raised, the surface reaction rate increases, and (2) because the partial pressure of the reactant increases as the gas is heated by the substrate, more reactant evaporates into the gas phase. The relevant equation here is

$$\text{rate} = \upsilon_m C_r k_o \exp(-E_A/RT) = \upsilon_m C_T \exp(-\Delta H_{\text{vap}}/RT) k_o \exp(-E_A/RT)$$

$$\text{(surface-reaction-limited with precursor particle evaporation).} \tag{14.3}$$

This suggests that the observed activation energy cannot be greater than the sum of the activation energy for the surface reaction and the activation energy for the evaporation of the precursor. In practice, the situation is more complicated than this limiting case, because the particles above the substrate do not reach the same temperature as the substrate, due primarily to thermophoresis, which keeps the particles from approaching the substrate.

At higher temperatures, the rate becomes transport limited and is given by

$$\text{rate} = \frac{\upsilon_m D C_r}{L} = \left(\frac{\upsilon_m D_o}{L}\right)\left(\frac{p_o}{p}\right)\left(\frac{T}{T_o}\right)^{3/2}\left(\frac{p_r}{RT}\right), \tag{14.4}$$

where $D$ is the diffusion coefficient, $cm^2/sec$, $D_o$ is the diffusion coefficient at reference conditions, $cm^2/sec$; $L$ is the thickness of the concentration boundary layer, cm; $C_r$ is the reactant concentration outside the concentration boundary layer, $molecules/cm^3$; $T_o$ is the reference temperature for the gas above the substrate, K; $p$ is the total pressure, $dynes/cm^2$; $p_o$ is the total pressure at reference conditions, $dynes/cm^2$; and $p_r$ is the reactant partial pressure outside the concentration boundary, $dynes/cm^2$. Because the precursor particles can evaporate and increase the partial pressure of the precursor as the temperature of the substrate is increased, the precursor partial pressure can be written in terms of a temperature-dependent concentration, and the expression for the transport-limited rate becomes

$$\text{rate} = B_5\left(\frac{T}{T_o}\right)^{3/2} \exp(-\Delta H_{\text{vap}}/RT) \quad \text{(transport limited)}, \tag{14.5}$$

where $B_5$ is a constant. Because the dependence on $T^{3/2}$ is not easily detectable on a logarithmic scale, the exponential term dominates, and the apparent activation energy is given by the enthalpy of evaporation.

*This discussion of cases A, B, and C suggests that the activation energy for the surface reaction can be observed only when the precursor is completely evaporated and no feed rate limitation is encountered (case A).* Otherwise the observed activation energy either is the enthalpy of evaporation of the precursor (C) or is less than or equal to the sum of the activation energy for the surface reaction and the enthalpy of evaporation of the precursor (B). The situations discussed are idealized cases in which the precursor evaporates rapidly and is able to keep the vapor saturated. When this does not occur, a weaker dependence on temperature will be observed.

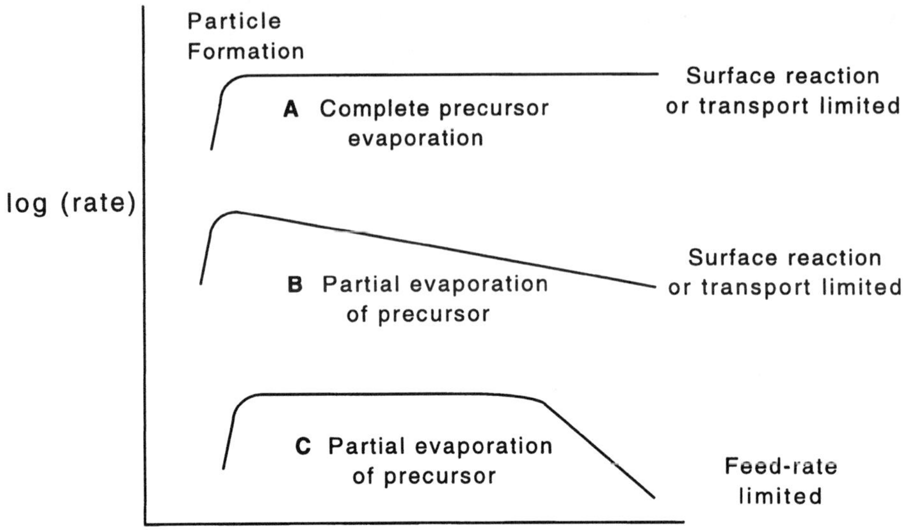


**Fig. 14.11.** Plots of log (rate) vs. $1/T$, showing possible dependencies of deposition rate on preheater temperature in AACVD where the preheater temperature is varied.

Similar behavior occurs in AACVD when the temperature of the preheater (the section of heated tubing before the reactor) is varied, as shown in Figure 14.11. The behavior described here occurs only when the evaporation time for the particles is long relative to the time they spend in the hot gas above the substrate in the reactor. Otherwise a more complex situation will arise in which the overall system of combined preheater and reactor must be considered. This is the opposite of the cases discussed where the evaporation rate was assumed rapid.

In case C of Figure 14.11 (not Fig. 14.10), only small amounts of precursor are fed into the system. This can occur for precursors with low solubilities in the solvent, small droplet sizes, low droplet concentrations, low vapor pressures, and low carrier gas flows. As the preheater temperature is increased, the deposition rate increases because the partial pressure of the precursor vapor increases, which assumes that some unevaporated precursor is present. At sufficiently high preheater temperatures, the precursor evaporates completely before entering the reactor, leading to no dependence of the deposition rate on the preheater temperature. At even higher preheater temperatures, particle formation can occur in the gas phase due to homogeneous gas-phase reactions, which then leads to a drop in deposition rate. An example of this is shown in Figure 14.12, where varying the preheater temperature leads to an increase in deposition rate consistent with the temperature dependence of the vapor pressure of the precursor, $\Delta H_{vap} = 6.7$ Kcal/mol.

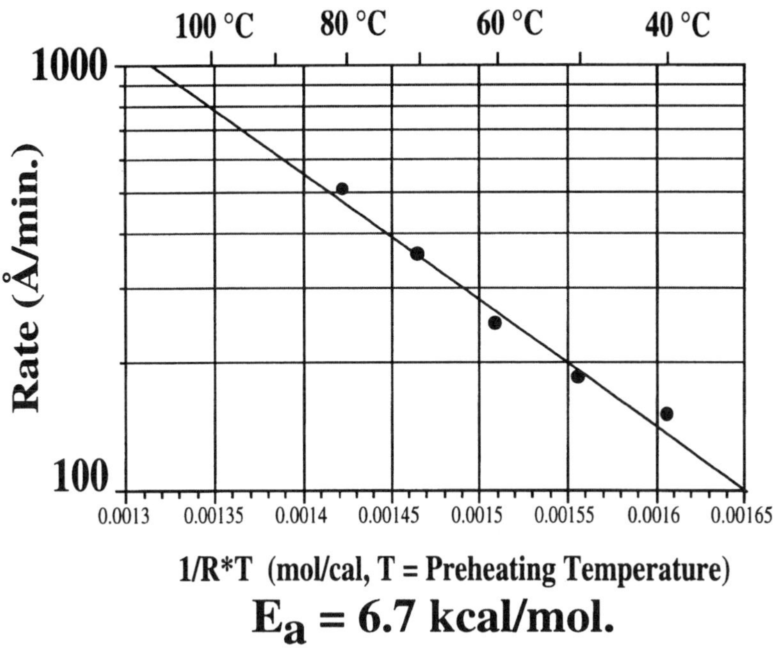


**Fig. 14.12.** Plot of log (rate) vs. $1/T$ for copper deposition by AACVD. Varying the preheater temperature resulted in an activation energy of 6.7 kcal/mol, corresponding to the enthalpy of evaporation. (Courtesy of Roger et al.)[9]

In case A, the feed rate of the precursor does not limit the deposition rate, and the precursor particles evaporate completely in the preheater. This corresponds to a situation indistinguishable from ordinary CVD. Because the substrate temperature is held constant, the rate does not depend on the preheater temperature.

In case B, the feed rate does not limit the rate, but increasing the preheater temperature increases the partial pressure of the precursor in the gas phase, in turn increasing the partial pressure at the substrate surface. If the reactor itself is operated at a constant temperature corresponding to surface-reaction-limited conditions, the rate is given by

$$\begin{aligned} \text{rate} &= v_m C_T \exp(-\Delta H_{\text{vap}}/RT_{\text{pre}}) k_o \exp(-E_A/RT_{\text{sub}}) \\ &= B_6 \exp(-\Delta H_{\text{vap}}/RT_{\text{pre}}) \qquad \text{(surface reaction)}, \end{aligned} \tag{14.6}$$

where $B_6$ is a constant and the temperature $T_{\text{pre}}$ that is being varied is the preheater temperature. The apparent activation energy is equal to the enthalpy of evaporation of the precursor.

If the rate is transport limited in the reactor, it is given by

$$\text{rate} = B_7 \exp(-\Delta H_{\text{vap}}/RT_{\text{pre}}), \tag{14.7}$$

where $B_7$ is a constant and the temperature $T_{pre}$ that is being varied is the preheater temperature. Again, the apparent activation energy is equal to the enthalpy of evaporation of the precursor.

The preceding discussion suggests that if the rate of precursor evaporation is slow and the precursor partial pressure in the reactor is determined by the precursor partial pressure at the exit of the preheater, the activation energy associated with the surface reaction can be observed only under some conditions. The activation energy for precursor evaporation can be observed, however, when the precursor is not fully evaporated in the preheater.

The variation of the deposition rate with the concentration of precursor in the solution is also an important consideration. The concentration of the precursor in the solution and the volume of solution per volume of carrier gas, along with the carrier gas flow rate, determine the partial pressure of the precursor. Note that this situation is an idealized equilibrium case and that if the evaporation rate is slow and the residence time is short, the desired precursor partial pressure may not be obtained. This information can be used to calculate the partial pressure of the precursor, considering that the partial pressure can be no higher than the vapor pressure. If the number calculated is less than the vapor pressure, the precursor can evaporate completely if sufficient time is available. However, if the value is larger than the vapor pressure, then the precursor particles cannot evaporate completely, regardless of how much time is available, and both vapor and particles will be present. *Thus, changing the precursor concentration in the solution can change the temperature dependence of the rate by providing either pure vapor or both vapor and precursor particles above the substrate surface.*

## 14.6 DEPOSITION USING DROPLETS OR SOLID PARTICLES FORMED IN SITU

For routes involving deposition with droplets or particles formed in situ, the particle formation process is coupled to the problem of depositing an aerosol onto a surface, as shown in Figure 14.13. The particles are almost always formed by gas-to-particle conversion, and particle sizes are often less than 100 nm, allowing the use of convective diffusion equations to describe particle transport and deposition. This class of routes encompasses a wide variety of technologies, including optical fiber production, particle formation by expansion of a supercritical fluid followed by deposition, ionized cluster deposition, neutral cluster deposition, and aerosol jet etching. In these processes, the particles are formed in the gas approaching or passing by the surface. Some fraction of the particles is not deposited and corresponds to reduced yields.

Models have been developed for many of these approaches. The wide variety of system geometries and particle formation and deposition mechanisms, however, has precluded the development of a single model that describes all the approaches. The most important technology from the

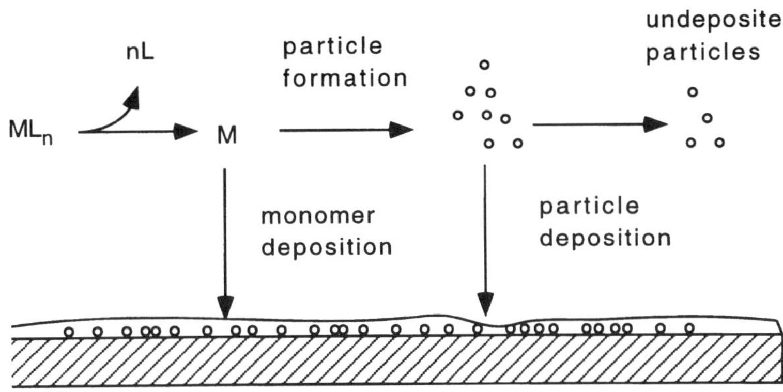


**Fig. 14.13.** Coupled vapor and particle deposition where particles are formed in situ by gas-to-particle conversion.

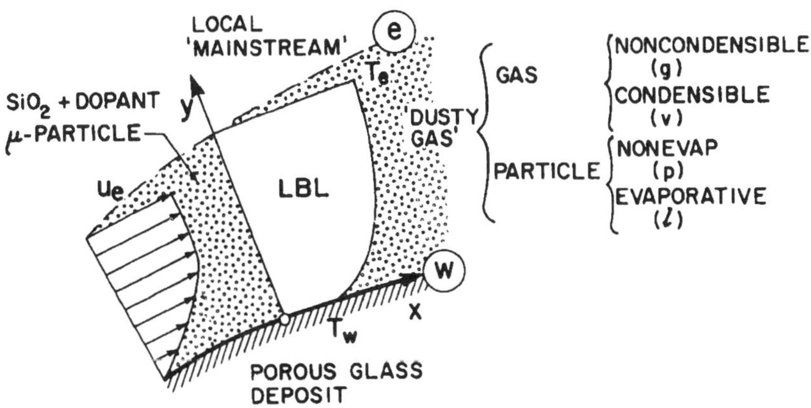


**Fig. 14.14.** Velocity and temperature profiles during thermophoretic deposition of $SiO_2$–$GeO_2$ particles. Gas velocity, temperature, and species concentration profiles all vary with position from the surface. LBL is the laminar boundary layer, *w* denotes the wall, and *e* refers the edge of the LBL. Reprinted from *Chemical Engineering Science 44* (3), Park and Rosner,[12] "Multiphase continuum theory of dopant redistribution across aerosol-laden laminar nonisothermal boundary layers," pp. 603–617. Copyright 1989, with permission of Elsevier Science Ltd., Kidlington OX5 1GB, UK.

industrial perspective is optical fiber production, an example that illustrates many of the features of this class of processes.

Figure 14.14 shows particles of $SiO_2$ doped with species which include $GeO_2$, $P_2O_5$, and $B_2O_3$, formed in a hot gas flowing past a cooler surface that consists of a porous glass deposit.[12] This is the situation encountered in what is known as outside vapor deposition (OVD) (a misleading term because the deposition occurs through particles, not vapor). The gas flowing over the substrate is shown in laminar flow, in which the velocity of the gas varies from nearly zero at the surface to some value $u_e$ farther away that is related to the total flow rate of the gas through the system. The temperature varies from $T_w$ at the surface to $T_e$ far from the surface. This temperature gradient drives the deposition of particles onto the substrate by thermophoresis. In the example presented in the figure, the dusty gas consists of noncondensible carrier gases such as nitrogen, condensible species such as $GeO_2$, nonevaporating components in the particles ($SiO_2$), and evaporative components in the particles ($GeO_2$). This distinction is made because $GeO_2$ is much more volatile than $SiO_2$, leading to the presence of $GeO_2$ vapor under conditions where no $SiO_2$ vapor exists.

An interesting but practical problem is the composition of the deposit formed during OVD as a function of operating parameters. Figure 14.15

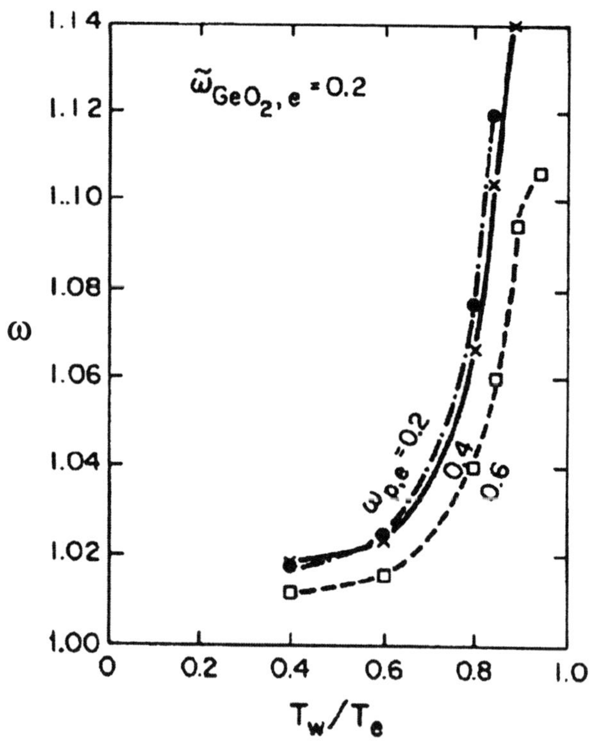


**Fig. 14.15.** Transport-induced differences between actual $GeO_2$ content of deposited glass and that expected based on chemical thermodynamics and steady-flow element balances. The quantity $\omega$ represents actual Ge content relative to the content expected on the basis of thermodynamics. Reprinted from *Chemical Engineering Science 44*(3), Park and Rosner,[12] "Multiphase continuum theory of dopant redistribution across aerosol-laden laminar nonisothermal boundary layers," pp. 603–617. Copyright 1989, with permission of Elsevier Science Ltd., Kidlington OX5 1GB, UK.

shows that the composition of the deposited glass, consisting of Ge and Si oxides, can be very different from that predicted ignoring vapor and particle transport effects.[12] The quantity $\omega$ plotted is the ratio of the actual Ge content in the glass relative to that expected solely on the basis of chemical thermodynamics and steady-flow element balances. The actual fraction of Ge can be 10% higher than that predicted when transport effects are ignored. Deviations of this magnitude are more than large enough to change optical properties and optical fiber performance.

Another important aspect of optical fiber production is estimating the fraction of the particles that deposit. A commonly encountered geometry in addition to OVD is flow through a tube wherein particle deposition occurs on the inside of the tube. Here, the deposition efficiency is important because particles that do not deposit lead to reduced yields and long processing times. Figure 14.16 shows the deposition efficiency as a function of the parameter $\beta_{\text{th}}$.[13] The term Pr is the Prandtl number, $\mu C_p/k$, where $\mu$ is the viscosity of the gas, $C_p$ is the heat capacity at constant pressure, and $k$ is the thermal conductivity of the gas. The term $\theta^*$ is $T_w/(T_e - T_w)$, where $T_w$ is the wall temperature and $T_e$ is the gas temperature at the entrance to the region of the tube with the cooled walls. The parameter $K_{\text{th}}$ is the thermophoretic coefficient.[14] This plot is also useful for estimating the fraction of particles lost when a hot gas containing particles exits a tubular flow reactor. An important point is that the deposition efficiency depends relatively weakly on particle size and characteristics through $K_{\text{th}}$ and Pr, but

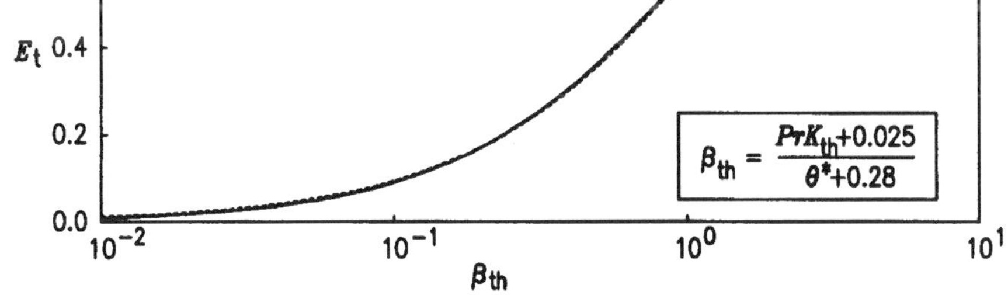


**Fig. 14.16.** Plot of thermophoretic deposition efficiency $E_t$ of aerosol particles flowing through a tube with cool walls as a function of the dimensionless quantity $\beta_{\text{th}}$. Symbols are defined in the text. Reprinted from *Journal of Aerosol Science*, *25*, Stratmann, F., Otto, E., and Fissan, H., 1305–1319, Copyright 1994 with permission of Elsevier Science Ltd., Kidlington OX5 1GB UK.

depends strongly on the temperature difference between the walls and hot gas through the term $\theta^*$. This term is roughly 0.33 for a wall temperature near room temperature and a gas temperature of 1000 K corresponding to deposition efficiencies of roughly 0.4. The plot illustrates the difficulty in depositing most of the particles by thermophoresis. It also demonstrates that appreciable fractions of particles will be lost when exiting hot-wall tubular flow systems designed for powder production.

## 14.7 SIMULTANEOUS PARTICLE AND VAPOR DEPOSITION

Coupled particle and vapor deposition can utilize particles formed ex situ by spray routes, gas-to-particle conversion, and suspension of powders or in situ by gas-to-particle conversion. In some cases, the particles are formed in situ by gas-to-particle conversion, and the deposition of particles is then coupled to the CVD process because the same precursor forms both the particles and the film. This latter situation is encountered in particle-precipitation-aided CVD in which thermophoretic deposition is exploited. Other methods of particle deposition can also be employed, including electrophoresis, gravitational settling, impaction, diffusion, etc. This class of aerosol routes to films is just beginning to be developed, and relatively few demonstrations and models exist for the full process including both CVD and particle deposition. Hurt and Allendorf present one of the few models for a subset of this class of processes.[15] The overall qualitative features, however, have been identified.

In general, for ex situ particle formation, simultaneous particle and vapor deposition may occur as illustrated in Figure 14.17. Particle deposition may occur by any of the mechanisms identified and described in Chapter 3. The

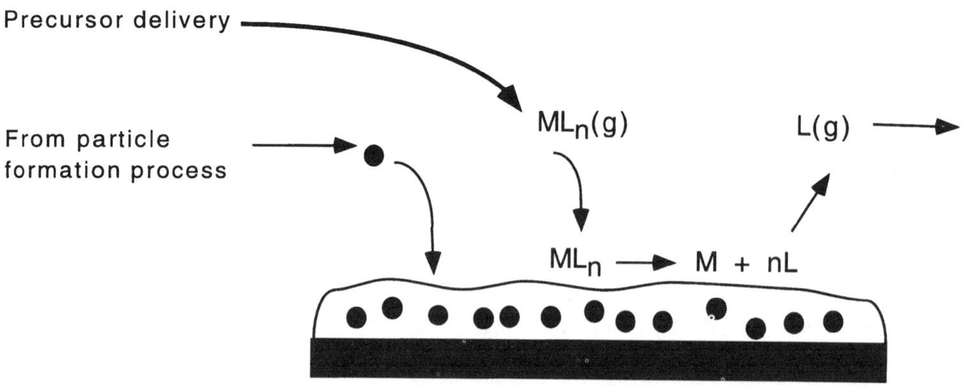


**Fig. 14.17.** Simultaneous particle and vapor deposition to form composite films. $ML_n$ is the molecular precursor, where M is the metal and L the ligand.

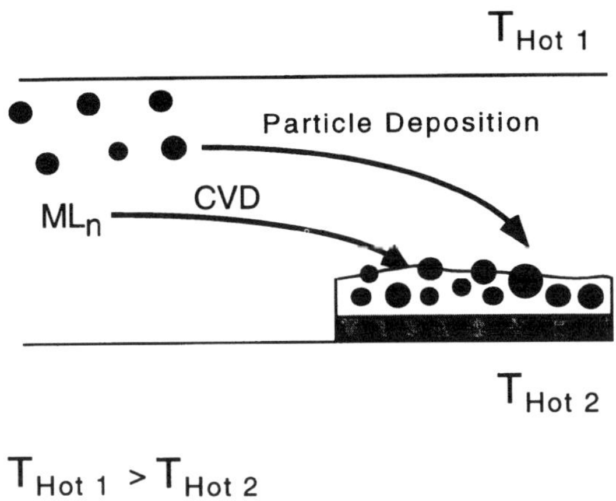


**Fig. 14.18.** Simultaneous particle and vapor deposition by thermophoresis. The substrate is cooler than the upper surface, providing a temperature gradient that drives thermophoretic particle deposition.

deposited particles can lead to a higher film surface area than in the absence of the particles, resulting in increased CVD rates. In addition, particle deposition contributes to the growth rate of the film. If the deposited film is porous, CVD can occur in the pores, producing what is known as chemical vapor infiltration (CVI), a process used to impregnate porous materials by CVD.

Figure 14.18 shows a system in which particles formed ex situ are co-deposited with vapor. The particles are forced to deposit by keeping the upper surface hotter than the substrate. In this case, the precursor must be introduced close to the substrate to avoid contact with the hotter upper surface, where CVD could also take place. The particles are shown as being formed ex situ, but can be formed in situ. The latter situation occurs during particle-precipitation-aided CVD. Although this route has been used for the fabrication of homogeneous materials, it has not been used extensively for composite film generation.

Figure 14.19 shows how particles can be co-deposited with vapor by charging them and then applying an electrical field. The particles shown are formed ex situ; however, they can also be formed in situ. A corona discharge obtained by applying a high field between a surface with a sharp radius of curvature (the tip of a wire) and the substrate can be used to charge the particles. As an example, Danek et al. have used this approach, with particles formed by spraying generating composite films containing quantum dots.[16, 17]

Particles can also be formed in situ, as shown in Figure 14.20. The figure illustrates deposition by diffusion, but other mechanisms can be used as well. This situation has not been investigated in detail, but probably occurs during particle formation in CVD where small particles with high diffusivities are formed immediately above the substrate. When a single precursor is used,

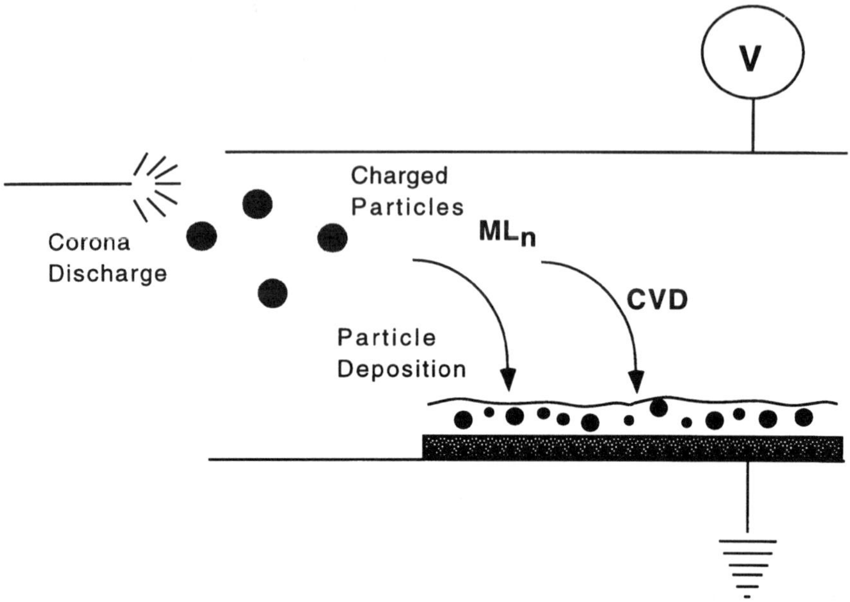


**Fig. 14.19.** Simultaneous particle and vapor deposition by charged particles in an electric field.

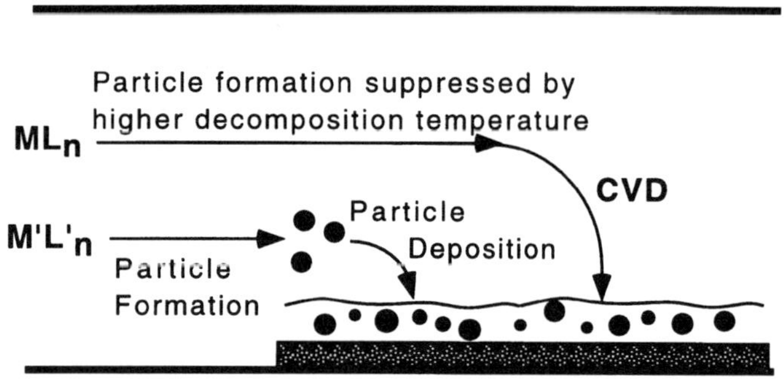


**Fig. 14.20.** In situ particle formation and co-deposition with vapor to form a composite film.

the process corresponds to particle formation during CVD, which is often undesirable. On the other hand, when two different precursors are used and only one forms particles by gas-phase reaction, composite films can be produced. In this case, the two precursors must have substantially different rates of gas-phase reaction. The use of this approach for composite film fabrication has not been investigated in detail.

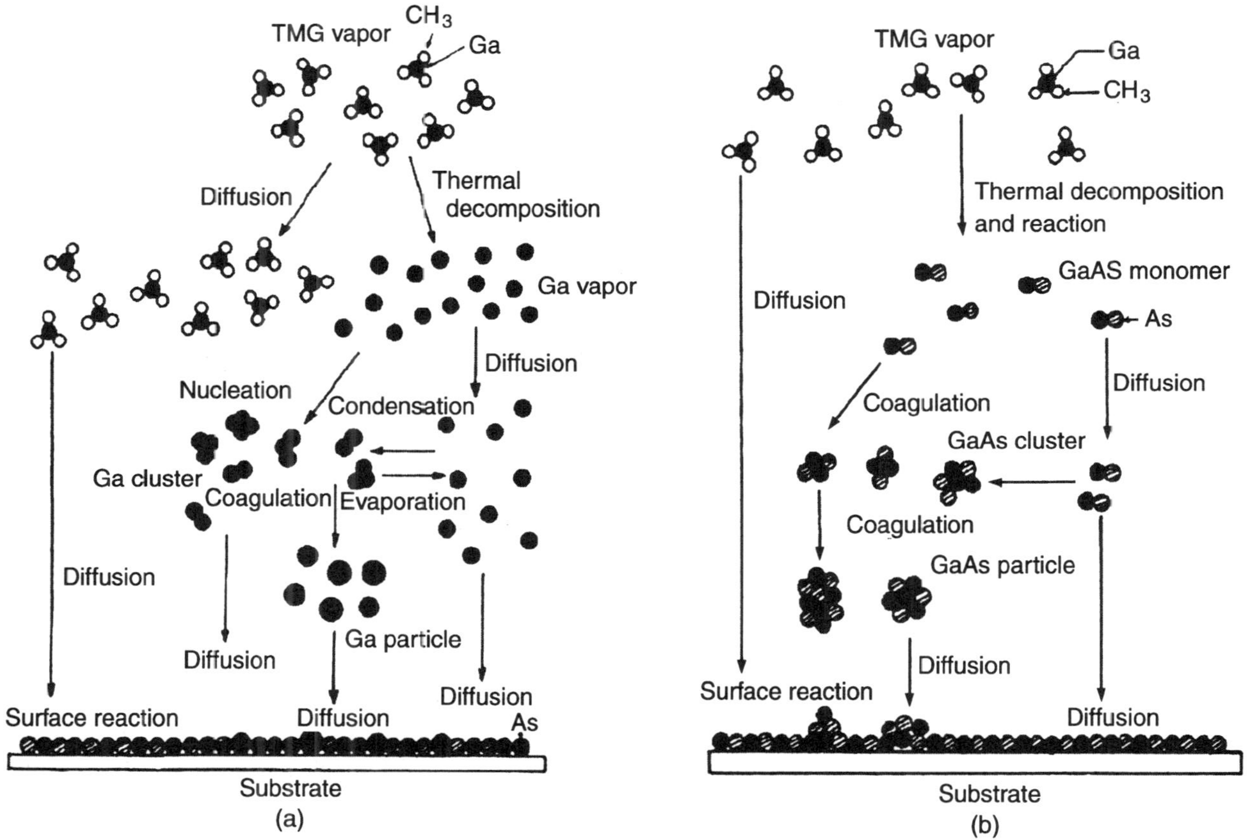


**Fig. 14.21.** GaAs film formation in which gas-phase reactions can form Ga particles and GaAs particles that can diffuse to the surface and deposit along with vapor. Okuyama et al.[18] Reproduced with permission from the *Japanese Journal of Applied Physics*.

## 14.8 PARTICLE FORMATION DURING CVD

The formation of particles during CVD is often undesirable because it can lead to degraded film properties. This is a complex problem in which CVD is coupled to the particle formation and deposition processes. The homogeneous reactions leading to the formation of particles in the gas phase are in many cases the same reactions exploited for particle formation. For systems in which particle formation must be avoided, it is useful to understand the qualitative features of the process. Specifically, the influence of reactant concentration, total pressure, and system geometry on the rate of particle formation and film properties must be understood.

Figure 14.21 shows the steps that can occur during the formation of Ga and GaAs clusters in the gas phase in CVD.[18] Gas-phase reactions lead to the formation of condensable species such as Ga and GaAs. These species can be transported to the walls of the system, thereby reducing their gas-phase concentration, which is determined by the balance between formation and loss by transport. If the gas-phase concentration exceeds the critical value for particle formation, nucleation occurs. Once formed by condensation and coagulation, the nucleated particles can grow. This qualitative picture suggests several strategies for reducing particle formation. Reducing the total pressure and the dimensions of the system can inhibit particle formation by increasing the rate of removal of the condensable species by diffusion, thereby reducing the gas-phase concentration of those species. Reducing the reactant concentration or substrate temperature has a similar effect.

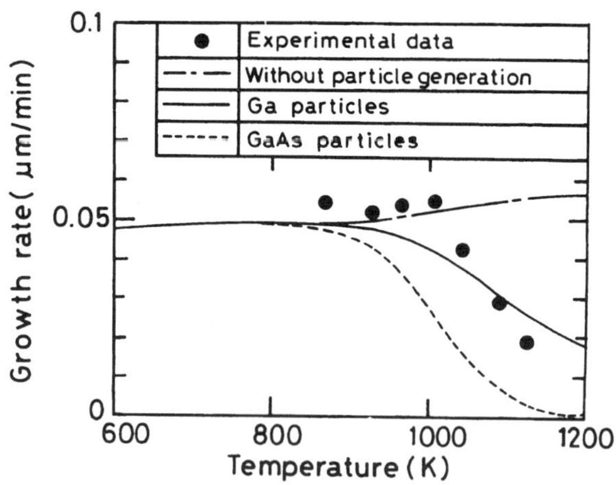


**Fig. 14.22.** Plot of growth rate at the axial position $x = 13$ cm as a function of substrate temperature with both modes of particle generation for GaAs deposition. Okuyama et al.[18] Reproduced with permission from the *Japanese Journal of Applied Physics*.

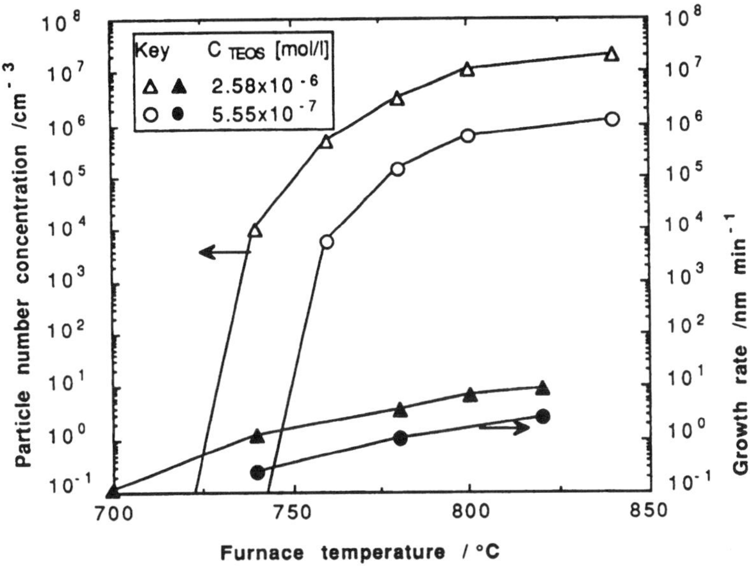


**Fig. 14.23.** Plot of gas-phase particle concentration and deposition rate in a TEOS-Atmospheric Pressure CVD reactor as a function of furnace temperature for two TEOS concentrations. TEOS = tetraethylortheosilicate. Adachi et al.[19] Reproduced with permission from the *Japanese Journal of Applied Physics*.

An indicator of particle formation in the gas phase (shown in Figure 14.22) is a reduction in the deposition rate as the growth rate is increased by increasing the substrate temperature.[18] The data points are measurements, while the curves are obtained from a sophisticated model taking into account CVD with Ga and GaAs particle formation in the gas phase.[18] Although the deposition rate is reduced once sufficient numbers of particles are produced in the gas phase, particles can also be produced under conditions where the deposition rate is unaffected. This occurs when the mass of particles formed is a small fraction of the mass deposited by CVD. Such a situation is shown in Figure 14.23 for the CVD of $SiO_2$ using tetraethylorthosilicate (TEOS) at atmospheric pressure.[19] Particles at concentrations up to $10^7$ number/$cm^3$ are formed at deposition rates of 10 nm/min without a reduction in deposition rate.

One possible outcome of particle formation in the gas phase is the deposition of particles onto the substrate, leading to degraded film properties. Figure 14.24 shows regions on a substrate where particles were deposited during CVD of $TiO_2$ in a tubular flow reactor using titanium tetraisopropoxide and oxygen for the conditions outlined in Table 14.1.[20] A comparison of the results presented in the figure with those shown in the table indicate that particle formation is favored by higher total pressures (a, b), higher reactant concentration (a, g, h), and higher temperature (d, e).

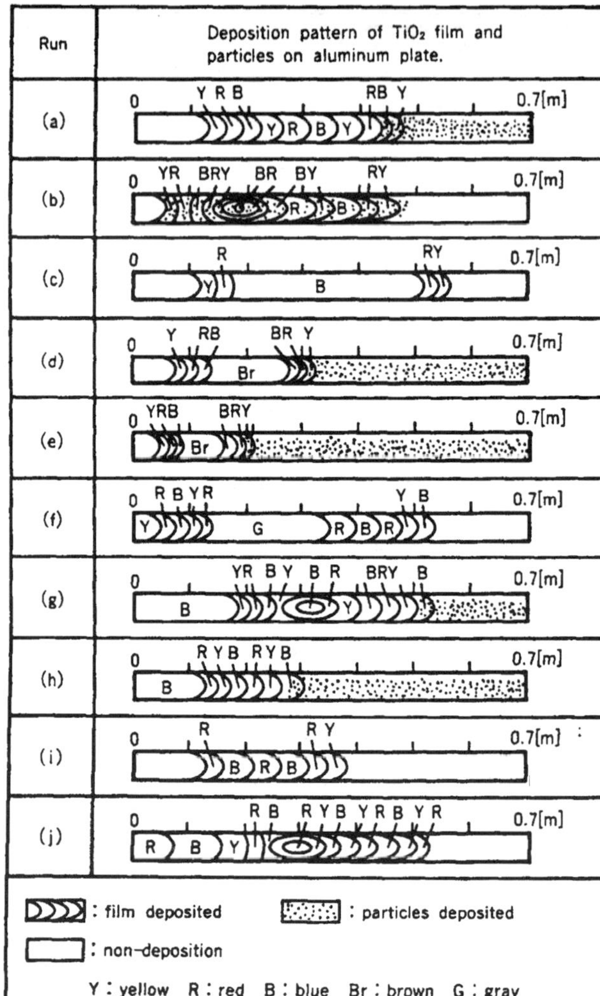


**Fig. 14.24.** Patterns of $TiO_2$ vapor and particle deposition on substrates for conditions of Table 14.1. (Courtesy of Oshima et al.)[20]

These trends agree with those predicted by the qualitative picture presented in the text and also show that extensive particle deposition can occur, yielding films with nonuniform properties.

Particle formation in the gas phase has a strong impact on deposition rates and film morphology. The occurrence of particle formation can be inferred from the dependence of the deposition rate on operating parameters, including reactant concentrations, reactor diameter, and total pressure. Figure 14.25 shows how particle formation can occur during CVD of copper oxides from CuI in a tubular flow system.[21] The CuI can react in the gas phase to

**TABLE 14.1. Conditions used for films shown in Figure 14.24. TTIP = titanium tetraisopropoxide. Courtesy of Ohshima et al.[20]**

| Run | Pressure, *P* [Pa] | Volume flow, *Q* [$m^3$/sec] | TTIP concentration, *C* [%] | Furnace temperature, *T* [K] |
|---|---|---|---|---|
| a | $1.01 \times 10^5$ | — | — | — |
| b | $1.13 \times 10^4$ | — | — | 523 |
| c | $1.13 \times 10^3$ | — | 0.1 | — |
| d | $1.01 \times 10^5$ | — | — | 623 |
| e | — | $1.67 \times 10^{-5}$ | — | 723 |
| f | $1.13 \times 10^3$ | — | — | 623 |
| g | $1.01 \times 10^3$ | — | 0.025 | — |
| h | — | — | 0.4 | — |
| i | $1.13 \times 10^3$ | — | 0.025 | 523 |
| j | — | — | 0.4 | — |

dashes indicate same as next value below in column.

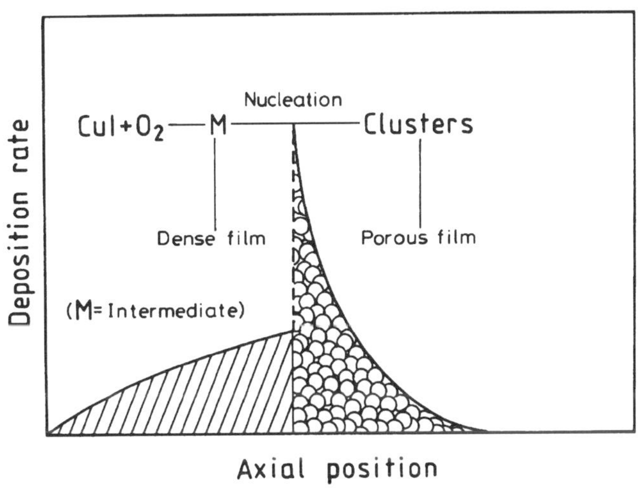


**Fig. 14.25.** Schematic diagram showing film formation mechanism of CuI + $O_2$ system. (Courtesy of Hong & Komiyama.[21] Reprinted with permission of the American Ceramic Society, Westerville, OH. Copyright 1991 by the American Ceramic Society.

form an intermediate (possibly $Cu_2O$) that can diffuse to the film while, at the same time, its concentration builds up as it flows down the reactor. At some point, the concentration of the intermediate builds up to a level at which clusters form. The clusters can deposit, resulting in a porous film with a greater thickness than that obtained in the region where particles do not deposit. The influence of operating parameters on particle formation and deposition has been quantified for this system and provides useful insights.

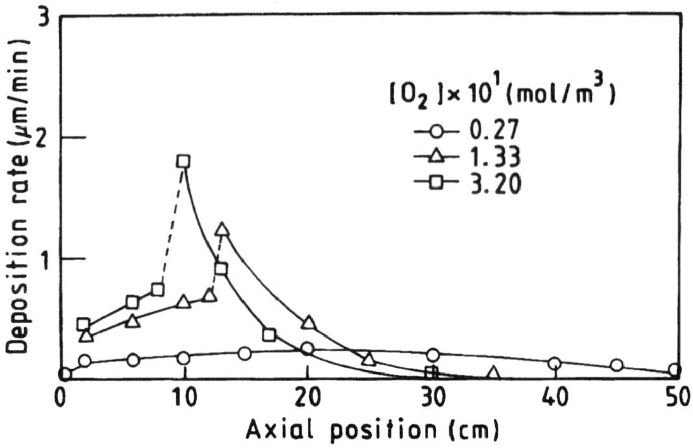


**Fig. 14.26.** Axial variation of copper-oxide film deposition rate at various oxygen concentrations. Reaction conditions: [CuI] = 0.94%, $P_t = 5.33 \times 10^3$ N/m$^2$, $d = 6.7$ mm, $t = 19$ ms. (Courtesy of Hong & Komiyama.[21] Reprinted with permission of the American Ceramic Society, Westerville, OH. Copyright 1991 by the American Ceramic Society.

Figure 14.26 shows how the oxygen concentration influences the rate of $CuO_x$ CVD.[21] Increasing the oxygen concentration increases the rate of the gas-phase reaction, which moves the point at which particle formation and deposition occur upstream. Increasing the oxygen concentration also increases the amount of particulate deposited on the substrate, resulting in thicker films. At sufficiently low concentrations of oxygen, particle formation and deposition do not occur. This corresponds to the situation where any monomer generated in the gas phase is able to diffuse to the surface faster than it is completely formed, thereby keeping the monomer concentration in the gas phase below the critical value required for particle formation and growth.

The qualitative picture of particle formation in CVD systems suggests that reducing the dimensions of the system will favor a reduction in particle formation. Experimental results for the $CuO_x$ system are in agreement with this hypothesis, as shown in Figure 14.27.[21] When the reactor diameter was reduced from 6.7 to 3.2 mm, particles did not form. This increased the rate of transport of the monomer (formed by gas-phase reaction) to the substrate surface, in turn reducing its gas-phase concentration to a value below the critical value required for particle formation and growth.

Particle formation should be reduced by reducing the reactor pressure. The reduced pressure should result in increased rates of transport of the reactant and intermediate to the substrate, thereby reducing the concentration of the intermediate in the gas phase. Figure 14.28 shows that reducing the pressure reduced the size of the region over which particle formation and

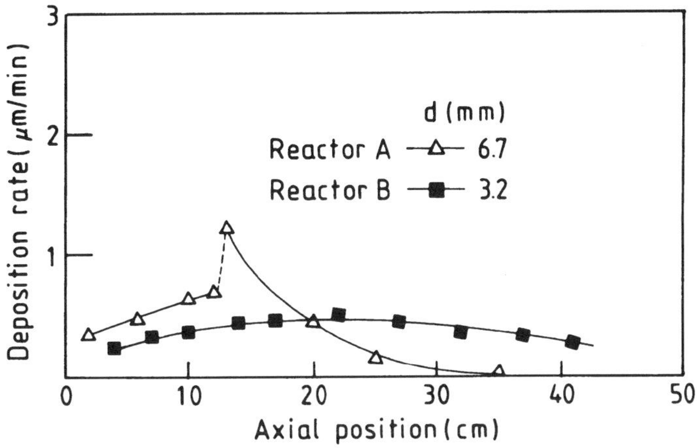


**Fig. 14.27.** Axial variation of copper-oxide deposition rate in reactors with different diameters. Reaction conditions: [CuI] = 0.94%, $[O_2]$ = 20.3%, [He] = 78.8%; $P_t$ = $5.33 \times 10^3$ $N/m^2$, $t$ = 19 ms. (Courtesy of Hong & Komiyama.[21] Reprinted with permission of the American Ceramic Society, Westerville, OH. Copyright 1991 by the American Ceramic Society.

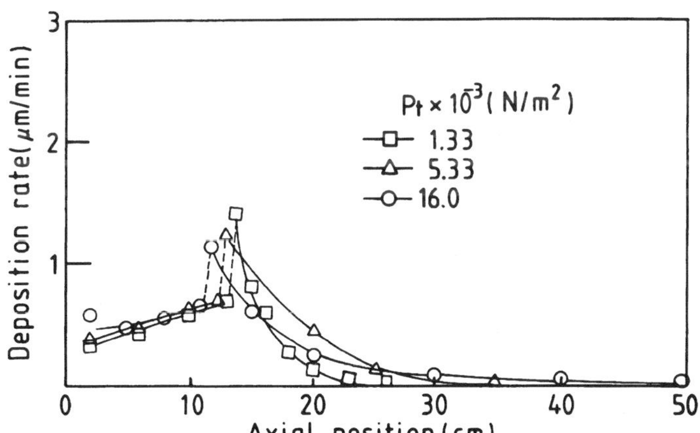


**Fig. 14.28.** Axial variation of copper-oxide deposition rate at various total pressures. Reaction conditions: [CuI] = 0.94%, $[O_2]$ = 20.3%, [He] = 78.8%, $d$ = 6.7 mm, $t$ = 19 ms. Hong and Komiyama.[21] Reprinted with permission of the American Ceramic Society, Westerville, OH. Copyright 1991 by the American Ceramic Society.

deposition occurred.[21] However, the effect was less dramatic than that for changes in system geometry or reaction kinetics.

The previous examples demonstrate the effect of gas-phase particle formation on the deposition rate profiles in a tubular flow reactor. The size and number of particles formed during CVD is also important, because particles that are formed can deposit onto the substrate, thereby degrading its properties. Knowledge of particle characteristics can suggest strategies for reducing particle formation and deposition. As an example, particle formation was studied during CVD of W using silane reduction of $WF_6$. The equations suggested were

$$WF_6 + \tfrac{3}{2}SiH_4 \longrightarrow W + \tfrac{3}{2}SiF_4 + 3H_2 > 250°C \qquad (14.8)$$

and

$$WF_6 + \tfrac{3}{4}SiH_4 \longrightarrow W + \tfrac{3}{4}SiF_4 + 3HF > 600°C. \qquad (14.9)$$

Figures 14.29, 14.30 and 14.31 show the effects of carrier gas ($H_2$ versus Ar), substrate temperature, $SiH_4$–$WF_6$ ratio, and total pressure on particle con-

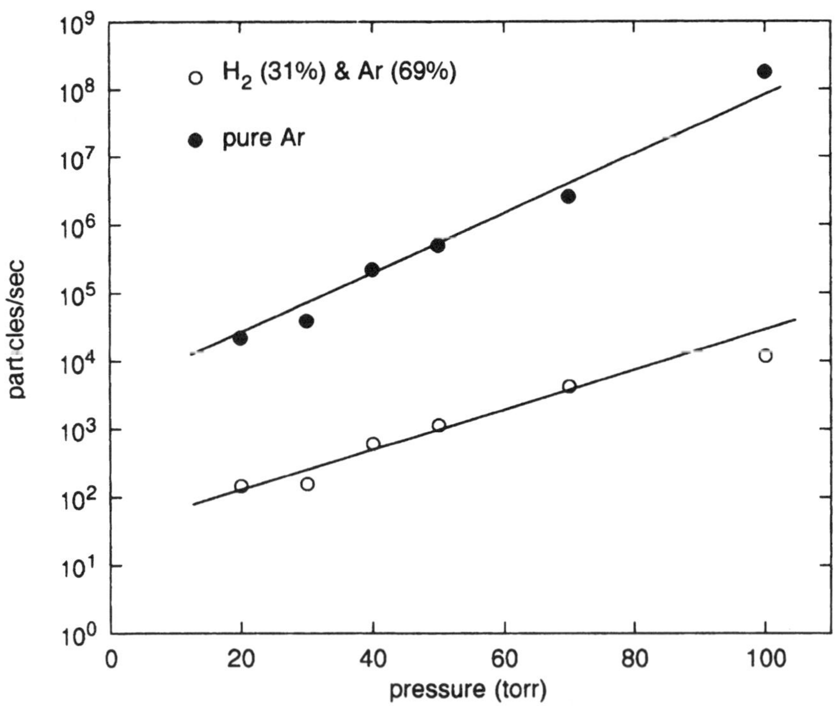


**Fig. 14.29.** Addition of hydrogen suppresses particle formation over a wide range of pressures. $T$ = 450°C, $SiH_4/WF_6$ = 1.2. Reprinted with permission from McInerney, Mountsier, Chin, and Broadbent.[22] Copyright 1993 American Vacuum Society.

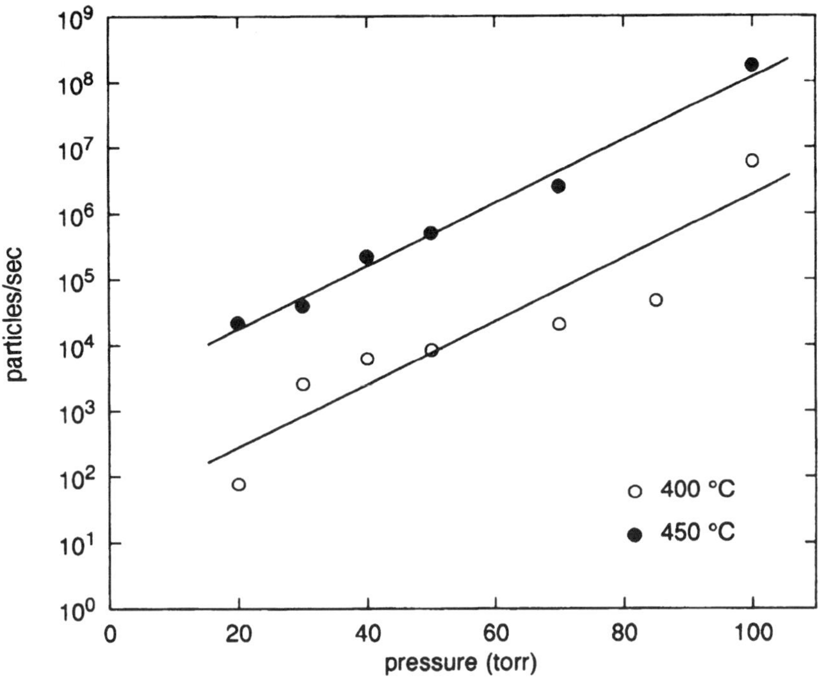


**Fig. 14.30.** Particle formation rate increases with increasing temperature and pressure. $SiH_4/WF_6 = 1.2$. Reprinted with permission from McInerney, Mountsier, Chin, and Broadbent.[22] Copyright 1993 American Vacuum Society.

centration.[22] The method used to count the particles detected only particles larger than roughly 0.1 $\mu$m. Because smaller particles may have been present, it is possible that higher concentrations of particles were obtained. Figure 14.29 shows that changing from argon to hydrogen carrier gas reduced particle formation. The same effect was found using He instead of $H_2$, suggesting that the influence of the carrier gas was to increase the diffusion coefficient of the gas-phase species. This in turn increased the rate of loss of the condensible species to the walls, thereby reducing its concentration in the gas phase. Increasing pressure increased particle concentrations because the higher pressure reduced the rate of diffusion of particle-forming species to the walls.

Figure 14.30 shows that increasing the temperature increased the particle formation rate. The increased temperature increased the rate of the gas-phase reaction, leading to the formation of the condensable product. Increasing pressure in general increased the rate of particle formation. This occurred because higher pressures gave lower diffusion coefficients for the gas-phase species, corresponding to reduced transport rates to the walls and higher gas-phase concentrations of the condensable species.

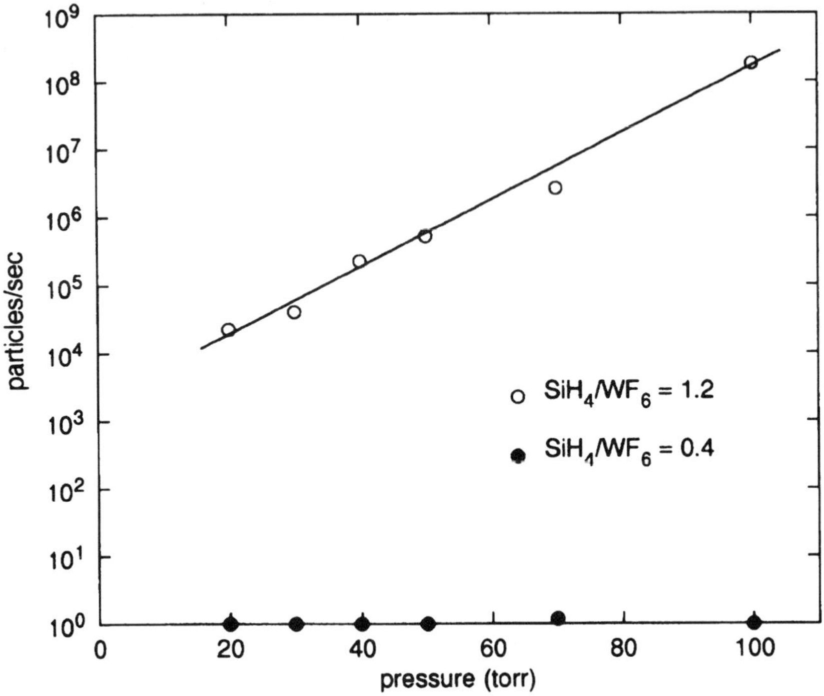


**Fig. 14.31.** Particle formation rate as a function of increasing pressure for two different $SiH_4$–$WF_6$ flow ratios. $T$ = 450°C. Reprinted with permission from McInerney, Mountsier, Chin, and Broadbent.[22] Copyright 1993 American Vacuum Society.)

Figure 14.31 shows that the gas-phase reaction chemistry played a major role in determining the number of particles formed. Decreasing the ratio of $SiH_4$ to $WF_6$ from 1.2 to 0.4 completely eliminated particle formation. This was likely due to a decrease in the rate of the gas-phase reaction forming those species which led to particle formation.

## 14.9 PARTICLE FORMATION DURING LOW-PRESSURE PLASMA DEPOSITION AND ETCHING

The formation of particles in low-pressure plasmas is an extremely complicated phenomenon. Although particles were observed in plasmas as early as 1924 by Langmuir and others,[23] only relatively recently have laser light scattering diagnostics demonstrated that many plasma reactors produce and trap large quantities of particles. This is a major problem for the microelectronics industry because the particles that are formed can deposit onto substrates, thereby destroying the devices being fabricated on the substrate.

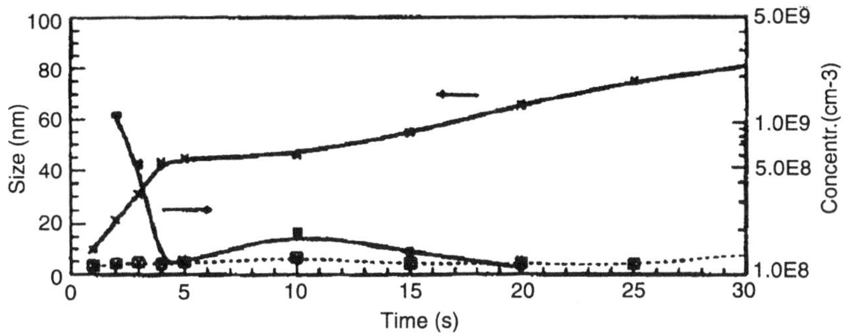


**Fig. 14.32.** Evolution of particle sizes (x), concentration, (■) and standard deviation (⊠) up to 30 s for particle formation in low-temperature plasma. (Courtesy of Boufendi & Bouchoule).[24]

Particle formation in these plasmas is an unsteady phenomenon. Figure 14.32 shows the change in the size and concentration of particles in a low-pressure plasma.[24] After a period of time, the volume in which the particles are observed has sharp boundaries, and the cloud of particles behaves collectively like a liquid droplet. This situation, along with the evolution of the cloud, is shown in Figure 14.33.[25] The spatial distribution of the particles that is shown is determined by the forces that act on them, as summarized by Barnes et al.[26] The forces include gravitational, neutral and ion drag, electrostatic, thermophoretic, and pressure gradient forces.

Although particles are formed in dc discharges, they are usually observed in larger quantities in radiofrequency plasmas. Numerous coupled physical and chemical phenomena occur just for particles in plasmas. (See also Chapter 8.) For particle growth, atomic and molecular quantities or mechanisms to be considered include radical and ion fluxes, bond energies between atoms in molecules, temperature, desorption rates, surface charges, and sputtering. A variety of aspects associated with charged species must also be taken into account: the floating potential (resulting from having to match the fluxes of charged species to and from a particle), electron and ion fluxes, electron affinity and work function, electrostriction, field and thermionic emission, and photoelectric charging. For calculating the position of the particles, the electrostatic–gravitational balance, collisional drag from ions and neutrals, ensemble polarizability, and particle mass have to be considered. The temperature of the particles, which is not necessarily the same as that of the gas or electrons, is influenced by surface radical recombination, surface electron–ion recombination, surface quenching of energetic species, thermionic emission, pyrolysis, radiation, and Knudsen or continuum cooling. This is a formidable list of considerations and illustrates the complexities involved.

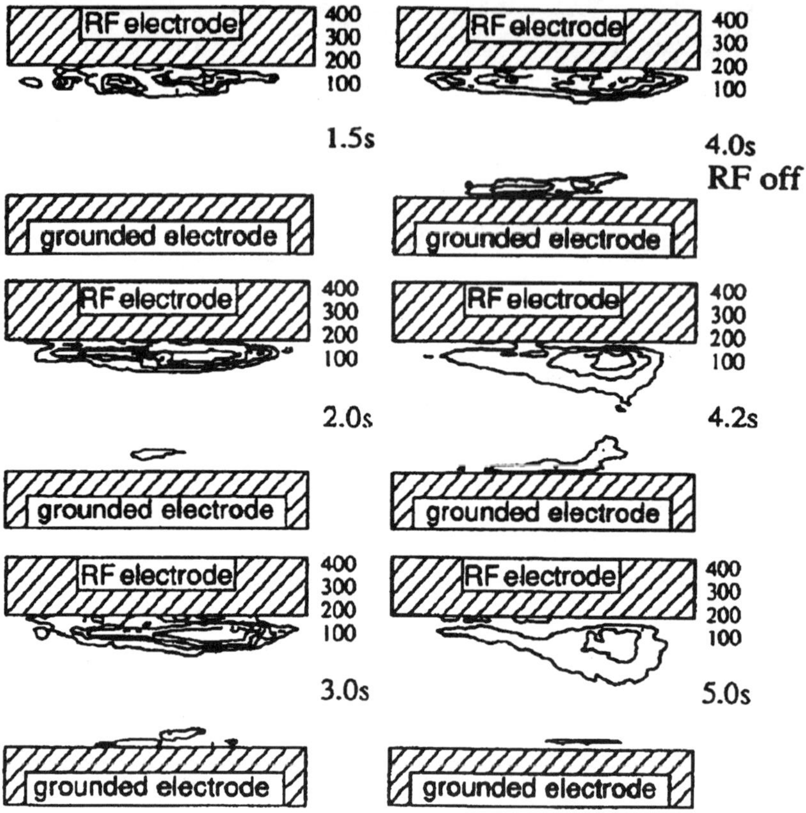


**Fig. 14.33.** Evolution of laser light scattering, shown as contours of intensity distribution. Particle formation is localized. (Courtesy of Watanabe & Shiratani.)[25]

During film formation, particles can originate from gas-phase chemistry or from sputtering of the electrodes. As an example, if a gas containing carbon atoms at a typical pressure in a plasma can dissociate, then the concentration of carbon atoms can be near $10^{15}$ number/cm$^3$. Because the equilibrium vapor pressure of carbon is nearly zero at room temperature, dissociation of gaseous species can produce a highly supersaturated vapor. In this case, the critical cluster size can be nearly the same as that of a monomer (atom), leading to rapid growth by collision and coalescence.

Low-pressure plasmas can also be used for etching. Here, particles can also form as the result of sputtering, an approach that has been used intentionally to form nanophase powders. Sputtering occurs when atoms are dislodged from a surface as a result of impinging energetic species. Sputtering can give gas-phase atom densities in the $10^7$ number/cm$^3$ range, which is sufficient to account for the observed particle growth kinetics.[27]

Ion-induced nucleation has been suggested to be the mechanism by which particles are formed. Several groups have examined the criteria for the size and charge at which an ion grows to be a stable particle.[28, 29] The small particles must not gain too many charges, or they can undergo Coulomb explosion—the disintegration of a particle because of the repulsive forces between like charges. Nucleation and growth rates in plasmas can be extremely high. Watanabe et al. reported $10^8$ number/cm$^3$ of 100-nm particles after 1 second.[30] In plasmas without gaseous reactants, where the particles can come only from materials in the electrodes, such particles are observed after a few minutes. Particles with sizes up to the micrometer range have been observed in a wide variety of low-pressure plasma etching and deposition systems.

The particles in plasmas derived from the walls of the system and from the gas-phase chemistry have different morphologies, which can be used to distinguish the particle formation mechanism. Large nonspherical particles have been observed by Selwyn et al., Stoffels et al., and others, suggesting a heterogeneous mechanism.[31, 32] Particles formed during the sputtering of carbon often have monodisperse size distributions and frequently display a morphology similar to cauliflowers. The particles are roughly spherical, but are not dense. The internal morphology often consists of columnar structures, which are consistent with a ballistic deposition process[33] for growth rather than a diffusion-limited process.[34]

A large number of other issues associated with this very complicated problem are discussed in a special issue of *Plasma Sources Science and Technology* (volume 3, 1994).

## 14.10 REFERENCES

1. Kodas, T.T. & Hampden-Smith, M.J. (Eds.). *The Chemistry of Metal CVD*. Weinheim, West Germany: VCH, Verlagsgesellschaft, 1994.
2. Friedlander, S.K. *Smoke, Dust & Haze*. New York: Wiley Interscience, 1977.
3. Rader, D.J. & Geller, A.S. *Plasma Sources Science and Technology 3*(3), 1994, 426.
4. Willeke, K. & Baron, P.A. (Eds.). *Handbook of Aerosol Measurement: Principles, Techniques and Applications*. New York: Van Nostrand Reinhold, 1993.
5. Kodas, T.T. *Angew. Chemie: Internat. Ed. Eng., 28*, 1989, 794–807.
6. Siefert, W. *Thin Solid Films 120*, 1984, 267–274.
7. Chung, G.Y., Kim, H.D., Ahn, B.T., & Im, H.B. *Thin Solid Films 232*, 1993, 28–33.
8. Hitchman, M.L. & Jensen, K.F. *Chemical Vapor Deposition: Principles and Applications*. New York: Academic, 1993.
9. Roger, C., Corbitt, T., Kodas, T.T., & Hampden-Smith, M.J. *Appl. Phys. Lett. 65*(8), 1994, 1021–1023.
10. Roger, C., Corbitt, T., Xu, C., Zeng, D., Powell, Q., Chandler, C.D., Nyman, M., Hampden-Smith, M.J., & Kodas, T.T. *J. NanoSt. Mater. 4*(5), 1994, 529–535.
11. Xu, C., Hampden-Smith, M.J., & Kodas, T.T. *Adv. Mater. 6*(10), 1994, 746.

12. Park, H.M. & Rosner, D.E. *Chemical Engineering Science 44*(3), 1989, 603–617.
13. Stratmann, F., Otto, E., & Fissan, H. *J. Aerosol Sci. 25*(7), 1994, 1305–1319.
14. Talbot, L., Cheng, R.K., Schefer, R.W., & Willis, D.R. *J. Fluid Mech. 101*(4), 1980, 737–758.
15. Hurt, R.H. & Allendorf, M.D. *AIChE J. 37*(10), 1991, 1485–1496.
16. Danek, M., Jensen, K.F., Murray, C.B., & Bawendi, M.G. *J. Crystal Growth 145*, 1994, 714–720.
17. Danek, M., Jensen, K.F., Murray, C.B., & Bawendi, M.G. *Appl. Phys. Lett. 65*(22), 1994, 2795–2797.
18. Okuyama, K., Huang, D.D., Seinfeld, J.H., Tani, N., & Matsui, I. *Jpn. J. Appl. Phys. 31*(1), 1992, 1–11.
19. Adachi, M., Okuyama, K., Tohge, N., Shimada, M., Satoh, J.-I., & Muroyama, M. *Jpn. J. Appl. Phys. pt. 2 31*, 1992, L1439–L1442.
20. Ohshima, K., Torimoto, Y., Tsuto, K., Okuyama, K., Ushio, R., & Kousaka, Y. *Kona 9*, 1991, 59–71.
21. Hong, L.-S. & Komiyama, H. *J. Am. Ceram. Soc. 74*(7), 1991, 1597–1604.
22. McInerney, E.J., Mountsier, T.W., Chin, B.L., & Broadbent, E.K. *J. Vac. Sci. Technol. B 11*, 1993, 734.
23. Langmuir, I., Found, G., & Dittmer, A.F. *Science 60*, 1924, 392.
24. Boufendi, L. & Bouchoule, A. *Plasma Sources Science and Technology 3*(3), 1994, 262–267.
25. Watanabe, Y. & Shiratani, M. *Plasma Sources Science and Technology 3*(3), 1994, 286–291.
26. Barnes, M.S., Keller, J.H., Forster, J.C., O'Neill, J.A., & Coultas, D.K. *Phys. Rev. Lett. 68*, 1992, 313–316.
27. Ganguly, B.N., Garscadden, A., Williams, J., & Haaland, P.D. *J. Vac. Sci. Technol. A 11*, 1993, 1119.
28. Gibson, E.G. *Phys. Fluids 9*, 1966, 2389.
29. Breslin, A.C. & Emeleus, K.G. *Int. J. Electron 31*, 1971, 189.
30. Wantanabe, Y., Shiritani, M., & Yamashita, M. *Appl. Phys. Lett. 61*, 1992, 1510.
31. Selwyn, G.S., Singh, J., & Bennett, R.S. *J. Vac. Sci. Technol. A 7*(4), 1989, 2758–2765.
32. Stoffels, W.W., Stoffels, E.C., Kroesen, G.M.W., Haverlag, M., denBoer, J.H.W.G., & deHoog, F.J. *Plasma Sources Sci. Technol. 3*(3), 1994, 320–324.
33. Ramanlal, P. & Sander, L.M. *Phys. Rev. Lett. 54*, 1985, 1828.
34. Weitz, D.A. & Oliveria, M. *Phys. Rev. Lett. 52*, 1984, 1433–1436.

# 15 Film Generation Technologies

## 15.1 INTRODUCTION

A wide variety of aerosol or spray technologies have been developed for fabrication and modification of films. The most extensively studied processes are spray pyrolysis (droplet deposition), AACVD (aerosol delivery of volatile precursors for CVD), optical fiber production (thermophoretic deposition), and spraying of plasma and molten material. These processes, which have been used to deposit materials such as metal oxides, nonoxide ceramics, metals, organics, and composites at high rates at various temperatures, can be divided into vapor, solution droplet, molten droplet, and solid-particle deposition processes. (See Table 15.1 at the end of the chapter.) In some cases a clear distinction is not possible, as when both solid particles and vapors deposit (particle-precipitation-aided CVD) or when droplets and solid particles (a variation of spray pyrolysis) deposit simultaneously. Another way of categorizing the various processes is in terms of the precursors to the films, which may be liquid-phase precursors or solid particles formed in a variety of ways; this approach is used here.

In this chapter, we examine the technologies of film formation. We provide a basic description of each; an overview of the reactants, products, and chemistry specific to the technology; examples of the implementation of the approach; and a discussion of any theory that pertains to the technology but was not covered in Chapter 14. Section 15.2 covers liquid-phase precursors, including spray pyrolysis, AACVD, supercritical fluid spraying, molten droplet deposition, and aerosol jet etching and cleaning. Section 15.3 discusses solid-particle deposition: ionized cluster beams, neutral cluster beams, inertial deposition, thermophoretic deposition (with optical fibers as a separate section), electrophoretic deposition, simultaneous particle deposition and CVD, and particle formation during conventional CVD, plasma-assisted CVD, and plasma etching.

## 15.2 LIQUID-PHASE PRECURSORS

### 15.2.1 Droplet Deposition (Film Pyrolysis)

***15.2.1.1 Introduction*** A variety of methods are available for applying precursors in solutions onto surfaces with the goal of forming a film. Perhaps the simplest and most well-known method is dip or spin coating. In this process, a substrate is coated with a thin layer of a solution containing one or more precursors. The surface is allowed to dry, and then the substrate is heated, which results in thermal decomposition of the precursor(s) to form a film. In order to control the thickness of the film, this coating process is repeated several times. However, not all solutions form uniform films after the solvent evaporates, which leads to a nonuniform coating. Nonuniform film formation is often caused by the use of precursors that crystallize from solution to form crystalline nonuniform layers of the dry precursor. The problem can be avoided by using precursors that tend to form amorphous or very fine-grained films, but this approach is not always feasible. The ceramic or metallic yield —the volume fraction of the product relative to the precursor—must also be high so that porous films will not be formed.

Spray pyrolysis has been used extensively for applying precursor solutions to a surface. Several review articles discuss the technique.[1–9] An aerosol of a solution is formed, and the droplets are deposited onto a surface, where solvent evaporation and chemical reactions take place, resulting in a film. The primary advantage over dip and spin coating is that a uniform film can be obtained with a greater variety of precursors because drying takes place during deposition. Even precursors that tend to form nonuniform films during solvent evaporation using dip or spin coating can be applied. An example is metal nitrates, which tend to crystallize while the solution is drying, to form discontinuous films. Still, these metallic compounds can be applied uniformly to surfaces by spray routes. Film generation by droplet deposition requires nonvolatile, soluble precursors. Typical precursors are low-cost metal nitrates, acetates, and chlorides, an advantage over higher cost precursors such as those used for sol-gel approaches. The solution is then atomized to form droplets. In the simplest case, the droplets are deposited directly onto a substrate heated to 300 to 500°C, where the solvent evaporates and the reactants decompose to form the product or to form precursors to the product, which are in turn converted to the final product by further thermal processing.

Film properties depend strongly on the characteristics of the aerosol droplets and the deposition system. The primary problems encountered in spray pyrolysis are due to nonuniform droplet sizes. Large (1000-$\mu$m) droplets and precursors with low solubilities ($< 1$ g/liter) can result in the formation of rough films because of the large amount of solvent that must evaporate from the surface. Using very small droplets (1 $\mu$m) may result in solvent evaporation and the formation of solid particles before droplet impaction, which can also result in rough films. Smaller droplets can follow the gas stream and avoid striking the surface, leading to reduced deposition rates and poor conversion of precursor to film. Thus, an optimum droplet size exists for the deposition of solution droplets. However, these problems have not been studied extensively.

The quality of the films also depends strongly on the substrate temperature, as well as the geometry of the reactor. The dependence on substrate temperature is complex: Low temperatures lead to incomplete solvent evaporation from the surface, while very high temperatures can result in rough films. The latter problem occurs when solid particles are formed in the gas phase and then deposit onto the substrate. The deposition system geometry influences the droplet deposition efficiency, which determines how uniform the film is. Deposition of only very large droplets can lead to porous films.

Some of the advantages of film formation by spray pyrolysis (droplet deposition) include:

- Simplicity and low cost (only a substrate heater and an atomizer are needed; operation is generally at atmospheric pressure)
- Many choices for liquid-phase precursors (nitrates, chlorides, acetates)
- A wide variety of multicomponent materials can be formed
- Produces oriented films (ZnO, $SnO_2$, $YBa_2Cu_3O_7$, Pb–Bi–Sr–Ca–Cu–O), $CuInSe_2$, and more complex materials can be formed
- High deposition rates (0.1–1 $\mu$m/min)
- Demonstrated scale-up

The primary disadvantages are:

- Formation of porous films
- Low purity in some cases
- Limited products: oxides, sulfides, selenides, and other oxidation-resistant materials (nonoxide ceramics and most metals not produced)

***15.2.1.2 Precursors and Products*** A variety of materials have been produced by droplet deposition. Tables 15.2 through 15.6 at the end of the chapter show some of the materials, the precursors, and the conditions under which they have been deposited. A particularly large amount of work has been done with the CdS (Table 15.3), metal selenide (Table 15.4), ZnO (Table 15.5), and $YBa_2Cu_3O_{7-y}$ and Pb–Bi–Sr–Ca–Cu–O (Table 15.6) systems. Indium oxide, cadmium stannate, and other films have also been used as transparent electrodes and as oxide semiconductors in solar cells. Other applications of the technique are for photochemical cell electrodes, gas-sensing elements, antireflective coatings, and thermal coatings. Solar cell applications have driven the use of this approach because of the need for low-cost methods of forming film over large areas. The types of precursors that have been used most extensively are metal chlorides and nitrates. Metal acetates and alkoxides have been used to a lesser extent.

Although much work has been accomplished using spray pyrolysis with various metal halide precursors to produce metal oxides, the high volatility of some forms of these species suggests that the process is AACVD, rather than droplet deposition. For example, the temperatures required for a vapor pressure of 10 torr for various metal chlorides are as follows: $SnCl_4$, 10°C; $SbCl_3$, 85°C; $SbCl_5$, 62°C; $BiCl_3$, 264°C; $GaCl_3$, 76°C, $TiCl_4$ 21°C. These temperatures are far lower than those used for typical droplet deposition conditions, suggesting that the various species can easily vaporize and carry out CVD. In contrast, other metal halides have low vapor pressures and require higher temperatures. For example, the temperatures required for a vapor pressure of 10 torr are quite high for many metal chlorides: $ZnCl_2$, 508°C; $CdCl_2$, 656°C; CuCl, 702°C; and AgCl, 1074°C. A number of values of vapor pressures were given in Chapter 8; the values presented here illustrate that it is important to choose a precursor with the appropriate volatility for the film formation process.

A further consideration in choosing precursors arises from a problem that occurs in multicomponent systems: It is common for the film to have a different composition from that of the starting solution. There are many possible causes of this problem, including the use of one volatile precursor along with one or more involatile precursors, the formation of a volatile intermediate on the surface followed by desorption, and the splattering of droplets from which one or more components have already precipitated onto the surface.

*15.2.1.3 Experimental Details* Droplet deposition systems typically rely on the formation and deposition of droplets in the 1 to 1000 $\mu$m range. Deposition rates as high as 1 $\mu$m/min have been reported, and film thicknesses of 1 to 10 $\mu$m are common. The largest number of applications have been for films with complex compositions, such as ceramic superconductors, as well as for solar cells, for which low cost and the ability to deposit onto large areas are important.

Figure 15.1 shows a typical experimental system.[10] The most commonly used droplet formation apparatus is a two-fluid atomizer, through which the solution and a carrier gas are passed under pressure to form the droplets. These atomizers typically produce droplets from 10 to 1000 $\mu$m in diameter. Ultrasonic atomizers with submerged transducers have also been used, but they produce droplets on the order of 1 to 5 $\mu$m, which cannot be deposited as easily as larger droplets can. Substrate heating is typically carried out by placing the substrate on a 200 to 500°C surface, often with a medium to promote heat transfer between the substrate and the heater. A common problem is that the temperature of the substrate decreases over time due to evaporative cooling. This is alleviated by cycling the aerosol flow on and off in order to minimize the temperature range over which deposition takes place. The substrate is moved relative to the spray, to ensure that a uniform coating is obtained. This simple process is usually carried out at atmospheric pressure, often without any enclosure, and the films have to be annealed in most cases to optimize properties.

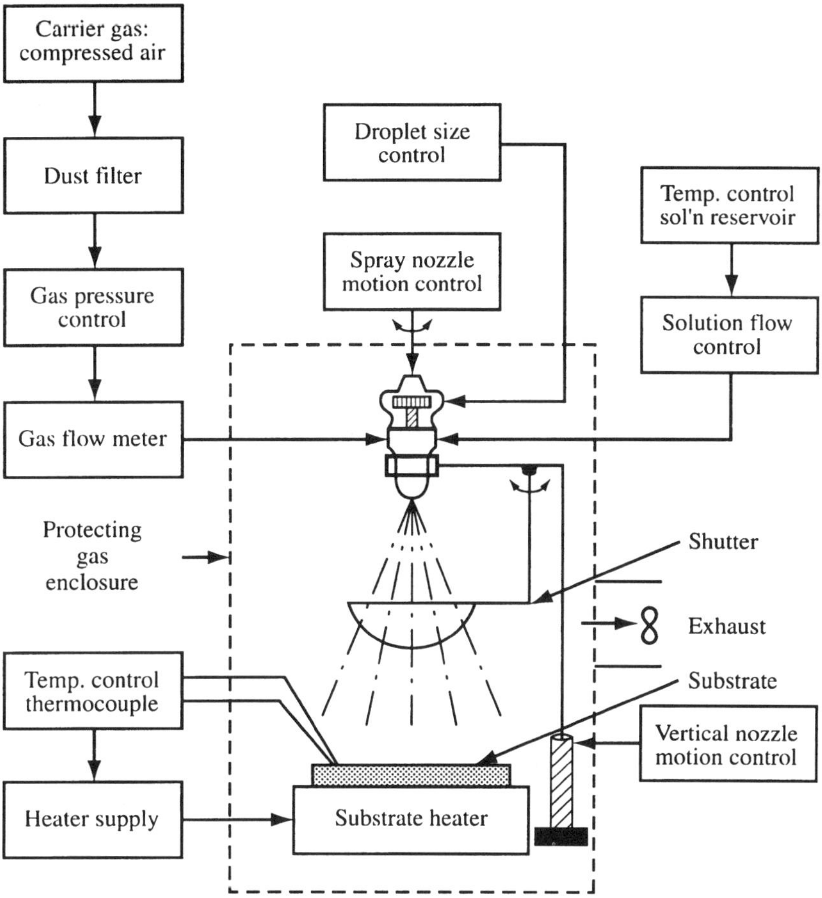


**Fig. 15.1.** Typical droplet deposition (spray pyrolysis) system used at laboratory scale. (Reprinted from Omar et al.[10] *Journal of Material Science: Materials in Electronics*, *1*, (2), 1990, Chapman & Hall.)

***15.2.1.4 Theory*** The basic theory of droplet deposition was covered in Chapter 14. Although some theoretical studies have examined a number of aspects of the approach,[11] no comprehensive models have been devised to describe the operation of this specific type of film generation system.

### 15.2.2 Aerosol-Assisted Chemical Vapor Deposition

***15.2.2.1 Introduction*** Deposition of high-quality thin films of electronic materials is required for a number of applications. Spray pyrolysis (droplet deposition) cannot in many cases produce dense, oriented films because of problems such as solvent evaporation from the surface. Physical vapor deposition (PVD) techniques such as sputtering and evaporation require high vacuum conditions and cannot coat substrates with complex topographies because they are ballistic processes in which atoms travel in straight lines on their way to the surface.[12] These methods often are not feasible or are too costly for industrial applications, which require large areas to be coated. For these reasons, and because they allow conformal deposition over large areas at high rates with relatively inexpensive equipment, CVD processes are often employed.

A number of problems arise with CVD in the case of multicomponent films produced with low-vapor-pressure precursors. One problem is the difficulty in introducing several precursors into the reactor at controlled rates that are constant with time. The precursors are metal-containing solids with relatively low vapor pressures that must be heated to high temperatures, often several hundred degrees Celsius to obtain useful vapor pressures. High temperatures can cause precursors in the solid state to react and form volatile products. An example is the use of $Ba_4(tmhd)_8$ (where tmhd = tetramethylheptanedionate) for the generation of $YBa_2Cu_3O_{7-x}$ films, which can decompose during sublimation. (See Chapter 8.) These volatile by-products add to the vapor pressure of the precursor, leading to reduced precursor transport rates when the total gas flow rate is held constant. A second problem is that the surface areas of solid precursors change with time, often resulting in reduced evaporation rates as the surface area of the solid precursor decreases with time. A third problem is that the low vapor pressures of the precursors lead to difficulties in controlling mass flow rates using conventional mass flow controllers; complex systems are usually required. All of these problems result in precursor evaporation rates that vary with time in different manners for the different precursors and also lead to changes in the stoichiometry of the films.

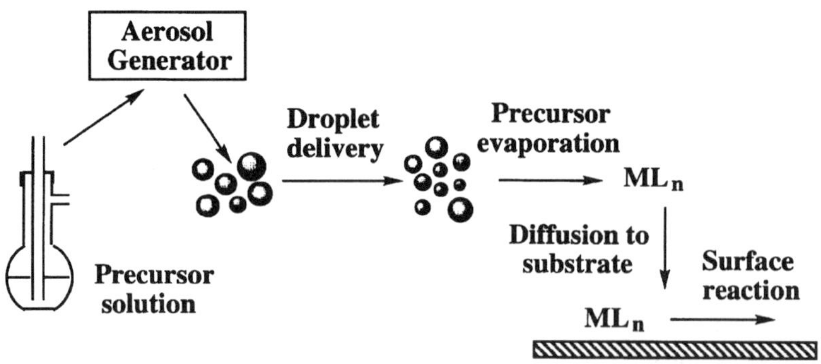


**Fig. 15.2.** Schematic representation of aerosol-assisted chemical vapor deposition process.

An alternative method of depositing thin films is AACVD, the key steps of which are shown in Figure 15.2.[12–14] In this process, a solution containing volatile precursors is converted into an aerosol. Before or near the substrate, the precursors evaporate into the gas phase. This is followed by thermal decomposition of the gaseous molecular precursors on the heated surface to deposit a thin film. In another example, the solvent and precursor(s) can be vaporized in a hot-wall system before reaching the substrate. Still another variation is to dry the precursor solution and use a powder feeder to introduce dry particles into the reactor.[12]

The AACVD approach has several advantages when thermally sensitive precursors are used to generate multicomponent films. Because the precursors are held outside the reactor at room temperature or even at a reduced temperature, they do not thermally degrade. Also, the precursor delivery rates are constant with time because the ratio and concentration of precursors in the solution and the droplet size and delivery rate do not change with time. This allows the reproducible deposition of multicomponent films with constant composition as a function of deposition time. Conformal films are also possible because deposition occurs by CVD.

AACVD has several advantages:

- Deposition of multicomponent films
- Controlled stoichiometry
- Simplicity
- Wide variety of materials
- Oriented or epitaxial films
- High deposition rates
- Many choices for precursors
- Extended range of thermal stability for volatile precursors
- Possibility of doping
- Conformal coverage
- Reproducible, constant delivery rates for multicomponent systems

The approach, however, also has several disadvantages:

- It requires total pressures near atmospheric pressure or high enough that the droplets (particles) do not settle in the reactor before evaporating; low-pressure AACVD, however, has been demonstrated.
- A solvent is present in most cases and may be incompatible with the precursors or the chemistry of film formation. This problem is relatively easily overcome by a suitable choice of solvent.
- Precursors can react in solution to form involatile intermediates.
- Residue particles always remain in the gas phase after solvent evaporation because of the limited purity of available solvents and reactants.

*15.2.2.2* ***Precursors and Products*** A number of materials have been fabricated into thin films using AACVD. Tables 15.7 and 15.8 at the end of the chapter show representative precursors, products, and conditions from the literature. The materials that have been studied most extensively are $SnO_2$ and $YBa_2Cu_3O_7$. The use of organometallic precursors has allowed substrate temperatures to be as low as 300 to 500°C, although higher deposition temperatures, approaching 1000°C, have also been used. The most commonly used precursors have been metal $\beta$-diketonates such as metal acetylacetonates (also known as pentanedionates) and tetramethylheptanedionates (also known as dipivaloylmethanoates), while a typical solvent is toluene. A note of caution is that the use of flammable solvents with oxygen-containing atmospheres requires careful consideration of the flammability limits for the mixture.[15]

*15.2.2.3* ***Experimental Details*** AACVD has been carried out in vertical and horizontal flow geometries in both hot- and warm-wall reactor configurations. Many studies have relied on preheating sections for precursor evaporation. Deposition rates as high as 5 $\mu$m/min have been obtained during the production of high-quality films. The droplet generator used in most cases is an ultrasonic atomizer. This choice has been motivated by the need to form droplets with diameters near 1 $\mu$m for which solvent and precursor evaporation can occur within the short residence times ($<1$ sec) found in typical systems.

Figure 15.3 shows a laboratory-scale AACVD system in which the gas flow is directed at the substrate, with the flow in the vertical downward direction.[15] This approach is often used because of its simplicity and the relatively well-defined mass transfer characteristics associated with stagnation point flow (the flow of a gas stream at a surface). The vertically upward flow is the more desirable configuration, because it minimizes free convection, but it has has been studied in much detail.

Figure 15.4 shows a horizontal-flow hot-wall system with ultrasonic generation of the aerosol.[16] Like the vertical-flow system, this system is also used because of its simplicity, but the mass transfer characteristics of its geometry are more complicated than those of the vertical system. In general, it is more difficult to obtain uniform deposition on substrates using a horizontal configuration than a vertical one.[17]

As an example of AACVD, a system similar to that shown in Figure 15.3 has been used for generation of copper films.[13, 14, 18] Crystalline copper was deposited onto $SiO_2$ substrates at low temperatures (120 to 200°C), resulting in higher deposition rates compared with traditional CVD that took place for similar conditions. The precursor was (hfac)Cu(1,5-COD) (hfac = hexafluoroacetylacetonate, 1,5-COD = 1,5-cyclooctadiene) in a toluene solution, which was atomized through a TSI-3076 aerosol generator using nitrogen and preheated (40 to 100°C) to partially vaporize the precursor. This atomizer provided submicron-sized droplets, which ensured rapid solvent and

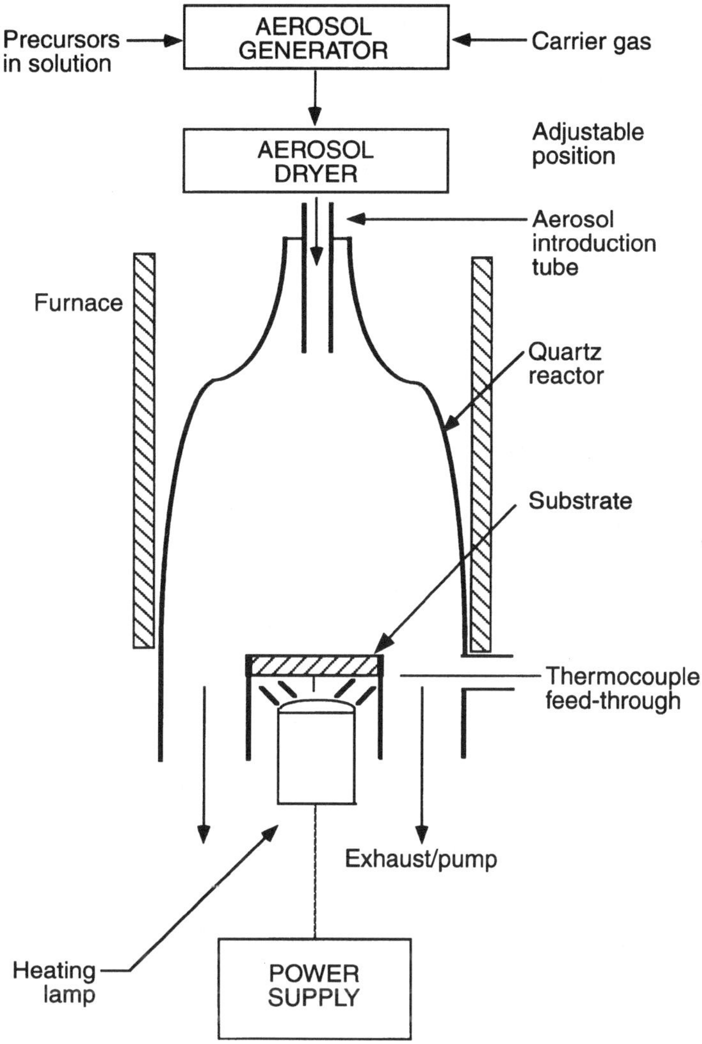


**Fig. 15.3.** Vertical-flow cold-wall system for aerosol-assisted chemical vapor deposition.

precursor evaporation. The precursor and solvent vapors and the partially evaporated precursor particles exiting the preheater were directed towards a substrate, where more precursor vaporized as the gas was heated in the thermal boundary layer above the substrate, and then reacted on the substrate at temperatures of 120 to 200°C. Copper was deposited at atmospheric pressure at rates of 10 to 100 nm/min. The activation energy was 6.7 $kcal mol^{-1}$, which corresponds to the enthalpy of vaporization of (hfac)Cu(1,5-COD) (6.4 $kcal mol^{-1}$).[13, 14, 18] This correspondence suggested that as the

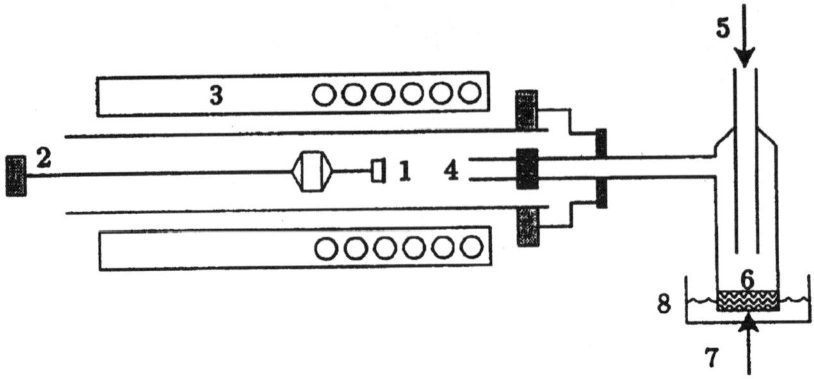


1 Substrate
2 Substrate rotation motor
3 Furnace
4 Spray nozzle
5 Carrier gas
6 Solution
7 Membrane
8 Ultrasonic Humidifier

**Fig. 15.4.** Horizontal-flow warm-wall system for aerosol-assisted chemical vapor deposition. (Reprinted from Cui et al., "Photocatalytic properties of titanium (IV) oxide thin films prepared by spin coating and spray pyrolysis," *Materials Research Bulletin*, *28*, pp. 195–201. Copyright 1993, with permission from Elsevier Science Ltd., The Boulevard, Langford Lane, Kidlington OX5 1GB, UK.)

temperature was varied, more precursor evaporated, resulting in a higher partial pressure and higher deposition rate. At preheating temperatures above 80°C the deposition rate decreased dramatically due to gas-phase reactions of the precursor. These results are typical for the operation of this type of system.

***15.2.2.4 Theory*** The theory describing AACVD is not well developed. A full model would require a description of the gas velocity profiles; the gas temperature profiles; aerosol particle transport by convection, diffusion, gravitational settling, thermophoresis, inertia, and electrophoresis (for charged particles in electric fields); and the aerosol dynamics, including particle evaporation, vapor concentration profiles, gas-phase reactions, and surface reactions. The problem can be simplified by first calculating the gas velocity and temperature profiles separately. This still leaves, however, a large number of coupled physical and chemical phenomena. Although no full quantitative models have been developed for AACVD, the qualitative features are well understood. (See Chapter 14.)

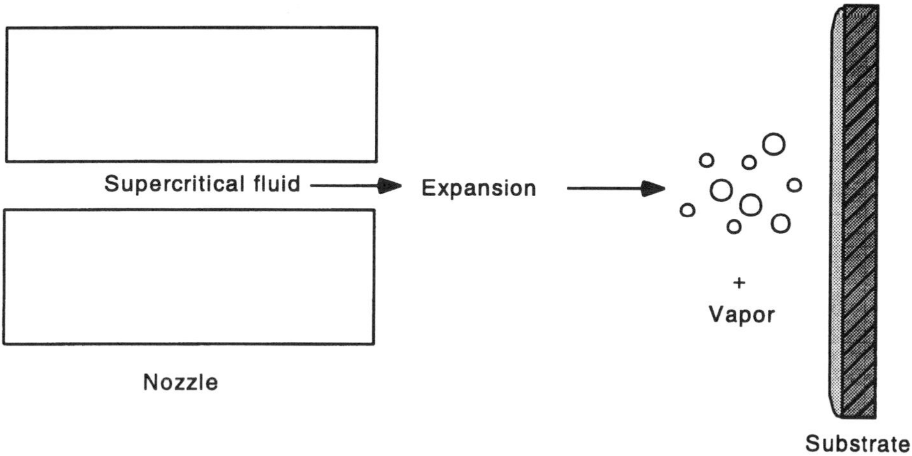


**Fig. 15.5.** Film formation by supercritical expansion of solutions. Both vapor and particles can be deposited.

### 15.2.3 Supercritical Expansion of Solutions

***15.2.3.1 Introduction*** Supercritical expansion of solutions relies on the expansion of a solution of precursors from the supercritical state to the gaseous state, sometimes with accompanying formation and deposition of particles, droplets, and/or vapor onto a surface. (See Fig. 15.5.) Dissolving volatile precursors into the solution can be used to introduce such precursors into the gas phase for CVD, as is done in AACVD. Thus, droplet deposition, particle deposition, physical vapor deposition, and CVD are all possible with supercritical expansion processes.

Supercritical expansion of solutions has several attractive features:

- Species that are not soluble in common solvents may be used.
- Thermally sensitive precursors and/or materials may be used.
- Unique phases may be formed.
- The process takes place at low temperatures.
- The solvent can be gaseous at ambient conditions.

The primary disadvantages are that the process is complex, few studies of it have been carried out, it is not thoroughly understood, a limited number of combinations of solute and solvents are available, and solute often gets incorporated into the film.

***15.2.3.2 Precursors and Products*** Table 15.9 at the end of the chapter shows some of the few types of films that have been produced by this

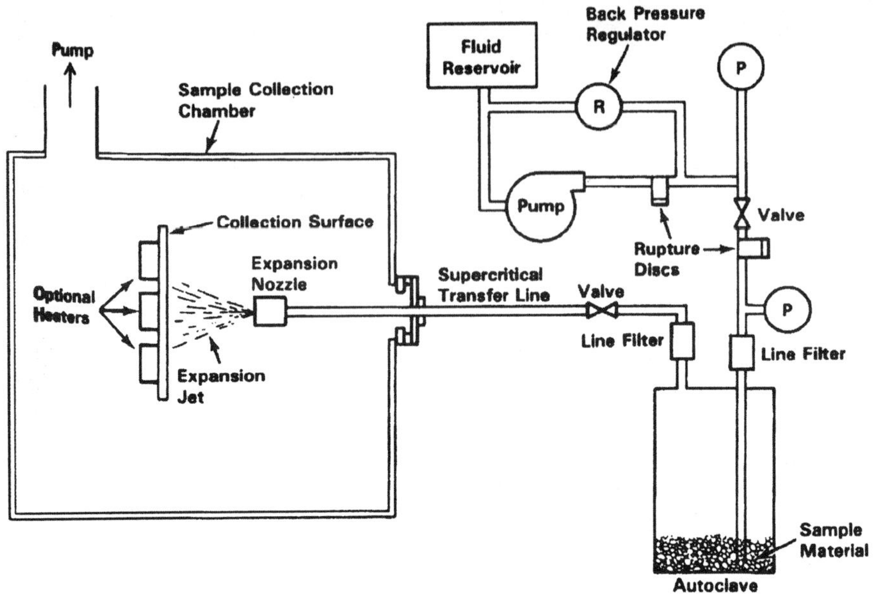


**Fig. 15.6.** Schematic diagram of experimental system for film formation by supercritical expansion of solutions. (Reprinted from Matson, D.W., Petersen, R.C., and Smith, R.D. *Journal of Material Science*, *22*, 1987, 1919–1928. Chapman & Hall.)

approach. Most studies have examined silicon-dioxide films, which can be formed by dissolving $SiO_2$ into supercritical $H_2O$ and spraying the fluid at a substrate. Deposition rates of 1 $\mu$m/min have been obtained. Numerous organic materials have also been produced in the form of particulates, but not films.

*15.2.3.3* ***Experimental Details*** A schematic representation of a typical supercritical fluid expansion system is shown in Figure 15.6.[19] Solvent is introduced through the use of a pump capable of reaching pressures many times that of the critical pressure of the fluid. A high-pressure chamber containing the solute is kept at the required pressure, and temperature is usually the parameter used to control solubility. The fluid may be kept in the chamber under supercritical conditions, or supercritical conditions may be reached in the nozzle before the fluid expands into the lower pressure environment.

The nozzle is a key component because it controls the fluid expansion characteristics, which influence the properties of the product. Reaching supercritical conditions is often difficult for many reasons, forcing the use of short lengths (<10 mm) of tubing with small diameters (<100 $\mu$m, inside diameter). Because of the corrosive nature of many of the fluids, stainless steel or more exotic tubing materials must be used.

An example that has been studied by Matson et al. is the $SiO_2$–$H_2O$ solute–solvent system.[19] Near the critical point, the density changes rapidly with small pressure and temperature variations, corresponding to large changes in the solubility of silica near the critical point of water (374°C and 991 bar). To avoid complications resulting from the complex solubility behavior near the critical point, most studies are conducted at pressures several times those of the critical pressure. Typical conditions for this system were 200 to 1000 ppm of $SiO_2$ in $H_2O$, a fluid temperature of 500°C, a fluid pressure of 600 bar and a background pressure of 10 torr. Various forms of $SiO_2$ were deposited, depending on the conditions. In general, the films did not have unique properties that could justify using the approach commercially.

***15.2.3.4 Theory*** Only a brief overview of the theory is given here, because of the complexity of the process and the limited number of existing applications. A supercritical fluid is a substance above its critical temperature such that the fluid remains as a single phase regardless of the applied pressure. The density of a supercritical fluid may approach that of liquids.[20–22] The key property of supercritical fluids that is exploited for materials production is their enhanced solvating capabilities compared with those of liquids and gases under normal conditions. This capability allows many materials, such as $SiO_2$, to be dissolved for which ordinary solvents are unavailable. Passing a supercritical fluid solution through a nozzle to reduce the pressure to subcritical values results in the precipitation of the solutes. A further advantage is that the supercritical solvent may be a gas at ambient conditions, allowing the process to be carried out without the use of conventional solvents that are liquids at ambient conditions.

Few detailed models of supercritical expansion and nucleation processes for the generation of particles or films exist. (See Chapter 14.) However, a qualitative description of the process involves nucleation and condensation of the solutes, which are initially placed in the solution. Eliminating the nucleation and condensation of the solute and solvent results in a molecular stream that can be used for film generation. When nucleation of the solute species occurs to form particles, film formation can occur by particle deposition. The particle size can be controlled by varying the operating parameters. In extreme cases, large particles are formed and the method can be used for powder production.

### 15.2.4 Molten Droplet Deposition

***15.2.4.1 Introduction*** Plasma spraying, flame spraying, and molten-metal atomization and spraying are specialized forms of droplet deposition in which particles of a solid precursor are melted and deposited, after which they solidify and form a film (Fig. 15.7).[23] Plasma and flame reactor particle formation have been discussed in Chapters 10 and 11.

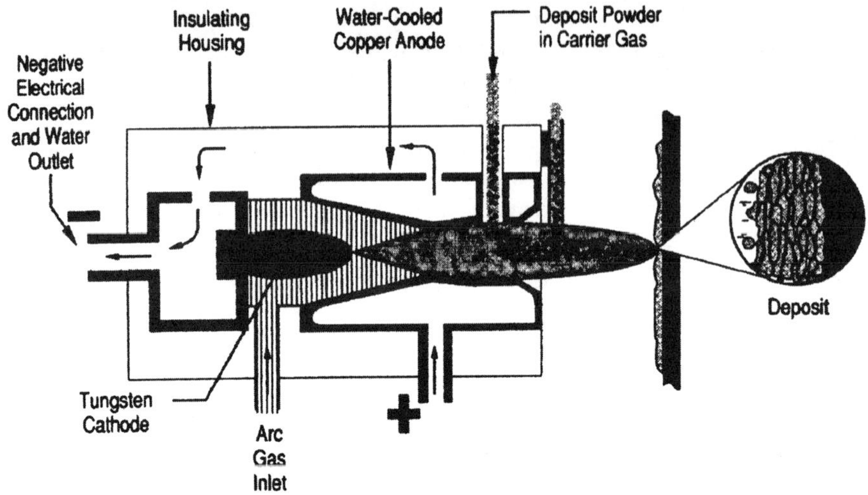


**Fig. 15.7.** Schematic of plasma spray gun. (Adapted from Sampath and Herman.[23] Reproduced with permission from the Minerals, Metals and Materials Society.)

The rapid impact deposition of plasma-sprayed or flame-sprayed molten particles can yield metastable structures of a wide range of alloys and ceramics. Plasma and flame spray methods are variations on molten droplet deposition and combine the steps of melting, rapid solidification, and consolidation into a single process. Plasma and flame spraying allow the production of bulk, near-net-shaped forms with the benefits of rapid solidification (homogeneity, small grain size, formation of metastable phases). Plasma and flame spraying have the advantage of medium to high throughputs (5 to 50 kg/h) and the ability to coat or form complex objects. Conventional plasma and flame spraying in air cannot handle many metals and intermetallics because of defects including unmelted particles, oxide phases, and porosity.

Vacuum plasma spraying and low-pressure plasma spraying allow the formation of metal coatings consisting of dense oxide-free and pore-free deposits with nearly theoretical density and low levels of residual stresses, because the high temperature of the process leads to annealing. The high velocities that are attained in low-pressure approaches can cause the particles to flatten, leading to increases in efficiency and in the density of deposits. For these reasons, low-pressure plasma spraying is an effective means for consolidating reactive metals and for the production of free-standing forms.

A variety of nonplasma spray methods allow spray formation of coatings and objects. One such approach is the Osprey process, which can form near-net-shaped disks, billets, and tubes from a variety of metallic alloys and composites. In this method, molten metals or other materials are atomized and sprayed onto a mandrel. The throughput of these atomizers is hundreds

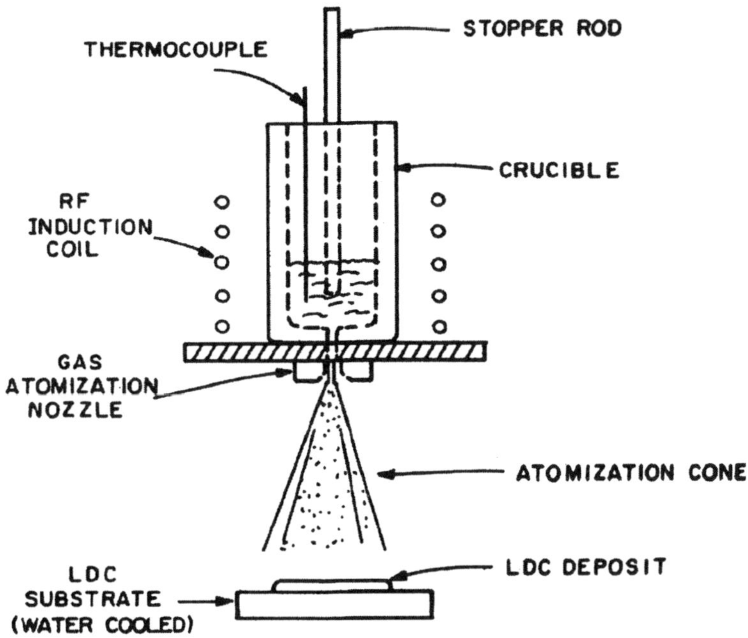


**Fig. 15.8.** Apparatus for generation and deposition of molten metal droplets onto surfaces. (Reprinted with permission from Chin, T.S., Hara, Y., Lavernia, E.J., O'Handley, R.C., and Grant, N.J. *J. Appl. Phys.*, *59*(4), 1297–1300. Copyright 1986 American Institute of Physics.)

of kilograms per hour, but the particle velocity is not high, leading to coatings with low densities. Plasma and flame spraying provide higher velocities because the flame allows high-density deposits with useful metastable phases to be formed. In addition, it is possible to spray a wide range of materials, such as refractory metals, ceramics, and their composites.

Gas-atomized molten metal deposition is used to manufacture tubes, disks, and sheet preforms. A typical system is shown in Figure 15.8.[24] Two types of microstructures are commonly found in the coatings. The first is consolidated splats, which consist of columnar grains growing perpendicular to the substrate with boundaries between the splats and a large amount of porosity in the deposit. Such spray deposits require further consolidation to remove the pores before the product can be used. The second microstructure consists of a continuous equiaxed grain structure with only a small amount of porosity. Boundaries between sites of droplet deposition are not observed.

Some of the advantages of this class of processes are:

- Simplicity and low cost
- Capability of processing a wide variety of materials

- High deposition rates
- Demonstrated scale-up (used industrially for coating objects and forming pieces)

The primary disadvantages are:

- Porous films can be formed
- Low purity in some cases
- Limited to materials that can be easily melted

***15.2.4.2 Precursors and Products*** Plasma spraying has been used to deposit a variety of materials onto surfaces. Oxides, nonoxides, metals, and composites have been deposited with very high deposition rates. However, porous films frequently result. No tabulated results are given here because of the wide variety of materials that have been deposited and because the precursors in almost all cases are powders with the same composition as the film.[23] For example, some of the metal compositions that are sprayed are Al–V alloys, oxide-dispersion-strengthened materials, Al–Cu alloys, Fe–C alloys, Fe–C–Cr alloys, and Fe–C–Si alloys.

***15.2.4.3 Experimental Details*** Numata et al. have demonstrated a system for melting powders in a plasma and depositing them onto surfaces.[25] Solidification rates are often on the order of $10^6$ K/sec. Particle sizes are frequently 10 to 100 $\mu$m in diameter, and high throughputs of up to 50 kg/h are possible.

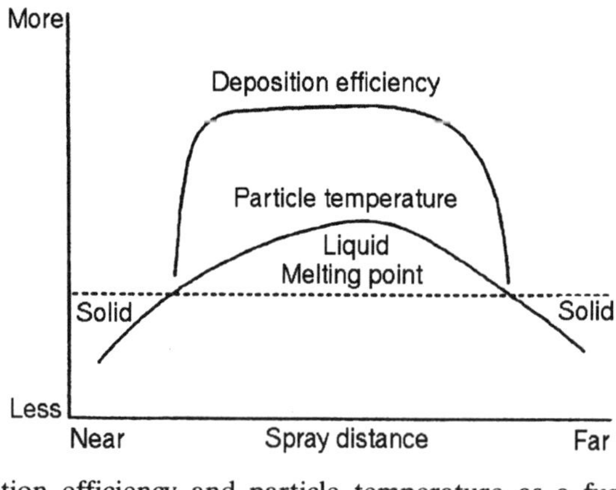


**Fig. 15.9.** Deposition efficiency and particle temperature as a function of distance from substrate for plasma spraying. (Adapted from Pober & Abbott.[26] Reprinted with permission of the American Ceramic Society, Post Office Box 6136, Westerville, Ohio 43086-6136. Copyright 1994 by the American Ceramic Society. All rights reserved.)

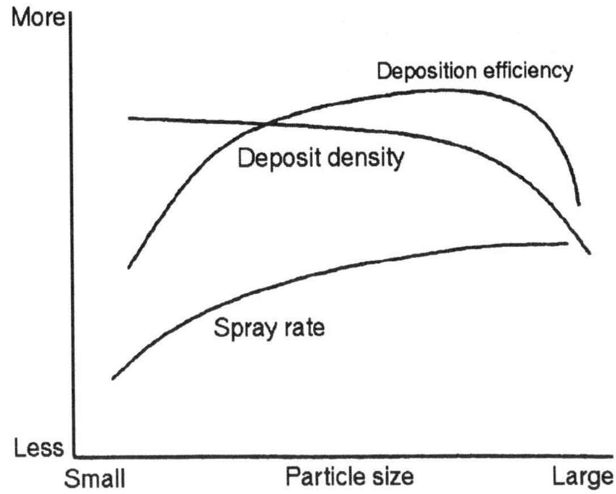


**Fig. 15.10.** Deposition efficiency, deposit density, and spray rates as a function of particle size during plasma spraying. (Adapted from Pober & Abbott.[26] Reprinted with permission of the American Ceramic Society, Post Office Box 6136, Westerville, Ohio 43086-6136. Copyright 1994 by the American Ceramic Society. All rights reserved.)

***15.2.4.4 Theory*** Some theoretical studies have examined plasma and molten metal spraying. The qualitative features are understood mainly from extensive experience. Figures 15.9 and 15.10 show qualitatively the dependence of deposition efficiency and other parameters on the spray distance and particle size.[26] An optimum spraying distance occurs where the particles are molten and the deposition efficiency is high. A large spraying distance leads to particle solidification before deposition. Similarly, large droplet sizes lead to lower deposit densities and lower efficiencies.

Two regimes exist under which spray deposition can occur. In one regime, the depositing droplets are liquid or partially liquid, and the droplet deposition rate is low enough to allow the heat from each particle to be transferred into the substrate. Because the droplets contain liquid when they deposit, they form a splat. Each splat solidifies before the arrival of subsequent molten droplets.

In the other regime, the depositing droplets are liquid or partially liquid, but the droplet deposition rate is high enough that the rate of heat supply associated with the droplets is greater than the rate at which the heat can be transferred into the substrate. As a result, each droplet does not solidify before another droplet deposits. This gives rise to a pool of liquid on the surface, which acts to reduce the porosity in the deposit.

The first regime can provide fine-grained structures, but the grains are often columnar and the coating is frequently highly porous. The second regime provides lower porosities, but the grain size may be large because of the slow cooling.

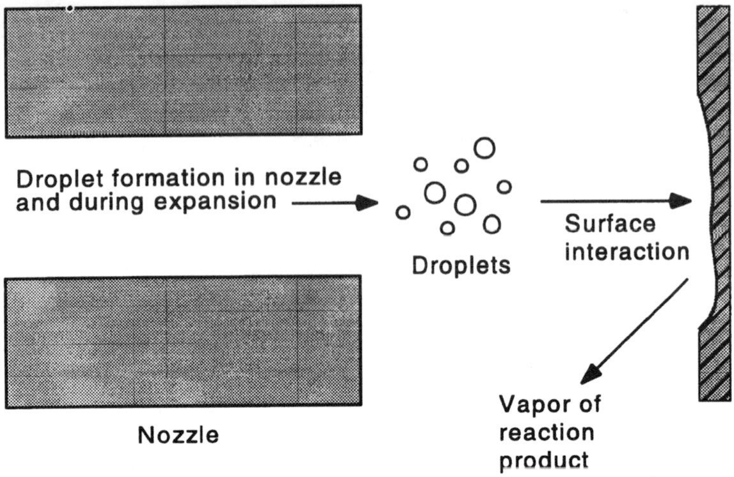


**Fig. 15.11.** Etching and cleaning of surfaces by deposition of particles formed with a nozzle.

### 15.2.5 Aerosol Jet Etching, Aerosol Cleaning, and Related Processes

***15.2.5.1 Introduction*** The formation and deposition of droplets onto surfaces can also be employed to remove materials from surfaces (Fig. 15.11). If material is removed, the process is called etching; however, if molecular and particulate contaminants are removed, the process is termed cleaning. In practice, however, etching and cleaning sometimes rely on the same physical and chemical phenomena. Both processes play important roles in the fabrication of microelectronic circuits. This section emphasizes aerosol jet etching.

Etching is a key class of processes used to form patterns in microelectronic circuits. Etching and cleaning are, in a general sense, techniques that result in removal of material from a surface and can be divided into wet and dry methods. Wet etching or cleaning usually involves the conversion of a solid material on a substrate into a soluble species that dissolves into the liquid. Dry etching and cleaning involve the conversion of a solid species into a volatile product that desorbs into the gas phase. To generate patterns with features on the order of microns, etching processes must be both anisotropic and selective towards certain materials without degrading the properties of the substrate. Anisotropy is not required for cleaning processes, but a high degree of selectivity is necessary so that other materials are not also removed. Wet processes etch isotropically and are not useful for submicron-sized features, but are used extensively to clean surfaces. Dry etching methods can be anisotropic and reasonably selective, but can result in damage, surface contamination, and uneven etching across the wafer. Also, these methods are not used extensively for cleaning, primarily because of the lack of suitable chemical approaches. Aerosol jet etching, or reactive cluster-beam etching, has been developed as an alternative that provides cleaning and selective, anisotropic etching. (See Table 15.10 at the end of the chapter.)

Aerosol jet etching begins with a vapor that is expanded out of a nozzle into a region of lower pressure, where a supersonic flow field is formed.[27] The expanding gas becomes supersaturated, and droplets or particles form, resulting in a condensed molecular beam, or cluster beam. This beam contains clustered atoms or molecules with sizes from atomic dimensions up to droplets of millions of atoms or more. The aerosol droplets contain molecular species such as HF and HCl, which can react with the substrate to form volatile species. The substrate is usually heated, which results in rapid reaction and desorption of volatile products containing the atoms that compose the substrate. The droplets are usually deposited by inertial impaction.

Some of the advantages of jet etching are:

- Many choices for reactants
- Etching or cleaning of a wide variety of multicomponent materials
- Possibility of high etch rates (0.1−1 $\mu$m demonstrated)

The primary disadvantages are:

- Complex process
- Demonstrated only for a few materials
- No demonstrated scale-up
- Difficulty achieving wafer-scale uniformity

***15.2.5.2 Precursors and Products*** The primary materials that have been etched are $SiO_2$ and $Hg_{1-x}Cd_xTe$. Most studies have used HF as the reactant because of its ability to react with $SiO_2$ at reasonable rates to form volatile products (e.g., $SiF_4$). (See Table 15.10.)

Aerosol jet cleaning involves directing a clean liquid or solid particles of a frozen volatile species such as $CO_2$ onto a surface at high velocities. The droplets or particles strike the surface, and the contaminants on the surface are removed. This approach has been used with solid $CO_2$ particles, which have the benefit of being soft and being gases at room temperature. (See Table 15.10.) The mechanism is physical and not chemical: The impinging particles dislodge contaminants from the surface by mechanical means.

***15.2.5.3 Experimental Details*** Figures 15.12 and 15.13 show an apparatus used to etch $SiO_2$.[27, 28] Anisotropic etching is attained because the droplets travel only in the vertical direction. Thus, vertical sidewalls can be obtained. The materials that have been etched include $SiO_2$ and $Hg_{1-x}Cd_xTe$. Etch rates of up to 1 $\mu$m/min have been achieved for $SiO_2$ at temperatures below 50°C. These etch rates are of the same magnitude as those obtained in conventional plasma etchers, which require vacuum systems to operate.

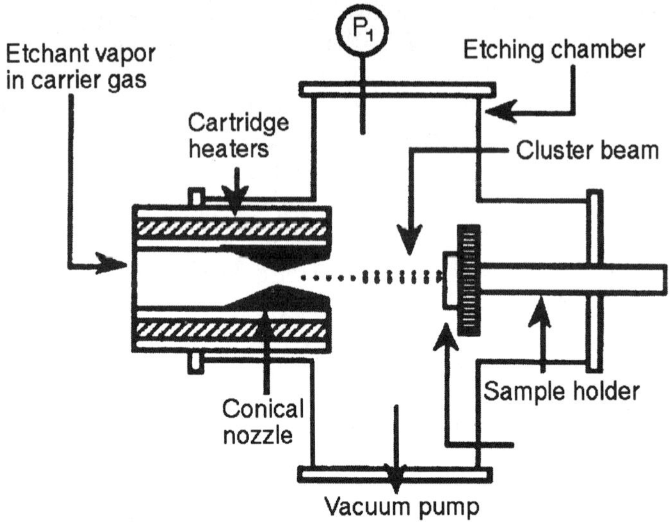


**Fig. 15.12.** Experimental apparatus for aerosol jet etching. (Reprinted with permission from Chung, C.W., Brock, J.R., and Trachtenberg, I. *Appl. Phys. Lett.*, *63*(244), 1993, 3341–3343. Copyright 1995 American Institute of Physics.)

A mixture of an inert gas and an etchant vapor (HF) is expanded through a nozzle into a vacuum ($10^{-4}$ torr), where nucleation occurs to form a beam of ultrafine droplets. The upstream pressure can range from tens to thousands of torr. The expanding supersonic jet has an isentropic region in which the Mach number is greater than unity. The substrate is placed in this region, and etching is carried out by particles colliding with the surface. Silicon-dioxide films formed by CVD on silicon have been etched with particles consisting of 38.2 wt.% of hydrofluoric acid with $H_2O$. The nozzle diameter was 0.1 mm, with a length from throat to exit of the nozzle of 5.0 mm and a total cone angle of 20°. The vapor of the droplets did not etch the silicon dioxide; etching occurred only by impaction and reaction of the liquid particles. The etch rate was varied by the process variables up to rates of 1 $\mu$m/min. Anisotropic etching was demonstrated, with little or no lateral etching under the mask. In addition, the etching was selective, with little or no etching of the aluminum masking layer and the underlying silicon substrate. Submicrometer patterns were produced using this approach.

*15.2.5.4* ***Theory*** The theory describing aerosol jet etching and related processes involves calculating the gas dynamic behavior of the jet coupled to the aerosol dynamics. Because particle or droplet formation takes place in the jet, the theory is the same as for particle formation in nozzles. The gas dynamics and aerosol dynamics have been studied by numerical simulation, using the continuity, momentum, and energy equations along with binary

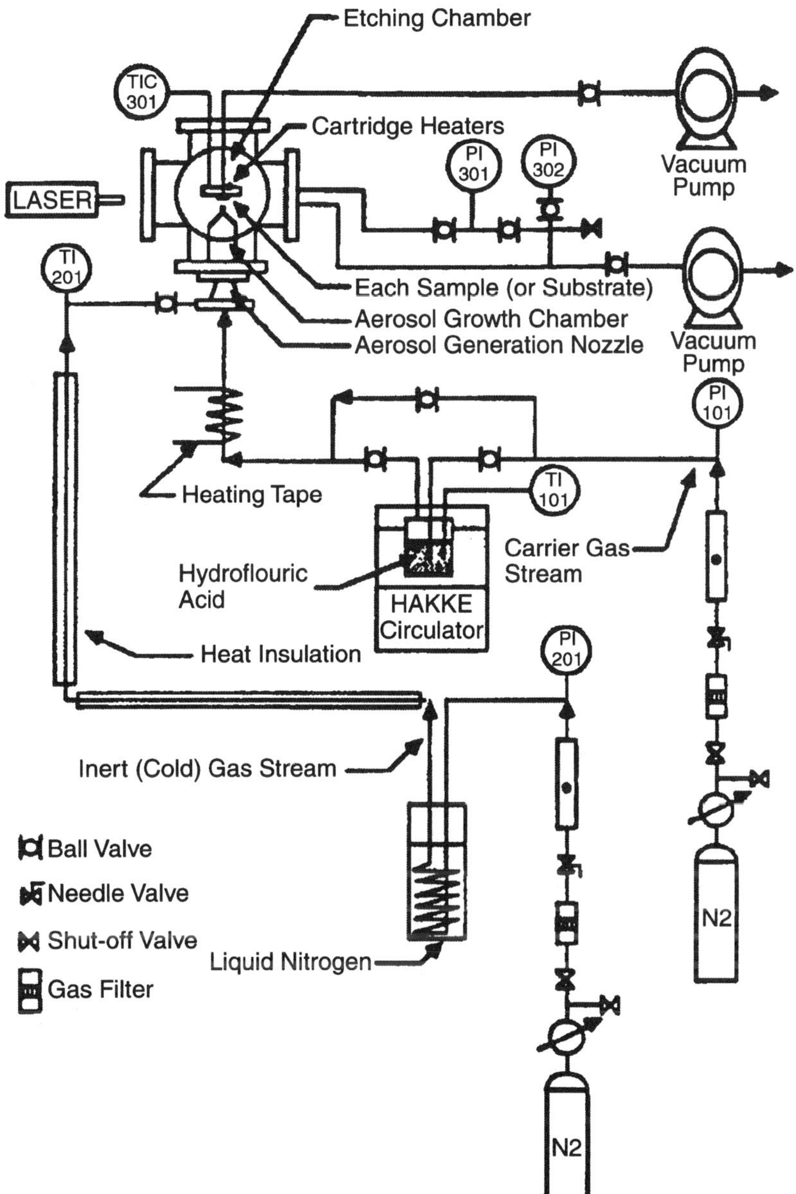


**Fig. 15.13.** Overall systems for aerosol jet etching. (Adapted from Chen et al.)[28]

homogeneous nucleation theory.[29] The particle size was typically 10 to 100 nm and increased with increasing upstream pressure, increasing etchant vapor concentration, increasing nozzle throat diameter, and decreasing source temperature. The calculations suggested that the particle size could be controlled by adjusting the upstream pressure, upstream temperature, etchant vapor concentration, and nozzle geometry. The nozzle geometry is also of critical importance for the aerosol dynamics. Converging–diverging supersonic nozzles are generally the most desirable type of nozzle because they provide higher cluster concentrations and bigger clusters.[30, 31]

## 15.3 SOLID-PARTICLE DEPOSITION

### 15.3.1 Introduction

Previous sections discussed the deposition of droplets onto surfaces and the use of droplets to transport volatile precursors to a substrate, where precursor evaporation and CVD occurred to form films. This section discusses techniques for generating films by the deposition of solid particles onto substrates.

A variety of vacuum processes exist for film formation. Expansion techniques rely on passing particles in a gas through an orifice or nozzle, after which expansion takes place to give a beam of particles that can be deposited onto surfaces. A variation on this basic scheme is ionized cluster-beam (ICB) deposition, in which particles are formed by expansion of a vapor into a vacuum, followed by cluster formation and charging of some fraction of the particles. The particles are then accelerated towards the substrate using an electric field in a vacuum chamber. Finally, the particles are deposited on the substrate. Neutral cluster-beam deposition, in which the particles are created and formed into a beam without being charged, can also be used.

Nonvacuum techniques rely on deposition by diffusion, impaction, sedimentation, electrophoresis, and thermophoresis at or near atmospheric pressure. Thermophoretic deposition techniques are based on the formation of particles in the gas phase, followed by deposition in a temperature gradient to form the film. The most well-known example is optical fiber preform fabrication by particle formation and deposition by thermophoresis (also called modified chemical vapor deposition, or MCVD). Particle-precipitation-aided CVD is a lower temperature version of the process used to produce optical fibers. In particle-precipitation aided CVD, particle deposition and CVD occur simultaneously. The sections that follow discuss a variety of vacuum and non-vacuum techniques for generating films.

### 15.3.2 Ionized Cluster Beam Deposition

***15.3.2.1 Introduction*** Ionized cluster beam (ICB) deposition involves the use of ionized particles (clusters) that are accelerated in an electric field in a vacuum towards a substrate (Fig. 15.14).[32] The acceleration imparts a high

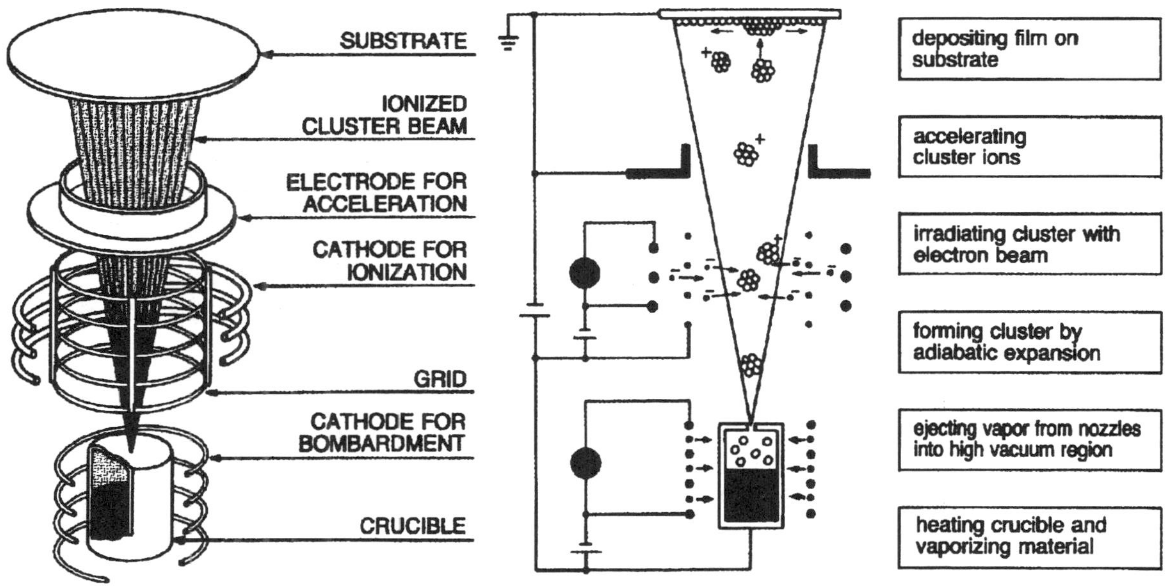


**Fig. 15.14.** Schematic depiction of the sequence of events occurring during ionized cluster-beam film formation. (Adapted from Mitsubishi advertisement.)[32]

kinetic energy to the clusters, which leads to extensive atomic migration after cluster deposition. This migration allows the formation of high-quality films of a variety of materials, including metals, metal oxides, and organics. The clusters are usually formed by evaporation of a volatile source from a crucible through a nozzle, although other methods of cluster formation are possible as well.

The two most important advantages of ICB deposition are the low charge-to-mass ratio of the ionized clusters, which allows the formation of a beam with a high loading of clusters with suitable energy, and control over the kinetic energy of the ionized clusters along with the efficient transfer of the energy to the film. A third advantage is that the energy associated with the clusters is transferred to the migrating adatoms as the cluster breaks apart upon impact. These advantages enable low-temperature epitaxial deposition. Ionized cluster beam deposition is different from other deposition methods, such as molecular-beam epitaxy, vapor-phase epitaxy, and liquid-phase epitaxy, which usually involve neutral species.

Some of the advantages of ICB deposition are:

- Possibility of a wide variety of multicomponent materials
- Oriented and epitaxial films
- High deposition rates (0.1 to 1 $\mu$m demonstrated)
- Demonstrated scale-up
- Commercial availability of process instrument

The primary disadvantage of the process is that complex and expensive high-vacuum equipment is required.

***15.3.2.2 Precursors and Products*** Table 15.11 at the end of the chapter shows some of the materials that have been deposited by the ICB deposition technique. These include metals, ceramics, and organic materials. Low tem perature epitaxy has been demonstrated for systems in which large lattice misfit conditions exist.[33] Aluminum could be deposited on Si and $CaF_2$ while accommodating the lattice misfit at the initial stages of the deposition. Monolithic Al–insulator–semiconductor single-crystal structures could be formed. Large-area deposition of GaAs epitaxial films, composition control in $Cd_{1-x}Mn_xTe$ films, and deposition of smooth, high-quality CdTe/PbTe multilayers have been demonstrated.

***15.3.2.3 Experimental Details*** Figure 15.15 shows a typical system used for ICB deposition.[34] The main features of this vacuum system are a heated source, an ionizing region, an accelerating region, and the substrate. The basic operation of ICB deposition systems involves vaporizing the precursor through a nozzle into a high-vacuum chamber. The operating conditions are chosen so that the vapor exiting the nozzle undergoes an adiabatic expansion

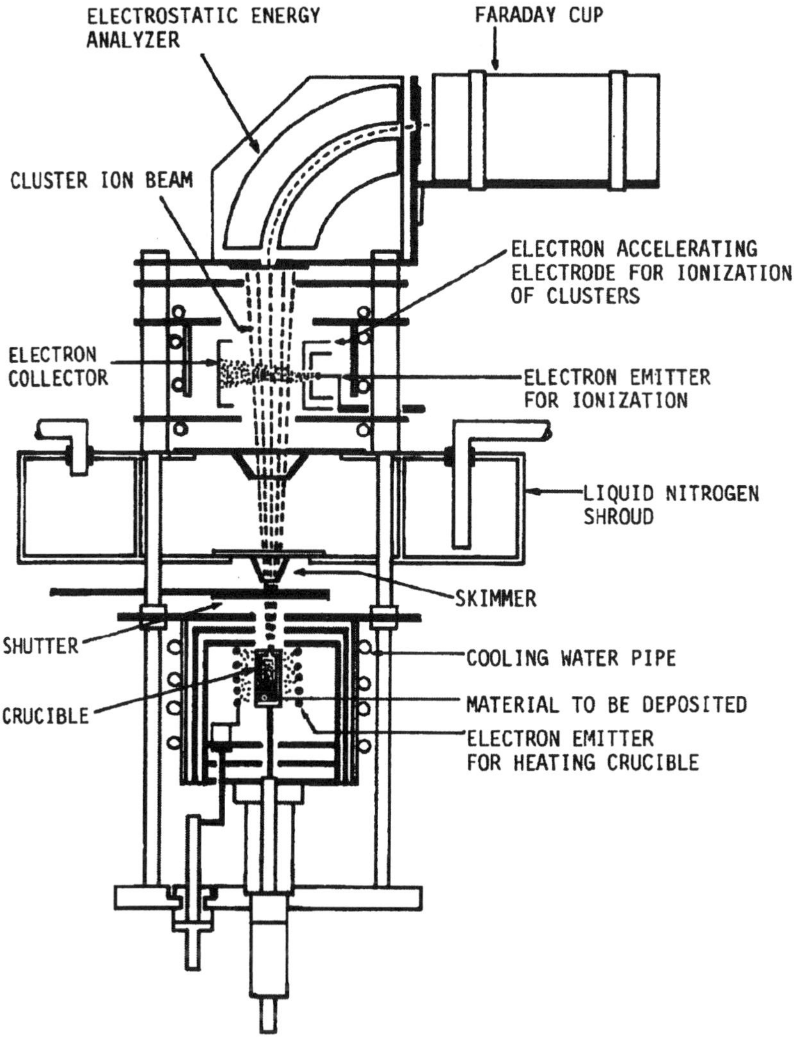


**Fig. 15.15.** Schematic of ionized cluster-beam apparatus showing complexity of process. (Adapted from Yamada & Takagi.[34] Reprinted from Yamada and Takagi, Vaporized-metal cluster formation and ionized-cluster beam deposition and epitaxy, *Thin Solid Films*, pp. 105–115. Copyright 1981, with permission from Elsevier Science Ltd., The Boulevard, Langford Lane, Kidlington OX5 1GB, UK.)

and subsequent cooling to a supersaturated state, which leads to the formation of clusters. Once formed, the clusters are subjected to bombardment by energetic electrons, which causes ionization of some of the clusters. The charged clusters can then be accelerated toward the substrate by an applied potential. Some of the clusters are not ionized, but also strike the surface and contribute to deposition.

***15.3.2.4 Theory*** One of the major theoretical considerations in ICB deposition, from an aerosol perspective, is determining the size distribution of the clusters impinging on the substrate. Calculations using classical nucleation theory have shown that the nucleation rates for metals are similar to those for water and inert gases, even though metals have high surface energies. A typical model involves a description of the gas expansion in the nozzle, coupled to equations describing the aerosol dynamics. Experimental evaluations of cluster formation have been made.[34–37] Typical time-of-flight spectra have shown from several hundred to several thousand atoms per cluster.

Figure 15.16 shows the Ag cluster size distribution exiting a nozzle as a function of the stagnation pressure calculated using the procedure just outlined.[38] Increasing the stagnation pressure increased the cluster size because of the larger amount of vapor available for cluster formation.

A second aspect to the modeling of ICB deposition is the dynamic processes associated with film formation, including the breaking of clusters

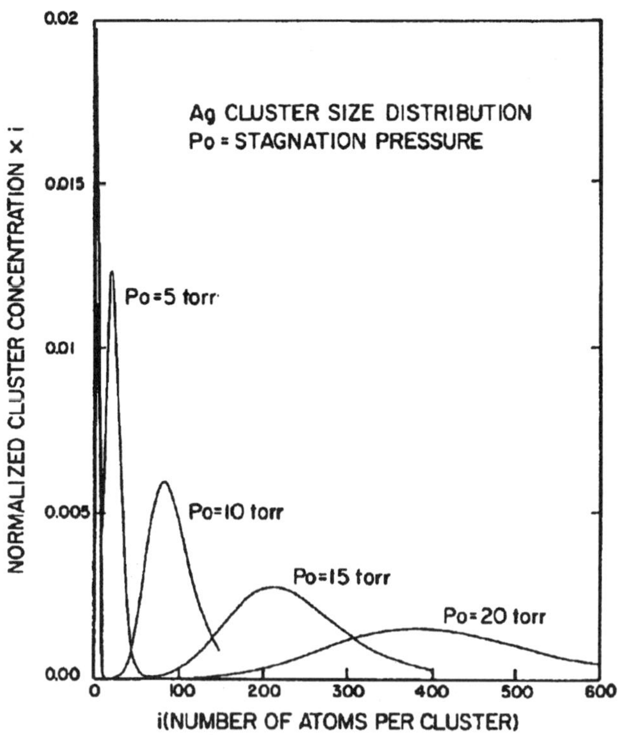


**Fig. 15.16.** Cluster size distribution for silver particles formed in a nozzle used for ionized cluster-beam deposition. (Reprinted with permission from Yang, S.-N. and Lu, T.-M. *J. Vac. Sci. Technol. B*, 5(1), 355–358. Copyright 1987 American Vacuum Society.)

into atoms upon striking the surface, sputtering of impurities from the substrate surface, formation of nucleation sites, and migration of adatoms. In practice, these processes can be controlled by changing the acceleration of the ionized clusters and the fraction of ionized clusters in the beam. As a result, the film properties can be controlled as well.

### 15.3.3 Neutral Cluster Beam

Neutral cluster-beam deposition is similar to ICB deposition, except that the clusters are not ionized and accelerated before deposition. A variety of approaches have been used to produce the particles or clusters used for deposition. The most well-studied utilizes evaporation of a volatile species into a carrier gas, where cooling results in particle formation.

A typical approach involves the formation of a metal vapor by evaporation from a heated crucible.[39] The evaporation rate is controlled by the gas pressure, crucible temperature, and nozzle or hole size in the crucible. The metal vapor is carried in an inert gas such as He or Ar and expands through an orifice or nozzle. The unsaturated vapor can be mixed with cool inert gas, resulting in a supersaturated stream in which clusters are formed. The clusters grow by condensation of metal atoms and can be deposited using a gas cell, which decelerates the clusters before they are deposited and removes any uncondensed metal atoms from the cluster beam.

Two of the advantages of this process are the wide variety of multicomponent materials that are possible and the ability of the process to produce oriented and epitaxial films. The primary disadvantage is the complex and expensive high-vacuum equipment required.

The neutral cluster-beam approach has been used to form films of several materials. (See Table 15.11 at the end of the chapter.) For example, gold clusters with diameters of roughly 1 nm were prepared in the gas phase, expanded into a vacuum chamber to form a neutral cluster beam, and deposited at room temperature onto substrates.[39] Several monolayers of these clusters were deposited onto nitrocellulose, glass, mica, and NaCl substrates, which resulted in the formation of smooth, ultrathin films with a quality similar to that obtained by ICB deposition. The films were extremely uniform and smooth, with a surface height that varied less than 1 nm across the entire film.

An example of a neutral cluster-beam deposition system is discussed by Hayashi.[40] Particles or clusters are formed by heating a metal, which is drawn into a nozzle, where the particles are formed into a beam by expansion through several stages of a vacuum system. The particles can be modified before being deposited onto a surface. These systems are typically expensive and complicated and require high pumping capacities.

The theory for neutral cluster beams is the same as for ionized cluster beams and for nozzle expansions used to form particles.

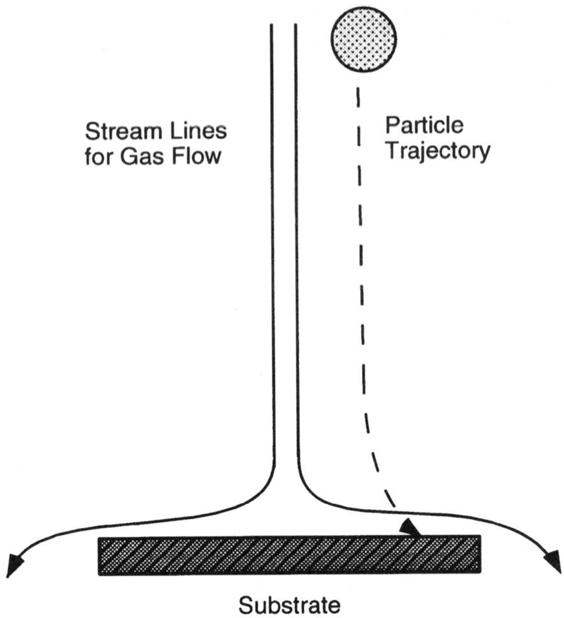


**Fig. 15.17.** Particle deposition by impaction for film formation.

### 15.3.4 Inertial Particle Deposition

***15.3.4.1 Introduction*** Inertial particle deposition has been used extensively to form a variety of products, including ceramic filters, sensors, and hard coatings; Figure 15.17 shows a schematic of the deposition process. Typically, a carrier gas stream is directed at a surface, resulting in a flow field approximating stagnation point flow. In the absence of temperature gradients, particle deposition takes place by impaction. As the gas flows by the substrate, the inertia of the particles causes them to travel across flow streamlines and collide with the surface. In this sense, the technology is very similar to that of multistage impactors used to collect particles classified by size. Larger areas are coated by moving the substrate relative to the nozzle.

The primary strength of inertial particle deposition is its generation of thick films. Typically, postprocessing is required to densify the porous films. The structure of these films before and after densification has not been studied extensively.

Some of the advantages of inertial particle deposition are as follows:

- Simplicity and low cost
- Many choices for a method to produce particles
- Deposition of any multicomponent material
- High deposition rates
- Commercially available

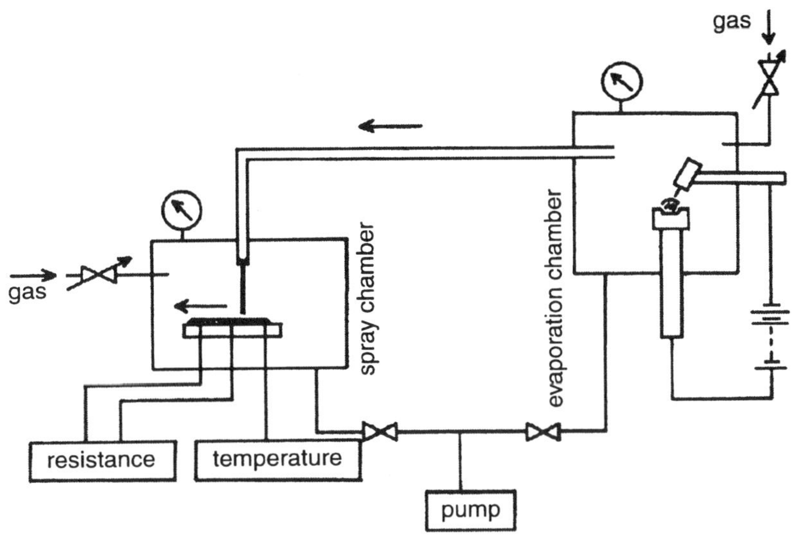


**Fig. 15.18.** Experimental apparatus for inertial particle deposition. (Adapted from Kashu et al.[41] Reproduced with permission from the *Japanese Journal of Applied Physics*.)

The primary disadvantages are that porous films can be formed and that it is difficult to coat large areas uniformly in short times.

***15.3.4.2 Precursors and Products*** Solid particles consisting of virtually any material can be used to form coatings (see Table 15.12 at the end of the chapter), including thin films of ceramic superconductors ($YBa_2Cu_3O_7$, $YBa_2Cu_4O_8$, Bi–Pb–Sr–Ca–Cu–O), metals (Fe–Ag, Ag), and composites ($Ag–ZrO_2$). Such particles have been formed by virtually all available methods, from liquid-phase to solid-state and gas-phase routes.

***15.3.4.3 Experimental Details*** Figure 15.18 shows a typical deposition chamber used for the fabrication of films.[41] In one such system, tin-dioxide particles were introduced through a nozzle 0.8 mm in diameter into the deposition chamber, which was pumped down to 30 torr. The particles were accelerated to about 200 m/sec and then deposited on the glass substrate to form a film. Also shown in the figure is the overall system for particle generation and deposition.[41] Particles are formed by evaporation of a metal and are carried in an inert gas and passed through a nozzle, after which they are deposited onto a substrate. Reduced pressures are required for deposition of particles smaller than 0.1 $\mu$m.

***15.3.4.4 Theory*** The mechanism of inertial deposition has been studied extensively, and useful correlations are available in the literature.[42] The

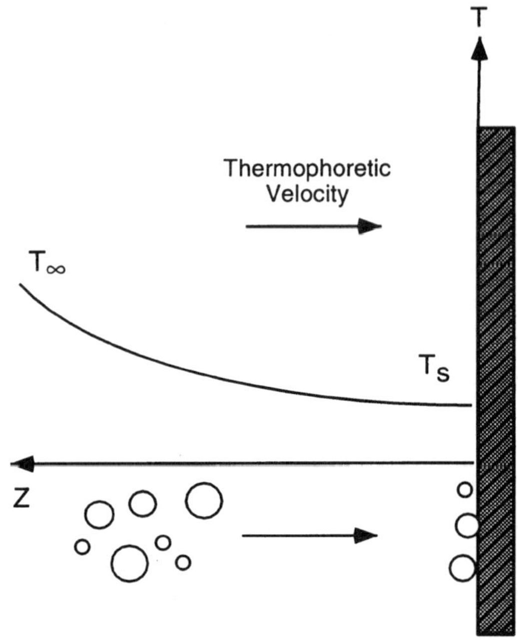


**Fig. 15.19.** Temperature profile and coordinate system for thermophoretic particle deposition. $T_s$ is the substrate temperature, and $T_\infty$ is the temperature far from the substrate.

theory is identical to that of inertial impactors. A key problem with this approach is particle bounce, which has been studied extensively.[43-49]

### 15.3.5 Thermophoretic Aerosol Deposition

*15.3.5.1* ***Introduction*** Thermophoretic particle deposition relies on the thermophoretic velocity of particles in a temperature gradient.[50] This velocity results from the differences in the kinetic energies of molecules striking a particle; in a simple sense, molecules on the higher temperature side of the particle have higher energy and cause particles they collide with to be driven in the direction of lower temperature. Figure 15.19 shows a schematic of this approach.

Some of the advantages of thermophoretic deposition are:

- Simplicity and low cost
- Ability to deposit any particles suspended in a gas
- Ability to deposit a wide variety of multicomponent materials
- High deposition rates (>1 $\mu$m/min demonstrated)
- Demonstrated scale-up (used industrially for producing optical fibers)

The primary disadvantages are:

- Formation of porous films that require annealing
- Difficulty in obtaining uniform deposition over large areas
- Low yields

*15.3.5.2* ***Precursors and Products*** Table 15.13 at the end of the chapter lists some of the materials that have been formed as films using this approach. An example is the formation of thick films of superconducting ceramics by deposition of ceramic particles formed by aerosol decomposition.[51] A modification of this process has been used to produce Bi–Sr–Ca–Cu–O films.[52] Because of the wide variety of conditions employed, it is not possible to give typical numbers for thermophoretic deposition systems.

*15.3.5.3* ***Experimental Details*** Figure 15.20 shows a system used to deposit films of Y–Ba–Cu–O.[53] Precursors for Y, Ba, and Cu were placed into a solution that was atomized to form droplets. The droplets were introduced into a flame reactor, where reaction and particle formation occurred and the particles were deposited by thermophoresis onto the cooler substrate. The maximum substrate temperature was 940°C. The substrate was moved relative to the flame to provide uniform coatings.

*15.3.5.4* ***Theory*** The theory of thermophoretic deposition is well developed for explaining coating of the inside and outside of a tube. The development of this theory has been motivated by the importance of the method for the

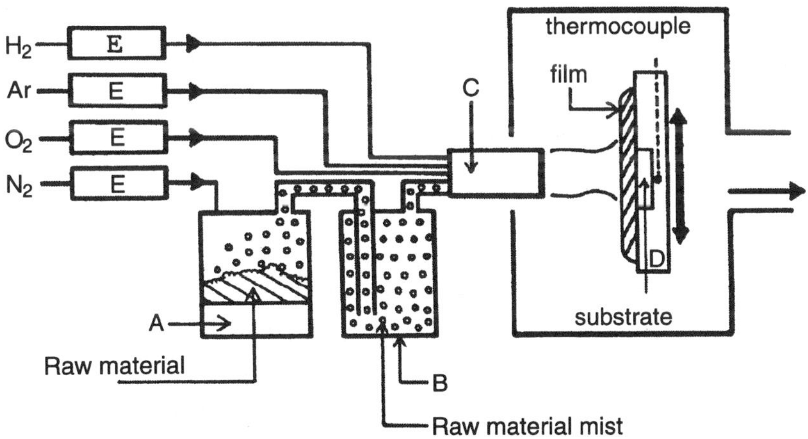


**Fig. 15.20.** Apparatus for thermophoretic deposition of particles formed by spray pyrolysis. (Adapted from Koguchi et al.[53] Reproduced with permission from the *Japanese Journal of Applied Physics*.)

production of optical fibers. The theory is discussed in Chapter 7, in Section 15.3.6, and in Chapter 3.

### 15.3.6 Thermophoretic Aerosol Deposition: Optical Fibers

***15.3.6.1 Introduction*** Optical fiber generation is a subset of thermophoretic deposition that involves the formation of porous cylindrical preforms of doped $SiO_2$ particles that are eventually drawn into optical fibers. Two manufacturing techniques are used: "inside" and "outside" processes. In both approaches, $SiO_2$ particles are formed by the reaction of $SiCl_4$ vapor with oxygen and/or water. Various dopant species are also added to control the refractive index. The particles are then deposited by thermophoresis onto the inside or outside of preforms, which are later consolidated and drawn to form the fiber. A schematic representation of an inside deposition process is shown in Figure 15.21,[54] and the aerosol dynamics are shown in Figure 15.22.[55]

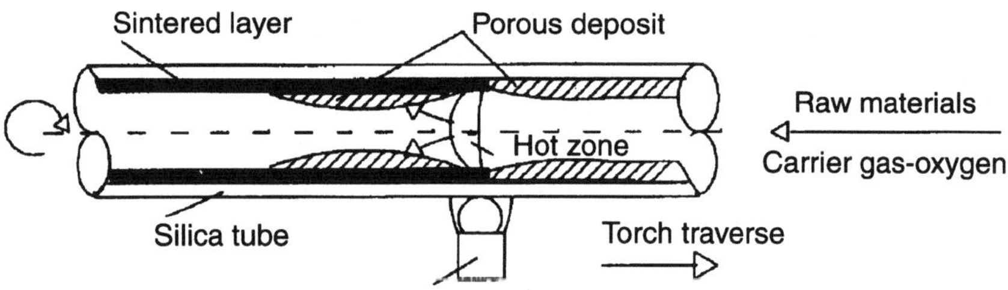


**Fig. 15.21.** Schematic of optical fiber formation process. (Reprinted from *Journal of Aerosol Science*, *26*, Kasik, I. and Matejec, V., 399–406, Copyright 1995, with kind permission from Elsevier Science Ltd., Kidlington OX5 1GB UK.)

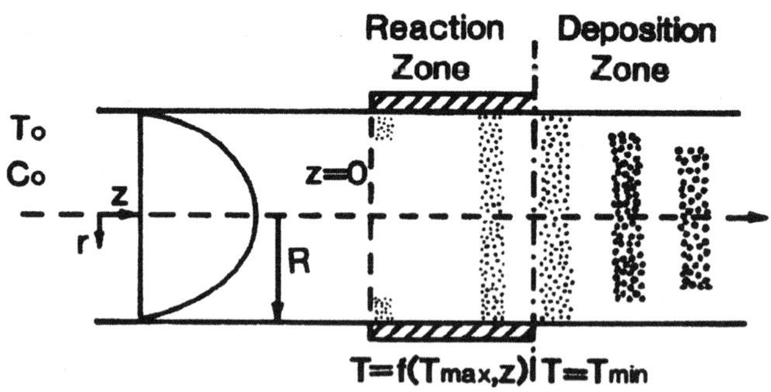


**Fig. 15.22.** Aerosol phenomena involved in optical fiber production by thermophoretic particle deposition. (Reprinted from *Chemical Engineering Science*, *44*, (11), Kim, S.E. and Pratsinis, S.E., pp. 2475–2482, Copyright 1989, with kind permission from Elsevier Science Ltd., The Boulevard, Langford Lane, Kidlington OX5 1GB UK. and Reproduced with permission of the American Institute of Chemical Engineers. Copyright © 1988 AIChE. All rights reserved.)

Modified chemical vapor deposition (MCVD) is an example of an inside deposition process. A precursor such as $SiCl_4$ in oxygen is passed through a rotating preform tube heated by a flame that is moved back and forth along the outside of the tube to drive the reaction. Heating of the tube also provides the temperature gradient necessary for thermophoretic deposition. The particles formed are not large enough for inertial impaction, and the residence time is too short for diffusion to the tube walls. The silica particles are deposited downstream of the heated section by thermophoresis and are sintered to a clear glass as the torch moves back and forth. By changing the composition of the reactant stream to include other elements, one can vary the composition of the coating to change the refractive index. Typical precursors for dopants are $GeCl_4$ and $SiF_4$.

Outside deposition processes rely on thermophoretic deposition of particles generated in a flame reactor. The particles are formed in flames in which $SiCl_4$ and reactants for dopants are reacted with oxygen. Vapor axial deposition (VAD) and outside vapor deposition (OVD) are examples of this class of approaches.

In the OVD process, the flame reactor used to produce the particles consists of concentric rings of holes (Fig. 15.23).[56] The VAD flame is produced using a series of concentric tubes. In the VAD process, the preform is pulled away from the flame as material is deposited.

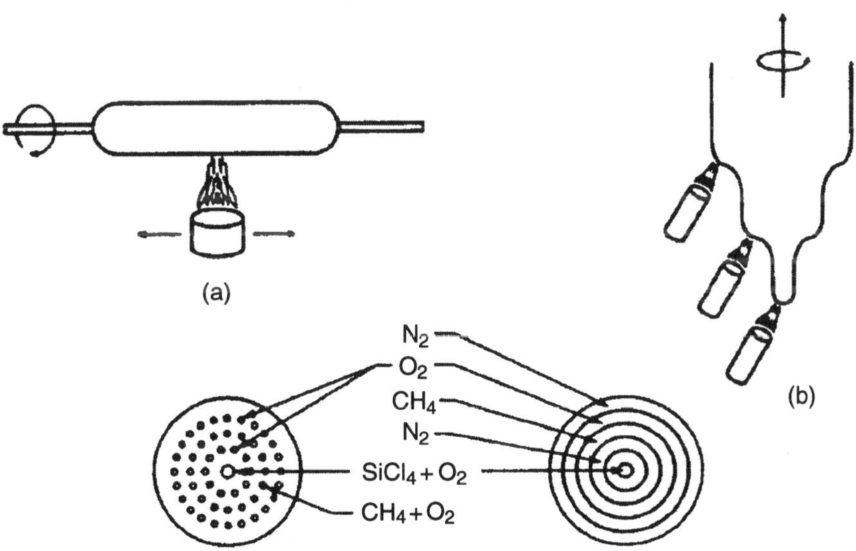


**Fig. 15.23.** Outside vapor deposition process for optical fiber production. (Reprinted from *Journal of Aerosol Science*, *22*, Bautista, J.R. and Atkins, R.M., 667–675. Copyright 1991 with kind permission from Elsevier Science Ltd., The Boulevard, Langford Lane, Kidlington OX5 1GB UK.)

After deposition of the required amount of material, the soot preforms are dehydrated and sintered in a furnace, thereby producing a clear glass preform. The resulting preforms are typically 1 m long and roughly 5 cm in diameter and are drawn at speeds in excess of 5 m/sec to yield approximately 100 km of fiber, roughly 100 $\mu$m in diameter.

***15.3.6.2 Precursors and Products*** Precursors and products for optical fiber production are shown in Table 15.13 at the end of the chapter. The reactants are usually metal halides, while the products are metal oxides that usually contain appreciable amounts of water. Typically, this water must be removed, and the porous preform must be consolidated before being drawn into a fiber.

***15.3.6.3 Experimental Details*** The specific operating conditions used to form optical fibers are proprietary, but several discussions of the conditions in experimental systems are available. See Table 15.13 at the end of the chapter.

***15.3.6.4 Theory*** The inside deposition process has been modeled extensively. The treatment was initiated by Walker et al.;[57] later, Kim and Pratsinis gave a more detailed account.[58]

Figures 15.24 and 15.25 show how particle concentrations vary in the radial and axial directions for inside deposition.[58] The tube can be divided

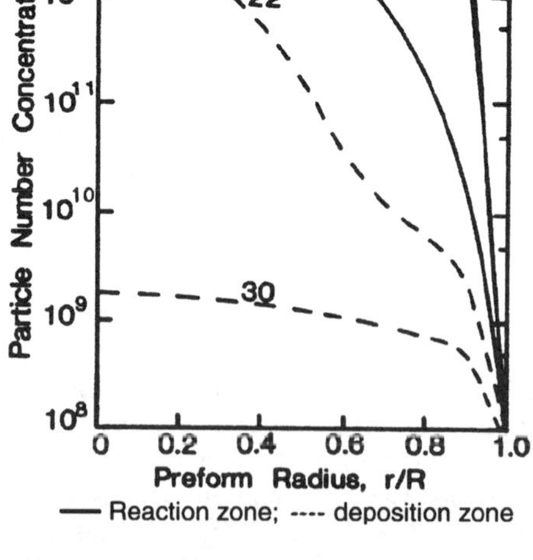


**Fig. 15.24.** Particle concentrations as a function of radial position for different distances along the tube. At short distances, only particles near the wall have deposited. (Adapted from Kim & Pratsinis.[58] Reproduced with permission of the American Institute of Chemical Engineers. Copyright © 1988 AIChE. All rights reserved.)

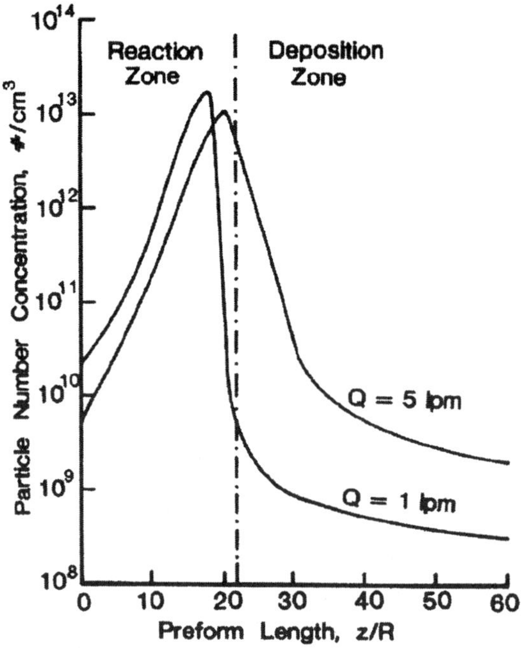


**Fig. 15.25.** Particle number concentration along a preform, showing how the particle concentration decreases in a section with cool walls (the deposition zone). (Adapted from Kim & Pratsinis.[58] Reproduced with permission of the American Institute of Chemical Engineers. Copyright © 1988 AIChE. All rights reserved.)

into the reaction zone, where the $SiCl_4$ is converted into particles and the particle concentration increases, and the deposition zone, where the particles deposit by thermophoresis and the particle concentration decreases along the tube. At the same time, the particle concentration near the walls decreases because of the loss of particles by thermophoresis. Higher flow rates result in lower deposition efficiencies because of the reduced residence time in the region where deposition is desired.

***15.3.7 Deposition of Charged Solid Particles in Electric Fields*** The deposition of charged solid particles has been used in a variety of applications. A typical process relies on a method of forming particles in the gas phase or of introducing particles into the gas phase from the powder or liquid form. The particles are then charged and brought into a region with an electric field that drives them to a surface, where they deposit. This approach has also been used to deposit liquid droplets onto surfaces, as discussed in Section 15.2.4. For both liquids and solids, the main motivation for using charged particles is the high deposition efficiency that can be obtained.

Some of the advantages of this process are:

- Simplicity and low cost
- Ability to deposit any particles suspended in a gas
- Ability to deposit a wide variety of multicomponent materials
- High deposition rates (>1 $\mu$m/min demonstrated)

The primary disadvantages are that porous films are formed if thick coatings are made without simultaneous CVD and that uniform deposition over large areas is difficult to obtain.

Virtually any type of particle can be deposited by this route; however, few examples exist perhaps because of the complexity of the process. One example is the deposition of charged Ag particles onto InP substrates for applications involving quantum dots (Fig. 15.26).[59] These zero-dimensional quantum structures have potential applications in lasers. First, a polydisperse stream of silver particles was produced by evaporation–condensation of silver using a tube furnace and silver powder. Monodisperse Ag aerosol particles (40 nm) were then obtained by means of a differential mobility analyzer. This instrument selects particles of a given size from a polydisperse aerosol stream by exploiting the differences in the electrical mobility of charged particles of different sizes. The charged silver particles were then deposited by passing

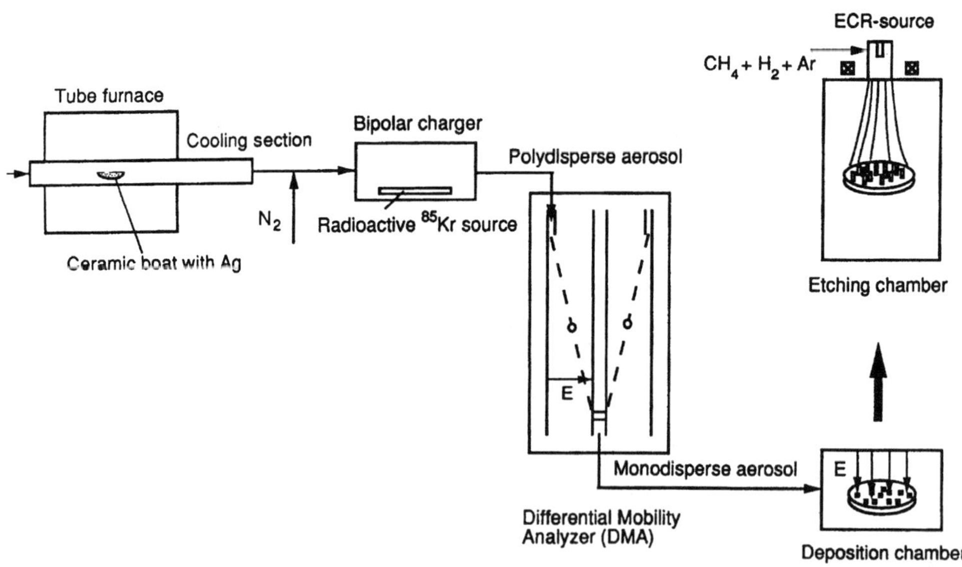


**Fig. 15.26.** Electrophoretic deposition system relying on classification of particles before deposition to provide a narrow particle size distribution. (Adapted from Wiedensohler et al.[59] Reprinted from *Journal of Aerosol Science*, *24*, Wiedensohler, A., Maximov, I., Samuelson, L. and Hansson, H.C., 555–558. Copyright 1993, with kind permission from Elsevier Science Ltd., The Boulevard, Langford Lane, Kidlington OX5 1GB UK.)

them through a deposition chamber to which a potential was applied. The quantum dots were produced on the InP surface by dry etching, with the Ag particles used as masks. The InP columns produced by this approach had average diameters of 60 nm.

The basic theory of electrophoretic deposition is well established,[50] but little has been done in the way of examining the modeling of this class of deposition reactors.

## 15.4 SIMULTANEOUS PARTICLE DEPOSITION AND CVD

Many approaches have been developed in which particles and vapor are deposited simultaneously. These approaches can combine thermophoretic, diffusional, gravitational, and electrophoretic deposition with CVD to produce composite films or to provide films at high rates. Figure 15.27 shows how the process can produce films.[60]

Numerous examples exist of the different deposition mechanisms. Particle-precipitation-aided CVD involves forming particles in a CVD system and depositing the particles onto a substrate in parallel with deposition by CVD. (The method is discussed in Chapter 14.) The system is designed to provide a temperature gradient towards the substrate to drive the particles onto the substrate. Charged CdSe particles have been deposited simultaneously with CVD of ZnSe for the generation of CdSe particles in a ZnSe matrix.[61, 62] Composite diamond–metal films have been formed by deposition of metal powder and diamond CVD.[63] Silica fibers have been coated with ultrafine $SiO_2$ particles by diffusional deposition.[64] Gravitational deposition has been used to produce films of SiC.[60]

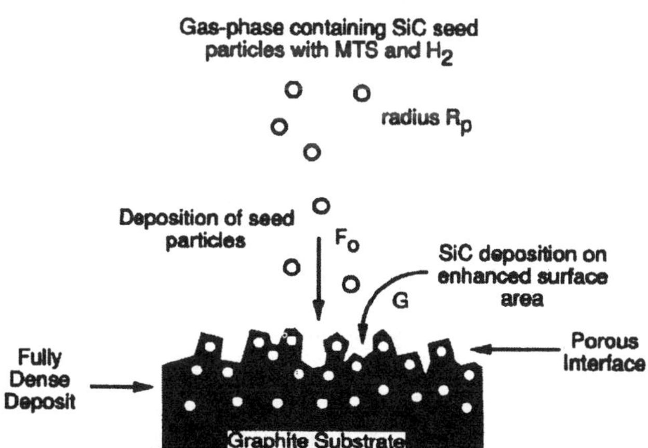


**Fig. 15.27.** Simultaneous particle deposition and chemical vapor deposition. (Adapted from Allendorf et al.[60] Reprinted with permission from Allendorf, M.D., Hurt, R.H., and Yang, N. *J. Mater. Res.*, *8*(7), 1993, 1651–1665.)

Some of the advantages of this process are:

- Simplicity and low cost
- Ability to deposit any particles suspended in a gas
- Ability to deposit a wide variety of composite materials
- High deposition rates (>1 $\mu m/min$ demonstrated)
- Numerous deposition mechanisms

The primary disadvantages are:

- Formation of porous films if CVD rate is too low
- Difficulty in obtaining uniform deposition over large areas
- Not extensively studied

Table 15.14 at the end of the chapter gives some of the materials that have been formed by these routes. The most extensively studied approach has been thermophoretic deposition with CVD. AlN and $TiO_2$ films have been formed at rates of 10 to 1000 $\mu m/hr$. These rates are much higher than those obtained with conventional CVD processes.

Figure 15.28 shows a system used for simultaneous thermophoretic deposition and CVD.[65] The CVD precursor is introduced separately to avoid CVD on the walls before the substrate and to avoid particle formation except above the substrate. The walls are kept at a higher temperature than the substrate, resulting in thermophoretic deposition.

Figure 15.29 shows a system for the simultaneous deposition of diamond and metal.[63] The metal powder is introduced into the tail flame of the plasma gun, along with the precursor for diamond CVD. The substrate is cooled to provide thermophoretic deposition of particles.

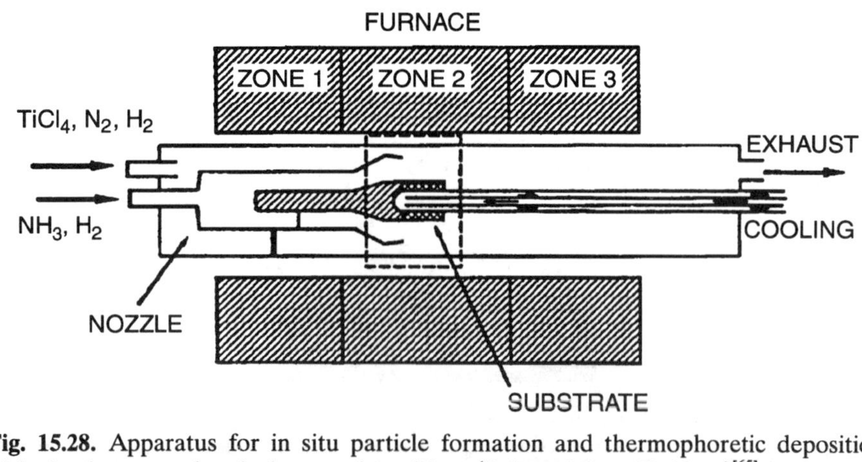


**Fig. 15.28.** Apparatus for in situ particle formation and thermophoretic deposition with simultaneous chemical vapor deposition. (Adapted from Dekker.)[65]

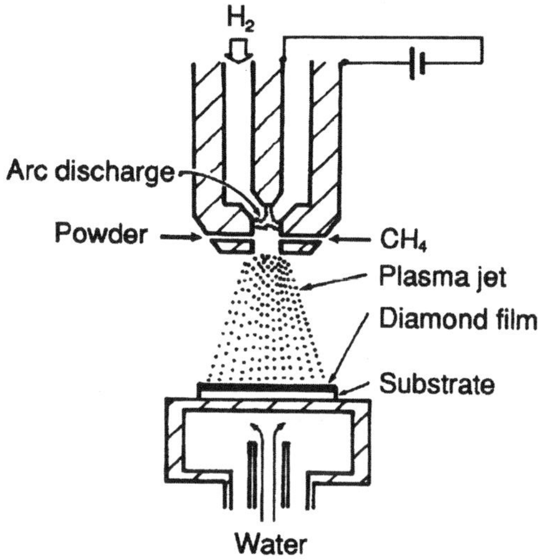


**Fig. 15.29.** Flame spray process for simultaneous diamond particle and chemical vapor deposition. (Adapted from Kawarada et al.[63] Reprinted with permission from Kawarada, M., Kurihara, K., and Sasaki, K. *J. Appl. Phys.*, *71*(3), 1442–1445. Copyright 1992 American Institute of Physics.)

The theory describing the individual physical phenomena used for particle formation in these processes is well established and is discussed in Chapter 14. Few full models, however, accounting for simultaneous particle deposition and vapor deposition have been developed.[60]

## 15.5 PARTICLE FORMATION DURING CVD

### 15.5.1 Introduction

Particle formation during CVD has been reported by a wide variety of investigators. This phenomenon is of interest because the same precursors used to form thin films by CVD can be used to form particles by gas-to-particle conversion. Simultaneous particle deposition and CVD can rely on in situ particle formation. The qualitative features of this problem are well understood. Chemical vapor deposition involves passing gaseous reactants either through a hot tube (hot-wall CVD; Fig. 15.30) or over a hot substrate contained in a cold-wall system.[12] The former is usually termed a hot-wall reactor, while the latter is called a cold-wall or, often, a single-wafer reactor. In the hot-wall case, the system is identical to the hot-wall reactor described in Chapter 11, which is used to produce particles. The only difference is the lower pressure needed for CVD and the presence of wafers to be coated. In the cold-wall system, the gas above the substrate is heated. This results in a smaller volume of heated gas compared with that of the hot-wall system, but qualitatively, the situation is the same as for the hot-wall method. The gas

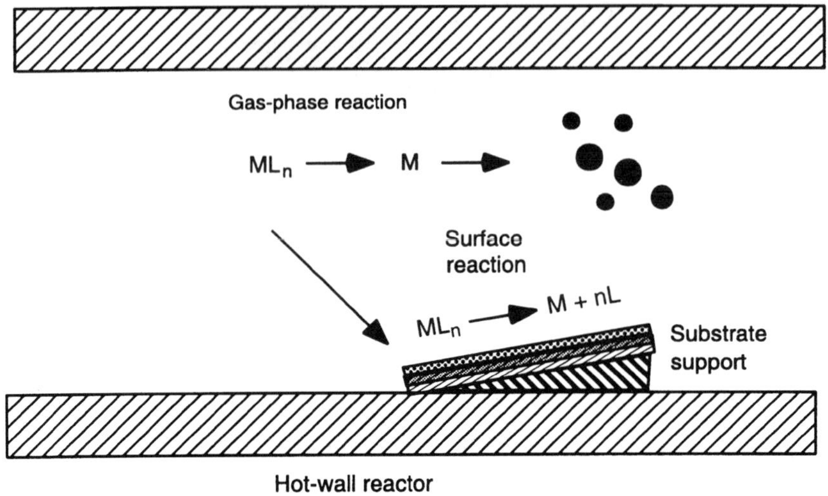


**Fig. 15.30.** Particle formation during hot-wall chemical vapor deposition.

stream containing the reactive precursor is heated by the walls of the hot-wall reactor or by the substrate in the cold-wall reactor, leading to homogeneous reactions of the precursors. These gas-phase reactions result in particle formation under certain conditions, as outlined in Section 15.5.4.

### 15.5.2 Precursors and Products

Particle formation has been observed for a wide variety of materials and is a general phenomenon. At the same time, particle formation is also a major problem during the coating of large sheets of window glass using CVD. Although industry has extensive experience with this problem, few discussions have been published. The most well-studied systems are particle formation during CVD using $SiH_4$, CuI, and $Ti(O\text{-}i\text{-}Pr)_4$. Table 15.15 at the end of the chapter gives examples of CVD systems in which particle formation has occurred.

### 15.5.3 Experimental Details

An example of a CVD system set up to study particle formation is shown in Figure 15.31.[66] The precursor dimethylethylaminealane was introduced into a low-pressure hot-wall CVD system in which particle formation occurred. The particles were collected on a plate and examined by scanning electron microscopy. Particle formation has also been examined using optical particle counters designed to operate at reduced pressures.

### 15.5.4 Theory

As the reactants flow through a CVD reactor, the heating of the gas drives gas-phase reactions of the precursors, which can result in particle formation.

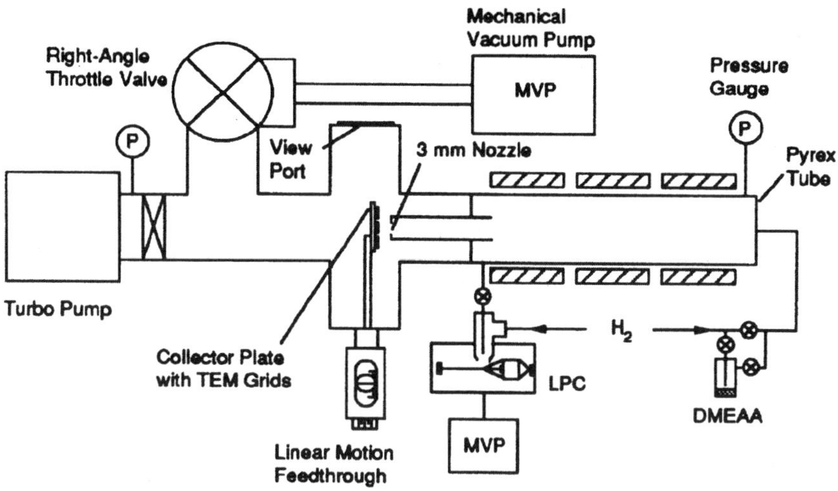


**Fig. 15.31.** System for studies of particle formation during chemical vapor deposition. (Adapted from Simmonds et al.)[66]

Particle formation is favored by higher pressures and temperatures and by larger reactors. Consider flow through a tube under isothermal conditions. A reaction takes place to form a species in the gas phase, which may then nucleate to form particles. The product also may be transported to the walls, where it is lost. This results in a balance between formation and loss of the reaction product, which determines the vapor concentration and, in turn, the rate of particle formation. A theoretical analysis[67–69] has shown that the occurrence of particle formation depends on the magnitude of the dimensionless number

$$E = \frac{R_t^2}{\tau_p D}, \tag{15.1}$$

where $\tau_p$ is the characteristic time for particle formation, $R_t$ is the radius of the reactor tube, and $D$ is the diffusion coefficient of the reaction product in the carrier gas.

Particle formation is favored by large values of $E$. Larger reactor diameters mean that the species has to diffuse a larger distance to be lost to the walls. Large values of $E$ correspond to a small diffusion coefficient, which occurs at higher pressure and hinders losses to the walls. Finally, large $E$ corresponds to a time for particle formation that is small.

Particle formation in annular hot-wall reactors with vapor deposition has been modeled, taking into account the dynamics of particle formation and growth. Figure 15.32 shows how particle concentrations evolve.[67–69] The particle concentration in the radial direction drops to zero near the walls because the product vapor concentration there is low. Most of the particle

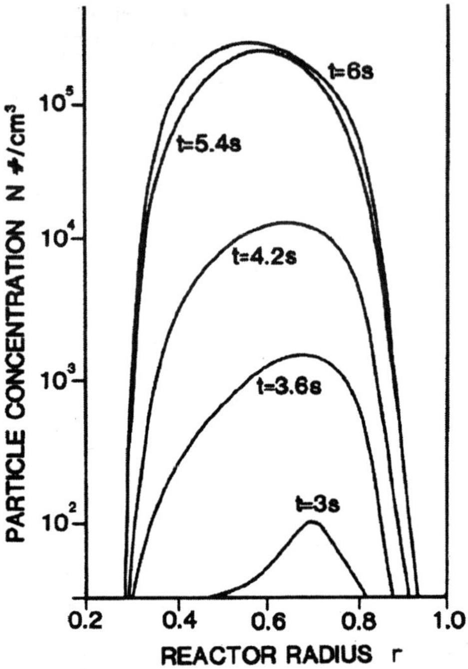


**Fig. 15.32.** Particle concentrations in tubular flow reactor with simultaneous vapor deposition and particle formation. (Adapted from Pratsinis et al.)[69]

formation occurs in the region farthest from the walls. Reducing the dimensions of the reactor in the direction transverse to the flow results in reduced particle concentrations and a lower yield of particles. The reason is that, according to the preceding criterion, as the dimensions of the system shrink, the reaction product can diffuse more rapidly to the walls thereby reducing its concentration in the gas phase. This results in a lower rate of particle formation. Larger particles are formed as the dimensions of the reactor shrink because fewer particles are formed and each particle consumes a larger fraction of the gas-phase reaction product.

## 15.6 REFERENCES

1. Biswas, D.R. *J. Mater. Sci. 21*, 1986, 2217–2223.
2. Kodas, T.T. *Advanced Mater. 6*, 1989, 180–192.
3. Orban de Xivry, E., Streydio, J.-M., & Berote, G. Chemical spray technology adapted to thin film semiconductor growth. In Vincenzini, P. (Ed.). *High Tech Ceramics*. Amsterdam: Elsevier, 1987, 1709–1718.
4. Tomar, M.S. & Garcia, F.J. *Prog. Cryst. Growth Charact. 4*, 1981, 221–248.
5. Krishnakumar, R., Subramanian, V., Ramprakash, Y., & Lakshmanan, A.S. *Materials Chem. Phys. 15*, 1987, 385–395.

6. Lampkin, C.M. *Prog. Crystal Growth Charact. 1*, 1979, 405–416.
7. Oktik, S. *Prog. Crystal Growth Charact. 17*, 1988, 171–240.
8. Pamplin, B.R. *Prog. Cryst. Growth Charact. 1*, 1979, 395–403.
9. Sears, W.M. & Gee, M.A. *Thin Solid Films 165*, 1988, 265–277.
10. Omar, O.A., Ragaie, H.F., & Fikry, W.F. *J. Mater. Sci.*, 1990, 79–83.
11. Biswas, P., Jones, C.L., & Flagan, R.C. *Aerosol Sci. Tech. 7*(2), 1987, 231–246.
12. Kodas, T.T. & Hampden-Smith, M.J. (Eds.). *The Chemistry of Metal CVD*. Weinheim, West Germany: VCH, Verlagsgesellschaft, 1994.
13. Roger, C., Corbitt, T., Xu, C., Zeng, D., Powell, Q., Chandler, C.D., Nyman, M., Hampden-Smith, M.J., & Kodas, T.T. *J. NanoSt. Mater. 4*(5), 1994, 529–535.
14. Roger, C., Corbitt, T., Kodas, T.T., & Hampden-Smith, M.J. *Appl. Phys. Lett. 65*(8), 1994, 1021–1023.
15. Salazar, K.V., Ott, K.C., Rye, R.C., Hubbard, K.M., Peterson, E.J., Coulter, J.Y., & Kodas, T.T. *Physica C. 198*(3-4), 1992, 303–308.
16. Cui, H., Shen, H.-S., Gao, Y.-M., Dwight, K., & Wold, A. *Mat. Res. Bull. 28*, 1993, 195–201.
17. Hess, D.W. & Jensen, K.F. (Eds.). *Microelectronics Processing*. Washington: American Chemical Society, 1989.
18. Roger, C., Corbitt, T., Kodas, T.T., & Hampden-Smith, M.J. *J. Aerosol Sci. 25*, 1994, S221.
19. Matson, D.W., Petersen, R.C., & Smith, R.D. *J. Mater. Sci. 22*, 1987, 1919–1928.
20. Paulaitis, M.E., Penninger, J.M.L., Gray, R.D., & Davidson, P. (Eds.). *Chemical Engineering at Supercritical Conditions*. Ann Arbor, MI: Ann Arbor, Science, 1983.
21. Schneider, G.M., Stahl, E., & Wilke, G. *Extraction with Supercritical Gases*. Deerfield Beach, FL: Verlag Chemie, 1980.
22. Van Wassen, U., Swaid, I., & Schneider, G.M. *Agnew. Chem. Int. Ed. Engng. 19*, 1980, 575.
23. Sampath, S. & Herman, H. *J. Minerals Metals & Mater. Soc. 45*(7), 1993, 42–49.
24. Chin, T.S., Hara, Y., Lavernia, E.J., O'Handley, R.C., & Grant, N.J. *J. Appl. Phys. 59*(4), 1986, 1297–1300.
25. Numata, K., Hoshino, K., Yoshida, Y., Inoue, K., Maeda, H., Kamekawa, Y., & Suzuki, M. *American Institute of Physics*, 1992, 418–424.
26. Pober, R. & Abbott, R. *Am. Ceram. Soc. Bull. 73*(2), 1994, 45–50.
27. Chung, C.W., Brock, J.R., & Trachtenberg, I. *Appl. Phys. Lett. 63*(244), 1993, 3341–3343.
28. Chen, Y.L., Brock, J.R., & Trachtenberg, I. *Aerosol Science Technol. 12*, 1990, 842–855.
29. Mirabel, P. & Katz, J.L. *J. Chem. Phys. 60*, 1974, 1138.
30. Hagena, O.F. & Obert, W. *J. Chem. Phys. 56*(5), 1972, 1793–1802.
31. Jonkman, H.T., Even, U., & Kommandeur, J. *J. Phys. Chem. 89*, 1985, 4240.
32. Mitsubishi, Japan, external document.
33. Yamada, I., Takaoka, H., Usui, H., & Takagi, T. *J. Vac. Sci. Technol. A 4*(3), 1986, 722–726.
34. Yamada, I. & Takagi, T. *Thin Solid Films 80*, 1981, 105–115.

35. Ueda, A., Usui, H., & Takagi, T. *Proceedings of 9th Symposium on Ion Sources and Ion Assisted Technology*. Tokyo: IEEJ, 1985, 45.
36. Usui, H., Ueda, A., Yamada, I., & Takagi, T. *Proceedings of 9th Symposium on Ion Sources and Ion Assisted Technology*. Tokyo: IEEJ, 1985, 39.
37. Yamada, I., Stein, G.D., Usui, H., & Takagi, T. *Proceedings of 10th International Symposium on Molecular Beams*. Cannes, France, 1985, 47.
38. Yang, S.-N. & Lu, T.-M. *J. Vac. Sci. Technol. B 5*(1), 1987, 355–358.
39. Ramachandra, A., Vaziri, M., & Andres, R.P. *Mat. Res. Soc. Sym. Proc. 206*, 1991, 397–401.
40. Hayashi, C. *Physics Today 40*(12), 1987, 44–51.
41. Kashu, S., Fuchita, E., & Manabe, T. *Jpn. J. Appl. Phys. 23*(12), 1984, L910–L912.
42. Friedlander, S.K., *Smoke, Dust and Haze*. New York: Wiley Interscience, 1977.
43. Wall, S., John, W., Wang, H.-C., & Goren, S.L. *Aerosol Sci. Technol. 12*, 1990, 926–946.
44. Brenner, S.S., Wrient, H.A., & Oriani, R.A. *Wear 68*, 1981, 169–190.
45. Dahneke, B. *J. Colloid Interface Sci. 37*(2), 1971, 342–353.
46. Dahneke, B. *J. Colloid Interface Sci. 40*(1), 1972, 1–13.
47. Dahneke, B. *J. Colloid Interface Sci. 45*(3), 1973, 584–590.
48. Dahneke, B. *J. Colloid Interface Sci. 51*(1), 1975, 58–65.
49. Rogers, L.N. & Reed, J. *J. Phys. D. Appl. Phys. 17*, 1984, 677–689.
50. Hinds, W.C. *Aerosol Technology*. New York: Wiley Interscience, 1982.
51. Kodas, T.T., Engler, E., & Lee, V. *Appl. Phys. Lett. 54*, 1989, 1923–1925.
52. Matsuno, S., Egawa, K., Yoshizaki, K., & Watarai, H. *Preparation of* $YBa_2Cu_3O_{7-y}$ *and Bi–Sr–Ca–Cu–O superconducting thin films by fine particle deposition method*. Tokyo: MRS Intern. Meeting on Advanced Materials, 1988.
53. Koguchi, M., Matsuda, Y., Kinoshita, E., & Hirabayashi, K. *J. Appl. Phys. Japan 29*(1), 1990, L33–L35.
54. Kasik, I. & Matejec, V. *J. Aerosol Sci. 26*(3), 1995, 399–406.
55. Kim, K.-S. & Pratsinis, S.E. *Chem. Eng. Sci. 44*(11), 1989, 2475–2482.
56. Bautista, J.R. & Atkins, R.M. *J. Aerosol Sci. 22*(5), 1991, 667–675.
57. Walker, K.L., Geyling, F.T., & Nagel, S.R. *J. Am. Ceram. Soc. 63*(9–10), 1980, 552–558.
58. Kim, K.-S. & Pratsinis, S.E. *AIChE J. 34*(6), 1988, 912–921.
59. Wiedensohler, A., Maximov, I., Samuelson, L., & Hansson, H.-C. *J. Aerosol Sci. 24*(4), 1993, 555–558.
60. Allendorf, M.D., Hurt, R.H., & Yang, N. *J. Mater. Res. 8*(7), 1993, 1651–1665.
61. Danek, M., Jensen, K.F., Murray, C.B., & Bawendi, M.G. *J. Crystal Growth 145*, 1994, 714–720.
62. Danek, M., Jensen, K.F., Murray, C.B., & Bawendi, M.G. *Appl. Phys. Lett. 65*(22), 1994, 2795–2797.
63. Kawarada, M., Kurihara, K., & Sasaki, K. *J. Appl. Phys. 71*(3), 1992, 1442–1445.
64. Fotou, G.P. & Pratsinis, S.E. *Chem. Eng. Sci. 49*(10), 1994, 1651–1662.
65. Dekker, J.P. *CVD techniques for the synthesis or modification of porous ceramics*. Ph.D. dissertation, Delft University, 1994.

66. Simmonds, M.G., Gladfelter, W.L., Li, H., & McMurry, P.H. *Mat. Res. Soc. Symp. Proc. 282*, 1993, 317–322.

67. Pratsinis, S.E., Kodas, T.T., Dudukovic, M.P., & Friedlander, S.K. *Chem. Eng. Sci. 41*, 1986, 693–700.

68. Pratsinis, S.E., Kodas, T.T., Dudukovic, M.P., & Friedlander, S.K. *Ind. Eng. Chem. Process Des. Dev. 25*(3), 1986, 634–642.

69. Pratsinis, S.E., Kodas, T.T., Dudukovic, M.P., & Friedlander, S.K. *Powder Tech. 47*, 1986, 17–23.

## 15.7 TABLES

**TABLE 15.1. Definition of methods for film generation**

| Name | Process |
|---|---|
| Spray Pyrolysis | Solution droplet deposition onto heated substrate |
| Aerosol-Assisted CVD (also called AACVD, spray pyrolysis, pyrosol process, aerosol CVD, spray MOCVD) | Chemical vapor deposition onto heated substrate with aerosol delivery of precursors, either in solution or solid-particle form |
| Aerosol Jet Etching and Cleaning | Deposition of solution particles or droplets onto heated substrate to form volatile products after reaction or to dislodge particles from the surface |
| Ionized Cluster Beam | Formation of ionized particles in vacuum system that are accelerated towards a substrate, where deposition occurs |
| Neutral Cluster Beam | Formation of particles in vacuum system that are deposited onto substrates |
| Supercritical Expansion of Solutions | Formation of particles by expansion of supercritical solution, followed by deposition |
| Inertial Deposition of Solid Particles | Formation of particles by assorted routes, followed by inertial deposition onto substrates |
| Thermophoretic Deposition | Deposition of particles in a temperature gradient, as with optical fibers |
| Particle-Precipitation-Aided CVD | Simultaneous deposition of vapor by CVD and of particles by thermophoresis |
| Electrophoretic Deposition | Deposition of charged particles in an electric field |
| Gravitational Deposition | Deposition of large ($\gg 1\ \mu m$) particles by sedimentation |
| Particle Formation During CVD | Particle formation due to homogeneous gas-phase reactions during CVD |

**TABLE 15.2. Assorted precursors, products, and conditions for film generation by droplet deposition for simple metal oxides**

| Material | Precursors | Conditions, °C | References |
|---|---|---|---|
| Sn(Sb) oxide | Sn and Sb butoxides | $T_s$ = 300–600 high-pressure atomizer | [1] |
| Ni–Zn Ferrite | $FeCl_2$, $NiCl_2$ $ZnCl_2$ | $T_s$ = 400 | [2] |
| $Al_2O_3$ doped with Ce | $AlCl_3$, $CeCl_3$ | $T_s$ = 280–486 | [3] |
| CdO | $Cd(NO_3)_2$ | $T_s$ = 550 | [4] |
| $In_2O_3$ doped with Sn | $InCl_3$, $SnCl_4$ | $T_s$ = 450 | [5] |
| $BaTiO_3$ | $Ba(O_2CCH_3)_2$ $Ti(O\text{-}i\text{-}Pr)_4$ + $H_2O$ | Formed sol, which was sprayed $T_s$ = 450 | [6] |
| $Ga_2O_3$ | $Ga(NO_3)_3$ | $T_s$ = 450, ultrasonic atomizer | [7] |
| $SnO_2$ | di-n-butyl tin diacetate | $T_s$ = 280–310 | [8] |
| $Ga_2O_3$ $Ga_2O_3$:Co | $GaCl_3$ $CoCl_2$ | $T_s$ = 350 | [9] |
| Mixed Al, Cr, Mo, W gels | $Al[M(CO)_3C_5H_5]_3$ M = Cr, Mo, W | $T_s$ = 165 | [10] |
| $TiO_2$ | $Ti(OCH_2CH_3)_4$ | $T_s$ = 300 supercritical spraying | [11] |
| $In_2O_3$:F | $InCl_3$ | $T_s$ = 310 | [12] |
| $Co_3S_4$, $Co_4S_3$, $Co_6S_5$, $Co_2O_3$ $Co_3O_4$ mixtures | $Co(NO_3)_x$ $(NH_2)_2CS$ | $T_s$ = 130–180 | [13] |
| $CrO_3$, $Cr_2O_3$ | $Cr(NO_3)_3 \cdot 9H_2O$ | $T_s$ = 500 | [14] |
| $Fe_2O_3$ | Iron citrate | $T_s$ = 25–225 | [15] |
| $\alpha$-$Fe_2O_3$ | Iron oxalate | $T_s$ = 400 | [16] |
| $Y_{3-x}Gd_xFe_5O_{12}$ | Metal citrate-nitrate gel particles in liquid | $T_s$ = 550 | [17] |
| $Fe_2O_3$ | $FeCl_3$ | $T_s$ = 500 | [18] |
| $H_3PO_4(WO_3)_{12} \cdot 29H_2O$ | $Na_2WO_4 \cdot 2HFe(NO_3)_2$ | $T_s$ = 25 | [19] |
| $Fe_2O_3$ | Iron citrate | $T_s$ = 400 | [20] |
| $CoO_x$, $FeO_x$ | $Co(NO_3)_2$, $Fe(NO_3)_2$ | $T_s$ = 170 | [21] |

*Note*: Unless otherwise stated, oxygen in the film is derived from precursor, air, or water present during deposition.

## References to Table 15.2


1. Gonzalez-Oliver, C.J.R. *J. Non-Crystal. Solids 82*, 1986, 400–410.
2. Talisa, S.H., Yoo, K.C., Abe, M., & Itoh, T. *J. Appl. Phys. 64*(10), 1988, 5819–5821.
3. Falcony, C., Garcia, M., Ortiz, A., Miranda, O., Gradilla, I., Soto, G., Cota-Araiza, L., Farias, M.H., & Alonso, J.C. *J. Electrochem. Soc. 141*(10), 1994, 2860–2863.
4. Gurumurugan, K., Mangalaraj, D., & Narayandass, S.K. *Thin Solid Films 251*, 1994, 7–9.
5. Ishida, T., Kouno, H., Kobayashi, H., & Nakato, Y. *J. Electrochem. Soc. 141*(5), 1994, 1357–1361.
6. Tripathi, A.K., Chariar, V., Goel, T.C., & Pillai, P.K.C. *Materials Sci. Eng. B25*, 1994, 34–38.
7. Wu, P., Gao, Y.-M., Kershaw, R., Dwight, K., & Wold, A. *Mat. Res. Bull. 25*, 1990, 357–363.
8. Yagi, I., Ikeda, E., & Kuniya, Y. *J. Mater. Res. 9*(3), 1994, 663–668.
9. Kim, H.-G. & Kim, W.-T. *J. Appl. Phys. 62*(5), 1987, 2000–2002.
10. Burlitch, J.M., DeMott, G.J., & Kohlstedt, D.L. *Mat. Res. Soc. Symp. Proc. 131*, 1989, 453–458.
11. Bocquet, J.F., Chhor, K., & Pommier, C. *Surface Coatings Technol. 70*, 1994, 73–78.
12. Singh, S.P., Raza, A., Sharma, A.K., Agnihotri, O.P., & Tewari, L.M. *Thin Solid Films 105*, 1983, 131–138.
13. Choudhury, C. & Sehgal, H.K. *Solar Energy 28*(1), 1982, 25–31.
14. Misho, R.H., Murad, W.A., & Fattahallah, G.H. *Thin Solid Films 169*, 1989, 235–239.
15. Pawar, S.K. & Pawar, S.H. *Mat. Res. Bull. 18*, 1983, 211–215.
16. Murthy, A.S.N. & Reddy, K.S. *Mat. Res. Bull. 19*, 1984, 241–246.
17. Bahadur, D., Roy, D., & Ram, S. Characterization of amorphous films of gadolinium iron garnet synthesized by spray pyrolysis. *Fourth Intrnl. Conf. on Ferrites, Part* 2 *16*, San Francisco, 1984, 651–655.
18. Pawar, S.H., Patil, P.S., Madhale, R.D., & Lokhande, C.D. *Indian J. Pure Appl. Phys. 27*, 1989, 227–230.
19. Lee, T.J., Fey, G.T.K., Lin, W.J., Yao, P.C., & Hsu, S.E. *Mater. Res. Soc. Symp. Proc. 135*, 1989, 597–602.
20. Sharon, M. & Prasad, B.M. *Solar Energy Materials 8*(4), 1983, 457–469.
21. Uma, C.S., Malhotra, L.K., & Chopra, K.L. *Bull. Mater. Sci. 8*(3), 1986, 385–389.

**TABLE 15.3. Precursors, products, and conditions for droplet deposition of metal sulfides (mainly CdS)**

| Material | Precursors | Conditions, °C | References |
|---|---|---|---|
| CdS solar cells | Cd propionate, formate, acetate, chloride, nitrate | $T_s = 320–380$ two-fluid atomizer | [1] |
| CdS, CuS solar cells | CuI, $CdCl_2$ | $T_s < 470$ electrostatic, two-fluid atomizer | [2] |
| CdS solar cells | $CdCl_2$ | $T_s = 380–420$ | [3] |
| CdS solar cells | $CdCl_2$ + $(NH_2)_2CS$ | $T_s = 500$ two-fluid atomizer | [4] |
| $Sb_2S_3$ | $SbCl_3$, thioacetamide | $T_s = 300$ | [5] |
| $CuBiS_2$ | CuCl, $BiCl_3$, $(NH_2)_2CS$ | $T_s = 150–400$ glass nozzle | [6] |
| $Cd_{1-x}Zn_xS$–$CuInSe_2$ fabricated solar cell | CuCl, $InCl_3$, $(NH_2)_2CSe$ | $T_s = 300$ | [7] |
| CdS–CdTe fabricated solar cell | $CdCl_2$, $(NH_2)_2CS$ | $T_s = 375–400$ large-area conveyor belt | [8] |
| $SnS_2$ | $SnCl_4$, $(NH_2)_2CS$ | $T_s = 300$ double-nozzle sprayer | [9] |
| CdS doped with oxygen | $CdCl_2$, $Cd(CH_3COO)_2$, $(NH_2)_2CS$ | $T_s = 200–300$ nozzle translated for coating uniformity | [10] |
| $Co_xIn_2S_{3+x}$ | $CoCl_2$, $InCl_3$, $(NH_2)_2CS$ | $T_s = 270$ | [11] |
| CdS:In | $CdCl_2$, $InCl_3$, $NH_2CS$ | $T_s = 325$ nozzle translated | [12] |
| $CuInS_2$ | $Cu(CH_3COO)_2$, $InCl_3$, $(NH_2)_2CS$ | $T_s = 200–450$ | [13] |
| CdS | $CdCl_2$, $(NH_2)_2CS$ | $T_s = 250$ studied intermediates | [14] |
| CdS | $CdCl_2$, $(NH_2)_2CS$ | $T_s = 235–450$ studied nucleation and growth on surface | [15] |
| $CuInS_2$ | $Cu(CH_3COO)_2 \cdot H_2O$, $InCl_3$, $(NH_2)_2CS$ | $T_s = 300$ | [16] |

**TABLE 15.3.** (*Continued*)

| Material | Precursors | Conditions, °C | References |
|---|---|---|---|
| $Zn_xCd_{1-x}S$ $x = 0–0.35$ | $ZnCl_2$, $CdCl_2$, $(NH_2)_2CS$ | $T_s = 450$ | [17] |
| $CdS_xSe_{1-x}$ $Zn_xCd_{1-x}S$ $0 \leq x \leq 1$ | $CdCl_2$, $(NH_2)_2CS$, $(NH_2)_2CSe$ | $T_s = 450$ | [18] |
| CdS | $CdCl_2$, $(NH_2)_2CS$ | $T_s = 340$ studied purity as function of $O_2$ partial pressure | [19] |
| $Bi_2CdS_4$ | $BiCl_3$, $CdCl_2$, $(NH_2)_2CS$ | $T_s = 250–450$ examined effect of substrate temperature | [20, 21] |
| $AgInS_2$ $AgIn_5S_8$ | AgCl, $InCl_3$ $(NH_2)_2CS$ | $T_s = 200–250$ | [22] |
| $CdS{:}Al_2O_3$ $CdS{:}SnO_2$ | $AlCl_3$, $CdCl_2$, $SnCl_4$, $(NH_2)_2CS$ | $T_s = 380$ | [23] |
| ZnS:Tb ZnS:Sm | $Zn(CH_3COO)_2$, $TbCl_3$, $SmCl_3$ $(NHCH_3)_2CS$ | $T_s = 375$ phosphor material | [24] |
| ZnS:In, Al, Cu doped | $ZnCl_2$, $InCl_3$, $AlCl_3$, $CuCl_2$, $(NHCH_3)_2CS$ | $T_s = 220–330$ | [25] |

## References to Table 15.3

1. Chamberlin, R.R. & Skarman, J.S. *Electrochem. Soc 113*(1), 1966, 86–89.
2. Cumberbatch, T.J., Barden, P.E., & Durrant, J. *Thin Solid Films 167*, 1988, 169–174.
3. Ebothé, J., Nguyen Cong, H., & Chartier, P. Preparation atmosphere effects on the texture and spectral properties of thin sprayed CdS films. In Vincenzini, P. (Ed.), *High Tech Ceramics*. Amsterdam: Elsevier, 1987, 1727–1736.
4. Gupta, B.K. & Agnihotri, O.P. *Solid State Communications 23*, 1977, 295–300.
5. Bhosale, C.H., Uplane, M.D., Patil, P.S., & Lockhande, C.D. *Thin Solid Films 248*, 1994, 137–139.
6. Pawar, S.H., Pawar, A.J., & Bhosale, P.N. *Bull. Mater. Sci. 8*(3), 1986, 423–426.
7. Raja Ram, P., Thangaraj, R., & Agnihotri, O.P. *Bull. Mater. Sci. 8*(4), 1986, 279–284.
8. Jordan, J.F. & Albright, S.P. *Solar Cells 23*(1–2), 1988, 107–113.
9. Abass, A.K., Majeid, K.J., Jassim, H.A., & Murad, W.A. *Solid State Comm. 57*(10), 1986, 805–808.

10. Richards, D., El-Korashy, A.M., & Stirn, R.J. *J. Vac. Sci. Tech. A. 2*(2), 1984, 332–335.

11. Kim, W.-T., Yun, C.-S., Jeong, H.-M., & Kim, C.-D. *J. Appl. Phys. 60*(7), 1986, 2357–2360.

12. Uekita, M., Nelson, K.F., Bates, C.W., Jr., Mooney, J.B., & Recktenwald, J.M. *Thin Solid Films 88*, 1982, 275–278.

13. Tiwari, A.N., Pandya, D.K., & Chopra, K.L. *Thin Solid Films 130*, 1985, 217–230.

14. Krunks, M., Mellikov, E., & Sork, E. *Thin Solid Films 145*, 1986, 105–109.

15. Albin, D.S. & Risbud, S.H. *Thin Solid Films 147*, 1987, 203–212.

16. Onnagawa, H. & Miyashita, K. *Jpn. J. Appl. Phys. 23*(8), 1984, 965–969.

17. Chynoweth, T.A. & Bube, R.H. *J. Appl. Phys. 51*(3), 1980, 1844–1846.

18. Feigelson, R.S., N'Diaye, A., Yin, S.-Y., & Bube, R. *J. Appl. Phys. 48*(7), 1977, 3162–3164.

19. Shirreffs, B.F., Cheng, C.H., Geib, K., & Jones, K.A. *J. Electrochem. Soc. Solid State Tech. 131*(2), 1984, 440–443.

20. Pawar, S.H., Tamhankar, S.P., & Lokhande, C.D. *J. Mater. Sci. Lett. 3*, 1984, 427–430.

21. Pawar, S.H., Tamhankar, S.P., & Lokhande, C.D. *Phys. Status Solidi A. 82*(2), 1984, K195–K198.

22. Gorska, M., Beaulieu, R., Loferski, J.J., & Roessler, B. *Thin Solid Films 67*, 1980, 341–345.

23. Thakoor, A.P., Raj, B., Pandya, D.K., & Chopra, K.L. *Thin Solid Films 83*, 1981, 231–237.

24. Ortiz, A., Falcony, C., Garcia, M., & Sanchez, A. *Semicond. Sci. Tech. 3*, 1988, 537–541.

25. Ortiz, A., Garcia, M., & Sanchez, A. *J. Electrochem. Soc. 136*(4), 1989, 1232–1235.

**TABLE 15.4. Precursors, products, and conditions for droplet deposition for selenides and tellurides**

| Material | Precursors | Conditions, °C | References |
|---|---|---|---|
| $CuInSe_2$ solar cells | CuCl, $(CH_3NH)_2CSe_2$ | $T_s$ = 250 two-fluid atomizer | [1] |
| $CuInSe_2$ | $CuCl_2$, $InCl_3$, $(NH_2)_2CSe$ | $T_s$ = 200–250 studied solution chemistry | [2] |
| $CuInSe_2$ | $CuCl_2$, $InCl_3$, $(NH_2)_2CSe$ | $T_s$ = 180–310 correlated properties with fabrication conditions | [3] |
| $CuInSe_2$ | $CuCl_2$ $InCl_3$, $(CH_3NH)_2CSe$ | $T_s$ = 150–400 | [4] |
| $CuInSe_2$ $CuGaSe_2$ $CuGaS_2$ $CuInS_2$ | $CuCl_2$, $InCl_3$, $GaCl_3$, $(NH_2)_2CSe$ $(NH_2)_2CS$ | $T_s$ = 200–600 poor film above 450°C | [5] |
| $CuInSe_2$:$In_2O_3$ | — | $T_s$ = 250 correlated film properties with $In_2O_3$ content | [2] |
| CdTe | $CdCl_2$, $TeCl_4$ | $T_s$ = 425–500 | [6] |
| CdSe | $CdCl_2$, $(NH_2)_2CSe$ | $T_s$ = 200–300 photoelectrochemical cell applications | [7] |

## References to Table 15.4


1. Mooney, J.B. & Lamoreaux, R.H. *Electrochem. Soc. Proc. 83*(7), 1983, 89–93.
2. Brown, B.J., Haba, B., & Bates, C.W., Jr. *J. Electrochem. Soc. 135*(6), 1988, 1559–1561.
3. Bougnot, J., Duchemin, S., & Savelli, M. *Solar Cells 16*(1–4), 1986, 221–236.
4. Bates, C.W., Jr., Varner, E.B., Alshvang, S., & Summers, K. *Mat. Res. Soc. Proc. 282*, 1993, 653–658.
5. Pamplin, B. & Feigelson, R.S. *Thin Solid Films 60*, 1979, 141–146.
6. Boone, J.L., vanDoren, P., & Berry, A.K. *Thin Solid Films 87*, 1982, 259–264.
7. Mangalhara, J.P., Thangaraj, R., & Pagnihotri, O.P. *Bull. Mater. Sci. 10*(4), 1988, 333–340.

**TABLE 15.5. Precursors, products, and conditions for droplet deposition for zinc oxide**

| Material | Precursors | Conditions, °C | References |
|---|---|---|---|
| In-doped ZnO | Zinc acetate<br>$InCl_3$ | $T_s$ = 300–450<br>two-fluid atomizer | [1] |
| ZnO | Zinc acetate | $T_s$ = 400<br>two-fluid atomizer | [2] |
| ZnO solar cells | $Zn(NO_3)_2$<br>$ZnCl_2$<br>Zn acetate | $T_s$ = 340–480<br>two-fluid atomizer | [3] |
| ZnO | $ZnCl_2$ | $T_s$ = 375–525<br>studied effect of substrate temperature | [4] |
| ZnO | $Zn(CH_3COO)_2$<br>$ZnCl_2$ | $T_s$ = 300–450<br>no deposition onto InP using zinc chloride; possibly vapor deposition | [5] |
| ZnO:Tb | $Zn(CH_3COO)_2$<br>$TbCl_3$ | $T_s$ = 270–400 | [6] |

## References to Table 15.5

1. Major, S., Banerjee, A., & Chopra, K.L. *Thin Solid Films 108*, 1983, 333–340.
2. Major, S., Banerjee, A., & Chopra, K.L. *Thin Solid Films 125*, 1985, 179–185.
3. Aranovich, J., Ortiz, A., & Bube, R.H. *J. Vac. Sci. Technol. 16*(4), 1979, 994–1003.
4. Cossement, D. & Streydio, J.-M. *J. Crystal Growth 72*, 1985, 57–60.
5. Eberspacher, C., Fahrenbruch, A.L., & Bube, R.H. *Thin Solid Films 136*, 1986, 1–10.
6. Ortiz, A., Falcony, C., Garcia, M., & Sanchez, A. *J. Phys. D 20*(5), 1987, 670–671.

**TABLE 15.6. Precursors, products, and conditions for droplet deposition for ceramic superconductors**

| Material | Precursors | Conditions, °C | References |
|---|---|---|---|
| Y–Ba–Cu–O<br>2–7 μm thick | Nitrates | $T_s$ = 200–400 | [1] |
| Tl–Ca–Ba–Cu | Nitrates | $T_s$ = 300<br>ultrasonic atomizer | [2] |
| Y–Ba-Cu-O<br>1.0 μm thick | Nitrates | $T_s$ = 250–300<br>two-fluid atomizer | [3] |
| $YBa_2Cu_3O_7$<br>(no rate) | Nitrates | $T_s$ = 250–850<br>two-fluid atomizer | [4] |
| $YBa_2Cu_3O_7$<br>$YBa_2Cu_3O_7$–Ag<br>Bi–Sr–Ca–Cu–O | Nitrates | $T_s$ = 325–350<br>two-fluid atomizer | [5] |
| Y–Ba–Cu–O | Nitrates | $T_s$ = 400 | [6] |
| $YBa_2Cu_3O_{7-x}$<br>0.2–1 μm/hr | Nitrates | $T_s$ = 200–400<br>ultrasonic | [7] |
| $YBa_2Cu_3O_{7-x}$ | Nitrates | $T_s$ = 150–190<br>ultrasonic | [8] |
| Y–Ba–Cu–O | Nitrates | $T_s$ = 400<br>two-fluid atomizer | [9, 10] |
| $YBa_2Cu_3O_7$<br>~ 1 μm/hr | Nitrates | $T_s$ = 250 | [11] |
| Y–Ba–Cu–O | Nitrates | $T_s$ = 200–450<br><20-μm droplets | [12] |
| Y–Er–Ba–Cu–O | Nitrates | $T_s$ = 400 | [13] |
| Y–Ba–Cu–O | Nitrates | $T_s$ = 650 | [14] |
| $Bi_{2.15}Sr_{1.68}Ca_{1.17}$<br>$Cu_2O_x$ | Nitrates | $T_s$ = 500–550 | [15] |
| Bi–Sr–Ca–Cu–O | Nitrates | $T_s$ = 600 | [16] |
| Bi–Pb–Sr–Ca–Cu–O | Nitrates | $T_s$ = 400–500 | [17] |
| Bi–Sr–Ca–Cu–O | Nitrates | $T_s$ = 400 | [18] |
| Bi–Sr–Ca–Cu–O | Nitrates | $T_s$ = 350 | [19] |
| Bi–Pb–Sr–Co–Cu–O | Nitrates | $T_s$ = 400 | [20] |
| Bi–Ca–Sr–Cu–O | Acetates | $T_s$ = 25 | [21] |

## References to Table 15.6


1. Chu, J.J., Liu, R.S., Wu, P.T., & Chen, L.J. *Physica C 153–155*, 1988, 804–805.
2. DeLuca, J.A., Garbauskas, M.F., Bolon, R.B., McMullen, J.G., Balz, W.E., & Karas, P.L. *J. Mater. Res. 6*(7), 1991, 1415–1424.
3. Derraa, A. & Sayer, M. *J. Appl. Phys. 68*(3), 1990, 1401–1403.
4. Henry, R.L., Lessoff, H., Swiggard, E.M., & Qadri, S.B. *J. Crystal Growth 85*, 1987, 615–618.
5. Henry, R.L., Cukauskas, E.J., Qadri, S.B., Singer, A.H., & Campisi, G.G. *IEEE Trans. Magnetics 25*(2), 1989, 2352–2355.

6. Kawai, M., Kawai, T., Masuhira, H., & Takahasi, M. *Jpn. J. Appl. Phys.* *26*(10), 1987, L1740–L1742.

7. Langlet, M., Senet, E., Deschanvres, J.L., Delabouglise, G., Weiss, F., & Joubert, J.C. *J. Less-Common Metals 151*, 1989, 399–406.

8. Peng, J., Danzinger, M., Haubner, R., & Lux, B. *J. de Physique IV C2*, 1991, CZ721–CZ728.

9. Saxena, A.K., Arya, S.P.S., Das, B., Singh, A.K., Tiwari, R.S., & Srivastava, O.N. *Solid State Comm.* *66*(10), 1988, 1063–1065.

10. Saxena, A.K., Das, B., Arya, S.P.S., Mandal, P., Tripathi, C.C., Singh, A.K., Tiwari, R.S., & Srivastava, O.N. *Reviews of Solid State Science* *2*(2, 3), 1988, 397–402.

11. Wu, N.L., Wang, Y.K., & Lin, S.H. Preparation of $YBa_2Cu_3O_{7-x}$ silver composite ribbons by spray pyrolysis. Am. Institute of Physics, 1992, 303–311.

12. Risbud, S.H., Bania, W.R., & McIntyre, L.C., Jr. *Physica C 153–155*, 1988, 1679–1680.

13. Ban, E., Matsuoka, Y., & Ogawa, H. *J. Phys. D.: Appl. Phys. 22*, 1989, 1235–1236.

14. Nobumasa, H., Shimizu, K., Kitano, Y., Tanaka, M., & Kawai, T. *Jpn. J. Appl., Phys.* *27*(8), 1988, 1544–1545.

15. Cooper, E.I., Giess, E.A., & Gupta, A. *Materials Lett.* *7*(1, 2), 1988, 5–8.

16. Hsu, H.M., Yee, I., DeLuca, J., Hilbert, C., Miracky, R.F., and Smith, L.N. *J. Appl. Phys. Lett.* *54*(10), 1989, 957–959.

17. Koo, H.S., Tu, G.C., Chu, J.J., Lee, W.H., & Wu, P.T. *J. Mater. Sci. Lett. 9*, 1990, 1326–1328.

18. Walia, D.K., Gupta, A.K., Reddy, G.S.N., Tomar, V.S., Kataria, N.D., Ojha, V.N., & Khare, N. *Solid State Comm.* *71*(11), 1989, 987–990.

19. Roul, B.K., Pramanik, P., & Chopra, K.L. *Physica C 160*, 1989, 439–442.

20. Nobumasa, H., Shimizu, K., Kitano, Y., & Kawai, T. *Jpn. J. Appl. Phys. Part 2* *27*(9), 1988, L1669–L1671.

21. Vaslow, D.F., Dieckmann, G.H., Elli, D.D., Ellis, A.B., Holmes, D.S., Lefkow, A., MacGregor, M., Nordman, J.E., Petras, M.F., & Yang, Y. *Appl. Phys. Lett.* *53*(4), 1988, 324–336.

**TABLE 15.7. Chemical vapor deposition with aerosol delivery of precursors**

| Material | Precursors | Conditions, °C | References |
|---|---|---|---|
| Cu, Ag | ($\beta$-diketonate)CuL<br>($\beta$-diketonate)$AgL_n$<br>$Pd(hfac)_2$ | $T_s$ = 100–250<br>spray nozzle | [1, 2] |
| $Al_2O_3$ | $Al(acac)_3$ | $T_s$ = 430<br>ultrasonic atomizer | [3] |
| $Al_2O_3$ | $Al(acac)_3$ | $T_s$ = 400–600<br>spray nozzle | [4] |
| $Al_2O_3$<br>$Al_2O_3$–$ZrO_2$ | Al(sec-butoxide)$_3$<br>Zr(n-butoxide)$_4$ | $T_s$ = 550 | [5] |
| CuO | $Cu(O_2CCH_3)_2$ | $T_s$ = 375<br>ultrasonic atomizer | [6] |
| $Cr_2O_3$ | $Cr(acac)_3$ | $T_s$ = 450<br>ultrasonic atomizer | [7] |
| $\gamma$-$Fe_2O_3$–$Fe_3O_4$ | $Fe(acac)_3$ | $T_s$ = 420–540<br>ultrasonic atomizer | [8] |
| $Nd_2O_3$ | $Nd(tfac)_3$ | $T_s$ = 600 | [8] |
| $SnO_2$ doped with Sb | $SnCl_4$ | $T_s$ = 575–670<br>two-stage: evaporation and deposition | [9] |
| $SnO_2$ | $SnCl_4$ | $T_s$ = 300–500<br>ultrasonic atomizer | [10] |
| $SnO_2$ | $SnCl_4$ | $T_s$ = 250–400<br>conventional spray nozzle | [11] |
| $SnO_2$ | $SnCl_4$ | $T_s$ = 450<br>two-fluid atomizer | [12] |
| $SnO_2$<br>(Cd, Zn)S | $SnCl_4$<br>$CdCl_2$<br>$ZnCl_2$ | $T_s$ = 400–550<br>airless, pneumatic, electrostatic, ultrasonic | [13] |
| $SnO_2$ | $SnCl_4$ | $T_s$ = 250–300 | [14] |
| $SnO_2$<br>solar cells | $SnCl_4$,<br>Sb chloride | $T_s$ = 300–600<br>two-fluid atomizer | [15] |
| $In_2O_3$:$SnO_2$<br>(ITO) | $SnCl_4$, $In(NO_3)_3 \cdot 5H_2O$ | $T_s$ = 400–500<br>radio frequency plasma above substrate | [16] |
| $SnO_2$ | $SnCl_4$, $NH_4F$<br>alcohol solvent | $T_s$ = 410<br>sprayed upward and downward; upward gave better film properties | [17] |
| $SnO_2$<br>375–480 nm/min | $SnCl_4$ | $T_s$ = 200–500<br>two-fluid atomizer | [18] |

**TABLE 15.7.** (*Continued*)

| Material | Precursors | Conditions, °C | References |
|---|---|---|---|
| $SnO_2$ | $SnCl_4$ | $T_s = 500$ | [19] |
| $SnO_2$ | $SnCl_4$ | $T_s = 425$ ultrasonic | [20] |
| $TiO_2$ | $TiCl_4$ | $T_s = 292–462$ | [21] |
| $TiO_2$ | $Ti(O\text{-}i\text{-}Pr)_2(acac)_2$ | $T_s = 500$ ultrasonic atomizer | [22] |
| $TiO_2$ | $Ti(acac)_4$ | $T_s = 380$ ultrasonic atomizer | [23] |
| $ZrO_2$ | $Zr(acac)_4$ | $T_s = 450$ ultrasonic atomizer | [24] |
| $In_2O_3$, $SnO_2$ | $InCl_3$, $SnCl_4$ | $T_s = 300–450$ ultrasonic atomizer | [25] |
| $In_2O_3$, $SnO_2$, $Fe_2O_3$, $Cr_2O_3$, $Al_2O_3$, $Y_2O_3$, $V_2O_3$, $VO_2$, $TiO_2$, $ZrO_2$ | Fe, Cr, In, Y, V acetylacetonates, $Al(O\text{-}i\text{-}Pr)_3$, $SnCl_4$, $Zr(OR)_4$, and $Ti(OR)_4$ | $T_s = 300–650$ ultrasonic atomizer | [26] |
| $Fe_2O_3$, $SnO_2$, $In_2O_3$, $Cr_2O_3$,$V_2O_3$, Pd, Ru | Fe, In, Cr, V, Pd, and Ru acetylacetonates, $SnCl_4$ | $T_s = 380–560$ ultrasonic atomizer | [27] |
| Pb-doped $SnO_2$ | $(n\text{-}Bu)_2Sn(CH_3COO)_2$ | $T_s = 460–560$ ultrasonic atomizer | [28] |
| $Y_2O_3$, $La_2O_3$, $La_2CuO_4$ | $Y(acac)_3$, $La(acac)_3$, $Cu(acac)_2$ | $T_s = 400–550$ ultrasonic atomizer | [29] |
| Pb–La–Zr–Ti–O | Metal nitrates | $T_s = 165$ | [30] |
| $LiTaO_3$ | $LiTa(O\text{-}t\text{-}Bu)_6$ | Pulsed liquid injection $T_s = 550–800$ | [31] |
| $LiNbO_3$ | $LiNb(OEt)_6$ | $T_s = 463$ ultrasonic atomizer | [32] |
| $MgF_2$ | $Mg(tfac)_2$, $Mg(hfac)_2$ | $T_s = 400–700$ ultrasonic atomizer | [33] |
| CdS | $CdCl_2$ $(NH_2)_2CS$ | $T_s = 250–450$ ultrasonic atomizer | [34] |
| ZnS | $ZnCl_2$ and $(NH_2)_2CS$ | $T_s = 500$ ultrasonic atomizer | [35] |
| ZnS, CdS $Zn_{1-x}Cd_xS$ | thiocarboxylates | $T_s = 125–250$ | [36, 37] |
| $Ga_2S_3$ | thiocarboxylates | $T_s = 275–310$ | [38] |
| $In_2S_3$ | thiocarboxylates | $T_s = 210$ | [39] |
| CaS, SrS $Ca_{1-x}Sr_xS$ | thiocarboxylates | $T_s = 300–400$ | [40, 41] |

*Note*: Unless otherwise stated, oxygen in the film is derived from precursors, air, or water present during deposition.

## References to Table 15.7

1. Roger, C., Corbitt, T., Kodas, T.T., & Hampden-Smith, M.J. *Appl. Phys. Lett* *65*(8), 1994, 1021–1023.
2. Roger, C., Corbitt, T., Kodas, T.T., & Hampden-Smith, M.J. *J. Aerosol Sci. 25*, 1994, S221.
3. DeSisto, W.J., Qian, Y.-T., Hannigan, C., Edwards, J.O., Kershaw, R., Dwight, K., & Wold, A. *Mat. Res. Bull. 25*(2), 1990, 183–189.
4. Lambeck, P.V., Hilderink, L., & Popma, T.J.A. Thin-layer formation by chemical aerosol deposition. *Aerosols: Formation and Reactivity, 2nd Int. Aerosol Conf. Berlin.* Oxford, U.K.: Pergamon, 1986, 964.
5. Nguyen, D., van Roode, M., & Johar, A.S. *Thin Solid Films 135*, 1986, L19–L21.
6. DeSisto, W., Sosnowski, M., Smith, F., Deluca, J., Kershaw, R., Dwight, K., & Wold, A. *Mat. Res. Bull. 24*(6), 1989, 753–760.
7. Qian, Y.-T., Kershaw, R., Dwight, K., & Wold, A. *Mat. Res. Bull. 25*(10), 1990, 1243–1250.
8. Langlet, M. & Joubert, J.C. The pyrosol process or the pyrolysis of an ultrasonically generated aerosol. In Rao, C.N.R. (Ed.), *Chemistry of Advanced Materials*, Oxford, U.K.: Blackwell Scientific Publications, 1993, 55.
9. Parker, E.H. & Piazza, G.J. *Method for depositing thin, transparent metal oxide films.* United States Patent 4,571,350, Feb. 18, 1986.
10. Sears, W.M. & Gee, M.A. *Thin Solid Films 165*, 1988, 265–277.
11. Miki-Yoshida, M. & Andrade, E. *Thin Solid Films 224*, 1993, 87–96.
12. Afify, H.H., Momtaz, R.S., Badawy, W.A., & Nasser, S.A. *J. Mater. Sci. 2*, 1991, 40–45.
13. Orban de Xivry, E., Streydio, J.-M., & Berote, G. Chemical spray technology adapted to thin film semiconductor growth. In Vincenzini, P. (Ed.) *High Tech Ceramics.* Amsterdam: Elsevier, 1987, 1709–1718.
14. Sharma, T.P. & Pandey, C.P. *Bull. Mater. Sci. 72*, 1985, 131–135.
15. Mulla, I.S., Soni, H.S., Rao, V.J., & Sinha, A.P.B. *J. Mater. Sci. 21*, 1986, 1280–1288.
16. Lee, D.H., Vuong, K.D., Williams, J.A.A., Fagan, J., Condrate, R.A., Sr., & Wang, X.W. *J. Mater. Res. 11*(4), 1996, 895–903.
17. Arya, S.P.S. *Cryst. Res. Technol. 21*(3), 1986, K38–K42.
18. Omar, O.A., Ragaie, H.F., & Fikry, W.F. *J. Mater. Sci.*, 1990, 79–83.
19. Garcia, F.J., Muci, J., & Tomar, M.S. *Thin Solid Films 97*, 1982, 47–51.
20. Yoon, W.H. & Song, J.S. *Thin Solid Films 224*, 1993, 203–208.
21. Zhang, S., Zhu, Y.F., & Brodie, D.E. *Thin Solid Films 213*, 1992, 265–270.
22. Cui, H., Shen, H.-S., Gao, Y.-M., Dwight, K., & Wold, A. *Mat. Res. Bull. 28*, 1993, 195–201.
23. Xu, W.W., Kershaw, R., Dwight, K., & Wold, A. *25*, 1990, 1385–1392.
24. Gao, Y.-M., Wu, P., Kershaw, R., Dwight, K., & Wold, A. *Mat. Res. Bull. 25*, 1990, 871–876.
25. Siefert, W. *Thin Solid Films 121*, 1984, 275–282.

26. Blandenet, G., Court, M., & Legard, Y. *Thin Solid Films 77*, 1981, 81.

27. Viguié, J.C. & Spitz, J. *J. Electrochem. Soc. 122*(44), 1975, 585–588.

28. Labeau, M., Gautheron, B., Delabouglise, G., Peña, J., Ragel, V., Varela, A., Román, J., Martinez, J., González-Calbet, J.M., & Vallet-Regí, M. *Sensors and Actuators B 15–16*(1–3), 1993, 379–383.

29. Gao, Y.-M., Lee, W., Trehan, R., Kershaw, R., Dwight, K., & Wold, A. *Mat. Res. Bull. 26*, 1991, 1247–1254.

30. Delikouras, E.A. & Perlmutter, D.D. *J. Am. Ceram. Soc. 77*(12), 1994, 3142–3144.

31. Xie, H. & Raj, R. *Appl. Phys. Lett 63*(23), 1993, 3146–3148.

32. Weinberg, M.C. *J. Am. Ceram. Soc. 66*(6), 1983, 439–443.

33. Joosten, P.H., Heller, P., Nabben, H.J.P., VanHal, H.A.M., Popma, T.J.A., & Haisma, J. *Applied Optics 24*(16), 1985, 2674–2678.

34. Chung, G.Y., Kim, H.D., Ahn, B.T., & Im, H.B. *Thin Solid Films 232*, 1993, 28–33.

35. Maatman, D., Gruisinga, W., Lambeck, P.V., & Popma, T.J.A. *J. Aerosol Sci. 19*(7), 1988, 1368–1372.

36. Nyman, M., Hampden-Smith, M.J., & Duesler, E. *Adv. Mater., CVD 5*, 1996, 171–174.

37. Nyman, M., Hampden-Smith, M.J., Kodas, T.T., Yap, G., & Rheingold, A.L. *Chem. Mater. 10*(3), 1998, 913–921.

38. Shang, G., Hampden-Smith, M.J., & Duesler, E.N. *J. Chem. Soc. Chem. Commun.*, 1996, 1773–1774.

39. Shang, G., Kunze, K., & Hampden-Smith, M.J. *Adv. Mater., CVD Comm. 2*(9), 1996, 242–244.

40. Kunze, K., Bihry, L., Anatassova, P., Hampden-Smith, M.J., & Duesler, E.N. *Adv. Mater., CVD 2*(3), 1996, 105–108.

41. Kunze, K., Hampden-Smith, M.J., Kodas, T.T., & Duesler, E.N. *Inorg. Chem.*, forthcoming.

**TABLE 15.8. Chemical vapor deposition with aerosol delivery of precursors for ceramic superconductors**

| Material | Precursors | Conditions, °C | References |
|---|---|---|---|
| $Y_1Ba_2Cu_3O_{7-\delta}$ | Y, Ba, Cu tmhd | $T_s$ = 400–560<br>ultrasonic atomizer | [1] |
| $Y_1Ba_2Cu_3O_{7-\delta}$ | Y, Ba, Cu tmhd | $T_s$ = 500–970<br>solid particles | [2, 3] |
| $Y_1Ba_2Cu_3O_{7-\delta}$ | Y, Ba, Cu tmhd | $T_s$ = 700<br>ultrasonic atomizer | [4] |
| $Y_1Ba_2Cu_3O_{7-\delta}$ | Y, Ba, Cu tmhd | $T_s$ = 450<br>ultrasonic atomizer | [5, 6] |
| $Y_1Ba_2Cu_3O_{7-\delta}$ | Y, Ba, Cu tmhd | $T_s$ = 500–900<br>warm-wall, two-fluid nozzle | [7] |
| $Y_1Ba_2Cu_3O_{7-\delta}$ | Metal nitrates | $T_s$ = 800–900<br>ultrasonic atomizer and plasma to vaporize involatile species | [8] |
| $Y_1Ba_2Cu_3O_{7-\delta}$<br>Bi–Sr–Ca–Cu–O | Bi, Sr, Ca, Y, Ba, Cu tmhd | $T_s$ = 700–830<br>ultrasonic atomizer | [9] |

## References to Table 15.8


1. Driessen, A., Tang, Q., Hilderink, L., & Popma, T.J.A. *Mat. Res. Soc. Symp. Proc. 169*. 1989, 601.
2. Lackey, W.C., Hanigofsky, J.A., Shapiro, M.J., Carter, W.B., Hill, D.N., Barefield, E.K., Judson, E.A., O'Brien, D.F., Chung, Y.S., Moss, T.S., & More, K.L. Preparation of superconducting wire by deposition of $YBa_2Cu_3O_x$ onto fibers. 11th International CVD Conf., Seattle, 1990.
3. Lackey, W.J., Carter, W.B., Hanigofsky, J.A., Hill, D.N., Bearefield, E.K., Neumeier, G., O'Brien, D.F., Shapiro, M.J., Thompson, J.R., Green, A.J., Moss, II., T.S., Jake, R.A., & Efferson, K.R. *Appl. Phys. Lett. 56*(12), 1990, 1175–1177.
4. Matsuno, S., Uchikawa, F., Utsunomiya, S., & Nakabayashi, S. *Appl. Phys. Lett. 60*(19), 1992, 2427–2429.
5. Tang, Q., Driessen, A., Hoekstra, P., Hilderink, L.T., Van Silfhout, A., & Popma, T.J.A. *J. Less-Common Metals 164–165*, 1990, 1587–1594.
6. Tang, Q., Albers, H., Driessen, A., Hilderink, L.T.H., Lambeck, P.V., & Popma, T.J.A. *J. Aerosol Sci. 27*(S1), 1990, S737–S740.
7. Salazar, K.V., Ott, K.C., Rye, R.C., Hubbard, K.M., Peterson, E.J., Coulter, J.Y., & Kodas, T.T. *Physica C. 198*(3–4), 1992, 303–308.
8. Koukitu, A., Hasegawa, Y., Seki, H., Komijama, H., Tanaka, I., & Kamioka, Y. *J. Appl. Phys. Japan 28*(7), 1989, L1212–L1213.
9. Gorbenko, O.Y., Fuflyigin, V.N., Erokhin, Y., Graboy, I.E., Kaul, A.R., & Tretyakov, Y.D. *J. Mater. Chem. 4*, 1994, 1585–1589.

**TABLE 15.9. Materials produced by supercritical nozzle expansion**

| Material | Precursors | Conditions | References |
|---|---|---|---|
| $SiO_2$, $GeO_2$ 1 $\mu$m/min | $SiO_2$, $GeO_2$ | Supercritical $H_2O$ production of powders | [1] |
| $SiO_2$ + polymers | $SiO_2$ | $H_2O$, $NH_3$, $H_2O$ ethanol, n-pentane as solvents, rapid expansion | [2] |
| $SiO_2$ | $SiO_2$ | $H_2O$, $NH_3$, $H_2O$ ethanol, n-pentane as solvents | [3] |

## References to Table 15.9


1. Matson, D.W., Petersen, R.C., & Smith, R.D. *J. Mater. Sci. 22*, 1987, 1919–1928.
2. Matson, D.W., Fulton, J.L., Petersen, R.C., & Smith, R.D. *Ind. Eng. Chem. Res. 26*, 1987, 2298–2306.
3. Petersen, R.C., Matson, D.W., & Smith, R.D. *J. Amer. Chem. Soc. 108*(8), 1986, 2100–2102.


**TABLE 15.10. Aerosol jet etching and cleaning**

| Material | Precursors | Conditions | References |
|---|---|---|---|
| $SiO_2$ | HF | Droplets 0.05–0.2 $\mu$m $T_s$ = 30–50°C | [1, 2] |
| $SiO_2$ | HF | Droplets 0.05–0.2 $\mu$m $T_s$ = 30–50°C | [3] |
| Fe–Ag, frozen $CO_2$ | $CO_2$ | Cleaning substrate with $CO_2$ particles | [4] |
| Frozen $CO_2$ | $CO_2$ | Cleaning substrate with $CO_2$ particles | [5] |

## References to Table 5.10


1. Chen, Y.L., Brock, J.R., & Trachtenberg, I. *Aerosol Science Technol. 12*, 1990, 842–855.
2. Chen, Y.L., Brock, J.R., & Trachtenberg, I. *Appl. Phys. Lett. 51*(26), 1987, 2203–2205.
3. Chung, C.W., Brock, J.R., & Trachtenberg, I. *Appl. Phys. Lett. 63*(244), 1993, 3341–3343.
4. Kashu, S., Fuchita, E., & Manabe, T. *Jpn. J. Appl. Phys. 23*(12), 1984, L910–L912.
5. Sherman, R. & Whitlock, W. *J. Vac. Sci. Technol. B8*(3), 1990, 563–567.

**TABLE 15.11. Neutral and ionized cluster-beam materials and conditions**

| Material | Precursors | Conditions | References |
|---|---|---|---|
| Au, Ag | Au, Ag | ICB | [1] |
| Au | Au | $d_p = 1$ nm<br>ICB | [2] |
| Ag, Ag–$C_2H_2$ | Ag, $C_2H_2$ | NCB<br>4–5-nm particles | [3] |
| In, Pb, Ag | In, Pd, Ag | ICB and NCB | [4] |
| $Hg_{1-x}Cd_xTe$ | CdTe, HgTe,<br>Te ingots | ICB | [5] |
| Cu on glass<br>Si on Si<br>ZnS on NaCl | Au, Si, ZnS | ICB | [6] |
| BeO on glass | Be, $O_2$ | Reactive ICB | [6] |
| Al on various surfaces<br>GaAs,<br>$Cd_{1-x}$ $Mn_xTe$<br>CdTe–PbTe<br>multilayers | Elements | Epitaxial ICB,<br>separate beams for<br>each element | [7] |
| Anthracene<br>polyethylene | Anthracene<br>polyethylene | ICB, good adhesion at<br>highest accelerations | [8] |
| Al on Si(111)<br>and Si(100) | Al | Epitaxial ICB | [9] |
| Ag, Au, Si | Elements | NCB<br>5–10-nm particles | [10] |

## References to Table 15.11

1. Knauer, W. & Poeschel, R.L. *J. Vac. Sci. Technol. B 6*(1), 1988, 456–460.
2. Ramachandra, A., Vaziri, M., & Andres, R.P. *Mat. Res. Soc. Sym. Proc. 206*, 1991, 397–401.
3. Bielefeld, J.D. & Andres, R.P. *Mater. Res. Soc. Symp. Proc. 317*, 1994, 155–159.
4. Stein, G.D. Design and use of metal cluster beam sources: Implications for thin film devices. *Proc. Int'l. Ion Engineering Cong.* Kyoto, Japan: the Institute, 1983, 1165–1176B.
5. Takaoka, G.H., Murakami, S., Ishikawa, J., & Takagi, T. *Appl. Phys. Lett. 54*(25), 1989, 2550–2552.
6. Takagi, T., Matsubara, I., & Takaoka, H. *J. Appl. Phys. 51*(10), 1980, 5419–5424.
7. Yamada, I., Takaoka, H., Usui, H., & Takagi, T. *J. Vac. Sci. Technol. A 4*(3), 1986, 722–726.
8. Usui, H., Yamada, I., & Takagi, T. *J. Vac. Sci. Technol. A 4*(1), 1986, 52–60.
9. Yamada, I., Inokawa, H., & Takagi, T. *J. Appl. Phys. 56*(10), 1984, 2746–2750.
10. Ichihashi, T. *Jpn. J. Appl. Phys. 25*(8), 1986, 1247–1251.

**TABLE 15.12. Materials deposited by inertial deposition and conditions used to produce films**

| Material | Precursors | Conditions, °C | References |
|---|---|---|---|
| $SnO_2$ | $Sn + O_2$ | $T_s = 25$<br>0.01–0.1-$\mu$m particles | [1] |
| Ag–$ZrO_2$, $ZrO_2$ | Zr(t-butoxide)$_4$ | $T_s = 25$<br>0.03–0.15-$\mu$m<br>17–33 $\mu$m thick | [2] |
| $Bi_4Sr_3Ca_3Cu_4O$ | $Bi_4Sr_3Ca_3Cu_4O$ | $T_s = 850$ | [3] |
| $YBa_2Cu_3O_7$ | Y–Ba–Cu–O particles | $T_s = 400$ | [4] |
| Pb–Bi–Sr–Ca–Cu–O 5–200 $\mu$m thick | 1-$\mu$m particles of Pb–Bi–Sr–Ca–Cu–O | $T_s = 400$ | [5] |
| Diamond-metal composites, W, Mo, Si, Cu, Pt, Ag, Mi, Co, Fe | Metals, diamond | $T_s = 1000$–$1200$<br>180 $\mu$m/hr<br>50–150 $\mu$m thick | [6] |
| Ag, Au, Cu, Fe, Ni, Pt, Ta, W | Metals | dc electric arc to form particles | [7] |
| $Si_3N_4$ | $(CH_3SiHNH)_x$ | Aerosol pyrolysis by $CO_2$ laser | [8] |
| $YBa_2Cu_3O_{7-x}$ $YBa_2Cu_4O_8$ 25 $\mu$m thick | Powder from liquid-phase route entrained | $T_s = 400$ | [9] |
| $Fe_2O_3$ | $Fe_2(C_2O_4)_3$ | Spray pyrolysis for particle formation | [10] |

## References to Table 15.12


1. Adachi, M., Okuyama, K., Kousaka, Y., & Tanaka, H. *J. Aerosol Sci. 19*(2), 1988, 253–263.
2. Adachi, M., Okuyama, K., Moon, S., Tohge, N., & Kousaka, Y. *J. Mater. Sci. 24*(6), 1989, 2275–2280.
3. Baker, R.C., Hurng, W.M., & Steinfink, H. *Appl. Phys. Lett. 54*(4), 1988–89, 371–373.
4. Hatanaka, K., Kaitou, M., Umehara, M., Kashu, S., & Hayashi, C. Preparation of superconducting thick films of Y–Ba–Cu–O by gas deposition of ultrafine powder. *Proc. 1st Intern. Symp. on Superconductivity.* Nagoya, Japan. 1988, Tokyo: Springer–Verlag, 1989, 341–345.
5. Kashu, S., Matsuzaki, Y., Kaito, M., Toyokawa, M., Hatanaka, K., & Hayashi, C. Preparation of superconducting thick films of Bi–Pb–Sr–Ca–Cu–O by gas deposition of fine powder. *Proc. 2nd Intern. Symp. on Superconductivity.* Tsukuba, Japan. 1989, Reston, VA: Foreign Broadcast Information Service, 1990, 413–419.
6. Kawarada, M., Kurihara, K., & Sasaki, K. *J. Appl. Phys. 71*(3), 1992, 1442–1445.

7. Kizaki, Y., Osada, H., & Aoki, H. *Jpn. J. Appl. Phys. 32*(1–12A), 1993, 5771–5777.
8. Magee, A.P., Strutt, P.R., & Gonsalves, K.E. *Chem. Mater. 2*, 1990, 232–235.
9. Murakami, H., Nishino, J., Yaegashi, S., Shiohara, Y., & Tanaka, S. *Jpn. J. Appl. Phys. 30*(2A), 1991, L185–L187.
10. Althainz, P., Schuy, L., Goschnick, J., & Ache, H.J. *Thin Solid Films 24*, 1994, 366–369.

**TABLE 15.13. Materials, precursors, and conditions for thermophoretic particle deposition to form films**

| Material | Precursors | Conditions | References |
|---|---|---|---|
| $SiO_2$ | $SiCl_4$ | Measured particle size distribution | [1] |
| Y–Ba–Cu–O<br>Bi–Si–Ca–Cu–O<br>Tl–Ca–Ba–Cu–O | Metal nitrates | Particles made by spray pyrolysis | [2] |
| $SiO_2$ | $SiCl_4$ | Measured particle size | [3] |
| $YBa_2Cu_3O_x$ | Y, Ba, Cu acetates | $T_s$ = 940°C large particles | [4] |
| $YBa_2Cu_3O_{7-x}$<br>Bi–Sr–Ca–Cu–O | Nitrates and acetates | $T_s$ = 200–300°C, thermophoretic deposition, particles produced by spray pyrolysis | [5] |
| $SiO_2$ doped optical fiber | $SiCl_4$, $POCl_3$, $GeCl_4$, $C_2Cl_3F_3$, $O_2$ | Measured particle deposition rates | [6] |
| $SiO_2$ doped optical fiber | $SiCl_4$ | Studied role of HCl in process | [7, 8] |
| $SiO_2$ doped optical fiber | $SiCl_4$ | Used laser to enhance deposition | [9] |
| $SiO_2$ doped optical fiber | $SiCl_4$ | Studied particle deposition | [10] |
| $SiO_2$ | $SiCl_4$ | Studied particle deposition | [11] |
| $SiO_2$ | $SiCl_4$ | Enhanced deposition with electric field | [12] |

*Note*: Unless otherwise stated, oxygen in the film is derived from precursor, air, or water present during deposition.

## References to Table 15.13


1. Bautista, J.R. & Atkins, R.M. *J. Aerosol Sci. 22*(5), 1991, 667–675.
2. Kodas, T.T., Engler, E., & Lee, V. *Appl. Phys. Lett. 54*, 1989, 1923–1925.
3. Bautista, J.R., Potkay, E., & Scatton, D.L. *Mat. Res. Soc. Symp. Proc. 117*, 1988, 151–156.
4. Koguchi, M., Matsuda, Y., Kinoshita, E., & Hirabayashi, K. *J. Appl. Phys. Japan 29*(1), 1990, L33–L35.
5. Matsuno, S., Egawa, K., Yoshizaki, K., & Watarai, H. *Preparation of* $YBa_2Cu_3O_{7-y}$ *and Bi–Sr–Ca–Cu–O superconducting thin films by fine particle deposition method.* MRS Intern. Meeting on Advanced Materials, Tokyo, 1988.
6. Kasik, I. & Matejec, V. *J. Aerosol Sci. 26*(3), 1995, 399–406.
7. Wood, D.L. & Shirk, J.S. *J. Am. Ceram. Soc. 64*(6), 1981, 325–327.
8. Wood, D.L., Kometani, T.Y., Luongo, J.P., & Saifi, M.A. *J. Am Ceram. Soc. 62*(11–12), 1979, 638–639.
9. Cipolla, J.J., Wang, C.Y., & Morse, T.F. Laser induced thermophoresis and particulate deposition efficiency. *ASME/AIChE National Heat Transfer Conf.* Seattle, New York: Am. Soc. Mech. Eng., 1983, 21–28.
10. Park, K.S. & Choi, M. *Int. J. Heat Mass Transfer 37*(11), 1994, 1593–1603.
11. Graham, G.M. & Alam, M.K. *Aerosol Sci. Tech. 15*, 1991, 69–76.
12. Hwang, J. & Daily, J.W. *J. Aerosol Sci. 26*(1), 1995, 5–18.

**TABLE 15.14. Simultaneous particle deposition and chemical vapor deposition**

| Material | Precursors | Conditions, °C | References |
|---|---|---|---|
| SiC | $MeSiCl_3 + H_2$ | $T_s = 1287–1347$<br>10 μm/min | [1–3] |
| TiN | $TiCl_4$, $H_2$, $NH_3$ | $T_s = 650–900$<br>10–1000 μm/hr | [4] |
| AlN, $TiO_2$, $ZrO_2$ | $Zr(C_8H_7O)_4$,<br>$AlCl_3$,<br>Ti(O–*i*–Pr)$_4$ | $T_s = 427–1027$ | [5, 6] |
| AlN | $AlCl_3$, $NH_3$ | $T_s = 725–888$<br>100 nm/sec | [5, 6] |
| $TiO_2$ | Ti(O–*i*–Pr)$_4$ | $T_s = 350$<br>30 nm/sec | [7] |
| Ag, $BaTiO_3$,<br>yttrium aluminum garnet<br>$YBa_2Cu_3O_7$ | Assorted metallorganics such as β-diketonates | Flame reactor<br>$T_s$ not specified | [8] |
| CdSe–ZnSe | CdSe particles and $H_2Se$, $ZnEt_2$, $H_2$ | Cold-wall reactor | [9, 10] |
| $TiO_2$–polymer | $Ti[OCH(CH_3)_2]_4$ | Plasma-enhanced CVD | [11] |
| Nb particles in $Nb_7Al_3$ | $Nb_3Al$ target | Pulsed laser deposition | [12] |
| $TiO_2$ | Ti(O–*i*–Pr)$_4$ | Impinging jet reactor | [13] |
| SiC with powders and fibers | $CH_3SiCl_3$ | Powders entrained with gaseous reactant | [14] |
| SiC–$TiB_2$<br>SiC–$Si_3N_4$<br>SiC–$B_4C$ | $CH_3SiCl_3$ | Feed powders of $TiB_2$, $Si_3N_4$, $B_4C$;<br>SiC by CVD | [15] |
| Bi–Pb–Sr–Ca–Cu–O | Bi–Pb–O and Sr–Ca–Cu–O particles | Particles made by spray pyrolysis; particles deposited by filtration; no CVD | [16] |
| $Si_3N_4$ | $SiCl_4$, $NH_3$ | Particles deposited electrostatically; no CVD | [17] |
| $SiO_2$ on fibers | $SiCl_4$ | Fibers in flame reactor | [18] |

## References to Table 15.14


1. Hurt, R.H. & Allendorf, M.D. *AIChE J. 37*(10), 1991, 1485–1496.
2. Hurt, R.H. & Allendorf, M.D. *Mat. Res. Soc. Symp. Proc.* Mat. Res. Soc. *250*, 1992, 59–64.
3. Allendorf, M.D., Hurt, R.H., & Yang, N. *J. Mater. Res. 8*(7), 1993, 1651–1665.
4. Dekker, J.P., van der Put, P.J., Veringa, H.J., & Schoonman, J. *International Ceramic Conference*, Melbourne: Commonwealth Sci. & Ind. Res. Org. 1992.
5. Komiyama, H., Osawa, T., Kazi, H., & Konno, T. Rapid growth of AlN films by particle precipitation aided chemical vapor deposition. In Vincenzini, P. (Ed.), *High Tech Ceramics*. Amsterdam: Elsevier, 1987, 667–676.
6. Komiyama, H., Osawa, T., Shimogaki, Y., Wakita, N., Minamiyama, M., & Ueoka, T. Particle precipitation aided chemical vapor deposition for rapid growth of ceramic films—preparation of 1mm-thick AlN, $TiO_2$ and $ZrO_2$ films. *Proc. 10th Int. Conf. Chemical Vapor Deposition*. Electrochemical Society, Pennington, NJ: 1987, 1119–1128.
7. Shimogaki, Y. & Komiyama, H. *J. Ceram. Soc. Jpn. Inter. ed 95*, 1987, 65–69.
8. Hunt, A.T. & Carter, W.B. *Appl. Phys. Lett. 63*, 1993, 266.
9. Danek, M., Jensen, K.F., Murray, C.B., & Bawendi, M.G. *Appl. Phys. Lett. 65*(22), 1994, 2795–2797.
10. Danek, M., Jensen, K.F., Murray, C.B., & Bawendi, M.G. *J. Crystal Growth 145*, 1994, 714–720.
11. Ratcliffe, P.J., Hopkins, J., Fitzpatrick, A.D., Barker, C.P., & Badyal, J.P.S. *J. Mater. Chem. 4*(7), 1994, 1055–1060.
12. Chen, L.-C., Hall, E.L., & Lou, K.A. *Mat. Res. Soc. Symp. Proc.* Mat. Res. Soc. *273*, 1992, 377–383.
13. Gokoglu, S.A., Stewart, G.D., Collins, J., & Rosner, D.E. *Mat. Res. Soc. Symp. Proc.* Mat. Res. Soc. *335*, 1994, 171–176.
14. Reagan, P. *J. de Physique IV 3*(C3), 1993, 541–548.
15. Zhao, G.Y., Thiart, J.J., Orlicki, D., & Hlavacek, V. *Chem. Eng. Sci. 49*(24B), 1994, 4917–4938.
16. Takao, Y., Awano, M., Kuwahara, Y., & Murase, Y. *J. Mater. Res. 9*(8), 1994, 2128–2132.
17. Yamamoto, H. & Masuda, S. *IEEE Trans. Industry Applications 27*(2), 1991, 307–310.
18. Fotou, G.P. & Pratsinis, S.E. *Chem. Eng. Sci. 49*(10), 1994, 1651–1662.

**TABLE 15.15. Particle formation during chemical vapor deposition**

| Material | Precursors | Notes | References |
|---|---|---|---|
| Al | $Me_2EtNAlH_3$ | $T_s$ = 180°C; trace amounts of CO, $H_2O$, and $O_2$ increased particle formation | [1] |
| Co | $Co(C_5H_5)_2$<br>$Co_2(CO)_8$<br>$Co(C_5H_5)(CO)_2$<br>$CoCF_3(CO_4)_4$ | All precursors formed particles, thereby reducing deposition rate; lower pressure reduced particle formation | [2] |
| $Al_2O_3$ | $AlCl_3$, $H_2$, $CO_2$ | Hot-wall reactor | [3] |
| $TiO_2$ | $Ti(O-i-Pr)_4$ | Increased temperature led to particle formation and decreased deposition rate when particle formation occurred | [4] |
| Si | $SiCl_4$ | HCl depressed particle formation up to rates of 40 $\mu m/min$ | [5] |
| Si | $SiH_4$ | Particles formed above critical concentration of $SiH_4$ when $T$ = 1050°C | [6] |
| Si | $SiH_4$ | Particles formed above critical $SiH_4$ concentration at constant temperature; low-pressure CVD reactor; $T$ = 1100°C | [7, 8] |
| Si | $SiH_4$ | Particles formed above 725°C and critical concentration of $SiH_4$; low-pressure CVD reactor | [9] |
| $SiO_2$ | $N_2O-SiH_4$ (He diluted) | Particle formation at partial pressure of reactants of 130 Pa. $T_s$ = 230°C; Parallel-plate Plasma-enhanced CVD reactor | [10] |
| Ga | $Ga(CH_3)_3$ | Droplets formed in bubbler used for vaporizing precursor | [11] |

**TABLE 15.15.** (*Continued*)

| Material | Precursors | Notes | References |
|---|---|---|---|
| $SiO_2$ | $SiH_4$ with $O_2$ and Ar | Total pressure below 100 mtorr; gas-cooled system; $T_s$ = 40–50°C; post-type radio frequency magnetron Plasma-enhanced CVD reactor | [12] |
| $SiO_2$ | TEOS (tetraethylortho-silicate) | Particles formed above 740°C; amorphous particles formed from 740–800°C; hot-wall atmospheric pressure CVD reactor | [13] |
| GaAs | $GaMe_3$, $AsH_3$ | Particles formed above 750°C; deposition rate decreased | [14] |
| GaAs | $AsH_3$, $Ga(CH_3)_3$ | Formation of GaAs particles suggested | [15] |
| GaAs | $AsH_3$, $Ga(CH_3)_3$ | Formation of Ga or GaAs particles suggested | [16] |
| $Cd_xHg_{1-x}Te$ | $TeEt_2$, $CdEt_2$ Hg | Particles formed with $H_2$ carrier gas; photo-CVD | [17] |
| TiC | $TiCl_4$, $C_3H_8$, $H_2$ | Particles formed above 950°C; deposition rate decreased | [18] |
| BN | $B(C_2H_5)_3$, $NH_3$ | Particles formed above 1000°C; deposition rate decreased | [19] |
| AlN | $AlCl_3$, $NH_3$ | Atmospheric pressure, hot wall, $T_s$ = 600–1100°C | [20] |
| $Mg_3N_2$ | Mg vapor, $NH_3$, $N_2$ | $T_s$ = 800°C; hot-wall reactor | [21] |
| ZnSe | $ZnMe_2 + H_2Se$ | Lower pressures eliminated particle formation | [22] |
| W–Si | $WF_6$, $SiH_4$ | Thorough study of pressure, flow, temperature dependence of particle formation | [23] |
| $Cu_2O$ | CuI, $O_2$ | Thorough study | [24] |
| $TiO_2$ | $Ti(O{-}i{-}Pr)_4$ | Thorough study | [25] |

## References to Table 15.15

1. Simmonds, M.G., Gladfelter, W.L., Rao, N., Szymanski, W.W., Ahn, K.-H., & McMurry, P.H. *J. Vac. Sci. Technol. A 9*(5), 1991, 2782–2784.
2. Dormans, G.J.M., Meekes, G.J.B.M., & Staring, E.G.J. *Journal of Crystal Growth 114*, 1991, 364–372.
3. Lux, B., Colombier, C., Altena, H., & Stjernberg, K. *Thin Solid Films 138*, 1986, 49–64.
4. Lee, H.-Y. & Kim, H.-G. *Thin Solid Films 229*(2), 1993, 187–191.
5. Bloem, J. *J. Crystal Growth 18*, 1973, 70–76.
6. Eversteijn, F.C. *Gas-phase decomposition of silane in a horizontal epitaxial reactor*. Philips Res. Repts., 1971.
7. Murthy, T.U.M.S., Miyamoto, N., Shimbo, M., & Nishizawa, J. *J. Crystal Growth 33*, 1976, 1–7.
8. Sarma, K.R. & Rice, Jr., M.J. *J. Electrochem. Soc. Solid-State Sci. Tech. 128*(12), 1981, 2647–2655.
9. Van den Brekel, C.H.J. & Bollen, L.J.M. *J. Crystal Growth 54*, 1981, 310–322.
10. Smith, D. & Alimonda, A.S. *J. Electrochem. Soc. 140*, 1993, 1496–1503.
11. Deppert, K. & Wiedensohler, A. *Cryst. Res. Technol. 29*(5), 1994, 613–616.
12. Chu, J.H. & Lin, I. *J. Appl. Phys. 74*, 1993, 4741.
13. Adachi, M., Okuyama, K., & Tohoge, N. Formation of silicon dioxide films and gas-phase nucleation in an APCVD process using the TEOS/ozone/helium system. *Proc. 12th Inter. Symp. CVD*, Pennington, NJ: Electrochem. Soc., *93–2*. 1993, 306–312.
14. Leys, M.R. & Veenvliet, H. *J. Crystal Growth 55*, 1981, 145–153.
15. Frolov, I.A., Boldyrevskii, P.B., Druz, B.L., & Sokolov, E.B. *Inorganic Materials 13*(1), 1977, 632–634.
16. Nishizawa, J. & Kurobayashi, T. *J. Electrochem. Soc. 130*(2), 1983, 413–417.
17. Irvine, S.J.C., Giess, C.J., Mullin, J.B., Blackmore, G.W., & Dosser, O.D. *J. Vac. Sci. Technol. B 3*(5), 1985, 1450.
18. Kim, D.G., Yoo, J.S., & Chun, J.S. *J. Vac. Sci. Technol. A 4*(2), 1986, 219–221.
19. Nakamura, K. *J. Electrochem. Soc. 133*(6), 1986, 1120–1123.
20. Egashira, Y., Kim, H.-J., & Komiyama, H. *J. Am. Ceram. Soc. 77*(8), 1993, 2009–2016.
21. Murata, T., Itatani, K., Howell, F.S., Kishioka, A., & Kinoshita, M. *J. Am. Ceram. Soc. 76*(11), 1993, 2909–2911.
22. Stutius, W. *Appl. Phys. Lett. 33*(7), 1978, 656–658.
23. McInerney, E.J., Mountsier, T.W., Chin, B.L., & Broadbent, E.K. *J. Vac. Sci. Technol. B 11*, 1993, 734.
24. Hong, L.-S. & Komiyama, H. *J. Am. Ceram. Soc. 74*(7), 1991, 1597–1604.
25. Oshima, K., Torimoto, Y., Tsuto, K., Okuyama, K., Ushio, R., and Kousaka, Y. *KONA 9*, 1991, 59–71.

# 16 Aerosol Reactor Components

## 16.1 INTRODUCTION

Powder or film generation by aerosol processes involves a number of steps, or unit operations. For example, powder generation by spray pyrolysis involves atomizing a solution to form droplets, entraining the droplets in a gas stream to form an aerosol, transporting the aerosol to a heat source, heating the droplets to drive the reactions to form the products, collecting the powder, and cleaning up gaseous by-products. Similarly, powder generation by gas-to-particle conversion involves delivering gaseous reactants to a heat source (the reactor), inducing the desired chemicals to react, collecting the powder that forms, and cleaning up gases. Film formation requires either the generation and deposition of droplets or the delivery of gaseous materials. In this chapter, we discuss the separate operations that together form an aerosol powder or film generation system. Droplet generators, virtual and conventional impactors, cyclones, volatile precursor delivery systems, and powder collection by electrostatic precipitation and filtration are examined. For background, the reader is referred to Bayvel and Orzechowski[1] and LeFebvre[2] for sprays; to Kodas and Hampden-Smith[3] for delivery of reactants; and to Willeke and Baron[4] for all other topics.

## 16.2 ATOMIZATION

### 16.2.1 Introduction

The devices used to convert bulk liquids into droplets are usually called atomizers or nebulizers. Atomization is used in a variety of applications involving the generation and processing of fine particles. Liquid metal atomization is widely used to produce metal powders with diameters of 1 to 5000 $\mu$m.[5–7] Powders obtained by this route are used for the production of sintered machine parts, soldering materials, and metal pastes used to form patterns on circuit boards. Two starting points for information are the texts by Bayvel and Orzechowski[1] and Lefebvre.[2]

Atomization is also used to deposit anticorrosion coatings, plastics, resins, adhesives, and ceramic and stoneware glazing. It is utilized to wash and clean parts in order to prepare them for subsequent physical or chemical processes in the microelectronics industry. Atomization is the basis for spray-drying, in which liquids are sprayed into a heated gas stream to remove the moisture. The water or other liquid evaporates quickly because of the large surface area of the drops.

Atomization also plays a key role in spray pyrolysis (Chapters 12 and 13) for powder production. In this case, droplet sizes are usually 0.1 to 100 $\mu$m, significantly smaller than those used for spray-drying. Various routes to films (Chapters 14 and 15) also exploit atomization. Droplet deposition relies on the generation of droplets of involatile reactants. AACVD utilizes droplets of solutions of volatile precursors.

The requirements for the droplet size distribution and, therefore, the atomizer, vary, depending on the application and scale, laboratory or industrial. The discussion that follows outlines the differences among atomizers, describes their relevant characteristics, and explains which types of atomizers are suitable for the various aerosol routes to materials.

### 16.2.2 Classification of Atomizers

Atomizers can be classified by means of various approaches, such as the type of energy used for atomization. The simplest class of atomizers uses the energy of the liquid itself to drive the breakup of the liquid; the pressure of the fluid is converted into kinetic energy, which atomizes the liquid. These atomizers can be classified into three types: *jet atomizers, swirl atomizers, and jet–swirl atomizers*.[1,2] Jet atomizers themselves can be grouped into two types: continuous and intermittent. Continuous atomizers are the simplest of all atomizers, while intermittent atomizers are the most complex. Jet atomizers produce jets that break apart as they leave the atomizer. High injection pressures are required to produce the kinetic energy necessary to break up the fluid into sufficiently small droplets. Swirl atomizers are the most widely used of all types of atomizers because they require lower pressures than jet atomizers. Swirl atomizers exploit the swirling of a liquid inside an atomizer to give a conical sheet of liquid that breaks up more easily than a jet. Jet–swirl atomizers combine the features of jet and swirl atomizers.

*Pneumatic* or *gas-assist atomizers* use the energy of a carrier gas to break up sheets or jets of a liquid. Pneumatic atomizers are the most complex of the atomizers because the liquid and gas phases interact in a complex manner. A common mode of interaction is the transverse introduction of a liquid stream into a high-velocity gas stream.

*Rotary atomizers* impinge a liquid onto a rapidly rotating surface, usually driven by electric motors, that transmit its rotational energy to the liquid, causing it to leave the atomizer with high kinetic energy and break apart. This type of atomizer can provide narrow size distributions under the correct operating conditions. Rotary atomizers are often used in the atomization of solutions, because they can be fed with a liquid at low pressure. They are common in spray-drying systems.

Other types of atomizers rely on more exotic phenomena to convert the liquid into droplets. Examples are *ultrasonic* and *electrospray atomizers*. These atomizers have been used extensively at the laboratory scale for aerosol routes to materials because of their ability to produce smaller droplets than can be obtained by other methods.

One of the major challenges for atomization technology is the production of high concentrations of droplets with small sizes. Most atomizer technologies—e.g., jet, gas-assist, and rotary atomizers—when operated at industrial rates, produce droplets with diameters greater than 10 $\mu$m and concentrations lower than $10^7$ droplets/$cm^3$. A variety of applications, including especially spray pyrolysis for powder generation, require droplets of 1 to 10 $\mu$m in size at concentrations greater than $10^7$ droplets/$cm^3$ (but below the coagulation limit of roughly $10^9$ droplets/$cm^3$). This has led to an emphasis on ultrasonic, electrostatic, and other technologies capable of producing droplets with sizes less than 10 $\mu$m at high concentrations.

The conventional jet, gas-assist, rotary, etc., atomizer technologies have been discussed in detail in various books.[1, 2] For this reason, we focus our discussion on less conventional approaches, including ultrasonic and electrostatic atomizers.

### 16.2.3 Pressure and Two-Fluid Atomization

In general, pressure and two-fluid atomizers cannot easily produce high concentrations of droplets smaller than 1 $\mu$m. Several designs are available for droplets in the 1 to 10 $\mu$m range,[8] and a massive number of atomizers are available for droplets greater than 10 $\mu$m. We shall concentrate on the discussion of atomizers capable of producing droplets less than 10 $\mu$m in size.

The physical phenomena that are exploited to produce droplets by these routes vary. *Pressure atomizers* work by discharging a liquid through a small aperture under high pressure. They have the advantage of not requiring any associated gas flow (therefore, high droplet concentrations can be achieved), but have the disadvantage of producing large droplet sizes. *Rotary atomizers* work by centrifugal atomization: A liquid is directed onto a rotating element that converts the liquid stream to droplets. These atomizers, too, have the advantage of not requiring any associated gas flow, but have the disadvantage of providing large droplet sizes. *Air-assist atomizers* expose the liquid to a stream of air flowing at high velocity. They require a gas flow to produce droplets, but have the advantage of providing somewhat smaller droplet sizes than are produced by pressure and rotary atomizers. *Air-blast atomizers* are similar to air-assist nozzles, but use much higher quantities of air or steam. These atomizers are discussed by Lefebvre and will not be examined here.[2]

Many variations of the *Collison* (not collision) design of two-fluid atomizers have been used for spray pyrolysis at the laboratory scale to produce droplets with sizes on the order of 1 $\mu$m.[9, 10] The Collison design relies on drawing a liquid stream into a high-velocity gas stream that atomizes the liquid. The droplets that are formed are directed towards a surface where the gas stream makes a sharp bend, and the larger droplets impact on the surface because of their greater inertia, while the smaller particles (with less inertia) follow the gas stream, thereby narrowing the size distribution and reducing the average droplet size. A typical design is shown in Figure 16.1. This particular atomizer produces droplets less than 1 $\mu$m in size. Several designs (BGI, TSI) are available that produce droplet concentrations of $10^6$ to $10^7/\text{cm}^3$ and sizes of 1 $\mu$m or less.

*Effervescent atomization* has been used for spray pyrolysis applications at the laboratory scale. This technique is illustrated in Figure 16.2.[11] The fluid to be atomized is pressurized to sufficient levels to dissolve significant quantities of the carrier gas in the liquid. The liquid is atomized into a region at atmospheric pressure in which the dissolved gas comes out of solution, resulting in further breakdown of the droplets. Droplet sizes are usually greater than 1 $\mu$m, but high throughputs are obtained relative to the Collison design.

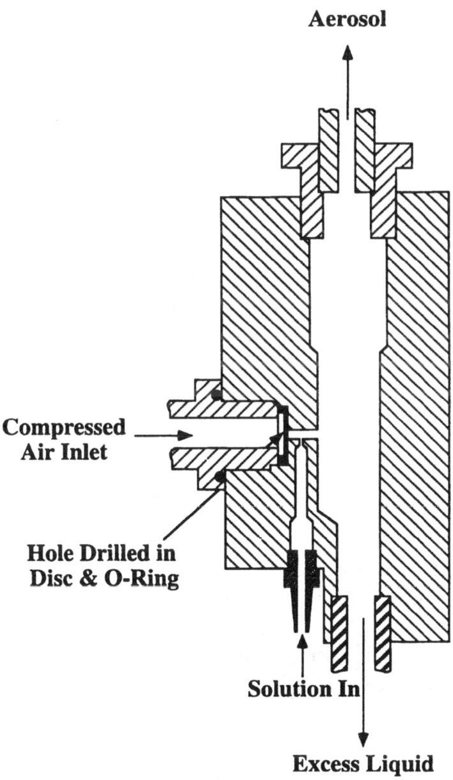


**Fig. 16.1.** Collison-type aerosol generator capable of producing high concentrations of submicron-sized droplets.

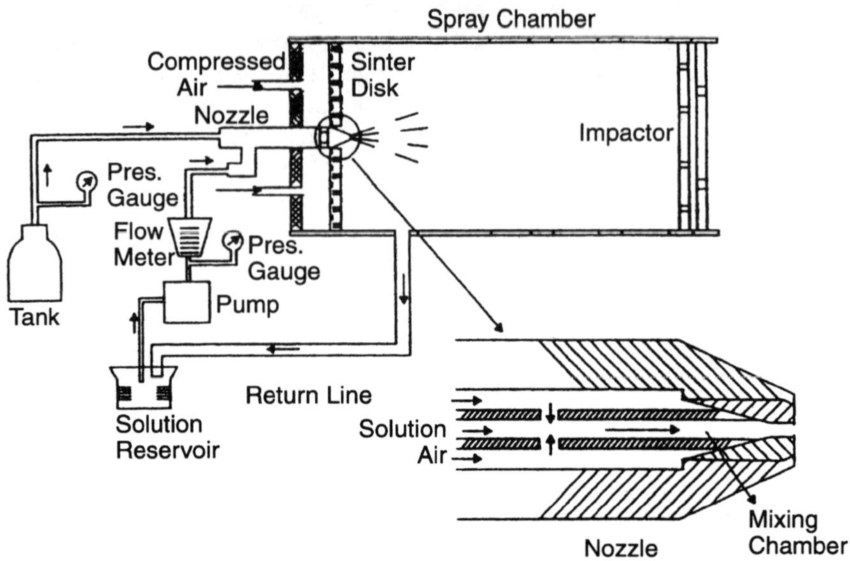


**Fig. 16.2.** Effervescent spray system in which gas is dissolved in the liquid to be atomized and then forms bubbles in the gas after exiting the nozzle, thereby breaking up the liquid stream into small droplets. (Reproduction with permission of Messing.)[11]

### 16.2.4 Ultrasonic Atomizers

Ultrasonic atomizers use an ultrasonic transducer or horn that vibrates at ultrasonic frequencies (typically, 50 kHz–2.4 MHz) to produce the short wavelengths required for liquid atomization.[12–19] The most useful characteristics of ultrasonic atomizers for materials processing by aerosol techniques are their low spray velocity, the small or nonexistent amount of carrier gas required to operate them, and the small droplet size produced, which can be less than 10 $\mu$m. This makes it easy to carry small droplets at a high concentration into a flow reactor or towards a surface without using a high gas flow. Because of these useful characteristics, ultrasonic atomizers have been employed extensively in aerosol methods for generating both films and powders at the laboratory and on industrial scales. These types of atomizers are also used to spray acidic or basic solutions onto semiconductor wafers to etch features into them.

There are three classes of ultrasonic atomizers. One version passes the liquid through a tube that is driven ultrasonically (Fig. 16.3).[20] In the other versions, the liquid comes into contact with an ultrasonic horn, or the ultrasonic transducer is submerged in the liquid (Fig. 16.4).[21] The principle of operation is similar for these examples of ultrasonic atomizers. When a liquid comes into contact with an ultrasonically driven surface, a wave pattern appears on the surface of the liquid. When the amplitude of the

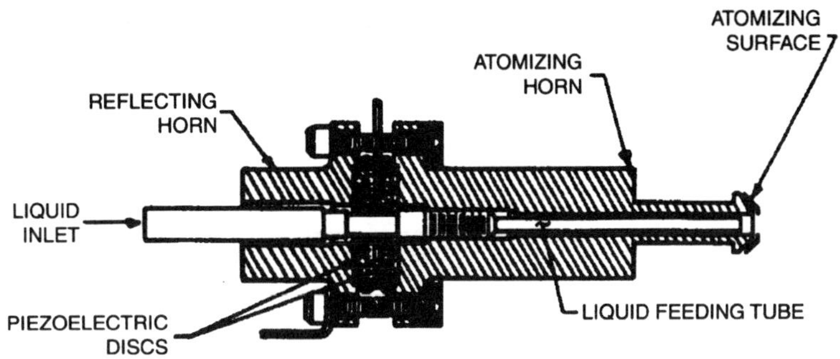


**Fig. 16.3.** Ultrasonic horn atomizer in which a liquid stream flows through a tube and then comes into contact with a transducer. (Courtesy of Berger.)[20]

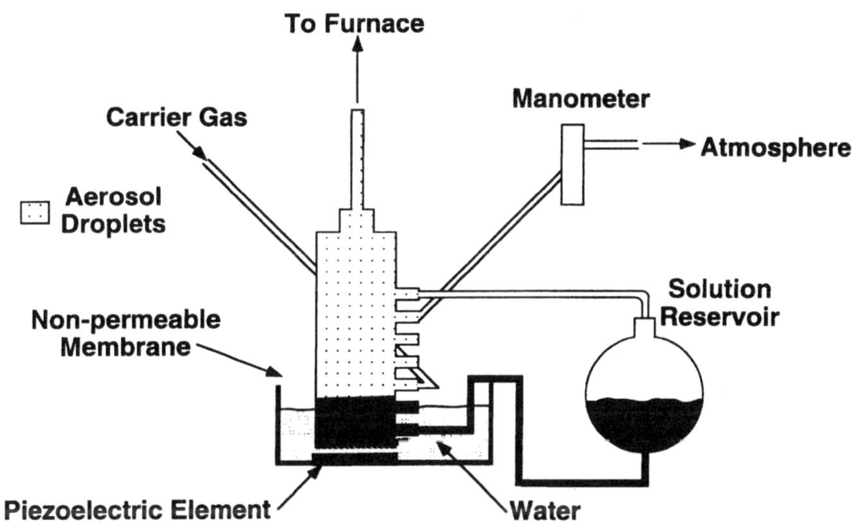


**Fig. 16.4.** Submerged ultrasonic transducer for generation of droplets. The system shown can produce roughly 1 gm/hr of submicron-sized powder with a gas-phase particle concentration of $10^6$ to $10^7/cm^3$. (Adapted from Lyons.)[21]

vibration is sufficient, the wave height[14] is sufficient for the wave crests in the film to become unstable. This instability drives the formation of droplets, which are ejected from the surface. A carrier gas can then be used to sweep the droplets away from the surface.

The class of ultrasonic atomizers relying on a capillary tube has been described by Berger.[22] As shown in Figure 16.3, the nozzle consists of piezoelectric transducers placed between a pair of sections of the horn. The surface on which the droplets are formed is at the end of the stem. Liquid is delivered through a tube to the transducers. These atomizers which rely on making a liquid come into contact with an exposed transducer operate in a similar manner. The maximum flow rate is typically 10 liters/h at 55 kHz, with a droplet or particle size range of 30 to 60 $\mu$m. The output drops dramatically as the frequency increases.

Ultrasonic transducers that are submerged in a liquid provide some of the smallest droplets of all ultrasonic atomizers. An example is shown in Figure 16.4. The submerged ultrasonic transducer can provide 1 to 5 $\mu$m droplets for frequencies of 1.5 to 2.4 MHz with low gas flow rates that correspond to high droplet concentrations. In medical applications, these droplets, when inhaled, can penetrate deep into the lungs to deliver medication.[19] The major disadvantage of this type of atomizer is the low liquid flow rates per transducer.

Submerged ultrasonic transducers have been used in a large number of studies of spray pyrolysis. (See Chapter 13.) Much smaller particles can be obtained with these ultrasonic atomizers (typically, 2 to 4 $\mu$m) than with the other ultrasonic types, which produce droplets with mass median diameters of 20 $\mu$m and larger. The droplet size depends on the surface tension, density, and frequency of the ultrasonic source, according to the equation

$$d_p = \left( \frac{\pi \sigma}{4 \rho f_t^2} \right)^{1/3}, \tag{16.1}$$

where $d_p$ is the droplet diameter (cm), $\sigma$ is the surface tension (erg/cm$^2$), $\rho$ is the fluid density (g/cm$^3$), and $f_t$ is the transducer frequency (1/sec).

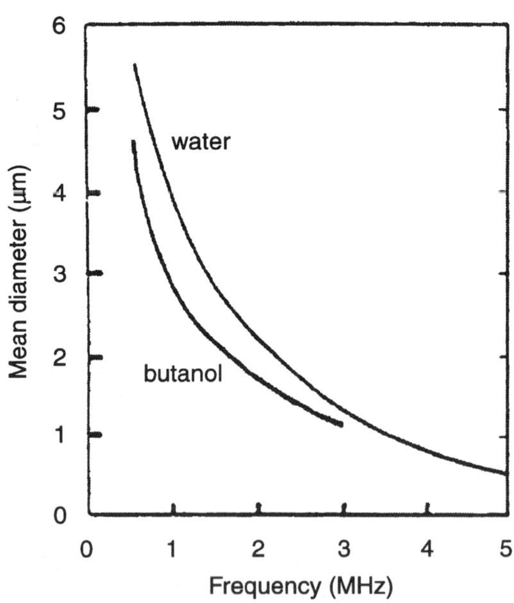


**Fig. 16.5.** Calculated dependence of droplet size on transducer frequency for generation with submerged ultrasonic transducer. (Courtesy of Tang.)[23]

Figure 16.5 show how median droplet diameter depends on frequency.[23] Transducers with frequencies ranging from 1.6 to 2.4 MHz are commercially available and produce droplets of 2 to 3 $\mu$m in median diameter.

The primary advantages of ultrasonic generators with submerged transducers are:

- narrow droplet size distribution
- simplicity
- high concentration of droplets
- droplet size appropriate for spray pyrolysis
- ability to vary droplet size by varying frequency

Several studies have measured droplet size distributions.[24, 25] Figure 16.6 shows droplet size distributions produced using this type of ultrasonic generator.[26] The figure demonstrates that the droplet sizes required for most spray pyrolysis applications for powder generation correspond to frequencies near 2 MHz. Droplet concentrations depend on the details of operation of the reactor, but typical values are $10^6$ to $10^7$ droplets/cm$^3$.

Ultrasonic atomizers such as that shown in Figure 16.4 exhibit a dependence of the aerosol output on the level of the liquid above the transducer, as shown in Figure 16.7.[26] This requires careful design of these atomizers to maintain a constant liquid level. Such generators also exhibit a strong dependence of the aerosol output on the ultrasonic power and gas flow. Typically, a certain gas flow is required to optimize the output in the sense that a maximum in aerosol concentration is obtained as the gas flow rate is varied. This situation is illustrated in Figure 16.8.[26]

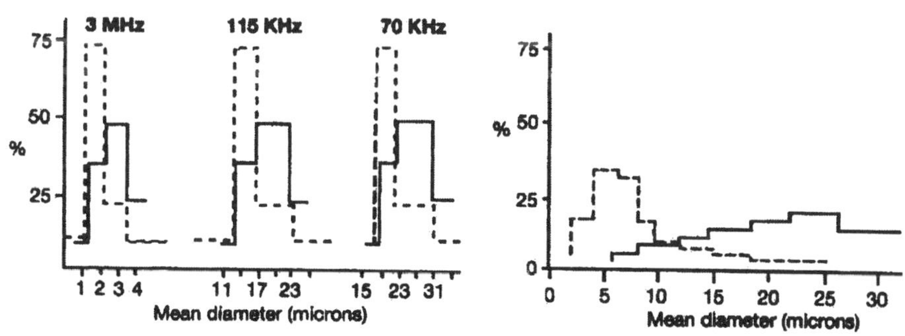


**Fig. 16.6.** Number (---) and volume (—) distribution of water droplets obtained by ultrasonic atomization (left) compared to number (---) and volume (—) distribution obtained by pneumatic atomization (right). Reprinted with permission from Langlet, M., Joubert, J.C. *In* Rao, C.N.R. (Ed.), *Chemistry of Advanced Materials*, 1993, Oxford: Blackwell Scientific, 55.[26]

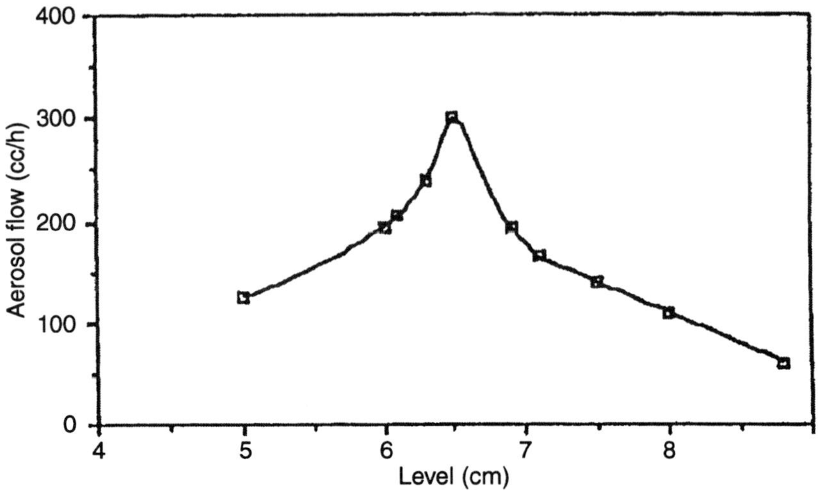


**Fig. 16.7.** Measured dependence of aerosol flow as a function of level of liquid above transducer in system similar to that shown in Figure 16.4. Reprinted with permission from Langlet, M., Joubert, J.C. *In* Rao, C.N.R. (Ed.), *Chemistry of Advanced Materials*, 1993, Oxford: Blackwell Scientific, 55.[26]

### 16.2.5 Electrostatic Atomizers

Electrostatic atomizers exploit charges for generating and modifying droplets and their sizes. There are two types of electrostatic atomization. In the first type, the electrostatic forces are insufficient to influence the stability of the droplet or liquid stream. However, once produced by conventional means, the drops are given electric charges by corona or induction charging. These atomizers are also termed hybrid atomizers because the liquid is atomized using pressure, pneumatic, or rotary atomizers and the drops are charged after being formed. A typical median diameter for the drops they produce is in the range 30 to 50 $\mu$m for a liquid flow rate of 50 mL/min. These atomizers find applications where high droplet deposition efficiencies are needed.

In the other type of electrostatic atomization, which has been studied extensively by aerosol scientists, the electrostatic forces by themselves are strong enough to atomize the liquid. Here the transfer of the charge to the liquid proceeds through contact (i.e., contact of the liquid with the source of electrons). This type of electrostatic atomization depends on the electrical conductivity of the liquid. The repulsion between charges on the liquid causes the stream to disintegrate into fine droplets. The expansion in the droplets due to the charges on the surface is opposed by surface tension forces. Typical droplet sizes are submicrons to millimeters, and the flow rates reported in the literature range from $10^{-7}$ to 1 mL/sec. Although the method is capable of producing submicron-sized droplets, it does not yet

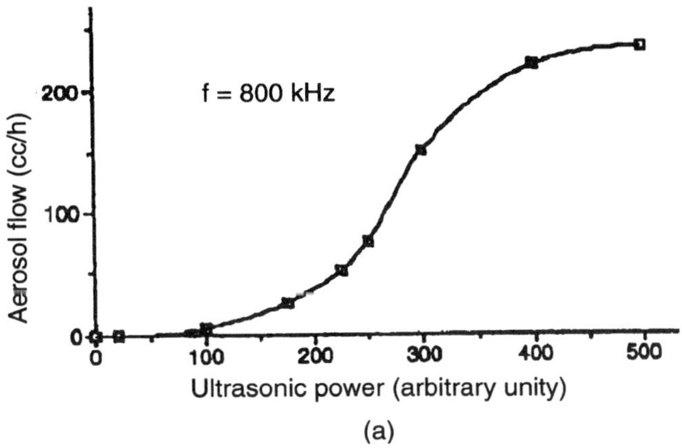


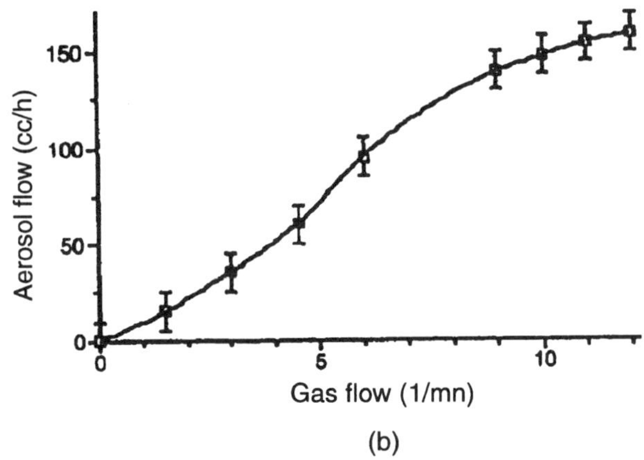


**Fig. 16.8.** Effect of (a) ultrasonic power level and (b) carrier gas flow rate on ultrasonic atomization of butanol. Reprinted with permission from Langlet, M., Joubert, J.C. *In* Rao, C.N.R. (Ed.), *Chemistry of Advanced Materials*, 1993, Oxford: Blackwell Scientific, 55.[26]

allow simultaneous high liquid flow rates and droplets of that size.[27–29] As a result, it has been used primarily on the laboratory scale. A typical system for droplet generation is shown in Figure 16.9.[30] This type of atomizer has been used extensively to form droplets that are deposited onto surfaces. A measured size distribution is shown in Figure 16.10.[31]

Electrostatic atomizers have the advantage that the charged droplets can be manipulated using electrical fields. This allows droplets to be deposited onto surfaces even in the presence of thermophoretic forces, which drive the particles away from a hot surface.

Submicron-sized metal powders have been produced by electrostatic atomization of a molten metal. Particles as small as 0.01 $\mu$m and as large as 10 to

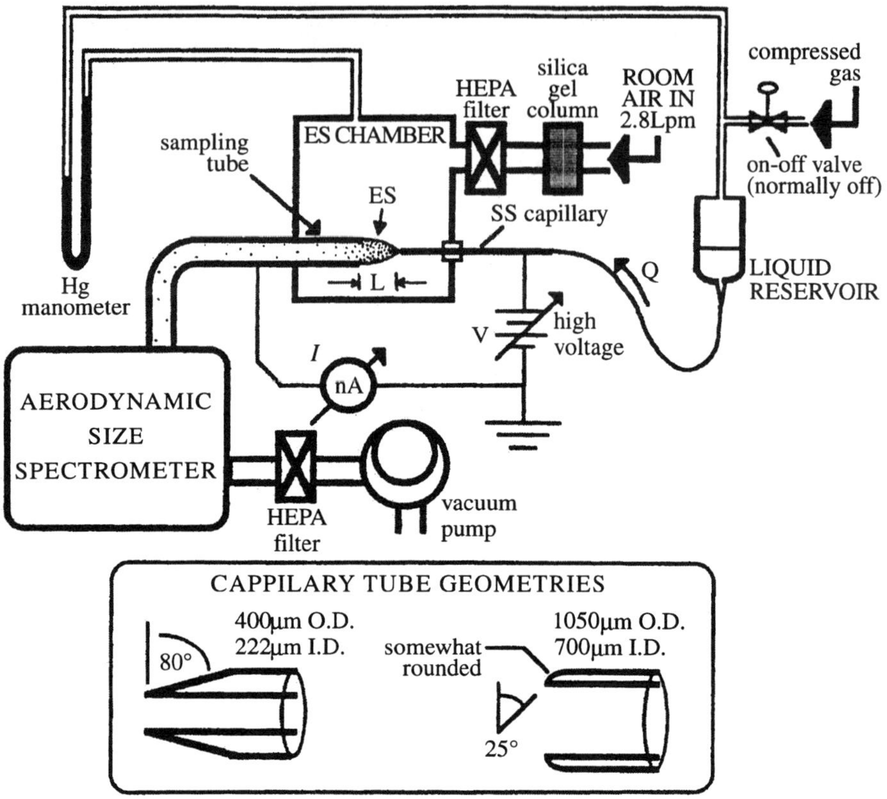


**Fig. 16.9.** Electrospray system for droplet generation. Reprinted from *Journal of Aerosol Science 25*, Rosell-Llompart, J. and Fernandez de la Mora, J., 1093–119. Copyright 1994, with kind permission from Elsevier Science Ltd., Kidlington OX5 1GB UK.[30]

100 $\mu$m have been generated. A facility with a large number of such atomizers can produce 1000 kg of powder per day.[32]

### 16.2.6 Summary of Atomizer Characteristics

Table 16.1 gives a summary of the characteristics of the various atomizers. An important conclusion is that only the submerged ultrasonic, Collison-type, and electrospray atomizers are capable of producing droplets below roughly 10 $\mu$m. Other atomizers with similar droplet sizes, based on other designs, exist but are not commercially available.

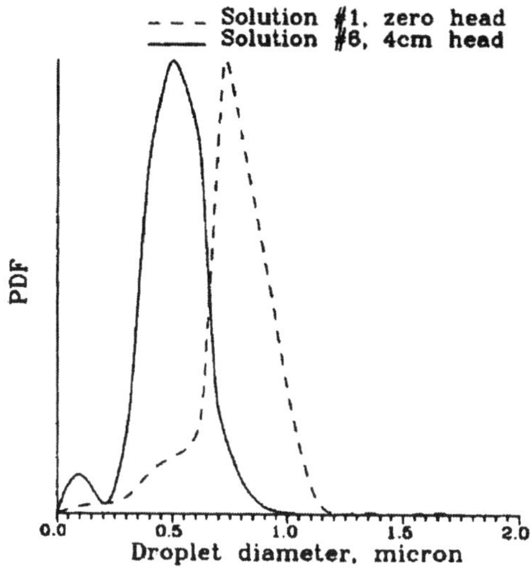


**Fig. 16.10.** Droplet size distribution obtained from electrospraying. Reprinted from *Journal of Aerosol Science 25*, Naqwi, A.A., 1201–1211. Copyright 1994, with kind permission from Elsevier Science Ltd., Kidlington OX5 1GB UK.[31]

**TABLE 16.1. Atomizer characteristics**

| Type | Average Droplet Size | Spread of Distribution | Gas Flow Rate | Liquid Delivery Rate |
|---|---|---|---|---|
| Ultrasonic, submerged transducer | 1–10 μm | Narrow | Low | Low |
| Ultrasonic tube or horn | 10–1000 μm | Medium | Low | Medium |
| Electrospray | < 1–1000 μm | Narrow | Low | Low–high |
| Collison | < 1–1000 μm | Broad | High | Low |
| Rotary | 10–1000 μm | Broad | Low | High |
| Air blast | 10–1000 μm | Broad | High | High |
| Air assist | 10–1000 μm | Broad | High | High |
| Effervescent | 10–1000 μm | Broad | High | High |
| Pressure | 10–1000 μm | Broad | Low | High |

## 16.3 DRY POWDER FEEDERS

The dispersion of dry bulk powders into an aerosol stream has many applications, including inhalation studies, instrument calibration, aerosol characterization, drug delivery, and materials processing such as the classification of powders by particle size. The aerosolization of powders from bulk materials has been accomplished using fluidized beds and variations on powder feeders in combination with jet mills and venturi-type powder dispersion devices.[33–35]

The optimum aerosol stream generated with these devices is one that contains unagglomerated particles with a consistent and controllable instantaneous and time-integrated particle mass concentration and throughput. These stream characteristics depend primarily on the attributes of the powder that is generated, the device used to generate the powders, and the operating parameters of the generating device. The dispersion characteristics of dry powders depend largely on the size, shape, and composition of the particles, the charging characteristics of the powder in the device, the extent of agglomeration of the particles in the powders, and the water absorption characteristics of the powder. In general, hydrophobic powders that are unagglomerated in the bulk disperse better into the gas than do agglomerated and hydrophilic powders, and dispersibility increases with particle size.

*Fluidized-bed particle generators* use the mechanical energy of fluidized beads to disperse small, micron-sized particles. These generators consist of beds of beads hundreds of microns in size that are loaded with the micron-sized particles to be generated. The mixture of beads and small particles is fluidized, and the small particles are released from the bed and carried off in the fluidizing gas. For a given powder, the mass output of the generator and the size of the particles generated are dependent upon the fluidizing gas velocity, the bed height, and the mass of particles relative to the mass of the beads.[33]

*Powder-feeder–air-dispersion* combinations use a mechanical dust-feeding mechanism such as a screw feeder or a scraper to feed the powder to a jet mill or venturi, where the particles are dispersed in a gas. For a given powder, the degree of dispersion and the mass output of the devices are dependent primarily upon the feed rate to the gas dispersion device and the gas flow characteristics in the device.

## 16.4 DROPLET IMPACTORS AND VIRTUAL IMPACTORS

Inertial classification is a technique that can be used to remove larger droplets or particles from a gas stream by impaction on a surface or to remove smaller droplets or particles while concentrating the remaining droplets or particles. For this reason, it is a powerful technique that has numerous applications in the processing of both films and powders. Inertial classifiers, which include impactors, virtual impactors, and cyclones are widely used in the aerosol field, but also have seen limited applications in materials processing.

*Droplet impactors* (Fig. 16.11) rely on directing a gas stream towards a surface near which the gas stream turns, and the inertia of the larger particles or droplets causes them to collide with the surface.[36] This is also the arrange-

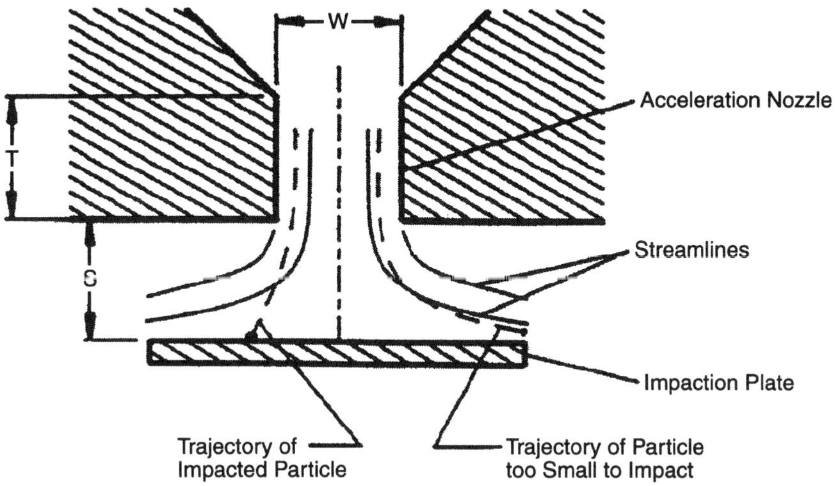


**Fig. 16.11.** Impactor relying on inertial deposition of particles. *Aerosol Measurement: Principles, Techniques and Applications*, Willeke, K., Copyright 1993, Van Nostrand Reinhold. Reprinted by permission of John Wiley & Sons, Inc.[36]

ment used for fabrication of films by droplet deposition (spray pyrolysis of films). Droplet impactors have been used and studied both theoretically and experimentally.[37–44] The studies provide excellent and useful guidelines for the design of systems for film generation by droplet deposition. Many versions of these devices are available commercially for collecting particles, and many more have been designed, built, and used in special studies of materials processing by aerosol routes. Particles as small as 1 nm can be collected using this technique.

*Virtual impactors* (Fig. 16.12) direct a gas stream towards an opening while drawing off most of the gas stream to the sides.[45–58] The larger particles are unable to follow the gas drawn off to the sides and therefore travel into the opening. As a result, the larger particles remain airborne and are concentrated in the small amount of remaining gas. The feature of keeping the larger particles in the gas phase is important for further processing of the droplets or particles. Cutoff points (the diameter above or below which particles are retained or removed) can be as small as 1 $\mu$m in practice.

*Cyclones* (Fig. 16.13) are also widely used to remove larger particles from a gas stream.[4, 59] Cyclones rely on passing a gas tangentially into a cylindrical-shaped container, resulting in a gas flow in the radial direction. The inertia of the particles causes them to collide with the walls, where they are captured. Cyclones can collect much larger quantities of particles than impactors can and, for this reason, are often used to collect particles from gas streams.

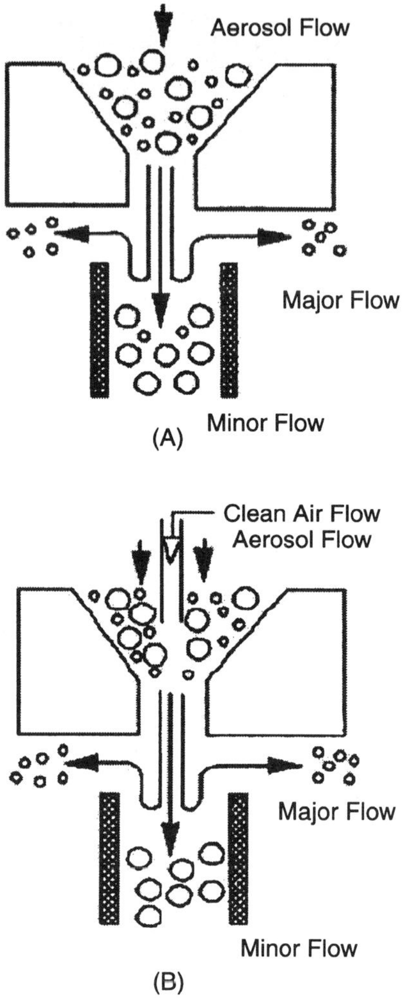


**Fig. 16.12.** Virtual impactor relying on inertial separation of particles. Reprinted with permission of Elsevier Science, Inc. from "A virtual impactor with clean air core for the generation of aerosols with narrow size distributions," by Chein, H. and Lundgren, D.A. *Aerosol Science and Technology*, 18, 376–388, 1993 by American Association of Aerosol Research.[47]

Numerous models are available commercially. Cyclones are most effective for particles larger than a few micrometers in size.

The inertial collection of particles relies on the momentum of the particles. Collection occurs by forcing the gas flow to make a sharp turn and then capturing the particles with sufficient inertia to cross gas streamlines that collide with the surface. Particles with less inertia remain in the gas flow and do not cross streamlines. A good example of inertial collection is an automobile moving down the highway. Insects do not usually hit the windshield at low speeds (relative to each other), but do so at high speeds. Large insects hit the windshield, while smaller insects do not. Finally, the bugs tend to strike regions of the car where the flow of air has to turn more sharply. This situation is quantified in terms of the Stokes number, discussed in Chapter 3.

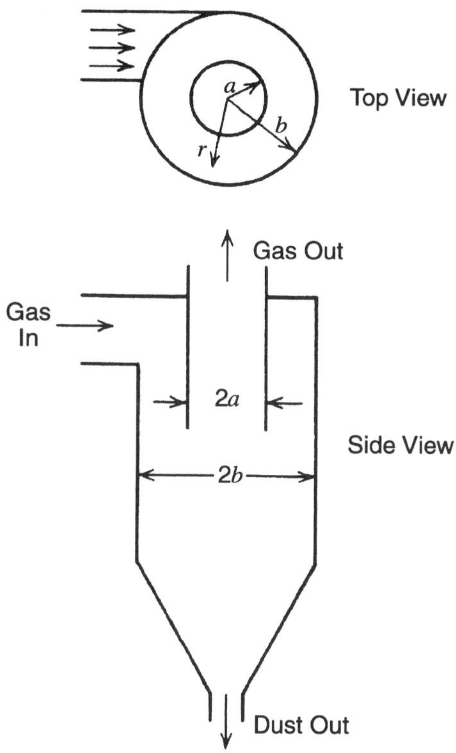


**Fig. 16.13.** Cyclone relying on inertial classification of particles. S.K. Friedlander, *Smoke, Dust, and Haze*, VCH Publishers, Inc. New York, 1994. Reprinted by permission of John Wiley & Sons, Inc.[59]

It is useful to compare the operating principles of the different types of inertial devices. A conventional impactor (Fig. 16.11) consists of a gas jet directed at a flat plate. Variations include either round or rectangular nozzles, single or multiple nozzles, and flat or cylindrical impaction surfaces. In a virtual impactor (Fig. 16.12), the impaction plate is replaced by a collection tube, which is chosen to be larger than the jet. The larger particles travel nearly in a straight line into the collection tube, while the smaller particles leave with the majority of the gas flow. Only a small fraction (often just 10%) of the original amount of gas passes through the collection tube, thereby concentrating the particles. In a cyclone, the aerosol stream is passed through an inlet and enters a cylindrical region tangentially at the outer edge of the cylinder, which is connected to a conical-shaped region. The gas spirals down the outer edge of the cylinder and cone walls and then reverses direction, after which it spirals upward along the center of the cyclone and exits through the top. The larger particles are collected on the cylindrical and conical walls by means of their inertial forces. Agglomerates of impacted particles fall off the walls to the bottom, where they are collected.

A major difference between virtual and conventional impactors is that particles smaller than the cutoff size of the virtual impactor remain in both the major and minor flows. The small particles in the large-particle fraction can be eliminated by providing a central core of clean air in the nozzle of the impactor. Besides being an inertial classifier, a virtual impactor can be used to concentrate particles larger than the cutoff size. Because these particles are concentrated in the minor flow, the concentration factor is equal to the ratio of the total flow rate to the minor flow rate. Concentration factors of 20 can be obtained.

## 16.5 POWDER COLLECTION: FILTRATION AND ELECTROSTATIC PRECIPITATORS

Separation of particles from gases is used for a variety of applications that consist primarily of the cleaning of gas for use (compressed air), the removal of particles in a process as waste or product, or the cleaning of the air in an enclosure (e.g., the removal of cigarette smoke from a room through filtration and air circulation units). The discussion in this section will concentrate on the removal of particles as a powder product from an aerosol process stream. The reader is referred to Willeke and Baron for an in-depth discussion.[4]

Aerosol particles can be separated from a gas using sedimentation, centrifugal force, inertia, interception, electrostatic precipitation, diffusion, and

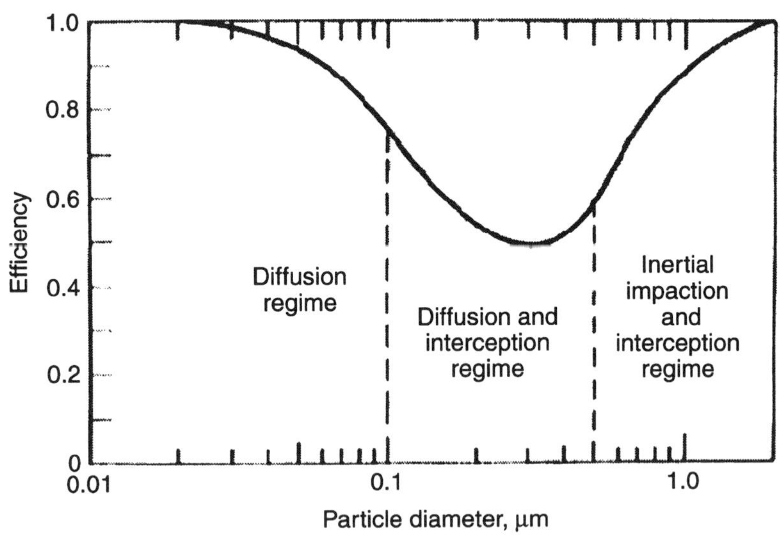


**Fig. 16.14.** Filtration efficiency curve for particles in a gas. Note that the efficiency is high for small particles because of diffusional collection, in contrast to particles in liquids, where the collection efficiency is lower for smaller particles. *Aerosol Measurement: Principles, Techniques and Applications*, Willeke, K., Copyright 1993, Van Nostrand Reinhold. Reprinted by permission of John Wiley & Sons, Inc.[60]

thermophoresis (Fig. 16.14).[60] Although a large number of devices are available that use one or more of these mechanisms to collect particles from an aerosol stream, the most effective mechanism of powder collection depends primarily on the particle size distribution, the particle density, and the gas mean free path of the aerosol stream.

In selecting a method for collecting powder, one must consider many aspects of the aerosol stream, because it is these features that will determine the most efficient method of particle collection. Among such features are the average particle size, particle size distribution, particle density, gas mean free path, gas composition, chemical and physical properties of the powders, particle morphology, and end use of the powder. Other matters to be considered in selecting a powder collection device are the collection temperature, the mass loading of the aerosol stream, the stream flow rate, the desired collection efficiency, whether the particles or gas are corrosive, and even the value of the powder relative to the cost of the unit, all of which are related to the economics of the process.

These factors determine the type, size, and materials of construction of the powder collection device, which, in turn, determine the cost of the unit. In general, the smaller the particles, the higher the collection efficiency, the more corrosive the environment, the higher the collection temperature, mass loading, and gas flow rate, the more the powder collection unit will cost. Of course, special consideration should be given to each application and device.

Devices used for particle collection include filters, cyclones, settling chambers, electrostatic precipitators, thermophoretic collectors, scrubbers, impactors, and combinations of these devices. The most common type of collection devices are filters, cyclones, settling chambers, and electrostatic collection devices, which are discussed in the remainder of this section.

### 16.5.1 Filtration

Filters can be designed for use in almost any application. For large-scale processes, filtration is generally more expensive than cyclones or settling chambers. Filters are commonly used, however, to collect particles with aerodynamic diameters less than 10 microns because these particles are difficult to collect using cyclones or settling chambers.

During filtration, particles are collected primarily by impaction, interception, and diffusion onto the filter substrate (Fig. 16.14). A common misconception is that small particles are difficult to collect from gases. This is true only for liquids; Figure 16.14 shows that smaller aerosol particles are collected more efficiently than larger particles once the size of the particles is below a few tenths of a micrometer. This occurs because of the relatively high diffusion coefficients of particles in gases. Depending upon the type of filter and the aerosol used, particles can penetrate the top surface of the filter and enter into the filter body. For the collection of a product powder, this is undesirable, as powder that enters the body of the filter cannot be recovered easily. For this reason, high surface-loading filters are desirable when high recovery rates are required.

Filters can be used to process streams with flow rates ranging from less than 1 mL/min to many thousands of cubic feet per minute. The size of the filter unit required depends on the system requirements, such as the flow rate and the maximum allowable pressure drop across the filter, the pressure drop per surface area of filter per gas flow rate, the mass loading of the aerosol, and the physical properties of the powder. The permeability of the filter depends upon the type of filter used and the pore size of the filter. For a given type of filter, the smaller the pore size, the lower is the permeability of the filter, and the larger is the surface area of the filter required for a given gas flow rate and pressure drop across the filter. Normally, filters with higher efficiency have lower permeabilities; therefore, the more stringent the collection efficiency requirements, the larger the filter unit will be.

Small laboratory processes normally use plate filters, small cylindrical filters, or small bag filters. Large industrial-scale processes use baghouse filters, which are large housings that contain multiple banks of bag or cylindrical filters. Many types of filters have different efficiencies, permeabilities, corrosion resistances, temperature resistances, and surface-loading properties. Some of the most commonly used filters for powder recovery on the laboratory scale are nylon, polysulfone, polycarbonate, polypropylene, polyethylene, metallic, ceramic and Teflon-based filters. A large number of baghouse filters also are available, ranging from simple polyester fabrics to stainless-steel mesh filters and Teflon filters. In the removal of product powder, contaminates are undesirable; therefore, filters that do not flake off into the powder are desired.

As the filter collects particles on the surface, the permeability of the filter decreases; thus, the pressure drop across the filter increases. As different powders load up in different configurations on the filter, the amount of material that can build up on the filter before the powder is removed depends on the powder's packing properties. In small-scale systems, once the filters are loaded with powder, they are normally removed and scraped to remove the powder, and then they are discarded. In large-scale processes, when the filters have been loaded with powder, the powder is normally discharged from the surface of the filter cartridges by shaking or purging the filters from the inside out with gas that releases the powder, which then drops from the filters to a hopper, where it is collected in a bucket.

### 16.5.2 Cyclones

Cyclones operate on the principal of spiraling an aerosol gas through a cylinder, thereby imposing an artificial gravity on the particles, in a manner similar to a centrifuge, so that the particles are forced in an outward direction to be deposited on the inner wall of the cylinder.[4] Following deposition, the particles fall by gravity into the base of the cyclone, where they are collected in a pot. After the gas is spiraled through the centrifuge and the particles are collected, the gas exits the cyclone through the top of the cylinder.

Both droplets and dry particles can be collected using cyclones. These devices are most efficient, however, at collecting particles that have aerodynamic diameters greater than 5 microns, although specialty cyclones can collect particles as small as 1 or 2 microns in diameter. At the same time, good collection efficiencies in this size range are difficult to obtain, especially in larger cyclones.

Cyclones are commercially available and can be designed to handle virtually any flow rate. They are perhaps the most economical and widely used powder collection devices available. Cyclones are normally constructed of metal and can be used at temperatures as high as 1000 °C. Advantages of cyclones over bag filters include less of a powder holdup, little or no increase in the pressure drop through the cyclone as powder is collected, simplicity in design and construction, the ability to process high-temperature gas streams, and relatively low cost. However, cyclone collection efficiencies for particles less than about 5 microns in diameter are often lower than those obtained using filters. Also, filters are usually much more forgiving in collection efficiencies for variations in flow rate.

Cyclones can be placed in series or in front of other collection devices to reduce the material loading of the gas. When cyclones are used in series, they act as size classifiers by removing smaller and smaller particles in each successive step.

### 16.5.3 Settling Chambers

Settling chambers use gravitational settling of particles as a means of collecting them. The aerosol stream is sent into the base of a relatively large chamber, which is sized such that the upward velocity of the gas is smaller than the settling velocity of the particles. The particles thus settle out of the gas. Due to their mode of collection, settling chambers are used for particles with significant settling velocities or particles generally greater than 25 to 50 microns. The size of settling chambers depends on the settling velocity of the particles and the flow rate of the gas being processed. Settling chambers can be used for both large- and small-scale processes.

### 16.5.4 Electrostatic Precipitators

In electrostatic precipitators, charged particles are collected on an oppositely charged surface (Fig. 16.15).[34] This is accomplished by flowing an aerosol stream through an electrical field that ionizes the gas and creates a small electrical current from one plate to another. The current exchanged between the oppositely charged surfaces depends primarily upon the composition of the gas, voltage potential between the charged surfaces, and the distance

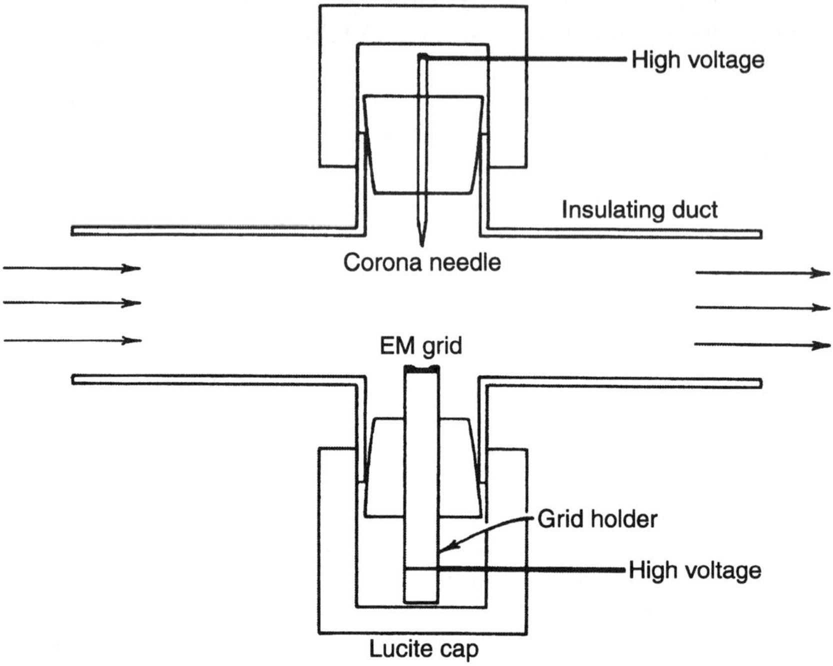


**Fig. 16.15.** Electrostatic precipitation for sampling particles. *Aerosol Technology*, Hinds, W.C., Copyright 1982 Wiley Interscience, Reprinted by permission of John Wiley & Sons, Inc.[34]

between the surfaces. Normally the amperage between the plates is on the order of microamperes, and the voltage between the plates is tens of thousands of volts. If the gas breaks down, arcing between the charged surfaces will occur; this is not wanted because spikes in the amperage discharge are hard to deal with electrically, and the side receiving the discharge can develop pits. Also, the particle collection efficiency decreases drastically during arcing. For this reason, the operation characteristics of electrostatic precipitators are strongly dependent on the ionization potential of the gas.

One of the main advantages of the electrostatic precipitator is that the collection efficiency of the particles actually increases with a decrease in particle size, due to an increase in particle mobility with decreased size.[4] Electrostatic precipitators are commercially available at the scale of most processes.

## 16.6 VOLATILE REACTANT DELIVERY SYSTEMS

The delivery of volatile precursors to a reactor can be accomplished by a variety of methods, which are primarily determined by the volatility and the chemical and physical properties of the precursor. This topic is discussed in detail elsewhere.[3] In general, gaseous reactants are easier to deliver than liquid precursors which, in turn, are easier to deliver than solid reactants. This is due to the volatility of the gaseous reactants and the associated simplicity of the vaporization equipment. Generally, a continuous and consistent precursor output is desired, and controlled precursor mass loading and mass flow rates are a necessity. Additionally, ease and accuracy of determining and controlling the mass output of the precursor is beneficial.

Special considerations need to be taken for each reactant. For example, reactants that are sensitive to certain environments (such as air or moisture) must be stored and delivered in environments that do not degrade the precursor. Also, the thermal stability of precursors can limit the time and temperature at which they can be heated. As an example, Figure 16.16 shows the evaporation rates of a solid precursor as a function of time at different temperatures.[61] The rate drops with time because of changes in precursor characteristics such as surface area and degree of degradation by solid-state reactions. In some instances, special techniques such as spraying are required to deliver thermally unstable precursors.[61]

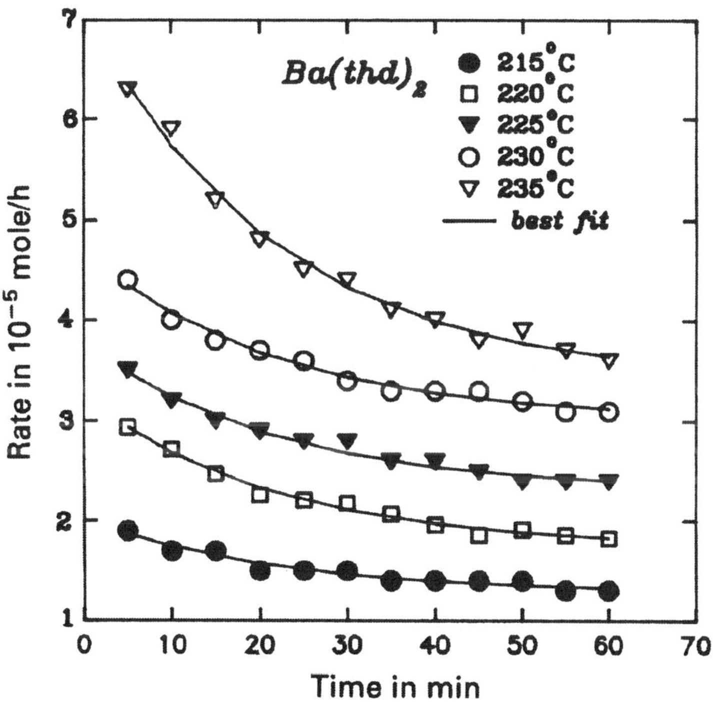


**Fig. 16.16.** Evaporation rates of $Ba(thd)_2$ as a function of time; $P = 0$ torr, $Q = 34.65$ sccm. Curves are drawn on the basis of fitted empirical equations. Reprinted from *Thermochim Acta 240*, July 15, Chou, K.S. and Tsai, G.J., Dynamic Evaporation Behavior of Diketonate Compounds of Yttrium, Copper and Barium, 129–139, 1994, with kind permission of Elsevier Science - NL, 1055 KV Amsterdam, The Netherlands.[61]

Reactants that are already in a gaseous state, such as silane or ammonia, can be delivered and regulated relatively easily by using a pressurized gas cylinder, a pressure regulator, and a flowmeter assembly that is resistant to chemical attack by the gas. Gases are the most convenient form of precursor to deliver, because they are easily delivered in a continuous and consistent stream that can be controlled and monitored.

Liquid reactants are normally generated using bubblers, evaporation pots, below-saturation-level evaporation chambers, specialized liquid delivery systems, or spraying equipment such as nozzles or nebulizers. In standard bubblers, a carrier gas is bubbled through the liquid, thereby facilitating the evaporation of the precursor into the carrier gas. For a given precursor, the amount of precursor evaporated is dependent upon the temperature of the bubbler (which controls the vapor pressure of the precursor), the gas flow rate through the bubbler, the pressure of the bubbler (vacuum or pressurized), and the saturation level of the gas exiting the bubbler. Bubblers are perhaps the easiest and most common method of delivering volatile liquid precursors on a laboratory scale. The vapor production characteristics of bubblers are usually very good, and the mass output of bubblers can be easily monitored through changes in their volume or weight.

Evaporators are containers of the volatile liquid or solid precursors that are open to the system. Most often, evaporators are used on processes that operate under vacuum, and no carrier gases are used. However, evaporators can also be used in high-pressure systems if the precursor is volatile enough to deliver the reactant at the required rate. In evaporators, the precursors simply evaporate and expand into the system. The evaporation rate is a function of the precursor vapor pressure, the system pressure, and the evacuation rate of the system. These parameters can be controlled by varying the temperature of the evaporator and the system pressure. Quite often, the vapor pressure of the precursor sets the pressure of the system.

Liquid delivery systems are similar to evaporators, except that the precursor (usually a liquid) is delivered to the evaporation chamber in a controlled fashion. (The mass throughput of the liquid is controlled.) The evaporation chamber is operated at a temperature such that the precursor immediately evaporates. With this method, the precursor delivery rate is controlled by the feed rate to the evaporation chamber. A liquid delivery system can be used for pure liquids, solutions, or solids, although the delivery of liquids is much easier than that of solid reactants. Liquid reactants can be delivered to the evaporation chamber by a pump or a spray. A spray is often better because it facilitates more efficient evaporation. Solids can be delivered with a dry-powder dispenser or by dissolving the precursor in an inert solution and spraying the solution into a chamber, where it evaporates. Spraying solid precursor solutions is not always possible, because few solvents are compatible with both the precursor and the process itself. Still, the approach is valuable for precursors that are thermally unstable: The precursor does not have to be heated for extended periods of time and is heated and evaporated only as needed.

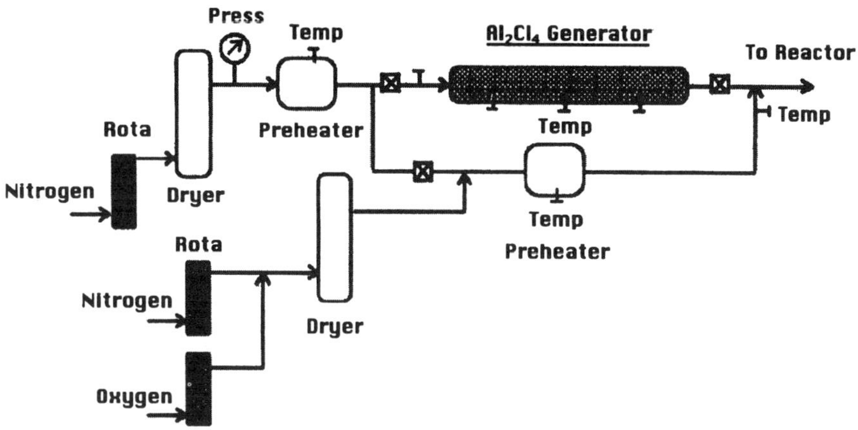


**Fig. 16.17.** System for generating $AlCl_3$ vapor for introduction into a gas-to-particle conversion reactor. Reprinted by permission of Powell.[62]

Solid reactants are usually the most difficult precursors to deliver at a consistent vapor feed rate. The simplest method for delivering solids is to flow gas over a heated bed of the solid reactant. However, the vaporization rate of the precursor can vary greatly, as the saturation level of the gas is dependent upon the surface area of the powder, which changes as it evaporates (Fig. 16.16). One method of getting around this problem is to provide a large amount of powder surface area, so that the carrier gas always gets saturated in the evaporator or the surface area of the precursor powder changes very little with time. A method for accomplishing this is to provide a packed bed of the precursor and a high substrate surface area, such as beads, and then flow the carrier gas through the packed bed as in Figure 16.17.[62] For this case, channeling through the bed must avoided.

As discussed previously, solid reactants can also be delivered by spraying solutions or by using evaporators. Dissolving the solid reactant in a liquid and spraying the solution into an evaporation chamber is advantageous because the delivery rate is controlled and steady. However, as mentioned before, solvents that are compatible with the process and the precursor are not always available.

Finally, reaction-based vapor generators that take advantage of the reaction of two or more bulk materials to generate the desired volatile reactant are sometimes used. Usually, a solid and a gas are reacted to give off a volatile precursor, as is done in the production of $TiO_2$, wherein raw rutile ore or titanium metal is reacted with chlorine gas to form a gaseous $TiCl_4$, which is purified and delivered to be processed. In using this type of vapor generator, it is often advantageous to have the gas feed stream rate be limiting such that the delivery of the vapor precursor is regulated by the amount of gas fed into the generator.

## 16.7 REFERENCES


1. Bayvel, L. & Orzechowski, Z. *Liquid Atomization.* Washington, DC: Taylor and Francis, 1993.
2. Lefebvre, A.H. *Atomization and Sprays*, New York: Hemisphere Publishing Co., 1989.
3. Kodas, T.T. & Hampden-Smith, M.J. (Eds.). *The Chemistry of Metal CVD*, Weinheim, West Germany: VCH, Verlagsgesellschaft, 1994.
4. Willeke, K. & Baron, P.A. (Eds.). *Handbook of Aerosol Measurement: Principles, Techniques and Applications*. New York: Van Nostrand Reinhold, 1993.
5. See, J.B. & Johnston, G.H. *Powder Technol. 21*, 1978, 119–133.
6. Roylance, T.F., Pickering, S.J., Hay, N., & Thomas, G.H. A rotary cup air blast atomizer for molten slag—an investigation of particle size and spray trajectory. Institute of Energy, ICLASS-85, 3rd International Conference on Liquid Atomisation and Spray Systems at Imperial College, London, 8/9/10 July 1985. *LP/VB 2*. London: 1985, 1–16.
7. Smith, N.P., Norris, J.O.W., Hemsley, D.J., & Yeoman, M.L., Measurement of the size and velocity of liquid metal drops during atomization. 3rd Int. Conf. on Liquid Atomisation and Spray Systems, Imperial College, London, 1985. *LP/IVB(b)/1*, London: 1985, 1–9.
8. Salisbury, C.L., Tuncel, G., & Ondov, J.M. *Aerosol Sci. Technol. 15*, 1991, 156–169.
9. May, K.R. *J. Aerosol Sci. 4*, 1973, 235–243.
10. Mercer, T.T. *Aerosol Technology in Hazard Evaluation.* New York: Academic Press, 1973, Chapter 9.
11. Messing, G. Nanoparticulates Workshop, Monterey, CA, 1994.
12. Denton, M.B. & Schwartz, D.B. *Rev. Sci. Instrum. 45*(1), 1974, 81–83.
13. Goddard, R.F., Mercer, T.T., O'Neill, P.X.F., Flores, R.L., & Sanchez, R. *J. Asthma Research 5*(4), 1968, 355–368.
14. Lang, R.J. *J. Acoustical Soc. Am. 34*(1), 1962, 6–8.
15. Lierke, E.G. *Ultrasonics*, October 1967, 214–218.
16. Perron, R.R., *IEEE Trans. Sonics and Ultrasonics SU-14*(4), 1967, 149–153.
17. Peskin, R.L. & Raco, R.J. *J. Accoustical Soc. Am. 35*(9), 1963, 1378–1381.
18. Pointon, A.J. *IEEE Proc. 129 pt. A*, 1982, 285–307.
19. Topp, M.N. & Eisenklam, P. *Ultrasonics*, 1972, 95–101.
20. Berger, H.L., Characterization of a Class of Widely Applicable Ultrasonic Nozzles, *Proceedings of the 3rd International Conference on Liquid Atomization and Spray Systems*, London, July 1985, 1A/2/1–13
21. Lyons, S.W. *Aerosol Dynamics of Superconducting Ceramic Powder Generation by Gas-Phase Processes*. Master's thesis, University of New Mexico, 1992.
22. Berger, H.L. *Proceedings of the 3rd Int. Conf. on Liquid Atomisation and Spray Systems A2*. London: 1985, 1–13.
23. Tang, Q. *Aerosol-deposited YBaCuO thin films and superconducting optical detectors*. Ph.D. dissertation, Twente University, Enschede, the Netherlands, 1992.

24. Rodes, C., Smith, T., Crouse, R., & Ramachandran, G. *Aerosol Sci. Technol. 13*, 1990, 220–229.
25. Tarr, M.A., Zhy, G., & Browner, R.F. *Applied Spectroscopy 45*(9), 1991, 1424–1432.
26. Langlet, M. & Joubert, J.C. The pyrosol process or the pyrolysis of an ultrasonically generated aerosol. In Rao, C.N.R. (Ed.), *Chemistry of Advanced Materials*. Oxford, U.K., Blackwell Scientific Publications, 1993, 55.
27. Graf, P.E., Breakup of small liquid volume by electrical charging. *Proceedings of API Research Conference on Distillate Fuel Combustion*, New York: Am. Petroleum Institute, *1701*. 1962, *CP*62–4.
28. Luther, F.E. Electrostatic automization of No. 2 heating oil. *Proceedings of API Research Conference on Distillate Fuel Combustion*, New York: Am. Petroleum Institute, *1701*. 1962, *CP*62–3.
29. Meesters, G.M.H., Vercoulen, P.H.W., Marijnissen, J.C.M., & Scarlett, B. *J. Aerosol Sci. 23*(1), 1992, 37–49.
30. Rosell-Llompart, J. & Fernandez de la Mora, J. *J. Aerosol Sci. 25*(6), 1994, 1093–1119.
31. Naqwi, A.A. *J. Aerosol Sci. 25*(6), 1994, 1201–1211.
32. Bailey, A.G. *Atomization Spray Technol. 2*, 1986, 45.
33. Carpenter, R.L. & Yerks, K. *Am. Ind. Hyg. Assoc. J. 41*, 1980, 449–454.
34. Hinds, W.C. *Aerosol Technology*, New York: Wiley Interscience, 1982.
35. Koch, W., Lodding, H., Oenning, G., & Muhle, H. *J. Aerosol Sci. 17*(3), 1986, 499–504.
36. Marple, V., Rubow, K., & Olson, B. In Willeke, K. & Baron, P.A. (Eds.), *Handbook of Aerosol Measurement: Principles, Techniques and Applications*. New York: Van Nostrand Reinhold, 1993, 206.
37. Biswas, P., Jones, C.L., & Flagan, R.C. *Aerosol Sci. Tech. 7*(2), 1987, 231–246.
38. Burton, R.M. & Lundgren, D.A. *Aerosol Sci. Tech. 6*, 1987, 289–301.
39. Her, J.Y. & Kim, S.S. *Aerosol Sci. Technol. 14*, 1991, 157–164.
40. Hillamo, R.E. & Kauppinen, E.I. *Aerosol Sci. Technol. 14* (1), 1991, 33–47.
41. Marple, V.A. & Olson, B.A., *Aerosol Sci. Technol. 22*, 1995, 124–134.
42. Mercer, T.T. & Stafford, R.G. *Am. Occup. Hyg. 12*, 1969, 41–48.
43. Ranz, W.E. & Marshall, W.R., Jr. *Chemical Engineering Progress 48* (4), 1952, 173–180.
44. Regtuit, H.E., DeRuiter, C.J., Vrins, E.L.M., Hofschreuder, P., Oeseburg, F., & Benschop, F.M. *J. Aerosol Sci. 21* (7), 1990, 919–933.
45. Anderson, T.L., Charlson, R.J., & Covert, D.S. *Aerosol Sci. Technol. 19*, 1993, 317–329.
46. Biswas, P. & Flagan, R.C. *J. Aerosol Sci. 19*(1), 1988, 113–121.
47. Chein, H. & Lundgren, D.A. *Aerosol Sci. Technol. 18*, 1993, 376–388.
48. Chen, B.T., Yeh, H.C., & Cheng, Y.S. *J. Aerosol Sci. 16*(4), 1985, 343–354.
49. Chen, B.T., Yeh, H.C., & Johnson, N.F. *J. Aerosol Sci. 27*(1), 1996, 83–94.
50. Forney, L.J., Ravenhall, D.G., & Lee, S.S. *Environ. Sci. Technol. 16*(8), 1982, 492ff.
51. Friedman, W. & Willeke, K. *Aerosol Sci. Technol. 5*, 1986, 447–457.

52. Keskinen, J., Janka, K., & Lehtimaki, M. *Aerosol Sci. Technol. 6*, 1987, 79–83.
53. Novick, V.J. & Alvarez, J.L. *Aerosol Sci. Technol. 6*, 1987, 63–70.
54. Pillai, R.S., Yeates, D.B., Eljamal, M., Miller, I.F., & Hickey, A.J. *J. Aerosol Sci. 25*(1), 1994, 187–197.
55. Rader, D.J. & Marple, V.A. *Aerosol Sci. Technol. 4*, 1985, 141–156.
56. Ravenhall, D.G., Forney, L.J., & Hubbard, A.L. *J. Colloid Interface Sci. 85*(2), 1982, 508–520.
57. Sioutas, C., Koutrakis, P., & Burton, R.M. *J. Aerosol Sci. 25*(7), 1994, 1321–1330.
58. Sioutas, C., Koutrakis, P., & Olson, B.A. *Aerosol Sci. Technol. 21*, 1994, 223–235.
59. Friedlander, S.K. *Smoke, Dust & Haze*. New York: Wiley Interscience, 1977.
60. Lee, K.W. & Ramamurthi, M. In Willeke, K. & Baron, P.A. (Eds.), *Handbook of Aerosol Measurement: Principles, Techniques and Applications*. New York: Van Nostrand Reinhold, 1993, 179.
61. Chou, K.-S. & Tsai, G.-J. *Thermochim. Acta 240*, 1994, 129.
62. Powell, Q.H. *Gas-phase production and coating of titania in tubular, hot-wall reactors*. Master's thesis, University of New Mexico, 1995.

# 17 Measurement Techniques

## 17.1 INTRODUCTION

In previous chapters, we discussed the characteristics used to evaluate the quality of powders and films in the context of aerosol routes to powder and film production. Characteristics for powders include size distribution, morphology, phase composition and distribution, crystallinity, chemical composition, thermo-physical and chemical properties, surface properties, and other more specialized properties. For films, the characteristics are similar, but film thickness replaces size. The measurement of physical and chemical characteristics of materials produced by aerosol processes involves many techniques, which are summarized in Tables. 17.1 and 17.2. Because several books already discuss the characterization of film properties, we do not discuss the subject in detail. Rather, we focus on the advantages and disadvantages of techniques for characterizing powders produced by aerosol routes. These techniques are discussed only at sufficient depth to give a feeling for their applications to aerosol-derived materials. The reader is referred to Willeke and Baron, Vincent, Reed, and Feldman and Meyer for discussions of general aerosol measurements, sampling, powder characterization, and surface science techniques, respectively.[1-4]

## 17.2 PARTICLE COLLECTION METHODS

Aerosol particles can be characterized by collecting them onto a surface and then analyzing them, or by measuring some property of the particles while they are in the gas phase. The former approach is particularly useful for materials applications and is discussed here. The collection surface is often a filter that allows the gas flow to pass but captures the particles. In other cases, the particles may be collected by relying on their inertia as is done in impactors and cyclones or in electric, thermal, or magnetic force fields.

Filtration is the most common method for collecting particles.[5, 6] The collection of particles in a filter occurs by inertial impaction, diffusion, interception, gravitational settling, and electrostatic attraction.[7] Which mechanism dominates depends on the size, density, and shape of the particle, its electrical charge, the filtration flow velocity, and the mechanical and electrical properties of the filter. Filtration is the standard approach in most aerosol processes.

**TABLE 17.1. Techniques for measuring characteristics of collected particles**

| Physical characteristics | Chemical characteristics |
|---|---|
| *Size* | *Phase Composition* |
| **Liquid Phase** | X-ray diffraction |
| Chromatography and field flow fractionation | Analytical transmission electron microscopy |
| Electronic methods | |
| Sedimentation velocity | *Elemental Composition* |
| Light scattering | *Bulk* |
| | X-ray fluorescence |
| **Solid Phase** | Emission and absorption spectroscopy |
| Sieving | |
| Image analysis and microscopy | Fourier-transform infrared spectroscopy |
| **Gas Phase** | Proton-induced X-ray emission spectroscopy |
| Optical particle counter | |
| Electrical aerosol analyzer | Nuclear reaction analysis |
| Differential-mobility particle sizer | Gas chromatography with mass spectroscopy |
| Cascade impactor | |
| Aerodynamic particle sizer | Nuclear magnetic resonance |
| Diffusion battery | Electron paramagnetic resonance |
| | Raman spectroscopy |
| *Shape* | *Surface Sensitive* |
| Scanning electron microscopy | Auger electron spectroscopy |
| Transmission electron microscopy | Energy-dispersive spectroscopy |
| Optical microscopy | X-ray photoelectron spectroscopy |
| Scanning tunneling microscopy | Secondary-ion mass spectroscopy |
| Atomic force microscopy | |
| Tap density | *Thermochemical and Thermophysical* |
| *Microstructure: Porosity* | |
| Mercury porosimetry | Differential scanning calorimetry |
| Surface area | Differential thermal analysis |
| He density | Thermogravimetric analysis |
| Pellet green density | Thermogravimetric analysis with mass spectroscopy |
| *Microstructure: Grain Size* | |
| Scanning electron microscopy | |
| Transmission electron microscopy | |
| Scanning tunneling microscopy | |
| Atomic force microscopy | |
| *Assorted* | |
| Small-angle neutron scattering | |
| Small-angle X-ray scattering | |

**TABLE 17.2. Particle size analysis techniques**

| Method | Size | Sample Size |
|---|---|---|
| Scanning electron microscopy | 0.01–50 μm | mg level of dry powder |
| Transmission electron microscopy | 0.001–1 μm | |
| Optical microscopy | 0.2–400 μm | |
| Sieving | 20–5000 μm | 5–20 g dry powder |
| Electronic methods (Coulter counter) | 0.1–400 μm | <1 g in liquid |
| Chromatography and field flow fractionation | 1 nm–200 μm (depending on technique) | <1 g in liquid |
| Sedimentation velocity | | |
| (gravity) | 0.2–100 μm | <1 g in liquid |
| (centrifuge) | 0.02–100 μm | <1g in liquid |
| Optical particle counter | > 0.1 μm | In sampled gas |
| Condensation particle counter | > 0.01 μm or > 0.005 μm | In sampled gas |
| Electrical aerosol analyzer or differential mobility particle sizer | 0.01–1 μm | In sampled gas |
| Impactor | 0.01–25 μm | In sampled gas, mg level |
| Aerodynamic particle sizer | 0.1–25 μm | In sampled gas |
| Single-particle counters | > 0.1 μm | In sampled gas or in liquid |
| Ensemble-particle counters (DLS)* | 0.01–1 μm | In situ or in liquid |

*Dynamic light scattering

The inertia of particles can also be exploited to remove the particles from a gas stream. If the stream is directed at a surface, the inertia of the particles in the gas forces them to deviate from the gas flow streamlines. In single and multistage impactors, the gas stream is passed through a nozzle and directed at an impaction surface. Particles with sufficient inertia reach the collection surface and impact upon it.[8] The particles can be removed by impaction stage to give powder samples that have been segregated according to size.

The diffusional motion of particles can be exploited to collect small particles in diffusion batteries. Deposition by diffusion occurs because of the Brownian motion of the particles, which causes them to move away from a gas streamline. The instrument used to separate particles according to their diffusion coefficients is called a diffusion battery. The most common instruments use screens and can size particles in the 0.001 to 0.2 μm range.[9] Once the particles are deposited on the screens, they can be observed by microscopy.

Thermophoretic collection is often used to sample particles by means of microscope grids.[10] This is accomplished in what is known as a thermal precipitator, which uses a heated wire to create a temperature gradient in a flow channel. The particle is acted upon by a force in the direction of decreasing temperature. Because the thermophoretic velocity does not depend strongly on particle size, these devices work over large size ranges. A drawback is that low gas flow rates are required. The grids can be placed in various optical and electron microscopes.

Electrostatic deposition is also used to collect submicron-sized particles onto microscope grids.[10, 11] A typical device consists of a needle electrode and a flat surface. The high voltage between the needle and surface generates a high concentration of unipolar ions in a corona discharge. The ions charge the particles, and the electric field drives the charged particles towards the collection surface.

Weighing how much material is collected on a surface or filter in a specific period for a known gas flow is one of the simplest and most important measurement techniques. The filter is weighed before collection, particles are collected, and the filter is reweighed. The same approach is used for determining the mass deposited on the collection plates of impactors. Several types of instruments are available for measuring the mass in real time. With a known gas flow rate and sampling time, the aerosol mass and particle number concentrations can be calculated in a simple manner.

Finally, always remember that proper sampling is necessary regardless of the measurement technique, to ensure that the measured size distribution and other characteristics are not distorted by sampling. This is discussed in detail by Vincent.[2]

## 17.3 PHYSICAL PROPERTIES: SIZE AND MORPHOLOGY

Particle size and morphology play critical roles in materials applications. Powders are often formed into bulk materials by procedures in which the powder is formed into a shape (green body) and then sintered to form a dense body. Particle size plays a crucial role because the rate of densification depends on the particle size to a high power, often as $d_p^3$ or $d_p^4$.[12, 13] Thus, small changes in particle size make a large difference in densification rate. The particle size distribution is also crucial. Very narrow or very broad size distributions lead to difficulties in densifying the powder to form a dense part. Further, particle morphology plays a major role in densification. Hollow particles require more shrinkage of a green body to densify to theoretical densities. This requirement usually dictates the use of longer sintering time or higher sintering temperatures, which leads to degraded properties, often due to excessive grain growth. Similarly, powders that consist of smaller particles fused together to form an agglomerate do not pack well into a green body and also require higher temperatures and longer times to reach their theoretical densities. Techniques for quantifying particle size and morphology are discussed next. The reader is referred to Willeke and Baron[1] and Arai[14] for in-depth discussion.

### 17.3.1 Dry Methods: Sieving

Particles produced by aerosol routes are typically collected by dry processes such as filtration. After or during collection, it is often desirable to measure the size distribution of the particles without placing them in a liquid or reentraining them into a gas, both of which can result in changes in the distribution.

Sieving is a simple and easy method for sorting particles ranging from 20 $\mu$m to much greater than 50 $\mu$m.[3] For particles less than or near 20 $\mu$m, however, extensive agglomeration can occur. Smaller particle sizes can be determined by using liquids, but these methods are often difficult to reproduce in the laboratory. Thus, sieving is not routinely used for characterizing powders produced by gas-phase routes. However, sieving can be used to size powders formed by spray-drying processes, where the average particle size is much greater than 10 micrometers.

### 17.3.2 Dry Methods: Scanning and Transmission Electron Microscopy

Scanning and transmission electron microscopes are used to size particles, to examine their microstructure, to quantify phase composition and distribution, and, together with energy-dispersive spectroscopy, to identify the elements present in a sample. Here we discuss the use of these microscopes to study particle size and morphology; later sections discuss other aspects of the instruments. Scanning electron microscopy and transmission electron microscopy differ in that electrons must pass through the sample in the latter, while this constraint is absent in the former. More specialized techniques based on transmission electron microscopy, such as electron energy-loss spectroscopy, can provide information on particle composition.[15] Computer analysis of images of particles from both types of device is a convenient and viable method for directly determining the size, size distribution, and shape characterization of fine particles. The most common image analysis system consists of a microscope, a video camera, and an image-analyzing computer; images of particles are collected with optical, scanning electron, or transmission electron microscopes, depending on the size range of interest. For example, transmission electron microscopy allows sizing of particles from the nanometer up to the micrometer range. Only small amounts of materials are needed, but a good dispersion of particles is required so that single particles are present on the surface. Poor dispersion leads to problems in distinguishing between particles that touch or that are actually fused (Fig. 17.1).[16] Great care must be taken to obtain a representative sample, whereupon the surface of the filter can be examined with a microscope to analyze the size

**Fig. 17.1.** Scanning electron micrograph of as-synthesized $TiO_2$ particles partially fused together and made by spray calcination at 1150°C. (Courtesy of Gurav and Kodas, personal communication.)[16]

distribution. Even with computer sizing of the particles, however, this method remains time intensive because the sample must be prepared on the surface, an image must be chosen, and the image must be analyzed. Another problem is that of obtaining good statistics: Usually, a typical image does not contain enough particles to give a statistically significant result. Further, because electron microscope analysis is carried out in vacuum, it can be used only with materials that do not evaporate or degrade when exposed to a vacuum and an electron beam. Even for inorganic materials, annealing or other changes in sample characteristics can be a problem. Still, scanning electron microscopy and transmission electron microscopy are used extensively by the $TiO_2$ pigment industry.[17]

### 17.3.3 Liquid-Phase Methods: Light Scattering

Many routes for the processing of powders to form bulk objects rely on first dispersing the powder in a liquid. The behavior of the powder in the liquid determines to a large extent how well the powder can be formed into a high-quality green body. Thick coatings are also made, by applying a paste or ink containing particles onto a surface, and here again, the particle behavior in liquids is of critical importance. For this reason, many liquid-phase routes are used to characterize powders produced by aerosol processes. Note that complete dispersion must be obtained if the goal is to measure the particle size distribution; otherwise, false readings will be obtained. These techniques rely on various physical characteristics, such as optical and electrical properties or sedimentation velocity. As a result, they usually give different average sizes with different physical significance. Miller and Lines[18] and Sheppard[19] have summarized recent advances in methods for determining the size of particles that have been collected and placed in liquids. Observe that the particle size and the spread of the size distribution obtained by dispersing a solid in a liquid requires that *all* the powder be dispersed to obtain a true representation.

Light scattering is a powerful method that can be applied to particles that have been collected and suspended in liquids. A wide variety of instruments are available. The main advantage of light scattering in the analysis of collected aerosol powers is that the measurements give the behavior of the particles as dispersed in the liquid. This means that agglomeration of particles on a collection filter, for example, will be observed, provided that the agglomerate is not broken up by the liquid. The main disadvantage is that if agglomeration does occur, the measurements do not provide any direct insight into the aerosol process itself. Furthermore, the particles must be spherical with a known refractive index in order to obtain quantitative results. Particles as small as 0.1 $\mu$m can be detected in single-particle instruments, while much smaller particles can be detected by using dynamic light scattering. Even smaller particles can be examined by X-ray or Neutron scattering.

### 17.3.4 Liquid-Phase Methods: Sedimentation Velocity

Particle size distributions can be obtained from measurements of the sedimentation velocity of particles dispersed in liquids. The particle settling velocity is measured by letting the particles settle a known distance and detecting them using visible light or X-rays. Particle size distributions are calculated under the assumption that the particles are solid and spherical and that Stokes's law can be used to relate velocity to diameter. Particles greater than 0.1 $\mu$m can be sized, while particles below 0.1 $\mu$m are influenced too strongly by Brownian motion to allow measurements, unless centrifuges are used. The upper limit for this technique is roughly 50 $\mu$m, the

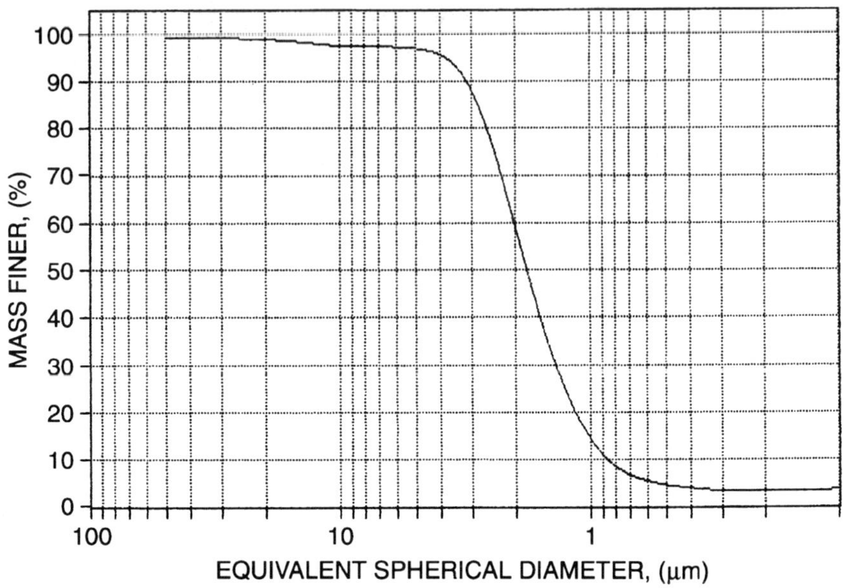


**Fig. 17.2.** Particle size distribution as measured by sedigraph of $BaTiO_3$ powders made by flame spray pyrolysis at an adiabatic temperature of approximately 1900°C. (Courtesy of Brewster.)[20]

size above which particles settle too rapidly and do not follow Stokes's law. Care must be taken to fully disperse the particles in the liquid. For this reason, surfactants are often used to aid dispersion. Figure 17.2 shows a particle size distribution obtained from sedimentation velocity measurements for $BaTiO_3$ particles produced by flame spray pyrolysis.[20]

### 17.3.5 Liquid Phase Methods: Assorted

Chromatography and field flow fractionation can be used to size particles. The particles are placed in a liquid, classified into size fractions by various separation methods, and then counted to determine their size distribution.[18] To separate particles according to size within a flow channel, conventional chromatography is often employed. In field flow fractionation, however, the retention of particles is induced by an electrical, centrifugal, or thermal gradient.

In hydrodynamic chromatography, particles are passed through a packed bed of nonporous columnar material. A particle's residence time in the column depends on the size of the particles, the size of the spheres in the bed, and the flow rate and ionic strength of the eluting solution. This technique is useful for particles from 20 nm to 2 $\mu$m. Size exclusion chromatography is similar to hydrodynamic chromatography, but utilizes a porous bed. The flow through the porous material separates particles by steric exclusion. The upper limit of this technique is nearly 500 nm, and the lower limit is molecular size. Capillary particle chromatography is an extension of hydrodynamic chromatography that is used to separate particles from 0.2 to 200 $\mu$m by employing a long capillary tube instead of a packed bed.

For particles from 10 nm to 1 $\mu$m, field flow fractionation methods can be used. Fractionation with centrifugal sedimentation occurs within closely spaced parallel plates. The particles roll or tumble along the outer wall to produce the segregation.

In electrical sensing methods (e.g., the Coulter counter), the particles to be sized are suspended at a low concentration in a highly diluted electrolyte solution in order to avoid coincidence errors. The particles are then drawn through a small aperture in an insulating wall, across which a current also flows.[18] Each particle passing through the sensing zone displaces some of the electrolyte solution, resulting in a transient change in the electrical impedance across the aperture. The volume of solution displaced determines the amplitude of the electrical pulse produced as the particle passes through the sensing zone. A variety of apertures may be employed to classify particles from 0.1 to roughly 400 $\mu$m.

Because these techniques require the resuspension of small collected particles in a liquid, they are not broadly used to characterizate particles produced by aerosol processes. The resuspension of submicrometer-sized particles into a liquid is a complex problem and can lead to significant errors. However, because so many particle processing schemes rely on suspensions of particles, liquid-phase measurements aid in understanding the behavior of particles in subsequent processing steps.

### 17.3.6 Gas-Phase Methods: Condensation Particle Counter

Gas-phase measurements can provide in situ or ex situ gas-phase particle size distributions. The primary advantage of these methods in aerosol processing is that measurements of particles in the gas phase provide the best measure of the physical and chemical events occurring in a system. Collection of particles followed by solid-state or liquid-phase measurements always has the problem of correctly relating the measured information back to the gas-phase process.

Most gas-phase techniques provide particle size distributions for a gas that is sampled from the system, but the size distribution is still measured while the particles are in the gas phase. Some examples of the instruments employed are the *electrical aerosol analyzer*, *differential mobility particle sizer*, a combination of the *differential mobility analyzer* and the *condensation particle counter*, and the *aerodynamic particle sizer*. Some instruments rely on separating particles in the gas phase, followed by collecting and subsequently measuring them. Among these instruments are cascade and virtual impactors.

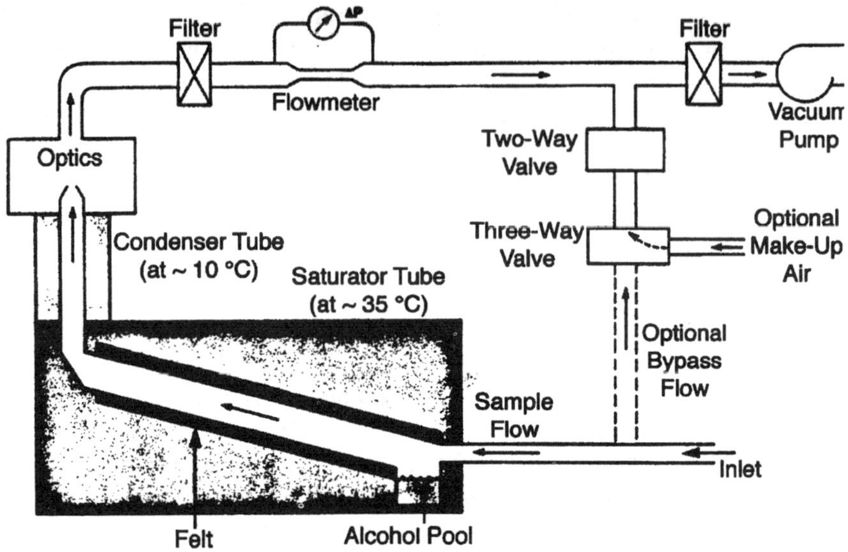


**Fig. 17.3.** Schematic of a condensation particle counter. (Adapted from Instruction Manual, Model 3022 Condensation Particle Counter, TSI, Inc., St. Paul, MN, 1988.)

This section discusses all of the foregoing instruments, which are commercially available from several sources.

A condensation particle counter (Fig. 17.3) measures the total particle concentration above a certain limit, near 0.005 $\mu$m in the most advanced instruments. The device operates by drawing in an aerosol stream, saturating the stream with an organic vapor at higher temperatures, cooling the gas stream to provide particle growth by vapor condensation, and counting the particles by light scattering. Limitations on the instrument are that the particle concentrations must be below $10^7/\text{cm}^3$ and the gas stream must be at room temperature when it enters the instrument. The condensation particle counter does not distinguish between particles with different sizes, shapes, and compositions. It is often used together with other aerosol techniques that classify particles by size, in which case it is employed to count particles in each size range. Detailed studies of these instruments have been carried out.[21–25]

### 17.3.7 Gas-Phase Methods: Differential-Mobility Particle Sizer and Electrical Aerosol Analyzer

A differential-mobility particle sizer is a combination of a differential-mobility analyzer and a condensation particle counter. An electrical aerosol analyzer is a differential-mobility analyzer in conjunction with an electrometer (a device that measures the current associated with the collection of charged particles) (Fig. 17.4). The differential-mobility analyzer is an instrument that separates particles according to their electrical mobilities.[7] Particles are charged with a Boltzmann equilibrium charge distribution and then sent in laminar flow along the outside edge of an annular region. A potential is applied across the region, causing charged particles to move in the radial direction at different velocities, depending on their size. At the bottom, a single (differential) size cut is taken out of the differential-mobility analyzer to give nearly monodisperse particles of known size. A condensation particle counter or an electrometer is then used to count the particles.[7] By changing

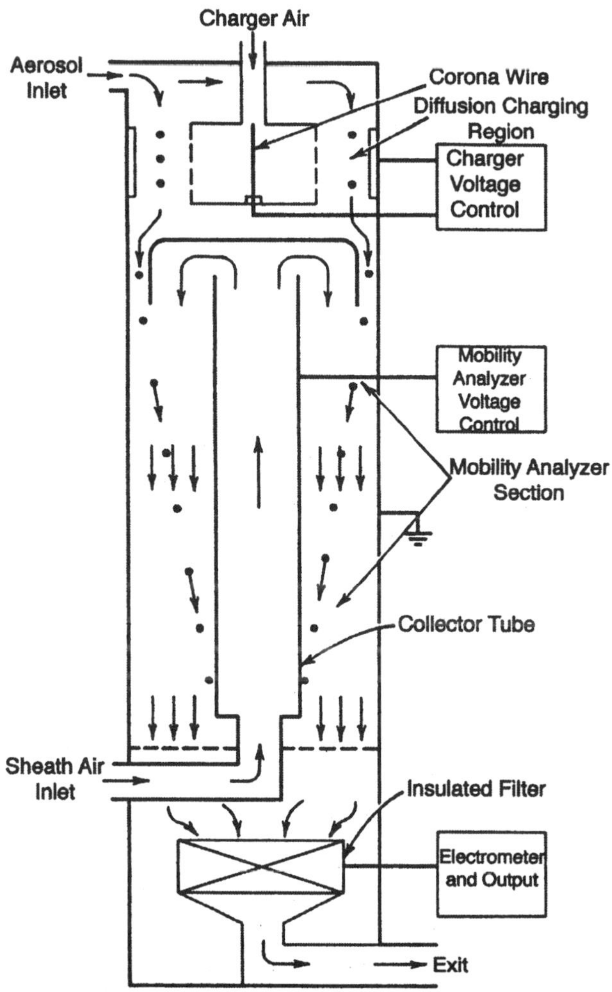


**Fig. 17.4.** Schematic of an electrical aerosol analyzer. (*Aerosol Technology*, Hinds, W.C., Copyright 1982. Reprinted by permission of John Wiley & Sons, Inc.[7])

the voltage and scanning over the entire size range, one can obtain a full particle size distribution over the range 0.005 to 1 $\mu$m. Among the disadvantages of the technique are that the particle stream must be at room temperature and that difficulties exist in determining the size of agglomerated particles. Advantages are that the particles remain in the gas phase, particle concentrations greater than $10^7/\text{cm}^3$ can be handled, and nanometer-size particles can be detected. Numerous detailed studies of these instruments have been carried out.[26–29]

### 17.3.8 Gas-Phase Methods: Single-Particle Optical Particle Counter

Single-particle optical particle counters size and count particles greater than 0.1 $\mu$m as they pass through a sample volume defined by a conventional light or laser beam. As the particles pass through this volume, they scatter light, which is collected and focused onto a detector. The particle size is obtained from the intensity of the scattered light. Several reviews of the technique are available.[30] Limitations include coincidence errors at particle concentrations over roughly $10^3$ particles/cm$^3$, and the necessity of keeping the aerosol stream at room temperature for instruments that withdraw the gas from the system. Furthermore, calibration of the instrument relies on knowing the refractive index of the particles that are sized. Single-particle optical particle counters are often used in conjunction with an aerosol dilution system that reduces particle concentration to below $10^3/\text{cm}^3$. These instruments are rarely used for aerosol routes to powders because of the restriction to low particle concentrations; however, they have been utilized for particles of different chemical compositions.[31]

Single-particle phase-based instruments are also available. These instruments are based on the phase Doppler technique and are combined with laser Doppler velocimetry to give simultaneous measurements of the size and velocity of single particles. The technique can be used in situ and does not depend on light intensity, so beam attenuation and window fouling effects are minimized. The system works by measuring the spatial and temporal frequency of the Doppler-shifted light that is scattered as single particles pass through the viewing volume. Measurements are performed with the use of two detectors. The phase shift between the signals from the detectors is related to the particle diameter, refractive index, and receiving geometry of spherical particles. Limitations on the technique include its inability to size particles below 0.1 $\mu$m and the restriction to concentrations on the order of $10^3$ particles/cm$^3$. The advantage is the in situ capability. These instruments are used often for obtaining droplet size distributions in sprays. Low-pressure systems are available for examining particle formation in CVD, etching, and other systems.[32]

Imaging single-particle optical counters work by providing a direct image of the particle. The accuracy is limited by diffraction and the depth of field effects. The lower limit is usually more than 10 $\mu$m in commercial instruments. These devices are rarely used for aerosol processes, except for spray processes, for which they can provide information about droplet size distributions.

### 17.3.9 Gas-Phase Methods: Dynamic Light Scattering

Dynamic light scattering is one of the few techniques for in situ gas-phase measurements of particle size in the submicron range. The approach, also known as *diffusion broadening spectroscopy* or *quasi-elastic light scattering*, can size particles in the 0.01 to 1 $\mu$m range and does not require knowledge of particle properties (i.e., density and refractive index), so no calibration of the instrument is necessary. The method works by measuring the degree of spectral broadening induced in an incident laser beam by scattering from particles in Brownian motion (random thermal motion). The average particle diffusion coefficient, which is related to particle size, is measured. Several versions of this instrument are available commercially. The main advantages of the technique are its in situ capability and the relatively small particle size that is capable of being detected. The main disadvantages are the complexity of the method and high cost of the instrument.

### 17.3.10 Gas-Phase Methods: Assorted

Particle field imaging relies on producing images of ensembles of particles which can then be analyzed by various software packages. The technique requires particles much larger than 1 $\mu$m and is rarely used for aerosol routes. Fraunhofer diffraction determines a size distribution from a measurement of the diffraction patterns that result from viewing many particles simultaneously. Numerous commercial instruments are available, as are reviews of this approach.[33] The lower detection limit is roughly 1 $\mu$m, the upper limit several hundred micrometers. As with some of the other light scattering instruments, these are rarely used for sizing particles in processes that generate materials, except for spray processes.

### 17.3.11 Gas-Phase Methods: Multistage Impactors

Multistage or cascade impactors allow particles to be collected from a gas in accordance with their size. In these impactors, each stage corresponds to a certain range of particle sizes. The different size ranges are usually obtained by decreasing the nozzle diameter and increasing the gas velocity. Each stage is assumed to collect particles with diameters larger than a certain cutoff size. The result is a collection of particles sorted into groups by size.

Impactors have several practical problems, especially with regard to solid particles. The particles may bounce or be reentrained from the surface and then be collected at subsequent stages.[34, 35] This problem is usually mitigated by coating the impaction surfaces with a grease to improve particle collection. However, depositing particles build up, and subsequent particles may bounce because they strike particles that have already deposited. This limited loading capacity is one of the drawbacks of impactors. The main advantages of the instrument are its ability to segregate particles by size, sample directly from the gas at high temperatures, and take the samples from the stages for subsequent analysis by other techniques. Particles as small as 0.01 $\mu$m can be routinely analyzed with the use of numerous commercially available instruments. Hypersonic impactors have been developed that can separate particles down to 1 nm.[36]

### 17.3.12 Gas-Phase Methods: Virtual Impactors

Virtual impactors (described in more detail in Chapter 16) exploit the inertia of particles without allowing them to impact onto a solid surface. Instead, the particles are separated into two gas streams, composed of the majority and the minority flows. Particles with high inertia are collected in the minority stream of gas, while smaller particles with lower inertia follow the majority gas stream. The particles in both flows can be collected on a filter and analyzed. Alternatively, they can be analyzed with other techniques that do not rely on collection.[37–39] Virtual impactors are not as convenient as cascade impactors and therefore are not used often for aerosol routes to materials.

### 17.3.13 Gas-Phase Methods: Cyclones

Cyclones remove particles from gas streams by exploiting centrifugal forces.[40, 41] The sampled gas stream moves in the radial direction; particles with high inertia cannot make such a sharp turn and strike the walls. Cyclones do not yield cutoff particle sizes as sharp as those of impactors, but they can handle large volumes of gas and particles. The use of several cyclones in series allows the collection and classification of particles by size.[42] The devices are used most often for a collection, rather than classification, function. Particles as small as 1 $\mu$m can be sized with a cascade of cyclones.

### 17.3.14 Gas-Phase Methods: Aerodynamic Particle Sizer

Aerodynamic particle sizers size particles on the basis of their time of flight. The particles are usually detected by light scattering. The major difference between this instrument and an optical single-particle counter is that, instead of measuring the intensities of the scattered light pulses, the aerodynamic particle sizer measures the velocities of particles exiting from an acceleration nozzle. These velocities are then related to particle size.[43, 44] The major disadvantage of the technique is that only low particle concentrations ($<10^3/cm^3$) can be handled, due to coincidence errors. Because of this limitation, these instruments are rarely used for aerosol routes to materials.

Aerodynamic particle sizers have been used to determine shape factors of nonspherical particles.[45]

## 17.4 PHYSICAL PROPERTIES: MICROSTRUCTURE

The microstructure (crystallite size, pore size, etc.) of particles plays a critical role in determining the properties and the rate of densification of the particles. To characterize microstructure, it is necessary to measure porosity, surface area, density, and grain size of particles. If the particles are to be used in catalytic applications, it is important to know their total surface area.

### 17.4.1 Mercury Porosimetry

Mercury porosimetry utilizes mercury intrusion to measure pore size and pore size distributions.[3] A sample of powder is placed into a chamber where mercury under a controlled pressure is forced into the pores of the material. As the pressure is increased, the mercury enters smaller pores. The accessible pore size corresponding to a given pressure is obtained from the Washburn equation, which relates the applied pressure to the intruded radius of a cylindrical pore (Fig. 17.5).[46] This technique is often used to look for

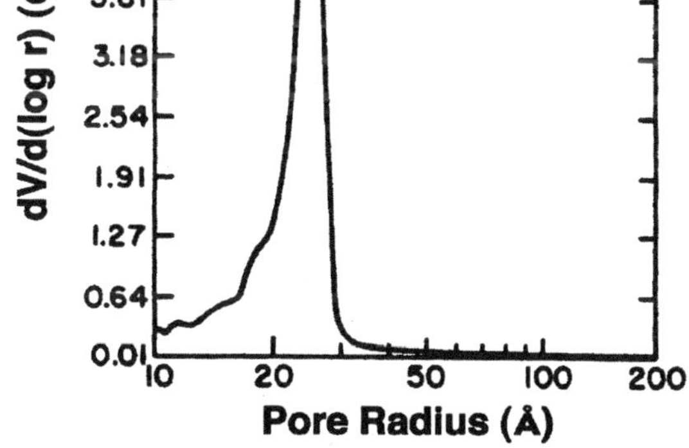


**Fig. 17.5.** Pore size distribution of silica xerogel, calculated from a nitrogen desorption isotherm. (Reprinted with permission of The American Ceramic Society, Westerville, Ohio. Copyright 1991 by The American Ceramic Society. All rights reserved.)[46]

porosity in particles produced by spray routes. Sample quantities of one gram are frequently sufficient for analysis. Pores as small as 1 nm can be detected.

### 17.4.2 Surface Area

The surface area of a powder is frequently measured by nitrogen, carbon dioxide, hydrogen, or helium adsorption with data analysis that uses the Brunauer–Emmet–Teller isotherm method,[3] which provides information about the porosity of particles and the primary particle size for those particles consisting of agglomerates of smaller particles. In general, adsorption measurements can give pore size distributions and total surface areas. Different types of information can be obtained with different types of adsorbing probe molecules or atoms. For example, $N_2$ adsorbs on all surfaces, while $H_2$ selectively adsorbs on metal particle surfaces. Therefore, this combination can be used to determine the total surface area and accessible surface area of a metal in a heterogeneous catalyst. If the particles are smooth, monodisperse spheres, the primary particle size can be related to surface area. Specific surface areas can range from a few square meters per gram for micrometer-sized solid particles to hundreds of square meters per gram for highly porous or very small particles. Figure 17.6 shows how measured surface areas can depend on particle formation conditions for

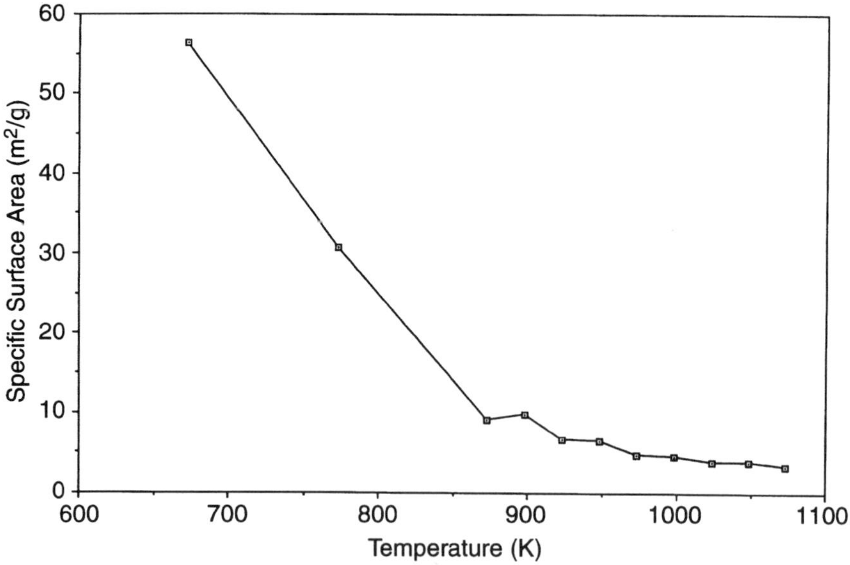


**Fig. 17.6.** Specific surface area of PdO measured by Brunauer-Emmett-Teller nitrogen adsorption analysis as a function of the reactor temperature during spray pyrolysis of palladium nitrate. (Adapted from Gurav et al.)[47]

spray pyrolysis.[47] This is a critical measurement and is almost always carried out for aerosol-derived materials. Sample quantities of 1 gram are often sufficient. An example for aerosol-derived Si powders is given by Sheldon and Haggerty.[48]

### 17.4.3 Density

The apparent density of individual particles can vary from that of the bulk material to much lower values if the particles contain trapped pores or are hollow. When no closed porosity is present, helium pycnometry measurements provide the density of a material.[3] A sample is weighed and placed in a vessel into which helium is introduced. The volume of the sample is obtained by computing the volume of gas displaced by the powder. The density is predicted by noting the volume and weight of the sample. Hollow particles with no porosity provide densities lower than that of the bulk material. If the bulk density is known, such a measurement aids in determining the amount of closed porosity in a sample. This measurement is often used for powders derived by spray routes. Sample sizes are typically 1 gram with no restrictions on particle size.

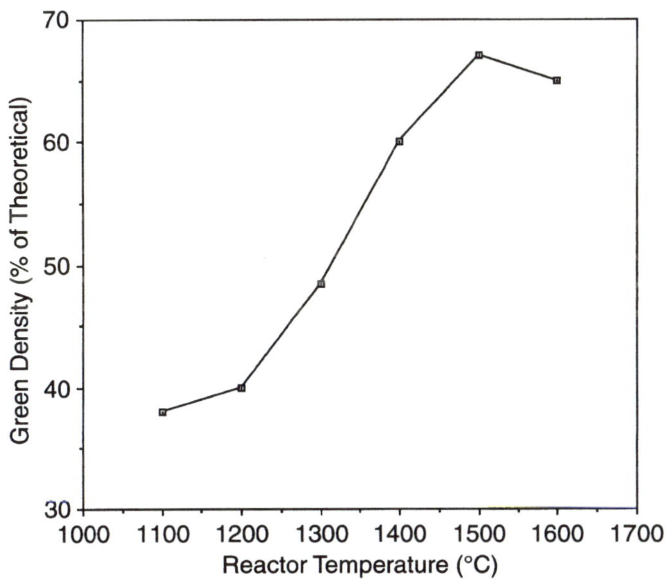


**Fig. 17.7.** Variation of the green density of pellets of Ba–Ca–Ti–O as a function of reactor temperature during powder synthesis by spray pyrolysis. (Courtesy of Ortega et al.)[49]

### 17.4.4 Pellet Green Density

Various powder-processing techniques such as dry pressing produce green pellets—unfired compacts of powders. The green density—that is, the apparent density of the pellet—is an indirect, but practical measure of the quality of a powder. Although it is well known that pellet-processing variations can result in large variations in green density, the method still provides a way of quantifying the packing characteristics of powders from an applications point of view. Typically, compacts of hollow particles will have low green densities, on the order of less than 20 to 40%. In contrast, solid particles can provide green densities of 50 to 60% (Fig. 17.7).[49] This approach is used routinely when the aerosol-derived powder is to be sintered to form bulk materials in subsequent processing stages.

### 17.4.5 Tap Density

Tap density is a technique that measures the apparent density of powders (as opposed to particles). A powder is placed into a cylinder, and then the cylinder is tapped a prescribed number of times. After this tapping, the volume of the powder is measured. Using this volume and the weight of the powder, one can calculate the tap density. The tap-density technique is used extensively by industry to quantify the performance of powders. Qualitatively, a higher tap density corresponds to less agglomeration in the powder. Industry has found that tap density correlates well with performance during powder processing to form films and bulk materials for powders composed of particles with sizes greater than or equal to 1 $\mu$m. Often, several grams are required for this measurement.

### 17.4.6 Transmission Electron Microscopy

Transmission electron microscopy is frequently used to observe and determine grain sizes ranging from the nanometer to the micrometer scale. Milligram amounts are needed, and the sample must be placed on a special grid for analysis. This can be accomplished by dipping the grid into the powder or by sampling from the gas onto the grid. The crystallinity (microstructure) of materials is important for several reasons. Processing of ceramic or metallic powders involves particle sintering. At the same time, sintering rates and the grain-size distribution of densified ceramics depend on the initial grain-size distribution (crystallinity) within each particle, as well as the size distribution of the particles. Edelson and Glaeser[50] have shown that sintering of spherical particles consisting of agglomerates of smaller particles can result in broad grain-size distributions, which are undesirable.[51] Particles with smaller grain sizes also have larger grain surface areas, providing a large driving force for increases in grain size.[13] Therefore, particles consisting of nanocrystalline grains usually densify at lower temperatures than single-crystal particles do. Transmission electron microscopy is used routinely to observe crystallite size in aerosol-derived powders.

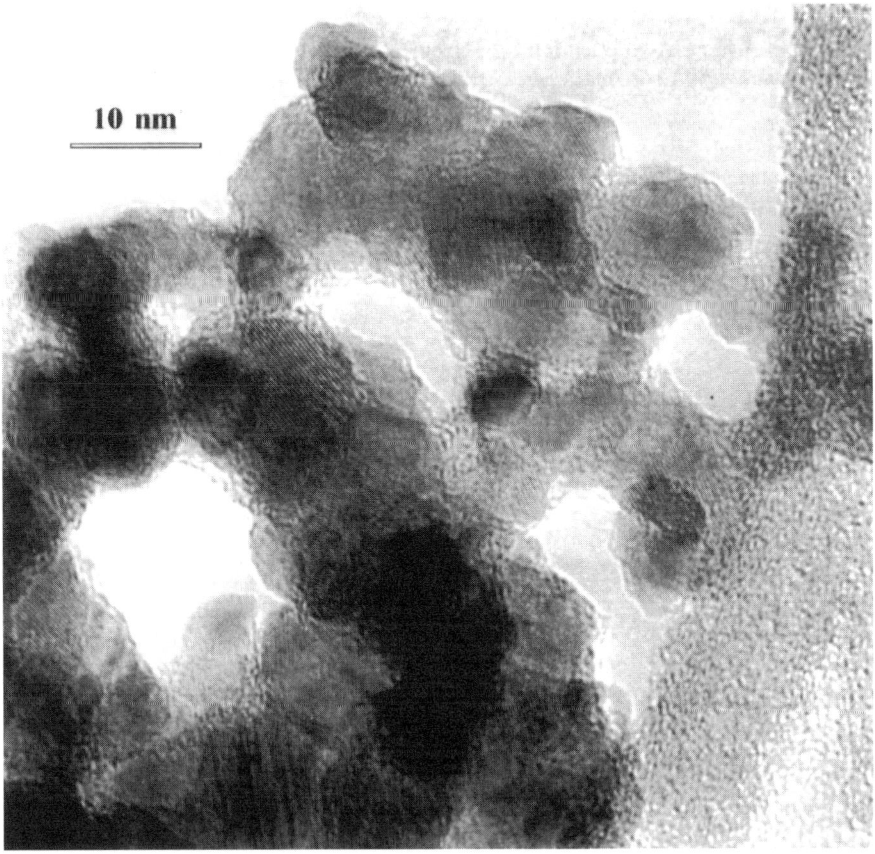


**Fig. 17.8.** Transmission electron micrograph of ZnS particles made at 700°C by gas-phase reaction of a single-source organometallic precursor. The crystallite size is 3 to 10 nm. (Courtesy of Dericotte.)[52]

Figure 17.8[52] shows how transmission electron microscopy can be used to examine microstructure and determine whether particles are polycrystalline or single crystals.[4] The approach requires that the particles be small enough to allow penetration of the electron beam through the sample. This is not possible for many powders made by spray pyrolysis, which are too large (often, roughly 1 $\mu$m).

## 17.5 PHYSICAL PROPERTIES: PHASE COMPOSITION AND DISTRIBUTION

The phase composition of a sample is a measure of the type and number of phases present in that sample. Even in the case of relatively simple systems such as metals or simple oxides, several different phases may be present (e.g., for $TiO_2$, brookite, rutile, and anatase). For some complex materials, such as ceramic superconductors, tens of phases may be formed in one sample. These phases have profoundly different attributes that influence the properties of the powder and its behavior in subsequent processing steps. For these reasons, determining phase content is a critical step in the production of any powder or film. The two main approaches to quantifying phase composition and distribution are X-ray diffraction and electron diffraction in a transmission electron microscope.

### 17.5.1 X-ray Diffraction

X-ray diffraction is a standard quantitative tool for identifying crystalline phases in powder or film samples.[4] The crystalline phases diffract X-rays according to Bragg's equation, which relates lattice spacing to X-ray wavelength. Only a few milligrams of sample are needed. Crystalline phases are detected when present at levels of about 1 percent or greater. The Scherrer equation can be used to relate the widths of the peaks to the crystallite size.[14] The unit cell dimensions $(a, b, c)$ of the crystal can also be determined and therefore provide more information about materials with a complex structure. Amorphous materials, however, are not observed. In routine work, X-ray diffraction relies on the availability of standards that allow the identification of peaks in the diffraction pattern. The orientation of films can be determined by examining the intensity of the diffraction peaks. Stress can also be measured.

### 17.5.2 Transmission Electron Microscopy and Electron Diffraction

Microdiffraction with transmission electron microscopy can provide information about the crystalline structure of phases present in particles.[4] Single particles can be examined, so only very small amounts of a sample are needed. The electron diffraction pattern shows whether the particles are polycrystalline (ring patterns), a single crystal (spot patterns), or amorphous. Different phases in a particle can be identified by electron diffraction, which gives patterns characteristic of a given phase of a material. Coatings can usually be easily distinguished because of differences in the contrast between the coating and core material.

## 17.6 CHEMICAL COMPOSITION: BULK MATERIALS

Aerosol routes and other routes to powders and films often result in multicomponent materials with varying chemical composition. Multicomponent materials formed from several precursors may have different compositions at different reactor operating conditions because of vapor wall losses.

For these reasons, determining the chemical composition of a multicomponent particle or film is of great importance. Even in single-component systems, the presence of chemical impurities may be detrimental and must be evaluated.

There are a variety of techniques for determining the chemical composition of aerosol particles. An outline of these methods is presented in this section. The discussion will include X-ray fluorescence, proton-induced X-ray emission, nuclear reaction analysis, and emission and absorption spectroscopy. Several reviews and books examine the chemical characterization of aerosols.[1, 53]

### 17.6.1 X-ray Fluorescence

X-ray fluorescence is used to determine the chemical composition of liquid and solid samples.[4] The sample is irradiated with X rays, producing photons (fluorescence). The wavelength of the emitted light is determined by the elements in the sample. This technique can be used to examine multicomponent solid samples when it is desirable to determine the ratio of the elements present in the sample. Elements with $Z > 11$ are detectable down to roughly 10 ppm. Only 10-to-100-nanogram amounts of particles are required in order to determine chemical composition. Both energy- and wavelength-dispersive instruments are available. Knapp and Bennett[54] discuss the use of X-ray fluorescence for sized aerosol particles.

### 17.6.2 Fourier-Transform Infrared Spectroscopy

Fourier-transform infrared spectroscopy measures the amount of infrared radiation absorbed by a sample as a function of the reciprocal of the wavelength (wavenumber) of the sample.[3] This absorption can be correlated with the different types of bonds present in the sample. In this sense, the technique does not provide a measure of composition directly, but gives only an indirect measure similar to the way solid-state nuclear magnetic resonance works. The mass of sample required depends on the nature of the sample, but 10-mg amounts are usually adequate. An important application for aerosol-derived materials is characterization of the surfaces of $SiO_2$ particles produced by flame processes.

### 17.6.3 Proton-induced X-ray Emission Spectroscopy

Feldman and Mayer discuss the use of ions to induce the emission of X-rays from a sample.[4] Elements from Na to U can be detected. Ten-micrometer beams can be used with proton-induced X-ray emission for high-sensitivity detection of trace elements. In addition, the method can simultaneously identify several elements in complex samples.

### 17.6.4 Nuclear Reaction Analysis

Nuclear reaction analysis probes samples with an ion beam and examines the light-ion reaction products.[4] The technique is also known as instrumental neutron activation analysis. The probe beam induces radioactivity (gamma rays) in the powder or film, which is measured to determine the identity of the elements in the sample. One advantage is that light elements such as H, Be, B, C, O, and F can be detected. The detection limit for thermal nuclear reaction analysis is $10^{-8}$ to $10^{-10}$ gram. Because this method has poor spatial resolution, it is commonly used for samples of particles collected on filters of impactors. Although the technique requires a nuclear reactor, several laboratories provide services on a routine basis. An example of an application to aerosols is given by Podzimek et al.[55]

### 17.6.5 Emission and Absorption Spectroscopy

According to Reed, emission spectroscopy and absorption spectroscopy are extremely sensitive methods that detect trace amounts (ppm) of elements in samples of 5-mg powders.[3] An electric arc or a laser flash is employed to provide information about the wavelengths and intensities of emitted or absorbed light. This technique is often used for quick qualitative surveys of samples.

Alkali and boron are analyzed to the ppm level (with ppb detectability possible for some elements) with flame emission spectroscopy. The technique is commonly used for liquid samples that are sprayed into flames.

Inductively coupled plasma emission spectroscopy also utilizes liquid samples. This technique is a reliable, fast, and multielement method that can analyze about 40 elements.

Atomic absorption spectroscopy, the industry standard for quantitative analysis to ppm levels, can analyze 30 to 40 elements, one element at a time, present at concentrations of $< 0.1\%$. The sample, in solution form, is sprayed into a flame through which the light from a lamp is passed. Dissociated atoms in the flame absorb the light and reduce the intensity transmitted. The species that are present are identified from the wavelengths at which absorption takes place. The major advantage of this method is its high sensitivity.

Emission and absorption spectroscopy methods require that the sample be able to be dissolved completely in a liquid (often a mixture of concentrated acids).

### 17.6.6 Combustion Elemental Analysis

Combustion elemental analysis is a method that is used primarily for the detection of light elements in samples containing organic constituents. The technique involves combusting the sample to convert the carbon, hydrogen, and nitrogen present to $CO_2$, $H_2O$, $NO_x$ and, by gas chromatography, measuring the amounts of these products that are evolved. This technique is accurate to within approximately ±0.3 wt.%, inexpensive, and requires only about 1 to 5 mg of sample. Problems can occur when metals are present, due to incomplete oxidation (e.g., metal carbides may be formed). These problems are generally resolved by adding oxidizing agents such as $WO_3$.

## 17.7 CHEMICAL COMPOSITION: SURFACE-SENSITIVE METHODS

### 17.7.1 Auger Electron Spectroscopy

One of several surface sensitive methods for determining the chemical composition of materials is Auger electron spectroscopy.[4] This technique can be used on both powders and films. The environment is an ultrahigh-vacuum chamber in which an electron beam is directed at a sample and results in the emission of electrons (the so-called Auger electrons) with energies that are characteristic of the elements in the sample. Depth profiling can be obtained by sputtering for solid samples of both powders and films. In this way, these techniques can give information about the bulk composition of a powder or film. Auger electron spectroscopy is surface sensitive because the Auger electrons can travel only 0.5 to 2 nm after being formed. The technique can detect elements from Li through U with up to 0.3% sensitivity. Specialized scanning Auger electron spectroscopy instruments with a spatial resolution of 0.1 $\mu$m are available. Even single particles can be examined, so the amount of material required is very small. This technique is particularly useful for aerosol processes in which coated particles are produced. Surface charging is a problem for insulating materials such as $SiO_2$. Figure 17.9 shows how the technique can be used to analyze $TiO_2$ particles coated with $SiO_2$.[56] Scanning Auger electron microscopy can be utilized to generate compositional maps of solids.

### 17.7.2 Energy-Dispersive Spectroscopy

Energy-dispersive spectroscopy is a technique that probes samples with electrons and results in the emission of X rays with energies characteristic of the elements present in the sample.[4] The technique can be used on both powders and films. Because an electron beam is used as a probe, the composition of particles or films in the nm range can be examined. Even single particles can be examined by this technique. The probing depth is approximately 1 $\mu$m. Energy-dispersive spectroscopy has a detection limit of 0.1% for $Z > 11$. Thus, elements with atomic weights greater than sodium can usually be detected (Fig. 17.10). Resolution down to less than 1 $\mu$m is available with combined energy-dispersive spectroscopy and scanning electron microscopy and down to 50 nm or less with combined energy-dispersive spectroscopy and transmission electron microscopy. Maps of a film or particle can be used to quantify the chemical homogeneity of multicomponent films and particles. These maps show the distribution of elements as a

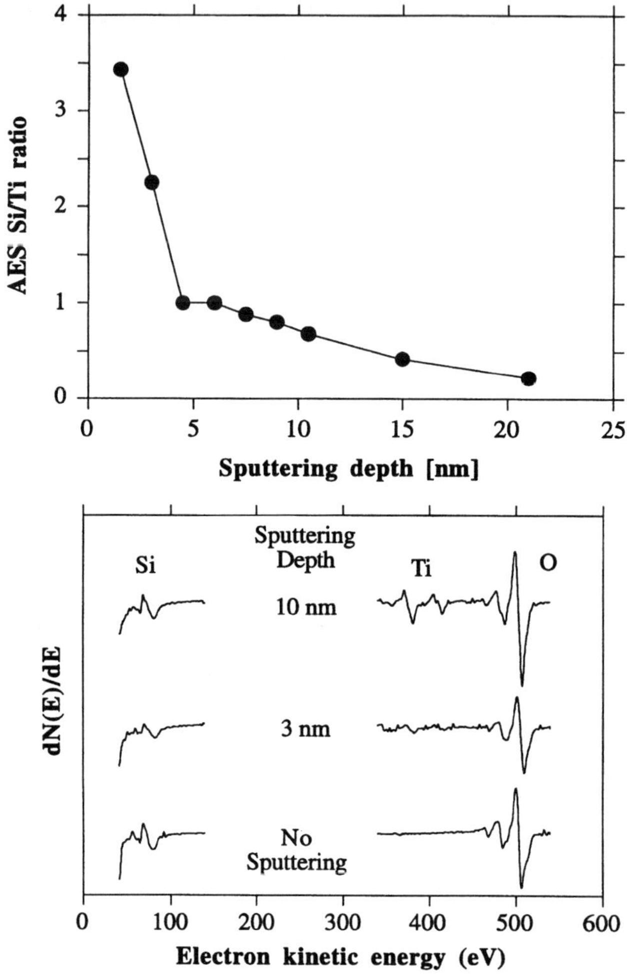


**Fig. 17.9.** Top: Relative concentration of Si vs. Ti as a function of sputtering depth of a $TiO_2$ particle coated with $SiO_2$. Bottom: Auger electron spectra from which relative concentration in top panel was calculated. (Atanasova, personal communication.)[56]

function of position in the sample. Energy-dispersive spectroscopy can also be used to examine a large number of particles and look for variations in composition from particle to particle. Quantitative measurements require flat surfaces and calibration standards, but semiquantitative information can be obtained for particles without standards. This approach has been used, for example, to examine the composition of thin coatings on particles (Fig. 17.10).[57]

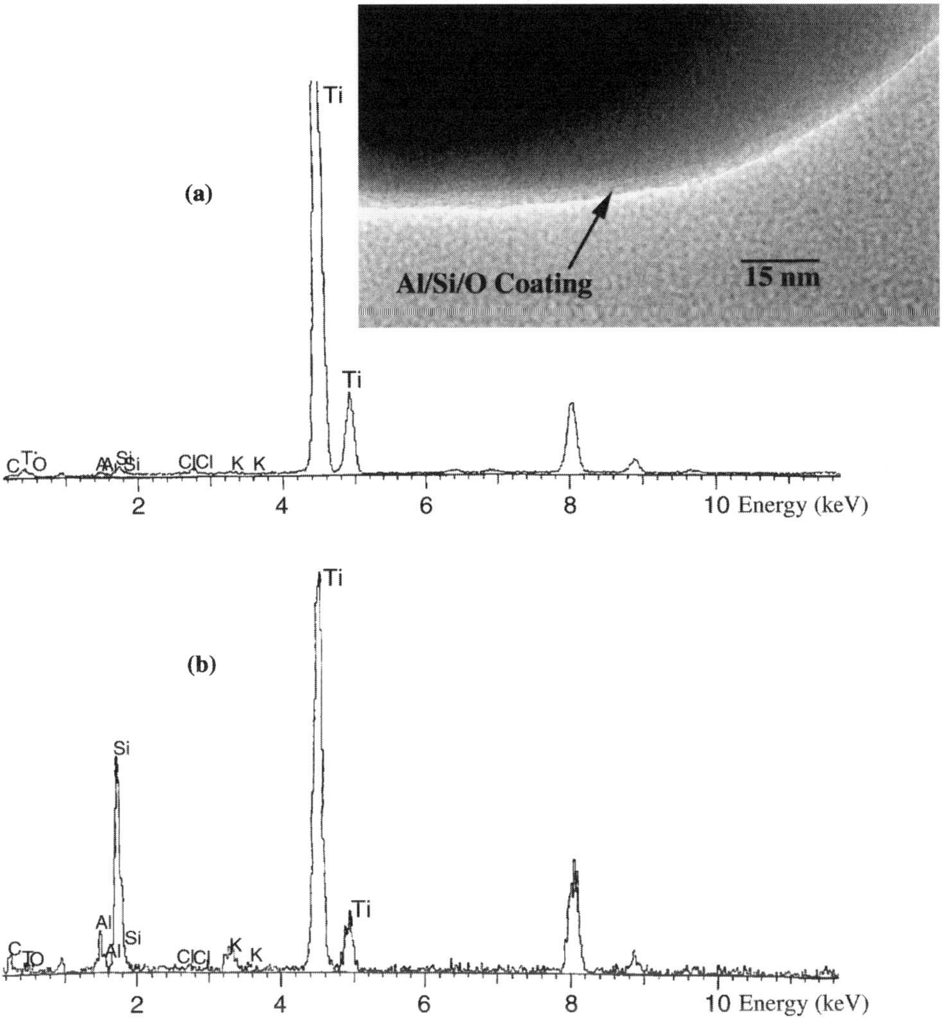


**Fig. 17.10.** Energy-dispersive spectra of a $TiO_2$ particle with a Al–Si–O coating: (a) center or bulk of the particle and (b) edge of the particle. Inset shows the corresponding transmission electron micrograph of the particle and the coating on its surface. (Courtesy of Powell.)[57]

### 17.7.3 X-ray Photoelectron Spectroscopy

X-ray photoelectron spectroscopy can detect elements from Li to U[4] and allows both powders and films to be analyzed. The technique is also known as electron spectroscopy for chemical analysis (ESCA). X-ray photoelectron spectroscopy irradiates particles or films with X-rays, resulting in the emission of photoelectrons with energies characteristic of the elements in the sample and their chemical bonding. The probing depth is approximately

3 nm, with 0.1-monolayer sensitivity. A spatial resolution of 10 $\mu$m can be obtained in state-of-the-art instruments, which limits the use of the technique to relatively large particles. Since submonolayer amounts are detected, small amounts of sample are needed. The primary advantage of X-ray photoelectron spectroscopy over Auger electron spectroscopy and energy-dispersive spectroscopy is the information it yields about the chemical bonding of the elements (the oxidation state). The technique is especially useful for characterizing coated particles; information similar to that shown in Figure 17.9 can be obtained routinely.[58]

### 17.7.4 Mass Spectrometric Methods

Mass spectrometers are sensitive instruments used for the detection of elements from H to U. They can detect species at ppm levels.[4] Various mass spectrometric methods differ by the means they use to introduce the sample. Atoms or clusters can be vaporized from surfaces with the use of pulsed lasers, or they can be sputtered by means of ions, or they can be introduced directly from the gas phase into the mass spectrometer.[59–65] Mass spectrometric methods are usually not used for aerosols on a routine basis, because of their complexity. Secondary ion mass spectroscopy, however, is a standard technique for analyzing films and powders.

## 17.8 THERMOPHYSICAL AND THERMOCHEMICAL ANALYSIS

Powder samples often contain various species that are adsorbed onto the surfaces of the particles in them. A powder may contain unreacted precursors that, when heated, will react to form gaseous species. The powder also may contain certain phases that, when heated, will undergo phase transitions, resulting in changes in properties of the powder. All of these are examples of phenomena that can be examined by heating the sample and measuring either its weight loss or gain or its heat release or take-up. An example of the importance of this kind of analysis is the incorporation of $NO_x$ in $YBa_2Cu_3O_7$ and Pb–Bi–Sr–Ca–Cu–O powders. When heated, these materials release NO and $NO_2$, which can degrade properties in powder-processing steps.

### 17.8.1 Thermogravimetric Analysis

Thermogravimetric analysis measures the change in the weight of a sample as it is heated at a known rate.[66] The primary use of the technique for materials produced by aerosol processes is to determine the extent of reaction and to detect species such as water. To determine the *identity* of the evolved species, thermogravimetric analysis can be used together with mass spectrometry. This variation identifies the species released at a certain temperature. From an examination of the rate of weight loss, kinetic parameters can also be determined.[67] Thermogravimetric analysis typically requires

10-mg samples. As a note of caution, the data is often sample-mass sensitive because of the necessity for species to diffuse into and out of the bed of particles. This potential problem must always be considered in interpreting results of the technique.

### 17.8.2 Differential Thermal Analysis and Differential Scanning Calorimetry

Differential thermal analysis measures the amount of heat released or absorbed by a sample as it is heated at a known rate.[66] When the change in enthalpy is determined, the method is called differential scanning calorimetry. The presence of exothermic or endothermic processes at certain temperatures provides information about the nature of phase changes and chemical reactions occurring in the material as it is heated. Differential thermal analysis and differential scanning calorimetry can often be used as a sensitive method for establishing the presence or absence of secondary phases in samples if these phases undergo transformations at known temperatures. Like thermogravimetric analysis, differential thermal analysis requires 10-mg samples.

### 17.8.3 Dilatometry

The use of dilatometry begins with a powder formed into a pellet.[3] The pellet is heated at a controlled temperature ramp, and its size is measured. When densification begins, the size of the pellet decreases. This decrease continues until the densification process stops at some higher temperature. Dilatometry measurements are critical for powders, because they provide information about relative densification temperatures.

## 17.9 SURFACE PROPERTIES

The surface properties of powders suspended in liquids can play a critical role in powder processing.[14] For example, the surface chemistry determines how well a powder disperses into a liquid. For this reason, it is often important to understand the behavior of particles in solution. A variety of measurements are available to characterize surface properties in solution: electrophoresis, sedimentation potential, electro-osmosis, streaming potential, and zeta potential. All of these measurements are related and can provide information about the nature of ionized functional groups on the surfaces of particles.

## 17.10 ASSORTED METHODS

A variety of other methods can characterize particles produced by aerosol processes, including Raman spectroscopy, electron spin resonance, small-angle X-ray scattering, small-angle neutron scattering, extended X-ray absorption fine-structure analysis, and other methods. These are relatively specialized techniques that are expensive and are not usually used on a routine basis.

A class of techniques that is being used on an increasing basis includes scanning tunneling microscopy and atomic force microscopy. These techniques provide information about structure down to the atomic level for both conductors and insulators.[68] This level of resolution is important in the case of nanometer-size clusters on surfaces, which may be difficult to image by techniques such as transmission electron microscopy.

Finally, numerous characterization techniques exist which are specific to a property or application. Examples are photoluminescence, cathodoluminescence, electroluminescence, electrophoretic mobility,[69] UV–visible absorption, magnetometry, electrical measurements, and many others.

## 17.11 REFERENCES

1. Willeke, K. & Baron, P.A. (Eds.) *Handbook of Aerosol Measurement: Principles, Techniques and Applications*. New York: Van Nostrand Reinhold, 1993.
2. Vincent, J.H. *Aerosol Sampling*. New York: John Wiley and Sons, 1989.
3. Reed, J.S. *Introduction to the Principles of Ceramic Processing*, New York: John Wiley and Sons, 1988.
4. Feldman, L. & Mayer, J. *Fundamentals of Surface and Thin Film Analysis*. New York: North-Holland 1986, 352.
5. Lippmann, M. *Size-selective health hazard sampling*. Cincinnati: Air Sampling Instruments, 1989, 163–198.
6. Lippmann, M. *Sampling aerosols by filtration*. Cincinnati: Air Sampling Instruments, 1989, 305–336.
7. Hinds, W.C. *Aerosol Technology*. New York: Wiley Interscience, 1982.
8. Marple, V.A. & Willeke, K. Inertial impactors. In Lundgren, D.A. (Ed.) *Aerosol Measurement*. Gainesville, FL: University Presses of Florida, 1979, 90–107.
9. Cheng, Y.S. & Yeh, H.C. Performance of a screen-type diffusion battery. In Marple, V.A., & Liu, B.Y.H. (Eds.) *Aerosols in the Mining and Industrial Work Environments*. Ann Arbor, MI: Ann Arbor Science Publishers, 1983, 1077–1094.
10. Swift, D.L. & Lippmann, M. *Electrostatic and thermal precipitators*. Cincinnati: *Air Sampling Instruments*, 1989, 387–402.
11. Enliang, L., Yingmin, W., & Raper, J.A. *Aerosol Sci. Technol. 12*, 1990, 947–952.
12. Kingery, W.D., & Berg, M. *J. Appl. Phys. 26*(10), 1955, 1205–1212.
13. Kingery, W.D., Bowen, A.K. & Uhlmann, D.R. *Introduction to Ceramics*, New York: John Wiley and Sons, 1976.
14. Arai, Y. *Chemistry of Powder Production*. New York: Chapman and Hall, 1996.
15. Maynard, A.D. *J. Aerosol Sci. 26*(5), 1995 757–777.
16. Gurav, A. & Kodas, T.T. personal communication, 1997.
17. Judin, V.P.S. *Chemistry in Britain 28*(6), 1993, 503–505.

18. Miller, B.V. & Lines, R.W. *CRC Critical Reviews in Analytical Chemistry 20*(2), 1988, 75–116.
19. Sheppard, L.M. *Ceram. Bull. 67*(5), 1988, 878–883.
20. Brewster, J. *Aerosol Generation of* $BaTiO_3$. Master's thesis, University of New Mexico, 1997.
21. Niessner, R., Daeumer, B., & Klockow, D. *Aerosol Sci. Technol. 12*, 1990, 953–963.
22. Su, Y.F., Cheng, Y.S., Newton, G.J., & Yeh, H.C. *Aerosol Sci. Technol. 12*, 1990, 1050–1054.
23. Zhang, Z.Q. & Liu, B.Y.H. *Aerosol Sci. Technol. 13*, 1990, 493–504.
24. Peters, T.M., Chein, H., Lundgren, D.A., & Keady, P.B. *Aerosol Science and Technology 19*, 1993, 396–405.
25. Mertes, S., Schröder, F., & Wiedensohler, A. *Aerosol Sci. Technol. 23*, 1995, 257–261.
26. Wang, S.C. & Flagan, R.C. *Aerosol Sci. Technol. 13*, 1990, 230–240.
27. Hering, S.V., & Stolzenburg, M.R. *Aerosol Sci. Technol. 23*, 1995, 155–173.
28. Zhang, S.-H., Akutsu, Y., Russell, L.M., Flagan, R.C., & Seinfeld, J.H. *Aerosol Sci. Technol. 23*, 1995, 357–372.
29. Fissan, H., Hummes, D., Stratmann, F., Buscher, P., Neumann, S., Pui, D.Y.H., & Chen, D. *Aerosol Sci. Technol. 24*, 1996, 1–13.
30. Knollenberg, R.G. & Luehr, R. Open cavity laser "active" scattering particle spectrometry from 0.05 to 5 microns. In Liu, B.Y.H. (Ed.), *Measurement, Sampling and Analysis*. New York: Academic Press, 1976, 669–696.
31. Kim, Y.J. *Aerosol Sci. Technol. 22*, 1995, 33–42.
32. Caldow, R., Pui, D.Y.H., Szymanski, W.W., & Liu, B.Y.H. *Aerosol Sci. Technol. 12*, 1990, 981–991.
33. Felton, C.C., Sheppard, J.C., & Campbell, M.J. *Environ. Sci. Technol. 24*, 1990, 1841–1847.
34. Rao, A.K. & Whitby, K.T. *J. Aerosol Sci. 9*, 1978, 77–86.
35. Rao, A.K. & Whitby, K.T. *J. Aerosol Sci. 9*, 1978, 87–100.
36. Olawoyin, O.O., Raunemaa, T.M., & Hopke, P.K. *Aerosol Sci. Technol. 23*, 1995, 121–130.
37. Loo, B.W., Jaklevic, J.M., & Goulding, F.S. Dichotomous virtual impactors for large scale monitoring of airborne particulate matter. In Liu, B.Y.H. (Ed.), *Fine Particles: Aerosol Generation, Measurement, Sampling and Analysis*. New York: Academic Press, 1976, 311–350.
38. Chen, B.T., Yeh, H.C., & Cheng, Y.S. *J. Aerosol Sci. 16*(4), 1985, 343–354.
39. Biswas, P. & Flagan, R.C. *J. Aerosol Sci. 19*(1), 1988, 113–121.
40. DeOtte, R.E., Jr. *Aerosol Sci. Technol. 12*, 1990, 1055–1066.
41. Kim, J.C. & Lee, K.W. *Aerosol Science and Technology 12*, 1990, 1003–1015.
42. Smith, W.B., Wilson, R.R., & Harris, D.B. *Environ. Sci. Technol. 13*, 1979, 1387–1392.
43. Wilson, J.C. & Liu, B.Y.H. *J. Aerosol Sci. 11*, 1980, 139–150.

44. Remiarz, R.J., Agarwal, J.K., Quant, F.R., & Sem, G.J. A real-time aerodynamic particle size analyzer. In Marple, V.A. & Liu, B.Y.H. (Eds.), *Aerosols in the Mining and Industrial Work Environments*. Ann Arbor, MI: Ann Arbor Science Publishers, 1983, 879–895.
45. Brockmann, J.E. & Rader, D.J. *Aerosol Sci. Technol. 13*, 1990, 162–172.
46. Malla, P.B., Komarneni, S., Taguchi, H., & Kido, H. *J. Amer. Ceram. Soc. 74*(12), 1991, 2988–2995.
47. Gurav, A., Pluym, T., Kodas, T.T., & Pratsinis, S.E. *Chem. Eng. Commun. 151*, 1996, 211–226.
48. Sheldon, B.W. & Haggerty, J.S. *J. Am. Ceram. Soc. 74*(6), 1991, 1417–1424.
49. Ortega, J., Kodas, T.T., Chadda, S., Smith, D.M., Ciftcioglu, M., & Brennan, J. *Chem. Mater. 3*, 1991, 746–751.
50. Edelson, L.H. & Glaeser, A.M. *J. Amer. Ceram. Soc. 71*(4), 1988, 225–235.
51. Barringer, E.A. & Bowen, H.K. *Effects of particle packing on the sintered mirostructure*. Cambridge, MA: MIT Industrial Liaison Program, 1984.
52. Dericotte, D. *Formation of Nanoscale Single and Multicomponent Particles via Two Gas-Phase Routes: Matched Reactivities and Single-Source Precursors*. Master's thesis, University of New Mexico, 1997.
53. Niessner, R. *J. Amer. Ceram. Soc. 337*, 1990, 565–576.
54. Knapp, K.T. & Bennett, R.L. *Aerosol Sci. Technol. 12*, 1990, 1067–1074.
55. Podzimek, J., Wojnar, V., & Straka, M. *Aerosol Sci. Technol. 13*, 1990, 173–182.
56. Atanasova, P., personal communication, 1997.
57. Powell, Q.H. *Gas-phase production and coating of titania in tubular, hot-wall reactors*. Master's thesis, University of New Mexico, 1995.
58. Day, R.E. & Egerton, T.A. *Colloids and Surfaces 23*, 1987, 137–155.
59. Carson, P.G., Neubauer, K.R., Johnston, M.V., & Wexler, A.S. *J. Aerosol Sci. 26*(4), 1995, 535–545.
60. Giggy, C.L., Friedlander, S.K., & Sinha, M.P. *Atmos. Env. 23*(10), 1989, 2223–2229.
61. Hinz, K.-P., Kauffmann, R., & Spengler, B. *Anal. Chem. 66*, 1994, 2071–2076.
62. Noble, C.A., Nordmeyer, T., Salt, K., Morrical, B., & Prather, K.A. *Trend. Anal. Chem. 13*(5), 1994, 218–222.
63. Nordmeyer, T. & Prather, K.A. *Anal. Chem. 66*, 1994, 3540–3542.
64. Murphy, D.M. & Thomson, D.S. *Aerosol Sci. Technol. 22*, 1995, 237–249.
65. Reents, W.D.J., Downey, S.W., Emerson, A.B., Mujsce, A.M., Muller, A.J., Siconolfi, D.J., Sinclair, J.D., & Swanson, A.G. *Aerosol Sci. Technol. 23*, 1995, 263–270.
66. Hench, L.L. & Gould, R.W. *Characterization of Ceramics*. New York: Marcel Dekker, 1971.
67. Biswas, P., Zhou, D., Zitkovsky, I., Blue, C., & Boolchand, P. *Mat. Lett. 8*(6, 7), 1989, 233–237.
68. Neddermeyer, H. *Critical Reviews in Solid State and Materials Science 16*, 1990, 309–335.
69. Kagawa, M., Syono, Y., Imamura, Y., & Usui, S. *Comm. Am. Ceram. Soc. 69*(3), 1986, C50–C51.

# INDEX

In this index, page numbers followed by "t" designate tables; *See also* designates related topics. Chemicals are primarily indexed by symbols.